Linear Algebra and Optimization with Applications to Machine Learning

Volume II

Fundamentals of Optimization Theory with Applications to Machine Learning

Linear Algebra and Optimization with Applications to Machine Learning

Volume II

Fundamentals of Optimization Theory with Applications to Machine Learning

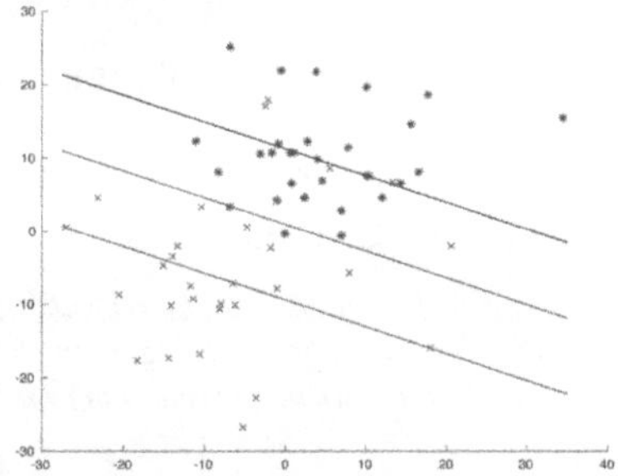

Jean Gallier

Jocelyn Quaintance

University of Pennsylvania, USA

World Scientific

NEW JERSEY • LONDON • SINGAPORE • BEIJING • SHANGHAI • HONG KONG • TAIPEI • CHENNAI • TOKYO

Published by

World Scientific Publishing Co. Pte. Ltd.
5 Toh Tuck Link, Singapore 596224
USA office: 27 Warren Street, Suite 401-402, Hackensack, NJ 07601
UK office: 57 Shelton Street, Covent Garden, London WC2H 9HE

British Library Cataloguing-in-Publication Data
A catalogue record for this book is available from the British Library.

LINEAR ALGEBRA AND OPTIMIZATION WITH APPLICATIONS TO MACHINE LEARNING
Volume II: Fundamentals of Optimization Theory with Applications to Machine Learning

ISBN 978-981-121-656-5 (hardcover)
ISBN 978-981-121-657-2 (ebook for institutions)
ISBN 978-981-121-658-9 (ebook for individuals)

For any available supplementary material, please visit
https://www.worldscientific.com/worldscibooks/10.1142/11722#t=suppl

Desk Editor: Liu Yumeng

Preface

In recent years, computer vision, robotics, machine learning, and data science have been some of the key areas that have contributed to major advances in technology. Anyone who looks at papers or books in the above areas will be baffled by a strange jargon involving exotic terms such as kernel PCA, ridge regression, lasso regression, support vector machines (SVM), Lagrange multipliers, KKT conditions, *etc.* Do support vector machines chase cattle to catch them with some kind of super lasso? No! But one will quickly discover that behind the jargon which always comes with a new field (perhaps to keep the outsiders out of the club), lies a lot of "classical" linear algebra and techniques from optimization theory. And there comes the main challenge: in order to understand and use tools from machine learning, computer vision, and so on, one needs to have a firm background in linear algebra and optimization theory. To be honest, some probability theory and statistics should also be included, but we already have enough to contend with.

Many books on machine learning struggle with the above problem. How can one understand what are the dual variables of a ridge regression problem if one doesn't know about the Lagrangian duality framework? Similarly, how is it possible to discuss the dual formulation of SVM without a firm understanding of the Lagrangian framework?

The easy way out is to sweep these difficulties under the rug. If one is just a consumer of the techniques we mentioned above, the cookbook recipe approach is probably adequate. But this approach doesn't work for someone who really wants to do serious research and make significant contributions. To do so, we believe that one must have a solid background in linear algebra and optimization theory.

This is a problem because it means investing a great deal of time and energy studying these fields, but we believe that perseverance will be amply rewarded.

This second volume covers some elements of optimization theory and applications, especially to machine learning. This volume is divided in five parts:

(1) Preliminaries of Optimization Theory.
(2) Linear Optimization.
(3) Nonlinear Optimization.
(4) Applications to Machine Learning.
(5) An appendix consisting of two chapters; one on Hilbert bases and the Riesz–Fischer theorem, the other one containing `Matlab` code.

Part 1 is devoted to some preliminaries of optimization theory. The goal of most optimization problems is to minimize (or maximize) some objective function J subject to equality or inequality constraints. Therefore it is important to understand when a function J has a minimum or a maximum (an optimum). In most optimization problems, we need to find necessary conditions for a function $J\colon \Omega \to \mathbb{R}$ to have a local extremum with respect to a subset U of Ω (where Ω is open). This can be done in two cases:

(1) The set U is defined by a set of equations,
$$U = \{x \in \Omega \mid \varphi_i(x) = 0,\ \ 1 \leq i \leq m\},$$
where the functions $\varphi_i\colon \Omega \to \mathbb{R}$ are continuous (and usually differentiable).
(2) The set U is defined by a set of inequalities,
$$U = \{x \in \Omega \mid \varphi_i(x) \leq 0,\ \ 1 \leq i \leq m\},$$
where the functions $\varphi_i\colon \Omega \to \mathbb{R}$ are continuous (and usually differentiable).

The case of equality constraints is much easier to deal with and is treated in Chapter 4.

In the case of equality constraints, a necessary condition for a local extremum with respect to U can be given in terms of *Lagrange multipliers*.

Part 2 deals with the special case where the objective function is a linear form and the constraints are affine inequality and equality constraints. This subject is known as *linear programming*, and the next four chapters give an introduction to the subject.

Part 3 is devoted to nonlinear optimization, which is the case where the objective function J is not linear and the constraints are inequality constraints. Since it is practically impossible to say anything interesting if the constraints are not convex, we quickly consider the convex case.

Chapter 13 is devoted to some general results of optimization theory. A main theme is to find sufficient conditions that ensure that an objective function has a minimum which is achieved. We define gradient descent methods (including Newton's method), and discuss their convergence.

Chapter 14 contains the most important results of nonlinear optimization theory. Theorem 14.2 gives necessary conditions for a function J to have a minimum on a subset U defined by convex inequality constraints in terms of the Karush–Kuhn–Tucker conditions. Furthermore, if J is also convex and if the KKT conditions hold, then J has a global minimum.

We illustrate the KKT conditions on an interesting example from machine learning the so-called *hard margin support vector machine*; see Sections 14.5 and 14.6. The problem is to separate two disjoint sets of points, $\{u_i\}_{i=1}^p$ and $\{v_j\}_{j=1}^q$, using a hyperplane satisfying some optimality property (to maximize the margin).

Section 14.7 contains the most important results of the chapter. The notion of Lagrangian duality is presented and we discuss *weak duality* and *strong duality*.

In Chapter 15, we consider some deeper aspects of the the theory of convex functions that are not necessarily differentiable at every point of their domain. Some substitute for the gradient is needed. Fortunately, for convex functions, there is such a notion, namely *subgradients*. A major motivation for developing this more sophisticated theory of differentiation of convex functions is to extend the Lagrangian framework to convex functions that are not necessarily differentiable.

Chapter 16 is devoted to the presentation of one of the best methods known at the present for solving optimization problems involving equality constraints, called ADMM (alternating direction method of multipliers). In fact, this method can also handle more general constraints, namely, membership in a convex set. It can also be used to solve *lasso minimization*.

In Section 16.4, we prove the convergence of ADMM under exactly the same assumptions as in Boyd et al. [Boyd *et al.* (2010)]. It turns out that Assumption (2) in Boyd et al. [Boyd *et al.* (2010)] implies that the matrices $A^\top A$ and $B^\top B$ are invertible (as we show after the proof of Theorem 16.1). This allows us to prove a convergence result stronger than the convergence result proven in Boyd et al. [Boyd *et al.* (2010)].

The next four chapters constitute Part 4, which covers some applications of optimization theory (in particular Lagrangian duality) to machine learning.

Chapter 17 is an introduction to positive definite kernels and the use of kernel functions in machine learning called a *kernel function.*

We illustrate the kernel methods on kernel PCA.

In Chapter 18 we return to the problem of separating two disjoint sets of points, $\{u_i\}_{i=1}^p$ and $\{v_j\}_{j=1}^q$, but this time we do not assume that these two sets are separable. To cope with nonseparability, we allow points to invade the safety zone around the separating hyperplane, and even points on the wrong side of the hyperplane. Such a method is called *soft margin support vector machine (SVM)*. We discuss variations of this method, including ν-SV classification. In each case we present a careful derivation of the dual. We prove rigorous results about the existence of support vectors.

In Chapter 19, we discuss *linear regression*, *ridge regression*, *lasso regression* and *elastic net regression*.

In Chapter 20 we present *ν-SV Regression*. This method is designed in the same spirit as soft margin SVM, in the sense that it allows a margin of error. Here the errors are penalized in the ℓ^1-sense. We present a careful derivation of the dual and discuss the existence of support vectors.

The methods presented in Chapters 18, 19 and 20 have all been implemented in `Matlab`, and much of this code is given in Appendix B. Remarkably, ADMM emerges as the main engine for solving most of these optimization problems. Thus it is nice to see the continuum spanning from theoretical considerations of convergence and correctness to practical matters of implementation. It is fun to see how these abstract Lagrange multipliers yield concrete results such as the weight vector w defining the desired hyperplane in regression or SVM.

Except for a few exceptions we provide complete proofs. We did so to make this book self-contained, but also because we believe that no deep knowledge of this material can be acquired without working out some proofs. However, our advice is to skip some of the proofs upon first reading, especially if they are long and intricate.

The chapters or sections marked with the symbol ⊛ contain material that is typically more specialized or more advanced, and they can be omitted upon first (or second) reading.

Acknowledgement: We would like to thank Christine Allen-Blanchette, Kostas Daniilidis, Carlos Esteves, Spyridon Leonardos, Stephen Phillips, João Sedoc, Stephen Shatz, Jianbo Shi, and Marcelo Siqueira, for reporting typos and for helpful comments. Thanks to Gilbert Strang. We learned much from his books which have been a major source of inspiration. Special thanks to Steven Boyd. We learned a lot from his remarkable book on convex optimization and his papers, and Part III of our book is significantly inspired by his writings. The first author also wishes to express his deepest gratitute to Philippe G. Ciarlet who was his teacher and mentor in 1970–1972 while he was a student at ENPC in Paris. Professor Ciarlet was by far his best teacher. He also knew how to instill in his students the importance of intellectual rigor, honesty, and modesty. He still has his typewritten notes on measure theory and integration, and on numerical linear algebra. The latter became his wonderful book Ciarlet [Ciarlet (1989)], from which we have borrowed heavily.

Contents

NonLinear Optimization 313

Chapter 1

Introduction

This second volume covers some elements of optimization theory and applications, especially to machine learning. This volume is divided in five parts:

(1) Preliminaries of Optimization Theory.
(2) Linear Optimization.
(3) Nonlinear Optimization.
(4) Applications to Machine Learning.
(5) An appendix consisting of two chapters; one on Hilbert bases and the Riesz–Fischer theorem, the other one containing `Matlab` code.

Part 1 is devoted to some preliminaries of optimization theory. The goal of most optimization problems is to minimize (or maximize) some objective function J subject to equality or inequality constraints. Therefore it is important to understand when a function J has a minimum or a maximum (an optimum). If the function J is sufficiently differentiable, then a necessary condition for a function to have an optimum typically involves the derivative of the function J, and if J is real-valued, its gradient ∇J.

Thus it is desirable to review some basic notions of topology and calculus, in particular, to have a firm grasp of the notion of derivative of a function between normed vector spaces. Partial derivatives $\partial f / \partial A$ of functions whose range and domain are spaces of matrices tend to be used casually, even though in most cases a correct definition is never provided. It is possible, and simple, to define rigorously derivatives, gradients, and directional derivatives of functions defined on matrices and to avoid these nonsensical partial derivatives.

Chapter 2 contains a review of basic topological notions used in analysis.

We pay particular attention to complete metric spaces and complete normed vector spaces. In fact, we provide a detailed construction of the completion of a metric space (and of a normed vector space) using equivalence classes of Cauchy sequences. Chapter 3 is devoted to some notions of differential calculus, in particular, directional derivatives, total derivatives, gradients, Hessians, and the inverse function theorem.

Chapter 4 deals with extrema of real-valued functions. In most optimization problems, we need to find necessary conditions for a function $J\colon \Omega \to \mathbb{R}$ to have a local extremum with respect to a subset U of Ω (where Ω is open). This can be done in two cases:

(1) The set U is defined by a set of equations,

$$U = \{x \in \Omega \mid \varphi_i(x) = 0,\ \ 1 \leq i \leq m\},$$

where the functions $\varphi_i\colon \Omega \to \mathbb{R}$ are continuous (and usually differentiable).

(2) The set U is defined by a set of inequalities,

$$U = \{x \in \Omega \mid \varphi_i(x) \leq 0,\ \ 1 \leq i \leq m\},$$

where the functions $\varphi_i\colon \Omega \to \mathbb{R}$ are continuous (and usually differentiable).

In (1), the equations $\varphi_i(x) = 0$ are called *equality constraints*, and in (2), the inequalities $\varphi_i(x) \leq 0$ are called *inequality constraints*. The case of equality constraints is much easier to deal with and is treated in Chapter 4.

If the functions φ_i are convex and Ω is convex, then U is convex. This is a very important case that we will discuss later. In particular, if the functions φ_i are affine, then the equality constraints can be written as $Ax = b$, and the inequality constraints as $Ax \leq b$, for some $m \times n$ matrix A and some vector $b \in \mathbb{R}^m$. We will also discuss the case of affine constraints later.

In the case of equality constraints, a necessary condition for a local extremum with respect to U can be given in terms of *Lagrange multipliers*. In the case of inequality constraints, there is also a necessary condition for a local extremum with respect to U in terms of generalized Lagrange multipliers and the *Karush–Kuhn–Tucker* conditions. This will be discussed in Chapter 14.

In Chapter 5 we discuss Newton's method and some of its generalizations (the Newton–Kantorovich theorem). These are methods to find the zeros of a function.

Chapter 6 covers the special case of determining when a quadratic function has a minimum, subject to affine equality constraints. A complete answer is provided in terms of the notion of symmetric positive semidefinite matrices.

The Schur complement is introduced in Chapter 7. We give a complete proof of a criterion for a matrix to be positive definite (or positive semidefinite) stated in Boyd and Vandenberghe [Boyd and Vandenberghe (2004)] (Appendix B).

Part 2 deals with the special case where the objective function is a linear form and the constraints are affine inequality and equality constraints. This subject is known as linear programming, and the next four chapters give an introduction to the subject. Although linear programming has been supplanted by convex programming and its variants, it is still a great workhorse. It is also a great warm up for the general treatment of Lagrangian duality. We pay particular attention to versions of Farkas' lemma, which is at the heart of duality in linear programming.

Part 3 is devoted to nonlinear optimization, which is the case where the objective function J is not linear and the constraints are inequality constraints. Since it is practically impossible to say anything interesting if the constraints are not convex, we quickly consider the convex case.

In optimization theory one often deals with function spaces of infinite dimension. Typically, these spaces either are Hilbert spaces or can be completed as Hilbert spaces. Thus it is important to have some minimum knowledge about Hilbert spaces, and we feel that this minimum knowledge includes the projection lemma, the fact that a closed subset has an orthogonal complement, the Riesz representation theorem, and a version of the Farkas–Minkowski lemma. Chapter 12 covers these topics. A more detailed introduction to Hilbert spaces is given in Appendix A.

Chapter 13 is devoted to some general results of optimization theory. A main theme is to find sufficient conditions that ensure that an objective function has a minimum which is achieved. We define the notion of a coercive function. The most general result is Theorem 13.1, which applies to a coercive convex function on a convex subset of a separable

Hilbert space. In the special case of a coercive quadratic functional, we obtain the Lions–Stampacchia theorem (Theorem 13.4), and the Lax–Milgram theorem (Theorem 13.5). We define elliptic functionals, which generalize quadratic functions defined by symmetric positive definite matrices. We define gradient descent methods, and discuss their convergence. A gradient descent method looks for a descent direction and a stepsize parameter, which is obtained either using an exact line search or a backtracking line search. A popular technique to find the search direction is steepest descent. In addition to steepest descent for the Euclidean norm, we discuss steepest descent for an arbitrary norm. We also consider a special case of steepest descent, Newton's method. This method converges faster than the other gradient descent methods, but it is quite expensive since it requires computing and storing Hessians. We also present the method of conjugate gradients and prove its correctness. We briefly discuss the method of gradient projection and the penalty method in the case of constrained optima.

Chapter 14 contains the most important results of nonlinear optimization theory. We begin by defining the cone of feasible directions and then state a necessary condition for a function to have local minimum on a set U that is not necessarily convex in terms of the cone of feasible directions. The cone of feasible directions is not always convex, but it is if the constraints are inequality constraints. An inequality constraint $\varphi(u) \leq 0$ is said to be *active* is $\varphi(u) = 0$. One can also define the notion of *qualified constraint*. Theorem 14.1 gives necessary conditions for a function J to have a minimum on a subset U defined by qualified inequality constraints in terms of the Karush–Kuhn–Tucker conditions (for short KKT conditions), which involve nonnegative Lagrange multipliers. The proof relies on a version of the Farkas–Minkowski lemma. Some of the KTT conditions assert that $\lambda_i \varphi_i(u) = 0$, where $\lambda_i \geq 0$ is the Lagrange multiplier associated with the constraint $\varphi_i \leq 0$. To some extent, this implies that active constraints are more important than inactive constraints, since if $\varphi_i(u) < 0$ is an inactive constraint, then $\lambda_i = 0$. In general, the KKT conditions are useless unless the constraints are convex. In this case, there is a manageable notion of qualified constraint given by Slater's conditions. Theorem 14.2 gives necessary conditions for a function J to have a minimum on a subset U defined by convex inequality constraints in terms of the Karush–Kuhn–Tucker conditions. Furthermore, if J is also convex and if the KKT conditions hold, then J has a global minimum.

In Section 14.4, we apply Theorem 14.2 to the special case where the constraints are equality constraints, which can be expressed as $Ax = b$. In the special case where the convex objective function J is a convex quadratic functional of the form

$$J(x) = \frac{1}{2}x^\top Px + q^\top x + r,$$

where P is a $n \times n$ symmetric positive semidefinite matrix, the necessary and sufficient conditions for having a minimum are expressed by a linear system involving a matrix called the KKT matrix. We discuss conditions that guarantee that the KKT matrix is invertible, and how to solve the KKT system. We also briefly discuss variants of Newton's method dealing with equality constraints.

We illustrate the KKT conditions on an interesting example, the so-called hard margin support vector machine; see Sections 14.5 and 14.6. The problem is a classification problem, or more accurately a separation problem. Suppose we have two nonempty disjoint finite sets of p *blue* points $\{u_i\}_{i=1}^p$ and q *red* points $\{v_j\}_{j=1}^q$ in $\mathbb{R}^n$. Our goal is to find a hyperplane H of equation $w^\top x - b = 0$ (where $w \in \mathbb{R}^n$ is a nonzero vector and $b \in \mathbb{R}$), such that all the blue points u_i are in one of the two open half-spaces determined by H, and all the red points v_j are in the other open half-space determined by H.

If the two sets are indeed separable, then in general there are infinitely many hyperplanes separating them. Vapnik had the idea to find a hyperplane that maximizes the smallest distance between the points and the hyperplane. Such a hyperplane is indeed unique and is called a maximal hard margin hyperplane, or hard margin support vector machine. The support vectors are those for which the constraints are active.

Section 14.7 contains the most important results of the chapter. The notion of Lagrangian duality is presented. Given a primal optimization problem (P) consisting in minimizing an objective function $J(v)$ with respect to some inequality constraints $\varphi_i(v) \leq 0$, $i = 1, \ldots, m$, we define the *dual function* $G(\mu)$ as the result of minimizing the Lagrangian

$$L(v, \mu) = J(v) + \sum_{i=1}^{m} \mu_i \varphi_i(v)$$

with respect to v, with $\mu \in \mathbb{R}^m_+$. The dual program (D) is then to maximize $G(\mu)$ with respect to $\mu \in \mathbb{R}^m_+$. It turns out that G is a concave function, and the dual program is an unconstrained maximization. This is actually a misleading statement because G is generally a partial function, so maximizing $G(\mu)$ is equivalent to a constrained maximization problem in which the constraints specify the domain of G, but in many cases, we obtain a dual program simpler than the primal program. If d^* is the optimal value of the dual program and if p^* is the optimal value of the primal program, we always have

$$d^* \leq p^*,$$

which is known as *weak duality*. Under certain conditions, $d^* = p^*$, that is, the duality gap is zero, in which case we say that *strong duality* holds. Also, under certain conditions, a solution of the dual yields a solution of the primal, and if the primal has an optimal solution, then the dual has an optimal solution, but beware that the converse is generally false (see Theorem 14.5). We also show how to deal with equality constraints, and discuss the use of conjugate functions to find the dual function. Our coverage of Lagrangian duality is quite thorough, but we do not discuss more general orderings such as the semidefinite ordering. For these topics which belong to convex optimization, the reader is referred to Boyd and Vandenberghe [Boyd and Vandenberghe (2004)].

In Chapter 15, we consider some deeper aspects of the the theory of convex functions that are not necessarily differentiable at every point of their domain. Some substitute for the gradient is needed. Fortunately, for convex functions, there is such a notion, namely *subgradients*. Geometrically, given a (proper) convex function f, the subgradients at x are vectors normal to supporting hyperplanes to the epigraph of the function at $(x, f(x))$. The *subdifferential* $\partial f(x)$ to f at x is the set of all subgradients at x. A crucial property is that f is differentiable at x iff $\partial f(x) = \{\nabla f_x\}$, where ∇f_x is the gradient of f at x. Another important property is that a (proper) convex function f attains its minimum at x iff $0 \in \partial f(x)$. A major motivation for developing this more sophisticated theory of "differentiation" of convex functions is to extend the Lagrangian framework to convex functions that are not necessarily differentiable.

Experience shows that the applicability of convex optimization is significantly increased by considering extended real-valued functions, namely functions $f \colon S \to \mathbb{R} \cup \{-\infty, +\infty\}$, where S is some subset of $\mathbb{R}^n$ (usually convex). This is reminiscent of what happens in measure theory, where it is natural to consider functions that take the value $+\infty$.

In Section 15.1, we introduce extended real-valued functions, which are functions that may also take the values $\pm\infty$. In particular, we define proper convex functions, and the closure of a convex function. Subgradients and subdifferentials are defined in Section 15.2. We discuss some properties of subgradients in Section 15.3 and Section 15.4. In particular, we relate subgradients to one-sided directional derivatives. In Section 15.5, we discuss the problem of finding the minimum of a proper convex function and give some criteria in terms of subdifferentials. In Section 15.6, we sketch the generalization of the results presented in Chapter 14 about the Lagrangian framework to programs allowing an objective function and inequality constraints which are convex but not necessarily differentiable.

This chapter relies heavily on Rockafellar [Rockafellar (1970)]. We tried to distill the body of results needed to generalize the Lagrangian framework to convex but not necessarily differentiable functions. Some of the results in this chapter are also discussed in Bertsekas [Bertsekas (2009); Bertsekas *et al.* (2003); Bertsekas (2015)].

Chapter 16 is devoted to the presentation of one of the best methods known at the present for solving optimization problems involving equality constraints, called ADMM (alternating direction method of multipliers). In fact, this method can also handle more general constraints, namely, membership in a convex set. It can also be used to solve *lasso minimization*.

In this chapter, we consider the problem of minimizing a convex function J (not necessarily differentiable) under the equality constraints $Ax = b$. In Section 16.1, we discuss the dual ascent method. It is essentially gradient descent applied to the dual function G, but since G is maximized, gradient descent becomes gradient ascent.

In order to make the minimization step of the dual ascent method more robust, one can use the trick of adding the penalty term $(\rho/2)\left\|Au - b\right\|_2^2$ to the Lagrangian. We obtain the *augmented Lagrangian*

$$L_\rho(u, \lambda) = J(u) + \lambda^\top(Au - b) + (\rho/2)\left\|Au - b\right\|_2^2,$$

with $\lambda \in \mathbb{R}^m$, and where $\rho > 0$ is called the *penalty parameter*. We obtain the minimization Problem (P_ρ),

$$\begin{aligned} &\text{minimize} \quad J(u) + (\rho/2)\left\|Au - b\right\|_2^2 \\ &\text{subject to} \quad Au = b, \end{aligned}$$

which is equivalent to the original problem.

The benefit of adding the penalty term $(\rho/2)\left\|Au - b\right\|_2^2$ is that by Proposition 15.25, Problem (P_ρ) has a unique optimal solution under mild conditions on A. Dual ascent applied to the dual of (P_ρ) is called the *method of multipliers* and is discussed in Section 16.2.

The new twist in ADMM is to split the function J into two independent parts, as $J(x,z) = f(x) + g(z)$, and to consider the Minimization Problem (P_{admm}),

$$\begin{aligned} \text{minimize} \quad & f(x) + g(z) \\ \text{subject to} \quad & Ax + Bz = c, \end{aligned}$$

for some $p \times n$ matrix A, some $p \times m$ matrix B, and with $x \in \mathbb{R}^n$, $z \in \mathbb{R}^m$, and $c \in \mathbb{R}^p$. We also assume that f and g are convex.

As in the method of multipliers, we form the augmented Lagrangian

$$L_\rho(x,z,\lambda) = f(x) + g(z) + \lambda^\top(Ax + Bz - c) + (\rho/2)\left\|Ax + Bz - c\right\|_2^2,$$

with $\lambda \in \mathbb{R}^p$ and for some $\rho > 0$. The major difference with the method of multipliers is that instead of performing a minimization step jointly over x and z, ADMM first performs an x-minimization step and then a z-minimization step. Thus x and z are updated in an alternating or sequential fashion, which accounts for the term *alternating direction*. Because the Lagrangian is augmented, some mild conditions on A and B imply that these minimization steps are guaranteed to terminate. ADMM is presented in Section 16.3.

In Section 16.4, we prove the convergence of ADMM under exactly the same assumptions as in Boyd et al. [Boyd *et al.* (2010)]. It turns out that Assumption (2) in Boyd et al. [Boyd *et al.* (2010)] implies that the matrices $A^\top A$ and $B^\top B$ are invertible (as we show after the proof of Theorem 16.1). This allows us to prove a convergence result stronger than the convergence result proven in Boyd et al. [Boyd *et al.* (2010)]. In particular, we prove that *all* of the sequences (x^k), (z^k), and (λ^k) converge to optimal solutions $(\widetilde{x}, \widetilde{z})$, and $\widetilde{\lambda}$.

In Section 16.5, we discuss stopping criteria. In Section 16.6, we present some applications of ADMM, in particular, minimization of a proper closed convex function f over a closed convex set C in $\mathbb{R}^n$ and quadratic programming. The second example provides one of the best methods for solving quadratic problems, in particular, the SVM problems discussed in Chapter 18. Section 16.8 gives applications of ADMM to ℓ^1-norm problems, in particular, lasso regularization which plays an important role in machine learning.

The next four chapters constitute Part 4, which covers some applications of optimization theory (in particular Lagrangian duality) to machine learning.

Chapter 17 is an introduction to positive definite kernels and the use of kernel functions in machine learning.

Let X be a nonempty set. If the set X represents a set of highly nonlinear data, it may be advantageous to map X into a space F of much higher dimension called the *feature space*, using a function $\varphi\colon X \to F$ called a *feature map*. This idea is that φ "unwinds" the description of the objects in F in an attempt to make it linear. The space F is usually a vector space equipped with an inner product $\langle -, - \rangle$. If F is infinite dimensional, then we assume that it is a Hilbert space.

Many algorithms that analyze or classify data make use of the inner products $\langle \varphi(x), \varphi(y) \rangle$, where $x, y \in X$. These algorithms make use of the function $\kappa\colon X \times X \to \mathbb{C}$ given by

$$\kappa(x, y) = \langle \varphi(x), \varphi(y) \rangle, \qquad x, y \in X,$$

called a *kernel function*.

The kernel trick is to pretend that we have a feature embedding $\varphi\colon X \to F$ (actually unknown), but to only use inner products $\langle \varphi(x), \varphi(y) \rangle$ that can be evaluated using the original data through the known kernel function κ. It turns out that the functions of the form κ as above can be defined in terms of a condition which is reminiscent of positive semidefinite matrices (see Definition 17.2). Furthermore, every function satisfying Definition 17.2 arises from a suitable feature map into a Hilbert space; see Theorem 17.1.

We illustrate the kernel methods on kernel PCA (see Section 17.4).

In Chapter 18 we return to the problem of separating two disjoint sets of points, $\{u_i\}_{i=1}^p$ and $\{v_j\}_{j=1}^q$, but this time we do not assume that these two sets are separable. To cope with nonseparability, we allow points to invade the safety zone around the separating hyperplane, and even points on the wrong side of the hyperplane. Such a method is called soft margin support vector machine. We discuss variations of this method, including ν-SV classification. In each case we present a careful derivation of the dual and we explain how to solve it using ADMM. We prove rigorous results about the existence of support vectors.

In Chapter 19 we discuss linear regression. This problem can be cast as a learning problem. We observe a sequence of (distinct) pairs $((x_1, y_1), \ldots, (x_m, y_m))$ called a *set of training data*, where $x_i \in \mathbb{R}^n$ and $y_i \in \mathbb{R}$, viewed as input-output pairs of some unknown function f that we are trying to infer. The simplest kind of function is a linear function $f(x) = x^\top w$, where $w \in \mathbb{R}^n$ is a vector of coefficients usually called a *weight vector*. Since the problem is overdetermined and since our observations may be subject to errors, we can't solve for w exactly as the solution of the system $Xw = y$, so instead we solve the least-squares problem of minimizing $\|Xw - y\|_2^2$, where X is the $m \times n$ matrix whose *rows* are the row vectors $x_i^\top$. In general there are still infinitely many solutions so we add a regularizing term. If we add the term $K \|w\|_2^2$ to the objective function $J(w) = \|Xw - y\|_2^2$, then we have *ridge regression*. This problem is discussed in Section 19.1.

We derive the dual program. The dual has a unique solution which yields a solution of the primal. However, the solution of the dual is given in terms of the matrix $XX^\top$ (whereas the solution of the primal is given in terms of $X^\top X$), and since our data points x_i are represented by the rows of the matrix X, we see that this solution only involves inner products of the x_i. This observation is the core of the idea of kernel functions, which we introduce. We also explain how to solve the problem of learning an affine function $f(x) = x^\top w + b$.

In general the vectors w produced by ridge regression have few zero entries. In practice it is highly desirable to obtain sparse solutions, that is, vectors w with many components equal to zero. This can be achieved by replacing the regularizing term $K \|w\|_2^2$ by the regularizing term $K \|w\|_1$; that is, to use the ℓ^1-norm instead of the ℓ^2-norm; see Section 19.4. This method has the exotic name of *lasso regression*. This time there is no closed-form solution, but this is a convex optimization problem and there are efficient iterative methods to solve it. We show that ADMM provides an efficient solution.

Lasso has some undesirable properties, in particular when the dimension of the data is much larger than the number of data. In order to alleviate these problems, *elastic net regression* penalizes w with *both* an ℓ^2 regularizing term $K \|w\|_2^2$ and an ℓ^1 regularizing term $\tau \|w\|_1$. The method of elastic net blends ridge regression and lasso and attempts to retain their best properties; see Section 19.6. It can also be solved using ADMM but it appears to be much slower than lasso when K is small and the dimension of the data is much larger than the number of data.

In Chapter 20 we present ν-SV Regression. This method is designed in the same spirit as soft margin SVM, in the sense that it allows a margin of error. Here the errors are penalized in the ℓ^1-sense. We discuss several variations of the method and show how to solve them using ADMM. We present a careful derivation of the dual and discuss the existence of support vectors.

PART 1
Preliminaries for Optimization Theory

Chapter 2

Topology

This chapter contains a review of basic topological concepts. First metric spaces are defined. Next normed vector spaces are defined. Closed and open sets are defined, and their basic properties are stated. The general concept of a topological space is defined. The closure and the interior of a subset are defined. The subspace topology and the product topology are defined. Continuous maps and homeomorphisms are defined. Limits of sequences are defined. Continuous linear maps and multilinear maps are defined and studied briefly. Cauchy sequences and complete metric spaces are defined. We prove that every metric space can be embedded in a complete metric space called its completion. A complete normed vector space is called a *Banach space*. We prove that every normed vector space can be embedded in a complete normed vector space. We conclude with the contraction mapping theorem in a complete metric space.

2.1 Metric Spaces and Normed Vector Spaces

Most spaces considered in this book have a topological structure given by a metric or a norm, and we first review these notions. We begin with metric spaces. Recall that $\mathbb{R}_+ = \{x \in \mathbb{R} \mid x \geq 0\}$.

Definition 2.1. A *metric space* is a set E together with a function $d\colon E \times E \to \mathbb{R}_+$, called a *metric, or distance*, assigning a nonnegative real number $d(x, y)$ to any two points $x, y \in E$, and satisfying the following conditions for all $x, y, z \in E$:

(D1) $d(x, y) = d(y, x)$. (symmetry)
(D2) $d(x, y) \geq 0$, and $d(x, y) = 0$ iff $x = y$. (positivity)
(D3) $d(x, z) \leq d(x, y) + d(y, z)$. (triangle inequality)

Geometrically, Condition (D3) expresses the fact that in a triangle with vertices x, y, z, the length of any side is bounded by the sum of the lengths of the other two sides. From (D3), we immediately get

$$|d(x,\, y) - d(y,\, z)| \leq d(x,\, z).$$

Let us give some examples of metric spaces. Recall that the *absolute value* $|x|$ of a real number $x \in \mathbb{R}$ is defined such that $|x| = x$ if $x \geq 0$, $|x| = -x$ if $x < 0$, and for a complex number $x = a + ib$, by $|x| = \sqrt{a^2 + b^2}$.

Example 2.1.

(1) Let $E = \mathbb{R}$, and $d(x,\, y) = |x - y|$, the absolute value of $x - y$. This is the so-called natural metric on $\mathbb{R}$.

(2) Let $E = \mathbb{R}^n$ (or $E = \mathbb{C}^n$). We have the *Euclidean metric*

$$d_2(x,\, y) = \left(|x_1 - y_1|^2 + \cdots + |x_n - y_n|^2\right)^{\frac{1}{2}},$$

the distance between the points $(x_1, \ldots, x_n)$ and $(y_1, \ldots, y_n)$.

(3) For every set E, we can define the *discrete metric*, defined such that $d(x,\, y) = 1$ iff $x \neq y$, and $d(x,\, x) = 0$.

(4) For any $a, b \in \mathbb{R}$ such that $a < b$, we define the following sets:
$[a, b] = \{x \in \mathbb{R} \mid a \leq x \leq b\}$, (closed interval)
$(a, b) = \{x \in \mathbb{R} \mid a < x < b\}$, (open interval)
$[a, b) = \{x \in \mathbb{R} \mid a \leq x < b\}$, (interval closed on the left, open on the right)
$(a, b] = \{x \in \mathbb{R} \mid a < x \leq b\}$, (interval open on the left, closed on the right)
Let $E = [a, b]$, and $d(x,\, y) = |x - y|$. Then $([a, b], d)$ is a metric space.

We will need to define the notion of proximity in order to define convergence of limits and continuity of functions. For this we introduce some standard "small neighborhoods."

Definition 2.2. Given a metric space E with metric d, for every $a \in E$, for every $\rho \in \mathbb{R}$, with $\rho > 0$, the set

$$B(a, \rho) = \{x \in E \mid d(a,\, x) \leq \rho\}$$

is called the *closed ball of center a and radius ρ*, the set

$$B_0(a, \rho) = \{x \in E \mid d(a,\, x) < \rho\}$$

is called the *open ball of center a and radius ρ*, and the set

$$S(a,\, \rho) = \{x \in E \mid d(a,\, x) = \rho\}$$

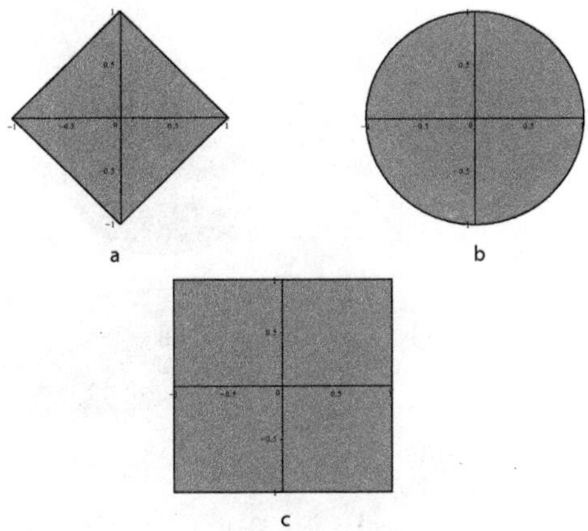

Fig. 2.1 Figure a shows the diamond shaped closed ball associated with $\| \|_1$. Figure b shows the closed unit disk associated with $\| \|_2$, while Figure c illustrates the closed unit ball associated with $\| \|_\infty$.

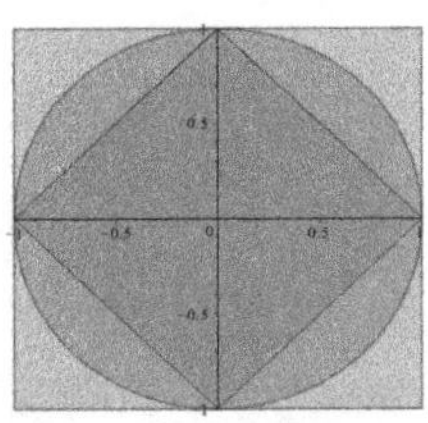

Fig. 2.2 The relationship between the closed unit balls centered at $(0, 0)$.

The set E itself is open, since for every $a \in E$, every open ball of center a is contained in E. In $E = \mathbb{R}^n$, given n intervals $[a_i, b_i]$, with $a_i < b_i$, it is easy to show that the open n-cube

$$\{(x_1, \ldots, x_n) \in E \mid a_i < x_i < b_i, \, 1 \leq i \leq n\}$$

is an open set. In fact, it is possible to find a metric for which such open n-cubes are open balls! Similarly, we can define the closed n-cube

$$\{(x_1, \ldots, x_n) \in E \mid a_i \leq x_i \leq b_i, \, 1 \leq i \leq n\},$$

which is a closed set.

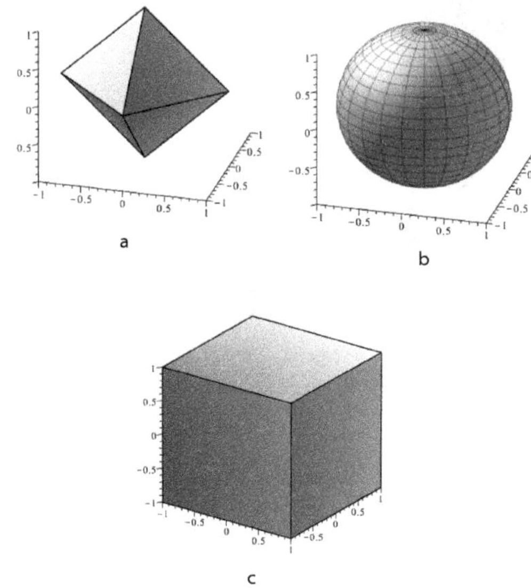

Fig. 2.3 Figure a shows the octahedral shaped closed ball associated with $\| \|_1$. Figure b shows the closed spherical associated with $\| \|_2$, while Figure c illustrates the closed unit ball associated with $\| \|_\infty$.

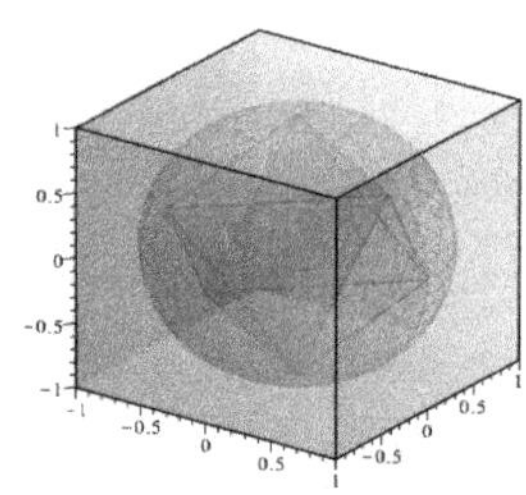

Fig. 2.4 The relationship between the closed unit balls centered at $(0, 0, 0)$.

The open sets satisfy some important properties that lead to the definition of a topological space.

Proposition 2.1. *Given a metric space E with metric d, the family $\mathcal{O}$ of all open sets defined in Definition 2.4 satisfies the following properties:*

(O1) For every finite family $(U_i)_{1 \leq i \leq n}$ of sets $U_i \in \mathcal{O}$, we have $U_1 \cap \cdots \cap$

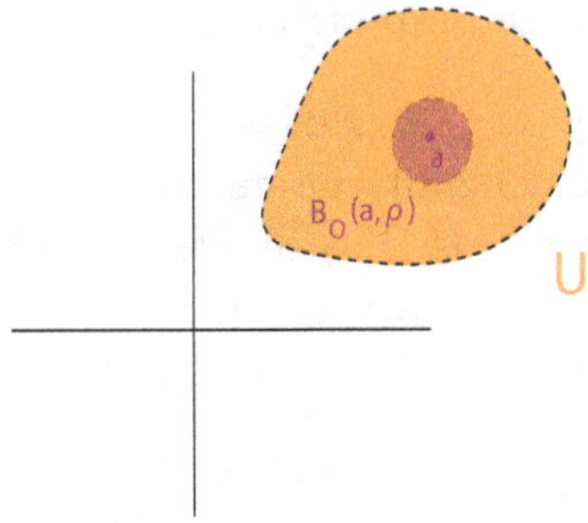

Fig. 2.5 An open set U in $E = \mathbb{R}^2$ under the standard Euclidean metric. Any point in the peach set U is surrounded by a small raspberry open set which lies within U.

$U_n \in \mathcal{O}$, i.e., $\mathcal{O}$ is closed under finite intersections.

(O2) For every arbitrary family $(U_i)_{i\in I}$ of sets $U_i \in \mathcal{O}$, we have $\bigcup_{i\in I} U_i \in \mathcal{O}$, i.e., $\mathcal{O}$ is closed under arbitrary unions.

(O3) $\emptyset \in \mathcal{O}$, and $E \in \mathcal{O}$, i.e., $\emptyset$ and E belong to $\mathcal{O}$.

Furthermore, for any two distinct points $a \neq b$ in E, there exist two open sets U_a and U_b such that, $a \in U_a$, $b \in U_b$, and $U_a \cap U_b = \emptyset$.

Proof. It is straightforward. For the last point, letting $\rho = d(a, b)/3$ (in fact $\rho = d(a, b)/2$ works too), we can pick $U_a = B_0(a, \rho)$ and $U_b = B_0(b, \rho)$. By the triangle inequality, we must have $U_a \cap U_b = \emptyset$. □

The above proposition leads to the very general concept of a topological space.

One should be careful that, in general, the family of open sets is not closed under infinite intersections. For example, in $\mathbb{R}$ under the metric $|x-y|$, letting $U_n = (-1/n, +1/n)$, each U_n is open, but $\bigcap_n U_n = \{0\}$, which is not open.

2.2 Topological Spaces

Motivated by Proposition 2.1, a topological space is defined in terms of a family of sets satisfying the properties of open sets stated in that proposition.

Definition 2.5. Given a set E, a *topology on E (or a topological structure on E)*, is defined as a family $\mathcal{O}$ of subsets of E called *open sets*, and satisfying the following three properties:

(1) For every finite family $(U_i)_{1 \leq i \leq n}$ of sets $U_i \in \mathcal{O}$, we have $U_1 \cap \cdots \cap U_n \in \mathcal{O}$, *i.e.*, $\mathcal{O}$ is closed under finite intersections.
(2) For every arbitrary family $(U_i)_{i \in I}$ of sets $U_i \in \mathcal{O}$, we have $\bigcup_{i \in I} U_i \in \mathcal{O}$, *i.e.*, $\mathcal{O}$ is closed under arbitrary unions.
(3) $\emptyset \in \mathcal{O}$, and $E \in \mathcal{O}$, i.e., $\emptyset$ and E belong to $\mathcal{O}$.

A set E together with a topology $\mathcal{O}$ on E is called a *topological space*. Given a topological space $(E, \mathcal{O})$, a subset F of E is a *closed set* if $F = E - U$ for some open set $U \in \mathcal{O}$, i.e., F is the complement of some open set.

It is possible that an open set is also a closed set. For example, $\emptyset$ and E are both open and closed.

Definition 2.6. When a topological space contains a proper nonempty subset U which is both open and closed, the space E is said to be *disconnected*.

By taking complements, we can state properties of the closed sets dual to those of Definition 2.5. If we denote the family of closed sets of E as $\mathcal{F} = \{F \subseteq E \mid E - F \in \mathcal{O}\}$, then the closed sets satisfy the following properties:

(1) For every finite family $(F_i)_{1 \leq i \leq n} \in \mathcal{F}$, we have $F_1 \cup \cdots \cup F_n \in \mathcal{F}$, *i.e.*, $\mathcal{F}$ is closed under finite unions.
(2) For every arbitrary family $(F_i)_{i \in I}$ of sets $F_i \in \mathcal{F}$, we have $\bigcap_{i \in I} F_i \in \mathcal{F}$, *i.e.*, $\mathcal{F}$ is closed under arbitrary intersections.
(3) $\emptyset \in \mathcal{F}$, and $E \in \mathcal{F}$, i.e., $\emptyset$ and E belong to $\mathcal{F}$.

One of the reasons why topological spaces are important is that the definition of a topology only involves a certain family $\mathcal{O}$ of sets, and not **how** such family is generated from a metric or a norm. For example, different metrics or different norms can define the same family of open sets. Many topological properties only depend on the family $\mathcal{O}$ and not on the specific metric or norm. But the fact that a topology is definable from a metric or a norm is important, because it usually implies nice properties of a space. *All our examples will be spaces whose topology is defined by a metric or a norm.*

Definition 2.7. A topological space $(E, \mathcal{O})$ is said to satisfy the *Hausdorff separation axiom (or T_2-separation axiom)* if for any two distinct points

$a \neq b$ in E, there exist two open sets U_a and U_b such that, $a \in U_a$, $b \in U_b$, and $U_a \cap U_b = \emptyset$. When the T_2-separation axiom is satisfied, we also say that $(E, \mathcal{O})$ is a *Hausdorff space*.

As shown by Proposition 2.1, any metric space is a topological Hausdorff space, the family of open sets being in fact the family of arbitrary unions of open balls. Similarly, any normed vector space is a topological Hausdorff space, the family of open sets being the family of arbitrary unions of open balls. The topology $\mathcal{O}$ consisting of all subsets of E is called the *discrete topology*.

Remark: Most (if not all) spaces used in analysis are Hausdorff spaces. Intuitively, the Hausdorff separation axiom says that there are enough "small" open sets. Without this axiom, some counter-intuitive behaviors may arise. For example, a sequence may have more than one limit point (or a compact set may not be closed). Nevertheless, non-Hausdorff topological spaces arise naturally in algebraic geometry. But even there, some substitute for separation is used.

It is also worth noting that the Hausdorff separation axiom implies the following property.

Proposition 2.2. *If a topological space $(E, \mathcal{O})$ is Hausdorff, then for every $a \in E$, the set $\{a\}$ is closed.*

Proof. If $x \in E - \{a\}$, then $x \neq a$, and so there exist open sets U_a and U_x such that $a \in U_a$, $x \in U_x$, and $U_a \cap U_x = \emptyset$. See Figure 2.6. Thus, for every $x \in E - \{a\}$, there is an open set U_x containing x and contained in $E - \{a\}$, showing by (O3) that $E - \{a\}$ is open, and thus that the set $\{a\}$ is closed. $\square$

Given a topological space $(E, \mathcal{O})$, given any subset A of E, since $E \in \mathcal{O}$ and E is a closed set, the family $\mathcal{C}_A = \{F \mid A \subseteq F,\ F \text{ a closed set}\}$ of closed sets containing A is nonempty, and since any arbitrary intersection of closed sets is a closed set, the intersection $\bigcap \mathcal{C}_A$ of the sets in the family $\mathcal{C}_A$ is the smallest closed set containing A. By a similar reasoning, the union of all the open subsets contained in A is the largest open set contained in A.

Definition 2.8. Given a topological space $(E, \mathcal{O})$, given any subset A of E, the smallest closed set containing A is denoted by $\overline{A}$, and is called the *closure, or adherence of A*. See Figure 2.7. A subset A of E is *dense in E* if $\overline{A} = E$. The largest open set contained in A is denoted by $\overset{\circ}{A}$, and is

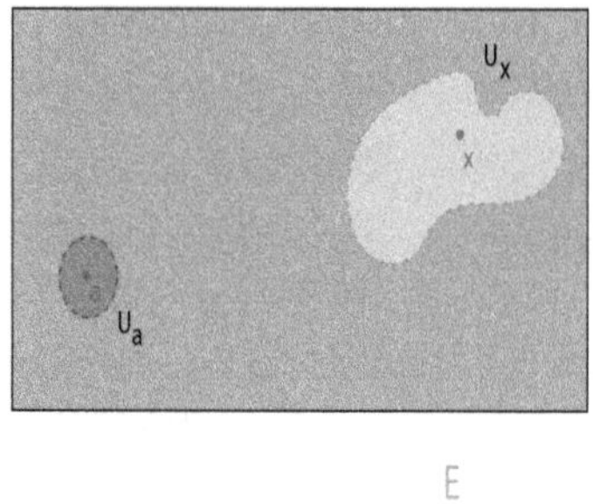

Fig. 2.6 A schematic illustration of the Hausdorff separation property.

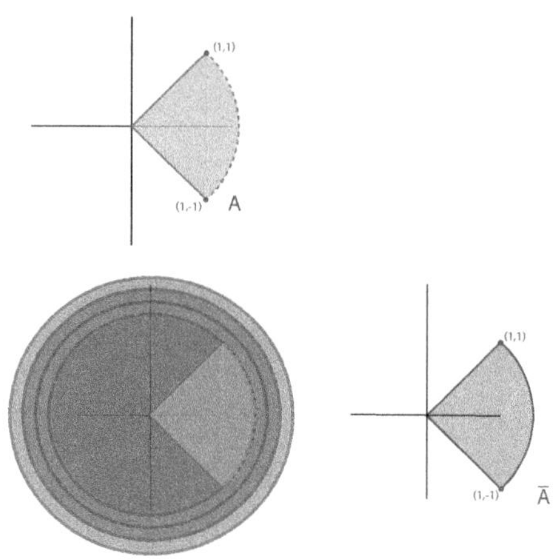

Fig. 2.7 The topological space $(E, \mathcal{O})$ is $\mathbb{R}^2$ with topology induced by the Euclidean metric. The subset A is the section $B_0(1)$ in the first and fourth quadrants bound by the lines $y = x$ and $y = -x$. The closure of A is obtained by the intersection of A with the closed unit ball.

called the *interior of* A. See Figure 2.8. The set $\operatorname{Fr} A = \overline{A} \cap \overline{E - A}$ is called the *boundary (or frontier) of* A. We also denote the boundary of A by ∂A. See Figure 2.9.

Remark: The notation $\overline{A}$ for the closure of a subset A of E is somewhat unfortunate, since $\overline{A}$ is often used to denote the set complement of A in E.

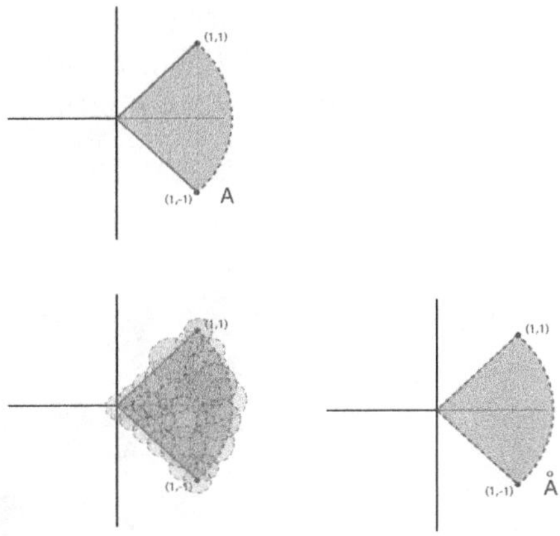

Fig. 2.8 The topological space $(E, \mathcal{O})$ is $\mathbb{R}^2$ with topology induced by the Euclidean metric. The subset A is the section $B_0(1)$ in the first and fourth quadrants bound by the lines $y = x$ and $y = -x$. The interior of A is obtained by the covering A with small open balls.

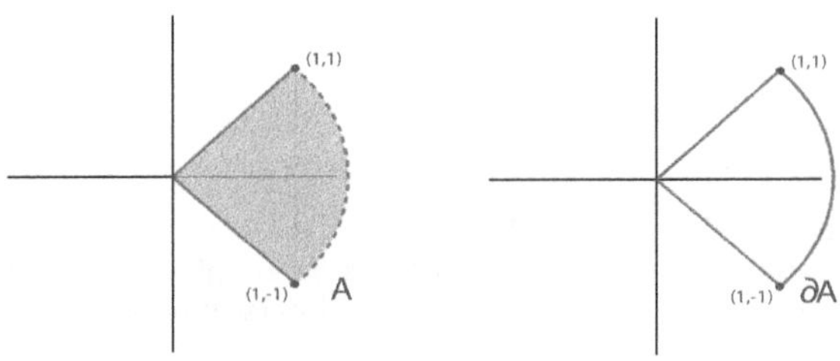

Fig. 2.9 The topological space $(E, \mathcal{O})$ is $\mathbb{R}^2$ with topology induced by the Euclidean metric. The subset A is the section $B_0(1)$ in the first and fourth quadrants bound by the lines $y = x$ and $y = -x$. The boundary of A is $\overline{A} - \mathring{A}$.

Still, we prefer it to more cumbersome notations such as clo(A), and we denote the complement of A in E by $E - A$ (or sometimes, A^c).

By definition, it is clear that a subset A of E is closed iff $A = \overline{A}$. The set $\mathbb{Q}$ of rationals is dense in $\mathbb{R}$. It is easily shown that $\overline{A} = \mathring{A} \cup \partial A$ and $\mathring{A} \cap \partial A = \emptyset$.

Another useful characterization of $\overline{A}$ is given by the following proposition.

Proposition 2.3. *Given a topological space $(E, \mathcal{O})$, given any subset A of E, the closure $\overline{A}$ of A is the set of all points $x \in E$ such that for every open set U containing x, then $U \cap A \neq \emptyset$. See Figure 2.10.*

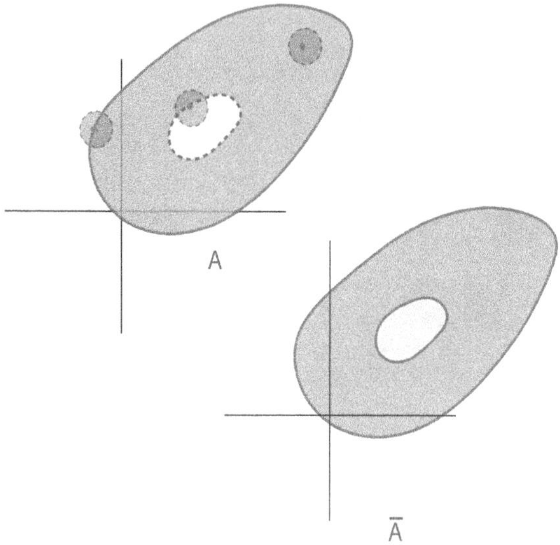

Fig. 2.10 The topological space $(E, \mathcal{O})$ is $\mathbb{R}^2$ with topology induced by the Euclidean metric. The purple subset A is illustrated with three red points, each in its closure since the open ball centered at each point has nontrivial intersection with A.

Proof. If $A = \emptyset$, since $\emptyset$ is closed, the proposition holds trivially. Thus assume that $A \neq \emptyset$. First assume that $x \in \overline{A}$. Let U be any open set such that $x \in U$. If $U \cap A = \emptyset$, since U is open, then $E - U$ is a closed set containing A, and since $\overline{A}$ is the intersection of all closed sets containing A, we must have $x \in E - U$, which is impossible. Conversely, assume that $x \in E$ is a point such that for every open set U containing x, $U \cap A \neq \emptyset$. Let F be any closed subset containing A. If $x \notin F$, since F is closed, then $U = E - F$ is an open set such that $x \in U$, and $U \cap A = \emptyset$, a contradiction. Thus, we have $x \in F$ for every closed set containing A, that is, $x \in \overline{A}$. $\square$

Often it is necessary to consider a subset A of a topological space E, and to view the subset A as a topological space.

2.3 Subspace and Product Topologies

The following proposition shows how to define a topology on a subset.

Proposition 2.4. *Given a topological space $(E, \mathcal{O})$, given any subset A of E, let*

$$\mathcal{U} = \{U \cap A \mid U \in \mathcal{O}\}$$

be the family of all subsets of A obtained as the intersection of any open set in $\mathcal{O}$ with A. The following properties hold.

(1) The space $(A, \mathcal{U})$ is a topological space.
(2) If E is a metric space with metric d, then the restriction $d_A \colon A \times A \to \mathbb{R}_+$ of the metric d to A defines a metric space. Furthermore, the topology induced by the metric d_A agrees with the topology defined by $\mathcal{U}$, as above.

Proof. Left as an exercise. □

Proposition 2.4 suggests the following definition.

Definition 2.9. Given a topological space $(E, \mathcal{O})$, given any subset A of E, the *subspace topology on A induced by $\mathcal{O}$* is the family $\mathcal{U}$ of open sets defined such that

$$\mathcal{U} = \{U \cap A \mid U \in \mathcal{O}\}$$

is the family of all subsets of A obtained as the intersection of any open set in $\mathcal{O}$ with A. We say that $(A, \mathcal{U})$ has the *subspace topology*. If (E, d) is a metric space, the restriction $d_A \colon A \times A \to \mathbb{R}_+$ of the metric d to A is called the *subspace metric*.

For example, if $E = \mathbb{R}^n$ and d is the Euclidean metric, we obtain the subspace topology on the closed n-cube

$$\{(x_1, \ldots, x_n) \in E \mid a_i \leq x_i \leq b_i, 1 \leq i \leq n\}.$$

See Figure 2.11.

One should realize that every open set $U \in \mathcal{O}$ which is entirely contained in A is also in the family $\mathcal{U}$, but $\mathcal{U}$ may contain open sets that are not in $\mathcal{O}$. For example, if $E = \mathbb{R}$ with $|x - y|$, and $A = [a, b]$, then sets of the form $[a, c)$, with $a < c < b$ belong to $\mathcal{U}$, but they are not open sets for $\mathbb{R}$ under $|x - y|$. However, there is agreement in the following situation.

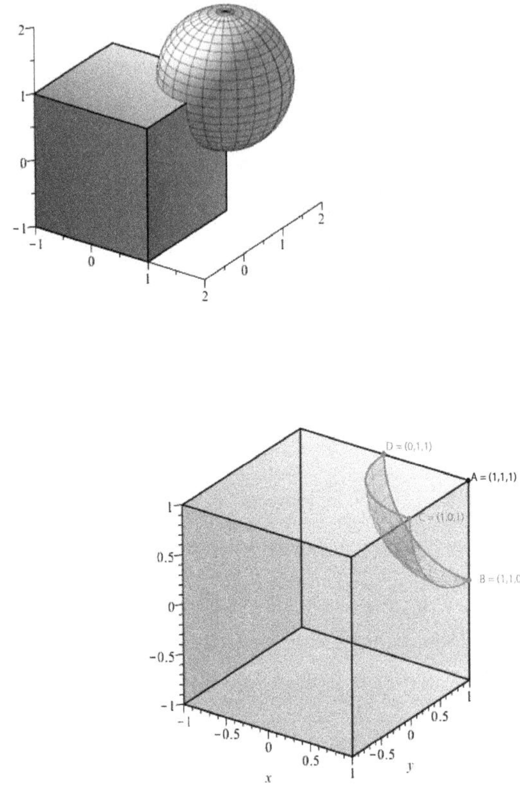

Fig. 2.11 An example of an open set in the subspace topology for $\{(x, y, z) \in \mathbb{R}^3 \mid -1 \leq x \leq 1, -1 \leq y \leq 1, -1 \leq z \leq 1\}$. The open set is the corner region $ABCD$ and is obtained by intersection the cube $B_0((1, 1, 1), 1)$.

Proposition 2.5. *Given a topological space $(E, \mathcal{O})$, given any subset A of E, if $\mathcal{U}$ is the subspace topology, then the following properties hold.*

(1) If A is an open set $A \in \mathcal{O}$, then every open set $U \in \mathcal{U}$ is an open set $U \in \mathcal{O}$.
(2) If A is a closed set in E, then every closed set w.r.t. the subspace topology is a closed set w.r.t. $\mathcal{O}$.

Proof. Left as an exercise. □

The concept of product topology is also useful. We have the following proposition.

Proposition 2.6. *Given n topological spaces $(E_i, \mathcal{O}_i)$, let $\mathcal{B}$ be the family of subsets of $E_1 \times \cdots \times E_n$ defined as follows:*

$$\mathcal{B} = \{U_1 \times \cdots \times U_n \mid U_i \in \mathcal{O}_i,\ 1 \leq i \leq n\},$$

and let $\mathcal{P}$ be the family consisting of arbitrary unions of sets in $\mathcal{B}$, including $\emptyset$. Then $\mathcal{P}$ is a topology on $E_1 \times \cdots \times E_n$.

Proof. Left as an exercise. □

Definition 2.10. Given n topological spaces $(E_i, \mathcal{O}_i)$, the *product topology on $E_1 \times \cdots \times E_n$* is the family $\mathcal{P}$ of subsets of $E_1 \times \cdots \times E_n$ defined as follows: if

$$\mathcal{B} = \{U_1 \times \cdots \times U_n \mid U_i \in \mathcal{O}_i,\ 1 \leq i \leq n\},$$

then $\mathcal{P}$ is the family consisting of arbitrary unions of sets in $\mathcal{B}$, including $\emptyset$. See Figure 2.12.

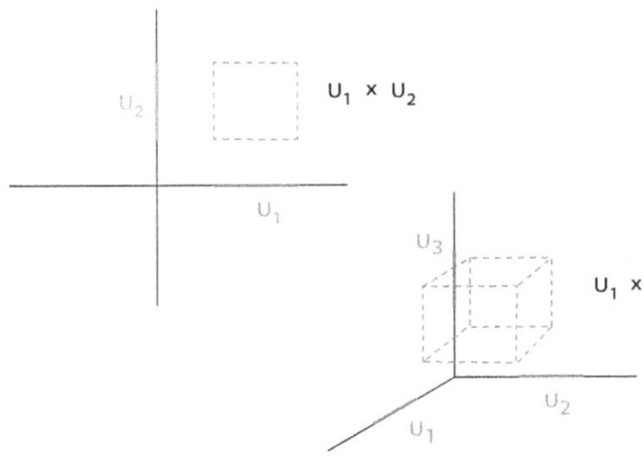

Fig. 2.12 Examples of open sets in the product topology for $\mathbb{R}^2$ and $\mathbb{R}^3$ induced by the Euclidean metric.

If each (E_i, d_{E_i}) is a metric space, there are three natural metrics that can be defined on $E_1 \times \cdots \times E_n$:

$$d_1((x_1, \ldots, x_n), (y_1, \ldots, y_n)) = d_{E_1}(x_1, y_1) + \cdots + d_{E_n}(x_n, y_n),$$

$$d_2((x_1, \ldots, x_n), (y_1, \ldots, y_n)) = \left((d_{E_1}(x_1, y_1))^2 + \cdots + (d_{E_n}(x_n, y_n))^2\right)^{\frac{1}{2}},$$

$$d_\infty((x_1, \ldots, x_n), (y_1, \ldots, y_n)) = \max\{d_{E_1}(x_1, y_1), \ldots, d_{E_n}(x_n, y_n)\}.$$

Proposition 2.7. *The following inequalities hold:*

$$\begin{aligned} d_\infty((x_1,\ldots,x_n),(y_1,\ldots,y_n)) &\le d_2((x_1,\ldots,x_n),(y_1,\ldots,y_n)) \\ &\le d_1((x_1,\ldots,x_n),(y_1,\ldots,y_n)) \\ &\le n d_\infty((x_1,\ldots,x_n),(y_1,\ldots,y_n)), \end{aligned}$$

so these distances define the same topology, which is the product topology.

If each $(E_i, \| \|_{E_i})$ is a normed vector space, there are three natural norms that can be defined on $E_1 \times \cdots \times E_n$:

$$\begin{aligned} \|(x_1,\ldots,x_n)\|_1 &= \|x_1\|_{E_1} + \cdots + \|x_n\|_{E_n}, \\ \|(x_1,\ldots,x_n)\|_2 &= \left(\|x_1\|_{E_1}^2 + \cdots + \|x_n\|_{E_n}^2\right)^{\frac{1}{2}}, \\ \|(x_1,\ldots,x_n)\|_\infty &= \max\left\{\|x_1\|_{E_1},\ldots,\|x_n\|_{E_n}\right\}. \end{aligned}$$

Proposition 2.8. *The following inequalities hold:*

$$\|(x_1,\ldots,x_n)\|_\infty \le \|(x_1,\ldots,x_n)\|_2 \le \|(x_1,\ldots,x_n)\|_1 \le n\|(x_1,\ldots,x_n)\|_\infty,$$

so these norms define the same topology, which is the product topology.

It can also be verified that when $E_i = \mathbb{R}$, with the standard topology induced by $|x - y|$, the topology product on $\mathbb{R}^n$ is the standard topology induced by the Euclidean norm.

Definition 2.11. Two metrics d_1 and d_2 on a space E are *equivalent* if they induce the same topology $\mathcal{O}$ on E (i.e., they define the same family $\mathcal{O}$ of open sets). Similarly, two norms $\| \|_1$ and $\| \|_2$ on a space E are *equivalent* if they induce the same topology $\mathcal{O}$ on E.

Given a topological space $(E, \mathcal{O})$, it is often useful, as in Proposition 2.6, to define the topology $\mathcal{O}$ in terms of a subfamily $\mathcal{B}$ of subsets of E.

Definition 2.12. Given a topological space $(E, \mathcal{O})$, we say that a family $\mathcal{B}$ of subsets of E is a *basis for the topology* $\mathcal{O}$, if $\mathcal{B}$ is a subset of $\mathcal{O}$, and if every open set U in $\mathcal{O}$ can be obtained as some union (possibly infinite) of sets in $\mathcal{B}$ (agreeing that the empty union is the empty set).

For example, given any metric space (E, d), $\mathcal{B} = \{B_0(a,\rho) \mid a \in E, \rho > 0\}$ is a basis for the topology. In particular, if $d = \| \|_2$, the open intervals form a basis for $\mathbb{R}$, while the open disks form a basis for $\mathbb{R}^2$. The open rectangles also form a basis for $\mathbb{R}^2$ with the standard topology.

It is immediately verified that if a family $\mathcal{B} = (U_i)_{i\in I}$ is a basis for the topology of $(E, \mathcal{O})$, then $E = \bigcup_{i\in I} U_i$, and the intersection of any two sets $U_i, U_j \in \mathcal{B}$ is the union of some sets in the family $\mathcal{B}$ (again, agreeing that the empty union is the empty set). Conversely, a family $\mathcal{B}$ with these properties is the basis of the topology obtained by forming arbitrary unions of sets in $\mathcal{B}$.

Definition 2.13. Given a topological space $(E, \mathcal{O})$, a *subbasis for* $\mathcal{O}$ is a family $\mathcal{S}$ of subsets of E, such that the family $\mathcal{B}$ of all finite intersections of sets in $\mathcal{S}$ (including E itself, in case of the empty intersection) is a basis of $\mathcal{O}$. See Figure 2.13.

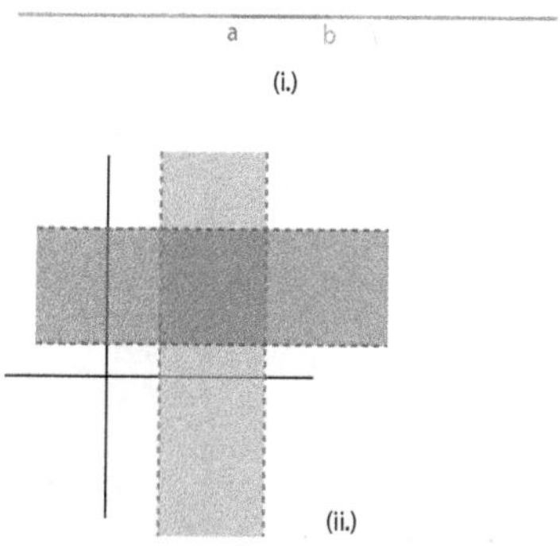

Fig. 2.13 Figure (*i.*) shows that the set of infinite open intervals forms a subbasis for $\mathbb{R}$. Figure (*ii.*) shows that the infinite open strips form a subbasis for $\mathbb{R}^2$.

The following proposition gives useful criteria for determining whether a family of open subsets is a basis of a topological space.

Proposition 2.9. *Given a topological space $(E, \mathcal{O})$ and a family $\mathcal{B}$ of open subsets in $\mathcal{O}$ the following properties hold:*

(1) *The family $\mathcal{B}$ is a basis for the topology $\mathcal{O}$ iff for every open set $U \in \mathcal{O}$ and every $x \in U$, there is some $B \in \mathcal{B}$ such that $x \in B$ and $B \subseteq U$. See Figure 2.14.*
(2) *The family $\mathcal{B}$ is a basis for the topology $\mathcal{O}$ iff*

(a) *For every $x \in E$, there is some $B \in \mathcal{B}$ such that $x \in B$.*

(b) *For any two open subsets, $B_1, B_2 \in \mathcal{B}$, for every $x \in E$, if $x \in B_1 \cap B_2$, then there is some $B_3 \in \mathcal{B}$ such that $x \in B_3$ and $B_3 \subseteq B_1 \cap B_2$. See Figure 2.15.*

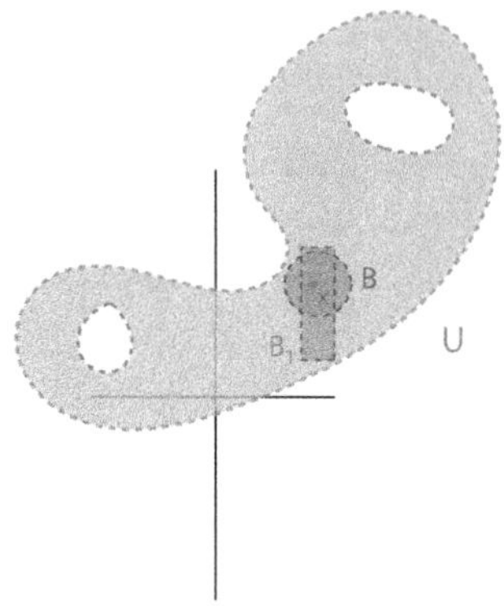

Fig. 2.14 Given an open subset U of $\mathbb{R}^2$ and $x \in U$, there exists an open ball B containing x with $B \subset U$. There also exists an open rectangle B_1 containing x with $B_1 \subset U$.

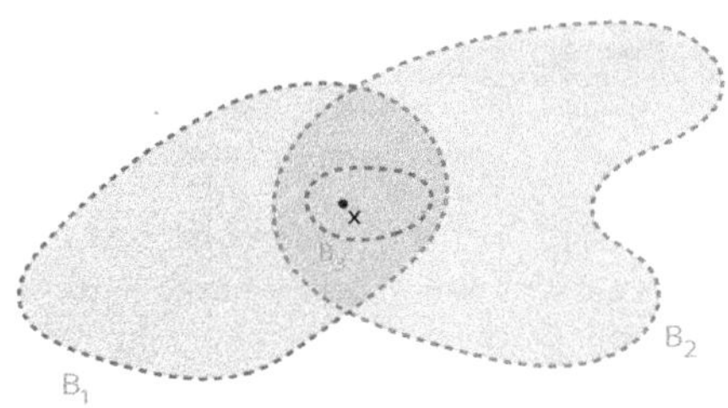

Fig. 2.15 A schematic illustration of Condition (b) in Proposition 2.9.

We now consider the fundamental property of continuity.

2.4 Continuous Functions

Definition 2.14. Let $(E, \mathcal{O}_E)$ and $(F, \mathcal{O}_F)$ be topological spaces, and let $f \colon E \to F$ be a function. For every $a \in E$, we say that f *is continuous at* a, if for every open set $V \in \mathcal{O}_F$ containing $f(a)$, there is some open set

$U \in \mathcal{O}_E$ containing a, such that, $f(U) \subseteq V$. See Figure 2.16. We say that f *is continuous* if it is continuous at every $a \in E$.

If $(E, \mathcal{O}_E)$ and $(F, \mathcal{O}_F)$ are topological spaces, and $f\colon E \to F$ is a function, for every nonempty subset $A \subseteq E$ of E, we say that f *is continuous on* A if the restriction of f to A is continuous with respect to $(A, \mathcal{U})$ and $(F, \mathcal{O}_F)$, where $\mathcal{U}$ is the subspace topology induced by $\mathcal{O}_E$ on A.

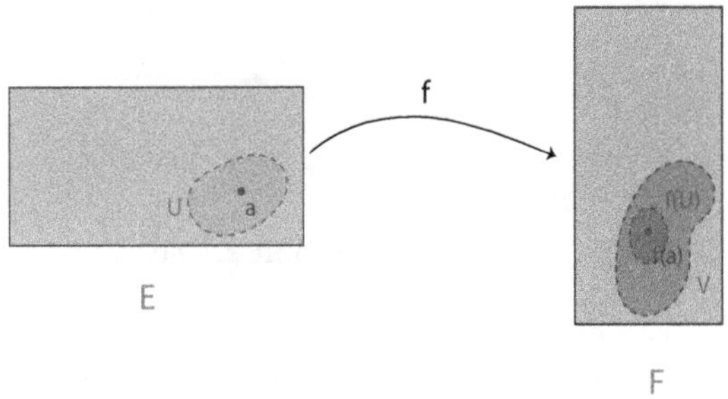

Fig. 2.16 A schematic illustration of Definition 2.14.

Definition 2.15. Let $(E, \mathcal{O}_E)$ be a topological space. Define a *neighborhood of* $a \in E$ as any subset N of E containing some open set $O \in \mathcal{O}$ such that $a \in O$.

Now if f is continuous at a and N is any neighborhood of $f(a)$, there is some open set $V \subseteq N$ containing $f(a)$, and since f is continuous at a, there is some open set U containing a, such that $f(U) \subseteq V$. Since $V \subseteq N$, the open set U is a subset of $f^{-1}(N)$ containing a, and $f^{-1}(N)$ is a neighborhood of a. Conversely, if $f^{-1}(N)$ is a neighborhood of a whenever N is any neighborhood of $f(a)$, it is immediate that f is continuous at a. See Figure 2.17.

It is easy to see that Definition 2.14 is equivalent to the following statements.

Proposition 2.10. *Let $(E, \mathcal{O}_E)$ and $(F, \mathcal{O}_F)$ be topological spaces, and let $f\colon E \to F$ be a function. For every $a \in E$, the function f is continuous at $a \in E$ iff for every neighborhood N of $f(a) \in F$, then $f^{-1}(N)$ is a neighborhood of a. The function f is continuous on E iff $f^{-1}(V)$ is an open set in $\mathcal{O}_E$ for every open set $V \in \mathcal{O}_F$.*

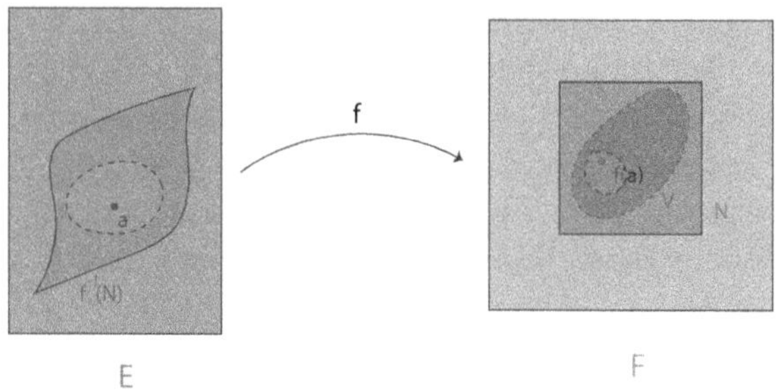

Fig. 2.17 A schematic illustration of the neighborhood condition.

If E and F are metric spaces, Proposition 2.10 can be restated as follows.

Proposition 2.11. *Let E and F be metric spaces defined by metrics d_1 and d_2. The function $f\colon E \to F$ is continuous at $a \in E$ iff for every $\epsilon > 0$, there is some $\eta > 0$ such that for every $x \in E$,*

$$\text{if } d_1(a, x) \leq \eta, \text{ then } d_2(f(a), f(x)) \leq \epsilon.$$

If E and F are normed vector spaces, Proposition 2.10 can be restated as follows.

Proposition 2.12. *Let E and F be normed vector spaces defined by norms $\| \, \|_1$ and $\| \, \|_2$. The function $f\colon E \to F$ is continuous at $a \in E$ iff for every $\epsilon > 0$, there is some $\eta > 0$ such that for every $x \in E$,*

$$\text{if } \|x - a\|_1 \leq \eta, \text{ then } \|f(x) - f(a)\|_2 \leq \epsilon.$$

It is worth noting that continuity is a topological notion, in the sense that equivalent metrics (or equivalent norms) define exactly the same notion of continuity.

An important example of a continuous function is the distance function in a metric space. One can show that in a metric space (E, d), the distance $d\colon E \times E \to \mathbb{R}$ is continuous, where $E \times E$ has the product topology. By the triangle inequality, we have

$$d(x, y) \leq d(x, x_0) + d(x_0, y_0) + d(y_0, y) = d(x_0, y_0) + d(x_0, x) + d(y_0, y)$$

and

$$d(x_0, y_0) \leq d(x_0, x) + d(x, y) + d(y, y_0) = d(x, y) + d(x_0, x) + d(y_0, y).$$

Consequently,

$$|d(x,y) - d(x_0,y_0)| \leq d(x_0,x) + d(y_0,y),$$

which proves that d is continuous at (x_0, y_0). In fact this shows that d is uniformly continuous; see Definition 2.21.

Similarly, for a normed vector space $(E, \| \|)$, the norm $\| \|: E \to \mathbb{R}$ is (uniformly) continuous.

Another important example of a continuous function is the projection of a product space. Given a product $E_1 \times \cdots \times E_n$ of topological spaces, as usual, we let $\pi_i: E_1 \times \cdots \times E_n \to E_i$ be the projection function such that, $\pi_i(x_1, \ldots, x_n) = x_i$. It is immediately verified that each π_i is continuous.

Definition 2.16. Given a topological space $(E, \mathcal{O})$, we say that a point $a \in E$ is *isolated* if $\{a\}$ is an open set in $\mathcal{O}$.

If $(E, \mathcal{O}_E)$ and $(F, \mathcal{O}_F)$ are topological spaces, any function $f: E \to F$ is continuous at every isolated point $a \in E$. In the discrete topology, every point is isolated.

As the following proposition shows, isolated points *do not* occur in nontrivial metric spaces.

Proposition 2.13. *In a nontrivial normed vector space* $(E, \| \|)$ *(with* $E \neq \{0\}$*), no point is isolated.*

Proof. To show this, we show that every open ball $B_0(u, \rho,)$ contains some vectors different from u. Indeed, since E is nontrivial, there is some $v \in E$ such that $v \neq 0$, and thus $\lambda = \|v\| > 0$ (by (N1)). Let

$$w = u + \frac{\rho}{\lambda + 1} v.$$

Since $v \neq 0$ and $\rho > 0$, we have $w \neq u$. Then,

$$\|w - u\| = \left\| \frac{\rho}{\lambda + 1} v \right\| = \frac{\rho \lambda}{\lambda + 1} < \rho,$$

which shows that $\|w - u\| < \rho$, for $w \neq u$. □

The following proposition shows that composition behaves well with respect to continuity.

Proposition 2.14. *Given topological spaces* $(E, \mathcal{O}_E)$, $(F, \mathcal{O}_F)$, *and* $(G, \mathcal{O}_G)$, *and two functions* $f: E \to F$ *and* $g: F \to G$, *if* f *is continuous at* $a \in E$ *and* g *is continuous at* $f(a) \in F$, *then* $g \circ f: E \to G$ *is continuous at* $a \in E$. *Given* n *topological spaces* $(F_i, \mathcal{O}_i)$, *for every function* $f: E \to F_1 \times \cdots \times F_n$, *then* f *is continuous at* $a \in E$ *iff every* $f_i: E \to F_i$ *is continuous at* a, *where* $f_i = \pi_i \circ f$.

Given a function $f\colon E_1\times\cdots\times E_n \to F$, we can fix $n-1$ of the arguments, say $a_1,\ldots,a_{i-1},a_{i+1},\ldots,a_n$, and view f as a function of the remaining argument,

$$x_i \mapsto f(a_1,\ldots,a_{i-1},x_i,a_{i+1},\ldots,a_n),$$

where $x_i \in E_i$. If f is continuous, it is clear that each f_i is continuous.

One should be careful that the converse is false! For example, consider the function $f\colon \mathbb{R}\times\mathbb{R}\to\mathbb{R}$, defined such that,

$$f(x,y)=\frac{xy}{x^2+y^2}\quad \text{if } (x,y)\neq(0,0), \quad\text{and}\quad f(0,0)=0.$$

The function f is continuous on $\mathbb{R}\times\mathbb{R}-\{(0,0)\}$, but on the line $y=mx$, with $m\neq 0$, we have $f(x,y)=\frac{m}{1+m^2}\neq 0$, and thus, on this line, $f(x,y)$ does not approach 0 when (x,y) approaches $(0,0)$. See Figure 2.18.

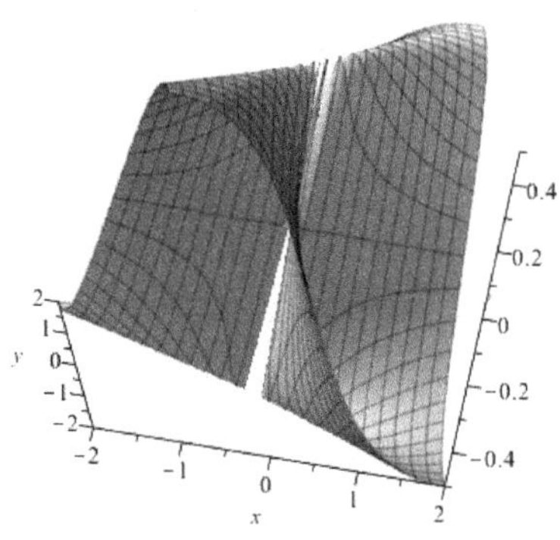

Fig. 2.18 The graph of $f(x,y)=\frac{xy}{x^2+y^2}$ for $(x,y)\neq(0,0)$. The bottom of this graph, which shows the approach along the line $y=-x$, does not have a z value of 0.

The following proposition is useful for showing that real-valued functions are continuous.

Proposition 2.15. *If E is a topological space, and $(\mathbb{R},|x-y|)$ the reals under the standard topology, for any two functions $f\colon E\to\mathbb{R}$ and $g\colon E\to\mathbb{R}$, for any $a\in E$, for any $\lambda\in\mathbb{R}$, if f and g are continuous at a, then $f+g$, λf, $f\cdot g$ are continuous at a, and f/g is continuous at a if $g(a)\neq 0$.*

Proof. Left as an exercise. ☐

Using Proposition 2.15, we can show easily that every real polynomial function is continuous.

The notion of isomorphism of topological spaces is defined as follows.

Definition 2.17. Let $(E, \mathcal{O}_E)$ and $(F, \mathcal{O}_F)$ be topological spaces, and let $f\colon E \to F$ be a function. We say that *f is a homeomorphism between E and F* if f is bijective, and both $f\colon E \to F$ and $f^{-1}\colon F \to E$ are continuous.

One should be careful that a bijective continuous function $f\colon E \to F$ is not necessarily a homeomorphism. For example, if $E = \mathbb{R}$ with the discrete topology, and $F = \mathbb{R}$ with the standard topology, the identity is not a homeomorphism. Another interesting example involving a parametric curve is given below. Let $L\colon \mathbb{R} \to \mathbb{R}^2$ be the function, defined such that

$$L_1(t) = \frac{t(1+t^2)}{1+t^4},$$
$$L_2(t) = \frac{t(1-t^2)}{1+t^4}.$$

If we think of $(x(t), y(t)) = (L_1(t), L_2(t))$ as a geometric point in $\mathbb{R}^2$, the set of points $(x(t), y(t))$ obtained by letting t vary in $\mathbb{R}$ from $-\infty$ to $+\infty$, defines a curve having the shape of a "figure eight," with self-intersection at the origin, called the "lemniscate of Bernoulli." See Figure 2.19. The map L is continuous, and in fact bijective, but its inverse L^{-1} is not continuous. Indeed, when we approach the origin on the branch of the curve in the upper left quadrant (i.e., points such that, $x \leq 0$, $y \geq 0$), then t goes to $-\infty$, and when we approach the origin on the branch of the curve in the lower right quadrant (i.e., points such that, $x \geq 0$, $y \leq 0$), then t goes to $+\infty$.

Fig. 2.19 The lemniscate of Bernoulli.

2.5 Limits and Continuity; Uniform Continuity

The definition of continuity utilizes open sets (or neighborhoods) to capture the notion of "closeness." Another way to quantify this notion of "closeness" is through the limit of a sequence.

Definition 2.18. Given any set E, a *sequence* is any function $x\colon \mathbb{N} \to E$, usually denoted by $(x_n)_{n\in\mathbb{N}}$, or $(x_n)_{n\geq 0}$, or even by (x_n).

Definition 2.19. Given a topological space $(E, \mathcal{O})$, we say that *a sequence* $(x_n)_{n\in\mathbb{N}}$ *converges to some* $a \in E$ if for every open set U containing a, there is some $n_0 \geq 0$, such that, $x_n \in U$, for all $n \geq n_0$. We also say that *a is a limit of* $(x_n)_{n\in\mathbb{N}}$. See Figure 2.20.

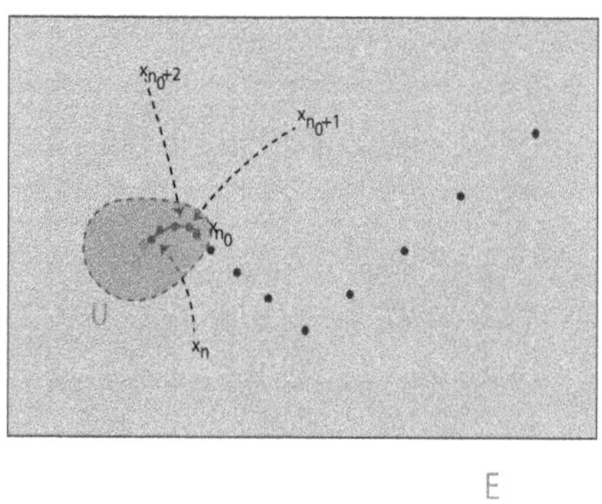

Fig. 2.20 A schematic illustration of Definition 2.19.

When E is a metric space, Definition 2.19 is equivalent to the following proposition.

Proposition 2.16. *Let E be a metric space with metric d. A sequence $(x_n)_{n\in\mathbb{N}} \subset E$ converges to some $a \in E$ iff*

for every $\epsilon > 0$, there is some $n_0 \geq 0$, such that, $d(x_n, a) \leq \epsilon$, for all $n \geq n_0$.

When E is a normed vector space, Definition 2.19 is equivalent to the following proposition.

Proposition 2.17. *Let E be a normed vector space with norm $\| \;\|$. A sequence $(x_n)_{n\in\mathbb{N}} \subset E$ converges to some $a \in E$ iff* *for every* $\epsilon > 0$, *there is some* $n_0 \geq 0$, *such that,* $\|x_n - a\| \leq \epsilon$, *for all* $n \geq n_0$.

The following proposition shows the importance of the Hausdorff separation axiom.

Proposition 2.18. *Given a topological space* $(E, \mathcal{O})$, *if the Hausdorff separation axiom holds, then every sequence has at most one limit.*

Proof. Left as an exercise. □

It is worth noting that the notion of limit is topological, in the sense that a sequence converge to a limit b iff it converges to the same limit b in any equivalent metric (and similarly for equivalent norms).

If E is a metric space and if A is a subset of E, there is a convenient way of showing that a point $x \in E$ belongs to the closure $\overline{A}$ of A in terms of sequences.

Proposition 2.19. *Given any metric space* (E, d), *for any subset* A *of* E *and any point* $x \in E$, *we have* $x \in \overline{A}$ *iff there is a sequence* (a_n) *of points* $a_n \in A$ *converging to* x.

Proof. If the sequence (a_n) of points $a_n \in A$ converges to x, then for every open subset U of E containing x, there is some n_0 such that $a_n \in U$ for all $n \geq n_0$, so $U \cap A \neq \emptyset$, and Proposition 2.3 implies that $x \in \overline{A}$.

Conversely, assume that $x \in \overline{A}$. Then for every $n \geq 1$, consider the open ball $B_0(x, 1/n)$. By Proposition 2.3, we have $B_0(x, 1/n) \cap A \neq \emptyset$, so we can pick some $a_n \in B_0(x, 1/n) \cap A$. This way, we define a sequence (a_n) of points in A, and by construction $d(x, a_n) < 1/n$ for all $n \geq 1$, so the sequence (a_n) converges to x. □

Before stating continuity in terms of limits, we still need one more concept, that of limit for functions.

Definition 2.20. Let $(E, \mathcal{O}_E)$ and $(F, \mathcal{O}_F)$ be topological spaces, let A be some nonempty subset of E, and let $f\colon A \to F$ be a function. For any $a \in \overline{A}$ and any $b \in F$, we say that $f(x)$ *approaches* b *as* x *approaches* a *with values in* A if for every open set $V \in \mathcal{O}_F$ containing b, there is some open set $U \in \mathcal{O}_E$ containing a, such that, $f(U \cap A) \subseteq V$. See Figure 2.21. This is denoted by

$$\lim_{x \to a, x \in A} f(x) = b.$$

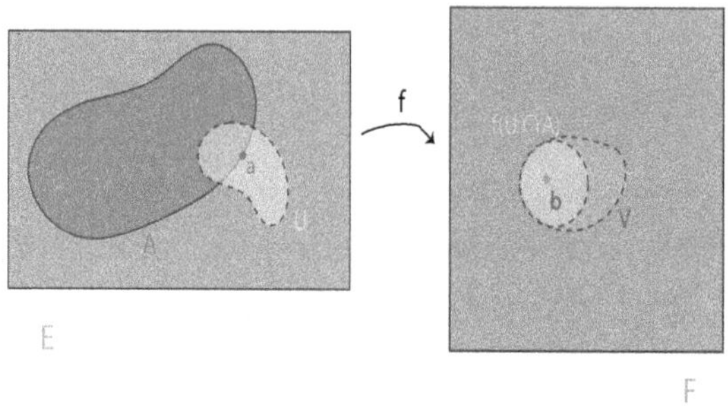

Fig. 2.21 A schematic illustration of Definition 2.20.

Note that by Proposition 2.3, since $a \in \overline{A}$, for every open set U containing a, we have $U \cap A \neq \emptyset$, and the definition is nontrivial. Also, even if $a \in A$, the value $f(a)$ of f at a plays no role in this definition.

When E and F are metric spaces, Definition 2.20 can be restated as follows.

Proposition 2.20. *Let E and F be metric spaces with metrics d_1 and d_2. Let A be some nonempty subset of E, and let $f\colon A \to F$ be a function. For any $a \in \overline{A}$ and any $b \in F$, $f(x)$ approaches b as x approaches a with values in A iff*

for every $\epsilon > 0$, there is some $\eta > 0$, such that, for every $x \in A$,

$$\text{if } d_1(x, a) \leq \eta, \text{ then } d_2(f(x), b) \leq \epsilon.$$

When E and F are normed vector spaces, Definition 2.20 can be restated as follows.

Proposition 2.21. *Let E and F be normed vector spaces with norms $\| \, \|_1$ and $\| \, \|_2$. Let A be some nonempty subset of E, and let $f\colon A \to F$ be a function. For any $a \in \overline{A}$ and any $b \in F$, $f(x)$ approaches b as x approaches a with values in A iff*

for every $\epsilon > 0$, there is some $\eta > 0$, such that, for every $x \in A$,

$$\text{if } \|x - a\|_1 \leq \eta, \text{ then } \|f(x) - b\|_2 \leq \epsilon.$$

We have the following result relating continuity at a point and the previous notion.

Proposition 2.22. *Let $(E, \mathcal{O}_E)$ and $(F, \mathcal{O}_F)$ be two topological spaces, and* *let $f: E \to F$ be a function. For any $a \in E$, the function f is continuous at a iff $f(x)$ approaches $f(a)$ when x approaches a (with values in E).*

Proof. Left as a trivial exercise. □

Another important proposition relating the notion of convergence of a sequence to continuity is stated without proof.

Proposition 2.23. *Let $(E, \mathcal{O}_E)$ and $(F, \mathcal{O}_F)$ be two topological spaces, and let $f: E \to F$ be a function.*

(1) If f is continuous, then for every sequence $(x_n)_{n \in \mathbb{N}}$ in E, if (x_n) converges to a, then $(f(x_n))$ converges to $f(a)$.
(2) If E is a metric space, and $(f(x_n))$ converges to $f(a)$ whenever (x_n) converges to a, for every sequence $(x_n)_{n \in \mathbb{N}}$ in E, then f is continuous.

A special case of Definition 2.20 will be used when E and F are (nontrivial) normed vector spaces with norms $\| \|_1$ and $\| \|_2$. Let U be any nonempty open subset of E. We showed earlier that E has no isolated points and that every set $\{v\}$ is closed, for every $v \in E$. Since E is nontrivial, for every $v \in U$, there is a nontrivial open ball contained in U (an open ball not reduced to its center). Then for every $v \in U$, $A = U - \{v\}$ is open and nonempty, and clearly, $v \in \overline{A}$. For any $v \in U$, if $f(x)$ approaches b when x approaches v with values in $A = U - \{v\}$, we say that *$f(x)$ approaches b when x approaches v with values $\neq v$ in U*. This is denoted by

$$\lim_{x \to v, x \in U, x \neq v} f(x) = b.$$

Remark: Variations of the above case show up in the following case: $E = \mathbb{R}$, and F is some arbitrary topological space. Let A be some nonempty subset of $\mathbb{R}$, and let $f: A \to F$ be some function. For any $a \in A$, we say that *f is continuous on the right at a* if

$$\lim_{x \to a, x \in A \cap [a, +\infty)} f(x) = f(a).$$

We can define *continuity on the left* at a in a similar fashion, namely

$$\lim_{x \to a, x \in A \cap (-\infty, a]} f(x) = f(a).$$

For example, the function $f: \mathbb{R} \to \mathbb{R}$

$$\begin{cases} f(x) = x & \text{if } x < 1 \\ f(x) = 2 & \text{if } x \geq 1, \end{cases}$$

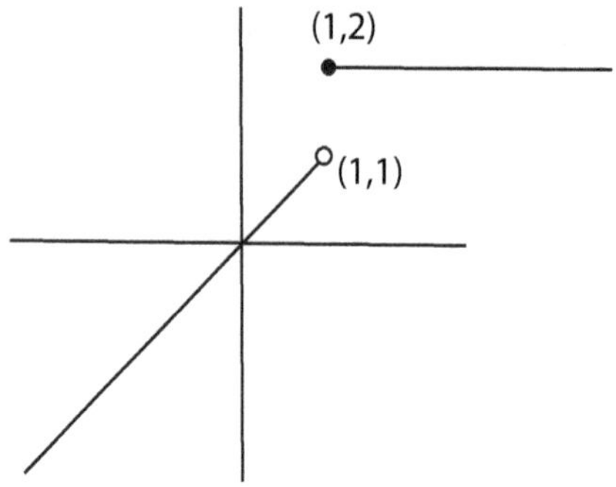

Fig. 2.22 The graph of the piecewise function $f(x) = x$ when $x < 1$ and $f(x) = 2$ when $x \geq 1$.

is continuous on the right at 1, but not continuous on the left at 1. See Figure 2.22.

Let us consider another variation. Let A be some nonempty subset of $\mathbb{R}$, and let $f\colon A \to F$ be some function. For any $a \in A$, we say that f *has a discontinuity of the first kind at* a if

$$\lim_{x \to a, x \in A \cap\, (-\infty, a)} f(x) = f(a_-)$$

and

$$\lim_{x \to a, x \in A \cap\, (a, +\infty)} f(x) = f(a_+)$$

both exist, and either $f(a_-) \neq f(a)$, or $f(a_+) \neq f(a)$. For example, the function $f\colon \mathbb{R} \to \mathbb{R}$

$$\begin{cases} f(x) = x & \text{if } x < 1 \\ f(x) = 2 & \text{if } x \geq 1, \end{cases}$$

has a discontinuity of the first kind at 1; both directional limits exits, namely $\lim_{x \to a, x \in A \cap\, (-\infty, a)} f(x) = 1$ and $\lim_{x \to a, x \in A \cap\, (a, +\infty)} f(x) = 2$, but $f(1_-) \neq f(1) = 2$. See Figure 2.22.

Note that it is possible that $f(a_-) = f(a_+)$, but f is still discontinuous at a if this common value differs from $f(a)$. Functions defined on a nonempty subset of $\mathbb{R}$, and that are continuous, except for some points of discontinuity of the first kind, play an important role in analysis.

In a metric space there is another important notion of continuity, namely uniform continuity.

Definition 2.21. Given two metric spaces, (E, d_E) and (F, d_F), a function, $f\colon E \to F$, is *uniformly continuous* if for every $\epsilon > 0$, there is some $\eta > 0$, such that for all $a, b \in E$,

$$\text{if} \quad d_E(a, b) \leq \eta \quad \text{then} \quad d_F(f(a), f(b)) \leq \epsilon.$$

See Figures 2.23 and 2.24.

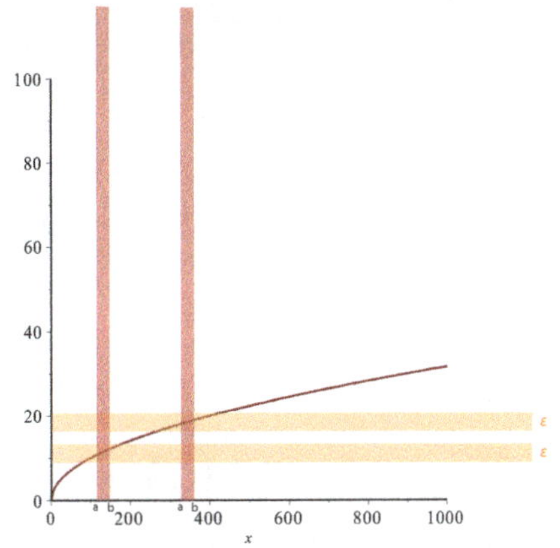

Fig. 2.23 The real valued function $f(x) = \sqrt{x}$ is uniformly continuous over $(0, \infty)$. Fix ϵ. If the x values lie within the rose colored η strip, the y values always lie within the peach ϵ strip.

As we saw earlier, the metric on a metric space is uniformly continuous, and the norm on a normed metric space is uniformly continuous.

Before considering differentials, we need to look at the continuity of linear maps.

2.6 Continuous Linear and Multilinear Maps

If E and F are normed vector spaces, we first characterize when a linear map $f\colon E \to F$ is continuous.

Proposition 2.24. *Given two normed vector spaces E and F, for any linear map $f\colon E \to F$, the following conditions are equivalent:*

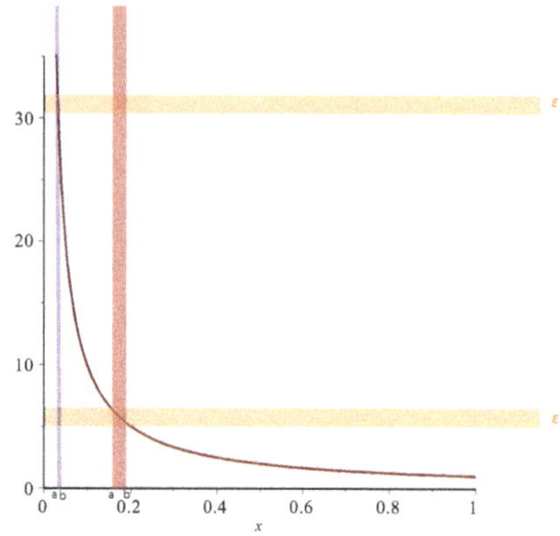

Fig. 2.24 The real valued function $f(x) = 1/x$ is not uniformly continuous over $(0, \infty)$. Fix ϵ. In order for the y values to lie within the peach epsilon strip, the widths of the eta strips decrease as $x \to 0$.

(1) The function f is continuous at 0.

(2) There is a constant $k \geq 0$ such that,

$$\|f(u)\| \leq k, \text{ for every } u \in E \text{ such that } \|u\| \leq 1.$$

(3) There is a constant $k \geq 0$ such that,

$$\|f(u)\| \leq k\|u\|, \text{ for every } u \in E.$$

(4) The function f is continuous at every point of E.

Proof. Assume (1). Then for every $\epsilon > 0$, there is some $\eta > 0$ such that, for every $u \in E$, if $\|u\| \leq \eta$, then $\|f(u)\| \leq \epsilon$. Pick $\epsilon = 1$, so that there is some $\eta > 0$ such that, if $\|u\| \leq \eta$, then $\|f(u)\| \leq 1$. If $\|u\| \leq 1$, then $\|\eta u\| \leq \eta\|u\| \leq \eta$, and so, $\|f(\eta u)\| \leq 1$, that is, $\eta\|f(u)\| \leq 1$, which implies $\|f(u)\| \leq \eta^{-1}$. Thus Condition (2) holds with $k = \eta^{-1}$.

Assume that (2) holds. If $u = 0$, then by linearity, $f(0) = 0$, and thus $\|f(0)\| \leq k\|0\|$ holds trivially for all $k \geq 0$. If $u \neq 0$, then $\|u\| > 0$, and since

$$\left\| \frac{u}{\|u\|} \right\| = 1,$$

we have

$$\left\| f\left(\frac{u}{\|u\|} \right) \right\| \leq k,$$

which implies that

$$\|f(u)\| \leq k\|u\|.$$

Thus Condition (3) holds.

If (3) holds, then for all $u, v \in E$, we have

$$\|f(v) - f(u)\| = \|f(v - u)\| \leq k\|v - u\|.$$

If $k = 0$, then f is the zero function, and continuity is obvious. Otherwise, if $k > 0$, for every $\epsilon > 0$, if $\|v - u\| \leq \frac{\epsilon}{k}$, then $\|f(v - u)\| \leq \epsilon$, which shows continuity at every $u \in E$. Finally it is obvious that (4) implies (1). $\square$

Among other things, Proposition 2.24 shows that a linear map is continuous iff the image of the unit (closed) ball is bounded. Since a continuous linear map satisfies the condition $\|f(u)\| \leq k\|u\|$ (for some $k \geq 0$), it is also uniformly continuous.

Definition 2.22. If E and F are normed vector spaces, the set of all continuous linear maps $f: E \to F$ is denoted by $\mathcal{L}(E; F)$.

Using Proposition 2.24, we can define a norm on $\mathcal{L}(E; F)$ which makes it into a normed vector space. This definition has already been given in Chapter 8 (Vol. I) (Definition 8.7 (Vol. I)) but for the reader's convenience, we repeat it here.

Definition 2.23. Given two normed vector spaces E and F, for every continuous linear map $f: E \to F$, we define the *norm* $\|f\|$ *of* f as

$$\|f\| = \inf \{k \geq 0 \mid \|f(x)\| \leq k\|x\|, \text{ for all } x \in E\} = \sup \{\|f(x)\| \mid \|x\| \leq 1\}.$$

From Definition 2.23, for every continuous linear map $f \in \mathcal{L}(E; F)$, we have

$$\|f(x)\| \leq \|f\|\|x\|,$$

for every $x \in E$. It is easy to verify that $\mathcal{L}(E; F)$ is a normed vector space under the norm of Definition 2.23. Furthermore, if E, F, G are normed vector spaces, and $f: E \to F$ and $g: F \to G$ are continuous linear maps, we have

$$\|g \circ f\| \leq \|g\|\|f\|.$$

We can now show that when $E = \mathbb{R}^n$ or $E = \mathbb{C}^n$, with any of the norms $\| \|_1$, $\| \|_2$, or $\| \|_\infty$, then every linear map $f: E \to F$ is continuous.

Proposition 2.25. *If $E = \mathbb{R}^n$ or $E = \mathbb{C}^n$, with any of the norms $\| \|_1$, $\| \|_2$, or $\| \|_\infty$, and F is any normed vector space, then every linear map $f: E \to F$ is continuous.*

Proof. Let $(e_1, \ldots, e_n)$ be the standard basis of $\mathbb{R}^n$ (a similar proof applies to $\mathbb{C}^n$). In view of Proposition 8.3 (Vol. I), it is enough to prove the proposition for the norm

$$\|x\|_\infty = \max\{|x_i| \mid 1 \le i \le n\}.$$

We have

$$\|f(v) - f(u)\| = \|f(v-u)\| = \left\| f\Big(\sum_{1\le i\le n} (v_i - u_i)e_i\Big)\right\|$$
$$= \left\|\sum_{1\le i\le n} (v_i - u_i)f(e_i)\right\|,$$

and so,

$$\|f(v)-f(u)\| \le \Big(\sum_{1\le i\le n} \|f(e_i)\|\Big) \max_{1\le i\le n} |v_i - u_i| = \Big(\sum_{1\le i\le n} \|f(e_i)\|\Big)\|v-u\|_\infty.$$

By the argument used in Proposition 2.24 to prove that (3) implies (4), f is continuous. □

Actually, we proved in Theorem 8.5 (Vol. I) that if E is a vector space of finite dimension, then any two norms are equivalent, so that they define the same topology. This fact together with Proposition 2.25 prove the following.

Theorem 2.1. *If E is a vector space of finite dimension (over $\mathbb{R}$ or $\mathbb{C}$), then all norms are equivalent (define the same topology). Furthermore, for any normed vector space F, every linear map $f\colon E \to F$ is continuous.*

If E is a normed vector space of infinite dimension, a linear map $f\colon E \to F$ may not be continuous.

As an example, let E be the infinite vector space of all polynomials over $\mathbb{R}$. Let

$$\|P(X)\| = \sup_{0\le x\le 1} |P(x)|.$$

We leave as an exercise to show that this is indeed a norm. Let $F = \mathbb{R}$, and let $f\colon E \to F$ be the map defined such that, $f(P(X)) = P(3)$. It is clear that f is linear. Consider the sequence of polynomials

$$P_n(X) = \left(\frac{X}{2}\right)^n.$$

It is clear that $\|P_n\| = \left(\frac{1}{2}\right)^n$, and thus, the sequence P_n has the null polynomial as a limit. However, we have

$$f(P_n(X)) = P_n(3) = \left(\frac{3}{2}\right)^n,$$

and the sequence $f(P_n(X))$ diverges to $+\infty$. Consequently, in view of Proposition 2.23 (1), f is not continuous.

We now consider the continuity of multilinear maps. We treat explicitly bilinear maps, the general case being a straightforward extension.

Proposition 2.26. *Given normed vector spaces E, F and G, for any bilinear map $f\colon E \times F \to G$, the following conditions are equivalent:*

(1) The function f is continuous at $\langle 0, 0\rangle$.
(2) There is a constant $k \geq 0$ such that,

$$\|f(u,v)\| \leq k, \quad \textit{for all } u \in E, v \in F \textit{ such that } \|u\|, \|v\| \leq 1.$$

(3) There is a constant $k \geq 0$ such that,

$$\|f(u,v)\| \leq k\|u\|\|v\|, \quad \textit{for all } u \in E, v \in F.$$

(4) The function f is continuous at every point of $E \times F$.

Proof. It is similar to that of Proposition 2.24, with a small subtlety in proving that (3) implies (4), namely that two different η's that are not independent are needed. ☐

In contrast to continuous linear maps, which must be uniformly continuous, nonzero continuous bilinear maps are **not** uniformly continuous. Let $f\colon E \times F \to G$ be a continuous bilinear map such that $f(a,b) \neq 0$ for some $a \in E$ and some $b \in F$. Consider the sequences (u_n) and (v_n) (with $n \geq 1$) given by

$$\begin{aligned} u_n &= (x_n, y_n) = (na, nb) \\ v_n &= (x'_n, y'_n) = \left(\left(n + \frac{1}{n}\right) a, \left(n + \frac{1}{n}\right) b\right). \end{aligned}$$

Obviously

$$\|v_n - u_n\| \leq \frac{1}{n}(\|a\| + \|b\|),$$

so $\lim_{n\mapsto\infty} \|v_n - u_n\| = 0$. On the other hand

$$f(x'_n, y'_n) - f(x_n, y_n) = \left(2 + \frac{1}{n^2}\right) f(a,b),$$

and thus $\lim_{n\mapsto\infty} \|f(x'_n, y'_n) - f(x_n, y_n)\| = 2\|f(a,b)\| \neq 0$, which shows that f is not uniformly continuous, because if this was the case, this limit would be zero.

Definition 2.24. If E, F, and G are normed vector spaces, we denote the set of all continuous bilinear maps $f\colon E \times F \to G$ by $\mathcal{L}_2(E, F; G)$.

Using Proposition 2.26, we can define a norm on $\mathcal{L}_2(E, F; G)$ which makes it into a normed vector space.

Definition 2.25. Given normed vector spaces E, F, and G, for every continuous bilinear map $f\colon E \times F \to G$, we define the *norm* $\|f\|$ *of* f as

$$\begin{aligned}\|f\| &= \inf\{k \geq 0 \mid \|f(x,y)\| \leq k\|x\|\|y\|, \text{ for all } x \in E, y \in F\}\\ &= \sup\{\|f(x,y)\| \mid \|x\|, \|y\| \leq 1\}.\end{aligned}$$

From Definition 2.23, we see that for every continuous bilinear map $f \in \mathcal{L}_2(E, F; G)$, we have

$$\|f(x,y)\| \leq \|f\|\|x\|\|y\|,$$

for all $x \in E, y \in F$. It is easy to verify that $\mathcal{L}_2(E, F; G)$ is a normed vector space under the norm of Definition 2.25.

Given a bilinear map $f\colon E \times F \to G$, for every $u \in E$, we obtain a linear map denoted $fu\colon F \to G$, defined such that, $fu(v) = f(u, v)$. Furthermore, since

$$\|f(x,y)\| \leq \|f\|\|x\|\|y\|,$$

it is clear that fu is continuous. We can then consider the map $\varphi\colon E \to \mathcal{L}(F; G)$, defined such that, $\varphi(u) = fu$, for any $u \in E$, or equivalently, such that,

$$\varphi(u)(v) = f(u, v).$$

Actually, it is easy to show that φ is linear and continuous, and that $\|\varphi\| = \|f\|$. Thus, $f \mapsto \varphi$ defines a map from $\mathcal{L}_2(E, F; G)$ to $\mathcal{L}(E; \mathcal{L}(F; G))$. We can also go back from $\mathcal{L}(E; \mathcal{L}(F; G))$ to $\mathcal{L}_2(E, F; G)$. We summarize all this in the following proposition.

Proposition 2.27. *Let E, F, G be three normed vector spaces. The map $f \mapsto \varphi$, from $\mathcal{L}_2(E, F; G)$ to $\mathcal{L}(E; \mathcal{L}(F; G))$, defined such that, for every $f \in \mathcal{L}_2(E, F; G)$,*

$$\varphi(u)(v) = f(u, v),$$

is an isomorphism of vector spaces, and furthermore, $\|\varphi\| = \|f\|$.

As a corollary of Proposition 2.27, we get the following proposition which will be useful when we define second-order derivatives.

Proposition 2.28. *Let E and F be normed vector spaces. The map app from $\mathcal{L}(E;F) \times E$ to F, defined such that, for every $f \in \mathcal{L}(E;F)$, for every $u \in E$,*

$$app(f,u) = f(u),$$

is a continuous bilinear map.

Remark: If E and F are nontrivial, it can be shown that $\|\text{app}\| = 1$. It can also be shown that composition

$$\circ \colon \mathcal{L}(E;F) \times \mathcal{L}(F;G) \to \mathcal{L}(E;G),$$

is bilinear and continuous.

The above propositions and definition generalize to arbitrary n-multilinear maps, with $n \geq 2$. Proposition 2.26 extends in the obvious way to any n-multilinear map $f \colon E_1 \times \cdots \times E_n \to F$, but condition (3) becomes:

There is a constant $k \geq 0$ such that,

$$\|f(u_1, \ldots, u_n)\| \leq k\|u_1\| \cdots \|u_n\|, \text{ for all } u_1 \in E_1, \ldots, u_n \in E_n.$$

Definition 2.25 also extends easily to

$$\begin{aligned}\|f\| = \inf \Big\{k \geq 0 \mid \|f(x_1, \ldots, x_n)\| \leq k\|x_1\| \cdots \|x_n\|, \\ \text{for all } x_i \in E_i, 1 \leq i \leq n\Big\} \\ = \sup \{\|f(x_1, \ldots, x_n)\| \mid \|x_n\|, \ldots, \|x_n\| \leq 1\}.\end{aligned}$$

Proposition 2.27 is also easily extended, and we get an isomorphism between continuous n-multilinear maps in $\mathcal{L}_n(E_1, \ldots, E_n; F)$, and continuous linear maps in

$$\mathcal{L}(E_1; \mathcal{L}(E_2; \ldots; \mathcal{L}(E_n; F))).$$

An obvious extension of Proposition 2.28 also holds.

Complete metric spaces and complete normed vector spaces are important tools in analysis and optimization theory, so we include some sections covering the basics.

2.7 Complete Metric Spaces and Banach Spaces

Definition 2.26. Given a metric space, (E, d), a sequence, $(x_n)_{n\in\mathbb{N}}$, in E is a *Cauchy sequence* if the following condition holds: for every $\epsilon > 0$, there is some $p \geq 0$, such that for all $m, n \geq p$, then $d(x_m, x_n) \leq \epsilon$.

If every Cauchy sequence in (E, d) converges we say that (E, d) is a *complete metric space*. A normed vector space $(E, \| \|)$ over $\mathbb{R}$ (or $\mathbb{C}$) which is a complete metric space for the distance $d(u, v) = \|v - u\|$, is called a *Banach space*.

The standard example of a complete metric space is the set $\mathbb{R}$ of real numbers. As a matter of fact, the set $\mathbb{R}$ can be defined as the "completion" of the set $\mathbb{Q}$ of rationals. The spaces $\mathbb{R}^n$ and $\mathbb{C}^n$ under their standard topology are complete metric spaces.

It can be shown that every normed vector space of finite dimension is a Banach space (is complete). It can also be shown that if E and F are normed vector spaces, and F is a Banach space, then $\mathcal{L}(E; F)$ is a Banach space. If E, F and G are normed vector spaces, and G is a Banach space, then $\mathcal{L}_2(E, F; G)$ is a Banach space.

An arbitrary metric space (E, d) is not necessarily complete, but there is a construction of a metric space $(\widehat{E}, \widehat{d})$ such that $\widehat{E}$ is complete, and there is a continuous (injective) distance-preserving map $\varphi\colon E \to \widehat{E}$ such that $\varphi(E)$ is dense in $\widehat{E}$. This is a generalization of the construction of the set $\mathbb{R}$ of real numbers from the set $\mathbb{Q}$ of rational numbers in terms of Cauchy sequences. This construction can be immediately adapted to a normed vector space $(E, \| \|)$ to embed $(E, \| \|)$ into a complete normed vector space $(\widehat{E}, \| \|_{\widehat{E}})$ (a Banach space). This construction is used heavily in integration theory where E is a set of functions.

2.8 Completion of a Metric Space

In order to prove a kind of uniqueness result for the completion $(\widehat{E}, \widehat{d})$ of a metric space (E, d), we need the following result about extending a uniformly continuous function.

Recall that E_0 is dense in E iff $\overline{E_0} = E$. Since E is a metric space, by Proposition 2.19, this means that for every $x \in E$, there is some sequence (x_n) converging to x, with $x_n \in E_0$.

Theorem 2.2. *Let E and F be two metric spaces, let E_0 be a dense subspace of E, and let $f_0\colon E_0 \to F$ be a continuous function. If f_0 is* *uniformly continuous and if F is complete, then there is a unique uniformly continuous function $f\colon E \to F$ extending f_0.*

Proof. We follow Schwartz's proof; see Schwartz [Schwartz (1980)] (Chapter XI, Section 3, Theorem 1).

Step 1. We begin by constructing a function $f\colon E \to F$ extending f_0. Since E_0 is dense in E, for every $x \in E$, there is some sequence (x_n) converging to x, with $x_n \in E_0$. Then the sequence (x_n) is a Cauchy sequence in E. We claim that $(f_0(x_n))$ is a Cauchy sequence in F.

Proof of the claim. For every $\epsilon > 0$, since f_0 is uniformly continuous, there is some $\eta > 0$ such that for all $(y, z) \in E_0$, if $d(y, z) \leq \eta$, then $d(f_0(y), f_0(z)) \leq \epsilon$. Since (x_n) is a Cauchy sequence with $x_n \in E_0$, there is some integer $p > 0$ such that if $m, n \geq p$, then $d(x_m, x_n) \leq \eta$, thus $d(f_0(x_m), f_0(x_n)) \leq \epsilon$, which proves that $(f_0(x_n))$ is a Cauchy sequence in F. ☐

Since F is complete and $(f_0(x_n))$ is a Cauchy sequence in F, the sequence $(f_0(x_n))$ converges to some element of F; denote this element by $f(x)$.

Step 2. Let us now show that $f(x)$ does not depend on the sequence (x_n) converging to x. Suppose that (x'_n) and (x''_n) are two sequences of elements in E_0 converging to x. Then the mixed sequence

$$x'_0, x''_0, x'_1, x''_1, \ldots, x'_n, x''_n, \ldots,$$

also converges to x. It follows that the sequence

$$f_0(x'_0), f_0(x''_0), f_0(x'_1), f_0(x''_1), \ldots, f_0(x'_n), f_0(x''_n), \ldots,$$

is a Cauchy sequence in F, and since F is complete, it converges to some element of F, which implies that the sequences $(f_0(x'_n))$ and $(f_0(x''_n))$ converge to the same limit.

As a summary, we have defined a function $f\colon E \to F$ by

$$f(x) = \lim_{n \mapsto \infty} f_0(x_n),$$

for any sequence (x_n) converging to x, with $x_n \in E_0$. See Figure 2.25.

Step 3. The function f extends f_0. Since every element $x \in E_0$ is the limit of the constant sequence (x_n) with $x_n = x$ for all $n \geq 0$, by definition $f(x)$ is the limit of the sequence $(f_0(x_n))$, which is the constant sequence with value $f_0(x)$, so $f(x) = f_0(x)$; that is, f extends f_0.

Step 4. We now prove that f is uniformly continuous. Since f_0 is uniformly continuous, for every $\epsilon > 0$, there is some $\eta > 0$ such that if

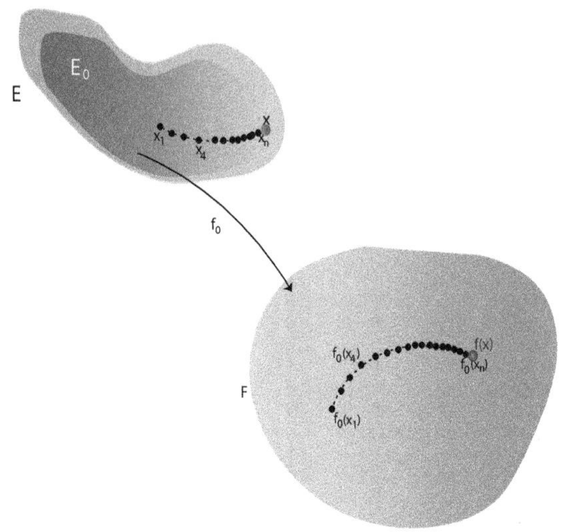

Fig. 2.25 A schematic illustration of the construction of $f\colon E \to F$ where $f(x) = \lim_{n\mapsto\infty} f_0(x_n)$ for any sequence (x_n) converging to x, with $x_n \in E_0$.

$a, b \in E_0$ and $d(a,b) \leq \eta$, then $d(f_0(a), f_0(b)) \leq \epsilon$. Consider any two points $x, y \in E$ such that $d(x,y) \leq \eta/2$. We claim that $d(f(x), f(y)) \leq \epsilon$, which shows that f is uniformly continuous.

Let (x_n) be a sequence of points in E_0 converging to x, and let (y_n) be a sequence of points in E_0 converging to y. By the triangle inequality,

$$d(x_n, y_n) \leq d(x_n, x) + d(x, y) + d(y, y_n) = d(x, y) + d(x_n, x) + d(y_n, y),$$

and since (x_n) converges to x and (y_n) converges to y, there is some integer $p > 0$ such that for all $n \geq p$, we have $d(x_n, x) \leq \eta/4$ and $d(y_n, y) \leq \eta/4$, and thus

$$d(x_n, y_n) \leq d(x, y) + \frac{\eta}{2}.$$

Since we assumed that $d(x,y) \leq \eta/2$, we get $d(x_n, y_n) \leq \eta$ for all $n \geq p$, and by uniform continuity of f_0, we get

$$d(f_0(x_n), f_0(y_n)) \leq \epsilon$$

for all $n \geq p$. Since the distance function on F is also continuous, and since $(f_0(x_n))$ converges to $f(x)$ and $(f_0(y_n))$ converges to $f(y)$, we deduce that the sequence $(d(f_0(x_n), f_0(y_n)))$ converges to $d(f(x), f(y))$. This implies that $d(f(x), f(y)) \leq \epsilon$, as desired.

Step 5. It remains to prove that f is unique. Since E_0 is dense in E, for every $x \in E$, there is some sequence (x_n) converging to x, with $x_n \in E_0$. Since f extends f_0 and since f is continuous, we get

$$f(x) = \lim_{n \mapsto \infty} f_0(x_n),$$

which only depends on f_0 and x and shows that f is unique. □

Remark: It can be shown that the theorem no longer holds if we either omit the hypothesis that F is complete or omit that f_0 is uniformly continuous.

For example, if $E_0 \neq E$ and if we let $F = E_0$ and f_0 be the identity function, it is easy to see that f_0 cannot be extended to a continuous function from E to E_0 (for any $x \in E - E_0$, any continuous extension f of f_0 would satisfy $f(x) = x$, which is absurd since $x \notin E_0$).

If f_0 is continuous but not uniformly continuous, a counter-example can be given by using $E = \overline{\mathbb{R}} = \mathbb{R} \cup \{\infty\}$ made into a metric space, $E_0 = \mathbb{R}$, $F = \mathbb{R}$, and f_0 the identity function; for details, see Schwartz [Schwartz (1980)] (Chapter XI, Section 3, page 134).

Definition 2.27. If (E, d_E) and (F, d_F) are two metric spaces, then a function $f: E \to F$ is *distance-preserving*, or an*isometry*, if

$$d_F(f(x), f(y)) = d_E(x, y), \quad \text{for all for all } x, y \in E.$$

Observe that an isometry must be injective, because if $f(x) = f(y)$, then $d_F(f(x), f(y)) = 0$, and since $d_F(f(x), f(y)) = d_E(x, y)$, we get $d_E(x, y) = 0$, but $d_E(x, y) = 0$ implies that $x = y$. Also, an isometry is uniformly continuous (since we can pick $\eta = \epsilon$ to satisfy the condition of uniform continuity). However, an isometry is not necessarily surjective.

We now give a construction of the completion of a metric space. This construction is just a generalization of the classical construction of $\mathbb{R}$ from $\mathbb{Q}$ using Cauchy sequences.

Theorem 2.3. *Let (E, d) be any metric space. There is a complete metric space $(\widehat{E}, \widehat{d})$ called a completion of (E, d), and a distance-preserving (uniformly continuous) map $\varphi: E \to \widehat{E}$ such that $\varphi(E)$ is dense in $\widehat{E}$, and the following extension property holds: for every complete metric space F and for every uniformly continuous function $f: E \to F$, there is a unique uniformly continuous function $\widehat{f}: \widehat{E} \to F$ such that*

$$f = \widehat{f} \circ \varphi,$$

as illustrated in the following diagram.

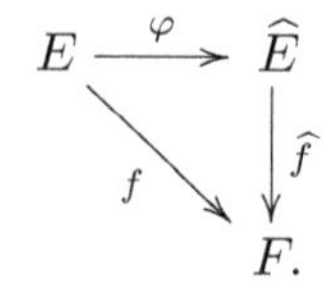

As a consequence, for any two completions $(\widehat{E}_1, \widehat{d}_1)$ *and* $(\widehat{E}_2, \widehat{d}_2)$ *of* (E, d)*, there is a unique bijective isometry between* $(\widehat{E}_1, \widehat{d}_1)$ *and* $(\widehat{E}_2, \widehat{d}_2)$*.*

Proof. Consider the set $\mathcal{E}$ of all Cauchy sequences (x_n) in E, and define the relation $\sim$ on $\mathcal{E}$ as follows:

$$(x_n) \sim (y_n) \quad \text{iff} \quad \lim_{n \mapsto \infty} d(x_n, y_n) = 0.$$

It is easy to check that $\sim$ is an equivalence relation on $\mathcal{E}$, and let $\widehat{E} = \mathcal{E}/\sim$ be the quotient set, that is, the set of equivalence classes modulo $\sim$. Our goal is to show that we can endow $\widehat{E}$ with a distance that makes it into a complete metric space satisfying the conditions of the theorem. We proceed in several steps.

Step 1. First let us construct the function $\varphi\colon E \to \widehat{E}$. For every $a \in E$, we have the constant sequence (a_n) such that $a_n = a$ for all $n \geq 0$, which is obviously a Cauchy sequence. Let $\varphi(a) \in \widehat{E}$ be the equivalence class $[(a_n)]$ of the constant sequence (a_n) with $a_n = a$ for all n. By definition of $\sim$, the equivalence class $\varphi(a)$ is also the equivalence class of all sequences converging to a. The map $a \mapsto \varphi(a)$ is injective because a metric space is Hausdorff, so if $a \neq b$, then a sequence converging to a does not converge to b. After having defined a distance on $\widehat{E}$, we will check that φ is an isometry.

Step 2. Let us now define a distance on $\widehat{E}$. Let $\alpha = [(a_n)]$ and $\beta = [(b_n)]$ be two equivalence classes of Cauchy sequences in E. The triangle inequality implies that

$$\begin{aligned} d(a_m, b_m) &\leq d(a_m, a_n) + d(a_n, b_n) + d(b_n, b_m) \\ &= d(a_n, b_n) + d(a_m, a_n) + d(b_m, b_n) \end{aligned}$$

and

$$\begin{aligned} d(a_n, b_n) &\leq d(a_n, a_m) + d(a_m, b_m) + d(b_m, b_n) \\ &= d(a_m, b_m) + d(a_m, a_n) + d(b_m, b_n), \end{aligned}$$

which implies that

$$|d(a_m, b_m) - d(a_n, b_n)| \leq d(a_m, a_n) + d(b_m, b_n).$$

Since (a_n) and (b_n) are Cauchy sequences, the above inequality shows that $(d(a_n, b_n))$ is a Cauchy sequence of nonnegative reals. Since $\mathbb{R}$ is complete, the sequence $(d(a_n, b_n))$ has a limit, which we denote by $\widehat{d}(\alpha, \beta)$; that is, we set

$$\widehat{d}(\alpha, \beta) = \lim_{n \mapsto \infty} d(a_n, b_n), \quad \alpha = [(a_n)],\ \beta = [(b_n)].$$

See Figure 2.26.

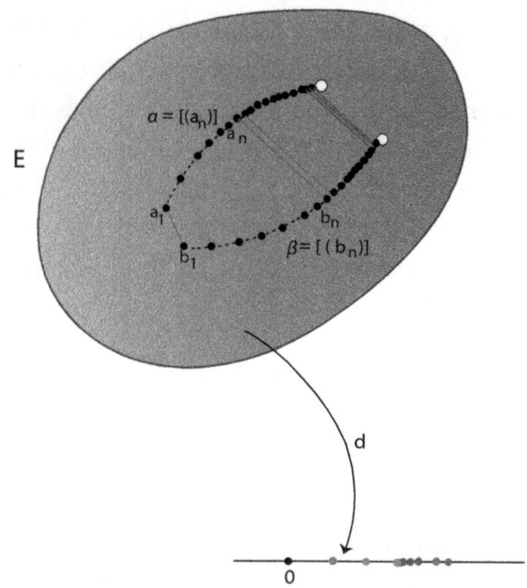

Fig. 2.26 A schematic illustration of $\widehat{d}(\alpha, \beta)$ from the Cauchy sequence $(d(a_n, b_n))$.

Step 3. Let us check that $\widehat{d}(\alpha, \beta)$ does not depend on the Cauchy sequences (a_n) and (b_n) chosen in the equivalence classes α and β.

If $(a_n) \sim (a'_n)$ and $(b_n) \sim (b'_n)$, then $\lim_{n \mapsto \infty} d(a_n, a'_n) = 0$ and $\lim_{n \mapsto \infty} d(b_n, b'_n) = 0$, and since

$$\begin{aligned} d(a'_n, b'_n) &\leq d(a'_n, a_n) + d(a_n, b_n) + d(b_n, b'_n) \\ &= d(a_n, b_n) + d(a_n, a'_n) + d(b_n, b'_n), \end{aligned}$$

and

$$\begin{aligned} d(a_n, b_n) &\leq d(a_n, a'_n) + d(a'_n, b'_n) + d(b'_n, b_n) \\ &= d(a'_n, b'_n) + d(a_n, a'_n) + d(b_n, b'_n), \end{aligned}$$

we have

$$|d(a_n, b_n) - d(a'_n, b'_n)| \leq d(a_n, a'_n) + d(b_n, b'_n),$$

so we have $\lim_{n\mapsto\infty} d(a'_n, b'_n) = \lim_{n\mapsto\infty} d(a_n, b_n) = \widehat{d}(\alpha, \beta)$. Therefore, $\widehat{d}(\alpha, \beta)$ is indeed well defined.

Step 4. Let us check that φ is indeed an isometry.

Given any two elements $\varphi(a)$ and $\varphi(b)$ in $\widehat{E}$, since they are the equivalence classes of the constant sequences (a_n) and (b_n) such that $a_n = a$ and $b_n = b$ for all n, the constant sequence $(d(a_n, b_n))$ with $d(a_n, b_n) = d(a, b)$ for all n converges to $d(a, b)$, so by definition $\widehat{d}(\varphi(a), \varphi(b)) = \lim_{n\mapsto\infty} d(a_n, b_n) = d(a, b)$, which shows that φ is an isometry.

Step 5. Let us verify that $\widehat{d}$ is a metric on $\widehat{E}$. By definition it is obvious that $\widehat{d}(\alpha, \beta) = \widehat{d}(\beta, \alpha)$. If α and β are two distinct equivalence classes, then for any Cauchy sequence (a_n) in the equivalence class α and for any Cauchy sequence (b_n) in the equivalence class β, the sequences (a_n) and (b_n) are inequivalent, which means that $\lim_{n\mapsto\infty} d(a_n, b_n) \neq 0$, that is, $\widehat{d}(\alpha, \beta) \neq 0$. Obviously, $\widehat{d}(\alpha, \alpha) = 0$.

For any equivalence classes $\alpha = [(a_n)]$, $\beta = [(b_n)]$, and $\gamma = [(c_n)]$, we have the triangle inequality

$$d(a_n, c_n) \leq d(a_n, b_n) + d(b_n, c_n),$$

so by continuity of the distance function, by passing to the limit, we obtain

$$\widehat{d}(\alpha, \gamma) \leq \widehat{d}(\alpha, \beta) + \widehat{d}(\beta, \gamma),$$

which is the triangle inequality for $\widehat{d}$. Therefore, $\widehat{d}$ is a distance on $\widehat{E}$.

Step 6. Let us prove that $\varphi(E)$ is dense in $\widehat{E}$. For any $\alpha = [(a_n)]$, let (x_n) be the constant sequence such that $x_k = a_n$ for all $k \geq 0$, so that $\varphi(a_n) = [(x_n)]$. Then we have

$$\widehat{d}(\alpha, \varphi(a_n)) = \lim_{m\mapsto\infty} d(a_m, a_n) \leq \sup_{p,q\geq n} d(a_p, a_q).$$

Since (a_n) is a Cauchy sequence, $\sup_{p,q\geq n} d(a_p, a_q)$ tends to 0 as n goes to infinity, so

$$\lim_{n\mapsto\infty} d(\alpha, \varphi(a_n)) = 0,$$

which means that the sequence $(\varphi(a_n))$ converge to α, and $\varphi(E)$ is indeed dense in $\widehat{E}$.

Step 7. Finally let us prove that the metric space $\widehat{E}$ is complete.

Let (α_n) be a Cauchy sequence in $\widehat{E}$. Since $\varphi(E)$ is dense in $\widehat{E}$, for every $n > 0$, there some $a_n \in E$ such that

$$\widehat{d}(\alpha_n, \varphi(a_n)) \leq \frac{1}{n}.$$

Since

$$\begin{aligned}\widehat{d}(\varphi(a_m), \varphi(a_n)) &\leq \widehat{d}(\varphi(a_m), \alpha_m) + \widehat{d}(\alpha_m, \alpha_n) + \widehat{d}(\alpha_n, \varphi(a_n)) \\ &\leq \widehat{d}(\alpha_m, \alpha_n) + \frac{1}{m} + \frac{1}{n},\end{aligned}$$

and since (α_m) is a Cauchy sequence, so is $(\varphi(a_n))$, and as φ is an isometry, the sequence (a_n) is a Cauchy sequence in E. Let $\alpha \in \widehat{E}$ be the equivalence class of (a_n). Since

$$\widehat{d}(\alpha, \varphi(a_n)) = \lim_{m \mapsto \infty} d(a_m, a_n)$$

and (a_n) is a Cauchy sequence, we deduce that the sequence $(\varphi(a_n))$ converges to α, and since $d(\alpha_n, \varphi(a_n)) \leq 1/n$ for all $n > 0$, the sequence (α_n) also converges to α.

Step 8. Let us prove the extension property. Let F be any complete metric space and let $f \colon E \to F$ be any uniformly continuous function. The function $\varphi \colon E \to \widehat{E}$ is an isometry and a bijection between E and its image $\varphi(E)$, so its inverse $\varphi^{-1} \colon \varphi(E) \to E$ is also an isometry, and thus is uniformly continuous. If we let $g = f \circ \varphi^{-1}$, then $g \colon \varphi(E) \to F$ is a uniformly continuous function, and $\varphi(E)$ is dense in $\widehat{E}$, so by Theorem 2.2 there is a unique uniformly continuous function $\widehat{f} \colon \widehat{E} \to F$ extending $g = f \circ \varphi^{-1}$; see the diagram below:

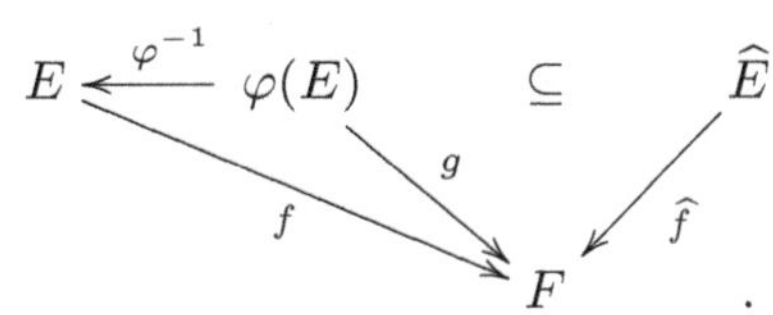

This means that

$$\widehat{f}|\varphi(E) = f \circ \varphi^{-1},$$

which implies that

$$(\widehat{f}|\varphi(E)) \circ \varphi = f,$$

that is, $f = \widehat{f} \circ \varphi$, as illustrated in the diagram below:

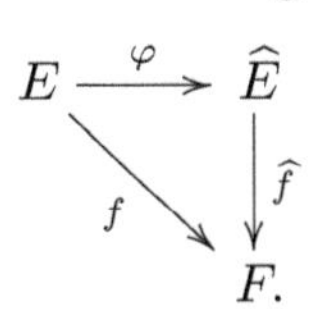

If $h \colon \widehat{E} \to F$ is any other uniformly continuous function such that $f = h \circ \varphi$, then $g = f \circ \varphi^{-1} = h|\varphi(E)$, so h is a uniformly continuous function extending g, and by Theorem 2.2, we have have $h = \widehat{f}$, so $\widehat{f}$ is indeed unique.

Step 9. Uniqueness of the completion $(\widehat{E}, \widehat{d})$ up to a bijective isometry.

Let $(\widehat{E}_1, \widehat{d}_1)$ and $(\widehat{E}_2, \widehat{d}_2)$ be any two completions of (E, d). Then we have two uniformly continuous isometries $\varphi_1 \colon E \to \widehat{E}_1$ and $\varphi_2 \colon E \to \widehat{E}_2$, so by the unique extension property, there exist unique uniformly continuous maps $\widehat{\varphi_2} \colon \widehat{E}_1 \to \widehat{E}_2$ and $\widehat{\varphi_1} \colon \widehat{E}_2 \to \widehat{E}_1$ such that the following diagrams commute:

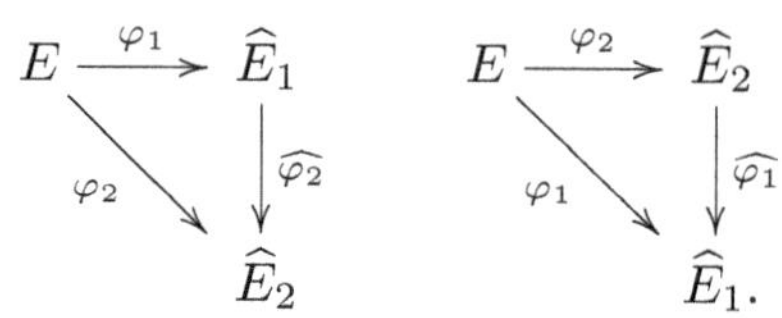

Consequently we have the following commutative diagrams:

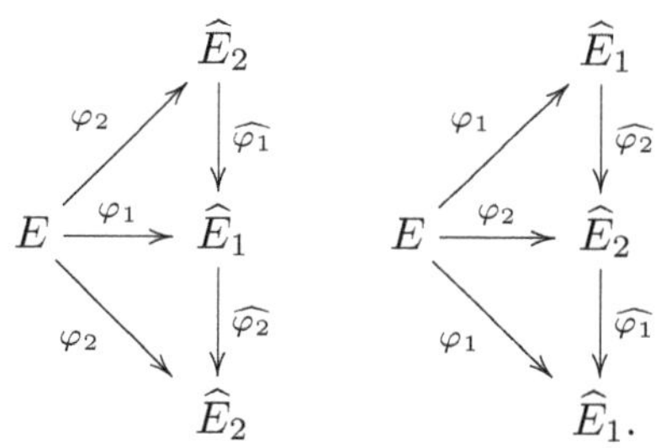

However, $\mathrm{id}_{\widehat{E}_1}$ and $\mathrm{id}_{\widehat{E}_2}$ are uniformly continuous functions making the following diagrams commute

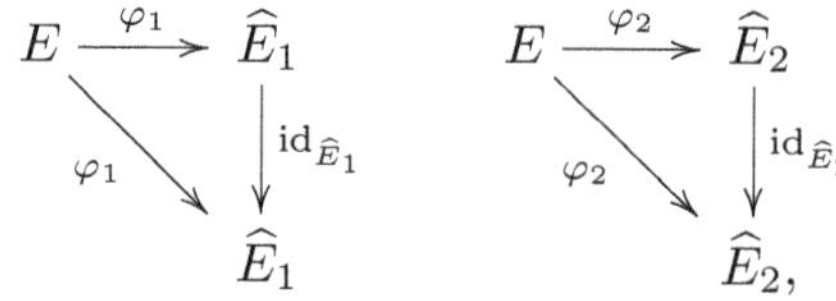

so by the uniqueness of extensions we must have

$$\widehat{\varphi_1} \circ \widehat{\varphi_2} = \mathrm{id}_{\widehat{E}_1} \quad \text{and} \quad \widehat{\varphi_2} \circ \widehat{\varphi_1} = \mathrm{id}_{\widehat{E}_2}.$$

This proves that $\widehat{\varphi_1}$ and $\widehat{\varphi_2}$ are mutual inverses. Now since $\varphi_2 = \widehat{\varphi_2} \circ \varphi_1$, we have

$$\widehat{\varphi_2} | \varphi_1(E) = \varphi_2 \circ \varphi_1^{-1},$$

and since φ_1^{-1} and φ_2 are isometries, so is $\widehat{\varphi_2} | \varphi_1(E)$. But we showed in Step 8 that $\widehat{\varphi_2}$ is the uniform continuous extension of $\widehat{\varphi_2} | \varphi_1(E)$ and $\varphi_1(E)$

is dense in $\widehat{E}_1$, so for any two elements $\alpha, \beta \in \widehat{E}_1$, if (a_n) and (b_n) are sequences in $\varphi_1(E)$ converging to α and β, we have

$$\widehat{d}_2((\widehat{\varphi_2}|\varphi_1(E))(a_n), ((\widehat{\varphi_2}|\varphi_1(E))(b_n)) = \widehat{d}_1(a_n, b_n),$$

and by passing to the limit we get

$$\widehat{d}_2(\widehat{\varphi_2}(\alpha), \widehat{\varphi_2}(\beta)) = \widehat{d}_1(\alpha, \beta),$$

which shows that $\widehat{\varphi_2}$ is an isometry (similarly, $\widehat{\varphi_1}$ is an isometry). □

Remarks:

(1) Except for Step 8 and Step 9, the proof of Theorem 2.3 is the proof given in Schwartz [Schwartz (1980)] (Chapter XI, Section 4, Theorem 1), and Kormogorov and Fomin [Kolmogorov and Fomin (1975)] (Chapter 2, Section 7, Theorem 4).

(2) The construction of $\widehat{E}$ relies on the completeness of $\mathbb{R}$, and so it cannot be used to construct $\mathbb{R}$ from $\mathbb{Q}$. However, this construction can be modified to yield a construction of $\mathbb{R}$ from $\mathbb{Q}$.

We show in Section 2.9 that Theorem 2.3 yields a construction of the completion of a normed vector space.

2.9 Completion of a Normed Vector Space

An easy corollary of Theorem 2.3 and Theorem 2.2 is that every normed vector space can be embedded in a complete normed vector space, that is, a Banach space.

Theorem 2.4. *If $(E, \| \|)$ is a normed vector space, then its completion $(\widehat{E}, \widehat{d})$ as a metric space (where E is given the metric $d(x, y) = \|x - y\|$) can be given a unique vector space structure extending the vector space structure on E, and a norm $\| \|_{\widehat{E}}$, so that $(\widehat{E}, \| \|_{\widehat{E}})$ is a Banach space, and the metric $\widehat{d}$ is associated with the norm $\| \|_{\widehat{E}}$. Furthermore, the isometry $\varphi \colon E \to \widehat{E}$ is a linear isometry.*

Proof. The addition operation $+\colon E \times E \to E$ is uniformly continuous because

$$\|(u' + v') - (u'' + v'')\| \leq \|u' - u''\| + \|v' - v''\|.$$

It is not hard to show that $\widehat{E} \times \widehat{E}$ is a complete metric space and that $E \times E$ is dense in $\widehat{E} \times \widehat{E}$. Then by Theorem 2.2, the uniformly continuous function $+$ has a unique continuous extension $+\colon \widehat{E} \times \widehat{E} \to \widehat{E}$.

The map $\cdot\colon \mathbb{R}\times E \to E$ is not uniformly continuous, but for any fixed $\lambda \in \mathbb{R}$, the map $L_\lambda\colon E \to E$ given by $L_\lambda(u) = \lambda\cdot u$ is uniformly continuous, so by Theorem 2.2 the function L_λ has a unique continuous extension $L_\lambda\colon \widehat{E} \to \widehat{E}$, which we use to define the scalar multiplication $\cdot\colon \mathbb{R} \times \widehat{E} \to \widehat{E}$. It is easily checked that with the above addition and scalar multiplication, $\widehat{E}$ is a vector space.

Since the norm $\| \;\|$ on E is uniformly continuous, it has a unique continuous extension $\| \;\|_{\widehat{E}}\colon \widehat{E} \to \mathbb{R}_+$. The identities $\|u + v\| \leq \|u\| + \|v\|$ and $\|\lambda u\| \leq |\lambda|\, \|u\|$ extend to $\widehat{E}$ by continuity. The equation

$$d(u, v) = \|u - v\|$$

also extends to $\widehat{E}$ by continuity and yields

$$\widehat{d}(\alpha, \beta) = \|\alpha - \beta\|_{\widehat{E}},$$

which shows that $\| \;\|_{\widehat{E}}$ is indeed a norm and that the metric $\widehat{d}$ is associated to it. Finally, it is easy to verify that the map φ is linear. The uniqueness of the structure of normed vector space follows from the uniqueness of continuous extensions in Theorem 2.2. □

Theorem 2.4 and Theorem 2.2 will be used to show that every Hermitian space can be embedded in a Hilbert space.

We refer the readers to the references cited at the end of this chapter for a discussion of the concepts of compactness and connectedness. They are important, but of less immediate concern.

2.10 The Contraction Mapping Theorem

If (E, d) is a nonempty complete metric space, every map $f\colon E \to E$, for which there is some k such that $0 \leq k < 1$ and

$$d(f(x), f(y)) \leq kd(x, y) \quad \text{for all } x, y \in E$$

has the very important property that it has a unique fixed point, that is, there is a unique, $a \in E$, such that $f(a) = a$.

Definition 2.28. Let (E, d) be a metric space. A map $f\colon E \to E$ is a *contraction* (or a *contraction mapping*) if there is some real number k such that $0 \leq k < 1$ and

$$d(f(u), f(v)) \leq kd(u, v) \quad \text{for all } u, v \in E.$$

The number k is often called a *Lipschitz constant*.

Furthermore, the fixed point of a contraction mapping can be computed as the limit of a fast converging sequence.

The fixed point property of contraction mappings is used to show some important theorems of analysis, such as the implicit function theorem and the existence of solutions to certain differential equations. It can also be used to show the existence of fractal sets defined in terms of iterated function systems. Since the proof is quite simple, we prove the fixed point property of contraction mappings. First observe that a contraction mapping is (uniformly) continuous.

Theorem 2.5. *(Contraction Mapping Theorem) If (E, d) is a nonempty complete metric space, every contraction mapping, $f\colon E \to E$, has a unique fixed point. Furthermore, for every $x_0 \in E$, if we define the sequence $(x_n)_{\geq 0}$ such that $x_{n+1} = f(x_n)$ for all $n \geq 0$, then $(x_n)_{n\geq 0}$ converges to the unique fixed point of f.*

Proof. First we prove that f has at most one fixed point. Indeed, if $f(a) = a$ and $f(b) = b$, since

$$d(a, b) = d(f(a), f(b)) \leq kd(a, b)$$

and $0 \leq k < 1$, we must have $d(a, b) = 0$, that is, $a = b$.

Next we prove that (x_n) is a Cauchy sequence. Observe that

$$\begin{aligned}
d(x_2, x_1) &\leq kd(x_1, x_0),\\
d(x_3, x_2) &\leq kd(x_2, x_1) \leq k^2 d(x_1, x_0),\\
\vdots \quad &\quad \vdots\\
d(x_{n+1}, x_n) &\leq kd(x_n, x_{n-1}) \leq \cdots \leq k^n d(x_1, x_0).
\end{aligned}$$

Thus, we have

$$\begin{aligned}
d(x_{n+p}, x_n) &\leq d(x_{n+p}, x_{n+p-1}) + d(x_{n+p-1}, x_{n+p-2}) + \cdots + d(x_{n+1}, x_n)\\
&\leq (k^{p-1} + k^{p-2} + \cdots + k + 1)k^n d(x_1, x_0)\\
&\leq \frac{k^n}{1-k} d(x_1, x_0).
\end{aligned}$$

We conclude that $d(x_{n+p}, x_n)$ converges to 0 when n goes to infinity, which shows that (x_n) is a Cauchy sequence. Since E is complete, the sequence (x_n) has a limit, a. Since f is continuous, the sequence $(f(x_n))$ converges to $f(a)$. But $x_{n+1} = f(x_n)$ converges to a and so $f(a) = a$, the unique fixed point of f. ☐

The above theorem is also called the *Banach fixed point theorem*. Note that no matter how the starting point x_0 of the sequence (x_n) is chosen, (x_n) converges to the unique fixed point of f. Also, the convergence is fast, since

$$d(x_n, a) \leq \frac{k^n}{1-k}\, d(x_1, x_0).$$

2.11 Further Readings

A thorough treatment of general topology can be found in Munkres [Munkres (2000, 1991)], Dixmier [Dixmier (1984)], Lang [Lang (1997)], Schwartz [Schwartz (1991, 1980)], Bredon [Bredon (1993)], and the classic, Seifert and Threlfall [Seifert and Threlfall (1980)].

2.12 Summary

The main concepts and results of this chapter are listed below:

- Metric space, distance, metric.
- Euclidean metric, discrete metric.
- Closed ball, open ball, sphere, bounded subset.
- Normed vector space, norm.
- Open and closed sets.
- Topology, topological space.
- Hausdorff separation axiom, Hausdorff space.
- Discrete topology.
- Closure, dense subset, interior, frontier or boundary.
- Subspace topology.
- Product topology.
- Basis of a topology, subbasis of a topology.
- Continuous functions.
- Neighborhood of a point.
- Homeomorphisms.
- Limits of sequences.
- Continuous linear maps.
- The norm of a continuous linear map.
- Continuous bilinear maps.
- The norm of a continuous bilinear map.
- The isomorphism between $\mathcal{L}(E, F; G)$ and $\mathcal{L}(E, \mathcal{L}(F; G))$.
- Cauchy sequences.

- Complete metric spaces and Banach spaces.
- Completion of a metric space or of a normed vector space.
- Contractions.
- The contraction mapping theorem.

2.13 Problems

Problem 2.1. Prove Proposition 2.1.

Problem 2.2. Give an example of a countably infinite family of closed sets whose union is not closed.

Problem 2.3. Prove Proposition 2.4.

Problem 2.4. Prove Proposition 2.5.

Problem 2.5. Prove Proposition 2.6.

Problem 2.6. Prove Proposition 2.7.

Problem 2.7. Prove Proposition 2.8.

Problem 2.8. Prove Proposition 2.9.

Problem 2.9. Prove Proposition 2.10.

Problem 2.10. Prove Proposition 2.11 and Proposition 2.12.

Problem 2.11. Prove Proposition 2.14.

Problem 2.12. Prove Proposition 2.15.

Problem 2.13. Prove Proposition 2.16 and Proposition 2.17.

Problem 2.14. Prove Proposition 2.18.

Problem 2.15. Prove Proposition 2.20 and Proposition 2.21.

Problem 2.16. Prove Proposition 2.22.

Problem 2.17. Prove Proposition 2.23.

Chapter 3

Differential Calculus

This chapter contains a review of basic notions of differential calculus. First we review the definition of the derivative of a function $f\colon \mathbb{R} \to \mathbb{R}$. Next we define directional derivatives and the total derivative of a function $f\colon E \to F$ between normed vector spaces. Basic properties of derivatives are shown, including the chain rule. We show how derivatives are represented by Jacobian matrices. The mean value theorem is stated, as well as the implicit function theorem and the inverse function theorem. Diffeomorphisms and local diffeomorphisms are defined. Higher-order derivatives are defined, as well as the Hessian. Schwarz's lemma (about the commutativity of partials) is stated. Several versions of Taylor's formula are stated, and a famous formula due to Faà di Bruno's is given.

3.1 Directional Derivatives, Total Derivatives

We first review the notion of the derivative of a real-valued function whose domain is an open subset of $\mathbb{R}$.

Let $f\colon A \to \mathbb{R}$, where A is a nonempty open subset of $\mathbb{R}$, and consider any $a \in A$. The main idea behind the concept of the derivative of f at a, denoted by $f'(a)$, is that locally around a (that is, in some small open set $U \subseteq A$ containing a), the function f is approximated linearly[1] by the map

$$x \mapsto f(a) + f'(a)(x - a).$$

As pointed out by Dieudonné in the early 1960s, it is an "unfortunate accident" that if V is vector space of dimension one, then there is a bijection between the space V^* of linear forms defined on V and the field of scalars. As a consequence, the derivative of a real-valued function f defined on an

[1]Actually, the approximation is affine, but everybody commits this abuse of language.

open subset A of the reals can be defined as the scalar $f'(a)$ (for any $a \in A$). But as soon as f is a function of several arguments, the scalar interpretation of the derivative breaks down.

Part of the difficulty in extending the idea of derivative to more complex spaces is to give an adequate notion of linear approximation. The key idea is to use linear maps. This could be carried out in terms of matrices but it turns out that this neither shortens nor simplifies proofs. In fact, this is often the opposite.

We admit that the more intrinsic definition of the notion of derivative f'_a at a point a of a function $f\colon E \to F$ between two normed vector spaces E and F as a linear map requires a greater effort to be grasped, but we feel that the advantages of this definition outweigh its degree of abstraction. In particular, it yields a clear notion of the derivative of a function $f\colon \mathrm{M}_m(\mathbb{R}) \to \mathrm{M}_n(\mathbb{R})$ defined from $m \times m$ matrices to $n \times n$ matrices (many definitions make use of partial derivatives with respect to matrices that do not make any sense). But more importantly, the definition of the derivative as a linear map makes it clear that whether the space E or the space F is infinite dimensional does not matter. This is important in optimization theory where the natural space of solutions of the problem is often an infinite dimensional function space. Of course, to carry out computations one need to pick finite bases and to use Jacobian matrices, but this is a different matter.

Let us now review the formal definition of the derivative of a real-valued function.

Definition 3.1. Let A be any nonempty open subset of $\mathbb{R}$, and let $a \in A$. For any function $f\colon A \to \mathbb{R}$, the *derivative of f at $a \in A$* is the limit (if it exists)

$$\lim_{h \to 0,\, h \in U} \frac{f(a+h) - f(a)}{h},$$

where $U = \{h \in \mathbb{R} \mid a + h \in A,\, h \neq 0\}$. This limit is denoted by $f'(a)$, or $\mathrm{D}f(a)$, or $\frac{df}{dx}(a)$. If $f'(a)$ exists for every $a \in A$, we say that f *is differentiable on* A. In this case, the map $a \mapsto f'(a)$ is denoted by f', or $\mathrm{D}f$, or $\frac{df}{dx}$.

Note that since A is assumed to be open, $A - \{a\}$ is also open, and since the function $h \mapsto a + h$ is continuous and U is the inverse image of $A - \{a\}$ under this function, U is indeed open and the definition makes sense.

We can also define $f'(a)$ as follows: there is some function ϵ, such that,

$$f(a+h) = f(a) + f'(a) \cdot h + \epsilon(h)h,$$

whenever $a + h \in A$, where $\epsilon(h)$ is defined for all h such that $a + h \in A$, and

$$\lim_{h \to 0,\, h \in U} \epsilon(h) = 0.$$

Remark: We can also define the notion of *derivative of f at a on the left*, and *derivative of f at a on the right*. For example, we say that the *derivative of f at a on the left* is the limit $f'(a_-)$ (if it exists)

$$f'(a_-) = \lim_{h \to 0,\, h \in U} \frac{f(a+h) - f(a)}{h},$$

where $U = \{h \in \mathbb{R} \mid a + h \in A,\, h < 0\}$.

If a function f as in Definition 3.1 has a derivative $f'(a)$ at a, then it is continuous at a. If f is differentiable on A, then f is continuous on A. The composition of differentiable functions is differentiable.

Remark: A function f has a derivative $f'(a)$ at a iff the derivative of f on the left at a and the derivative of f on the right at a exist and if they are equal. Also, if the derivative of f on the left at a exists, then f is continuous on the left at a (and similarly on the right).

We would like to extend the notion of derivative to functions $f \colon A \to F$, where E and F are normed vector spaces, and A is some nonempty open subset of E. The first difficulty is to make sense of the quotient

$$\frac{f(a+h) - f(a)}{h}.$$

Since F is a normed vector space, $f(a+h) - f(a)$ makes sense. But how do we define the quotient by a vector? Well, we don't!

A first possibility is to consider the *directional derivative* with respect to a vector $u \neq 0$ in E. We can consider the vector $f(a+tu) - f(a)$, where $t \in \mathbb{R}$. Now,

$$\frac{f(a+tu) - f(a)}{t}$$

makes sense.

The idea is that in E, the points of the form $a + tu$ for t in some small interval $[-\epsilon, +\epsilon]$ in $\mathbb{R}$ form a line segment $[r, s]$ in A containing a, and that the image of this line segment defines a small curve segment on $f(A)$. This curve segment is defined by the map $t \mapsto f(a + tu)$, from $[r, s]$ to F, and the directional derivative $\mathrm{D}_u f(a)$ defines the direction of the tangent line at a to this curve; see Figure 3.1. This leads us to the following definition.

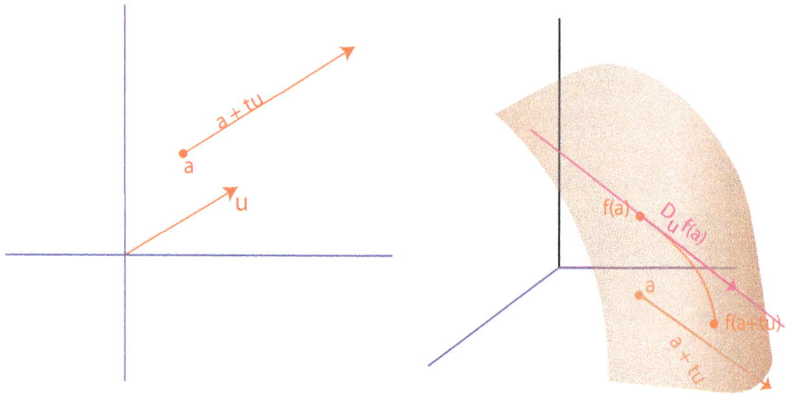

Fig. 3.1 Let $f\colon \mathbb{R}^2 \to \mathbb{R}$. The graph of f is the peach surface in $\mathbb{R}^3$, and $t \mapsto f(a+tu)$ is the embedded orange curve connecting $f(a)$ to $f(a+tu)$. Then $\mathrm{D}_u f(a)$ is the slope of the pink tangent line in the direction of u.

Definition 3.2. Let E and F be two normed vector spaces, let A be a nonempty open subset of E, and let $f\colon A \to F$ be any function. For any $a \in A$, for any $u \neq 0$ in E, the *directional derivative of f at a w.r.t. the vector u*, denoted by $\mathrm{D}_u f(a)$, is the limit (if it exists)

$$\mathrm{D}_u f(a) = \lim_{t\to 0,\, t\in U} \frac{f(a+tu) - f(a)}{t},$$

where $U = \{t \in \mathbb{R} \mid a+tu \in A,\, t \neq 0\}$ (or $U = \{t \in \mathbb{C} \mid a+tu \in A,\, t \neq 0\}$).

Since the map $t \mapsto a + tu$ is continuous, and since $A - \{a\}$ is open, the inverse image U of $A - \{a\}$ under the above map is open, and the definition of the limit in Definition 3.2 makes sense. The directional derivative is sometimes called the *Gâteaux derivative*.

Remark: Since the notion of limit is purely topological, the existence and value of a directional derivative is independent of the choice of norms in E and F, as long as they are equivalent norms.

In the special case where $E = \mathbb{R}$ and $F = \mathbb{R}$, and we let $u = 1$ (i.e., the real number 1, viewed as a vector), it is immediately verified that $\mathrm{D}_1 f(a) = f'(a)$, in the sense of Definition 3.1. When $E = \mathbb{R}$ (or $E = \mathbb{C}$) and F is any normed vector space, the derivative $\mathrm{D}_1 f(a)$, also denoted by $f'(a)$, provides a suitable generalization of the notion of derivative.

However, when E has dimension ≥ 2, directional derivatives present a serious problem, which is that their definition is not sufficiently uniform. Indeed, there is no reason to believe that the directional derivatives w.r.t. all nonnull vectors u share something in common. As a consequence, a function can have all directional derivatives at a, and yet not be continuous at a. Two functions may have all directional derivatives in some open sets, and yet their composition may not.

Example 3.1. Let $f\colon \mathbb{R}^2 \to \mathbb{R}$ be the function given by

$$f(x,y) = \begin{cases} \frac{x^2y}{x^4+y^2} & \text{if } (x,y) \neq (0,0) \\ 0 & \text{if } (x,y) = (0,0). \end{cases}$$

The graph of $f(x,y)$ is illustrated in Figure 3.2.

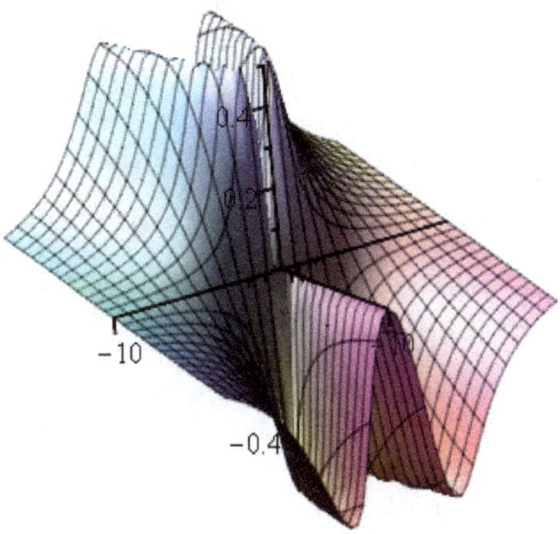

Fig. 3.2 The graph of the function from Example 3.1. Note that f is not continuous at $(0,0)$, despite the existence of $\mathrm{D}_u f(0,0)$ for all $u \neq 0$.

For any $u \neq 0$, letting $u = \begin{pmatrix} h \\ k \end{pmatrix}$, we have

$$\frac{f(0+tu) - f(0)}{t} = \frac{h^2k}{t^2h^4 + k^2},$$

so that

$$\mathrm{D}_u f(0,0) = \begin{cases} \frac{h^2}{k} & \text{if } k \neq 0 \\ 0 & \text{if } k = 0. \end{cases}$$

Thus, $D_u f(0,0)$ exists for all $u \neq 0$.

On the other hand, if $Df(0,0)$ existed, it would be a linear map $Df(0,0)\colon \mathbb{R}^2 \to \mathbb{R}$ represented by a row matrix $(\alpha\ \beta)$, and we would have $D_u f(0,0) = Df(0,0)(u) = \alpha h + \beta k$, but the explicit formula for $D_u f(0,0)$ is not linear. As a matter of fact, the function f is not continuous at $(0,0)$. For example, on the parabola $y = x^2$, $f(x,y) = \frac{1}{2}$, and when we approach the origin on this parabola, the limit is $\frac{1}{2}$, but $f(0,0) = 0$.

To avoid the problems arising with directional derivatives we introduce a more uniform notion.

Given two normed spaces E and F, recall that a linear map $f\colon E \to F$ is *continuous* iff there is some constant $C \geq 0$ such that

$$\|f(u)\| \leq C\,\|u\| \quad \text{for all } u \in E.$$

Definition 3.3. Let E and F be two normed vector spaces, let A be a nonempty open subset of E, and let $f\colon A \to F$ be any function. For any $a \in A$, we say that f is *differentiable at* $a \in A$ if there is a *continuous linear* map $L\colon E \to F$ and a function $h \mapsto \epsilon(h)$, such that

$$f(a+h) = f(a) + L(h) + \epsilon(h)\|h\|$$

for every $a + h \in A$, where $\epsilon(h)$ is defined for every h such that $a + h \in A$, and

$$\lim_{h \to 0,\, h \in U} \epsilon(h) = 0,$$

where $U = \{h \in E \mid a + h \in A,\ h \neq 0\}$. The linear map L is denoted by $Df(a)$, or Df_a, or $df(a)$, or df_a, or $f'(a)$, and it is called the *Fréchet derivative*, or *derivative*, or *total derivative*, or *total differential*, or *differential* of f at a; see Figure 3.3.

Since the map $h \mapsto a + h$ from E to E is continuous, and since A is open in E, the inverse image U of $A - \{a\}$ under the above map is open in E, and it makes sense to say that

$$\lim_{h \to 0,\, h \in U} \epsilon(h) = 0.$$

Note that for every $h \in U$, since $h \neq 0$, $\epsilon(h)$ is uniquely determined since

$$\epsilon(h) = \frac{f(a+h) - f(a) - L(h)}{\|h\|},$$

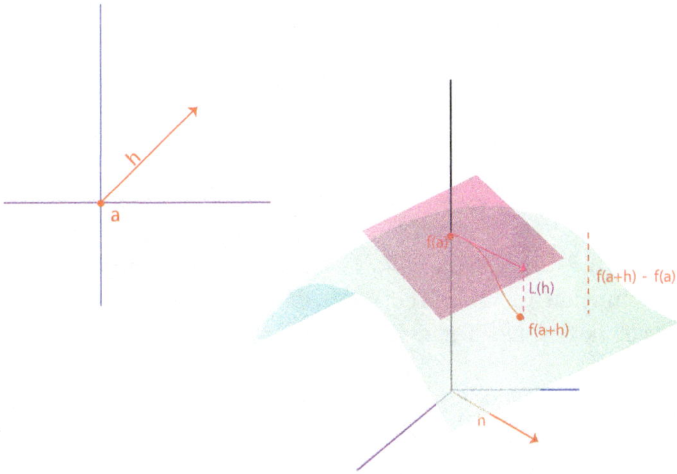

Fig. 3.3 Let $f\colon \mathbb{R}^2 \to \mathbb{R}$. The graph of f is the green surface in $\mathbb{R}^3$. The linear map $L = \mathrm{D}f(a)$ is the pink tangent plane. For any vector $h \in \mathbb{R}^2$, $L(h)$ is approximately equal to $f(a+h) - f(a)$. Note that $L(h)$ is also the direction tangent to the curve $t \mapsto f(a+th)$.

and that the value $\epsilon(0)$ plays absolutely no role in this definition. The condition for f to be differentiable at a amounts to the fact that

$$\lim_{h \mapsto 0} \frac{\|f(a+h) - f(a) - L(h)\|}{\|h\|} = 0 \tag{†}$$

as $h \neq 0$ approaches 0, when $a + h \in A$. However, it does no harm to assume that $\epsilon(0) = 0$, and we will assume this from now on.

Again, we note that the derivative $\mathrm{D}f(a)$ of f at a provides an affine approximation of f, locally around a.

Remarks:

(1) Since the notion of limit is purely topological, the existence and value of a derivative is independent of the choice of norms in E and F, as long as they are equivalent norms.

(2) If $h\colon (-a, a) \to \mathbb{R}$ is a real-valued function defined on some open interval containing 0, we say that h *is* $o(t)$ *for* $t \to 0$, and we write $h(t) = o(t)$, if

$$\lim_{t \mapsto 0,\ t \neq 0} \frac{h(t)}{t} = 0.$$

With this notation (the *little o notation*), the function f is differentiable at a iff

$$f(a+h) - f(a) - L(h) = o(\|h\|),$$

which is also written as

$$f(a+h) = f(a) + L(h) + o(\|h\|).$$

The following proposition shows that our new definition is consistent with the definition of the directional derivative and that *the continuous linear map L is unique*, if it exists.

Proposition 3.1. *Let E and F be two normed spaces, let A be a nonempty open subset of E, and let $f\colon A \to F$ be any function. For any $a \in A$, if $\mathrm{D}f(a)$ is defined, then f is continuous at a and f has a directional derivative $\mathrm{D}_u f(a)$ for every $u \neq 0$ in E. Furthermore,*

$$\mathrm{D}_u f(a) = \mathrm{D}f(a)(u)$$

and thus, $\mathrm{D}f(a)$ is uniquely defined.

Proof. If $L = \mathrm{D}f(a)$ exists, then for any nonzero vector $u \in E$, because A is open, for any $t \in \mathbb{R} - \{0\}$ (or $t \in \mathbb{C} - \{0\}$) small enough, $a + tu \in A$, so

$$\begin{aligned} f(a+tu) &= f(a) + L(tu) + \epsilon(tu)\|tu\| \\ &= f(a) + tL(u) + |t|\epsilon(tu)\|u\| \end{aligned}$$

which implies that

$$L(u) = \frac{f(a+tu) - f(a)}{t} - \frac{|t|}{t}\epsilon(tu)\|u\|,$$

and since $\lim_{t\mapsto 0} \epsilon(tu) = 0$, we deduce that

$$L(u) = \mathrm{D}f(a)(u) = \mathrm{D}_u f(a).$$

Because

$$f(a+h) = f(a) + L(h) + \epsilon(h)\|h\|$$

for all h such that $\|h\|$ is small enough, L is continuous, and $\lim_{h\mapsto 0} \epsilon(h)\|h\| = 0$, we have $\lim_{h\mapsto 0} f(a+h) = f(a)$, that is, f is continuous at a. ☐

When E is of finite dimension, every linear map is continuous (see Proposition 8.8 (Vol. I) or Theorem 2.1), and this assumption is then redundant.

Although this may not be immediately obvious, the reason for requiring the linear map $\mathrm{D}f_a$ to be continuous is to ensure that if a function f is differentiable at a, then it is continuous at a. This is certainly a desirable property of a differentiable function. In finite dimension this holds, but in infinite dimension this is not the case. The following proposition shows that if $\mathrm{D}f_a$ exists at a and if f is continuous at a, then $\mathrm{D}f_a$ must be a continuous map. So if a function is differentiable at a, then it is continuous iff the linear map $\mathrm{D}f_a$ is continuous. We chose to include the second condition rather that the first in the definition of a differentiable function.

Proposition 3.2. *Let E and F be two normed spaces, let A be a nonempty open subset of E, and let $f\colon A \to F$ be any function. For any $a \in A$, if $\mathrm{D}f_a$ is defined, then f is continuous at a iff $\mathrm{D}f_a$ is a continuous linear map.*

Proof. Proposition 3.1 shows that if $\mathrm{D}f_a$ is defined and continuous then f is continuous at a. Conversely, assume that $\mathrm{D}f_a$ exists and that f is continuous at a. Since f is continuous at a and since $\mathrm{D}f_a$ exists, for any $\eta > 0$ there is some ρ with $0 < \rho < 1$ such that if $\|h\| \leq \rho$ then

$$\|f(a+h) - f(a)\| \leq \frac{\eta}{2},$$

and

$$\|f(a+h) - f(a) - \mathrm{D}_a(h)\| \leq \frac{\eta}{2}\|h\| \leq \frac{\eta}{2},$$

so we have

$$\begin{aligned}\|\mathrm{D}_a(h)\| &= \|\mathrm{D}_a(h) - (f(a+h) - f(a)) + f(a+h) - f(a)\| \\ &\leq \|f(a+h) - f(a) - \mathrm{D}_a(h)\| + \|f(a+h) - f(a)\| \\ &\leq \frac{\eta}{2} + \frac{\eta}{2} = \eta,\end{aligned}$$

which proves that $\mathrm{D}f_a$ is continuous at 0. By Proposition 2.24, $\mathrm{D}f_a$ is a continuous linear map. $\square$

Example 3.2. Consider the map $f\colon \mathrm{M}_n(\mathbb{R}) \to \mathrm{M}_n(\mathbb{R})$ given by

$$f(A) = A^\top A - I,$$

where $\mathrm{M}_n(\mathbb{R})$ denotes the vector space of all $n \times n$ matrices with real entries equipped with any matrix norm, since they are all equivalent; for example, pick the Frobenius norm $\|A\|_F = \sqrt{\mathrm{tr}(A^\top A)}$. We claim that

$$Df(A)(H) = A^\top H + H^\top A, \quad \text{for all } A \text{ and } H \text{ in } \mathrm{M}_n(\mathbb{R}).$$

We have

$$\begin{aligned} f(A+H) - f(A) - (A^\top H + H^\top A) &= (A+H)^\top (A+H) - I \\ &\quad - (A^\top A - I) - A^\top H - H^\top A \\ &= A^\top A + A^\top H + H^\top A + H^\top H \\ &\quad - A^\top A - A^\top H - H^\top A \\ &= H^\top H. \end{aligned}$$

It follows that

$$\epsilon(H) = \frac{f(A+H) - f(A) - (A^\top H + H^\top A)}{\|H\|} = \frac{H^\top H}{\|H\|},$$

and since our norm is the Frobenius norm,

$$\|\epsilon(H)\| = \left\| \frac{H^\top H}{\|H\|} \right\| \leq \frac{\|H^\top\| \, \|H\|}{\|H\|} = \|H^\top\| = \|H\|,$$

so

$$\lim_{H \mapsto 0} \epsilon(H) = 0,$$

and we conclude that

$$Df(A)(H) = A^\top H + H^\top A.$$

If $\mathrm{D}f(a)$ exists for every $a \in A$, we get a map $\mathrm{D}f \colon A \to \mathcal{L}(E; F)$, called the *derivative of f on A*, and also denoted by *df*. Here $\mathcal{L}(E; F)$ denotes the vector space of continuous linear maps from E to F.

We now consider a number of standard results about derivatives.

3.2 Properties of Derivatives

A function $f \colon E \to F$ is said to be *affine* if there is some linear map $\overrightarrow{f} : E \to F$ and some fixed vector $c \in F$, such that

$$f(u) = \overrightarrow{f}(u) + c$$

for all $u \in E$. We call $\overrightarrow{f}$ the *linear map associated with f*.

Proposition 3.3. *Given two normed spaces E and F, if $f \colon E \to F$ is a constant function, then $\mathrm{D}f(a) = 0$, for every $a \in E$. If $f \colon E \to F$ is a continuous affine map, then $\mathrm{D}f(a) = \overrightarrow{f}$, for every $a \in E$, where $\overrightarrow{f}$ denotes the linear map associated with f.*

Proposition 3.4. *Given a normed space E and a normed vector space F, for any two functions $f, g\colon E \to F$, for every $a \in E$, if $\mathrm{D}f(a)$ and $\mathrm{D}g(a)$ exist, then $\mathrm{D}(f+g)(a)$ and $\mathrm{D}(\lambda f)(a)$ exist, and*

$$\begin{aligned}\mathrm{D}(f+g)(a) &= \mathrm{D}f(a) + \mathrm{D}g(a),\\ \mathrm{D}(\lambda f)(a) &= \lambda \mathrm{D}f(a).\end{aligned}$$

Given two normed vector spaces $(E_1, \| \|_1)$ and $(E_2, \| \|_2)$, there are three natural and equivalent norms that can be used to make $E_1 \times E_2$ into a normed vector space:

(1) $\|(u_1, u_2)\|_1 = \|u_1\|_1 + \|u_2\|_2$.
(2) $\|(u_1, u_2)\|_2 = (\|u_1\|_1^2 + \|u_2\|_2^2)^{1/2}$.
(3) $\|(u_1, u_2)\|_\infty = \max(\|u_1\|_1, \|u_2\|_2)$.

We usually pick the first norm. If E_1, E_2, and F are three normed vector spaces, recall that a bilinear map $f\colon E_1 \times E_2 \to F$ is *continuous* iff there is some constant $C \geq 0$ such that

$$\|f(u_1, u_2)\| \leq C \|u_1\|_1 \|u_2\|_2 \quad \text{for all } u_1 \in E_1 \text{ and all } u_2 \in E_2.$$

Proposition 3.5. *Given three normed vector spaces E_1, E_2, and F, for any continuous bilinear map $f\colon E_1 \times E_2 \to F$, for every $(a, b) \in E_1 \times E_2$, $\mathrm{D}f(a, b)$ exists, and for every $u \in E_1$ and $v \in E_2$,*

$$\mathrm{D}f(a,b)(u,v) = f(u,b) + f(a,v).$$

Proof. Since f is bilinear, a simple computation implies that

$$\begin{aligned}f((a,b)+(u,v)) - f(a,b) - (f(u,b)+f(a,v)) &= f(a+u, b+v) - f(a,b)\\ &\quad - f(u,b) - f(a,v)\\ &= f(a+u,b) + f(a+u,v) - f(a,b) - f(u,b) - f(a,v)\\ &= f(a,b) + f(u,b) + f(a,v) + f(u,v)\\ &\quad - f(a,b) - f(u,b) - f(a,v)\\ &= f(u,v).\end{aligned}$$

We define

$$\epsilon(u,v) = \frac{f((a,b)+(u,v)) - f(a,b) - (f(u,b)+f(a,v))}{\|(u,v)\|_1},$$

and observe that the continuity of f implies

$$\begin{aligned} \|f((a,b)+(u,v)) - f(a,b) - (f(u,b)+f(a,v))\| &= \|f(u,v)\| \\ &\leq C\,\|u\|_1\,\|v\|_2 \\ &\leq C\,(\|u\|_1 + \|v\|_2)^2 . \end{aligned}$$

Hence

$$\begin{aligned} \|\epsilon(u,v)\| = \left\| \frac{f(u,v)}{\|(u,v)\|_1} \right\| &= \frac{\|f(u,v)\|}{\|(u,v)\|_1} \leq \frac{C\,(\|u\|_1 + \|v\|_2)^2}{\|u\|_1 + \|v\|_2} \\ &= C\,(\|u\|_1 + \|v\|_2) = C\,\|(u,v)\|_1 , \end{aligned}$$

which in turn implies

$$\lim_{(u,v)\mapsto(0,0)} \epsilon(u,v) = 0. \qquad \square$$

We now state the very useful *chain rule*.

Theorem 3.1. *Given three normed spaces E, F, and G, let A be an open set in E, and let B an open set in F. For any functions $f\colon A \to F$ and $g\colon B \to G$, such that $f(A) \subseteq B$, for any $a \in A$, if $\mathrm{D}f(a)$ exists and $\mathrm{D}g(f(a))$ exists, then $\mathrm{D}(g \circ f)(a)$ exists, and*

$$\mathrm{D}(g \circ f)(a) = \mathrm{D}g(f(a)) \circ \mathrm{D}f(a).$$

Proof. Since f is differentiable at a and g is differentiable at $b = f(a)$, for every η such that $0 < \eta < 1$ there is some $\rho > 0$ such that for all s, t, if $\|s\| \leq \rho$ and $\|t\| \leq \rho$ then

$$\begin{aligned} f(a+s) &= f(a) + \mathrm{D}f_a(s) + \epsilon_1(s) \\ g(b+t) &= g(b) + \mathrm{D}g_b(t) + \epsilon_2(t), \end{aligned}$$

with $\|\epsilon_1(s)\| \leq \eta\,\|s\|$ and $\|\epsilon_2(t)\| \leq \eta\,\|t\|$. Since $\mathrm{D}f_a$ and $\mathrm{D}g_b$ are continuous, we have

$$\|\mathrm{D}f_a(s)\| \leq \|\mathrm{D}f_a\|\,\|s\| \quad \text{and} \quad \|\mathrm{D}g_b(t)\| \leq \|\mathrm{D}g_b\|\,\|t\| ,$$

which, since $\|\epsilon_1(s)\| \leq \eta\,\|s\|$ and $\eta < 1$, implies that

$$\begin{aligned} \|\mathrm{D}f_a(s) + \epsilon_1(s)\| &\leq \|\mathrm{D}f_a\|\,\|s\| + \|\epsilon_1(s)\| \\ &\leq \|\mathrm{D}f_a\|\,\|s\| + \eta\,\|s\| \leq (\|\mathrm{D}f_a\| + 1)\,\|s\| . \end{aligned}$$

Consequently, if $\|s\| < \rho/(\|\mathrm{D}f_a\| + 1)$, we have

$$\|\epsilon_2(\mathrm{D}f_a(s) + \epsilon_1(s))\| \leq \eta(\|\mathrm{D}f_a\| + 1)\,\|s\| , \qquad (*_1)$$

and

$$\|Dg_b(\epsilon_1(s))\| \leq \|Dg_b\| \, \|\epsilon_1(s)\| \leq \eta \, \|Dg_b\| \, \|s\| \, . \qquad (*_2)$$

Then since $b = f(a)$, using the above we have

$$\begin{aligned}(g \circ f)(a+s) &= g(f(a+s)) = g(b + Df_a(s) + \epsilon_1(s)) \\ &= g(b) + Dg_b(Df_a(s) + \epsilon_1(s)) + \epsilon_2(Df_a(s) + \epsilon_1(s)) \\ &= g(b) + (Dg_b \circ Df_a)(s) + Dg_b(\epsilon_1(s)) + \epsilon_2(Df_a(s) + \epsilon_1(s)).\end{aligned}$$

Now by $(*_1)$ and $(*_2)$ we have

$$\begin{aligned}\|Dg_b(\epsilon_1(s)) + \epsilon_2(Df_a(s) + \epsilon_1(s))\| &\leq \|Dg_b(\epsilon_1(s))\| + \|\epsilon_2(Df_a(s) + \epsilon_1(s))\| \\ &\leq \eta \, \|Dg_b\| \, \|s\| + \eta(\|Df_a\| + 1) \, \|s\| \\ &= \eta(\|Df_a\| + \|Dg_b\| + 1) \, \|s\| \, ,\end{aligned}$$

so if we write $\epsilon_3(s) = Dg_b(\epsilon_1(s)) + \epsilon_2(Df_a(s) + \epsilon_1(s))$ we proved that

$$(g \circ f)(a+s) = g(b) + (Dg_b \circ Df_a)(s) + \epsilon_3(s)$$

with $\epsilon_3(s) \leq \eta(\|Df_a\| + \|Dg_b\| + 1) \, \|s\|$, which proves that $Dg_b \circ Df_a$ is the derivative of $g \circ f$ at a. Since Df_a and Dg_b are continuous, so is $Dg_b \circ Df_a$, which proves our proposition. $\square$

Theorem 3.1 has many interesting consequences. We mention two corollaries.

Proposition 3.6. *Given three normed vector spaces E, F, and G, for any open subset A in E, for any $a \in A$, let $f : A \to F$ such that $Df(a)$ exists, and let $g : F \to G$ be a continuous affine map. Then $D(g \circ f)(a)$ exists, and*

$$D(g \circ f)(a) = \overrightarrow{g} \circ Df(a),$$

where $\overrightarrow{g}$ is the linear map associated with the affine map g.

Proposition 3.7. *Given two normed vector spaces E and F, let A be some open subset in E, let B be some open subset in F, let $f : A \to B$ be a bijection from A to B, and assume that Df exists on A and that Df^{-1} exists on B. Then for every $a \in A$,*

$$Df^{-1}(f(a)) = (Df(a))^{-1}.$$

Proposition 3.7 has the remarkable consequence that the two vector spaces E and F have the same dimension. In other words, a local property, the existence of a bijection f between an open set A of E and an open set B of F, such that f is differentiable on A and f^{-1} is differentiable on B, implies a global property, that the two vector spaces E and F have the same dimension.

Let us mention two more rules about derivatives that are used all the time.

Let $\iota\colon \mathbf{GL}(n, \mathbb{C}) \to \mathrm{M}_n(\mathbb{C})$ be the function (inversion) defined on invertible $n \times n$ matrices by

$$\iota(A) = A^{-1}.$$

Observe that $\mathbf{GL}(n, \mathbb{C})$ is indeed an open subset of the normed vector space $\mathrm{M}_n(\mathbb{C})$ of complex $n \times n$ matrices, since its complement is the closed set of matrices $A \in \mathrm{M}_n(\mathbb{C})$ satisfying $\det(A) = 0$. Then we have

$$d\iota_A(H) = -A^{-1}HA^{-1},$$

for all $A \in \mathbf{GL}(n, \mathbb{C})$ and for all $H \in \mathrm{M}_n(\mathbb{C})$.

To prove the preceding line observe that for H with sufficiently small norm, we have

$$\begin{aligned}\iota(A+H) - \iota(A) + A^{-1}HA^{-1} &= (A+H)^{-1} - A^{-1} + A^{-1}HA^{-1} \\ &= (A+H)^{-1}[I - (A+H)A^{-1} \\ &\quad + (A+H)A^{-1}HA^{-1}] \\ &= (A+H)^{-1}[I - I - HA^{-1} + HA^{-1} \\ &\quad + HA^{-1}HA^{-1}] \\ &= (A+H)^{-1}HA^{-1}HA^{-1}.\end{aligned}$$

Consequently, we get

$$\epsilon(H) = \frac{\iota(A+H) - \iota(A) + A^{-1}HA^{-1}}{\|H\|} = \frac{(A+H)^{-1}HA^{-1}HA^{-1}}{\|H\|},$$

and since

$$\left\|(A+H)^{-1}HA^{-1}HA^{-1}\right\| \leq \|H\|^2 \left\|A^{-1}\right\|^2 \left\|(A+H)^{-1}\right\|,$$

it is clear that $\lim_{H\mapsto 0} \epsilon(H) = 0$, which proves that

$$d\iota_A(H) = -A^{-1}HA^{-1}.$$

In particular, if $A = I$, then $d\iota_I(H) = -H$.

Next, if $f\colon \mathrm{M}_n(\mathbb{C}) \to \mathrm{M}_n(\mathbb{C})$ and $g\colon \mathrm{M}_n(\mathbb{C}) \to \mathrm{M}_n(\mathbb{C})$ are differentiable matrix functions, then

$$d(fg)_A(B) = df_A(B)g(A) + f(A)dg_A(B),$$

for all $A, B \in \mathbf{M}_n(\mathbb{C})$. This is known as the *product rule*.

In preparation for the next section on Jacobian matrices and the section on the implicit function theorem we need the following definitions.

When E is of finite dimension n, for any basis $(u_1, \ldots, u_n)$ of E, we can define the directional derivatives with respect to the vectors in the basis $(u_1, \ldots, u_n)$ (actually, we can also do it for an infinite basis). This way we obtain the definition of partial derivatives as follows:

Definition 3.4. For any two normed spaces E and F, if E is of finite dimension n, for every basis $(u_1, \ldots, u_n)$ for E, for every $a \in E$, for every function $f \colon E \to F$, the directional derivatives $\mathrm{D}_{u_j} f(a)$ (if they exist) are called the *partial derivatives of f with respect to the basis* $(u_1, \ldots, u_n)$. The partial derivative $\mathrm{D}_{u_j} f(a)$ is also denoted by $\partial_j f(a)$, or $\dfrac{\partial f}{\partial x_j}(a)$.

The notation $\dfrac{\partial f}{\partial x_j}(a)$ for a partial derivative, although customary and going back to Leibniz, is a "logical obscenity." Indeed, the variable x_j really has nothing to do with the formal definition. This is just another of these situations where tradition is just too hard to overthrow!

More generally we now consider the situation where E is a finite direct sum. Given a normed vector space $E = E_1 \oplus \cdots \oplus E_n$ and a normed vector space F, given any open subset A of E, for any $c = (c_1, \ldots, c_n) \in A$, we define the continuous functions $i_j^c \colon E_j \to E$, such that

$$i_j^c(x) = (c_1, \ldots, c_{j-1}, x, c_{j+1}, \ldots, c_n).$$

For any function $f \colon A \to F$, we have functions $f \circ i_j^c \colon E_j \to F$ defined on $(i_j^c)^{-1}(A)$, which contains c_j.

Definition 3.5. If $\mathrm{D}(f \circ i_j^c)(c_j)$ exists, we call it the *partial derivative of f w.r.t. its jth argument, at c*. We also denote this derivative by $\mathrm{D}_j f(c)$ of $\dfrac{\partial f}{\partial x_j}(c)$. Note that $\mathrm{D}_j f(c) \in \mathcal{L}(E_j; F)$.

This notion is a generalization of the notion defined in Definition 3.4. In fact, when E is of dimension n, and a basis $(u_1, \ldots, u_n)$ has been chosen, we can write $E = K u_1 \oplus \cdots \oplus K u_n$, (with $K = \mathbb{R}$ or $K = \mathbb{C}$), and then

$$\mathrm{D}_j f(c)(\lambda u_j) = \lambda \partial_j f(c),$$

and the two notions are consistent. We will use freely the notation $\dfrac{\partial f}{\partial x_j}(c)$ instead of $\mathrm{D}_j f(c)$.

The notion $\partial_j f(c)$ introduced in Definition 3.4 is really that of the vector derivative, whereas $\mathrm{D}_j f(c)$ $\left(= \dfrac{\partial f}{\partial x_j}(c)\right)$ is the corresponding linear map. The following proposition holds.

Proposition 3.8. *Given a normed vector space $E = E_1 \oplus \cdots \oplus E_n$, and a normed vector space F, given any open subset A of E, for any function $f\colon A \to F$, for every $c \in A$, if $\mathrm{D}f(c)$ exists, then each $\dfrac{\partial f}{\partial x_j}(c)$ exists, and*

$$\mathrm{D}f(c)(u_1, \ldots, u_n) = \frac{\partial f}{\partial x_j}(c)(u_1) + \cdots + \frac{\partial f}{\partial x_j}(c)(u_n),$$

for every $u_i \in E_i$, $1 \leq i \leq n$. The same result holds for the finite product $E_1 \times \cdots \times E_n$.

Proof. If $i_j\colon E_j \to E$ is the linear map given by

$$i_j(x) = (0, \ldots, 0, x, 0, \ldots, 0),$$

then

$$i_j^c(x) = (c_1, \ldots, c_{j-1}, 0, c_{j+1}, \ldots, c_n) + i_j(x),$$

which shows that i_j^c is affine, so $\mathrm{D}i_j^c(x) = i_j$. The proposition is then a simple application of Theorem 3.1. □

In the special case where F is a normed vector space of finite dimension m, for any basis $(v_1, \ldots, v_m)$ of F, every vector $x \in F$ can be expressed uniquely as

$$x = x_1 v_1 + \cdots + x_m v_m,$$

where $(x_1, \ldots, x_m) \in K^m$, the coordinates of x in the basis $(v_1, \ldots, v_m)$ (where $K = \mathbb{R}$ or $K = \mathbb{C}$). Thus, letting F_i be the standard normed vector space K with its natural structure, we note that F is isomorphic to the direct sum $F = K \oplus \cdots \oplus K$. Then every function $f\colon E \to F$ is represented by m functions $(f_1, \ldots, f_m)$, where $f_i\colon E \to K$ (where $K = \mathbb{R}$ or $K = \mathbb{C}$), and

$$f(x) = f_1(x)v_1 + \cdots + f_m(x)v_m,$$

for every $x \in E$. The following proposition is easily shown.

Proposition 3.9. *For any two normed vector spaces E and F, if F is of finite dimension m, for any basis $(v_1, \ldots, v_m)$ of F, a function $f\colon E \to F$ is differentiable at a iff each f_i is differentiable at a, and*

$$\mathrm{D}f(a)(u) = \mathrm{D}f_1(a)(u)v_1 + \cdots + \mathrm{D}f_m(a)(u)v_m,$$

for every $u \in E$.

3.3 Jacobian Matrices

If both E and F are of finite dimension, for any basis $(u_1, \ldots, u_n)$ of E and any basis $(v_1, \ldots, v_m)$ of F, every function $f\colon E \to F$ is determined by m functions $f_i\colon E \to \mathbb{R}$ (or $f_i\colon E \to \mathbb{C}$), where

$$f(x) = f_1(x)v_1 + \cdots + f_m(x)v_m,$$

for every $x \in E$. From Proposition 3.1, we have

$$\mathrm{D}f(a)(u_j) = \mathrm{D}_{u_j}f(a) = \partial_j f(a),$$

and from Proposition 3.9, we have

$$\mathrm{D}f(a)(u_j) = \mathrm{D}f_1(a)(u_j)v_1 + \cdots + \mathrm{D}f_i(a)(u_j)v_i + \cdots + \mathrm{D}f_m(a)(u_j)v_m,$$

that is,

$$\mathrm{D}f(a)(u_j) = \partial_j f_1(a)v_1 + \cdots + \partial_j f_i(a)v_i + \cdots + \partial_j f_m(a)v_m.$$

Since the j-th column of the $m \times n$-matrix representing $\mathrm{D}f(a)$ w.r.t. the bases $(u_1, \ldots, u_n)$ and $(v_1, \ldots, v_m)$ is equal to the components of the vector $\mathrm{D}f(a)(u_j)$ over the basis $(v_1, \ldots, v_m)$, the linear map $\mathrm{D}f(a)$ is determined by the $m \times n$-matrix $J(f)(a) = (\partial_j f_i(a))$, (or $J(f)(a) = (\partial f_i/\partial x_j)(a)$):

$$J(f)(a) = \begin{pmatrix} \partial_1 f_1(a) & \partial_2 f_1(a) & \ldots & \partial_n f_1(a) \\ \partial_1 f_2(a) & \partial_2 f_2(a) & \ldots & \partial_n f_2(a) \\ \vdots & \vdots & \ddots & \vdots \\ \partial_1 f_m(a) & \partial_2 f_m(a) & \ldots & \partial_n f_m(a) \end{pmatrix}$$

or

$$J(f)(a) = \begin{pmatrix} \dfrac{\partial f_1}{\partial x_1}(a) & \dfrac{\partial f_1}{\partial x_2}(a) & \ldots & \dfrac{\partial f_1}{\partial x_n}(a) \\ \dfrac{\partial f_2}{\partial x_1}(a) & \dfrac{\partial f_2}{\partial x_2}(a) & \ldots & \dfrac{\partial f_2}{\partial x_n}(a) \\ \vdots & \vdots & \ddots & \vdots \\ \dfrac{\partial f_m}{\partial x_1}(a) & \dfrac{\partial f_m}{\partial x_2}(a) & \ldots & \dfrac{\partial f_m}{\partial x_n}(a) \end{pmatrix}.$$

Definition 3.6. The matrix $J(f)(a)$ is called the *Jacobian matrix* of $\mathrm{D}f$ at a. When $m = n$, the determinant, $\det(J(f)(a))$, of $J(f)(a)$ is called the *Jacobian* of $\mathrm{D}f(a)$.

From a standard fact of linear algebra, we know that this determinant in fact only depends on $\mathrm{D}f(a)$, and not on specific bases. However, partial derivatives give a means for computing it.

When $E = \mathbb{R}^n$ and $F = \mathbb{R}^m$, for any function $f\colon \mathbb{R}^n \to \mathbb{R}^m$, it is easy to compute the partial derivatives $(\partial f_i/\partial x_j)(a)$. We simply treat the function $f_i\colon \mathbb{R}^n \to \mathbb{R}$ as a function of its j-th argument, leaving the others fixed, and compute the derivative as in Definition 3.1, that is, the usual derivative.

Example 3.3. For example, consider the function $f\colon \mathbb{R}^2 \to \mathbb{R}^2$, defined such that

$$f(r,\theta) = (r\cos(\theta), r\sin(\theta)).$$

Then we have

$$J(f)(r,\theta) = \begin{pmatrix} \cos(\theta) & -r\sin(\theta) \\ \sin(\theta) & r\cos(\theta) \end{pmatrix},$$

and the Jacobian (determinant) has value $\det(J(f)(r,\theta)) = r$.

In the case where $E = \mathbb{R}$ (or $E = \mathbb{C}$), for any function $f\colon \mathbb{R} \to F$ (or $f\colon \mathbb{C} \to F$), the Jacobian matrix of $\mathrm{D}f(a)$ is a column vector. In fact, this column vector is just $\mathrm{D}_1 f(a)$. Then for every $\lambda \in \mathbb{R}$ (or $\lambda \in \mathbb{C}$),

$$\mathrm{D}f(a)(\lambda) = \lambda \mathrm{D}_1 f(a).$$

This case is sufficiently important to warrant a definition.

Definition 3.7. Given a function $f\colon \mathbb{R} \to F$ (or $f\colon \mathbb{C} \to F$), where F is a normed vector space, the vector

$$\mathrm{D}f(a)(1) = \mathrm{D}_1 f(a)$$

is called the *vector derivative or velocity vector (in the real case)* at a. We usually identify $\mathrm{D}f(a)$ with its Jacobian matrix $\mathrm{D}_1 f(a)$, which is the column vector corresponding to $\mathrm{D}_1 f(a)$. By abuse of notation, we also let $\mathrm{D}f(a)$ denote the vector $\mathrm{D}f(a)(1) = \mathrm{D}_1 f(a)$.

When $E = \mathbb{R}$, the physical interpretation is that f defines a (parametric) curve that is the trajectory of some particle moving in $\mathbb{R}^m$ as a function of time, and the vector $\mathrm{D}_1 f(a)$ is the *velocity* of the moving particle $f(t)$ at $t = a$; see Figure 3.4.

It is often useful to consider functions $f\colon [a,b] \to F$ from a closed interval $[a,b] \subseteq \mathbb{R}$ to a normed vector space F, and its derivative $\mathrm{D}f(a)$ on $[a,b]$, even though $[a,b]$ is not open. In this case, as in the case of a real-valued function, we define the right derivative $\mathrm{D}_1 f(a_+)$ at a, and the left derivative $\mathrm{D}_1 f(b_-)$ at b, and we assume their existence.

Example 3.4.

(1) When $A = (0,1)$ and $F = \mathbb{R}^3$, a function $f\colon (0,1) \to \mathbb{R}^3$ defines a (parametric) curve in $\mathbb{R}^3$. If $f = (f_1, f_2, f_3)$, its Jacobian matrix at $a \in \mathbb{R}$ is

$$J(f)(a) = \begin{pmatrix} \dfrac{\partial f_1}{\partial t}(a) \\ \dfrac{\partial f_2}{\partial t}(a) \\ \dfrac{\partial f_3}{\partial t}(a) \end{pmatrix}.$$

See Figure 3.4.

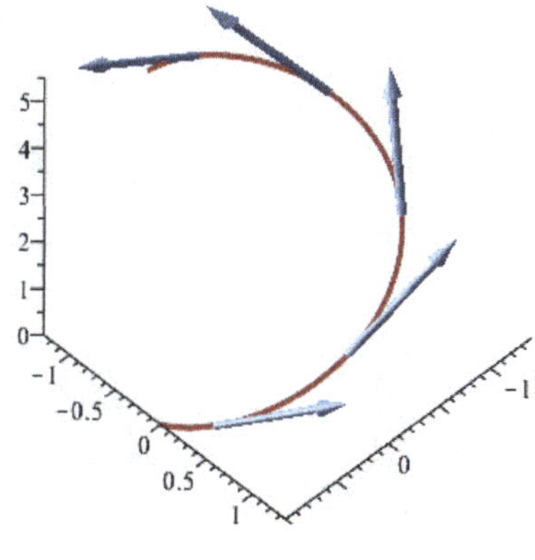

Fig. 3.4 The red space curve $f(t) = (\cos(t), \sin(t), t)$.

The velocity vectors $J(f)(a) = \begin{pmatrix} -\sin(t) \\ \cos(t) \\ 1 \end{pmatrix}$ are represented by the blue arrows.

(2) When $E = \mathbb{R}^2$ and $F = \mathbb{R}^3$, a function $\varphi\colon \mathbb{R}^2 \to \mathbb{R}^3$ defines a parametric surface. Letting $\varphi = (f, g, h)$, its Jacobian matrix at $a \in \mathbb{R}^2$ is

$$J(\varphi)(a) = \begin{pmatrix} \frac{\partial f}{\partial u}(a) & \frac{\partial f}{\partial v}(a) \\ \frac{\partial g}{\partial u}(a) & \frac{\partial g}{\partial v}(a) \\ \frac{\partial h}{\partial u}(a) & \frac{\partial h}{\partial v}(a) \end{pmatrix}.$$

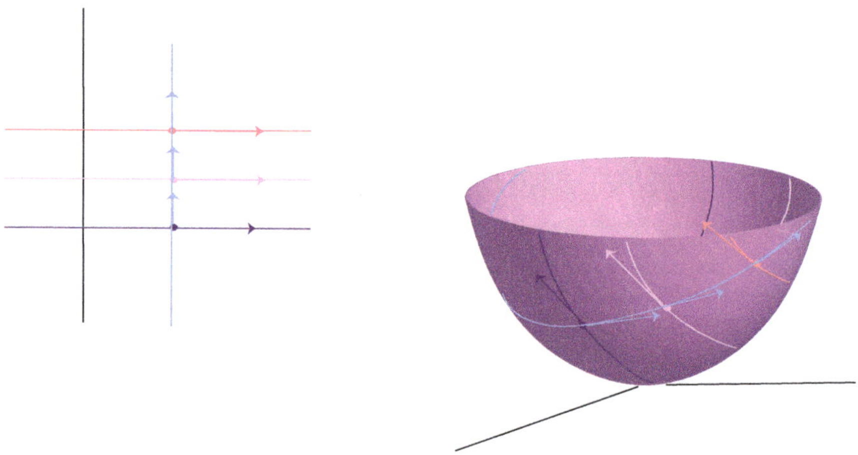

Fig. 3.5 The parametric surface $x = u, y = v, z = u^2 + v^2$.

See Figure 3.5. The Jacobian matrix is $J(f)(a) = \begin{pmatrix} 1 & 0 \\ 0 & 1 \\ 2u & 2v \end{pmatrix}$. The first column is the vector tangent to the pink u-direction curve, while the second column is the vector tangent to the blue v-direction curve.

(3) When $E = \mathbb{R}^3$ and $F = \mathbb{R}$, for a function $f\colon \mathbb{R}^3 \to \mathbb{R}$, the Jacobian matrix at $a \in \mathbb{R}^3$ is

$$J(f)(a) = \left(\frac{\partial f}{\partial x}(a) \; \frac{\partial f}{\partial y}(a) \; \frac{\partial f}{\partial z}(a) \right).$$

More generally, when $f\colon \mathbb{R}^n \to \mathbb{R}$, the Jacobian matrix at $a \in \mathbb{R}^n$ is the row vector

$$J(f)(a) = \left(\frac{\partial f}{\partial x_1}(a) \; \cdots \; \frac{\partial f}{\partial x_n}(a) \right).$$

Its transpose is a column vector called the *gradient* of f at a, denoted by $\text{grad} f(a)$ or $\nabla f(a)$. Then given any $v \in \mathbb{R}^n$, note that

$$\mathrm{D}f(a)(v) = \frac{\partial f}{\partial x_1}(a)\, v_1 + \cdots + \frac{\partial f}{\partial x_n}(a)\, v_n = \text{grad} f(a) \cdot v,$$

the scalar product of $\text{grad} f(a)$ and v.

Example 3.5. Consider the quadratic function $f \colon \mathbb{R}^n \to \mathbb{R}$ given by

$$f(x) = x^\top A x, \quad x \in \mathbb{R}^n,$$

where A is a real $n \times n$ symmetric matrix. We claim that

$$df_u(h) = 2u^\top A h \quad \text{for all } u, h \in \mathbb{R}^n.$$

Since A is symmetric, we have

$$\begin{aligned} f(u+h) &= (u^\top + h^\top)A(u+h) \\ &= u^\top A u + u^\top A h + h^\top A u + h^\top A h \\ &= u^\top A u + 2u^\top A h + h^\top A h, \end{aligned}$$

so we have

$$f(u+h) - f(u) - 2u^\top A h = h^\top A h.$$

If we write

$$\epsilon(h) = \frac{h^\top A h}{\|h\|}$$

for $h \notin 0$ where $\| \ \|$ is the 2-norm, by Cauchy–Schwarz we have

$$|\epsilon(h)| \leq \frac{\|h\| \, \|Ah\|}{\|h\|} \leq \frac{\|h\|^2 \, \|A\|}{\|h\|} = \|h\| \, \|A\|,$$

which shows that $\lim_{h \mapsto 0} \epsilon(h) = 0$. Therefore,

$$df_u(h) = 2u^\top A h \quad \text{for all } u, h \in \mathbb{R}^n,$$

as claimed. This formula shows that the gradient ∇f_u of f at u is given by

$$\nabla f_u = 2Au.$$

As a first corollary we obtain the gradient of a function of the form

$$f(x) = \frac{1}{2} x^\top A x - b^\top x,$$

where A is a symmetric $n \times n$ matrix and b is some vector $b \in \mathbb{R}^n$. Since the derivative of a linear function is itself, we obtain

$$df_u(h) = u^\top A h - b^\top h,$$

and the gradient of $f(x) = \frac{1}{2}x^\top Ax - b^\top x$, is given by

$$\nabla f_u = Au - b.$$

As a second corollary we obtain the gradient of the function

$$f(x) = \|Ax - b\|_2^2 = (Ax - b)^\top (Ax - b) = (x^\top A^\top - b^\top)(Ax - b)$$

which is the function to minimize in a least squares problem, where A is an $m \times n$ matrix. We have

$$f(x) = x^\top A^\top Ax - x^\top A^\top b - b^\top Ax + b^\top b = x^\top A^\top Ax - 2b^\top Ax + b^\top b,$$

and since the derivative of a constant function is 0 and the derivative of a linear function is itself, we get

$$df_u(h) = 2u^\top A^\top Ah - 2b^\top Ah.$$

Consequently, the gradient of $f(x) = \|Ax - b\|_2^2$ is given by

$$\nabla f_u = 2A^\top Au - 2A^\top b.$$

These two results will be heavily used in quadratic optimization.

When E, F, and G have finite dimensions, and $(u_1, \ldots, u_p)$ is a basis for E, $(v_1, \ldots, v_n)$ is a basis for F, and $(w_1, \ldots, w_m)$ is a basis for G, if A is an open subset of E, B is an open subset of F, for any functions $f\colon A \to F$ and $g\colon B \to G$, such that $f(A) \subseteq B$, for any $a \in A$, letting $b = f(a)$, and $h = g \circ f$, if $\mathrm{D}f(a)$ exists and $\mathrm{D}g(b)$ exists, by Theorem 3.1, the Jacobian matrix $J(h)(a) = J(g \circ f)(a)$ w.r.t. the bases $(u_1, \ldots, u_p)$ and $(w_1, \ldots, w_m)$ is the product of the Jacobian matrices $J(g)(b)$ w.r.t. the bases $(v_1, \ldots, v_n)$ and $(w_1, \ldots, w_m)$, and $J(f)(a)$ w.r.t. the bases $(u_1, \ldots, u_p)$ and $(v_1, \ldots, v_n)$:

$$J(h)(a) = \begin{pmatrix} \partial_1 g_1(b) & \partial_2 g_1(b) & \ldots & \partial_n g_1(b) \\ \partial_1 g_2(b) & \partial_2 g_2(b) & \ldots & \partial_n g_2(b) \\ \vdots & \vdots & \ddots & \vdots \\ \partial_1 g_m(b) & \partial_2 g_m(b) & \ldots & \partial_n g_m(b) \end{pmatrix} \begin{pmatrix} \partial_1 f_1(a) & \partial_2 f_1(a) & \ldots & \partial_p f_1(a) \\ \partial_1 f_2(a) & \partial_2 f_2(a) & \ldots & \partial_p f_2(a) \\ \vdots & \vdots & \ddots & \vdots \\ \partial_1 f_n(a) & \partial_2 f_n(a) & \ldots & \partial_p f_n(a) \end{pmatrix}$$

or

$$J(h)(a) = \begin{pmatrix} \frac{\partial g_1}{\partial y_1}(b) & \frac{\partial g_1}{\partial y_2}(b) & \ldots & \frac{\partial g_1}{\partial y_n}(b) \\ \frac{\partial g_2}{\partial y_1}(b) & \frac{\partial g_2}{\partial y_2}(b) & \ldots & \frac{\partial g_2}{\partial y_n}(b) \\ \vdots & \vdots & \ddots & \vdots \\ \frac{\partial g_m}{\partial y_1}(b) & \frac{\partial g_m}{\partial y_2}(b) & \ldots & \frac{\partial g_m}{\partial y_n}(b) \end{pmatrix} \begin{pmatrix} \frac{\partial f_1}{\partial x_1}(a) & \frac{\partial f_1}{\partial x_2}(a) & \ldots & \frac{\partial f_1}{\partial x_p}(a) \\ \frac{\partial f_2}{\partial x_1}(a) & \frac{\partial f_2}{\partial x_2}(a) & \ldots & \frac{\partial f_2}{\partial x_p}(a) \\ \vdots & \vdots & \ddots & \vdots \\ \frac{\partial f_n}{\partial x_1}(a) & \frac{\partial f_n}{\partial x_2}(a) & \ldots & \frac{\partial f_n}{\partial x_p}(a) \end{pmatrix}.$$

Thus, we have the familiar formula

$$\frac{\partial h_i}{\partial x_j}(a) = \sum_{k=1}^{k=n} \frac{\partial g_i}{\partial y_k}(b)\frac{\partial f_k}{\partial x_j}(a).$$

Given two normed vector spaces E and F of finite dimension, given an open subset A of E, if a function $f\colon A \to F$ is differentiable at $a \in A$, then its Jacobian matrix is well defined.

One should be warned that the converse is false. As evidenced by Example 3.1, there are functions such that all the partial derivatives exist at some $a \in A$, but yet, the function is not differentiable at a, and not even continuous at a. However, there are sufficient conditions on the partial derivatives for $\mathrm{D}f(a)$ to exist, namely, continuity of the partial derivatives.

If f is differentiable on A, then f defines a function $\mathrm{D}f\colon A \to \mathcal{L}(E; F)$. It turns out that the continuity of the partial derivatives on A is a necessary and sufficient condition for $\mathrm{D}f$ to exist and to be continuous on A.

If $f\colon [a, b] \to \mathbb{R}$ is a function which is continuous on $[a, b]$ and differentiable on $(a, b]$, then there is some c with $a < c < b$ such that

$$f(b) - f(a) = (b - a)f'(c).$$

This result is known as the *mean value theorem* and is a generalization of *Rolle's theorem*, which corresponds to the case where $f(a) = f(b)$.

Unfortunately, the mean value theorem fails for vector-valued functions. For example, the function $f\colon [0, 2\pi] \to \mathbb{R}^2$ given by

$$f(t) = (\cos t, \sin t)$$

is such that $f(2\pi) - f(0) = (0, 0)$, yet its derivative $f'(t) = (-\sin t, \cos t)$ does not vanish in $(0, 2\pi)$.

A suitable generalization of the mean value theorem to vector-valued functions is possible if we consider an inequality (an upper bound) instead of an equality. This generalized version of the mean value theorem plays an important role in the proof of several major results of differential calculus.

If E is an vector space (over $\mathbb{R}$ or $\mathbb{C}$), given any two points $a, b \in E$, the *closed segment* $[a, b]$ is the set of all points $a + \lambda(b - a)$, where $0 \leq \lambda \leq 1$, $\lambda \in \mathbb{R}$, and the *open segment* (a, b) is the set of all points $a + \lambda(b - a)$, where $0 < \lambda < 1$, $\lambda \in \mathbb{R}$.

Proposition 3.10. *Let E and F be two normed vector spaces, let A be an open subset of E, and let $f\colon A \to F$ be a continuous function on A.*

Given any $a \in A$ and any $h \neq 0$ in E, if the closed segment $[a, a+h]$ is contained in A, if $f\colon A \to F$ is differentiable at every point of the open segment $(a, a+h)$, and

$$\sup_{x \in (a,a+h)} \|\mathrm{D}f(x)\| \leq M,$$

for some $M \geq 0$, then

$$\|f(a+h) - f(a)\| \leq M\|h\|.$$

As a corollary, if $L\colon E \to F$ is a continuous linear map, then

$$\|f(a+h) - f(a) - L(h)\| \leq M\|h\|,$$

where $M = \sup_{x \in (a,a+h)} \|\mathrm{D}f(x) - L\|$.

The above proposition is sometimes called the "mean value theorem." Proposition 3.10 can be used to show the following important result.

Theorem 3.2. *Given two normed vector spaces E and F, where E is of finite dimension n, and where $(u_1, \ldots, u_n)$ is a basis of E, given any open subset A of E, given any function $f\colon A \to F$, the derivative $\mathrm{D}f\colon A \to \mathcal{L}(E; F)$ is defined and continuous on A iff every partial derivative $\partial_j f$ (or $\frac{\partial f}{\partial x_j}$) is defined and continuous on A, for all j, $1 \leq j \leq n$. As a corollary, if F is of finite dimension m, and $(v_1, \ldots, v_m)$ is a basis of F, the derivative $\mathrm{D}f\colon A \to \mathcal{L}(E; F)$ is defined and continuous on A iff every partial derivative $\partial_j f_i$ (or $\frac{\partial f_i}{\partial x_j}$) is defined and continuous on A, for all i, j, $1 \leq i \leq m$, $1 \leq j \leq n$.*

Theorem 3.2 gives a necessary and sufficient condition for the existence and continuity of the derivative of a function on an open set. It should be noted that a more general version of Theorem 3.2 holds, assuming that $E = E_1 \oplus \cdots \oplus E_n$, or $E = E_1 \times \cdots \times E_n$, and using the more general partial derivatives $\mathrm{D}_j f$ introduced before Proposition 3.8.

Definition 3.8. Given two normed vector spaces E and F, and an open subset A of E, we say that a function $f\colon A \to F$ is *of class C^0 on A or a C^0-function on A* if f is continuous on A. We say that $f\colon A \to F$ is *of class C^1 on A or a C^1-function on A* if $\mathrm{D}f$ exists and is continuous on A.

Since the existence of the derivative on an open set implies continuity, a C^1-function is of course a C^0-function. Theorem 3.2 gives a necessary and sufficient condition for a function f to be a C^1-function (when E is of finite dimension). It is easy to show that the composition of C^1-functions (on appropriate open sets) is a C^1-function.

3.4 The Implicit and the Inverse Function Theorems

Given three normed vector spaces E, F, and G, given a function $f: E \times F \to G$, given any $c \in G$, it may happen that the equation

$$f(x, y) = c$$

has the property that for some open sets $A \subseteq E$ and $B \subseteq F$, there is a function $g: A \to B$, such that

$$f(x, g(x)) = c,$$

for all $x \in A$. Such a situation is usually very rare, but if some solution $(a, b) \in E \times F$ such that $f(a, b) = c$ is known, under certain conditions, for some small open sets $A \subseteq E$ containing a and $B \subseteq F$ containing b, the existence of a unique $g: A \to B$ such that

$$f(x, g(x)) = c,$$

for all $x \in A$, can be shown. Under certain conditions, it can also be shown that g is continuous and differentiable. Such a theorem, known as the *implicit function theorem*, can be proven.

Example 3.6. Let $E = \mathbb{R}^2$, $F = G = \mathbb{R}$, $\Omega = \mathbb{R}^2 \times \mathbb{R} \cong \mathbb{R}^3$, $f: \mathbb{R}^2 \times \mathbb{R} \to \mathbb{R}$ given by

$$f((x_1, x_2), x_3) = x_1^2 + x_2^2 + x_3^2 - 1,$$

$a = (\sqrt{3}/(2\sqrt{2}), \sqrt{3}/(2\sqrt{2}))$, $b = 1/2$, and $c = 0$. The set of vectors $(x_1, x_2, x_3) \in \mathbb{R}^2$ such that

$$f((x_1, x_2), x_3) = x_1^2 + x_2^2 + x_3^2 - 1 = 0$$

is the unit sphere in $\mathbb{R}^3$. The vector (a, b) belongs to the unit sphere since $\|a\|_2^2 + b^2 - 1 = 0$. The function $g: \mathbb{R}^2 \to \mathbb{R}$ given by

$$g(x_1, x_2) = \sqrt{1 - x_1^2 - x_2^2}$$

satisfies the equation

$$f(x_1, x_2, g(x_1, x_2)) = 0$$

all for (x_1, x_2) in the open disk $\{(x_1, x_2) \in \mathbb{R}^2 \mid x_1^2 + x_2^2 < 1\}$, and $g(a) = b$. Observe that if we had picked $b = -1/2$, then we would need to consider the function

$$g(x_1, x_2) = -\sqrt{1 - x_1^2 - x_2^2}.$$

We now state a very general version of the implicit function theorem. The proof of this theorem is fairly involved and uses a fixed-point theorem for contracting mappings in complete metric spaces; it is given in Schwartz [Schwartz (1992)].

Theorem 3.3. *Let E, F, and G be normed vector spaces, let Ω be an open subset of $E \times F$, let $f\colon \Omega \to G$ be a function defined on Ω, let $(a, b) \in \Omega$, let $c \in G$, and assume that $f(a, b) = c$. If the following assumptions hold:*

(1) The function $f\colon \Omega \to G$ is continuous on Ω;
(2) F is a complete normed vector space;
(3) $\dfrac{\partial f}{\partial y}(x, y)$ exists for every $(x, y) \in \Omega$ and $\dfrac{\partial f}{\partial y}\colon \Omega \to \mathcal{L}(F; G)$ is continuous, where $\dfrac{\partial f}{\partial y}(x, y)$ is defined as in Definition 3.5;
(4) $\dfrac{\partial f}{\partial y}(a, b)$ is a bijection of $\mathcal{L}(F; G)$, and $\left(\dfrac{\partial f}{\partial y}(a, b)\right)^{-1} \in \mathcal{L}(G; F)$; this hypothesis implies that G is also a complete normed vector space;

then the following properties hold:

(a) There exist some open subset $A \subseteq E$ containing a and some open subset $B \subseteq F$ containing b, such that $A \times B \subseteq \Omega$, and for every $x \in A$, the equation $f(x, y) = c$ has a single solution $y = g(x)$, and thus there is a unique function $g\colon A \to B$ such that $f(x, g(x)) = c$, for all $x \in A$;
(b) The function $g\colon A \to B$ is continuous.

If we also assume that

(5) The derivative $\mathrm{D}f(a, b)$ exists;

then

(c) The derivative $\mathrm{D}g(a)$ exists, and

$$\mathrm{D}g(a) = -\left(\frac{\partial f}{\partial y}(a, b)\right)^{-1} \circ \frac{\partial f}{\partial x}(a, b);$$

and if in addition

(6) $\dfrac{\partial f}{\partial x}\colon \Omega \to \mathcal{L}(E; G)$ is also continuous (and thus, in view of (3), f is C^1 on Ω);

then

(d) *The derivative* $\mathrm{D}g\colon A \to \mathcal{L}(E;F)$ *is continuous, and*

$$\mathrm{D}g(x) = -\Big(\frac{\partial f}{\partial y}(x, g(x))\Big)^{-1} \circ \frac{\partial f}{\partial x}(x, g(x)),$$

for all $x \in A$.

Example 3.7. Going back to Example 3.6, write $x = (x_1, x_2)$ and $y = x_3$, so that the partial derivatives $\partial f/\partial x$ and $\partial f/\partial y$ are given in terms of their Jacobian matrices by

$$\begin{aligned}\frac{\partial f}{\partial x}(x, y) &= \begin{pmatrix}2x_1 & 2x_2\end{pmatrix}\\ \frac{\partial f}{\partial y}(x, y) &= 2x_3.\end{aligned}$$

If $0 < |b| \leq 1$ and $\|a\|_2^2 + b^2 - 1 = 0$, then Conditions (3) and (4) are satisfied. Conditions (1) and (2) obviously hold. Since $df_{(a,b)}$ is given by its Jacobian matrix as

$$df_{(a,b)} = \begin{pmatrix}2a_1 & 2a_2 & 2b\end{pmatrix},$$

Condition (5) holds, and clearly, Condition (6) also holds.

Theorem 3.3 implies that there is some open subset A of $\mathbb{R}^2$ containing a, some open subset B of $\mathbb{R}$ containing b, and a unique function $g\colon A \to B$ such that

$$f(x, g(x)) = 0$$

for all $x \in A$. In fact, we can pick A to be the open unit disk in $\mathbb{R}$, $B = (0, 2)$, and if $0 < b \leq 1$, then

$$g(x_1, x_2) = \sqrt{1 - x_1^2 - x_2^2},$$

else if $-1 \leq b < 0$, then

$$g(x_1, x_2) = -\sqrt{1 - x_1^2 - x_2^2}.$$

Assuming $0 < b \leq 1$, we have

$$\frac{\partial f}{\partial x}(x, g(x)) = (2x_1 \;\; 2x_2),$$

and

$$\Big(\frac{\partial f}{\partial y}(x, g(x))\Big)^{-1} = \frac{1}{2\sqrt{1 - x_1^2 - x_2^2}},$$

so according to the theorem,

$$dg_x = -\frac{1}{\sqrt{1 - x_1^2 - x_2^2}}(x_1 \;\; x_2),$$

which matches the derivative of g computed directly.

Observe that the functions $(x_1, x_2) \mapsto \sqrt{1 - x_1^2 - x_2^2}$ and $(x_1, x_2) \mapsto -\sqrt{1 - x_1^2 - x_2^2}$ are two differentiable parametrizations of the sphere, but the union of their ranges does not cover the entire sphere. Since $b \neq 0$, none of the points on the unit circle in the (x_1, x_2)-plane are covered. Our function f views b as lying on the x_3-axis. In order to cover the entire sphere using this method, we need four more maps, which correspond to b lying on the x_1-axis or on the x_2 axis. Then we get the additional (implicit) maps $(x_2, x_3) \mapsto \pm\sqrt{1 - x_2^2 - x_3^2}$ and $(x_1, x_3) \mapsto \pm\sqrt{1 - x_1^2 - x_3^2}$.

The implicit function theorem plays an important role in the calculus of variations.

We now consider another very important notion, that of a (local) diffeomorphism.

Definition 3.9. Given two topological spaces E and F and an open subset A of E, we say that a function $f\colon A \to F$ is a *local homeomorphism from A to F* if for every $a \in A$, there is an open set $U \subseteq A$ containing a and an open set V containing $f(a)$ such that f is a homeomorphism from U to $V = f(U)$. If B is an open subset of F, we say that $f\colon A \to F$ is a *(global) homeomorphism from A to B* if f is a homeomorphism from A to $B = f(A)$. If E and F are normed vector spaces, we say that $f\colon A \to F$ is a *local diffeomorphism from A to F* if for every $a \in A$, there is an open set $U \subseteq A$ containing a and an open set V containing $f(a)$ such that f is a bijection from U to V, f is a C^1-function on U, and f^{-1} is a C^1-function on $V = f(U)$. We say that $f\colon A \to F$ is a *(global) diffeomorphism from A to B* if f is a homeomorphism from A to $B = f(A)$, f is a C^1-function on A, and f^{-1} is a C^1-function on B.

Note that a local diffeomorphism is a local homeomorphism. Also, as a consequence of Proposition 3.7, if f is a diffeomorphism on A, then $\mathrm{D}f(a)$ is a bijection for every $a \in A$. The following theorem can be shown. In fact, there is a fairly simple proof using Theorem 3.3.

Theorem 3.4. *(Inverse Function Theorem) Let E and F be complete normed spaces, let A be an open subset of E, and let $f\colon A \to F$ be a C^1-function on A. The following properties hold:*

(1) For every $a \in A$, if $\mathrm{D}f(a)$ is a linear isomorphism (which means that both $\mathrm{D}f(a)$ and $(\mathrm{D}f(a))^{-1}$ are linear and continuous),[2] then there exist

[2]Actually, since E and F are Banach spaces, by the open mapping theorem, it is

some open subset $U \subseteq A$ containing a, and some open subset V of F containing $f(a)$, such that f is a diffeomorphism from U to $V = f(U)$. Furthermore,

$$\mathrm{D}f^{-1}(f(a)) = (\mathrm{D}f(a))^{-1}.$$

For every neighborhood N of a, the image $f(N)$ of N is a neighborhood of $f(a)$, and for every open ball $U \subseteq A$ of center a, the image $f(U)$ of U contains some open ball of center $f(a)$.

(2) *If $\mathrm{D}f(a)$ is invertible for every $a \in A$, then $B = f(A)$ is an open subset of F, and f is a local diffeomorphism from A to B. Furthermore, if f is injective, then f is a diffeomorphism from A to B.*

Proofs of the inverse function theorem can be found in Schwartz [Schwartz (1992)], Lang [Lang (1996)], Abraham and Marsden [Abraham and Marsden (1978)], and Cartan [Cartan (1990)].

The idea of Schwartz's proof is that if we define the function $f_1 \colon F \times \Omega \to F$ by

$$f_1(y, z) = f(z) - y,$$

then an inverse $g = f^{-1}$ of f is an implicit solution of the equation $f_1(y, z) = 0$, since $f_1(y, g(y)) = f(g(y)) - y = 0$. Observe that the roles of E and F are switched, but this is not a problem since F is complete. The proof consists in checking that the conditions of Theorem 3.3 apply.

Part (1) of Theorem 3.4 is often referred to as the "(local) inverse function theorem." It plays an important role in the study of manifolds and (ordinary) differential equations.

If E and F are both of finite dimension, and some bases have been chosen, the invertibility of $\mathrm{D}f(a)$ is equivalent to the fact that the Jacobian determinant $\det(J(f)(a))$ is nonnull. The case where $\mathrm{D}f(a)$ is just injective or just surjective is also important for defining manifolds, using implicit definitions.

Definition 3.10. Let E and F be normed vector spaces, where E and F are of finite dimension (or both E and F are complete), and let A be an open subset of E. For any $a \in A$, a C^1-function $f \colon A \to F$ is an *immersion at a* if $\mathrm{D}f(a)$ is injective. A C^1-function $f \colon A \to F$ is a *submersion at a* if $\mathrm{D}f(a)$ is surjective. A C^1-function $f \colon A \to F$ is an *immersion on A* (resp. *a submersion on A*) if $\mathrm{D}f(a)$ is injective (resp. surjective) for every $a \in A$.

---

sufficient to assume that $\mathrm{D}f(a)$ is continuous and bijective; see Lang [Lang (1996)].

When E and F are finite dimensional with $\dim(E) = n$ and $\dim(F) = m$, if $m \geq n$, then f is an immersion iff the Jacobian matrix, $J(f)(a)$, has full rank n for all $a \in E$, and if $n \geq m$, then f is a submersion iff the Jacobian matrix, $J(f)(a)$, has full rank m for all $a \in E$.

Example 3.8. For example, $f: \mathbb{R} \to \mathbb{R}^2$ defined by $f(t) = (\cos(t), \sin(t))$ is an immersion since $J(f)(t) = \begin{pmatrix} -\sin(t) \\ \cos(t) \end{pmatrix}$ has rank 1 for all t. On the other hand, $f: \mathbb{R} \to \mathbb{R}^2$ defined by $f(t) = (t^2, t^2)$ is not an immersion since $J(f)(t) = \begin{pmatrix} 2t \\ 2t \end{pmatrix}$ vanishes at $t = 0$. See Figure 3.6. An example of a submersion is given by the projection map $f: \mathbb{R}^2 \to \mathbb{R}$, where $f(x, y) = x$, since $J(f)(x, y) = \begin{pmatrix} 1 & 0 \end{pmatrix}$.

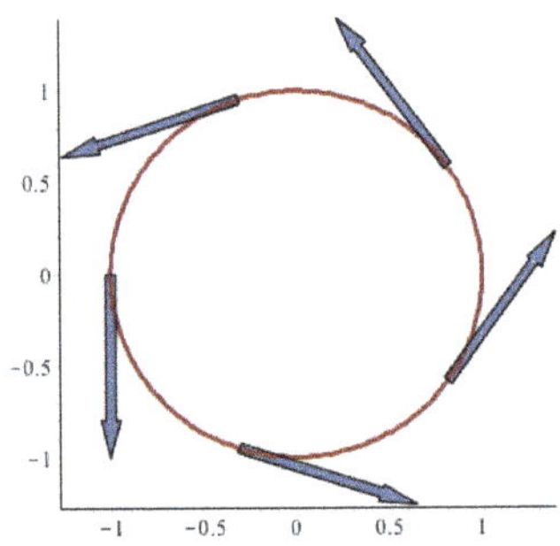

(i.)

(ii.)

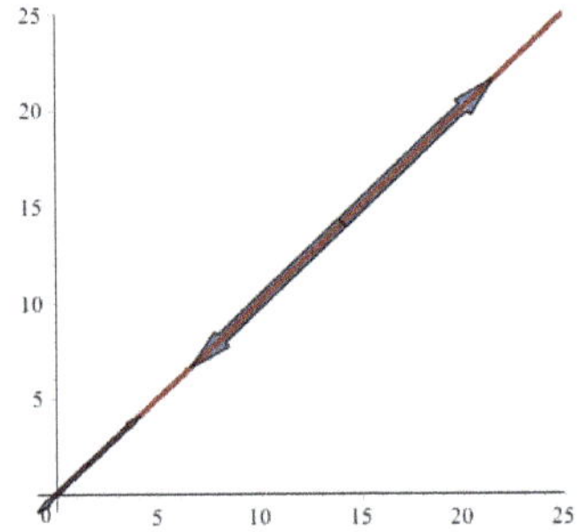

Fig. 3.6 Figure (*i.*) is the immersion of $\mathbb{R}$ into $\mathbb{R}^2$ given by $f(t) = (\cos(t), \sin(t))$. Figure (*ii.*), the parametric curve $f(t) = (t^2, t^2)$, is not an immersion since the tangent vanishes at the origin.

The following results can be shown.

Proposition 3.11. *Let A be an open subset of $\mathbb{R}^n$, and let $f\colon A \to \mathbb{R}^m$ be a function. For every $a \in A$, $f\colon A \to \mathbb{R}^m$ is a submersion at a iff there exists an open subset U of A containing a, an open subset $W \subseteq \mathbb{R}^{n-m}$, and a diffeomorphism $\varphi\colon U \to f(U) \times W$, such that,*

$$f = \pi_1 \circ \varphi,$$

where $\pi_1\colon f(U) \times W \to f(U)$ is the first projection. Equivalently,

$$(f \circ \varphi^{-1})(y_1, \ldots, y_m, \ldots, y_n) = (y_1, \ldots, y_m).$$

$$\begin{array}{ccc} U \subseteq A & \xrightarrow{\ \varphi\ } & f(U) \times W \\ & \searrow^{f} & \downarrow \pi_1 \\ & & f(U) \subseteq \mathbb{R}^m \end{array}$$

Furthermore, the image of every open subset of A under f is an open subset of F. (The same result holds for $\mathbb{C}^n$ and $\mathbb{C}^m$). See Figure 3.7.

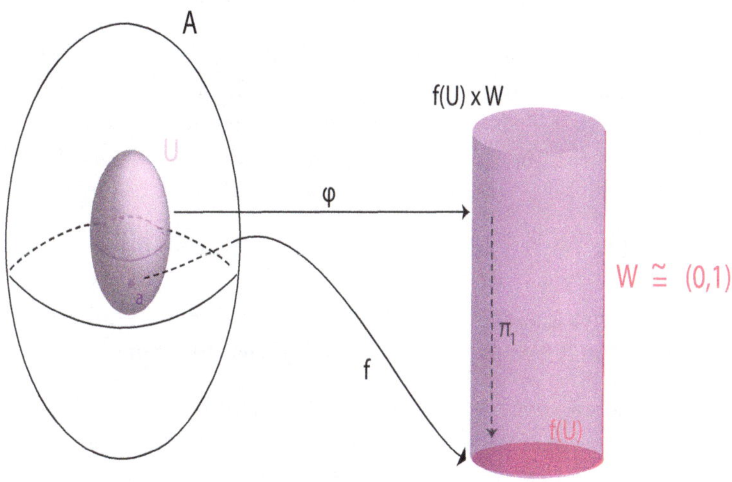

Fig. 3.7 Let $n = 3$ and $m = 2$. The submersion maps the solid lavender egg in $\mathbb{R}^3$ onto the bottom pink circular face of the solid cylinder $f(U) \times W$.

Proposition 3.12. *Let A be an open subset of $\mathbb{R}^n$, and let $f\colon A \to \mathbb{R}^m$ be a function. For every $a \in A$, $f\colon A \to \mathbb{R}^m$ is an immersion at a iff there exists* *an open subset U of A containing a, an open subset V containing $f(a)$ such that $f(U) \subseteq V$, an open subset W containing 0 such that $W \subseteq \mathbb{R}^{m-n}$, and a diffeomorphism $\varphi\colon V \to U \times W$, such that,*

$$\varphi \circ f = in_1,$$

where $in_1\colon U \to U \times W$ is the injection map such that $in_1(u) = (u, 0)$, or equivalently,

$$(\varphi \circ f)(x_1, \ldots, x_n) = (x_1, \ldots, x_n, 0, \ldots, 0).$$

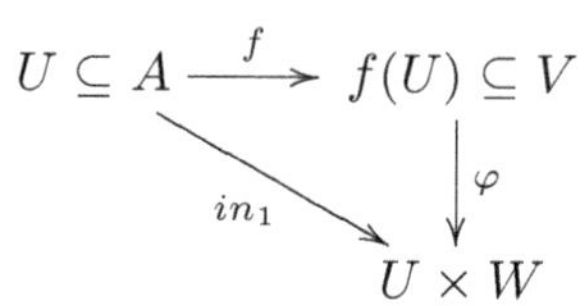

(The same result holds for $\mathbb{C}^n$ and $\mathbb{C}^m$.) See Figure 3.8.

We now briefly consider second-order and higher-order derivatives.

3.5 Second-Order and Higher-Order Derivatives

Given two normed vector spaces E and F, and some open subset A of E, if $\mathrm{D}f(a)$ is defined for every $a \in A$, then we have a mapping $\mathrm{D}f\colon A \to \mathcal{L}(E; F)$. Since $\mathcal{L}(E; F)$ is a normed vector space, if $\mathrm{D}f$ exists on an open subset U of A containing a, we can consider taking the derivative of $\mathrm{D}f$ at some $a \in A$.

Definition 3.11. Given a function $f\colon A \to F$ defined on some open subset A of E such that $\mathrm{D}f(a)$ is defined for every $a \in A$, if $\mathrm{D}(\mathrm{D}f)(a)$ exists for every $a \in A$, we get a mapping $\mathrm{D}^2 f\colon A \to \mathcal{L}(E; \mathcal{L}(E; F))$ called the *second derivative of f on A*, where $\mathrm{D}^2 f(a) = \mathrm{D}(\mathrm{D}f)(a)$, for every $a \in A$.

As in the case of the first derivative $\mathrm{D}f_a$ where $\mathrm{D}f_a(u) = \mathrm{D}_u f(a)$, where $\mathrm{D}_u f(a)$ is the directional derivative of f at a in the direction u, it would be useful to express $\mathrm{D}^2 f(a)(u)(v)$ in terms of two directional derivatives. This can indeed be done. If $\mathrm{D}^2 f(a)$ exists, then for every $u \in E$,

$$\mathrm{D}^2 f(a)(u) = \mathrm{D}(\mathrm{D}f)(a)(u) = \mathrm{D}_u(\mathrm{D}f)(a) \in \mathcal{L}(E; F).$$

We have the following result.

Proposition 3.13. *If $\mathrm{D}^2 f(a)$ exists, then $\mathrm{D}_u(\mathrm{D}_v f)(a)$ exists and*

$$\mathrm{D}^2 f(a)(u)(v) = \mathrm{D}_u(\mathrm{D}_v f)(a), \quad \textit{for all } u, v \in E.$$

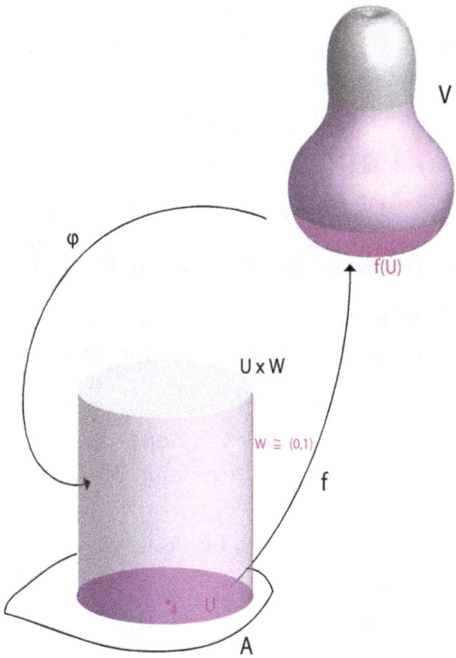

Fig. 3.8 Let $n = 2$ and $m = 3$. The immersion maps the purple circular base of the cylinder $U \times W$ to circular cup on the surface of the solid purple gourd.

Proof. Recall from Proposition 2.28, that the map app from $\mathcal{L}(E; F) \times E$ to F, defined such that for every $L \in \mathcal{L}(E; F)$, for every $v \in E$,

$$\mathrm{app}(L, v) = L(v),$$

is a continuous bilinear map. Thus, in particular, given a fixed $v \in E$, the linear map $\mathrm{app}_v \colon \mathcal{L}(E; F) \to F$, defined such that $\mathrm{app}_v(L) = L(v)$, is a continuous map.

Also recall from Proposition 3.6, that if $h \colon A \to G$ is a function such that $\mathrm{D}h(a)$ exits, and $k \colon G \to H$ is a continuous linear map, then, $\mathrm{D}(k \circ h)(a)$ exists, and

$$k(\mathrm{D}h(a)(u)) = \mathrm{D}(k \circ h)(a)(u),$$

that is,

$$k(\mathrm{D}_u h(a)) = \mathrm{D}_u(k \circ h)(a).$$

Applying these two facts to $h = \mathrm{D}f$, and to $k = \mathrm{app}_v$, we have

$$\mathrm{app}_v(D_u(Df)(a)) = \mathrm{D}_u(\mathrm{D}f)(a)(v) = \mathrm{D}_u(\mathrm{app}_v \circ \mathrm{D}f)(a).$$

But $(\mathrm{app}_v \circ \mathrm{D}f)(x) = \mathrm{D}f(x)(v) = \mathrm{D}_v f(x)$, for every $x \in A$, that is, $\mathrm{app}_v \circ \mathrm{D}f = \mathrm{D}_v f$ on A. So we have

$$\mathrm{D}_u(\mathrm{D}f)(a)(v) = \mathrm{D}_u(\mathrm{D}_v f)(a),$$

and since $\mathrm{D}^2 f(a)(u) = \mathrm{D}_u(\mathrm{D}f)(a)$, we get

$$\mathrm{D}^2 f(a)(u)(v) = \mathrm{D}_u(\mathrm{D}_v f)(a). \qquad \square$$

Definition 3.12. We denote $\mathrm{D}_u(\mathrm{D}_v f)(a)$ by $\mathrm{D}^2_{u,v} f(a)$ (or $\mathrm{D}_u \mathrm{D}_v f(a)$).

Recall from Proposition 2.27, that the map from $\mathcal{L}_2(E, E; F)$ to $\mathcal{L}(E; \mathcal{L}(E; F))$ defined such that $g \mapsto \varphi$ iff for every $g \in \mathcal{L}_2(E, E; F)$,

$$\varphi(u)(v) = g(u, v),$$

is an isomorphism of vector spaces. *Thus, we will consider* $\mathrm{D}^2 f(a) \in \mathcal{L}(E; \mathcal{L}(E; F))$ *as a continuous bilinear map in* $\mathcal{L}_2(E, E; F)$*, and we write* $\mathrm{D}^2 f(a)(u, v)$*, instead of* $\mathrm{D}^2 f(a)(u)(v)$.

Then the above discussion can be summarized by saying that when $\mathrm{D}^2 f(a)$ is defined, we have

$$\mathrm{D}^2 f(a)(u, v) = \mathrm{D}_u \mathrm{D}_v f(a).$$

Definition 3.13. When E has finite dimension and $(e_1, \ldots, e_n)$ is a basis for E, we denote $\mathrm{D}_{e_j} \mathrm{D}_{e_i} f(a)$ by $\dfrac{\partial^2 f}{\partial x_i \partial x_j}(a)$, when $i \neq j$, and we denote $\mathrm{D}_{e_i} \mathrm{D}_{e_i} f(a)$ by $\dfrac{\partial^2 f}{\partial x_i^2}(a)$.

The following important result attributed to Schwarz can be shown using Proposition 3.10. Given a bilinear map $h \colon E \times E \to F$, recall that h is *symmetric* if

$$h(u, v) = h(v, u),$$

for all $u, v \in E$.

Proposition 3.14. *(Schwarz's lemma) Given two normed vector spaces E and F, given any open subset A of E, given any $f \colon A \to F$, for every $a \in A$, if $\mathrm{D}^2 f(a)$ exists, then $\mathrm{D}^2 f(a) \in \mathcal{L}_2(E, E; F)$ is a continuous symmetric bilinear map. As a corollary, if E is of finite dimension n, and $(e_1, \ldots, e_n)$ is a basis for E, we have*

$$\frac{\partial^2 f}{\partial x_i \partial x_j}(a) = \frac{\partial^2 f}{\partial x_j \partial x_i}(a).$$

Remark: There is a variation of the above result which does not assume the existence of $\mathrm{D}^2 f(a)$, but instead assumes that $\mathrm{D}_u\mathrm{D}_v f$ and $\mathrm{D}_v\mathrm{D}_u f$ exist on an open subset containing a and are continuous at a, and concludes that $\mathrm{D}_u\mathrm{D}_v f(a) = \mathrm{D}_v\mathrm{D}_u f(a)$. This is a different result which does not imply Proposition 3.14 and is not a consequence of Proposition 3.14.

When $E = \mathbb{R}^2$, the existence of $\dfrac{\partial^2 f}{\partial x \partial y}(a)$ and $\dfrac{\partial^2 f}{\partial y \partial x}(a)$ is not sufficient to insure the existence of $\mathrm{D}^2 f(a)$.

When E is of finite dimension n and $(e_1, \ldots, e_n)$ is a basis for E, if $\mathrm{D}^2 f(a)$ exists, for every $u = u_1 e_1 + \cdots + u_n e_n$ and $v = v_1 e_1 + \cdots + v_n e_n$ in E, since $\mathrm{D}^2 f(a)$ is a symmetric bilinear form, we have

$$\begin{aligned}
\mathrm{D}^2 f(a)(u, v) &= \sum_{i=1,j=1}^{n} u_i v_j \mathrm{D}^2 f(a)(e_i, e_j) = \sum_{i=1,j=1}^{n} u_i v_j \mathrm{D}_{e_j}\mathrm{D}_{e_i} f(a) \\
&= \sum_{i=1,j=1}^{n} u_i v_j \frac{\partial^2 f}{\partial x_i \partial x_j}(a),
\end{aligned}$$

which can be written in matrix form as:

$$\mathrm{D}^2 f(a)(u, v) = U^\top \begin{pmatrix}
\dfrac{\partial^2 f}{\partial x_1^2}(a) & \dfrac{\partial^2 f}{\partial x_1 \partial x_2}(a) & \ldots & \dfrac{\partial^2 f}{\partial x_1 \partial x_n}(a) \\
\dfrac{\partial^2 f}{\partial x_1 \partial x_2}(a) & \dfrac{\partial^2 f}{\partial x_2^2}(a) & \ldots & \dfrac{\partial^2 f}{\partial x_2 \partial x_n}(a) \\
\vdots & \vdots & \ddots & \vdots \\
\dfrac{\partial^2 f}{\partial x_1 \partial x_n}(a) & \dfrac{\partial^2 f}{\partial x_2 \partial x_n}(a) & \ldots & \dfrac{\partial^2 f}{\partial x_n^2}(a)
\end{pmatrix} V,$$

where U is the column matrix representing u, and V is the column matrix representing v, over the basis $(e_1, \ldots, e_n)$. Note that the entries in this matrix are *vectors* in F, so the above expression is an abuse of notation, but since the u_i and v_j are scalars, the above expression makes sense since it is a bilinear combination. In the special case where $m = 1$, that is, $F = \mathbb{R}$ or $F = \mathbb{C}$, the Hessian matrix is an $n \times n$ matrix with scalar entries.

Definition 3.14. The above symmetric matrix is called the *Hessian of f at a*.

Example 3.9. Consider the function f defined on real invertible 2×2 matrices such that $ad - bc > 0$ given by

$$f(a, b, c, d) = \log(ad - bc).$$

We immediately verify that the Jacobian matrix of f is given by

$$df_{a,b,c,d} = \frac{1}{ad - bc} \begin{pmatrix} d & -c & -b & a \end{pmatrix}.$$

It is easily checked that if

$$A = \begin{pmatrix} a & b \\ c & d \end{pmatrix}, \quad X = \begin{pmatrix} x_1 & x_2 \\ x_3 & x_4 \end{pmatrix},$$

then

$$df_A(X) = \mathrm{tr}(A^{-1}X) = \frac{1}{ad - bc} \mathrm{tr}\left(\begin{pmatrix} d & -b \\ -c & a \end{pmatrix} \begin{pmatrix} x_1 & x_2 \\ x_3 & x_4 \end{pmatrix} \right).$$

Computing second-order derivatives, we find that the Hessian matrix of f is given by

$$Hf(A) = \frac{1}{(ad - bc)^2} \begin{pmatrix} -d^2 & cd & bd & -bc \\ cd & -c^2 & -ad & ac \\ bd & -ad & -b^2 & ab \\ -bc & ac & ab & -a^2 \end{pmatrix}.$$

Using the formula for the derivative of the inversion map and the chain rule we can show that

$$\mathrm{D}^2 f(A)(X_1, X_2) = -\mathrm{tr}(A^{-1}X_1A^{-1}X_2),$$

and so

$$Hf(A)(X_1, X_2) = -\mathrm{tr}(A^{-1}X_1A^{-1}X_2),$$

a formula which is far from obvious.

The function f can be generalized to matrices $A \in \mathbf{GL}^+(n, \mathbb{R})$, that is, $n \times n$ real invertible matrices of positive determinants, as

$$f(A) = \log\det(A).$$

It can be shown that the formulae

$$\begin{aligned} df_A(X) &= \mathrm{tr}(A^{-1}X) \\ \mathrm{D}^2 f(A)(X_1, X_2) &= -\mathrm{tr}(A^{-1}X_1A^{-1}X_2) \end{aligned}$$

also hold.

Example 3.10. If we restrict the function of Example 3.9 to symmetric positive definite matrices we obtain the function g defined by

$$g(a, b, c) = \log(ac - b^2).$$

We immediately verify that the Jacobian matrix of g is given by

$$dg_{a,b,c} = \frac{1}{ac - b^2} \begin{pmatrix} c & -2b & a \end{pmatrix}.$$

Computing second-order derivatives, we find that the Hessian matrix of g is given by

$$Hg(a,b,c) = \frac{1}{(ac - b^2)^2} \begin{pmatrix} -c^2 & 2bc & -b^2 \\ 2bc & -2(b^2 + ac) & 2ab \\ -b^2 & 2ab & -a^2 \end{pmatrix}.$$

Although this is not obvious, it can be shown that if $ac - b^2 > 0$ and $a, c > 0$, then the matrix $-Hg(a,b,c)$ is symmetric positive definite.

We now indicate briefly how higher-order derivatives are defined. Let $m \geq 2$. Given a function $f \colon A \to F$ as before, for any $a \in A$, if the derivatives $\mathrm{D}^i f$ exist on A for all i, $1 \leq i \leq m-1$, by induction, $\mathrm{D}^{m-1} f$ can be considered to be a continuous function $\mathrm{D}^{m-1} f \colon A \to \mathcal{L}_{m-1}(E^{m-1}; F)$.

Definition 3.15. Define $\mathrm{D}^m f(a)$, the *m-th derivative of f at a*, as

$$\mathrm{D}^m f(a) = \mathrm{D}(\mathrm{D}^{m-1} f)(a).$$

Then $\mathrm{D}^m f(a)$ *can be identified with a continuous m-multilinear map in* $\mathcal{L}_m(E^m; F)$. We can then show (as we did before) that if $\mathrm{D}^m f(a)$ is defined, then

$$\mathrm{D}^m f(a)(u_1, \ldots, u_m) = \mathrm{D}_{u_1} \ldots \mathrm{D}_{u_m} f(a).$$

Definition 3.16. When E if of finite dimension n and $(e_1, \ldots, e_n)$ is a basis for E, if $\mathrm{D}^m f(a)$ exists, for every $j_1, \ldots, j_m \in \{1, \ldots, n\}$, we denote $\mathrm{D}_{e_{j_m}} \ldots \mathrm{D}_{e_{j_1}} f(a)$ by

$$\frac{\partial^m f}{\partial x_{j_1} \ldots \partial x_{j_m}}(a).$$

Example 3.11. Going back to the function f of Example 3.9 given by $f(A) = \log\det(A)$, using the formula for the derivative of the inversion map, the chain rule and the product rule, we can show that

$$\mathrm{D}^m f(A)(X_1, \ldots, X_m) = (-1)^{m-1} \sum_{\sigma \in \mathfrak{S}_{m-1}} \mathrm{tr}(A^{-1} X_1 A^{-1} X_{\sigma(1)+1} \\ A^{-1} X_{\sigma(2)+1} \cdots A^{-1} X_{\sigma(m-1)+1})$$

for any $m \geq 1$, where $A \in \mathbf{GL}^+(n, \mathbb{R})$ and $X_1, \ldots X_m$ are any $n \times n$ real matrices.

Given a m-multilinear map $h \in \mathcal{L}_m(E^m; F)$, recall that h is *symmetric* if

$$h(u_{\pi(1)}, \ldots, u_{\pi(m)}) = h(u_1, \ldots, u_m),$$

for all $u_1, \ldots, u_m \in E$, and all permutations π on $\{1, \ldots, m\}$. Then the following generalization of Schwarz's lemma holds.

Proposition 3.15. *Given two normed vector spaces E and F, given any open subset A of E, given any $f\colon A \to F$, for every $a \in A$, for every $m \geq 1$, if $\mathrm{D}^m f(a)$ exists, then $\mathrm{D}^m f(a) \in \mathcal{L}_m(E^m; F)$ is a continuous symmetric m-multilinear map. As a corollary, if E is of finite dimension n, and $(e_1, \ldots, e_n)$ is a basis for E, we have*

$$\frac{\partial^m f}{\partial x_{j_1} \ldots \partial x_{j_m}}(a) = \frac{\partial^m f}{\partial x_{\pi(j_1)} \ldots \partial x_{\pi(j_m)}}(a),$$

for every $j_1, \ldots, j_m \in \{1, \ldots, n\}$, and for every permutation π on $\{1, \ldots, m\}$.

Because the trace function is invariant under permutation of its arguments ($\mathrm{tr}(XY) = \mathrm{tr}(YX)$), we see that the m-th derivatives in Example 3.11 are indeed symmetric multilinear maps.

If E is of finite dimension n, and $(e_1, \ldots, e_n)$ is a basis for E, $\mathrm{D}^m f(a)$ is a symmetric m-multilinear map, and we have

$$\mathrm{D}^m f(a)(u_1, \ldots, u_m) = \sum_j u_{1,j_1} \cdots u_{m,j_m} \frac{\partial^m f}{\partial x_{j_1} \ldots \partial x_{j_m}}(a),$$

where j ranges over all functions $j\colon \{1, \ldots, m\} \to \{1, \ldots, n\}$, for any m vectors

$$u_j = u_{j,1} e_1 + \cdots + u_{j,n} e_n.$$

The concept of C^1-function is generalized to the concept of C^m-function, and Theorem 3.2 can also be generalized.

Definition 3.17. Given two normed vector spaces E and F, and an open subset A of E, for any $m \geq 1$, we say that a function $f\colon A \to F$ is *of class C^m on A or a C^m-function on A* if $\mathrm{D}^k f$ exists and is continuous on A for every k, $1 \leq k \leq m$. We say that $f\colon A \to F$ is *of class C^∞ on A or a C^∞-function on A* if $\mathrm{D}^k f$ exists and is continuous on A for every $k \geq 1$. A C^∞-function (on A) is also called a *smooth function* (on A). A *C^m-diffeomorphism $f\colon A \to B$ between A and B* (where A is an open subset of E and B is an open subset of B) is a bijection between A and $B = f(A)$, such that both $f\colon A \to B$ and its inverse $f^{-1}\colon B \to A$ are C^m-functions.

Equivalently, f is a C^m-function on A if f is a C^1-function on A and $\mathrm{D}f$ is a C^{m-1}-function on A.

We have the following theorem giving a necessary and sufficient condition for f to a C^m-function on A.

Theorem 3.5. *Given two normed vector spaces E and F, where E is of finite dimension n, and where $(u_1, \ldots, u_n)$ is a basis of E, given any open subset A of E, given any function $f\colon A \to F$, for any $m \geq 1$, the derivative $\mathrm{D}^m f$ is a C^m-function on A iff every partial derivative $\mathrm{D}_{u_{j_k}} \ldots \mathrm{D}_{u_{j_1}} f$ (or $\dfrac{\partial^k f}{\partial x_{j_1} \ldots \partial x_{j_k}}(a)$) is defined and continuous on A, for all k, $1 \leq k \leq m$, and all $j_1, \ldots, j_k \in \{1, \ldots, n\}$. As a corollary, if F is of finite dimension p, and $(v_1, \ldots, v_p)$ is a basis of F, the derivative $\mathrm{D}^m f$ is defined and continuous on A iff every partial derivative $\mathrm{D}_{u_{j_k}} \ldots \mathrm{D}_{u_{j_1}} f_i$ (or $\dfrac{\partial^k f_i}{\partial x_{j_1} \ldots \partial x_{j_k}}(a)$) is defined and continuous on A, for all k, $1 \leq k \leq m$, for all i, $1 \leq i \leq p$, and all $j_1, \ldots, j_k \in \{1, \ldots, n\}$.*

Definition 3.18. When $E = \mathbb{R}$ (or $E = \mathbb{C}$), for any $a \in E$, $\mathrm{D}^m f(a)(1, \ldots, 1)$ is a vector in F, called the *mth-order vector derivative.* As in the case $m = 1$, we will usually identify the multilinear map $\mathrm{D}^m f(a)$ with the vector $\mathrm{D}^m f(a)(1, \ldots, 1)$.

Some notational conventions can also be introduced to simplify the notation of higher-order derivatives, and we discuss such conventions very briefly.

Recall that when E is of finite dimension n, and $(e_1, \ldots, e_n)$ is a basis for E, $\mathrm{D}^m f(a)$ is a symmetric m-multilinear map, and we have

$$\mathrm{D}^m f(a)(u_1, \ldots, u_m) = \sum_j u_{1,j_1} \cdots u_{m,j_m} \frac{\partial^m f}{\partial x_{j_1} \ldots \partial x_{j_m}}(a),$$

where j ranges over all functions $j\colon \{1, \ldots, m\} \to \{1, \ldots, n\}$, for any m vectors

$$u_j = u_{j,1} e_1 + \cdots + u_{j,n} e_n.$$

We can then group the various occurrences of ∂x_{j_k} corresponding to the same variable x_{j_k}, and this leads to the notation

$$\left(\frac{\partial}{\partial x_1}\right)^{\alpha_1} \left(\frac{\partial}{\partial x_2}\right)^{\alpha_2} \cdots \left(\frac{\partial}{\partial x_n}\right)^{\alpha_n} f(a),$$

where $\alpha_1 + \alpha_2 + \cdots + \alpha_n = m$.

If we denote $(\alpha_1, \ldots, \alpha_n)$ simply by α, then we denote

$$\left(\frac{\partial}{\partial x_1}\right)^{\alpha_1}\left(\frac{\partial}{\partial x_2}\right)^{\alpha_2}\cdots\left(\frac{\partial}{\partial x_n}\right)^{\alpha_n} f$$

by

$$\partial^\alpha f, \quad \text{or} \quad \left(\frac{\partial}{\partial x}\right)^\alpha f.$$

If $\alpha = (\alpha_1, \ldots, \alpha_n)$, we let $|\alpha| = \alpha_1 + \alpha_2 + \cdots + \alpha_n$, $\alpha! = \alpha_1! \cdots \alpha_n!$, and if $h = (h_1, \ldots, h_n)$, we denote $h_1^{\alpha_1} \cdots h_n^{\alpha_n}$ by h^α.

In the next section we survey various versions of Taylor's formula.

3.6 Taylor's Formula, Faà di Bruno's Formula

We discuss, without proofs, several versions of Taylor's formula. The hypotheses required in each version become increasingly stronger. The first version can be viewed as a generalization of the notion of derivative. Given an m-linear map $f\colon E^m \to F$, for any vector $h \in E$, we abbreviate

$$f(\underbrace{h, \ldots, h}_{m})$$

by $f(h^m)$. The version of Taylor's formula given next is sometimes referred to as the *formula of Taylor–Young*.

Theorem 3.6. *(Taylor–Young) Given two normed vector spaces E and F, for any open subset $A \subseteq E$, for any function $f\colon A \to F$, for any $a \in A$, if $\mathrm{D}^k f$ exists in A for all k, $1 \le k \le m-1$, and if $\mathrm{D}^m f(a)$ exists, then we have:*

$$f(a+h) = f(a) + \frac{1}{1!}\mathrm{D}^1 f(a)(h) + \cdots + \frac{1}{m!}\mathrm{D}^m f(a)(h^m) + \|h\|^m \epsilon(h),$$

for any h such that $a + h \in A$, and where $\lim_{h\to 0,\, h\neq 0} \epsilon(h) = 0$.

The above version of Taylor's formula has applications to the study of relative maxima (or minima) of real-valued functions. It is also used to study the local properties of curves and surfaces.

The next version of Taylor's formula can be viewed as a generalization of Proposition 3.10. It is sometimes called the *Taylor formula with Lagrange remainder* or *generalized mean value theorem*.

Theorem 3.7. *(Generalized mean value theorem) Let E and F be two normed vector spaces, let A be an open subset of E, and let $f\colon A \to F$ be a function on A. Given any $a \in A$ and any $h \neq 0$ in E, if the closed* *segment* $[a, a+h]$ *is contained in* A, $\mathrm{D}^k f$ *exists in* A *for all* k, $1 \leq k \leq m$, $\mathrm{D}^{m+1} f(x)$ *exists at every point* x *of the open segment* $(a, a+h)$, *and*

$$\max_{x \in (a,a+h)} \left\| \mathrm{D}^{m+1} f(x) \right\| \leq M,$$

for some $M \geq 0$, *then*

$$\left\| f(a+h) - f(a) - \left(\frac{1}{1!} \mathrm{D}^1 f(a)(h) + \cdots + \frac{1}{m!} \mathrm{D}^m f(a)(h^m) \right) \right\| \leq M \frac{\|h\|^{m+1}}{(m+1)!}.$$

As a corollary, if $L \colon E^{m+1} \to F$ *is a continuous* $(m+1)$*-linear map, then*

$$\left\| f(a+h) - f(a) - \left(\frac{1}{1!} \mathrm{D}^1 f(a)(h) + \cdots + \frac{1}{m!} \mathrm{D}^m f(a)(h^m) + \frac{L(h^{m+1})}{(m+1)!} \right) \right\| \leq M \frac{\|h\|^{m+1}}{(m+1)!},$$

where $M = \max_{x \in (a,a+h)} \left\| \mathrm{D}^{m+1} f(x) - L \right\|$.

The above theorem is sometimes stated under the slightly stronger assumption that f is a C^m-function on A. If $f \colon A \to \mathbb{R}$ is a real-valued function, Theorem 3.7 can be refined a little bit. This version is often called the *formula of Taylor–Maclaurin*.

Theorem 3.8. *(Taylor–Maclaurin) Let* E *be a normed vector space, let* A *be an open subset of* E, *and let* $f \colon A \to \mathbb{R}$ *be a real-valued function on* A. *Given any* $a \in A$ *and any* $h \neq 0$ *in* E, *if the closed segment* $[a, a+h]$ *is contained in* A, *if* $\mathrm{D}^k f$ *exists in* A *for all* k, $1 \leq k \leq m$, *and* $\mathrm{D}^{m+1} f(x)$ *exists at every point* x *of the open segment* $(a, a+h)$, *then there is some* $\theta \in \mathbb{R}$, *with* $0 < \theta < 1$, *such that*

$$f(a+h) = f(a) + \frac{1}{1!} \mathrm{D}^1 f(a)(h) + \cdots + \frac{1}{m!} \mathrm{D}^m f(a)(h^m) + \frac{1}{(m+1)!} \mathrm{D}^{m+1} f(a + \theta h)(h^{m+1}).$$

Example 3.12. Going back to the function f of Example 3.9 given by $f(A) = \log\det(A)$, we know from Example 3.11 that

$$\mathrm{D}^m f(A)(X_1, \ldots, X_m) = (-1)^{m-1} \sum_{\sigma \in \mathfrak{S}_{m-1}} \mathrm{tr}(A^{-1} X_1 A^{-1} X_{\sigma(1)+1} \cdots A^{-1} X_{\sigma(m-1)+1}) \quad (*)$$

for all $m \geq 1$, with $A \in \mathbf{GL}^+(n, \mathbb{R})$. If we make the stronger assumption that A is symmetric positive definite, then for any other symmetric positive definite matrix B, since the symmetric positive definite matrices form a convex set, the matrices $A + \theta(B - A) = (1-\theta)A + \theta B$ are also symmetric positive definite for $\theta \in [0, 1]$. Theorem 3.8 applies with $H = B - A$ (a symmetric matrix), and using $(*)$, we obtain

$$\begin{aligned}\log\det(A+H) = \log\det(A) + \operatorname{tr}\Bigg(A^{-1}H - \frac{1}{2}(A^{-1}H)^2 + \cdots \\ + \frac{(-1)^{m-1}}{m}(A^{-1}H)^m + \frac{(-1)^m}{m+1}((A+\theta H)^{-1}H)^{m+1}\Bigg),\end{aligned}$$

for some θ such that $0 < \theta < 1$. In particular, if $A = I$, for any symmetric matrix H such that $I + H$ is symmetric positive definite, we obtain

$$\begin{aligned}\log\det(I+H) = \operatorname{tr}\Bigg(H - \frac{1}{2}H^2 + \cdots \\ + \frac{(-1)^{m-1}}{m}H^m + \frac{(-1)^m}{m+1}((I+\theta H)^{-1}H)^{m+1}\Bigg),\end{aligned}$$

for some θ such that $0 < \theta < 1$. In the special case when $n = 1$, we have $I = 1$, H is a real such that $1 + H > 0$ and the trace function is the identity, so we recognize the partial sum of the series for $x \mapsto \log(1 + x)$,

$$\begin{aligned}\log(1+H) = H - \frac{1}{2}H^2 + \cdots + \frac{(-1)^{m-1}}{m}H^m \\ + \frac{(-1)^m}{m+1}(1+\theta H)^{-(m+1)}H^{m+1}.\end{aligned}$$

We also mention for "mathematical culture," a version with integral remainder, in the case of a real-valued function. This is usually called *Taylor's formula with integral remainder*.

Theorem 3.9. *(Taylor's formula with integral remainder) Let E be a normed vector space, let A be an open subset of E, and let $f\colon A \to \mathbb{R}$ be a real-valued function on A. Given any $a \in A$ and any $h \neq 0$ in E, if the closed segment $[a, a+h]$ is contained in A, and if f is a C^{m+1}-function on A, then we have*

$$\begin{aligned}f(a+h) = f(a) + \frac{1}{1!}\mathrm{D}^1 f(a)(h) + \cdots + \frac{1}{m!}\mathrm{D}^m f(a)(h^m) \\ + \int_0^1 \frac{(1-t)^m}{m!}\Big[\mathrm{D}^{m+1} f(a+th)(h^{m+1})\Big]\,dt.\end{aligned}$$

The advantage of the above formula is that it gives an explicit remainder.

We now examine briefly the situation where E is of finite dimension n, and $(e_1, \ldots, e_n)$ is a basis for E. In this case we get a more explicit expression for the expression

$$\sum_{i=0}^{k=m} \frac{1}{k!} \mathrm{D}^k f(a)(h^k)$$

involved in all versions of Taylor's formula, where by convention, $\mathrm{D}^0 f(a)(h^0) = f(a)$. If $h = h_1 e_1 + \cdots + h_n e_n$, then we have

$$\sum_{k=0}^{k=m} \frac{1}{k!} \mathrm{D}^k f(a)(h^k) = \sum_{k_1 + \cdots + k_n \leq m} \frac{h_1^{k_1} \cdots h_n^{k_n}}{k_1! \cdots k_n!} \left(\frac{\partial}{\partial x_1}\right)^{k_1} \cdots \left(\frac{\partial}{\partial x_n}\right)^{k_n} f(a),$$

which, using the abbreviated notation introduced at the end of Section 3.5, can also be written as

$$\sum_{k=0}^{k=m} \frac{1}{k!} \mathrm{D}^k f(a)(h^k) = \sum_{|\alpha| \leq m} \frac{h^\alpha}{\alpha!} \partial^\alpha f(a).$$

The advantage of the above notation is that it is the same as the notation used when $n = 1$, i.e., when $E = \mathbb{R}$ (or $E = \mathbb{C}$). Indeed, in this case, the Taylor–Maclaurin formula reads as:

$$f(a+h) = f(a) + \frac{h}{1!} \mathrm{D}^1 f(a) + \cdots + \frac{h^m}{m!} \mathrm{D}^m f(a) + \frac{h^{m+1}}{(m+1)!} \mathrm{D}^{m+1} f(a + \theta h),$$

for some $\theta \in \mathbb{R}$, with $0 < \theta < 1$, where $\mathrm{D}^k f(a)$ is the value of the k-th derivative of f at a (and thus, as we have already said several times, this is the kth-order vector derivative, which is just a scalar, since $F = \mathbb{R}$).

In the above formula, the assumptions are that $f \colon [a, a+h] \to \mathbb{R}$ is a C^m-function on $[a, a+h]$, and that $\mathrm{D}^{m+1} f(x)$ exists for every $x \in (a, a+h)$.

Taylor's formula is useful to study the local properties of curves and surfaces. In the case of a curve, we consider a function $f \colon [r, s] \to F$ from a closed interval $[r, s]$ of $\mathbb{R}$ to some vector space F, the derivatives $\mathrm{D}^k f(a)(h^k)$ correspond to vectors $h^k \mathrm{D}^k f(a)$, where $\mathrm{D}^k f(a)$ is the kth vector derivative of f at a (which is really $\mathrm{D}^k f(a)(1, \ldots, 1)$), and for any $a \in (r, s)$, Theorem 3.6 yields the following formula:

$$f(a+h) = f(a) + \frac{h}{1!} \mathrm{D}^1 f(a) + \cdots + \frac{h^m}{m!} \mathrm{D}^m f(a) + h^m \epsilon(h),$$

for any h such that $a + h \in (r, s)$, and where $\lim_{h \to 0,\, h \neq 0} \epsilon(h) = 0$.

In the case of functions $f \colon \mathbb{R}^n \to \mathbb{R}$, it is convenient to have formulae for the Taylor–Young formula and the Taylor–Maclaurin formula in terms of the gradient and the Hessian. Recall that the *gradient* $\nabla f(a)$ of f at $a \in \mathbb{R}^n$ is the column vector

$$\nabla f(a) = \begin{pmatrix} \frac{\partial f}{\partial x_1}(a) \\ \frac{\partial f}{\partial x_2}(a) \\ \vdots \\ \frac{\partial f}{\partial x_n}(a) \end{pmatrix},$$

and that

$$f'(a)(u) = \mathrm{D}f(a)(u) = \nabla f(a) \cdot u,$$

for any $u \in \mathbb{R}^n$ (where $\cdot$ means inner product). The above equation shows that *the direction of the gradient* $\nabla f(a)$ *is the direction of maximal increase of the function* f *at* a and that $\|\nabla f(a)\|$ *is the rate of change of* f *in its direction of maximal increase*. This is the reason why methods of "gradient descent" pick the direction *opposite* to the gradient (we are trying to minimize f).

The *Hessian matrix* $\nabla^2 f(a)$ of f at $a \in \mathbb{R}^n$ is the $n \times n$ symmetric matrix

$$\nabla^2 f(a) = \begin{pmatrix} \frac{\partial^2 f}{\partial x_1^2}(a) & \frac{\partial^2 f}{\partial x_1 \partial x_2}(a) & \dots & \frac{\partial^2 f}{\partial x_1 \partial x_n}(a) \\ \frac{\partial^2 f}{\partial x_1 \partial x_2}(a) & \frac{\partial^2 f}{\partial x_2^2}(a) & \dots & \frac{\partial^2 f}{\partial x_2 \partial x_n}(a) \\ \vdots & \vdots & \ddots & \vdots \\ \frac{\partial^2 f}{\partial x_1 \partial x_n}(a) & \frac{\partial^2 f}{\partial x_2 \partial x_n}(a) & \dots & \frac{\partial^2 f}{\partial x_n^2}(a) \end{pmatrix},$$

and we have

$$\mathrm{D}^2 f(a)(u, v) = u^\top \nabla^2 f(a)\, v = u \cdot \nabla^2 f(a) v = \nabla^2 f(a) u \cdot v,$$

for all $u, v \in \mathbb{R}^n$. This is the special case of Definition 3.14 where $E = \mathbb{R}^n$ and $F = \mathbb{R}$. Then we have the following three formulations of the formula of Taylor–Young of order 2:

$$\begin{aligned} f(a+h) &= f(a) + \mathrm{D}f(a)(h) + \frac{1}{2}\mathrm{D}^2 f(a)(h, h) + \|h\|^2 \epsilon(h) \\ f(a+h) &= f(a) + \nabla f(a) \cdot h + \frac{1}{2}(h \cdot \nabla^2 f(a) h) + (h \cdot h)\epsilon(h) \\ f(a+h) &= f(a) + (\nabla f(a))^\top h + \frac{1}{2}(h^\top \nabla^2 f(a)\, h) + (h^\top h)\epsilon(h), \end{aligned}$$

with $\lim_{h\mapsto 0}\epsilon(h) = 0$.

One should keep in mind that only the first formula is intrinsic (i.e., does not depend on the choice of a basis), whereas the other two depend on the basis and the inner product chosen on $\mathbb{R}^n$. As an exercise, the reader should write similar formulae for the Taylor–Maclaurin formula of order 2.

Another application of Taylor's formula is the derivation of a formula which gives the m-th derivative of the composition of two functions, usually known as "Faà di Bruno's formula." This formula is useful when dealing with geometric continuity of splines curves and surfaces.

Proposition 3.16. *Given any normed vector space E, for any function $f\colon \mathbb{R}\to\mathbb{R}$ and any function $g\colon \mathbb{R}\to E$, for any $a\in\mathbb{R}$, letting $b = f(a)$, $f^{(i)}(a) = \mathrm{D}^i f(a)$, and $g^{(i)}(b) = \mathrm{D}^i g(b)$, for any $m\geq 1$, if $f^{(i)}(a)$ and $g^{(i)}(b)$ exist for all i, $1\leq i\leq m$, then $(g\circ f)^{(m)}(a) = \mathrm{D}^m(g\circ f)(a)$ exists and is given by the following formula:*

$$
\begin{aligned}
&(g\circ f)^{(m)}(a)\\
&= \sum_{0\leq j\leq m}\ \sum_{\substack{i_1+i_2+\cdots+i_m=j\\ i_1+2i_2+\cdots+mi_m=m\\ i_1,i_2,\cdots,i_m\geq 0}} \frac{m!}{i_1!\cdots i_m!} g^{(j)}(b)\left(\frac{f^{(1)}(a)}{1!}\right)^{i_1}\cdots\left(\frac{f^{(m)}(a)}{m!}\right)^{i_m}.
\end{aligned}
$$

When $m = 1$, the above simplifies to the familiar formula

$$(g\circ f)'(a) = g'(b)f'(a),$$

and for $m = 2$, we have

$$(g\circ f)^{(2)}(a) = g^{(2)}(b)(f^{(1)}(a))^2 + g^{(1)}(b)f^{(2)}(a).$$

3.7 Further Readings

A thorough treatment of differential calculus can be found in Munkres [Munkres (1991)], Lang [Lang (1997)], Schwartz [Schwartz (1992)], Cartan [Cartan (1990)], and Avez [Avez (1991)]. The techniques of differential calculus have many applications, especially to the geometry of curves and surfaces and to differential geometry in general. For this, we recommend do Carmo [do Carmo (1976, 1992)] (two beautiful classics on the subject), Kreyszig [Kreyszig (1991)], Stoker [Stoker (1989)], Gray [Gray (1997)], Berger and Gostiaux [Berger and Gostiaux (1992)], Milnor [Milnor (1969)], Lang [Lang (1995)], Warner [Warner (1983)] and Choquet-Bruhat [Choquet-Bruhat *et al.* (1982)].

3.8 Summary

The main concepts and results of this chapter are listed below:

- Directional derivative ($\mathrm{D}_u f(a)$).
- Total derivative, Fréchet derivative, derivative, total differential, differential ($df(a), df_a$).
- Partial derivatives.
- Affine functions.
- The chain rule.
- Jacobian matrices ($J(f)(a)$), Jacobians.
- Gradient of a function ($\operatorname{grad} f(a)$, $\nabla f(a)$).
- Mean value theorem.
- C^0-functions, C^1-functions.
- The implicit function theorem.
- Local homeomorphisms, local diffeomorphisms, diffeomorphisms.
- The inverse function theorem.
- Immersions, submersions.
- Second-order derivatives.
- Schwarz's lemma.
- Hessian matrix.
- C^∞-functions, smooth functions.
- Taylor–Young's formula.
- Generalized mean value theorem.
- Taylor–MacLaurin's formula.
- Taylor's formula with integral remainder.
- Faà di Bruno's formula.

3.9 Problems

Problem 3.1. Let $f\colon \mathrm{M}_n(\mathbb{R}) \to \mathrm{M}_n(\mathbb{R})$ be the function defined on $n \times n$ matrices by

$$f(A) = A^2.$$

Prove that

$$\mathrm{D}f_A(H) = AH + HA,$$

for all $A, H \in \mathrm{M}_n(\mathbb{R})$.

Problem 3.2. Let $f\colon \mathrm{M}_n(\mathbb{R}) \to \mathrm{M}_n(\mathbb{R})$ be the function defined on $n \times n$ matrices by

$$f(A) = A^3.$$

Prove that

$$\mathrm{D}f_A(H) = A^2H + AHA + HA^2,$$

for all $A, H \in \mathrm{M}_n(\mathbb{R})$.

Problem 3.3. If $f\colon \mathrm{M}_n(\mathbb{R}) \to \mathrm{M}_n(\mathbb{R})$ and $g\colon \mathrm{M}_n(\mathbb{R}) \to \mathrm{M}_n(\mathbb{R})$ are differentiable matrix functions, prove that

$$d(fg)_A(B) = df_A(B)g(A) + f(A)dg_A(B),$$

for all $A, B \in \mathrm{M}_n(\mathbb{R})$.

Problem 3.4. Recall that $\mathfrak{so}(3)$ denotes the vector space of real skew-symmetric $n \times n$ matrices ($B^\top = -B$). Let $C\colon \mathfrak{so}(n) \to \mathrm{M}_n(\mathbb{R})$ be the function given by

$$C(B) = (I - B)(I + B)^{-1}.$$

(1) Prove that if B is skew-symmetric, then $I - B$ and $I + B$ are invertible, and so C is well-defined. Prove that

(2) Prove that

$$dC(B)(A) = -[I+(I-B)(I+B)^{-1}]A(I+B)^{-1} = -2(I+B)^{-1}A(I+B)^{-1}.$$

(3) Prove that $dC(B)$ is injective for every skew-symmetric matrix B.

Problem 3.5. Prove that

$$\begin{aligned} d^mC_B(H_1, \ldots, H_m) = 2(-1)^m \sum_{\pi \in \mathfrak{S}_m} &(I + B)^{-1}H_{\pi(1)}(I + B)^{-1} \\ &H_{\pi(2)}(I + B)^{-1} \cdots (I + B)^{-1}H_{\pi(m)}(I + B)^{-1}. \end{aligned}$$

Problem 3.6. Consider the function g defined for all $A \in \mathbf{GL}(n, \mathbb{R})$, that is, all $n \times n$ real invertible matrices, given by

$$g(A) = \det(A).$$

(1) Prove that

$$dg_A(X) = \det(A)\mathrm{tr}(A^{-1}X),$$

for all $n \times n$ real matrices X.

(2) Consider the function f defined for all $A \in \mathbf{GL}^+(n, \mathbb{R})$, that is, $n \times n$ real invertible matrices of positive determinants, given by

$$f(A) = \log g(A) = \log \det(A).$$

Prove that

$$\begin{aligned} df_A(X) &= \mathrm{tr}(A^{-1}X) \\ \mathrm{D}^2 f(A)(X_1, X_2) &= -\mathrm{tr}(A^{-1}X_1A^{-1}X_2), \end{aligned}$$

for all $n \times n$ real matrices X, X_1, X_2.

(3) Prove that

$$\mathrm{D}^m f(A)(X_1, \ldots, X_m) = (-1)^{m-1} \sum_{\sigma \in \mathfrak{S}_{m-1}} \mathrm{tr}(A^{-1}X_1A^{-1}X_{\sigma(1)+1} \\ A^{-1}X_{\sigma(2)+1} \cdots A^{-1}X_{\sigma(m-1)+1})$$

for any $m \geq 1$, where $X_1, \ldots X_m$ are any $n \times n$ real matrices.

Chapter 4

Extrema of Real-Valued Functions

This chapter deals with extrema of real-valued functions. In most optimization problems, we need to find necessary conditions for a function $J\colon \Omega \to \mathbb{R}$ to have a local extremum with respect to a subset U of Ω (where Ω is open). This can be done in two cases:

(1) The set U is defined by a set of equations,

$$U = \{x \in \Omega \mid \varphi_i(x) = 0,\ \ 1 \leq i \leq m\},$$

where the functions $\varphi_i\colon \Omega \to \mathbb{R}$ are continuous (and usually differentiable).

(2) The set U is defined by a set of inequalities,

$$U = \{x \in \Omega \mid \varphi_i(x) \leq 0,\ \ 1 \leq i \leq m\},$$

where the functions $\varphi_i\colon \Omega \to \mathbb{R}$ are continuous (and usually differentiable).

In (1), the equations $\varphi_i(x) = 0$ are called *equality constraints*, and in (2), the inequalities $\varphi_i(x) \leq 0$ are called *inequality constraints*. The case of equality constraints is much easier to deal with and is treated in this chapter.

If the functions φ_i are convex and Ω is convex, then U is convex. This is a very important case that we discuss later. In particular, if the functions φ_i are affine, then the equality constraints can be written as $Ax = b$, and the inequality constraints as $Ax \leq b$, for some $m \times n$ matrix A and some vector $b \in \mathbb{R}^m$. We will also discuss the case of affine constraints later.

In the case of equality constraints, a necessary condition for a local extremum with respect to U can be given in terms of *Lagrange multipliers*. In the case of inequality constraints, there is also a necessary condition for a local extremum with respect to U in terms of generalized Lagrange multipliers and the *Karush–Kuhn–Tucker* conditions. This will be discussed in Chapter 14.

4.1 Local Extrema, Constrained Local Extrema, and Lagrange Multipliers

Let $J: E \to \mathbb{R}$ be a real-valued function defined on a normed vector space E (or more generally, any topological space). Ideally we would like to find where the function J reaches a minimum or a maximum value, at least locally. In this chapter we will usually use the notations $dJ(u)$ or $J'(u)$ (or dJ_u or J'_u) for the derivative of J at u, instead of $\mathrm{D}J(u)$. Our presentation follows very closely that of Ciarlet [Ciarlet (1989)] (Chapter 7), which we find to be one of the clearest.

Definition 4.1. If $J: E \to \mathbb{R}$ is a real-valued function defined on a normed vector space E, we say that J has a *local minimum* (or *relative minimum*) at the point $u \in E$ if there is some open subset $W \subseteq E$ containing u such that

$$J(u) \leq J(w) \quad \text{for all } w \in W.$$

Similarly, we say that J has a *local maximum* (or *relative maximum*) at the point $u \in E$ if there is some open subset $W \subseteq E$ containing u such that

$$J(u) \geq J(w) \quad \text{for all } w \in W.$$

In either case, we say that J has a *local extremum* (or *relative extremum*) at u. We say that J has a *strict local minimum* (resp. *strict local maximum*) at the point $u \in E$ if there is some open subset $W \subseteq E$ containing u such that

$$J(u) < J(w) \quad \text{for all } w \in W - \{u\}$$

(resp.

$$J(u) > J(w) \quad \text{for all } w \in W - \{u\}).$$

By abuse of language, we often say that the point u itself "is a local minimum" or a "local maximum," even though, strictly speaking, this does not make sense.

We begin with a well-known necessary condition for a local extremum.

Proposition 4.1. *Let E be a normed vector space and let $J: \Omega \to \mathbb{R}$ be a function, with Ω some open subset of E. If the function J has a local extremum at some point $u \in \Omega$ and if J is differentiable at u, then*

$$dJ_u = J'(u) = 0.$$

Proof. Pick any $v \in E$. Since Ω is open, for t small enough we have $u + tv \in \Omega$, so there is an open interval $I \subseteq \mathbb{R}$ such that the function φ given by

$$\varphi(t) = J(u + tv)$$

for all $t \in I$ is well-defined. By applying the chain rule, we see that φ is differentiable at $t = 0$, and we get

$$\varphi'(0) = dJ_u(v).$$

Without loss of generality, assume that u is a local minimum. Then we have

$$\varphi'(0) = \lim_{t \mapsto 0_-} \frac{\varphi(t) - \varphi(0)}{t} \leq 0$$

and

$$\varphi'(0) = \lim_{t \mapsto 0_+} \frac{\varphi(t) - \varphi(0)}{t} \geq 0,$$

which shows that $\varphi'(0) = dJ_u(v) = 0$. As $v \in E$ is arbitrary, we conclude that $dJ_u = 0$. ☐

Definition 4.2. A point $u \in \Omega$ such that $J'(u) = 0$ is called a *critical point* of J.

If $E = \mathbb{R}^n$, then the condition $dJ_u = 0$ is equivalent to the system

$$\begin{aligned} \frac{\partial J}{\partial x_1}(u_1, \ldots, u_n) &= 0 \\ &\vdots \\ \frac{\partial J}{\partial x_n}(u_1, \ldots, u_n) &= 0. \end{aligned}$$

The condition of Proposition 4.1 is only a *necessary* condition for the existence of an extremum, but not a sufficient condition.

Here are some counter-examples. If $f \colon \mathbb{R} \to \mathbb{R}$ is the function given by $f(x) = x^3$, since $f'(x) = 3x^2$, we have $f'(0) = 0$, but 0 is neither a minimum nor a maximum of f as evidenced by the graph shown in Figure 4.1.

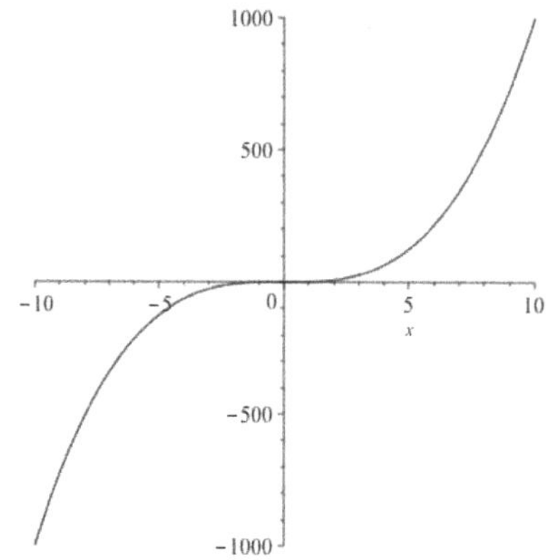

Fig. 4.1 The graph of $f(x) = x^3$. Note that $x = 0$ is a saddle point and not a local extremum.

If $g\colon \mathbb{R}^2 \to \mathbb{R}$ is the function given by $g(x, y) = x^2 - y^2$, then $g'_{(x,y)} = (2x \quad -2y)$, so $g'_{(0,0)} = (0 \quad 0)$, yet near $(0, 0)$ the function g takes negative and positive values. See Figure 4.2.

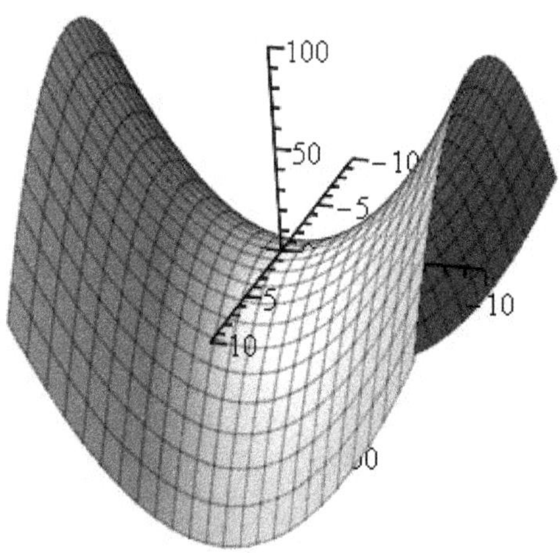

Fig. 4.2 The graph of $g(x, y) = x^2 - y^2$. Note that $(0, 0)$ is a saddle point and not a local extremum.

It is very important to note that the hypothesis that Ω *is open* is crucial for the validity of Proposition 4.1.

For example, if J is the identity function on $\mathbb{R}$ and $U = [0, 1]$, a closed subset, then $J'(x) = 1$ for all $x \in [0, 1]$, even though J has a minimum at $x = 0$ and a maximum at $x = 1$.

In many practical situations, we need to look for local extrema of a function J *under additional constraints*. This situation can be formalized conveniently as follows. We have a function $J\colon \Omega \to \mathbb{R}$ defined on some open subset Ω of a normed vector space, but we also have some subset U of Ω, and we are looking for the local extrema of J *with respect to the set* U.

The elements $u \in U$ are often called *feasible solutions* of the optimization problem consisting in finding the local extrema of some objective function J with respect to some subset U of Ω defined by a set of constraints. Note that in most cases, U is *not* open. In fact, U is usually closed.

Definition 4.3. If $J\colon \Omega \to \mathbb{R}$ is a real-valued function defined on some open subset Ω of a normed vector space E and if U is some subset of Ω, we say that J has a *local minimum* (or *relative minimum*) at the point $u \in U$ *with respect to* U if there is some open subset $W \subseteq \Omega$ containing u such that

$$J(u) \leq J(w) \quad \text{for all } w \in U \cap W.$$

Similarly, we say that J has a *local maximum* (or *relative maximum*) at the point $u \in U$ *with respect to* U if there is some open subset $W \subseteq \Omega$ containing u such that

$$J(u) \geq J(w) \quad \text{for all } w \in U \cap W.$$

In either case, we say that J has a *local extremum* at u *with respect to* U.

In order to find necessary conditions for a function $J\colon \Omega \to \mathbb{R}$ to have a local extremum with respect to a subset U of Ω (where Ω is open), we need to somehow incorporate the definition of U into these conditions. This can be done in two cases:

(1) The set U is defined by a set of equations,

$$U = \{x \in \Omega \mid \varphi_i(x) = 0, \ 1 \leq i \leq m\},$$

where the functions $\varphi_i\colon \Omega \to \mathbb{R}$ are continuous (and usually differentiable).

(2) The set U is defined by a set of inequalities,

$$U = \{x \in \Omega \mid \varphi_i(x) \leq 0,\ 1 \leq i \leq m\},$$

where the functions $\varphi_i \colon \Omega \to \mathbb{R}$ are continuous (and usually differentiable).

In (1), the equations $\varphi_i(x) = 0$ are called *equality constraints*, and in (2), the inequalities $\varphi_i(x) \leq 0$ are called *inequality constraints*.

An inequality constraint of the form $\varphi_i(x) \geq 0$ is equivalent to the inequality constraint $-\varphi_x(x) \leq 0$. An equality constraint $\varphi_i(x) = 0$ is equivalent to the conjunction of the two inequality constraints $\varphi_i(x) \leq 0$ and $-\varphi_i(x) \leq 0$, so the case of inequality constraints subsumes the case of equality constraints. However, the case of equality constraints is easier to deal with, and in this chapter we will restrict our attention to this case.

If the functions φ_i are convex and Ω is convex, then U is convex. This is a very important case that we will discuss later. In particular, if the functions φ_i are affine, then the equality constraints can be written as $Ax = b$, and the inequality constraints as $Ax \leq b$, for some $m \times n$ matrix A and some vector $b \in \mathbb{R}^m$. We will also discuss the case of affine constraints later.

In the case of equality constraints, a necessary condition for a local extremum with respect to U can be given in terms of *Lagrange multipliers*. In the case of inequality constraints, there is also a necessary condition for a local extremum with respect to U in terms of generalized Lagrange multipliers and the *Karush–Kuhn–Tucker* conditions. This will be discussed in Chapter 14.

We begin by considering the case where $\Omega \subseteq E_1 \times E_2$ is an open subset of a product of normed vector spaces and where U is the zero locus of some continuous function $\varphi \colon \Omega \to E_2$, which means that

$$U = \{(u_1, u_2) \in \Omega \mid \varphi(u_1, u_2) = 0\}.$$

For the sake of brevity, we say that J has a *constrained local extremum* at u instead of saying that J has a *local extremum* at the point $u \in U$ *with respect to* U.

In most applications, we have $E_1 = \mathbb{R}^{n-m}$ and $E_2 = \mathbb{R}^m$ for some integers m, n such that $1 \leq m < n$, Ω is an open subset of $\mathbb{R}^n$, $J \colon \Omega \to \mathbb{R}$, and we have m functions $\varphi_i \colon \Omega \to \mathbb{R}$ defining the subset

$$U = \{v \in \Omega \mid \varphi_i(v) = 0,\ 1 \leq i \leq m\}.$$

Fortunately, there is a necessary condition for constrained local extrema in terms of *Lagrange multipliers*.

Theorem 4.1. *(Necessary condition for a constrained extremum in terms of Lagrange multipliers) Let Ω be an open subset of $\mathbb{R}^n$, consider m C^1-functions $\varphi_i \colon \Omega \to \mathbb{R}$ (with $1 \leq m < n$), let*

$$U = \{v \in \Omega \mid \varphi_i(v) = 0,\ 1 \leq i \leq m\},$$

and let $u \in U$ be a point such that the derivatives $d\varphi_i(u) \in \mathcal{L}(\mathbb{R}^n;\mathbb{R})$ are linearly independent; equivalently, assume that the $m \times n$ matrix $\big((\partial\varphi_i/\partial x_j)(u)\big)$ has rank m. If $J \colon \Omega \to \mathbb{R}$ is a function which is differentiable at $u \in U$ and if J has a local constrained extremum at u, then there exist m numbers $\lambda_i(u) \in \mathbb{R}$, uniquely defined, such that

$$dJ(u) + \lambda_1(u)d\varphi_1(u) + \cdots + \lambda_m(u)d\varphi_m(u) = 0;$$

equivalently,

$$\nabla J(u) + \lambda_1(u)\nabla\varphi_1(u) + \cdots + \lambda_1(u)\nabla\varphi_m(u) = 0.$$

Theorem 4.1 will be proven as a corollary of Theorem 4.2, which gives a more general formulation that applies to the situation where E_1 is an infinite-dimensional Banach space. To simplify the exposition we postpone a discussion of this theorem until we have presented several examples illustrating the method of Lagrange multipliers.

Definition 4.4. The numbers $\lambda_i(u)$ involved in Theorem 4.1 are called the *Lagrange multipliers* associated with the constrained extremum u (again, with some minor abuse of language).

The linear independence of the linear forms $d\varphi_i(u)$ is equivalent to the fact that the Jacobian matrix $\big((\partial\varphi_i/\partial x_j)(u)\big)$ of $\varphi = (\varphi_1, \ldots, \varphi_m)$ at u has rank m. If $m = 1$, the linear independence of the $d\varphi_i(u)$ reduces to the condition $\nabla\varphi_1(u) \neq 0$.

A fruitful way to reformulate the use of Lagrange multipliers is to introduce the notion of the Lagrangian associated with our constrained extremum problem.

Definition 4.5. The *Lagrangian* associated with our constrained extremum problem is the function $L \colon \Omega \times \mathbb{R}^m \to \mathbb{R}$ given by

$$L(v, \lambda) = J(v) + \lambda_1\varphi_1(v) + \cdots + \lambda_m\varphi_m(v),$$

with $\lambda = (\lambda_1, \ldots, \lambda_m)$.

We have the following simple but important proposition.

Proposition 4.2. *There exists some* $\mu = (\mu_1, \ldots, \mu_m)$ *and some* $u \in U$ *such that*

$$dJ(u) + \mu_1 d\varphi_1(u) + \cdots + \mu_m d\varphi_m(u) = 0$$

if and only if

$$dL(u, \mu) = 0,$$

or equivalently

$$\nabla L(u, \mu) = 0;$$

that is, iff (u, λ) *is a critical point of the Lagrangian* L.

Proof. Indeed $dL(u, \mu) = 0$ if equivalent to

$$\begin{aligned}
\frac{\partial L}{\partial v}(u, \mu) &= 0 \\
\frac{\partial L}{\partial \lambda_1}(u, \mu) &= 0 \\
&\vdots \\
\frac{\partial L}{\partial \lambda_m}(u, \mu) &= 0,
\end{aligned}$$

and since

$$\frac{\partial L}{\partial v}(u, \mu) = dJ(u) + \mu_1 d\varphi_1(u) + \cdots + \mu_m d\varphi_m(u)$$

and

$$\frac{\partial L}{\partial \lambda_i}(u, \mu) = \varphi_i(u),$$

we get

$$dJ(u) + \mu_1 d\varphi_1(u) + \cdots + \mu_m d\varphi_m(u) = 0$$

and

$$\varphi_1(u) = \cdots = \varphi_m(u) = 0,$$

that is, $u \in U$. The converse is proven in a similar fashion (essentially by reversing the argument). ☐

If we write out explicitly the condition

$$dJ(u) + \mu_1 d\varphi_1(u) + \cdots + \mu_m d\varphi_m(u) = 0,$$

we get the $n \times m$ system

$$\frac{\partial J}{\partial x_1}(u) + \lambda_1 \frac{\partial \varphi_1}{\partial x_1}(u) + \cdots + \lambda_m \frac{\partial \varphi_m}{\partial x_1}(u) = 0$$

$$\vdots$$

$$\frac{\partial J}{\partial x_n}(u) + \lambda_1 \frac{\partial \varphi_1}{\partial x_n}(u) + \cdots + \lambda_m \frac{\partial \varphi_m}{\partial x_n}(u) = 0,$$

and it is important to note that the matrix of this system is the *transpose* of the Jacobian matrix of φ at u. If we write $\mathrm{Jac}(\varphi)(u) = \big((\partial\varphi_i/\partial x_j)(u)\big)$ for the Jacobian matrix of φ (at u), then the above system is written in matrix form as

$$\nabla J(u) + (\mathrm{Jac}(J)(u))^\top \lambda = 0,$$

where λ is viewed as a column vector, and the Lagrangian is equal to

$$L(u, \lambda) = J(u) + (\varphi_1(u), \ldots, \varphi_m(u))\lambda.$$

The beauty of the Lagrangian is that the constraints $\{\varphi_i(v) = 0\}$ have been incorporated into the function $L(v, \lambda)$, and that the necessary condition for the existence of a constrained local extremum of J is reduced to the necessary condition for the existence of a local extremum of the *unconstrained* L.

However, one should be careful to check that the assumptions of Theorem 4.1 are satisfied (in particular, the linear independence of the linear forms $d\varphi_i$).

Example 4.1. For example, let $J\colon \mathbb{R}^3 \to \mathbb{R}$ be given by

$$J(x, y, z) = x + y + z^2$$

and $g\colon \mathbb{R}^3 \to \mathbb{R}$ by

$$g(x, y, z) = x^2 + y^2.$$

Since $g(x, y, z) = 0$ iff $x = y = 0$, we have $U = \{(0, 0, z) \mid z \in \mathbb{R}\}$ and the restriction of J to U is given by

$$J(0, 0, z) = z^2,$$

which has a minimum for $z = 0$. However, a "blind" use of Lagrange multipliers would require that there is some λ so that

$$\frac{\partial J}{\partial x}(0,0,z) = \lambda \frac{\partial g}{\partial x}(0,0,z), \quad \frac{\partial J}{\partial y}(0,0,z) = \lambda \frac{\partial g}{\partial y}(0,0,z),$$
$$\frac{\partial J}{\partial z}(0,0,z) = \lambda \frac{\partial g}{\partial z}(0,0,z),$$

and since

$$\frac{\partial g}{\partial x}(x,y,z) = 2x, \quad \frac{\partial g}{\partial y}(x,y,z) = 2y, \quad \frac{\partial g}{\partial z}(0,0,z) = 0,$$

the partial derivatives above all vanish for $x = y = 0$, so at a local extremum we should also have

$$\frac{\partial J}{\partial x}(0,0,z) = 0, \quad \frac{\partial J}{\partial y}(0,0,z) = 0, \quad \frac{\partial J}{\partial z}(0,0,z) = 0,$$

but this is absurd since

$$\frac{\partial J}{\partial x}(x,y,z) = 1, \quad \frac{\partial J}{\partial y}(x,y,z) = 1, \quad \frac{\partial J}{\partial z}(x,y,z) = 2z.$$

The reader should enjoy finding the reason for the flaw in the argument.

One should also keep in mind that Theorem 4.1 gives only a necessary condition. The (u, λ) may *not* correspond to local extrema! Thus, it is always necessary to analyze the local behavior of J near a critical point u. This is generally difficult, but in the case where J is affine or quadratic and the constraints are affine or quadratic, this is possible (although not always easy).

Example 4.2. Let us apply the above method to the following example in which $E_1 = \mathbb{R}$, $E_2 = \mathbb{R}$, $\Omega = \mathbb{R}^2$, and

$$\begin{aligned} J(x_1, x_2) &= -x_2 \\ \varphi(x_1, x_2) &= x_1^2 + x_2^2 - 1. \end{aligned}$$

Observe that

$$U = \{(x_1, x_2) \in \mathbb{R}^2 \mid x_1^2 + x_2^2 = 1\}$$

is the unit circle, and since

$$\nabla\varphi(x_1, x_2) = \begin{pmatrix} 2x_1 \\ 2x_2 \end{pmatrix},$$

it is clear that $\nabla\varphi(x_1, x_2) \neq 0$ for every point $= (x_1, x_2)$ on the unit circle. If we form the Lagrangian

$$L(x_1, x_2, \lambda) = -x_2 + \lambda(x_1^2 + x_2^2 - 1),$$

Theorem 4.1 says that a necessary condition for J to have a constrained local extremum is that $\nabla L(x_1, x_2, \lambda) = 0$, so the following equations must hold:

$$\begin{aligned} 2\lambda x_1 &= 0 \\ -1 + 2\lambda x_2 &= 0 \\ x_1^2 + x_2^2 &= 1. \end{aligned}$$

The second equation implies that $\lambda \neq 0$, and then the first yields $x_1 = 0$, so the third yields $x_2 = \pm 1$, and we get two solutions:

$$\begin{aligned} \lambda &= \frac{1}{2}, & (x_1, x_2) &= (0, 1) \\ \lambda &= -\frac{1}{2}, & (x_1', x_2') &= (0, -1). \end{aligned}$$

We can check immediately that the first solution is a minimum and the second is a maximum. The reader should look for a geometric interpretation of this problem.

Example 4.3. Let us now consider the case in which J is a quadratic function of the form

$$J(v) = \frac{1}{2} v^\top A v - v^\top b,$$

where A is an $n \times n$ symmetric matrix, $b \in \mathbb{R}^n$, and the constraints are given by a linear system of the form

$$Cv = d,$$

where C is an $m \times n$ matrix with $m < n$ and $d \in \mathbb{R}^m$. We also assume that C has rank m. In this case the function φ is given by

$$\varphi(v) = (Cv - d)^\top,$$

because we view $\varphi(v)$ as a row vector (and v as a column vector), and since

$$d\varphi(v)(w) = C^\top w,$$

the condition that the Jacobian matrix of φ at u have rank m is satisfied. The Lagrangian of this problem is

$$L(v, \lambda) = \frac{1}{2} v^\top A v - v^\top b + (Cv - d)^\top \lambda = \frac{1}{2} v^\top A v - v^\top b + \lambda^\top (Cv - d),$$

where λ is viewed as a column vector. Now because A is a symmetric matrix, it is easy to show that

$$\nabla L(v, \lambda) = \begin{pmatrix} Av - b + C^\top \lambda \\ Cv - d \end{pmatrix}.$$

Therefore, the necessary condition for constrained local extrema is

$$\begin{aligned} Av + C^\top \lambda &= b \\ Cv &= d, \end{aligned}$$

which can be expressed in matrix form as

$$\begin{pmatrix} A & C^\top \\ C & 0 \end{pmatrix} \begin{pmatrix} v \\ \lambda \end{pmatrix} = \begin{pmatrix} b \\ d \end{pmatrix},$$

where the matrix of the system is a symmetric matrix. We should not be surprised to find the system discussed later in Chapter 6, except for some renaming of the matrices and vectors involved. As we will show in Section 6.2, the function J has a minimum iff A is positive definite, so in general, if A is only a symmetric matrix, the critical points of the Lagrangian do *not* correspond to extrema of J.

Remark: If the Jacobian matrix $\mathrm{Jac}(\varphi)(v) = \big((\partial\varphi_i/\partial x_j)(v)\big)$ has rank m for all $v \in U$ (which is equivalent to the linear independence of the linear forms $d\varphi_i(v)$), then we say that $0 \in \mathbb{R}^m$ is a *regular value* of φ. In this case, it is known that

$$U = \{v \in \Omega \mid \varphi(v) = 0\}$$

is a *smooth submanifold of dimension* $n - m$ *of* $\mathbb{R}^n$. Furthermore, the set

$$T_v U = \{w \in \mathbb{R}^n \mid d\varphi_i(v)(w) = 0,\ 1 \leq i \leq m\} = \bigcap_{i=1}^{m} \mathrm{Ker}\, d\varphi_i(v)$$

is the *tangent space* to U at v (a vector space of dimension $n - m$). Then, the condition

$$dJ(v) + \mu_1 d\varphi_1(v) + \cdots + \mu_m d\varphi_m(v) = 0$$

implies that $dJ(v)$ vanishes on the tangent space $T_v U$. Conversely, if $dJ(v)(w) = 0$ for all $w \in T_v U$, this means that $dJ(v)$ is orthogonal (in the sense of Definition 10.3 (Vol. I)) to $T_v U$. Since (by Theorem 10.4(b) (Vol. I)) the orthogonal of $T_v U$ is the space of linear forms spanned by $d\varphi_1(v), \ldots, d\varphi_m(v)$, it follows that $dJ(v)$ must be a linear combination of the $d\varphi_i(v)$. Therefore, when 0 is a regular value of φ, Theorem 4.1 asserts that if $u \in U$ is a local extremum of J, then $dJ(u)$ must vanish on the tangent space $T_u U$. We can say even more. The subset $Z(J)$ of Ω given by

$$Z(J) = \{v \in \Omega \mid J(v) = J(u)\}$$

(the *level set of level* $J(u)$) is a hypersurface in Ω, and if $dJ(u) \neq 0$, the zero locus of $dJ(u)$ is the tangent space $T_u Z(J)$ to $Z(J)$ at u (a vector space of dimension $n-1$), where

$$T_u Z(J) = \{w \in \mathbb{R}^n \mid dJ(u)(w) = 0\}.$$

Consequently, Theorem 4.1 asserts that

$$T_u U \subseteq T_u Z(J);$$

this is a geometric condition.

We now return to the general situation where E_1 and E_2 may be infinite-dimensional normed vector spaces (with E_1 a Banach space) and we state and prove the following general result about the method of Lagrange multipliers.

Theorem 4.2. *(Necessary condition for a constrained extremum) Let $\Omega \subseteq E_1 \times E_2$ be an open subset of a product of normed vector spaces, with E_1 a Banach space (E_1 is complete), let $\varphi\colon \Omega \to E_2$ be a C^1-function (which means that $d\varphi(\omega)$ exists and is continuous for all $\omega \in \Omega$), and let*

$$U = \{(u_1, u_2) \in \Omega \mid \varphi(u_1, u_2) = 0\}.$$

Moreover, let $u = (u_1, u_2) \in U$ be a point such that

$$\frac{\partial \varphi}{\partial x_2}(u_1, u_2) \in \mathcal{L}(E_2; E_2) \quad \text{and} \quad \left(\frac{\partial \varphi}{\partial x_2}(u_1, u_2)\right)^{-1} \in \mathcal{L}(E_2; E_2),$$

and let $J\colon \Omega \to \mathbb{R}$ be a function which is differentiable at u. If J has a constrained local extremum at u, then there is a continuous linear form $\Lambda(u) \in \mathcal{L}(E_2; \mathbb{R})$ such that

$$dJ(u) + \Lambda(u) \circ d\varphi(u) = 0.$$

Proof. The plan of attack is to use the implicit function theorem; Theorem 3.3. Observe that the assumptions of Theorem 3.3 are indeed met. Therefore, there exist some open subsets $U_1 \subseteq E_1$, $U_2 \subseteq E_2$, and a continuous function $g\colon U_1 \to U_2$ with $(u_1, u_2) \in U_1 \times U_2 \subseteq \Omega$ and such that

$$\varphi(v_1, g(v_1)) = 0$$

for all $v_1 \in U_1$. Moreover, g is differentiable at $u_1 \in U_1$ and

$$dg(u_1) = -\left(\frac{\partial \varphi}{\partial x_2}(u)\right)^{-1} \circ \frac{\partial \varphi}{\partial x_1}(u).$$

It follows that the restriction of J to $(U_1 \times U_2) \cap U$ yields a function G of a single variable, with

$$G(v_1) = J(v_1, g(v_1))$$

for all $v_1 \in U_1$. Now the function G is differentiable at u_1 and it has a local extremum at u_1 on U_1, so Proposition 4.1 implies that

$$dG(u_1) = 0.$$

By the chain rule,

$$\begin{aligned} dG(u_1) &= \frac{\partial J}{\partial x_1}(u) + \frac{\partial J}{\partial x_2}(u) \circ dg(u_1) \\ &= \frac{\partial J}{\partial x_1}(u) - \frac{\partial J}{\partial x_2}(u) \circ \left(\frac{\partial \varphi}{\partial x_2}(u)\right)^{-1} \circ \frac{\partial \varphi}{\partial x_1}(u). \end{aligned}$$

From $dG(u_1) = 0$, we deduce

$$\frac{\partial J}{\partial x_1}(u) = \frac{\partial J}{\partial x_2}(u) \circ \left(\frac{\partial \varphi}{\partial x_2}(u)\right)^{-1} \circ \frac{\partial \varphi}{\partial x_1}(u),$$

and since we also have

$$\frac{\partial J}{\partial x_2}(u) = \frac{\partial J}{\partial x_2}(u) \circ \left(\frac{\partial \varphi}{\partial x_2}(u)\right)^{-1} \circ \frac{\partial \varphi}{\partial x_2}(u),$$

if we let

$$\Lambda(u) = -\frac{\partial J}{\partial x_2}(u) \circ \left(\frac{\partial \varphi}{\partial x_2}(u)\right)^{-1},$$

then we get

$$\begin{aligned} dJ(u) &= \frac{\partial J}{\partial x_1}(u) + \frac{\partial J}{\partial x_2}(u) \\ &= \frac{\partial J}{\partial x_2}(u) \circ \left(\frac{\partial \varphi}{\partial x_2}(u)\right)^{-1} \circ \left(\frac{\partial \varphi}{\partial x_1}(u) + \frac{\partial \varphi}{\partial x_2}(u)\right) \\ &= -\Lambda(u) \circ d\varphi(u), \end{aligned}$$

which yields $dJ(u) + \Lambda(u) \circ d\varphi(u) = 0$, as claimed. □

Finally, we prove Theorem 4.1.

Proof of Theorem 4.1. The linear independence of the m linear forms $d\varphi_i(u)$ is equivalent to the fact that the $m \times n$ matrix $A = \big((\partial \varphi_i / \partial x_j)(u)\big)$ has rank m. By reordering the columns, we may assume that the first m columns are linearly independent. If we let $\varphi \colon \Omega \to \mathbb{R}^m$ be the function defined by

$$\varphi(v) = (\varphi_1(v), \ldots, \varphi_m(v))$$

for all $v \in \Omega$, then we see that $\partial\varphi/\partial x_2(u)$ is invertible and both $\partial\varphi/\partial x_2(u)$ and its inverse are continuous, so that Theorem 4.2 applies, and there is some (continuous) linear form $\Lambda(u) \in \mathcal{L}(\mathbb{R}^m; \mathbb{R})$ such that

$$dJ(u) + \Lambda(u) \circ d\varphi(u) = 0.$$

However, $\Lambda(u)$ is defined by some m-tuple $(\lambda_1(u), \ldots, \lambda_m(u)) \in \mathbb{R}^m$, and in view of the definition of φ, the above equation is equivalent to

$$dJ(u) + \lambda_1(u) d\varphi_1(u) + \cdots + \lambda_m(u) d\varphi_m(u) = 0.$$

The uniqueness of the $\lambda_i(u)$ is a consequence of the linear independence of the $d\varphi_i(u)$. □

We now investigate conditions for the existence of extrema involving the second derivative of J.

4.2 Using Second Derivatives to Find Extrema

For the sake of brevity, we consider only the case of local minima; analogous results are obtained for local maxima (replace J by $-J$, since $\max_u J(u) = -\min_u -J(u)$). We begin with a necessary condition for an unconstrained local minimum.

Proposition 4.3. *Let E be a normed vector space and let $J \colon \Omega \to \mathbb{R}$ be a function, with Ω some open subset of E. If the function J is differentiable in Ω, if J has a second derivative $\mathrm{D}^2 J(u)$ at some point $u \in \Omega$, and if J has a local minimum at u, then*

$$\mathrm{D}^2 J(u)(w, w) \geq 0 \quad \textit{for all } w \in E.$$

Proof. Pick any nonzero vector $w \in E$. Since Ω is open, for t small enough, $u + tw \in \Omega$ and $J(u + tw) \geq J(u)$, so there is some open interval $I \subseteq \mathbb{R}$ such that

$$u + tw \in \Omega \quad \text{and} \quad J(u + tw) \geq J(u)$$

for all $t \in I$. Using the Taylor–Young formula and the fact that we must have $dJ(u) = 0$ since J has a local minimum at u, we get

$$0 \leq J(u + tw) - J(u) = \frac{t^2}{2} \mathrm{D}^2 J(u)(w, w) + t^2 \left\|w\right\|^2 \epsilon(tw),$$

with $\lim_{t \mapsto 0} \epsilon(tw) = 0$, which implies that

$$\mathrm{D}^2 J(u)(w, w) \geq 0.$$

Since the argument holds for all $w \in E$ (trivially if $w = 0$), the proposition is proven. □

One should be cautioned that there is no converse to the previous proposition. For example, the function $f\colon x \mapsto x^3$ has no local minimum at 0, yet $df(0) = 0$ and $D^2 f(0)(u, v) = 0$. Similarly, the reader should check that the function $f\colon \mathbb{R}^2 \to \mathbb{R}$ given by

$$f(x, y) = x^2 - 3y^3$$

has no local minimum at $(0, 0)$; yet $df(0, 0) = 0$ since $df(x, y) = (2x, -9y^2)$, and for $u = (u_1, u_2)$, $D^2 f(0, 0)(u, u) = 2u_1^2 \geq 0$ since

$$D^2 f(x, y)(u, u) = \begin{pmatrix} u_1 & u_2 \end{pmatrix} \begin{pmatrix} 2 & 0 \\ 0 & -18y \end{pmatrix} \begin{pmatrix} u_1 \\ u_2 \end{pmatrix}.$$

See Figure 4.3.

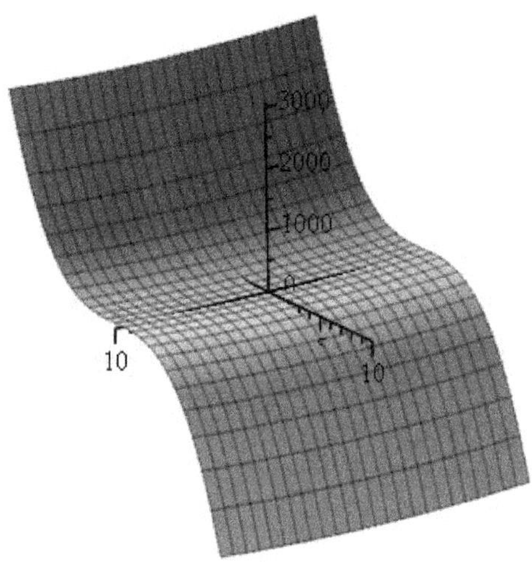

Fig. 4.3 The graph of $f(x, y) = x^2 - 3y^3$. Note that $(0, 0)$ not a local extremum despite the fact that $df(0, 0) = 0$.

When $E = \mathbb{R}^n$, Proposition 4.3 says that a necessary condition for having a local minimum is that the Hessian $\nabla^2 J(u)$ be positive semidefinite (it is always symmetric).

We now give sufficient conditions for the existence of a local minimum.

Theorem 4.3. *Let E be a normed vector space, let $J\colon \Omega \to \mathbb{R}$ be a function with Ω some open subset of E, and assume that J is differentiable in Ω and that $dJ(u) = 0$ at some point $u \in \Omega$. The following properties hold:*

(1) If $\mathrm{D}^2J(u)$ *exists and if there is some number* $\alpha \in \mathbb{R}$ *such that* $\alpha > 0$ *and*

$$\mathrm{D}^2J(u)(w,w) \geq \alpha \|w\|^2 \quad \textit{for all } w \in E,$$

then J *has a strict local minimum at* u.

(2) If $\mathrm{D}^2J(v)$ *exists for all* $v \in \Omega$ *and if there is a ball* $B \subseteq \Omega$ *centered at* v *such that*

$$\mathrm{D}^2J(v)(w,w) \geq 0 \quad \textit{for all } v \in B \textit{ and all } w \in E,$$

then J *has a local minimum at* u.

Proof. (1) Using the formula of Taylor–Young, for every vector w small enough, we can write

$$\begin{aligned} J(u+w) - J(u) &= \frac{1}{2}\mathrm{D}^2J(u)(w,w) + \|w\|^2 \epsilon(w) \\ &\geq \left(\frac{1}{2}\alpha + \epsilon(w)\right) \|w\|^2 \end{aligned}$$

with $\lim_{w \mapsto 0} \epsilon(w) = 0$. Consequently if we pick $r > 0$ small enough that $|\epsilon(w)| < \alpha$ for all w with $\|w\| < r$, then $J(u+w) > J(u)$ for all $u + w \in B$, where B is the open ball of center u and radius r. This proves that J has a local strict minimum at u.

(2) The formula of Taylor–Maclaurin shows that for all $u + w \in B$, we have

$$J(u+w) = J(u) + \frac{1}{2}\mathrm{D}^2J(v)(w,w) \geq J(u),$$

for some $v \in (u, w + w)$. □

There are no converses of the two assertions of Theorem 4.3. However, there is a condition on $\mathrm{D}^2J(u)$ that implies the condition of Part (1). Since this condition is easier to state when $E = \mathbb{R}^n$, we begin with this case.

Recall that a $n \times n$ symmetric matrix A is *positive definite* if $x^\top Ax > 0$ for all $x \in \mathbb{R}^n - \{0\}$. In particular, A must be invertible.

Proposition 4.4. *For any symmetric matrix* A*, if* A *is positive definite, then there is some* $\alpha > 0$ *such that*

$$x^\top Ax \geq \alpha \|x\|^2 \quad \textit{for all } x \in \mathbb{R}^n.$$

Proof. Pick any norm in $\mathbb{R}^n$ (recall that all norms on $\mathbb{R}^n$ are equivalent). Since the unit sphere $S^{n-1} = \{x \in \mathbb{R}^n \mid \|x\| = 1\}$ is compact and since the function $f(x) = x^\top Ax$ is never zero on S^{n-1}, the function f has a minimum $\alpha > 0$ on S^{n-1}. Using the usual trick that $x = \|x\| (x/\|x\|)$ for every nonzero vector $x \in \mathbb{R}^n$ and the fact that the inequality of the proposition is trivial for $x = 0$, from

$$x^\top Ax \geq \alpha \quad \text{for all } x \text{ with } \|x\| = 1,$$

we get

$$x^\top Ax \geq \alpha \|x\|^2 \quad \text{for all } x \in \mathbb{R}^n,$$

as claimed. □

We can combine Theorem 4.3 and Proposition 4.4 to obtain a useful sufficient condition for the existence of a strict local minimum. First let us introduce some terminology.

Definition 4.6. Given a function $J\colon \Omega \to \mathbb{R}$ as before, say that a point $u \in \Omega$ is a *nondegenerate critical point* if $dJ(u) = 0$ and if the Hessian matrix $\nabla^2 J(u)$ is invertible.

Proposition 4.5. *Let $J\colon \Omega \to \mathbb{R}$ be a function defined on some open subset $\Omega \subseteq \mathbb{R}^n$. If J is differentiable in Ω and if some point $u \in \Omega$ is a nondegenerate critical point such that $\nabla^2 J(u)$ is positive definite, then J has a strict local minimum at u.*

Remark: It is possible to generalize Proposition 4.5 to infinite-dimensional spaces by finding a suitable generalization of the notion of a nondegenerate critical point. Firstly, we assume that E is a Banach space (a complete normed vector space). Then we define the dual E' of E as the set of continuous linear forms on E, so that $E' = \mathcal{L}(E; \mathbb{R})$. Following Lang, we use the notation E' for the space of continuous linear forms to avoid confusion with the space $E^* = \mathrm{Hom}(E, \mathbb{R})$ of all linear maps from E to $\mathbb{R}$. A continuous bilinear map $\varphi\colon E \times E \to \mathbb{R}$ in $\mathcal{L}_2(E, E; \mathbb{R})$ yields a map Φ from E to E' given by

$$\Phi(u) = \varphi_u,$$

where $\varphi_u \in E'$ is the linear form defined by

$$\varphi_u(v) = \varphi(u, v).$$

It is easy to check that φ_u is continuous and that the map Φ is continuous. Then we say that φ is *nondegenerate* iff $\Phi\colon E \to E'$ is an isomorphism of Banach spaces, which means that Φ is invertible and that both Φ and Φ^{-1}

are continuous linear maps. Given a function $J\colon \Omega \to \mathbb{R}$ differentiable on Ω as before (where Ω is an open subset of E), if $\mathrm{D}^2J(u)$ exists for some $u \in \Omega$, we say that u is a *nondegenerate critical point* if $dJ(u) = 0$ and if $\mathrm{D}^2J(u)$ is nondegenerate. Of course, $\mathrm{D}^2J(u)$ is positive definite if $\mathrm{D}^2J(u)(w,w) > 0$ for all $w \in E - \{0\}$.

Using the above definition, Proposition 4.4 can be generalized to a nondegenerate positive definite bilinear form (on a Banach space) and Theorem 4.5 can also be generalized to the situation where $J\colon \Omega \to \mathbb{R}$ is defined on an open subset of a Banach space. For details and proofs, see Cartan [Cartan (1990)] (Part I Chapter 8) and Avez [Avez (1991)] (Chapter 8 and Chapter 10).

In the next section we make use of convexity; both on the domain Ω and on the function J itself.

4.3 Using Convexity to Find Extrema

We begin by reviewing the definition of a convex set and of a convex function.

Definition 4.7. Given any real vector space E, we say that a subset C of E is *convex* if either $C = \emptyset$ or if for every pair of points $u, v \in C$, the line segment connecting u and v is contained in C, i.e.,

$$(1-\lambda)u + \lambda v \in C \quad \text{for all } \lambda \in \mathbb{R} \text{ such that } 0 \leq \lambda \leq 1.$$

Given any two points $u, v \in E$, the *line segment* $[u, v]$ is the set

$$[u,v] = \{(1-\lambda)u + \lambda v \in E \mid \lambda \in \mathbb{R},\ 0 \leq \lambda \leq 1\}.$$

Clearly, a nonempty set C is convex iff $[u, v] \subseteq C$ whenever $u, v \in C$. See Figure 4.4 for an example of a convex set.

Definition 4.8. If C is a nonempty convex subset of E, a function $f\colon C \to \mathbb{R}$ is *convex* (on C) if for every pair of points $u, v \in C$, we have

$$f((1-\lambda)u+\lambda v) \leq (1-\lambda)f(u)+\lambda f(v) \quad \text{for all } \lambda \in \mathbb{R} \text{ such that } 0 \leq \lambda \leq 1.$$

The function f is *strictly convex* (on C) if for every pair of distinct points $u, v \in C$ ($u \neq v$), we have

$$f((1-\lambda)u+\lambda v) < (1-\lambda)f(u)+\lambda f(v) \quad \text{for all } \lambda \in \mathbb{R} \text{ such that } 0 < \lambda < 1;$$

see Figure 4.5.

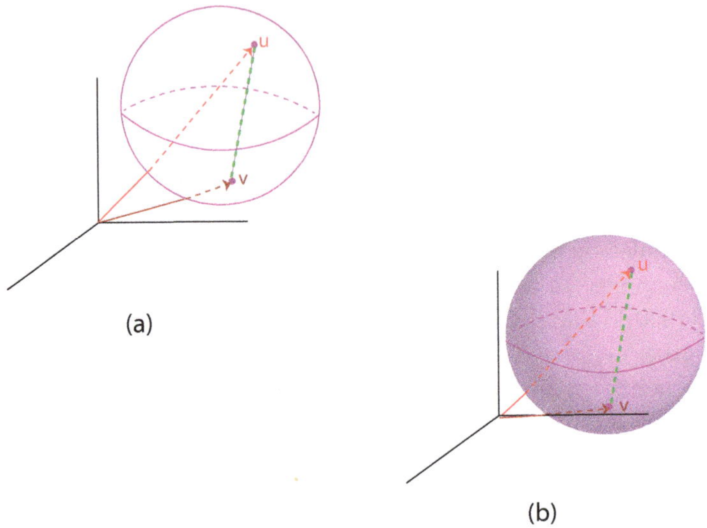

Fig. 4.4 Figure (a) shows that a sphere is not convex in $\mathbb{R}^3$ since the dashed green line does not lie on its surface. Figure (b) shows that a solid ball is convex in $\mathbb{R}^3$.

The *epigraph*[1] $\mathbf{epi}(f)$ of a function $f\colon A \to \mathbb{R}$ defined on some subset A of $\mathbb{R}^n$ is the subset of $\mathbb{R}^{n+1}$ defined as

$$\mathbf{epi}(f) = \{(x, y) \in \mathbb{R}^{n+1} \mid f(x) \leq y,\ x \in A\}.$$

A function $f\colon C \to \mathbb{R}$ defined on a convex subset C is *concave* (resp. *strictly concave*) if $(-f)$ is convex (resp. strictly convex).

It is obvious that a function f if convex iff its epigraph $\mathbf{epi}(f)$ is a convex subset of $\mathbb{R}^{n+1}$.

Example 4.4. Here are some common examples of convex sets.

- Subspaces $V \subseteq E$ of a vector space E are convex.
- *Affine subspaces*, that is, sets of the form $u + V$, where V is a subspace of E and $u \in E$, are convex.
- Balls (open or closed) are convex. Given any linear form $\varphi\colon E \to \mathbb{R}$, for any scalar $c \in \mathbb{R}$, the *closed half–spaces*

$$H^+_{\varphi,c} = \{u \in E \mid \varphi(u) \geq c\}, \qquad H^-_{\varphi,c} = \{u \in E \mid \varphi(u) \leq c\},$$

are convex.

[1] "Epi" means above.

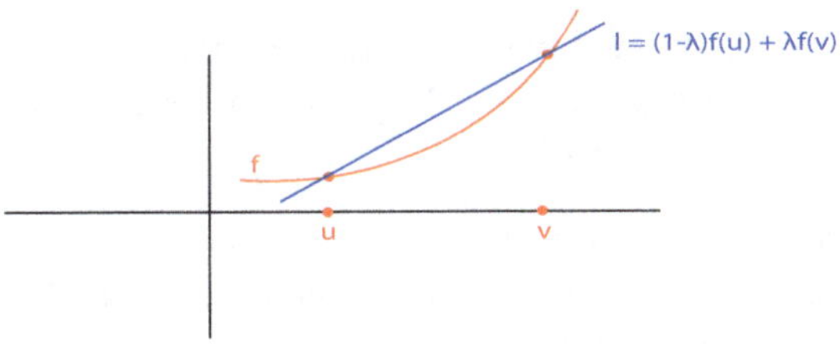

(a)

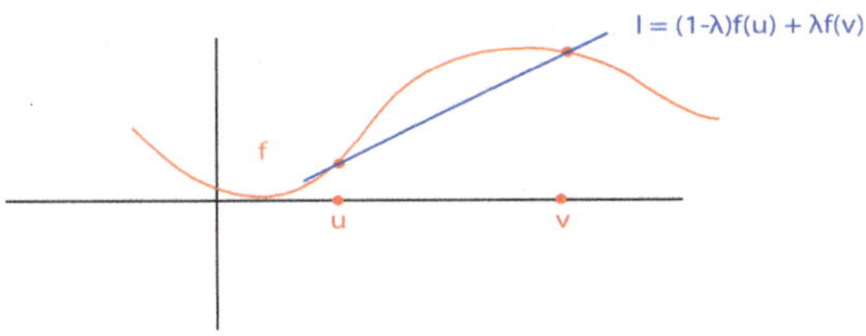

(b)

Fig. 4.5 Figures (a) and (b) are the graphs of real valued functions. Figure (a) is the graph of convex function since the blue line lies above the graph of f. Figure (b) shows the graph of a function which is not convex.

- Any intersection of half–spaces is convex.
- More generally, any intersection of convex sets is convex.

Example 4.5. Here are some common examples of convex and concave functions.

- Linear forms are convex functions (but not strictly convex).
- Any norm $\| \, \| : E \to \mathbb{R}_+$ is a convex function.
- The max function,

$$\max(x_1, \ldots, x_n) = \max\{x_1, \ldots, x_n\}$$

  is convex on $\mathbb{R}^n$.
- The exponential $x \mapsto e^{cx}$ is strictly convex for any $c \neq 0$ ($c \in \mathbb{R}$).
- The logarithm function is concave on $\mathbb{R}_+ - \{0\}$.
- The *log-determinant function* $\log\det$ is concave on the set of symmetric positive definite matrices. This function plays an important role in convex optimization.

An excellent exposition of convexity and its applications to optimization can be found in Boyd [Boyd and Vandenberghe (2004)].

Here is a necessary condition for a function to have a local minimum with respect to a convex subset U.

Theorem 4.4. *(Necessary condition for a local minimum on a convex subset) Let $J\colon \Omega \to \mathbb{R}$ be a function defined on some open subset Ω of a normed vector space E and let $U \subseteq \Omega$ be a nonempty convex subset. Given any $u \in U$, if $dJ(u)$ exists and if J has a local minimum in u with respect to U, then*

$$dJ(u)(v-u) \geq 0 \quad \textit{for all } v \in U.$$

Proof. Let $v = u + w$ be an arbitrary point in U. Since U is convex, we have $u + tw \in U$ for all t such that $0 \leq t \leq 1$. Since $dJ(u)$ exists, we can write

$$J(u+tw) - J(u) = dJ(u)(tw) + \|tw\| \, \epsilon(tw)$$

with $\lim_{t\mapsto 0} \epsilon(tw) = 0$. However, because $0 \leq t \leq 1$,

$$J(u+tw) - J(u) = t(dJ(u)(w) + \|w\| \, \epsilon(tw))$$

and since u is a local minimum with respect to U, we have $J(u+tw) - J(u) \geq 0$, so we get

$$t(dJ(u)(w) + \|w\| \, \epsilon(tw)) \geq 0.$$

The above implies that $dJ(u)(w) \geq 0$, because otherwise we could pick $t > 0$ small enough so that

$$dJ(u)(w) + \|w\| \, \epsilon(tw) < 0,$$

a contradiction. Since the argument holds for all $v = u + w \in U$, the theorem is proven. □

Observe that the convexity of U is a substitute for the use of Lagrange multipliers, but we now have to deal with an *inequality* instead of an equality.

In the special case where U is a subspace of E we have the following result.

Corollary 4.1. *With the same assumptions as in Theorem 4.4, if U is a subspace of E, if $dJ(u)$ exists and if J has a local minimum in u with respect to U, then*

$$dJ(u)(w) = 0 \quad \textit{for all } w \in U.$$

Proof. In this case since $u \in U$ we have $2u \in U$, and for any $u + w \in U$, we must have $2u - (u+w) = u - w \in U$. The previous theorem implies that $dJ(u)(w) \geq 0$ and $dJ(u)(-w) \geq 0$, that is, $dJ(u)(w) \leq 0$, so $dJ(u) = 0$. Since the argument holds for $w \in U$ (because U is a subspace, if $u, w \in U$, then $u + w \in U$), we conclude that

$$dJ(u)(w) = 0 \quad \text{for all } w \in U. \qquad \square$$

We will now characterize convex functions when they have a first derivative or a second derivative.

Proposition 4.6. *(Convexity and first derivative) Let $f \colon \Omega \to \mathbb{R}$ be a function differentiable on some open subset Ω of a normed vector space E and let $U \subseteq \Omega$ be a nonempty convex subset.*

(1) The function f is convex on U iff

$$f(v) \geq f(u) + df(u)(v - u) \quad \textit{for all } u, v \in U.$$

(2) The function f is strictly convex on U iff

$$f(v) > f(u) + df(u)(v - u) \quad \textit{for all } u, v \in U \textit{ with } u \neq v.$$

See Figure 4.6.

Proof. Let $u, v \in U$ be any two distinct points and pick $\lambda \in \mathbb{R}$ with $0 < \lambda < 1$. If the function f is convex, then

$$f((1-\lambda)u + \lambda v) \leq (1-\lambda)f(u) + \lambda f(v),$$

which yields

$$\frac{f((1-\lambda)u + \lambda v) - f(u)}{\lambda} \leq f(v) - f(u).$$

It follows that

$$df(u)(v-u) = \lim_{\lambda \mapsto 0} \frac{f((1-\lambda)u + \lambda v) - f(u)}{\lambda} \leq f(v) - f(u).$$

If f is strictly convex, the above reasoning does not work, because a strict inequality is not necessarily preserved by "passing to the limit." We have recourse to the following trick: for any ω such that $0 < \omega < 1$, observe that

$$(1-\lambda)u + \lambda v = u + \lambda(v - u) = \frac{\omega - \lambda}{\omega} u + \frac{\lambda}{\omega}(u + \omega(v - u)).$$

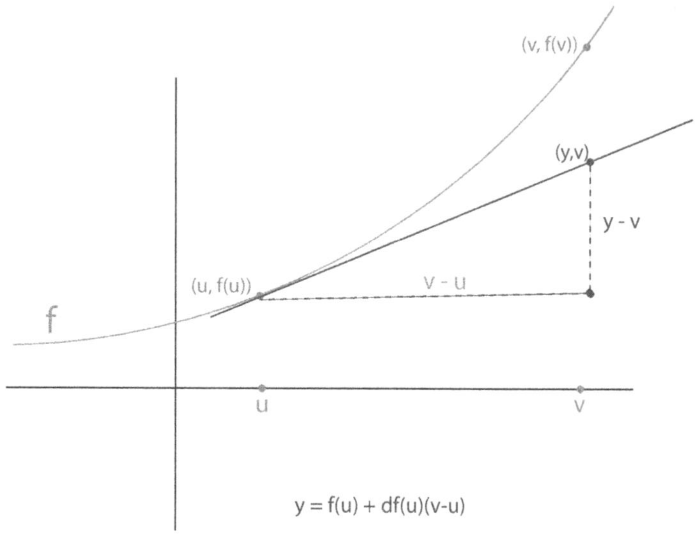

Fig. 4.6 An illustration of a convex valued function f. Since f is convex it always lies above its tangent line.

If we assume that $0 < \lambda \leq \omega$, the convexity of f yields

$$\begin{aligned} f(u+\lambda(v-u)) &= f\left(\left(1-\frac{\lambda}{\omega}\right)u + \frac{\lambda}{\omega}(u+\omega(v-u))\right) \\ &\leq \frac{\omega-\lambda}{\omega}f(u) + \frac{\lambda}{\omega}f(u+\omega(v-u)). \end{aligned}$$

If we subtract $f(u)$ to both sides, we get

$$\frac{f(u+\lambda(v-u)) - f(u)}{\lambda} \leq \frac{f(u+\omega(v-u)) - f(u)}{\omega}.$$

Now since $0 < \omega < 1$ and f is strictly convex,

$$f(u+\omega(v-u)) = f((1-\omega)u+\omega v) < (1-\omega)f(u) + \omega f(v),$$

which implies that

$$\frac{f(u+\omega(v-u)) - f(u)}{\omega} < f(v) - f(u),$$

and thus we get

$$\frac{f(u+\lambda(v-u)) - f(u)}{\lambda} \leq \frac{f(u+\omega(v-u)) - f(u)}{\omega} < f(v) - f(u).$$

If we let λ go to 0, by passing to the limit we get

$$df(u)(v-u) \leq \frac{f(u+\omega(v-u)) - f(u)}{\omega} < f(v) - f(u),$$

which yields the desired strict inequality.

Let us now consider the converse of (1); that is, assume that

$$f(v) \geq f(u) + df(u)(v-u) \quad \text{for all } u, v \in U.$$

For any two distinct points $u, v \in U$ and for any λ with $0 < \lambda < 1$, we get

$$\begin{aligned} f(v) &\geq f(v+\lambda(u-v)) - \lambda df(v+\lambda(u-v))(u-v) \\ f(u) &\geq f(v+\lambda(u-v)) + (1-\lambda)df(v+\lambda(u-v))(u-v), \end{aligned}$$

and if we multiply the first inequality by $1-\lambda$ and the second inequality by λ and them add up the resulting inequalities, we get

$$(1-\lambda)f(v) + \lambda f(u) \geq f(v+\lambda(u-v)) = f((1-\lambda)v + \lambda u),$$

which proves that f is convex.

The proof of the converse of (2) is similar, except that the inequalities are replaced by strict inequalities. □

We now establish a convexity criterion using the second derivative of f. This criterion is often easier to check than the previous one.

Proposition 4.7. *(Convexity and second derivative) Let $f\colon \Omega \to \mathbb{R}$ be a function twice differentiable on some open subset Ω of a normed vector space E and let $U \subseteq \Omega$ be a nonempty convex subset.*

(1) The function f is convex on U iff

$$D^2 f(u)(v-u, v-u) \geq 0 \quad \textit{for all } u, v \in U.$$

(2) If

$$D^2 f(u)(v-u, v-u) > 0 \quad \textit{for all } u, v \in U \textit{ with } u \neq v,$$

then f is strictly convex.

Proof. First assume that the inequality in Condition (1) is satisfied. For any two distinct points $u, v \in U$, the formula of Taylor–Maclaurin yields

$$\begin{aligned} f(v) - f(u) - df(u)(v-u) &= \frac{1}{2} D^2 f(w)(v-u, v-u) \\ &= \frac{\rho^2}{2} D^2 f(w)(v-w, v-w), \end{aligned}$$

for some $w = (1-\lambda)u + \lambda v = u + \lambda(v-u)$ with $0 < \lambda < 1$, and with $\rho = 1/(1-\lambda) > 0$, so that $v - u = \rho(v-w)$. Since $\mathrm{D}^2 f(w)(v-w, v-w) \geq 0$ for all $u, w \in U$, we conclude by applying Proposition 4.6(1).

Similarly, if (2) holds, the above reasoning and Proposition 4.6(2) imply that f is strictly convex.

To prove the necessary condition in (1), define $g\colon \Omega \to \mathbb{R}$ by

$$g(v) = f(v) - df(u)(v),$$

where $u \in U$ is any point considered fixed. If f is convex, since

$$g(v) - g(u) = f(v) - f(u) - df(u)(v-u),$$

Proposition 4.6 implies that $f(v) - f(u) - df(u)(v-u) \geq 0$, which implies that g has a local minimum at u with respect to all $v \in U$. Therefore, we have $dg(u) = 0$. Observe that g is twice differentiable in Ω and $\mathrm{D}^2 g(u) = \mathrm{D}^2 f(u)$, so the formula of Taylor–Young yields for every $v = u + w \in U$ and all t with $0 \leq t \leq 1$,

$$\begin{aligned} 0 \leq g(u+tw) - g(u) &= \frac{t^2}{2}\mathrm{D}^2 f(u)(tw, tw) + \|tw\|^2\, \epsilon(tw) \\ &= \frac{t^2}{2}(\mathrm{D}^2 f(u)(w,w) + 2\|w\|^2\, \epsilon(wt)), \end{aligned}$$

with $\lim_{t \mapsto 0} \epsilon(wt) = 0$, and for t small enough, we must have $\mathrm{D}^2 f(u)(w,w) \geq 0$, as claimed. $\square$

The converse of Proposition 4.7 (2) is false as we see by considering the strictly convex function f given by $f(x) = x^4$ and its second derivative at $x = 0$.

Example 4.6. On the other hand, if f is a quadratic function of the form

$$f(u) = \frac{1}{2}u^\top A u - u^\top b$$

where A is a symmetric matrix, we know that

$$df(u)(v) = v^\top(Au - b),$$

so

$$\begin{aligned} f(v) - f(u) - df(u)(v-u) &= \frac{1}{2}v^\top A v - v^\top b - \frac{1}{2}u^\top A u + u^\top b \\ &\quad - (v-u)^\top(Au - b) \\ &= \frac{1}{2}v^\top A v - \frac{1}{2}u^\top A u - (v-u)^\top A u \\ &= \frac{1}{2}v^\top A v + \frac{1}{2}u^\top A u - v^\top A u \\ &= \frac{1}{2}(v-u)^\top A(v-u). \end{aligned}$$

Therefore, Proposition 4.6 implies that if A is positive semidefinite, then f is convex and if A is positive definite, then f is strictly convex. The converse follows by Proposition 4.7.

We conclude this section by applying our previous theorems to convex functions defined on convex subsets. In this case local minima (resp. local maxima) are global minima (resp. global maxima). The next definition is the special case of Definition 4.1 in which $W = E$ but it does not hurt to state it explicitly.

Definition 4.9. Let $f\colon E \to \mathbb{R}$ be any function defined on some normed vector space (or more generally, any set). For any $u \in E$, we say that f has a *minimum* in u (resp. *maximum* in u) if

$$f(u) \leq f(v) \text{ (resp. } f(u) \geq f(v)) \quad \text{for all } v \in E.$$

We say that f has a *strict minimum* in u (resp. *strict maximum* in u) if

$$f(u) < f(v) \text{ (resp. } f(u) > f(v)) \quad \text{for all } v \in E - \{u\}.$$

If $U \subseteq E$ is a subset of E and $u \in U$, we say that f has a *minimum* in u (resp. *strict minimum* in u) *with respect to* U if

$$f(u) \leq f(v) \quad \text{for all } v \in U \quad \text{(resp. } f(u) < f(v) \quad \text{for all } v \in U - \{u\}),$$

and similarly for a *maximum* in u (resp. *strict maximum* in u) *with respect to* U with $\leq$ changed to $\geq$ and $<$ to $>$.

Sometimes, we say *global* maximum (or minimum) to stress that a maximum (or a minimum) is not simply a local maximum (or minimum).

Theorem 4.5. *Given any normed vector space E, let U be any nonempty convex subset of E.*

(1) For any convex function $J\colon U \to \mathbb{R}$, for any $u \in U$, if J has a local minimum at u in U, then J has a (global) minimum at u in U.
(2) Any strict convex function $J\colon U \to \mathbb{R}$ has at most one minimum (in U), and if it does, then it is a strict minimum (in U).
(3) Let $J\colon \Omega \to \mathbb{R}$ be any function defined on some open subset Ω of E with $U \subseteq \Omega$ and assume that J is convex on U. For any point $u \in U$, if $dJ(u)$ exists, then J has a minimum in u with respect to U iff

$$dJ(u)(v-u) \geq 0 \quad \textit{for all } v \in U.$$

(4) If the convex subset U in (3) is open, then the above condition is equivalent to

$$dJ(u) = 0.$$

Proof. (1) Let $v = u + w$ be any arbitrary point in U. Since J is convex, for all t with $0 \leq t \leq 1$, we have

$$J(u + tw) = J(u + t(v - u)) = J((1 - t)u + tv) \leq (1 - t)J(u) + tJ(v),$$

which yields

$$J(u + tw) - J(u) \leq t(J(v) - J(u)).$$

Because J has a local minimum in u, there is some t_0 with $0 < t_0 < 1$ such that

$$0 \leq J(u + t_0 w) - J(u) \leq t_0(J(v) - J(u)),$$

which implies that $J(v) - J(u) \geq 0$.

(2) If J is strictly convex, the above reasoning with $w \neq 0$ shows that there is some t_0 with $0 < t_0 < 1$ such that

$$0 \leq J(u + t_0 w) - J(u) < t_0(J(v) - J(u)),$$

which shows that u is a strict global minimum (in U), and thus that it is unique.

(3) We already know from Theorem 4.4 that the condition $dJ(u)(v - u) \geq 0$ for all $v \in U$ is necessary (even if J is not convex). Conversely, because J is convex, careful inspection of the proof of Part (1) of Proposition 4.6 shows that only the fact that $dJ(u)$ exists in needed to prove that

$$J(v) - J(u) \geq dJ(u)(v - u) \quad \text{for all } v \in U,$$

and if

$$dJ(u)(v - u) \geq 0 \quad \text{for all } v \in U,$$

then

$$J(v) - J(u) \geq 0 \quad \text{for all } v \in U,$$

as claimed.

(4) If U is open, then for every $u \in U$ we can find an open ball B centered at u of radius ϵ small enough so that $B \subseteq U$. Then for any $w \neq 0$ such that $\|w\| < \epsilon$, we have both $v = u + w \in B$ and $v' = u - w \in B$, so Condition (3) implies that

$$dJ(u)(w) \geq 0 \quad \text{and} \quad dJ(u)(-w) \geq 0,$$

which yields

$$dJ(u)(w) = 0.$$

Since the above holds for all $w \neq 0$ such such that $\|w\| < \epsilon$ and since $dJ(u)$ is linear, we leave it to the reader to fill in the details of the proof that $dJ(u) = 0$. ☐

Example 4.7. Theorem 4.5 can be used to rederive the fact that the least squares solutions of a linear system $Ax = b$ (where A is an $m \times n$ matrix) are given by the normal equation

$$A^\top A x = A^\top b.$$

For this, we consider the quadratic function

$$J(v) = \frac{1}{2} \|Av - b\|_2^2 - \frac{1}{2} \|b\|_2^2,$$

and our least squares problem is equivalent to finding the minima of J on $\mathbb{R}^n$. A computation reveals that

$$\begin{aligned} J(v) &= \frac{1}{2} \|Av - b\|_2^2 - \frac{1}{2} \|b\|_2^2 \\ &= \frac{1}{2}(Av - b)^\top (Av - b) - \frac{1}{2} b^\top b \\ &= \frac{1}{2}(v^\top A^\top - b^\top)(Av - b) - \frac{1}{2} b^\top b \\ &= \frac{1}{2} v^\top A^\top A v - v^\top A^\top b, \end{aligned}$$

and so

$$dJ(u) = A^\top A u - A^\top b.$$

Since $A^\top A$ is positive semidefinite, the function J is convex, and Theorem 4.5(4) implies that the minima of J are the solutions of the equation

$$A^\top A u - A^\top b = 0.$$

The considerations in this chapter reveal the need to find methods for finding the zeros of the derivative map

$$dJ \colon \Omega \to E',$$

where Ω is some open subset of a normed vector space E and E' is the space of all continuous linear forms on E (a subspace of E^*). Generalizations of *Newton's method* yield such methods and they are the object of the next chapter.

4.4 Summary

The main concepts and results of this chapter are listed below:

- Local minimum, local maximum, local extremum, strict local minimum, strict local maximum.

- Necessary condition for a local extremum involving the derivative; critical point.
- Local minimum with respect to a subset U, local maximum with respect to a subset U, local extremum with respect to a subset U.
- Constrained local extremum.
- Necessary condition for a constrained extremum.
- Necessary condition for a constrained extremum in terms of Lagrange multipliers.
- Lagrangian.
- Critical points of a Lagrangian.
- Necessary condition of an unconstrained local minimum involving the second-order derivative.
- Sufficient condition for a local minimum involving the second-order derivative.
- A sufficient condition involving nondegenerate critical points.
- Convex sets, convex functions, concave functions, strictly convex functions, strictly concave functions.
- Necessary condition for a local minimum on a convex set involving the derivative.
- Convexity of a function involving a condition on its first derivative.
- Convexity of a function involving a condition on its second derivative.
- Minima of convex functions on convex sets.

4.5 Problems

Problem 4.1. Find the extrema of the function $J(v_1, v_2) = v_2^2$ on the subset U given by

$$U = \{(v_1, v_2) \in \mathbb{R}^2 \mid v_1^2 + v_2^2 - 1 = 0\}.$$

Problem 4.2. Find the extrema of the function $J(v_1, v_2) = v_1 + (v_2 - 1)^2$ on the subset U given by

$$U = \{(v_1, v_2) \in \mathbb{R}^2 \mid v_1^2 = 0\}.$$

Problem 4.3. Let A be an $n \times n$ real symmetric matrix, B an $n \times n$ symmetric positive definite matrix, and let $b \in \mathbb{R}^n$.

(1) Prove that a necessary condition for the function J given by

$$J(v) = \frac{1}{2} v^\top A v - b^\top v$$

to have an extremum in $u \in U$, with U defined by

$$U = \{v \in \mathbb{R}^n \mid v^\top Bv = 1\},$$

is that there is some $\lambda \in \mathbb{R}$ such that

$$Au - b = \lambda Bu.$$

(2) Prove that there is a symmetric positive definite matrix S such that $B = S^2$. Prove that if $b = 0$, then λ is an eigenvalue of the symmetric matrix $S^{-1}AS^{-1}$.

(3) Prove that for all $(u, \lambda) \in U \times \mathbb{R}$, if $Au - b = \lambda Bu$, then

$$J(v) - J(u) = \frac{1}{2}(v-u)^\top (A - \lambda B)(v-u)$$

for all $v \in U$. Deduce that without additional assumptions, it is not possible to conclude that u is an extremum of J on U.

Problem 4.4. Let E be a normed vector space, and let U be a subset of E such that for all $u, v \in U$, we have $(1/2)(u+v) \in U$.

(1) Prove that if U is closed, then U is convex.

Hint. Every real $\theta \in (0,1)$ can be written in a unique way as

$$\theta = \sum_{n \geq 1} \alpha_n 2^{-n},$$

with $\alpha_n \in \{0,1\}$.

(2) Does the result in (1) holds if U is not closed?

Problem 4.5. Prove that the function f with domain $\mathrm{dom}(f) = \mathbb{R} - \{0\}$ given by $f(x) = 1/x^2$ has the property that $f''(x) > 0$ for all $x \in \mathrm{dom}(f)$, but it is not convex. Why isn't Proposition 4.7 applicable?

Problem 4.6. (1) Prove that the function $x \mapsto e^{ax}$ (on $\mathbb{R}$) is convex for any $a \in \mathbb{R}$.

(2) Prove that the function $x \mapsto x^a$ is convex on $\{x \in \mathbb{R} \mid x > 0\}$, for all $a \in \mathbb{R}$ such that $a \leq 0$ or $a \geq 1$.

Problem 4.7. (1) Prove that the function $x \mapsto |x|^p$ is convex on $\mathbb{R}$ for all $p \geq 1$.

(2) Prove that the function $x \mapsto \log x$ is concave on $\{x \in \mathbb{R} \mid x > 0\}$.

(3) Prove that the function $x \mapsto x \log x$ is convex on $\{x \in \mathbb{R} \mid x > 0\}$.

Problem 4.8. (1) Prove that the function f given by $f(x_1,\ldots,x_n) = \max\{x_1,\ldots,x_n\}$ is convex on $\mathbb{R}^n$.

(2) Prove that the function g given by $g(x_1,\ldots,x_n) = \log(e^{x_1}+\cdots+e^{x_n})$ is convex on $\mathbb{R}^n$.

Prove that

$$\max\{x_1,\ldots,x_n\} \leq g(x_1,\ldots,x_n) \leq \max\{x_1,\ldots,x_n\} + \log n.$$

Problem 4.9. In Problem 3.6, it was shown that

$$\begin{aligned} df_A(X) &= \mathrm{tr}(A^{-1}X) \\ \mathrm{D}^2 f(A)(X_1, X_2) &= -\mathrm{tr}(A^{-1}X_1A^{-1}X_2), \end{aligned}$$

for all $n \times n$ real matrices X, X_1, X_2, where f is the function defined on $\mathbf{GL}^+(n, \mathbb{R})$ (the $n \times n$ real invertible matrices of positive determinants), given by

$$f(A) = \log\det(A).$$

Assume that A is symmetric positive definite and that X is symmetric.

(1) Prove that the eigenvalues of $A^{-1}X$ are real (even though $A^{-1}X$ may **not** be symmetric).

Hint. Since A is symmetric positive definite, then so is A^{-1}, so we can write $A^{-1} = S^2$ for some symmetric positive definite matrix S, and then

$$A^{-1}X = S^2X = S(SXS)S^{-1}.$$

(2) Prove that the eigenvalues of $(A^{-1}X)^2$ are nonnegative. Deduce that

$$\mathrm{D}^2 f(A)(X, X) = -\mathrm{tr}((A^{-1}X)^2) < 0$$

for all nonzero symmetric matrices X and SPD matrices A. Conclude that the function $X \mapsto \log\det X$ is strictly concave on the set of symmetric positive definite matrices.

Chapter 5

Newton's Method and Its Generalizations

In Chapter 4 we investigated the problem of determining when a function $J: \Omega \to \mathbb{R}$ defined on some open subset Ω of a normed vector space E has a local extremum. Proposition 4.1 gives a necessary condition when J is differentiable: if J has a local extremum at $u \in \Omega$, then we must have

$$J'(u) = 0.$$

Thus we are led to the problem of finding the zeros of the derivative

$$J' : \Omega \to E',$$

where $E' = \mathcal{L}(E; \mathbb{R})$ is the set of linear continuous functions from E to $\mathbb{R}$; that is, the *dual* of E, as defined in the remark after Proposition 4.5.

This leads us to consider the problem in a more general form, namely, given a function $f: \Omega \to Y$ from an open subset Ω of a normed vector space X to a normed vector space Y, find

(i) Sufficient conditions which guarantee the *existence of a zero* of the function f; that is, an element $a \in \Omega$ such that $f(a) = 0$.
(ii) An *algorithm* for approximating such an a, that is, a sequence (x_k) of points of Ω whose limit is a.

In this chapter we discuss Newton's method and some of it generalizations to give (partial) answers to Problems (i) and (i).

5.1 Newton's Method for Real Functions of a Real Argument

When $X = Y = \mathbb{R}$, we can use *Newton's method* for find a zero of a function $f: \Omega \to \mathbb{R}$. We pick some initial element $x_0 \in \mathbb{R}$ "close enough" to a zero a

of f, and we define the sequence (x_k) by

$$x_{k+1} = x_k - \frac{f(x_k)}{f'(x_k)},$$

for all $k \geq 0$, provided that $f'(x_k) \neq 0$. The idea is to define x_{k+1} as the intersection of the x-axis with the tangent line to the graph of the function $x \mapsto f(x)$ at the point $(x_k, f(x_k))$. Indeed, the equation of this tangent line is

$$y - f(x_k) = f'(x_k)(x - x_k),$$

and its intersection with the x-axis is obtained for $y = 0$, which yields

$$x = x_k - \frac{f(x_k)}{f'(x_k)},$$

as claimed. See Figure 5.1.

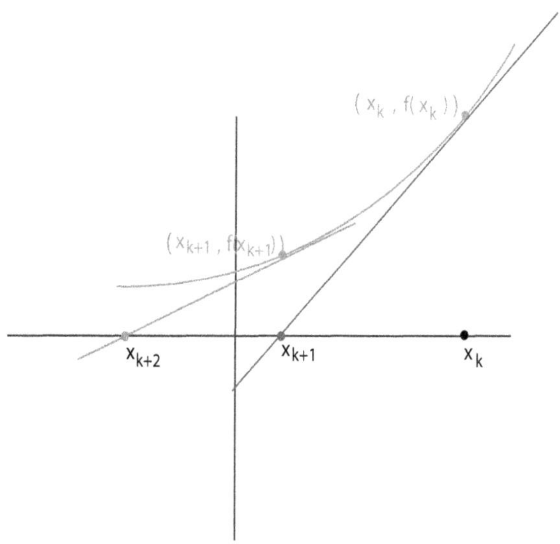

Fig. 5.1 The construction of two stages in Newton's method.

Example 5.1. If $\alpha > 0$ and $f(x) = x^2 - \alpha$, Newton's method yields the sequence

$$x_{k+1} = \frac{1}{2}\left(x_k + \frac{\alpha}{x_k}\right)$$

to compute the square root $\sqrt{\alpha}$ of α. It can be shown that the method converges to $\sqrt{\alpha}$ for any $x_0 > 0$; see Problem 5.1. Actually, the method also converges when $x_0 < 0$! Find out what is the limit.

The case of a real function suggests the following method for finding the zeros of a function $f\colon \Omega \to Y$, with $\Omega \subseteq X$: given a starting point $x_0 \in \Omega$, the sequence (x_k) is defined by

$$x_{k+1} = x_k - (f'(x_k))^{-1}(f(x_k)) \qquad (*)$$

for all $k \geq 0$.

For the above to make sense, it must be ensured that

(1) All the points x_k remain within Ω.
(2) The function f is differentiable within Ω.
(3) The derivative $f'(x)$ is a bijection from X to Y for all $x \in \Omega$.

These are rather demanding conditions but there are sufficient conditions that guarantee that they are met. Another practical issue is that it may be very costly to compute $(f'(x_k))^{-1}$ at every iteration step. In the next section we investigate generalizations of Newton's method which address the issues that we just discussed.

5.2 Generalizations of Newton's Method

Suppose that $f\colon \Omega \to \mathbb{R}^n$ is given by n functions $f_i\colon \Omega \to \mathbb{R}$, where $\Omega \subseteq \mathbb{R}^n$. In this case, finding a zero a of f is equivalent to solving the system

$$\begin{aligned} f_1(a_1 \ldots, a_n) &= 0 \\ f_2(a_1 \ldots, a_n) &= 0 \\ &\vdots \\ f_n(a_1 \ldots, a_n) &= 0. \end{aligned}$$

In the standard Newton method, the iteration step is given by $(*)$, namely

$$x_{k+1} = x_k - (f'(x_k))^{-1}(f(x_k)),$$

and if we define Δx_k as $\Delta x_k = x_{k+1} - x_k$, we see that $\Delta x_k = -(f'(x_k))^{-1}(f(x_k))$, so Δx_k is obtained by solving the equation

$$f'(x_k)\Delta x_k = -f(x_k),$$

and then we set $x_{k+1} = x_k + \Delta x_k$.

The generalization is as follows.

Variant 1. A single iteration of Newton's method consists in solving the linear system

$$(J(f)(x_k))\Delta x_k = -f(x_k),$$

and then setting

$$x_{k+1} = x_k + \Delta x_k,$$

where $J(f)(x_k) = \left(\frac{\partial f_i}{\partial x_j}(x_k)\right)$ is the Jacobian matrix of f at x_k.

In general it is very costly to compute $J(f)(x_k)$ at each iteration and then to solve the corresponding linear system. If the method converges, the consecutive vectors x_k should differ only a little, as also the corresponding matrices $J(f)(x_k)$. Thus, we are led to several variants of Newton's method.

Variant 2. This variant consists in keeping the same matrix for p consecutive steps (where p is some fixed integer ≥ 2):

$$\begin{aligned}
x_{k+1} &= x_k - (f'(x_0))^{-1}(f(x_k)), && 0 \leq k \leq p-1\\
x_{k+1} &= x_k - (f'(x_p))^{-1}(f(x_k)), && p \leq k \leq 2p-1\\
&\vdots\\
x_{k+1} &= x_k - (f'(x_{rp}))^{-1}(f(x_k)), && rp \leq k \leq (r+1)p-1\\
&\vdots
\end{aligned}$$

Variant 3. Set $p = \infty$, that is, use *the same matrix* $f'(x_0)$ *for all iterations*, which leads to iterations of the form

$$x_{k+1} = x_k - (f'(x_0))^{-1}(f(x_k)), \quad k \geq 0.$$

Variant 4. Replace $f'(x_0)$ by a particular matrix A_0 which is easy to invert:

$$x_{k+1} = x_k - A_0^{-1}f(x_k), \quad k \geq 0.$$

In the last two cases, if possible, we use an LU factorization of $f'(x_0)$ or A_0 to speed up the method. In some cases, it may even possible to set $A_0 = I$.

The above considerations lead us to the definition of a *generalized Newton method*, as in Ciarlet [Ciarlet (1989)] (Chapter 7). Recall that a linear map $f \in \mathcal{L}(E; F)$ is called an *isomorphism* iff f is continuous, bijective, and f^{-1} is also continuous.

Definition 5.1. If X and Y are two normed vector spaces and if $f\colon \Omega \to Y$ is a function from some open subset Ω of X, a *generalized Newton method* for finding zeros of f consists of

(1) A sequence of families $(A_k(x))$ of linear isomorphisms from X to Y, for all $x \in \Omega$ and all integers $k \geq 0$;

(2) Some starting point $x_0 \in \Omega$;

(3) A sequence (x_k) of points of Ω defined by

$$x_{k+1} = x_k - (A_k(x_\ell))^{-1}(f(x_k)), \quad k \geq 0, \tag{**}$$

where for every integer $k \geq 0$, the integer ℓ satisfies the condition

$$0 \leq \ell \leq k.$$

With $\Delta x_k = x_{k+1} - x_k$, Equation (**) is equivalent to solving the equation

$$A_k(x_\ell)(\Delta_k) = -f(x_k)$$

and setting $x_{k+1} = x_k + \Delta_k$. The function $A_k(x)$ usually depends on f'.

Definition 5.1 gives us enough flexibility to capture all the situations that we have previously discussed:

	Function	Index
Variant 1:	$A_k(x) = f'(x)$	$\ell = k$
Variant 2:	$A_k(x) = f'(x)$	$\ell = \min\{rp, k\}$, if $rp \leq k \leq (r+1)p - 1, r \geq 0$
Variant 3:	$A_k(x) = f'(x)$	$\ell = 0$
Variant 4:	$A_k(x) = A_0$	

where A_0 is a linear isomorphism from X to Y. The first case corresponds to Newton's original method and the others to the variants that we just discussed. We could also have $A_k(x) = A_k$, a fixed linear isomorphism independent of $x \in \Omega$.

Example 5.2. Consider the matrix function f given by

$$f(X) = A - X^{-1},$$

with A and X invertible $n \times n$ matrices. If we apply Variant 1 of Newton's method starting with any $n \times n$ matrix X_0, since the derivative of the function g given by $g(X) = X^{-1}$ is $dg_X(Y) = -X^{-1}YX^{-1}$, we have

$$f'_X(Y) = X^{-1}YX^{-1},$$

so

$$(f'_X)^{-1}(Y) = XYX$$

and the Newton step is

$$X_{k+1} = X_k - (f'_{X_k})^{-1}(f(X_k)) = X_k - X_k(A - X_k^{-1})X_k,$$

which yields the sequence (X_k) with

$$X_{k+1} = X_k(2I - AX_k), \quad k \geq 0.$$

In Problem 5.5, it is shown that Newton's method converges to A^{-1} iff the spectral radius of $I - X_0A$ is strictly smaller than 1, that is, $\rho(I - X_0A) < 1$.

The following theorem inspired by the *Newton–Kantorovich theorem* gives sufficient conditions that guarantee that the sequence (x_k) constructed by a generalized Newton method converges to a zero of f close to x_0. Although quite technical, these conditions are not very surprising.

Theorem 5.1. *Let X be a Banach space, let $f\colon \Omega \to Y$ be differentiable on the open subset $\Omega \subseteq X$, and assume that there are constants $r, M, \beta > 0$ such that if we let*

$$B = \{x \in X \mid \|x - x_0\| \leq r\} \subseteq \Omega,$$

then

(1)

$$\sup_{k\geq 0}\sup_{x\in B} \left\|A_k^{-1}(x)\right\|_{\mathcal{L}(Y;X)} \leq M,$$

(2) $\beta < 1$ and

$$\sup_{k\geq 0}\sup_{x,x'\in B} \|f'(x) - A_k(x')\|_{\mathcal{L}(X;Y)} \leq \frac{\beta}{M}$$

(3)

$$\|f(x_0)\| \leq \frac{r}{M}(1-\beta).$$

Then the sequence (x_k) defined by

$$x_{k+1} = x_k - A_k^{-1}(x_\ell)(f(x_k)), \quad 0 \leq \ell \leq k$$

is entirely contained within B and converges to a zero a of f, which is the only zero of f in B. Furthermore, the convergence is geometric, which means that

$$\|x_k - a\| \leq \frac{\|x_1 - x_0\|}{1-\beta}\beta^k.$$

Proof. We follow Ciarlet [Ciarlet (1989)] (Theorem 7.5.1, Section 7.5). The proof has three steps.

Step 1. We establish the following inequalities for all $k \geq 1$.

$$\|x_k - x_{k-1}\| \leq M\,\|f(x_{k-1})\| \tag{a}$$

$$\|x_k - x_0\| \leq r \quad (x_k \in B) \tag{b}$$

$$\|f(x_k)\| \leq \frac{\beta}{M}\,\|x_k - x_{k-1}\|. \tag{c}$$

We proceed by induction on k, starting with the base case $k = 1$. Since

$$x_1 = x_0 - A_0^{-1}(x_0)(f(x_0)),$$

we have $x_1 - x_0 = -A_0^{-1}(x_0)(f(x_0))$, so by (1) and (3) and since $0 < \beta < 1$, we have

$$\|x_1 - x_0\| \leq M \|f(x_0)\| \leq r(1-\beta) \leq r,$$

establishing (a) and (b) for $k = 1$. We also have $f(x_0) = -A_0(x_0)(x_1 - x_0)$, so

$-f(x_0) - A_0(x_0)(x_1 - x_0) = 0$ and thus

$$f(x_1) = f(x_1) - f(x_0) - A_0(x_0)(x_1 - x_0).$$

By the mean value theorem (Proposition 3.10) applied to the function $x \mapsto f(x) - A_0(x_0)(x)$, by (2), we get

$$\|f(x_1)\| \leq \sup_{x \in B} \|f'(x) - A_0(x_0)\| \, \|x_1 - x_0\| \leq \frac{\beta}{M} \|x_1 - x_0\|,$$

which is (c) for $k = 1$. We now establish the induction step.

Since by definition

$$x_k - x_{k-1} = -A_{k-1}^{-1}(x_\ell)(f(x_{k-1})), \quad 0 \leq \ell \leq k-1,$$

by (1) and the fact that by the induction hypothesis for (b), $x_\ell \in B$, we get

$$\|x_k - x_{k-1}\| \leq M \|f(x_{k-1})\|,$$

which proves (a) for k. As a consequence, since by the induction hypothesis for (c),

$$\|f(x_{k-1})\| \leq \frac{\beta}{M} \|x_{k-1} - x_{k-2}\|,$$

we get

$$\|x_k - x_{k-1}\| \leq M \|f(x_{k-1})\| \leq \beta \|x_{k-1} - x_{k-2}\|, \qquad (*_1)$$

and by repeating this step,

$$\|x_k - x_{k-1}\| \leq \beta^{k-1} \|x_1 - x_0\|. \qquad (*_2)$$

Using $(*_2)$ and (3), we obtain

$$\begin{aligned} \|x_k - x_0\| &\leq \sum_{j=1}^{k} \|x_j - x_{j-1}\| \leq \left(\sum_{j=1}^{k} \beta^{j-1}\right) \|x_1 - x_0\| \\ &\leq \frac{\|x_1 - x_0\|}{1-\beta} \leq \frac{M}{1-\beta} \|f(x_0)\| \leq r, \end{aligned}$$

which proves that $x_k \in B$, which is (b) for k.

Since

$$x_k - x_{k-1} = -A_{k-1}^{-1}(x_\ell)(f(x_{k-1}))$$

we also have $f(x_{k-1}) = -A_{k-1}(x_\ell)(x_k - x_{k-1})$, so we have

$$f(x_k) = f(x_k) - f(x_{k-1}) - A_{k-1}(x_\ell)(x_k - x_{k-1}),$$

and as in the base case, applying the mean value theorem (Proposition 3.10) to the function $x \mapsto f(x) - A_{k-1}(x_\ell)(x)$, by (2), we obtain

$$\|f(x_k)\| \leq \sup_{x \in B} \|f'(x) - A_{k-1}(x_\ell)\| \, \|x_k - x_{k-1}\| \leq \frac{\beta}{M} \|x_k - x_{k-1}\|,$$

proving (c) for k.

Step 2. Prove that f has a zero in B.

To do this we prove that (x_k) is a Cauchy sequence. This is because, using $(*_2)$, we have

$$\begin{aligned}\|x_{k+j} - x_k\| &\leq \sum_{i=0}^{j-1} \|x_{k+i+1} - x_{k+i}\| \leq \beta^k \left(\sum_{i=0}^{j-1} \beta^i\right) \|x_1 - x_0\| \\ &\leq \frac{\beta^k}{1-\beta} \|x_1 - x_0\|,\end{aligned}$$

for all $k \geq 0$ and all $j \geq 0$, proving that (x_k) is a Cauchy sequence. Since B is a closed ball in a complete normed vector space, B is complete and the Cauchy sequence (x_k) converges to a limit $a \in B$. Since f is continuous on Ω (because it is differentiable), by (c) we obtain

$$\|f(a)\| = \lim_{k \mapsto \infty} \|f(x_k)\| \leq \frac{\beta}{M} \lim_{k \mapsto \infty} \|x_k - x_{k-1}\| = 0,$$

which yields $f(a) = 0$.

Since

$$\|x_{k+j} - x_k\| \leq \frac{\beta^k}{1-\beta} \|x_1 - x_0\|,$$

if we let j tend to infinity, we obtain the inequality

$$\|x_k - a\| = \|a - x_k\| \leq \frac{\beta^k}{1-\beta} \|x_1 - x_0\|,$$

which is the last statement of the theorem.

Step 3. Prove that f has a unique zero in B.

Suppose $f(a) = f(b) = 0$ with $a, b \in B$. Since $A_0^{-1}(x_0)(A_0(x_0)(b-a)) = b - a$, we have

$$b - a = -A_0^{-1}(x_0)(f(b) - f(a) - A_0(x_0)(b-a)),$$

which by (1) and (2) and the mean value theorem implies that

$$\|b - a\| \leq \left\|A_0^{-1}(x_0)\right\| \sup_{x \in B} \|f'(x) - A_0(x_0)\| \, \|b - a\| \leq \beta \|b - a\|.$$

Since $0 < \beta < 1$, the inequality $\|b - a\| < \beta \|b - a\|$ is only possible if $a = b$. ☐

It should be observed that the conditions of Theorem 5.1 are typically quite stringent. It can be shown that Theorem 5.1 applies to the function f of Example 5.1 given by $f(x) = x^2 - \alpha$ with $\alpha > 0$, for any $x_0 > 0$ such that

$$\frac{6}{7}\alpha \leq x_0^2 \leq \frac{6}{5}\alpha,$$

with $\beta = 2/5$, $r = (1/6)x_0$, $M = 3/(5x_0)$. Such values of x_0 are quite close to $\sqrt{\alpha}$.

If we assume that we already know that some element $a \in \Omega$ is a zero of f, the next theorem gives sufficient conditions for a special version of a generalized Newton method to converge. For this special method the linear isomorphisms $A_k(x)$ are independent of $x \in \Omega$.

Theorem 5.2. *Let X be a Banach space and let $f\colon \Omega \to Y$ be differentiable on the open subset $\Omega \subseteq X$. If $a \in \Omega$ is a point such that $f(a) = 0$, if $f'(a)$ is a linear isomorphism, and if there is some λ with $0 < \lambda < 1/2$ such that*

$$\sup_{k \geq 0} \|A_k - f'(a)\|_{\mathcal{L}(X;Y)} \leq \frac{\lambda}{\|(f'(a))^{-1}\|_{\mathcal{L}(Y;X)}},$$

then there is a closed ball B of center a such that for every $x_0 \in B$, the sequence (x_k) defined by

$$x_{k+1} = x_k - A_k^{-1}(f(x_k)), \quad k \geq 0,$$

is entirely contained within B and converges to a, which is the only zero of f in B. Furthermore, the convergence is geometric, which means that

$$\|x_k - a\| \leq \beta^k \|x_0 - a\|,$$

for some $\beta < 1$.

A proof of Theorem 5.2 can be found in Ciarlet [Ciarlet (1989)] (Section 7.5).

For the sake of completeness, we state a version of the Newton–Kantorovich theorem which corresponds to the case where $A_k(x) = f'(x)$. In this instance, a stronger result can be obtained especially regarding upper bounds, and we state a version due to Gragg and Tapia which appears in Problem 7.5-4 of Ciarlet [Ciarlet (1989)].

Theorem 5.3. *(Newton–Kantorovich) Let X be a Banach space, and let $f\colon \Omega \to Y$ be differentiable on the open subset $\Omega \subseteq X$. Assume that there exist three positive constants λ, μ, ν and a point $x_0 \in \Omega$ such that*

$$0 < \lambda\mu\nu \leq \frac{1}{2},$$

and if we let

$$\rho^- = \frac{1-\sqrt{1-2\lambda\mu\nu}}{\mu\nu}$$
$$\rho^+ = \frac{1+\sqrt{1-2\lambda\mu\nu}}{\mu\nu}$$
$$B = \{x \in X \mid \|x - x_0\| < \rho^-\}$$
$$\Omega^+ = \{x \in \Omega \mid \|x - x_0\| < \rho^+\},$$

then $\overline{B} \subseteq \Omega$, $f'(x_0)$ *is an isomorphism of* $\mathcal{L}(X;Y)$, *and*

$$\left\|(f'(x_0))^{-1}\right\| \le \mu,$$
$$\left\|(f'(x_0))^{-1} f(x_0)\right\| \le \lambda,$$
$$\sup_{x,y\in\Omega^+} \|f'(x) - f'(y)\| \le \nu\, \|x - y\|\,.$$

Then $f'(x)$ *is isomorphism of* $\mathcal{L}(X;Y)$ *for all* $x \in B$, *and the sequence defined by*

$$x_{k+1} = x_k - (f'(x_k))^{-1}(f(x_k)), \quad k \ge 0$$

is entirely contained within the ball B *and converges to a zero* a *of* f *which is the only zero of* f *in* Ω^+. *Finally, if we write* $\theta = \rho^-/\rho^+$, *then we have the following bounds:*

$$\|x_k - a\| \le \frac{2\sqrt{1-2\lambda\mu\nu}}{\lambda\mu\nu}\frac{\theta^{2k}}{1-\theta^{2k}}\|x_1 - x_0\| \qquad \text{if } \lambda\mu\nu < \frac{1}{2}$$
$$\|x_k - a\| \le \frac{\|x_1 - x_0\|}{2^{k-1}} \qquad \text{if } \lambda\mu\nu = \frac{1}{2},$$

and

$$\frac{2\,\|x_{k+1} - x_k\|}{1+\sqrt{(1+4\theta^{2k}(1+\theta^{2k})^{-2})}} \le \|x_k - a\| \le \theta^{2k-1}\,\|x_k - x_{k-1}\|\,.$$

We can now specialize Theorems 5.1 and 5.2 to the search of zeros of the derivative $J'\colon \Omega \to E'$, of a function $J\colon \Omega \to \mathbb{R}$, with $\Omega \subseteq E$. The second derivative J'' of J is a continuous bilinear form $J''\colon E \times E \to \mathbb{R}$, but is is convenient to view it as a linear map in $\mathcal{L}(E, E')$; the continuous linear form $J''(u)$ is given by $J''(u)(v) = J''(u, v)$. In our next theorem, which follows immediately from Theorem 5.1, we assume that the $A_k(x)$ are isomorphisms in $\mathcal{L}(E, E')$.

Theorem 5.4. *Let* E *be a Banach space, let* $J\colon \Omega \to \mathbb{R}$ *be twice differentiable on the open subset* $\Omega \subseteq E$, *and assume that there are constants* $r, M, \beta > 0$ *such that if we let*

$$B = \{x \in E \mid \|x - x_0\| \le r\} \subseteq \Omega,$$

then

(1)

$$\sup_{k\geq 0}\sup_{x\in B}\left\|A_k^{-1}(x)\right\|_{\mathcal{L}(E';E)}\leq M,$$

(2) $\beta < 1$ *and*

$$\sup_{k\geq 0}\sup_{x,x'\in B}\|J''(x)-A_k(x')\|_{\mathcal{L}(E;E')}\leq\frac{\beta}{M}$$

(3)

$$\|J'(x_0)\|\leq\frac{r}{M}(1-\beta).$$

Then the sequence (x_k) *defined by*

$$x_{k+1}=x_k-A_k^{-1}(x_\ell)(J'(x_k)),\quad 0\leq\ell\leq k$$

is entirely contained within B *and converges to a zero* a *of* J'*, which is the only zero of* J' *in* B*. Furthermore, the convergence is geometric, which means that*

$$\|x_k-a\|\leq\frac{\|x_1-x_0\|}{1-\beta}\beta^k.$$

In the next theorem, which follows immediately from Theorem 5.2, we assume that the $A_k(x)$ are isomorphisms in $\mathcal{L}(E,E')$ that are independent of $x\in\Omega$.

Theorem 5.5. *Let* E *be a Banach space and let* $J\colon\Omega\to\mathbb{R}$ *be twice differentiable on the open subset* $\Omega\subseteq E$*. If* $a\in\Omega$ *is a point such that* $J'(a)=0$*, if* $J''(a)$ *is a linear isomorphism, and if there is some* λ *with* $0<\lambda<1/2$ *such that*

$$\sup_{k\geq 0}\|A_k-J''(a)\|_{\mathcal{L}(E;E')}\leq\frac{\lambda}{\|(J''(a))^{-1}\|_{\mathcal{L}(E';E)}},$$

then there is a closed ball B *of center* a *such that for every* $x_0\in B$*, the sequence* (x_k) *defined by*

$$x_{k+1}=x_k-A_k^{-1}(J'(x_k)),\quad k\geq 0,$$

is entirely contained within B *and converges to* a*, which is the only zero of* J' *in* B*. Furthermore, the convergence is geometric, which means that*

$$\|x_k-a\|\leq\beta^k\|x_0-a\|,$$

for some $\beta<1$*.*

When $E = \mathbb{R}^n$, the Newton method given by Theorem 5.4 yields an iteration step of the form

$$x_{k+1} = x_k - A_k^{-1}(x_\ell)\nabla J(x_k), \quad 0 \leq \ell \leq k,$$

where $\nabla J(x_k)$ is the gradient of J at x_k (here, we identify E' with $\mathbb{R}^n$). In particular, Newton's original method picks $A_k = J''$, and the iteration step is of the form

$$x_{k+1} = x_k - (\nabla^2 J(x_k))^{-1}\nabla J(x_k), \quad k \geq 0,$$

where $\nabla^2 J(x_k)$ is the Hessian of J at x_k.

Example 5.3. Let us apply Newton's original method to the function J given by $J(x) = \frac{1}{3}x^3 - 4x$. We have $J'(x) = x^2 - 4$ and $J''(x) = 2x$, so the Newton step is given by

$$x_{k+1} = x_k - \frac{x_k^2 - 4}{2x_k} = \frac{1}{2}\left(x_k + \frac{4}{x_k}\right).$$

This is the sequence of Example 5.1 to compute the square root of 4. Starting with any $x_0 > 0$ it converges very quickly to 2.

As remarked in Ciarlet [Ciarlet (1989)] (Section 7.5), generalized Newton methods have a very wide range of applicability. For example, various versions of gradient descent methods can be viewed as instances of Newton method. See Section 13.9 for an example.

Newton's method also plays an important role in convex optimization, in particular, interior-point methods. A variant of Newton's method dealing with equality constraints has been developed. We refer the reader to Boyd and Vandenberghe [Boyd and Vandenberghe (2004)], Chapters 10 and 11, for a comprehensive exposition of these topics.

5.3 Summary

The main concepts and results of this chapter are listed below:

- Newton's method for functions $f : \mathbb{R} \to \mathbb{R}$.
- Generalized Newton methods.
- The Newton-Kantorovich theorem.

5.4 Problems

Problem 5.1. If $\alpha > 0$ and $f(x) = x^2 - \alpha$, Newton's method yields the sequence

$$x_{k+1} = \frac{1}{2}\left(x_k + \frac{\alpha}{x_k}\right)$$

to compute the square root $\sqrt{\alpha}$ of α.

(1) Prove that if $x_0 > 0$, then $x_k > 0$ and

$$x_{k+1} - \sqrt{\alpha} = \frac{1}{2x_k}(x_k - \sqrt{\alpha})^2$$
$$x_{k+2} - x_{k+1} = \frac{1}{2x_{k+1}}(\alpha - x_{k+1}^2)$$

for all $k \geq 0$. Deduce that Newton's method converges to $\sqrt{\alpha}$ for any $x_0 > 0$.

(2) Prove that if $x_0 < 0$, then Newton's method converges to $-\sqrt{\alpha}$.

Problem 5.2. (1) If $\alpha > 0$ and $f(x) = x^2 - \alpha$, show that the conditions of Theorem 5.1 are satisfied by any $\beta \in (0, 1)$ and any x_0 such that

$$|x_0^2 - \alpha| \leq \frac{4\beta(1-\beta)}{(\beta+2)^2} x_0^2,$$

with

$$r = \frac{\beta}{\beta+2} x_0, \qquad M = \frac{\beta+2}{4x_0}.$$

(2) Prove that the maximum of the function defined on $[0, 1]$ by

$$\beta \mapsto \frac{4\beta(1-\beta)}{(\beta+2)^2}$$

has a maximum for $\beta = 2/5$. For this value of β, check that $r = (1/6)x_0$, $M = 3/(5x_0)$, and

$$\frac{6}{7}\alpha \leq x_0^2 \leq \frac{6}{5}\alpha.$$

Problem 5.3. Consider generalizing Problem 5.1 to the matrix function f given by $f(X) = X^2 - C$, where X and C are two real $n \times n$ matrices with C symmetric positive definite. The first step is to determine for which A does the inverse df_A^{-1} exist. Let g be the function given by $g(X) = X^2$. From Problem 3.1 we know that the derivative at A of the function g is $dg_A(X) = AX + XA$, and obviously $df_A = dg_A$. Thus we are led to figure out when the linear matrix map $X \mapsto AX + XA$ is invertible. This can be done using the Kronecker product.

Given an $m \times n$ matrix $A = (a_{ij})$ and a $p \times q$ matrix $B = (b_{ij})$, the *Kronecker product* (or *tensor product*) $A \otimes B$ of A and B is the $mp \times nq$ matrix

$$A \otimes B = \begin{pmatrix} a_{11}B & a_{12}B & \cdots & a_{1n}B \\ a_{21}B & a_{22}B & \cdots & a_{2n}B \\ \vdots & \vdots & \ddots & \vdots \\ a_{m1}B & a_{m2}B & \cdots & a_{mn}B \end{pmatrix}.$$

It can be shown (and you may use these facts without proof) that $\otimes$ is associative and that

$$\begin{aligned} (A \otimes B)(C \otimes D) &= AC \otimes BD \\ (A \otimes B)^\top &= A^\top \otimes B^\top, \end{aligned}$$

whenever AC and BD are well defined.

Given any $n \times n$ matrix X, let $\mathrm{vec}(X)$ be the vector in $\mathbb{R}^{n^2}$ obtained by concatenating the *rows* of X.

(1) Prove that $AX = Y$ iff

$$(A \otimes I_n)\mathrm{vec}(X) = \mathrm{vec}(Y)$$

and $XA = Y$ iff

$$(I_n \otimes A^\top)\mathrm{vec}(X) = \mathrm{vec}(Y).$$

Deduce that $AX + XA = Y$ iff

$$((A \otimes I_n) + (I_n \otimes A^\top))\mathrm{vec}(X) = \mathrm{vec}(Y).$$

In the case where $n = 2$ and if we write

$$A = \begin{pmatrix} a & b \\ c & d \end{pmatrix},$$

check that

$$A \otimes I_2 + I_2 \otimes A^\top = \begin{pmatrix} 2a & c & b & 0 \\ b & a+d & 0 & b \\ c & 0 & a+d & c \\ 0 & c & b & 2d \end{pmatrix}.$$

The problem is determine when the matrix $(A \otimes I_n) + (I_n \otimes A^\top)$ is invertible.

Remark: The equation $AX + XA = Y$ is a special case of the equation $AX + XB = C$ (sometimes written $AX - XB = C$), called the *Sylvester equation*, where A is an $m \times m$ matrix, B is an $n \times n$ matrix, and X, C are $m \times n$ matrices; see Higham [Higham (2008)] (Appendix B).

(2) In the case where $n = 2$, prove that

$$\det(A \otimes I_2 + I_2 \otimes A^\top) = 4(a+d)^2(ad - bc).$$

(3) Let A and B be any two $n \times n$ complex matrices. Use Schur factorizations $A = UT_1U^*$ and $B = VT_2V^*$ (where U and V are unitary and T_1, T_2 are upper-triangular) to prove that if $\lambda_1, \ldots, \lambda_n$ are the eigenvalues of A and $\mu_1, \ldots, \mu_n$ are the eigenvalues of B, then the scalars $\lambda_i\mu_j$ are the eigenvalues of $A \otimes B$, for $i, j = 1, \ldots, n$.
Hint. Check that $U \otimes V$ is unitary and that $T_1 \otimes T_2$ is upper triangular.

(4) Prove that the eigenvalues of $(A \otimes I_n) + (I_n \otimes B)$ are the scalars $\lambda_i + \mu_j$, for $i, j = 1, \ldots, n$. Deduce that the eigenvalues of $(A \otimes I_n) + (I_n \otimes A^\top)$ are $\lambda_i + \lambda_j$, for $i, j = 1, \ldots, n$. Thus $(A \otimes I_n) + (I_n \otimes A^\top)$ is invertible iff $\lambda_i + \lambda_j \neq 0$, for $i, j = 1, \ldots, n$. In particular, prove that if A is symmetric positive definite, then so is $(A \otimes I_n) + (I_n \otimes A^\top)$.
Hint. Use (3).

(5) Prove that if A and B are symmetric and $(A \otimes I_n) + (I_n \otimes A^\top)$ is invertible, then the unique solution X of the equation $AX + XA = B$ is symmetric.

(6) Starting with a symmetric positive definite matrix X_0, the general step of Newton's method is

$$X_{k+1} = X_k - (f'_{X_k})^{-1}(X_k^2 - C) = X_k - (g'_{X_k})^{-1}(X_k^2 - C),$$

and since g'_{X_k} is linear, this is equivalent to

$$X_{k+1} = X_k - (g'_{X_k})^{-1}(X_k^2) + (g'_{X_k})^{-1}(C).$$

But since X_k is SPD, $(g'_{X_k})^{-1}(X_k^2)$ is the unique solution of

$$X_kY + YX_k = X_k^2$$

whose solution is obviously $Y = (1/2)X_k$. Therefore the Newton step is

$$\begin{aligned} X_{k+1} &= X_k - (g'_{X_k})^{-1}(X_k^2) + (g'_{X_k})^{-1}(C) = X_k - \frac{1}{2}X_k + (g'_{X_k})^{-1}(C) \\ &= \frac{1}{2}X_k + (g'_{X_k})^{-1}(C), \end{aligned}$$

so we have

$$X_{k+1} = \frac{1}{2}X_k + (g'_{X_k})^{-1}(C) = (g'_{X_k})^{-1}(X_k^2 + C).$$

Prove that if X_k and C are symmetric positive definite, then $(g'_{X_k})^{-1}(C)$ is symmetric positive definite, and if C is symmetric positive semidefinite, then $(g'_{X_k})^{-1}(C)$ is symmetric positive semidefinite.
Hint. By (5) the unique solution Z of the equation $X_kZ+ZX_k = C$ (where C is symmetric) is symmetric so it can be diagonalized as $Z = QDQ^\top$ with Q orthogonal and D a real diagonal matrix. Prove that

$$Q^\top X_k QD + DQ^\top X_k Q = Q^\top CQ,$$

and solve the system using the diagonal elements.

Deduce that if X_k and C are SPD, then X_{k+1} is SPD.

Since $C = P\Sigma P^\top$ is SPD, it has an SPD square root (in fact unique) $C^{1/2} = P\Sigma^{1/2}P^\top$. Prove that

$$X_{k+1} - C^{1/2} = (g'_{X_k})^{-1}(X_k - C^{1/2})^2.$$

Prove that

$$\left\|(g'_{X_k})^{-1}\right\|_2 = \frac{1}{2\left\|X_k\right\|_2}.$$

Since

$$X_{k+1} - X_k = (g'_{X_k})^{-1}(C - X_k^2),$$

deduce that if $X_k \neq C^2$, then $X_k - X_{k+1}$ is SPD.

Open problem: Does Theorem 5.1 apply for some suitable r, M, β?

(7) Prove that if C and X_0 commute, provided that the equation $X_kZ+ZX_k = C$ has a unique solution for all k, then X_k and C commute for all k and Z is given by

$$Z = \frac{1}{2}X_k^{-1}C = \frac{1}{2}CX_k^{-1}.$$

Deduce that

$$X_{k+1} = \frac{1}{2}(X_k + X_k^{-1}C) = \frac{1}{2}(X_k + CX_k^{-1}).$$

This is the matrix analog of the formula given in Problem 5.1(1).

Prove that if C and X_0 have positive eigenvalues and C and X_0 commute, then X_{k+1} has positive eigenvalues for all $k \geq 0$ and thus the sequence (X_k) is defined.
Hint. Because X_k and C commute, X_k^{-1} and C commute, and obviously X_k and X_k^{-1} commute. By Proposition 22.15 of Vol. I, X_k, X_k^{-1}, and C are triangulable in a common basis, so there is some orthogonal matrix P and some upper-triangular matrices T_1, T_2 such that

$$X_k = PT_1P^\top, \quad X_k^{-1} = PT_1^{-1}P^\top, \quad C = PT_2P^\top.$$

It follows that

$$X_{k+1} = \frac{1}{2} P(T_1 + T_1^{-1} T_2) P^\top.$$

Also recall that the diagonal entries of an upper-triangular matrix are the eigenvalues of that matrix.

We conjecture that if C has positive eigenvalues, then the Newton sequence converges starting with any X_0 of the form $X_0 = \mu I_n$, with $\mu > 0$.

(8) Implement the above method in `Matlab` (there is a command `kron(A, B)` to form the Kronecker product of A and B). Test your program on diagonalizable and nondiagonalizable matrices, including

$$W = \begin{pmatrix} 10 & 7 & 8 & 7 \\ 7 & 5 & 6 & 5 \\ 8 & 6 & 10 & 9 \\ 7 & 5 & 9 & 10 \end{pmatrix}, \quad A_1 = \begin{pmatrix} 5 & 4 & 1 & 1 \\ 4 & 5 & 1 & 1 \\ 1 & 1 & 4 & 2 \\ 1 & 1 & 2 & 4 \end{pmatrix},$$

and

$$A_2 = \begin{pmatrix} 1 & 0 & 0 & 0 \\ -1 & 0.01 & 0 & 0 \\ -1 & -1 & 100 & 100 \\ -1 & -1 & -100 & 100 \end{pmatrix}, \quad A_3 = \begin{pmatrix} 1 & 1 & 1 & 1 \\ 0 & 1 & 1 & 1 \\ 0 & 0 & 1 & 1 \\ 0 & 0 & 0 & 1 \end{pmatrix}, \quad A_4 = \begin{pmatrix} 1 & -1 & 0 & 0 \\ 1 & 1 & 0 & 0 \\ 0 & 0 & 1 & -1 \\ 0 & 0 & 1 & 1 \end{pmatrix}.$$

What happens with

$$C = \begin{pmatrix} -1 & 0 \\ 0 & -1 \end{pmatrix}, \quad X_0 = \begin{pmatrix} 1 & -1 \\ 1 & 1 \end{pmatrix}.$$

The problem of determining when square roots of matrices exist and procedures for finding them are thoroughly investigated in Higham [Higham (2008)] (Chapter 6).

Problem 5.4. (1) Show that Newton's method applied to the function

$$f(x) = \alpha - \frac{1}{x}$$

with $\alpha \neq 0$ and $x \in \mathbb{R} - \{0\}$ yields the sequence (x_k) with

$$x_{k+1} = x_k(2 - \alpha x_k), \quad k \geq 0.$$

(2) If we let $r_k = 1 - \alpha x_k$, prove that $r_{k+1} = r_k^2$ for all $k \geq 0$. Deduce that Newton's method converges to $1/\alpha$ if $0 < \alpha x_0 < 2$.

Problem 5.5. (1) Show that Newton's method applied to the matrix function

$$f(X) = A - X^{-1}$$

with A and X invertible $n \times n$ matrices and started with any $n \times n$ matrix X_0 yields the sequence (X_k) with

$$X_{k+1} = X_k(2I - AX_k), \quad k \geq 0.$$

(2) If we let $R_k = I - AX_k$, prove that

$$R_{k+1} = I - (I - R_k)(I + R_k) = R_k^2$$

for all $k \geq 0$. Deduce that Newton's method converges to A^{-1} iff the spectral radius of $I - X_0A$ is strictly smaller than 1, that is, $\rho(I - X_0A) < 1$.

(3) Assume that A is symmetric positive definite and let $X_0 = \mu I$. Prove that the condition $\rho(I - X_0A) < 1$ is equivalent to

$$0 < \mu < \frac{2}{\rho(A)}.$$

(4) Write a `Matlab` program implementing Newton's method specified in (1). Test your program with the $n \times n$ matrix

$$A_n = \begin{pmatrix} 2 & -1 & 0 & \cdots & 0 \\ -1 & 2 & -1 & \ldots & 0 \\ \vdots & \ddots & \ddots & \ddots & \vdots \\ 0 & \cdots & -1 & 2 & -1 \\ 0 & \cdots & 0 & -1 & 2 \end{pmatrix},$$

and with $X_0 = \mu I_n$, for various values of n, including $n = 8, 10, 20$, and various values of μ such that $0 < \mu \leq 1/2$. Find some $\mu > 1/2$ causing divergence.

Problem 5.6. A method for computing the nth root $x^{1/n}$ of a positive real number $x \in \mathbb{R}$, with $n \in \mathbb{N}$ a positive integer $n \geq 2$, proceeds as follows: define the sequence (x_k), where x_0 is any chosen positive real, and

$$x_{k+1} = \frac{1}{n}\left((n-1)x_k + \frac{x}{x_k^{n-1}}\right), \quad k \geq 0.$$

(1) Implement the above method in `Matlab` and test it for various input values of x, x_0, and of $n \geq 2$, by running successively your program for $m = 2, 3, \ldots, 100$ iterations. Have your program plot the points (i, x_i) to watch how quickly the sequence converges.

Experiment with various choices of x_0. One of these choices should be $x_0 = x$. Compare your answers with the result of applying the of `Matlab` function $x \mapsto x^{1/n}$.

In some case, when x_0 is small, the number of iterations has to be at least 1000. Exhibit this behavior.

Problem 5.7. Refer to Problem 5.6 for the definition of the sequence (x_k).

(1) Define the *relative error* ϵ_k as

$$\epsilon_k = \frac{x_k}{x^{1/n}} - 1, \quad k \geq 0.$$

Prove that

$$\epsilon_{k+1} = \frac{x^{(1-1/n)}}{nx_k^{n-1}} \left(\frac{(n-1)x_k^n}{x} - \frac{nx_k^{n-1}}{x^{(1-1/n)}} + 1 \right),$$

and then that

$$\epsilon_{k+1} = \frac{1}{n(\epsilon_k + 1)^{n-1}} (\epsilon_k(\epsilon_k + 1)^{n-2}((n-1)\epsilon_k + (n-2)) + 1 - (\epsilon_k + 1)^{n-2}),$$

for all $k \geq 0$.

(2) Since

$$\epsilon_k + 1 = \frac{x_k}{x^{1/n}},$$

and since we assumed $x_0, x > 0$, we have $\epsilon_0 + 1 > 0$. We would like to prove that

$$\epsilon_k \geq 0, \quad \text{for all} \quad k \geq 1.$$

For this consider the variations of the function f given by

$$f(u) = (n-1)u^n - nx^{1/n}u^{n-1} + x,$$

for $u \in \mathbb{R}$.

Use the above to prove that $f(u) \geq 0$ for all $u \geq 0$. Conclude that

$$\epsilon_k \geq 0, \quad \text{for all} \quad k \geq 1.$$

(3) Prove that if $n = 2$, then

$$0 \leq \epsilon_{k+1} = \frac{\epsilon_k^2}{2(\epsilon_k + 1)}, \quad \text{for all} \quad k \geq 0,$$

else if $n \geq 3$, then

$$0 \leq \epsilon_{k+1} \leq \frac{(n-1)}{n}\epsilon_k, \quad \text{for all} \quad k \geq 1.$$

Prove that the sequence (x_k) converges to $x^{1/n}$ for every initial value $x_0 > 0$.

(4) When $n = 2$, we saw in Problem 5.7(3) that

$$0 \leq \epsilon_{k+1} = \frac{\epsilon_k^2}{2(\epsilon_k + 1)}, \quad \text{for all} \quad k \geq 0.$$

For $n = 3$, prove that

$$\epsilon_{k+1} = \frac{2\epsilon_k^2(3/2 + \epsilon_k)}{3(\epsilon_k + 1)^2}, \quad \text{for all} \quad k \geq 0,$$

and for $n = 4$, prove that

$$\epsilon_{k+1} = \frac{3\epsilon_k^2}{4(\epsilon_k + 1)^3}\left(2 + (8/3)\epsilon_k + \epsilon_k^2\right), \quad \text{for all} \quad k \geq 0.$$

Let μ_3 and μ_4 be the functions given by

$$\begin{aligned} \mu_3(a) &= \frac{3}{2} + a \\ \mu_4(a) &= 2 + \frac{8}{3}a + a^2, \end{aligned}$$

so that if $n = 3$, then

$$\epsilon_{k+1} = \frac{2\epsilon_k^2 \mu_3(\epsilon_k)}{3(\epsilon_k + 1)^2}, \quad \text{for all} \quad k \geq 0,$$

and if $n = 4$, then

$$\epsilon_{k+1} = \frac{3\epsilon_k^2 \mu_4(\epsilon_k)}{4(\epsilon_k + 1)^3}, \quad \text{for all} \quad k \geq 0.$$

Prove that

$$a\mu_3(a) \leq (a + 1)^2 - 1, \quad \text{for all} \quad a \geq 0,$$

and

$$a\mu_4(a) \leq (a + 1)^3 - 1, \quad \text{for all} \quad a \geq 0.$$

Let $\eta_{3,k} = \mu_3(\epsilon_1)\epsilon_k$ when $n = 3$, and $\eta_{4,k} = \mu_4(\epsilon_1)\epsilon_k$ when $n = 4$. Prove that

$$\eta_{3,k+1} \leq \frac{2}{3}\eta_{3,k}^2, \quad \text{for all} \quad k \geq 1,$$

and

$$\eta_{4,k+1} \leq \frac{3}{4}\eta_{4,k}^2, \quad \text{for all} \quad k \geq 1.$$

Deduce from the above that the rate of convergence of $\eta_{i,k}$ is very fast, for $i = 3, 4$ (and $k \geq 1$).

Remark: If we let $\mu_2(a) = a$ for all a and $\eta_{2,k} = \epsilon_k$, we then proved that

$$\eta_{2,k+1} \leq \frac{1}{2}\eta_{2,k}^2, \quad \text{for all} \quad k \geq 1.$$

Problem 5.8. This is a continuation of Problem 5.7.

(1) Prove that for all $n \geq 2$, we have

$$\epsilon_{k+1} = \left(\frac{n-1}{n}\right) \frac{\epsilon_k^2 \mu_n(\epsilon_k)}{(\epsilon_k + 1)^{n-1}}, \quad \text{for all} \quad k \geq 0,$$

where μ_n is given by

$$\mu_n(a) = \frac{1}{2}n + \sum_{j=1}^{n-4} \frac{1}{n-1}\left((n-1)\binom{n-2}{j} + (n-2)\binom{n-2}{j+1} - \binom{n-2}{j+2}\right)a^j + \frac{n(n-2)}{n-1}a^{n-3} + a^{n-2}.$$

Furthermore, prove that μ_n can be expressed as

$$\mu_n(a) = \frac{1}{2}n + \frac{n(n-2)}{3}a + \sum_{j=2}^{n-4} \frac{(j+1)n}{(j+2)(n-1)}\binom{n-1}{j+1}a^j + \frac{n(n-2)}{n-1}a^{n-3} + a^{n-2}.$$

(2) Prove that for every j, with $1 \leq j \leq n-1$, the coefficient of a^j in $a\mu_n(a)$ is less than or equal to the coefficient of a^j in $(a+1)^{n-1} - 1$, and thus

$$a\mu_n(a) \leq (a+1)^{n-1} - 1, \quad \text{for all} \quad a \geq 0,$$

with strict inequality if $n \geq 3$. In fact, prove that if $n \geq 3$, then for every j, with $3 \leq j \leq n-2$, the coefficient of a^j in $a\mu_n(a)$ is strictly less than the coefficient of a^j in $(a+1)^{n-1} - 1$, and if $n \geq 4$, this also holds for $j = 2$.

Let $\eta_{n,k} = \mu_n(\epsilon_1)\epsilon_k$ $(n \geq 2)$. Prove that

$$\eta_{n,k+1} \leq \left(\frac{n-1}{n}\right)\eta_{n,k}^2, \quad \text{for all} \quad k \geq 1.$$

Chapter 6

Quadratic Optimization Problems

In this chapter we consider two classes of quadratic optimization problems that appear frequently in engineering and in computer science (especially in computer vision):

(1) Minimizing
$$Q(x) = \frac{1}{2}x^\top Ax - x^\top b$$
over all $x \in \mathbb{R}^n$, or subject to linear or affine constraints.

(2) Minimizing
$$Q(x) = \frac{1}{2}x^\top Ax - x^\top b$$
over the unit sphere.

In both cases, A is a symmetric matrix. We also seek necessary and sufficient conditions for f to have a global minimum.

6.1 Quadratic Optimization: The Positive Definite Case

Many problems in physics and engineering can be stated as the minimization of some energy function, with or without constraints. Indeed, it is a fundamental principle of mechanics that nature acts so as to minimize energy. Furthermore, if a physical system is in a stable state of equilibrium, then the energy in that state should be minimal. For example, a small ball placed on top of a sphere is in an unstable equilibrium position. A small motion causes the ball to roll down. On the other hand, a ball placed inside and at the bottom of a sphere is in a stable equilibrium position because the potential energy is minimal.

The simplest kind of energy function is a quadratic function. Such functions can be conveniently defined in the form

$$Q(x) = x^\top Ax - x^\top b,$$

where A is a symmetric $n \times n$ matrix and x, b, are vectors in $\mathbb{R}^n$, viewed as column vectors. Actually, for reasons that will be clear shortly, it is preferable to put a factor $\frac{1}{2}$ in front of the quadratic term, so that

$$Q(x) = \frac{1}{2}x^\top Ax - x^\top b.$$

The question is, under what conditions (on A) does $Q(x)$ have a global minimum, preferably unique?

We give a complete answer to the above question in two stages:

(1) In this section we show that if A is symmetric positive definite, then $Q(x)$ has a unique global minimum precisely when

$$Ax = b.$$

(2) In Section 6.2 we give necessary and sufficient conditions in the general case, in terms of the pseudo-inverse of A.

We begin with the matrix version of Definition 20.2 (Vol. I).

Definition 6.1. A symmetric *positive definite matrix* is a matrix whose eigenvalues are strictly positive, and a symmetric *positive semidefinite matrix* is a matrix whose eigenvalues are nonnegative.

Equivalent criteria are given in the following proposition.

Proposition 6.1. *Given any Euclidean space E of dimension n, the following properties hold:*

(1) Every self-adjoint linear map $f\colon E \to E$ is positive definite iff

$$\langle f(x), x\rangle > 0$$

for all $x \in E$ with $x \neq 0$.

(2) Every self-adjoint linear map $f\colon E \to E$ is positive semidefinite iff

$$\langle f(x), x\rangle \geq 0$$

for all $x \in E$.

Proof. (1) First assume that f is positive definite. Recall that every self-adjoint linear map has an orthonormal basis $(e_1, \ldots, e_n)$ of eigenvectors, and let $\lambda_1, \ldots, \lambda_n$ be the corresponding eigenvalues. With respect to this basis, for every $x = x_1e_1 + \cdots + x_ne_n \neq 0$, we have

$$\langle f(x), x\rangle = \Big\langle f\Big(\sum_{i=1}^n x_ie_i\Big), \sum_{i=1}^n x_ie_i\Big\rangle = \Big\langle \sum_{i=1}^n \lambda_ix_ie_i, \sum_{i=1}^n x_ie_i\Big\rangle = \sum_{i=1}^n \lambda_ix_i^2,$$

which is strictly positive, since $\lambda_i > 0$ for $i = 1, \ldots, n$, and $x_i^2 > 0$ for some i, since $x \neq 0$.

Conversely, assume that

$$\langle f(x), x\rangle > 0$$

for all $x \neq 0$. Then for $x = e_i$, we get

$$\langle f(e_i), e_i\rangle = \langle \lambda_ie_i, e_i\rangle = \lambda_i,$$

and thus $\lambda_i > 0$ for all $i = 1, \ldots, n$.

(2) As in (1), we have

$$\langle f(x), x\rangle = \sum_{i=1}^n \lambda_ix_i^2,$$

and since $\lambda_i \geq 0$ for $i = 1, \ldots, n$ because f is positive semidefinite, we have $\langle f(x), x\rangle \geq 0$, as claimed. The converse is as in (1) except that we get only $\lambda_i \geq 0$ since $\langle f(e_i), e_i\rangle \geq 0$. ☐

Some special notation is customary (especially in the field of convex optimization) to express that a symmetric matrix is positive definite or positive semidefinite.

Definition 6.2. Given any $n \times n$ symmetric matrix A we write $A \succeq 0$ if A is positive semidefinite and we write $A \succ 0$ if A is positive definite.

Remark: It should be noted that we can define the relation

$$A \succeq B$$

between any two $n \times n$ matrices (symmetric or not) iff $A - B$ is symmetric positive semidefinite. It is easy to check that this relation is actually a partial order on matrices, called the *positive semidefinite cone ordering*; for details, see Boyd and Vandenberghe [Boyd and Vandenberghe (2004)], Section 2.4.

If A is symmetric positive definite, it is easily checked that A^{-1} is also symmetric positive definite. Also, if C is a symmetric positive definite $m \times m$ matrix and A is an $m \times n$ matrix of rank n (and so $m \geq n$ and the map $x \mapsto Ax$ is surjective onto $\mathbb{R}^m$), then $A^\top CA$ is symmetric positive definite.

We can now prove that

$$Q(x) = \frac{1}{2}x^\top Ax - x^\top b$$

has a global minimum when A is symmetric positive definite.

Proposition 6.2. *Given a quadratic function*

$$Q(x) = \frac{1}{2}x^\top Ax - x^\top b,$$

if A is symmetric positive definite, then $Q(x)$ has a unique global minimum for the solution of the linear system $Ax = b$. The minimum value of $Q(x)$ is

$$Q(A^{-1}b) = -\frac{1}{2}b^\top A^{-1}b.$$

Proof. Since A is positive definite, it is invertible since its eigenvalues are all strictly positive. Let $x = A^{-1}b$, and compute $Q(y) - Q(x)$ for any $y \in \mathbb{R}^n$. Since $Ax = b$, we get

$$\begin{aligned} Q(y) - Q(x) &= \frac{1}{2}y^\top Ay - y^\top b - \frac{1}{2}x^\top Ax + x^\top b \\ &= \frac{1}{2}y^\top Ay - y^\top Ax + \frac{1}{2}x^\top Ax \\ &= \frac{1}{2}(y - x)^\top A(y - x). \end{aligned}$$

Since A is positive definite, the last expression is nonnegative, and thus

$$Q(y) \geq Q(x)$$

for all $y \in \mathbb{R}^n$, which proves that $x = A^{-1}b$ is a global minimum of $Q(x)$. A simple computation yields

$$Q(A^{-1}b) = -\frac{1}{2}b^\top A^{-1}b. \qquad \square$$

Remarks:

(1) The quadratic function $Q(x)$ is also given by

$$Q(x) = \frac{1}{2}x^\top Ax - b^\top x,$$

but the definition using $x^\top b$ is more convenient for the proof of Proposition 6.2.

(2) If $Q(x)$ contains a constant term $c \in \mathbb{R}$, so that

$$Q(x) = \frac{1}{2}x^\top Ax - x^\top b + c,$$

the proof of Proposition 6.2 still shows that $Q(x)$ has a unique global minimum for $x = A^{-1}b$, but the minimal value is

$$Q(A^{-1}b) = -\frac{1}{2}b^\top A^{-1}b + c.$$

Thus when the energy function $Q(x)$ of a system is given by a quadratic function

$$Q(x) = \frac{1}{2}x^\top Ax - x^\top b,$$

where A is symmetric positive definite, finding the global minimum of $Q(x)$ is equivalent to solving the linear system $Ax = b$. Sometimes, it is useful to recast a linear problem $Ax = b$ as a variational problem (finding the minimum of some energy function). However, very often, a minimization problem comes with extra constraints that must be satisfied for all admissible solutions.

Example 6.1. For instance, we may want to minimize the quadratic function

$$Q(x_1, x_2) = \frac{1}{2}\left(x_1^2 + x_2^2\right)$$

subject to the constraint

$$2x_1 - x_2 = 5.$$

The solution for which $Q(x_1, x_2)$ is minimum is no longer $(x_1, x_2) = (0, 0)$, but instead, $(x_1, x_2) = (2, -1)$, as will be shown later.

Geometrically, the graph of the function defined by $z = Q(x_1, x_2)$ in $\mathbb{R}^3$ is a paraboloid of revolution P with axis of revolution Oz. The constraint

$$2x_1 - x_2 = 5$$

corresponds to the vertical plane H parallel to the z-axis and containing the line of equation $2x_1 - x_2 = 5$ in the xy-plane. Thus, as illustrated by Figure 6.1, the constrained minimum of Q is located on the parabola that is the intersection of the paraboloid P with the plane H.

A nice way to solve constrained minimization problems of the above kind is to use the method of *Lagrange multipliers* discussed in Section 4.1.

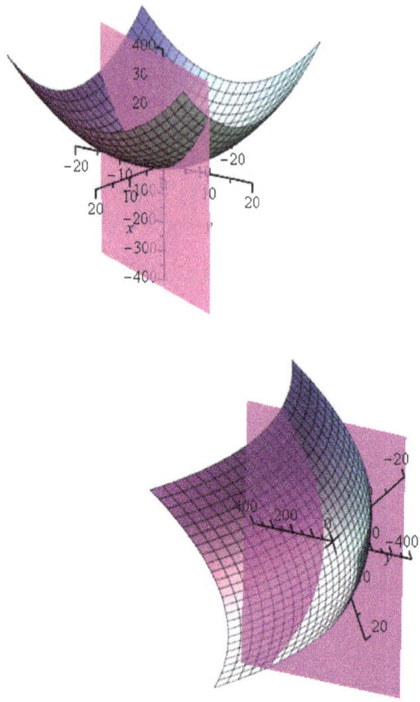

Fig. 6.1 Two views of the constrained optimization problem $Q(x_1, x_2) = \frac{1}{2}(x_1^2 + x_2^2)$ subject to the constraint $2x_1 - x_2 = 5$. The minimum $(x_1, x_2) = (2, -1)$ is the the vertex of the parabolic curve formed the intersection of the magenta planar constraint with the bowl shaped surface.

But first let us define precisely what kind of minimization problems we intend to solve.

Definition 6.3. The *quadratic constrained minimization problem* consists in minimizing a quadratic function

$$Q(x) = \frac{1}{2}x^\top A^{-1}x - b^\top x$$

subject to the linear constraints

$$B^\top x = f,$$

where A^{-1} is an $m \times m$ symmetric positive definite matrix, B is an $m \times n$ matrix of rank n (so that $m \geq n$), and where $b, x \in \mathbb{R}^m$ (viewed as column vectors), and $f \in \mathbb{R}^n$ (viewed as a column vector).

The reason for using A^{-1} instead of A is that the constrained minimization problem has an interpretation as a set of equilibrium equations in which the matrix that arises naturally is A (see Strang [Strang (1986)]). Since A and A^{-1} are both symmetric positive definite, this doesn't make any difference, but it seems preferable to stick to Strang's notation.

In Example 6.1 we have $m = 2$, $n = 1$,

$$A = \begin{pmatrix} 1 & 0 \\ 0 & 1 \end{pmatrix} = I_2, \quad b = \begin{pmatrix} 0 \\ 0 \end{pmatrix}, \quad B = \begin{pmatrix} 2 \\ -1 \end{pmatrix}, \quad f = 5.$$

As explained in Section 4.1, the method of Lagrange multipliers consists in incorporating the n constraints $B^\top x = f$ into the quadratic function $Q(x)$, by introducing extra variables $\lambda = (\lambda_1, \ldots, \lambda_n)$ called *Lagrange multipliers*, one for each constraint. We form the *Lagrangian*

$$L(x, \lambda) = Q(x) + \lambda^\top (B^\top x - f) = \frac{1}{2} x^\top A^{-1} x - (b - B\lambda)^\top x - \lambda^\top f.$$

We know from Theorem 4.1 that a necessary condition for our constrained optimization problem to have a solution is that $\nabla L(x, \lambda) = 0$. Since

$$\begin{aligned} \frac{\partial L}{\partial x}(x, \lambda) &= A^{-1} x - (b - B\lambda) \\ \frac{\partial L}{\partial \lambda}(x, \lambda) &= B^\top x - f, \end{aligned}$$

we obtain the system of linear equations

$$\begin{aligned} A^{-1} x + B\lambda &= b, \\ B^\top x &= f, \end{aligned}$$

which can be written in matrix form as

$$\begin{pmatrix} A^{-1} & B \\ B^\top & 0 \end{pmatrix} \begin{pmatrix} x \\ \lambda \end{pmatrix} = \begin{pmatrix} b \\ f \end{pmatrix}.$$

We shall prove in Proposition 6.3 below that our constrained minimization problem has a unique solution actually given by the above system.

Note that the matrix of this system is symmetric. We solve it as follows. Eliminating x from the first equation

$$A^{-1} x + B\lambda = b,$$

we get

$$x = A(b - B\lambda),$$

and substituting into the second equation, we get

$$B^\top A(b - B\lambda) = f,$$

that is,

$$B^\top AB\lambda = B^\top Ab - f.$$

However, by a previous remark, since A is symmetric positive definite and the columns of B are linearly independent, $B^\top AB$ is symmetric positive definite, and thus invertible. Thus we obtain the solution

$$\lambda = (B^\top AB)^{-1}(B^\top Ab - f), \qquad x = A(b - B\lambda).$$

Note that this way of solving the system requires solving for the Lagrange multipliers first.

Letting $e = b - B\lambda$, we also note that the system

$$\begin{pmatrix} A^{-1} & B \\ B^\top & 0 \end{pmatrix} \begin{pmatrix} x \\ \lambda \end{pmatrix} = \begin{pmatrix} b \\ f \end{pmatrix}$$

is equivalent to the system

$$\begin{aligned} e &= b - B\lambda, \\ x &= Ae, \\ B^\top x &= f. \end{aligned}$$

The latter system is called the *equilibrium equations* by Strang [Strang (1986)]. Indeed, Strang shows that the equilibrium equations of many physical systems can be put in the above form. This includes spring-mass systems, electrical networks and trusses, which are structures built from elastic bars. In each case, x, e, b, A, λ, f, and $K = B^\top AB$ have a physical interpretation. The matrix $K = B^\top AB$ is usually called the *stiffness matrix*. Again, the reader is referred to Strang [Strang (1986)].

In order to prove that our constrained minimization problem has a unique solution, we proceed to prove that the constrained minimization of $Q(x)$ subject to $B^\top x = f$ is equivalent to the unconstrained maximization of another function $-G(\lambda)$. We get $G(\lambda)$ by minimizing the Lagrangian $L(x, \lambda)$ treated as a function of x alone. The function $-G(\lambda)$ is the *dual function* of the Lagrangian $L(x, \lambda)$. Here we are encountering a special case of the notion of dual function defined in Section 14.7.

Since A^{-1} is symmetric positive definite and

$$L(x, \lambda) = \frac{1}{2} x^\top A^{-1} x - (b - B\lambda)^\top x - \lambda^\top f,$$

by Proposition 6.2 the global minimum (with respect to x) of $L(x, \lambda)$ is obtained for the solution x of

$$A^{-1}x = b - B\lambda,$$

that is, when

$$x = A(b - B\lambda),$$

and the minimum of $L(x, \lambda)$ is

$$\min_x L(x, \lambda) = -\frac{1}{2}(B\lambda - b)^\top A(B\lambda - b) - \lambda^\top f.$$

Letting

$$G(\lambda) = \frac{1}{2}(B\lambda - b)^\top A(B\lambda - b) + \lambda^\top f,$$

we will show in Proposition 6.3 that the solution of the constrained minimization of $Q(x)$ subject to $B^\top x = f$ is equivalent to the unconstrained maximization of $-G(\lambda)$. This is a special case of the duality discussed in Section 14.7.

Of course, since we minimized $L(x, \lambda)$ with respect to x, we have

$$L(x, \lambda) \geq -G(\lambda)$$

for all x and all λ. However, when the constraint $B^\top x = f$ holds, $L(x, \lambda) = Q(x)$, and thus for any admissible x, which means that $B^\top x = f$, we have

$$\min_x Q(x) \geq \max_\lambda -G(\lambda).$$

In order to prove that the unique minimum of the constrained problem $Q(x)$ subject to $B^\top x = f$ is the unique maximum of $-G(\lambda)$, we compute $Q(x) + G(\lambda)$.

Proposition 6.3. *The quadratic constrained minimization problem of Definition 6.3 has a unique solution (x, λ) given by the system*

$$\begin{pmatrix} A^{-1} & B \\ B^\top & 0 \end{pmatrix} \begin{pmatrix} x \\ \lambda \end{pmatrix} = \begin{pmatrix} b \\ f \end{pmatrix}.$$

Furthermore, the component λ of the above solution is the unique value for which $-G(\lambda)$ is maximum.

Proof. As we suggested earlier, let us compute $Q(x) + G(\lambda)$, assuming that the constraint $B^\top x = f$ holds. Eliminating f, since $b^\top x = x^\top b$ and $\lambda^\top B^\top x = x^\top B\lambda$, we get

$$\begin{aligned} Q(x) + G(\lambda) &= \frac{1}{2}x^\top A^{-1}x - b^\top x + \frac{1}{2}(B\lambda - b)^\top A(B\lambda - b) + \lambda^\top f \\ &= \frac{1}{2}(A^{-1}x + B\lambda - b)^\top A(A^{-1}x + B\lambda - b). \end{aligned}$$

Since A is positive definite, the last expression is nonnegative. In fact, it is null iff

$$A^{-1}x + B\lambda - b = 0,$$

that is,

$$A^{-1}x + B\lambda = b.$$

But then the unique constrained minimum of $Q(x)$ subject to $B^\top x = f$ is equal to the unique maximum of $-G(\lambda)$ exactly when $B^\top x = f$ and $A^{-1}x + B\lambda = b$, which proves the proposition. □

We can confirm that the maximum of $-G(\lambda)$, or equivalently the minimum of

$$G(\lambda) = \frac{1}{2}(B\lambda - b)^\top A(B\lambda - b) + \lambda^\top f,$$

corresponds to value of λ obtained by solving the system

$$\begin{pmatrix} A^{-1} & B \\ B^\top & 0 \end{pmatrix} \begin{pmatrix} x \\ \lambda \end{pmatrix} = \begin{pmatrix} b \\ f \end{pmatrix}.$$

Indeed, since

$$G(\lambda) = \frac{1}{2}\lambda^\top B^\top AB\lambda - \lambda^\top B^\top Ab + \lambda^\top f + \frac{1}{2}b^\top b,$$

and $B^\top AB$ is symmetric positive definite, by Proposition 6.2, the global minimum of $G(\lambda)$ is obtained when

$$B^\top AB\lambda - B^\top Ab + f = 0,$$

that is, $\lambda = (B^\top AB)^{-1}(B^\top Ab - f)$, as we found earlier.

Remarks:

(1) There is a form of duality going on in this situation. The constrained minimization of $Q(x)$ subject to $B^\top x = f$ is called the *primal problem*, and the unconstrained maximization of $-G(\lambda)$ is called the *dual problem*. Duality is the fact stated slightly loosely as

$$\min_x Q(x) = \max_\lambda -G(\lambda).$$

A general treatment of duality in constrained minimization problems is given in Section 14.7.

Recalling that $e = b - B\lambda$, since

$$G(\lambda) = \frac{1}{2}(B\lambda - b)^\top A(B\lambda - b) + \lambda^\top f,$$

we can also write

$$G(\lambda) = \frac{1}{2} e^\top A e + \lambda^\top f.$$

This expression often represents the total potential energy of a system. Again, the optimal solution is the one that minimizes the potential energy (and thus maximizes $-G(\lambda)$).

(2) It is immediately verified that the equations of Proposition 6.3 are equivalent to the equations stating that the partial derivatives of the Lagrangian $L(x, \lambda)$ are null:

$$\begin{aligned} \frac{\partial L}{\partial x_i} &= 0, \quad i = 1, \ldots, m, \\ \frac{\partial L}{\partial \lambda_j} &= 0, \quad j = 1, \ldots, n. \end{aligned}$$

Thus, the constrained minimum of $Q(x)$ subject to $B^\top x = f$ is an extremum of the Lagrangian $L(x, \lambda)$. As we showed in Proposition 6.3, this extremum corresponds to simultaneously minimizing $L(x, \lambda)$ with respect to x and maximizing $L(x, \lambda)$ with respect to λ. Geometrically, such a point is a *saddle point* for $L(x, \lambda)$. Saddle points are discussed in Section 14.7.

(3) The Lagrange multipliers sometimes have a natural physical meaning. For example, in the spring-mass system they correspond to node displacements. In some general sense, Lagrange multipliers are correction terms needed to satisfy equilibrium equations and the price paid for the constraints. For more details, see Strang [Strang (1986)].

Going back to the constrained minimization of $Q(x_1, x_2) = \frac{1}{2}(x_1^2 + x_2^2)$ subject to

$$2x_1 - x_2 = 5,$$

the Lagrangian is

$$L(x_1, x_2, \lambda) = \frac{1}{2}(x_1^2 + x_2^2) + \lambda(2x_1 - x_2 - 5),$$

and the equations stating that the Lagrangian has a saddle point are

$$\begin{aligned} x_1 + 2\lambda &= 0, \\ x_2 - \lambda &= 0, \\ 2x_1 - x_2 - 5 &= 0. \end{aligned}$$

We obtain the solution $(x_1, x_2, \lambda) = (2, -1, -1)$.

The use of Lagrange multipliers in optimization and variational problems is discussed extensively in Chapter 14.

Least squares methods and Lagrange multipliers are used to tackle many problems in computer graphics and computer vision; see Trucco and Verri [Trucco and Verri (1998)], Metaxas [Metaxas (1997)], Jain, Katsuri, and Schunck [Jain *et al.* (1995)], Faugeras [Faugeras (1996)], and Foley, van Dam, Feiner, and Hughes [Foley *et al.* (1993)].

6.2 Quadratic Optimization: The General Case

In this section we complete the study initiated in Section 6.1 and give necessary and sufficient conditions for the quadratic function $\frac{1}{2}x^\top Ax - x^\top b$ to have a global minimum. We begin with the following simple fact:

Proposition 6.4. *If A is an invertible symmetric matrix, then the function*

$$f(x) = \frac{1}{2}x^\top Ax - x^\top b$$

has a minimum value iff $A \succeq 0$, in which case this optimal value is obtained for a unique value of x, namely $x^* = A^{-1}b$, and with

$$f(A^{-1}b) = -\frac{1}{2}b^\top A^{-1}b.$$

Proof. Observe that

$$\frac{1}{2}(x - A^{-1}b)^\top A(x - A^{-1}b) = \frac{1}{2}x^\top Ax - x^\top b + \frac{1}{2}b^\top A^{-1}b.$$

Thus,

$$f(x) = \frac{1}{2}x^\top Ax - x^\top b = \frac{1}{2}(x - A^{-1}b)^\top A(x - A^{-1}b) - \frac{1}{2}b^\top A^{-1}b.$$

If A has some negative eigenvalue, say $-\lambda$ (with $\lambda > 0$), if we pick any eigenvector u of A associated with λ, then for any $\alpha \in \mathbb{R}$ with $\alpha \neq 0$, if we let $x = \alpha u + A^{-1}b$, then since $Au = -\lambda u$, we get

$$\begin{aligned} f(x) &= \frac{1}{2}(x - A^{-1}b)^\top A(x - A^{-1}b) - \frac{1}{2}b^\top A^{-1}b \\ &= \frac{1}{2}\alpha u^\top A\alpha u - \frac{1}{2}b^\top A^{-1}b \\ &= -\frac{1}{2}\alpha^2\lambda \left\|u\right\|_2^2 - \frac{1}{2}b^\top A^{-1}b, \end{aligned}$$

and since α can be made as large as we want and $\lambda > 0$, we see that f has no minimum. Consequently, in order for f to have a minimum, we must have $A \succeq 0$. If $A \succeq 0$, since A is invertible, it is positive definite, so $(x - A^{-1}b)^\top A(x - A^{-1}b) > 0$ iff $x - A^{-1}b \neq 0$, and it is clear that the minimum value of f is achieved when $x - A^{-1}b = 0$, that is, $x = A^{-1}b$. $\square$

Let us now consider the case of an arbitrary symmetric matrix A.

Proposition 6.5. *If A is an $n \times n$ symmetric matrix, then the function*

$$f(x) = \frac{1}{2} x^\top A x - x^\top b$$

has a minimum value iff $A \succeq 0$ and $(I - AA^+)b = 0$, in which case this minimum value is

$$p^* = -\frac{1}{2} b^\top A^+ b.$$

Furthermore, if A is diagonalized as $A = U^\top \Sigma U$ (with U orthogonal), then the optimal value is achieved by all $x \in \mathbb{R}^n$ of the form

$$x = A^+ b + U^\top \begin{pmatrix} 0 \\ z \end{pmatrix},$$

for any $z \in \mathbb{R}^{n-r}$, where r is the rank of A.

Proof. The case that A is invertible is taken care of by Proposition 6.4, so we may assume that A is singular. If A has rank $r < n$, then we can diagonalize A as

$$A = U^\top \begin{pmatrix} \Sigma_r & 0 \\ 0 & 0 \end{pmatrix} U,$$

where U is an orthogonal matrix and where Σ_r is an $r \times r$ diagonal invertible matrix. Then we have

$$\begin{aligned} f(x) &= \frac{1}{2} x^\top U^\top \begin{pmatrix} \Sigma_r & 0 \\ 0 & 0 \end{pmatrix} U x - x^\top U^\top U b \\ &= \frac{1}{2} (Ux)^\top \begin{pmatrix} \Sigma_r & 0 \\ 0 & 0 \end{pmatrix} U x - (Ux)^\top U b. \end{aligned}$$

If we write

$$Ux = \begin{pmatrix} y \\ z \end{pmatrix} \quad \text{and} \quad Ub = \begin{pmatrix} c \\ d \end{pmatrix},$$

with $y, c \in \mathbb{R}^r$ and $z, d \in \mathbb{R}^{n-r}$, we get

$$\begin{aligned} f(x) &= \frac{1}{2} (Ux)^\top \begin{pmatrix} \Sigma_r & 0 \\ 0 & 0 \end{pmatrix} U x - (Ux)^\top U b \\ &= \frac{1}{2} (y^\top \; z^\top) \begin{pmatrix} \Sigma_r & 0 \\ 0 & 0 \end{pmatrix} \begin{pmatrix} y \\ z \end{pmatrix} - (y^\top \; z^\top) \begin{pmatrix} c \\ d \end{pmatrix} \\ &= \frac{1}{2} y^\top \Sigma_r y - y^\top c - z^\top d. \end{aligned}$$

For $y = 0$, we get

$$f(x) = -z^\top d,$$

so if $d \neq 0$, the function f has no minimum. Therefore, if f has a minimum, then $d = 0$. However, $d = 0$ means that

$$Ub = \begin{pmatrix} c \\ 0 \end{pmatrix},$$

and we know from Proposition 21.5 (Vol. I) that b is in the range of A (here, U is $V^\top$), which is equivalent to $(I - AA^+)b = 0$. If $d = 0$, then

$$f(x) = \frac{1}{2} y^\top \Sigma_r y - y^\top c.$$

Consider the function $g\colon \mathbb{R}^r \to \mathbb{R}$ given by

$$g(y) = \frac{1}{2} y^\top \Sigma_r y - y^\top c, \quad y \in \mathbb{R}^r.$$

Since

$$\begin{pmatrix} y \\ z \end{pmatrix} = U^\top x$$

and $U^\top$ is invertible (with inverse U), when x ranges over $\mathbb{R}^n$, y ranges over the whole of $\mathbb{R}^r$, and since $f(x) = g(y)$, the function f has a minimum iff g has a minimum. Since Σ_r is invertible, by Proposition 6.4, the function g has a minimum iff $\Sigma_r \succeq 0$, which is equivalent to $A \succeq 0$.

Therefore, we have proven that if f has a minimum, then $(I - AA^+)b = 0$ and $A \succeq 0$. Conversely, if $(I - AA^+)b = 0$, then

$$\begin{aligned}
&\left(\begin{pmatrix} I_r & 0 \\ 0 & I_{n-r} \end{pmatrix} - U^\top \begin{pmatrix} \Sigma_r & 0 \\ 0 & 0 \end{pmatrix} U U^\top \begin{pmatrix} \Sigma_r^{-1} & 0 \\ 0 & 0 \end{pmatrix} U \right) b \\
&\quad = \left(\begin{pmatrix} I_r & 0 \\ 0 & I_{n-r} \end{pmatrix} - U^\top \begin{pmatrix} I_r & 0 \\ 0 & 0 \end{pmatrix} U \right) b = U^\top \begin{pmatrix} 0 & 0 \\ 0 & I_{n-r} \end{pmatrix} Ub = 0,
\end{aligned}$$

which implies that if

$$Ub = \begin{pmatrix} c \\ d \end{pmatrix},$$

then $d = 0$, so as above

$$f(x) = g(y) = \frac{1}{2} y^\top \Sigma_r y - y^\top c,$$

and because $A \succeq 0$, we also have $\Sigma_r \succeq 0$, so g and f have a minimum.

When the above conditions hold, since

$$A = U^\top \begin{pmatrix} \Sigma_r & 0 \\ 0 & 0 \end{pmatrix} U$$

is positive semidefinite, the pseudo-inverse A^+ of A is given by

$$A^+ = U^\top \begin{pmatrix} \Sigma_r^{-1} & 0 \\ 0 & 0 \end{pmatrix} U,$$

and since

$$f(x) = g(y) = \frac{1}{2} y^\top \Sigma_r y - y^\top c,$$

by Proposition 6.4 the minimum of g is achieved iff $y^* = \Sigma_r^{-1} c$. Since $f(x)$ is independent of z, we can choose $z = 0$, and since $d = 0$, for x^* given by

$$Ux^* = \begin{pmatrix} \Sigma_r^{-1} c \\ 0 \end{pmatrix} \quad \text{and} \quad Ub = \begin{pmatrix} c \\ 0 \end{pmatrix},$$

we deduce that

$$x^* = U^\top \begin{pmatrix} \Sigma_r^{-1} c \\ 0 \end{pmatrix} = U^\top \begin{pmatrix} \Sigma_r^{-1} & 0 \\ 0 & 0 \end{pmatrix} \begin{pmatrix} c \\ 0 \end{pmatrix} = U^\top \begin{pmatrix} \Sigma_r^{-1} & 0 \\ 0 & 0 \end{pmatrix} Ub = A^+ b, \quad (*)$$

and the minimum value of f is

$$f(x^*) = \frac{1}{2}(A^+ b)^\top A A^+ b - b^\top A^+ b = b^\top A^+ A A^+ b - b^\top A^+ b = -\frac{1}{2} b^\top A^+ b,$$

since A^+ is symmetric and $A^+ A A^+ = A^+$. For any $x \in \mathbb{R}^n$ of the form

$$x = A^+ b + U^\top \begin{pmatrix} 0 \\ z \end{pmatrix}, \quad z \in \mathbb{R}^{n-r},$$

since

$$x = A^+ b + U^\top \begin{pmatrix} 0 \\ z \end{pmatrix} = U^\top \begin{pmatrix} \Sigma_r^{-1} c \\ 0 \end{pmatrix} + U^\top \begin{pmatrix} 0 \\ z \end{pmatrix} = U^\top \begin{pmatrix} \Sigma_r^{-1} c \\ z \end{pmatrix},$$

and since $f(x)$ is independent of z (because $f(x) = g(y)$), we have

$$f(x) = f(x^*) = -\frac{1}{2} b^\top A^+ b. \qquad \square$$

The problem of minimizing the function

$$f(x) = \frac{1}{2} x^\top A x - x^\top b$$

in the case where we add either linear constraints of the form $C^\top x = 0$ or affine constraints of the form $C^\top x = t$ (where $t \in \mathbb{R}^m$ and $t \neq 0$) where C is an $n \times m$ matrix can be reduced to the unconstrained case using a

QR-decomposition of C. Let us show how to do this for linear constraints of the form $C^\top x = 0$.

If we use a QR decomposition of C, by permuting the columns of C to make sure that the first r columns of C are linearly independent (where $r = \mathrm{rank}(C)$), we may assume that

$$C = Q^\top \begin{pmatrix} R & S \\ 0 & 0 \end{pmatrix} \Pi,$$

where Q is an $n \times n$ orthogonal matrix, R is an $r \times r$ invertible upper triangular matrix, S is an $r \times (m - r)$ matrix, and Π is a permutation matrix (C has rank r). Then if we let

$$x = Q^\top \begin{pmatrix} y \\ z \end{pmatrix},$$

where $y \in \mathbb{R}^r$ and $z \in \mathbb{R}^{n-r}$, then $C^\top x = 0$ becomes

$$C^\top x = \Pi^\top \begin{pmatrix} R^\top & 0 \\ S^\top & 0 \end{pmatrix} Qx = \Pi^\top \begin{pmatrix} R^\top & 0 \\ S^\top & 0 \end{pmatrix} \begin{pmatrix} y \\ z \end{pmatrix} = 0,$$

which implies $y = 0$, and every solution of $C^\top x = 0$ is of the form

$$x = Q^\top \begin{pmatrix} 0 \\ z \end{pmatrix}.$$

Our original problem becomes

$$\begin{aligned} &\text{minimize} \quad \frac{1}{2}(y^\top \; z^\top)QAQ^\top \begin{pmatrix} y \\ z \end{pmatrix} + (y^\top \; z^\top)Qb \\ &\text{subject to} \;\; y = 0, \; y \in \mathbb{R}^r, \; z \in \mathbb{R}^{n-r}. \end{aligned}$$

Thus, the constraint $C^\top x = 0$ has been simplified to $y = 0$, and if we write

$$QAQ^\top = \begin{pmatrix} G_{11} & G_{12} \\ G_{21} & G_{22} \end{pmatrix},$$

where G_{11} is an $r \times r$ matrix and G_{22} is an $(n - r) \times (n - r)$ matrix and

$$Qb = \begin{pmatrix} b_1 \\ b_2 \end{pmatrix}, \quad b_1 \in \mathbb{R}^r, \; b_2 \in \mathbb{R}^{n-r},$$

our problem becomes

$$\text{minimize} \; \frac{1}{2} z^\top G_{22} z + z^\top b_2, \quad z \in \mathbb{R}^{n-r},$$

the problem solved in Proposition 6.5.

Constraints of the form $C^\top x = t$ (where $t \neq 0$) can be handled in a similar fashion. In this case, we may assume that C is an $n \times m$ matrix with full rank (so that $m \leq n$) and $t \in \mathbb{R}^m$. Then we use a QR-decomposition of the form

$$C = P \begin{pmatrix} R \\ 0 \end{pmatrix},$$

where P is an orthogonal $n \times n$ matrix and R is an $m \times m$ invertible upper triangular matrix. If we write

$$x = P \begin{pmatrix} y \\ z \end{pmatrix},$$

where $y \in \mathbb{R}^m$ and $z \in \mathbb{R}^{n-m}$, the equation $C^\top x = t$ becomes

$$(R^\top \; 0) P^\top x = t,$$

that is,

$$(R^\top \; 0) \begin{pmatrix} y \\ z \end{pmatrix} = t,$$

which yields

$$R^\top y = t.$$

Since R is invertible, we get $y = (R^\top)^{-1} t$, and then it is easy to see that our original problem reduces to an unconstrained problem in terms of the matrix $P^\top AP$; the details are left as an exercise.

6.3 Maximizing a Quadratic Function on the Unit Sphere

In this section we discuss various quadratic optimization problems mostly arising from computer vision (image segmentation and contour grouping). These problems can be reduced to the following basic optimization problem: given an $n \times n$ real symmetric matrix A

$$\begin{aligned} &\text{maximize} \quad x^\top Ax \\ &\text{subject to} \quad x^\top x = 1, \; x \in \mathbb{R}^n. \end{aligned}$$

In view of Proposition 21.10 (Vol. I), the maximum value of $x^\top Ax$ on the unit sphere is equal to the largest eigenvalue λ_1 of the matrix A, and it is achieved for any unit eigenvector u_1 associated with λ_1. Similarly, the minimum value of $x^\top Ax$ on the unit sphere is equal to the smallest eigenvalue λ_n of the matrix A, and it is achieved for any unit eigenvector u_n associated with λ_n.

A variant of the above problem often encountered in computer vision consists in minimizing $x^\top Ax$ on the ellipsoid given by an equation of the form

$$x^\top Bx = 1,$$

where B is a symmetric positive definite matrix. Since B is positive definite, it can be diagonalized as

$$B = QDQ^\top,$$

where Q is an orthogonal matrix and D is a diagonal matrix,

$$D = \mathrm{diag}(d_1, \ldots, d_n),$$

with $d_i > 0$, for $i = 1, \ldots, n$. If we define the matrices $B^{1/2}$ and $B^{-1/2}$ by

$$B^{1/2} = Q\,\mathrm{diag}\left(\sqrt{d_1}, \ldots, \sqrt{d_n}\right) Q^\top$$

and

$$B^{-1/2} = Q\,\mathrm{diag}\left(1/\sqrt{d_1}, \ldots, 1/\sqrt{d_n}\right) Q^\top,$$

it is clear that these matrices are symmetric, that $B^{-1/2}BB^{-1/2} = I$, and that $B^{1/2}$ and $B^{-1/2}$ are mutual inverses. Then if we make the change of variable

$$x = B^{-1/2}y,$$

the equation $x^\top Bx = 1$ becomes $y^\top y = 1$, and the optimization problem

$$\begin{aligned} &\text{minimize} \quad x^\top Ax \\ &\text{subject to} \quad x^\top Bx = 1,\ x \in \mathbb{R}^n, \end{aligned}$$

is equivalent to the problem

$$\begin{aligned} &\text{minimize} \quad y^\top B^{-1/2}AB^{-1/2}y \\ &\text{subject to} \quad y^\top y = 1,\ y \in \mathbb{R}^n, \end{aligned}$$

where $y = B^{1/2}x$ and $B^{-1/2}AB^{-1/2}$ are symmetric.

The complex version of our basic optimization problem in which A is a Hermitian matrix also arises in computer vision. Namely, given an $n \times n$ complex Hermitian matrix A,

$$\begin{aligned} &\text{maximize} \quad x^*Ax \\ &\text{subject to} \quad x^*x = 1,\ x \in \mathbb{C}^n. \end{aligned}$$

Again by Proposition 21.10 (Vol. I), the maximum value of $x^* Ax$ on the unit sphere is equal to the largest eigenvalue λ_1 of the matrix A, and it is achieved for any unit eigenvector u_1 associated with λ_1.

Remark: It is worth pointing out that if A is a *skew-Hermitian* matrix, that is, if $A^* = -A$, then $x^* Ax$ is *pure imaginary or zero.*

Indeed, since $z = x^* Ax$ is a scalar, we have $z^* = \overline{z}$ (the conjugate of z), so we have

$$\overline{x^* Ax} = (x^* Ax)^* = x^* A^* x = -x^* Ax,$$

so $\overline{x^* Ax} + x^* Ax = 2\mathrm{Re}(x^* Ax) = 0$, which means that $x^* Ax$ is pure imaginary or zero.

In particular, if A is a real matrix and if A is *skew-symmetric*, then

$$x^\top Ax = 0.$$

Thus, for any real matrix (symmetric or not),

$$x^\top Ax = x^\top H(A)x,$$

where $H(A) = (A + A^\top)/2$, the symmetric part of A.

There are situations in which it is necessary to add linear constraints to the problem of maximizing a quadratic function on the sphere. This problem was completely solved by Golub [Golub (1973)] (1973). The problem is the following: given an $n \times n$ real symmetric matrix A and an $n \times p$ matrix C,

$$\begin{aligned} &\text{minimize} \quad x^\top Ax \\ &\text{subject to} \quad x^\top x = 1,\ C^\top x = 0,\ x \in \mathbb{R}^n. \end{aligned}$$

As in Section 6.2, Golub shows that the linear constraint $C^\top x = 0$ can be eliminated as follows: if we use a QR decomposition of C, by permuting the columns, we may assume that

$$C = Q^\top \begin{pmatrix} R & S \\ 0 & 0 \end{pmatrix} \Pi,$$

where Q is an orthogonal $n \times n$ matrix, R is an $r \times r$ invertible upper triangular matrix, and S is an $r \times (p - r)$ matrix (assuming C has rank r). If we let

$$x = Q^\top \begin{pmatrix} y \\ z \end{pmatrix},$$

where $y \in \mathbb{R}^r$ and $z \in \mathbb{R}^{n-r}$, then $C^\top x = 0$ becomes

$$\Pi^\top \begin{pmatrix} R^\top & 0 \\ S^\top & 0 \end{pmatrix} Qx = \Pi^\top \begin{pmatrix} R^\top & 0 \\ S^\top & 0 \end{pmatrix} \begin{pmatrix} y \\ z \end{pmatrix} = 0,$$

which implies $y = 0$, and every solution of $C^\top x = 0$ is of the form

$$x = Q^\top \begin{pmatrix} 0 \\ z \end{pmatrix}.$$

Our original problem becomes

$$\begin{aligned} &\text{minimize} \quad (y^\top\ z^\top) QAQ^\top \begin{pmatrix} y \\ z \end{pmatrix} \\ &\text{subject to} \quad z^\top z = 1,\ z \in \mathbb{R}^{n-r}, \\ &\qquad\qquad\quad y = 0,\ y \in \mathbb{R}^r. \end{aligned}$$

Thus the constraint $C^\top x = 0$ has been simplified to $y = 0$, and if we write

$$QAQ^\top = \begin{pmatrix} G_{11} & G_{12} \\ G_{12}^\top & G_{22} \end{pmatrix},$$

our problem becomes

$$\begin{aligned} &\text{minimize} \quad z^\top G_{22} z \\ &\text{subject to} \quad z^\top z = 1,\ z \in \mathbb{R}^{n-r}, \end{aligned}$$

a standard eigenvalue problem.

Remark: There is a way of finding the eigenvalues of G_{22} which does not require the QR-factorization of C. Observe that if we let

$$J = \begin{pmatrix} 0 & 0 \\ 0 & I_{n-r} \end{pmatrix},$$

then

$$JQAQ^\top J = \begin{pmatrix} 0 & 0 \\ 0 & G_{22} \end{pmatrix},$$

and if we set

$$P = Q^\top JQ,$$

then

$$PAP = Q^\top JQAQ^\top JQ.$$

Now, $Q^\top JQAQ^\top JQ$ and $JQAQ^\top J$ have the same eigenvalues, so PAP and $JQAQ^\top J$ also have the same eigenvalues. It follows that the solutions of our optimization problem are among the eigenvalues of $K = PAP$, and at least r of those are 0. Using the fact that CC^+ is the projection onto the range of C, where C^+ is the pseudo-inverse of C, it can also be shown that

$$P = I - CC^+,$$

the projection onto the kernel of $C^\top$. So P can be computed directly in terms of C. In particular, when $n \geq p$ and C has full rank (the columns of C are linearly independent), then we know that $C^+ = (C^\top C)^{-1}C^\top$ and

$$P = I - C(C^\top C)^{-1}C^\top.$$

This fact is used by Cour and Shi [Cour and Shi (2007)] and implicitly by Yu and Shi [Yu and Shi (2001)].

The problem of adding affine constraints of the form $N^\top x = t$, where $t \neq 0$, also comes up in practice. At first glance, this problem may not seem harder than the linear problem in which $t = 0$, but it is. This problem was extensively studied in a paper by Gander, Golub, and von Matt [Gander *et al.* (1989)] (1989).

Gander, Golub, and von Matt consider the following problem: Given an $(n+m) \times (n+m)$ real symmetric matrix A (with $n > 0$), an $(n+m) \times m$ matrix N with full rank, and a nonzero vector $t \in \mathbb{R}^m$ with $\left\|(N^\top)^+ t\right\| < 1$ (where $(N^\top)^+$ denotes the pseudo-inverse of $N^\top$),

$$\begin{aligned} &\text{minimize} \quad x^\top A x \\ &\text{subject to} \quad x^\top x = 1,\ N^\top x = t,\ x \in \mathbb{R}^{n+m}. \end{aligned}$$

The condition $\left\|(N^\top)^+ t\right\| < 1$ ensures that the problem has a solution and is not trivial. The authors begin by proving that the affine constraint $N^\top x = t$ can be eliminated. One way to do so is to use a QR decomposition of N. If

$$N = P\begin{pmatrix} R \\ 0 \end{pmatrix},$$

where P is an orthogonal $(n+m) \times (n+m)$ matrix and R is an $m \times m$ invertible upper triangular matrix, then if we observe that

$$\begin{aligned} x^\top A x &= x^\top P P^\top A P P^\top x, \\ N^\top x &= (R^\top\ 0)P^\top x = t, \\ x^\top x &= x^\top P P^\top x = 1, \end{aligned}$$

and if we write

$$P^\top A P = \begin{pmatrix} B & \Gamma^\top \\ \Gamma & C \end{pmatrix},$$

where B is an $m \times m$ symmetric matrix, C is an $n \times n$ symmetric matrix, Γ is an $m \times n$ matrix, and

$$P^\top x = \begin{pmatrix} y \\ z \end{pmatrix},$$

with $y \in \mathbb{R}^m$ and $z \in \mathbb{R}^n$, we then get

$$\begin{aligned} x^\top Ax &= y^\top By + 2z^\top \Gamma y + z^\top Cz, \\ R^\top y &= t, \\ y^\top y + z^\top z &= 1. \end{aligned}$$

Thus

$$y = (R^\top)^{-1}t,$$

and if we write

$$s^2 = 1 - y^\top y > 0$$

and

$$b = \Gamma y,$$

we get the simplified problem

$$\begin{aligned} \text{minimize} \quad & z^\top Cz + 2z^\top b \\ \text{subject to} \quad & z^\top z = s^2, \ z \in \mathbb{R}^m. \end{aligned}$$

Unfortunately, if $b \neq 0$, Proposition 21.10 (Vol. I) is no longer applicable. It is still possible to find the minimum of the function $z^\top Cz + 2z^\top b$ using Lagrange multipliers, but such a solution is too involved to be presented here. Interested readers will find a thorough discussion in Gander, Golub, and von Matt [Gander *et al.* (1989)].

6.4 Summary

The main concepts and results of this chapter are listed below:

- Quadratic optimization problems; quadratic functions.
- Symmetric positive definite and positive semidefinite matrices.
- The positive semidefinite cone ordering.
- Existence of a global minimum when A is symmetric positive definite.
- Constrained quadratic optimization problems.
- Lagrange multipliers; Lagrangian.
- Primal and dual problems.

- Quadratic optimization problems: the case of a symmetric invertible matrix A.
- Quadratic optimization problems: the general case of a symmetric matrix A.
- Adding linear constraints of the form $C^\top x = 0$.
- Adding affine constraints of the form $C^\top x = t$, with $t \neq 0$.
- Maximizing a quadratic function over the unit sphere.
- Maximizing a quadratic function over an ellipsoid.
- Maximizing a Hermitian quadratic form.
- Adding linear constraints of the form $C^\top x = 0$.
- Adding affine constraints of the form $N^\top x = t$, with $t \neq 0$.

6.5 Problems

Problem 6.1. Prove that the relation

$$A \succeq B$$

between any two $n \times n$ matrices (symmetric or not) iff $A - B$ is symmetric positive semidefinite is indeed a partial order.

Problem 6.2. (1) Prove that if A is symmetric positive definite, then so is A^{-1}.

(2) Prove that if C is a symmetric positive definite $m \times m$ matrix and A is an $m \times n$ matrix of rank n (and so $m \geq n$ and the map $x \mapsto Ax$ is surjective onto $\mathbb{R}^m$), then $A^\top CA$ is symmetric positive definite.

Problem 6.3. Find the minimum of the function

$$Q(x_1, x_2) = \frac{1}{2}(2x_1^2 + x_2^2)$$

subject to the constraint

$$x_1 - x_2 = 3.$$

Problem 6.4. Consider the problem of minimizing the function

$$f(x) = \frac{1}{2}x^\top Ax - x^\top b$$

in the case where we add an affine constraint of the form $C^\top x = t$, with $t \in \mathbb{R}^m$ and $t \neq 0$, and where C is an $n \times m$ matrix of rank $m \leq n$. As in Section 6.2, use a QR-decomposition

$$C = P \begin{pmatrix} R \\ 0 \end{pmatrix},$$

where P is an orthogonal $n \times n$ matrix and R is an $m \times m$ invertible upper triangular matrix, and write

$$x = P \begin{pmatrix} y \\ z \end{pmatrix},$$

to deduce that

$$R^\top y = t.$$

Give the details of the reduction of this constrained minimization problem to an unconstrained minimization problem involving the matrix $P^\top AP$.

Problem 6.5. Find the maximum and the minimum of the function

$$Q(x, y) = (x\ y) \begin{pmatrix} 1 & 2 \\ 2 & 1 \end{pmatrix} \begin{pmatrix} x \\ y \end{pmatrix}$$

on the unit circle $x^2 + y^2 = 1$.

Chapter 7

Schur Complements and Applications

Schur complements arise naturally in the process of inverting block matrices of the form

$$M = \begin{pmatrix} A & B \\ C & D \end{pmatrix}$$

and in characterizing when symmetric versions of these matrices are positive definite or positive semidefinite. These characterizations come up in various quadratic optimization problems; see Boyd and Vandenberghe [Boyd and Vandenberghe (2004)], especially Appendix B. In the most general case, pseudo-inverses are also needed.

In this chapter we introduce Schur complements and describe several interesting ways in which they are used. Along the way we provide some details and proofs of some results from Appendix A.5 (especially Section A.5.5) of Boyd and Vandenberghe [Boyd and Vandenberghe (2004)].

7.1 Schur Complements

Let M be an $n \times n$ matrix written as a 2×2 block matrix

$$M = \begin{pmatrix} A & B \\ C & D \end{pmatrix},$$

where A is a $p \times p$ matrix and D is a $q \times q$ matrix, with $n = p + q$ (so B is a $p \times q$ matrix and C is a $q \times p$ matrix). We can try to solve the linear system

$$\begin{pmatrix} A & B \\ C & D \end{pmatrix} \begin{pmatrix} x \\ y \end{pmatrix} = \begin{pmatrix} c \\ d \end{pmatrix},$$

that is,

$$\begin{aligned} Ax + By &= c, \\ Cx + Dy &= d, \end{aligned}$$

by mimicking Gaussian elimination. If we assume that D is invertible, then we first solve for y, getting

$$y = D^{-1}(d - Cx),$$

and after substituting this expression for y in the first equation, we get

$$Ax + B(D^{-1}(d - Cx)) = c,$$

that is,

$$(A - BD^{-1}C)x = c - BD^{-1}d.$$

If the matrix $A - BD^{-1}C$ is invertible, then we obtain the solution to our system

$$\begin{aligned} x &= (A - BD^{-1}C)^{-1}(c - BD^{-1}d), \\ y &= D^{-1}(d - C(A - BD^{-1}C)^{-1}(c - BD^{-1}d)). \end{aligned}$$

If A is invertible, then by eliminating x first using the first equation, we obtain analogous formulas involving the matrix $D - CA^{-1}B$. The above formulas suggest that the matrices $A - BD^{-1}C$ and $D - CA^{-1}B$ play a special role and suggest the following definition:

Definition 7.1. Given any $n \times n$ block matrix of the form

$$M = \begin{pmatrix} A & B \\ C & D \end{pmatrix},$$

where A is a $p \times p$ matrix and D is a $q \times q$ matrix, with $n = p + q$ (so B is a $p \times q$ matrix and C is a $q \times p$ matrix), if D is invertible, then the matrix $A - BD^{-1}C$ is called the *Schur complement* of D in M. If A is invertible, then the matrix $D - CA^{-1}B$ is called the *Schur complement* of A in M.

The above equations written as

$$\begin{aligned} x &= (A - BD^{-1}C)^{-1}c - (A - BD^{-1}C)^{-1}BD^{-1}d, \\ y &= -D^{-1}C(A - BD^{-1}C)^{-1}c \\ &\quad + (D^{-1} + D^{-1}C(A - BD^{-1}C)^{-1}BD^{-1})d, \end{aligned}$$

yield a formula for the inverse of M in terms of the Schur complement of D in M, namely

$$\begin{aligned} &\begin{pmatrix} A & B \\ C & D \end{pmatrix}^{-1} \\ &= \begin{pmatrix} (A - BD^{-1}C)^{-1} & -(A - BD^{-1}C)^{-1}BD^{-1} \\ -D^{-1}C(A - BD^{-1}C)^{-1} & D^{-1} + D^{-1}C(A - BD^{-1}C)^{-1}BD^{-1} \end{pmatrix}. \end{aligned}$$

A moment of reflection reveals that

$$\begin{pmatrix} A & B \\ C & D \end{pmatrix}^{-1} = \begin{pmatrix} (A-BD^{-1}C)^{-1} & 0 \\ -D^{-1}C(A-BD^{-1}C)^{-1} & D^{-1} \end{pmatrix} \begin{pmatrix} I & -BD^{-1} \\ 0 & I \end{pmatrix},$$

and then

$$\begin{pmatrix} A & B \\ C & D \end{pmatrix}^{-1} = \begin{pmatrix} I & 0 \\ -D^{-1}C & I \end{pmatrix} \begin{pmatrix} (A-BD^{-1}C)^{-1} & 0 \\ 0 & D^{-1} \end{pmatrix} \begin{pmatrix} I & -BD^{-1} \\ 0 & I \end{pmatrix}.$$

By taking inverses, we obtain the following result.

Proposition 7.1. *If the matrix D is invertible, then*

$$\begin{pmatrix} A & B \\ C & D \end{pmatrix} = \begin{pmatrix} I & BD^{-1} \\ 0 & I \end{pmatrix} \begin{pmatrix} A-BD^{-1}C & 0 \\ 0 & D \end{pmatrix} \begin{pmatrix} I & 0 \\ D^{-1}C & I \end{pmatrix}.$$

The above expression can be checked directly and has the advantage of requiring only the invertibility of D.

Remark: If A is invertible, then we can use the Schur complement $D - CA^{-1}B$ of A to obtain the following factorization of M:

$$\begin{pmatrix} A & B \\ C & D \end{pmatrix} = \begin{pmatrix} I & 0 \\ CA^{-1} & I \end{pmatrix} \begin{pmatrix} A & 0 \\ 0 & D-CA^{-1}B \end{pmatrix} \begin{pmatrix} I & A^{-1}B \\ 0 & I \end{pmatrix}.$$

If $D - CA^{-1}B$ is invertible, we can invert all three matrices above, and we get another formula for the inverse of M in terms of $(D - CA^{-1}B)$, namely,

$$\begin{pmatrix} A & B \\ C & D \end{pmatrix}^{-1}$$

$$= \begin{pmatrix} A^{-1} + A^{-1}B(D-CA^{-1}B)^{-1}CA^{-1} & -A^{-1}B(D-CA^{-1}B)^{-1} \\ -(D-CA^{-1}B)^{-1}CA^{-1} & (D-CA^{-1}B)^{-1} \end{pmatrix}.$$

If A, D and both Schur complements $A - BD^{-1}C$ and $D - CA^{-1}B$ are all invertible, by comparing the two expressions for M^{-1}, we get the (non-obvious) formula

$$(A - BD^{-1}C)^{-1} = A^{-1} + A^{-1}B(D - CA^{-1}B)^{-1}CA^{-1}.$$

Using this formula, we obtain another expression for the inverse of M involving the Schur complements of A and D (see Horn and Johnson [Horn and Johnson (1990)]):

Proposition 7.2. *If A, D and both Schur complements $A - BD^{-1}C$ and $D - CA^{-1}B$ are all invertible, then*

$$\begin{pmatrix} A & B \\ C & D \end{pmatrix}^{-1} = \begin{pmatrix} (A-BD^{-1}C)^{-1} & -A^{-1}B(D-CA^{-1}B)^{-1} \\ -(D-CA^{-1}B)^{-1}CA^{-1} & (D-CA^{-1}B)^{-1} \end{pmatrix}.$$

If we set $D = I$ and change B to $-B$, we get

$$(A + BC)^{-1} = A^{-1} - A^{-1}B(I - CA^{-1}B)^{-1}CA^{-1},$$

a formula known as the *matrix inversion lemma* (see Boyd and Vandenberghe [Boyd and Vandenberghe (2004)], Appendix C.4, especially C.4.3).

7.2 Symmetric Positive Definite Matrices and Schur Complements

If we assume that our block matrix M is symmetric, so that A, D are symmetric and $C = B^\top$, then we see by Proposition 7.1 that M is expressed as

$$M = \begin{pmatrix} A & B \\ B^\top & D \end{pmatrix} = \begin{pmatrix} I & BD^{-1} \\ 0 & I \end{pmatrix} \begin{pmatrix} A - BD^{-1}B^\top & 0 \\ 0 & D \end{pmatrix} \begin{pmatrix} I & BD^{-1} \\ 0 & I \end{pmatrix}^\top,$$

which shows that M is similar to a block diagonal matrix (obviously, the Schur complement, $A - BD^{-1}B^\top$, is symmetric). As a consequence, we have the following version of "Schur's trick" to check whether $M \succ 0$ for a symmetric matrix.

Proposition 7.3. *For any symmetric matrix M of the form*

$$M = \begin{pmatrix} A & B \\ B^\top & C \end{pmatrix},$$

if C is invertible, then the following properties hold:

(1) $M \succ 0$ iff $C \succ 0$ and $A - BC^{-1}B^\top \succ 0$.
(2) If $C \succ 0$, then $M \succeq 0$ iff $A - BC^{-1}B^\top \succeq 0$.

Proof. (1) Since C is invertible, we have

$$M = \begin{pmatrix} A & B \\ B^\top & C \end{pmatrix} = \begin{pmatrix} I & BC^{-1} \\ 0 & I \end{pmatrix} \begin{pmatrix} A - BC^{-1}B^\top & 0 \\ 0 & C \end{pmatrix} \begin{pmatrix} I & BC^{-1} \\ 0 & I \end{pmatrix}^\top. \quad (*)$$

Observe that

$$\begin{pmatrix} I & BC^{-1} \\ 0 & I \end{pmatrix}^{-1} = \begin{pmatrix} I & -BC^{-1} \\ 0 & I \end{pmatrix},$$

so $(*)$ yields

$$\begin{pmatrix} I & -BC^{-1} \\ 0 & I \end{pmatrix} \begin{pmatrix} A & B \\ B^\top & C \end{pmatrix} \begin{pmatrix} I & -BC^{-1} \\ 0 & I \end{pmatrix}^\top = \begin{pmatrix} A - BC^{-1}B^\top & 0 \\ 0 & C \end{pmatrix},$$

and we know that for any symmetric matrix T, here $T = M$, and any invertible matrix N, here

$$N = \begin{pmatrix} I & -BC^{-1} \\ 0 & I \end{pmatrix},$$

the matrix T is positive definite ($T \succ 0$) iff $NTN^\top$ (which is obviously symmetric) is positive definite ($NTN^\top \succ 0$). But a block diagonal matrix is positive definite iff each diagonal block is positive definite, which concludes the proof.

(2) This is because for any symmetric matrix T and any invertible matrix N, we have $T \succeq 0$ iff $NTN^\top \succeq 0$. ☐

Another version of Proposition 7.3 using the Schur complement of A instead of the Schur complement of C also holds. The proof uses the factorization of M using the Schur complement of A (see Section 7.1).

Proposition 7.4. *For any symmetric matrix M of the form*

$$M = \begin{pmatrix} A & B \\ B^\top & C \end{pmatrix},$$

if A is invertible then the following properties hold:

(1) $M \succ 0$ iff $A \succ 0$ and $C - B^\top A^{-1} B \succ 0$.
(2) If $A \succ 0$, then $M \succeq 0$ iff $C - B^\top A^{-1} B \succeq 0$.

Here is an illustration of Proposition 7.4(2). Consider the nonlinear quadratic constraint

$$(Ax + b)^\top (Ax + b) \leq c^\top x + d,$$

were $A \in \mathrm{M}_n(\mathbb{R})$, $x, b, c \in \mathbb{R}^n$ and $d \in \mathbb{R}$. Since obviously $I = I_n$ is invertible and $I \succ 0$, we have

$$\begin{pmatrix} I & Ax + b \\ (Ax + b)^\top & c^\top x + d \end{pmatrix} \succeq 0$$

iff $c^\top x + d - (Ax + b)^\top (Ax + b) \succeq 0$ iff $(Ax + b)^\top (Ax + b) \leq c^\top x + d$, since the matrix (a scalar) $c^\top x + d - (Ax + b)^\top (Ax + b)$ is the Schur complement of I in the above matrix.

The trick of using Schur complements to convert nonlinear inequality constraints into linear constraints on symmetric matrices involving the semidefinire ordering $\succeq$ is used extensively to convert nonlinear problems into semidefinite programs; see Boyd and Vandenberghe [Boyd and Vandenberghe (2004)].

When C is singular (or A is singular), it is still possible to characterize when a symmetric matrix M as above is positive semidefinite, but this requires using a version of the Schur complement involving the pseudo-inverse of C, namely $A - BC^+B^\top$ (or the Schur complement, $C - B^\top A^+ B$, of A). We use the criterion of Proposition 6.5, which tells us when a quadratic function of the form $\frac{1}{2}x^\top Px - x^\top b$ has a minimum and what this optimum value is (where P is a symmetric matrix).

7.3 Symmetric Positive Semidefinite Matrices and Schur Complements

We now return to our original problem, characterizing when a symmetric matrix

$$M = \begin{pmatrix} A & B \\ B^\top & C \end{pmatrix}$$

is positive semidefinite. Thus, we want to know when the function

$$f(x, y) = (x^\top \ y^\top) \begin{pmatrix} A & B \\ B^\top & C \end{pmatrix} \begin{pmatrix} x \\ y \end{pmatrix} = x^\top Ax + 2x^\top By + y^\top Cy$$

has a minimum with respect to both x and y. If we hold y constant, Proposition 6.5 implies that $f(x, y)$ has a minimum iff $A \succeq 0$ and $(I - AA^+)By = 0$, and then the minimum value is

$$f(x^*, y) = -y^\top B^\top A^+ By + y^\top Cy = y^\top (C - B^\top A^+ B)y.$$

Since we want $f(x, y)$ to be uniformly bounded from below for all x, y, we must have $(I - AA^+)B = 0$. Now $f(x^*, y)$ has a minimum iff $C - B^\top A^+ B \succeq 0$. Therefore, we have established that $f(x, y)$ has a minimum over all x, y iff

$$A \succeq 0, \quad (I - AA^+)B = 0, \quad C - B^\top A^+ B \succeq 0.$$

Similar reasoning applies if we first minimize with respect to y and then with respect to x, but this time, the Schur complement $A - BC^+B^\top$ of C is involved. Putting all these facts together, we get our main result:

Theorem 7.1. *Given any symmetric matrix*

$$M = \begin{pmatrix} A & B \\ B^\top & C \end{pmatrix}$$

the following conditions are equivalent:

(1) $M \succeq 0$ *(M is positive semidefinite).*
(2) $A \succeq 0, \quad (I - AA^+)B = 0, \quad C - B^\top A^+ B \succeq 0.$
(3) $C \succeq 0, \quad (I - CC^+)B^\top = 0, \quad A - BC^+ B^\top \succeq 0.$

If $M \succeq 0$ as in Theorem 7.1, then it is easy to check that we have the following factorizations (using the fact that $A^+AA^+ = A^+$ and $C^+CC^+ = C^+$):

$$\begin{pmatrix} A & B \\ B^\top & C \end{pmatrix} = \begin{pmatrix} I & BC^+ \\ 0 & I \end{pmatrix} \begin{pmatrix} A - BC^+B^\top & 0 \\ 0 & C \end{pmatrix} \begin{pmatrix} I & 0 \\ C^+B^\top & I \end{pmatrix}$$

and

$$\begin{pmatrix} A & B \\ B^\top & C \end{pmatrix} = \begin{pmatrix} I & 0 \\ B^\top A^+ & I \end{pmatrix} \begin{pmatrix} A & 0 \\ 0 & C - B^\top A^+ B \end{pmatrix} \begin{pmatrix} I & A^+B \\ 0 & I \end{pmatrix}.$$

7.4 Summary

The main concepts and results of this chapter are listed below:

- Schur complements.
- The matrix inversion lemma.
- Symmetric positive definite matrices and Schur complements.
- Symmetric positive semidefinite matrices and Schur complements.

7.5 Problems

Problem 7.1. Prove that maximizing the function $g(\lambda)$ given by

$$g(\lambda) = c_0 + \lambda c_1 - (b_0 + \lambda b_1)^\top (A_0 + \lambda A_1)^+ (b_0 + \lambda b_1),$$

subject to

$$A_0 + \lambda A_1 \succeq 0, \quad b_0 + \lambda b_1 \in \mathrm{range}(A_0 + \lambda A_1),$$

with A_0, A_1 some $n \times n$ symmetric positive semidefinite matrices, $b_0, b_1 \in \mathbb{R}^n$, and $c_0, c_1 \in \mathbb{R}$, is equivalent to maximizing γ subject to the constraints

$$\lambda \geq 0$$
$$\begin{pmatrix} A_0 + \lambda A_1 & b_0 + \lambda b_1 \\ (b_0 + \lambda b_1)^\top & c_0 + \lambda c_1 - \gamma \end{pmatrix} \succeq 0.$$

Problem 7.2. Let $a_1, \ldots, a_m$ be m vectors in $\mathbb{R}^n$ and assume that they span $\mathbb{R}^n$.

(1) Prove that the matrix

$$\sum_{k=1}^{m} a_k a_k^\top$$

is symmetric positive definite.

(2) Define the matrix X by

$$X = \left(\sum_{k=1}^{m} a_k a_k^\top \right)^{-1}.$$

Prove that

$$\begin{pmatrix} \sum_{k=1}^{m} a_k a_k^\top & a_i \\ a_i^\top & 1 \end{pmatrix} \succeq 0, \quad i = 1, \ldots, m.$$

Deduce that

$$a_i^\top X a_i \leq 1, \quad 1 \leq i \leq m.$$

Problem 7.3. Consider the function g of Example 3.10 defined by

$$g(a, b, c) = \log(ac - b^2),$$

where $ac - b^2 > 0$. We found that the Hessian matrix of g is given by

$$Hg(a, b, c) = \frac{1}{(ac - b^2)^2} \begin{pmatrix} -c^2 & 2bc & -b^2 \\ 2bc & -2(b^2 + ac) & 2ab \\ -b^2 & 2ab & -a^2 \end{pmatrix}.$$

Use the Schur complement (of a^2) to prove that the matrix $-Hg(a, b, c)$ is symmetric positive definite if $ac - b^2 > 0$ and $a, c > 0$.

PART 2
Linear Optimization

Chapter 8

Convex Sets, Cones, $\mathcal{H}$-Polyhedra

8.1 What is Linear Programming?

What is *linear programming*? At first glance, one might think that this is some style of computer programming. After all, there is imperative programming, functional programming, object-oriented programming, *etc.* The term linear programming is somewhat misleading, because it really refers to a method for *planning* with linear constraints, or more accurately, an *optimization method* where both the objective function and the constraints are linear.[1]

Linear programming was created in the late 1940's, one of the key players being George Dantzing, who invented the simplex algorithm. Kantorovitch also did some pioneering work on linear programming as early as 1939. The term *linear programming* has a military connotation because in the early 1950's it was used as a synonym for plans or schedules for training troops, logistical supply, resource allocation, *etc.* Unfortunately the term linear programming is well established and we are stuck with it.

Interestingly, even though originally most applications of linear programming were in the field of economics and industrial engineering, linear programming has become an important tool in theoretical computer science and in the theory of algorithms. Indeed, linear programming is often an effective tool for designing approximation algorithms to solve hard problems (typically NP-hard problems). Linear programming is also the "baby version" of convex programming, a very effective methodology which has received much attention in recent years.

Our goal is to present the mathematical underpinnings of linear

[1]Again, we witness another unfortunate abuse of terminology; the constraints are in fact *affine*.

programming, in particular the existence of an optimal solution if a linear program is feasible and bounded, and the duality theorem in linear programming, one of the deepest results in this field. The duality theorem in linear programming also has significant algorithmic implications but we do not discuss this here. We present the simplex algorithm, the dual simplex algorithm, and the primal dual algorithm. We also describe the tableau formalism for running the simplex algorithm and its variants. A particularly nice feature of the tableau formalism is that the update of a tableau can be performed using elementary row operations identical to the operations used during the reduction of a matrix to row reduced echelon form (rref). What differs is the criterion for the choice of the pivot.

However, we do not discuss other methods such as the ellipsoid method or interior points methods. For these more algorithmic issues, we refer the reader to standard texts on linear programming. In our opinion, one of the clearest (and among the most concise!) is Matousek and Gardner [Matousek and Gartner (2007)]; Chvatal [Chvatal (1983)] and Schrijver [Schrijver (1999)] are classics. Papadimitriou and Steiglitz [Papadimitriou and Steiglitz (1998)] offers a very crisp presentation in the broader context of combinatorial optimization, and Bertsimas and Tsitsiklis [Bertsimas and Tsitsiklis (1997)] and Vanderbei [Vanderbei (2014)] are very complete.

Linear programming has to do with maximizing a linear cost function $c_1x_1 + \cdots + c_nx_n$ with respect to m "linear" inequalities of the form

$$a_{i1}x_1 + \cdots + a_{in}x_n \leq b_i.$$

These constraints can be put together into an $m \times n$ matrix $A = (a_{ij})$, and written more concisely as

$$Ax \leq b.$$

For technical reasons that will appear clearer later on, it is often preferable to add the nonnegativity constraints $x_i \geq 0$ for $i = 1, \ldots, n$. We write $x \geq 0$. It is easy to show that every linear program is equivalent to another one satisfying the constraints $x \geq 0$, at the expense of adding new variables that are also constrained to be nonnegative. Let $\mathcal{P}(A, b)$ be the set of *feasible solutions* of our linear program given by

$$\mathcal{P}(A, b) = \{x \in \mathbb{R}^n \mid Ax \leq b,\ x \geq 0\}.$$

Then there are two basic questions:

(1) Is $\mathcal{P}(A, b)$ nonempty, that is, does our linear program have a chance to have a solution?

(2) Does the objective function $c_1x_1 + \cdots + c_nx_n$ have a maximum value on $\mathcal{P}(A, b)$?

The answer to both questions can be **no**. But if $\mathcal{P}(A, b)$ is nonempty and if the objective function is bounded above (on $\mathcal{P}(A, b)$), then it can be shown that the maximum of $c_1x_1 + \cdots + c_nx_n$ is achieved by some $x \in \mathcal{P}(A, b)$. Such a solution is called an *optimal solution*. Perhaps surprisingly, this result is not so easy to prove (unless one has the simplex method at his disposal). We will prove this result in full detail (see Proposition 9.1).

The reason why linear constraints are so important is that the domain of potential optimal solutions $\mathcal{P}(A, b)$ is *convex*. In fact, $\mathcal{P}(A, b)$ is a convex polyhedron which is the intersection of half-spaces cut out by affine hyperplanes. The objective function being linear is convex, and this is also a crucial fact. Thus, we are led to study convex sets, in particular those that arise from solutions of inequalities defined by affine forms, but also convex cones.

We give a brief introduction to these topics. As a reward, we provide several criteria for testing whether a system of inequalities

$$Ax \leq b, \; x \geq 0$$

has a solution or not in terms of versions of the *Farkas lemma* (see Proposition 14.3 and Proposition 11.4). Then we give a complete proof of the strong duality theorem for linear programming (see Theorem 11.1). We also discuss the complementary slackness conditions and show that they can be exploited to design an algorithm for solving a linear program that uses both the primal problem and its dual. This algorithm known as the *primal dual algorithm*, although not used much nowadays, has been the source of inspiration for a whole class of approximation algorithms also known as primal dual algorithms.

We hope that this chapter and the next three will be a motivation for learning more about linear programming, convex optimization, but also convex geometry. The "bible" in convex optimization is Boyd and Vandenberghe [Boyd and Vandenberghe (2004)], and one of the best sources for convex geometry is Ziegler [Ziegler (1997)]. This is a rather advanced text, so the reader may want to begin with Gallier [Gallier (2016)].

8.2 Affine Subsets, Convex Sets, Affine Hyperplanes, Half-Spaces

We view $\mathbb{R}^n$ as consisting of *column vectors* ($n \times 1$ matrices). As usual, row vectors represent *linear forms*, that is linear maps $\varphi\colon \mathbb{R}^n \to \mathbb{R}$, in the sense that the row vector y (a $1 \times n$ matrix) represents the linear form φ if $\varphi(x) = yx$ for all $x \in \mathbb{R}^n$. We denote the space of linear forms (row vectors) by $(\mathbb{R}^n)^*$.

Recall that a *linear combination* of vectors in $\mathbb{R}^n$ is an expression

$$\lambda_1 x_1 + \cdots + \lambda_m x_m$$

where $x_1, \ldots, x_m \in \mathbb{R}^n$ and where $\lambda_1, \ldots, \lambda_m$ are *arbitrary* scalars in $\mathbb{R}$. Given a sequence of vectors $S = (x_1, \ldots, x_m)$ with $x_i \in \mathbb{R}^n$, the set of all linear combinations of the vectors in S is the smallest (linear) subspace containing S called the *linear span* of S, and denoted $\mathrm{span}(S)$. A *linear subspace* of $\mathbb{R}^n$ is any nonempty subset of $\mathbb{R}^n$ closed under linear combinations.

Definition 8.1. An *affine combination* of vectors in $\mathbb{R}^n$ is an expression

$$\lambda_1 x_1 + \cdots + \lambda_m x_m$$

where $x_1, \ldots, x_m \in \mathbb{R}^n$ and where $\lambda_1, \ldots, \lambda_m$ are scalars in $\mathbb{R}$ *satisfying the condition*

$$\lambda_1 + \cdots + \lambda_m = 1.$$

Given a sequence of vectors $S = (x_1, \ldots, x_m)$ with $x_i \in \mathbb{R}^n$, the set of all affine combinations of the vectors in S is the smallest affine subspace containing S called the *affine hull* of S and denoted $\mathrm{aff}(S)$.

Definition 8.2. An *affine subspace* A of $\mathbb{R}^n$ is any subset of $\mathbb{R}^n$ closed under affine combinations.

If A is a nonempty affine subspace of $\mathbb{R}^n$, then it can be shown that $V_A = \{a - b \mid a, b \in A\}$ is a linear subspace of $\mathbb{R}^n$ and that

$$A = a + V_A = \{a + v \mid v \in V_A\}$$

for *any* $a \in A$; see Gallier [Gallier (2011)] (Section 2.5).

Definition 8.3. Given an affine subspace A, the linear space $V_A = \{a - b \mid a, b \in A\}$ is called the *direction of* A. The *dimension* of the nonempty affine subspace A is the dimension of its direction V_A.

Definition 8.4. *Convex combinations* are affine combinations $\lambda_1 x_1 + \cdots + \lambda_m x_m$ satisfying the extra condition that $\lambda_i \geq 0$ for $i = 1, \ldots, m$.

A convex set is defined as follows.

Definition 8.5. A subset V of $\mathbb{R}^n$ is *convex* if for any two points $a, b \in V$, we have $c \in V$ for every point $c = (1 - \lambda)a + \lambda b$, with $0 \leq \lambda \leq 1$ ($\lambda \in \mathbb{R}$). Given any two points a, b, the notation $[a, b]$ is often used to denote the line segment between a and b, that is,

$$[a, b] = \{c \in \mathbb{R}^n \mid c = (1 - \lambda)a + \lambda b,\ 0 \leq \lambda \leq 1\},$$

and thus a set V is convex if $[a, b] \subseteq V$ for any two points $a, b \in V$ ($a = b$ is allowed). The *dimension* of a convex set V is the dimension of its affine hull $\text{aff}(A)$.

The empty set is trivially convex, every one-point set $\{a\}$ is convex, and the entire affine space $\mathbb{R}^n$ is convex.

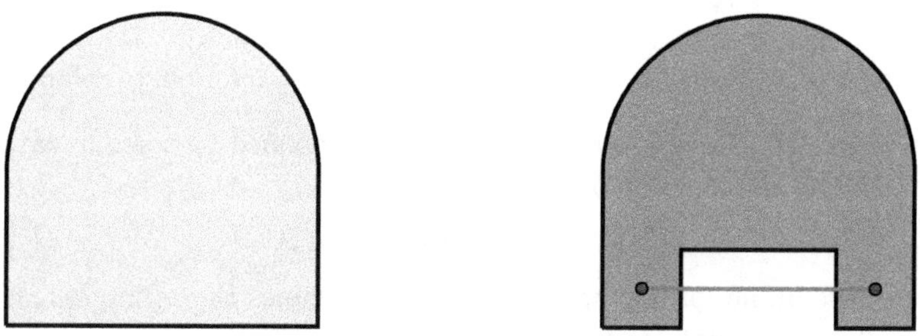

(a) (b)

Fig. 8.1 (a) A convex set; (b) A nonconvex set.

It is obvious that the intersection of any family (finite or infinite) of convex sets is convex.

Definition 8.6. Given any (nonempty) subset S of $\mathbb{R}^n$, the smallest convex set containing S is denoted by $\text{conv}(S)$ and called the *convex hull of S* (it is the intersection of all convex sets containing S).

It is essential not only to have a good understanding of $\text{conv}(S)$, but to also have good methods for computing it. We have the following simple but crucial result.

Proposition 8.1. *For any family $S = (a_i)_{i\in I}$ of points in $\mathbb{R}^n$, the set V of convex combinations $\sum_{i\in I} \lambda_i a_i$ (where $\sum_{i\in I} \lambda_i = 1$ and $\lambda_i \geq 0$) is the convex hull* conv(S) *of $S = (a_i)_{i\in I}$.*

It is natural to wonder whether Proposition 8.1 can be sharpened in two directions: (1) Is it possible to have a fixed bound on the number of points involved in the convex combinations? (2) Is it necessary to consider convex combinations of all points, or is it possible to consider only a subset with special properties?

The answer is yes in both cases. In Case 1, Carathéodory's theorem asserts that it is enough to consider convex combinations of $n+1$ points. For example, in the plane $\mathbb{R}^2$, the convex hull of a set S of points is the union of all triangles (interior points included) with vertices in S. In Case 2, the theorem of Krein and Milman asserts that a convex setthat is also compact is the convex hull of its extremal points (given a convex set S, a point $a \in S$ is *extremal* if $S - \{a\}$ is also convex).

We will not prove these theorems here, but we invite the reader to consult Gallier [Gallier (2016)] or Berger [Berger (1990b)].

Convex sets also arise as half-spaces cut out by affine hyperplanes.

Definition 8.7. An *affine form* $\varphi\colon \mathbb{R}^n \to \mathbb{R}$ is defined by some linear form $c \in (\mathbb{R}^n)^*$ and some scalar $\beta \in \mathbb{R}$ so that

$$\varphi(x) = cx + \beta \quad \text{for all } x \in \mathbb{R}^n.$$

If $c \neq 0$, the affine form φ specified by (c, β) defines the *affine hyperplane* (for short *hyperplane*) $H(\varphi)$ given by

$$H(\varphi) = \{x \in \mathbb{R}^n \mid \varphi(x) = 0\} = \{x \in \mathbb{R}^n \mid cx + \beta = 0\},$$

and the two *(closed) half-spaces*

$$H_+(\varphi) = \{x \in \mathbb{R}^n \mid \varphi(x) \geq 0\} = \{x \in \mathbb{R}^n \mid cx + \beta \geq 0\},$$
$$H_-(\varphi) = \{x \in \mathbb{R}^n \mid \varphi(x) \leq 0\} = \{x \in \mathbb{R}^n \mid cx + \beta \leq 0\}.$$

When $\beta = 0$, we call H a *linear hyperplane*.

Both $H_+(\varphi)$ and $H_-(\varphi)$ are convex and $H = H_+(\varphi) \cap H_-(\varphi)$.

For example, $\varphi\colon \mathbb{R}^2 \to \mathbb{R}$ with $\varphi(x, y) = 2x + y + 3$ is an affine form defining the line given by the equation $y = -2x - 3$. Another example of an affine form is $\varphi\colon \mathbb{R}^3 \to \mathbb{R}$ with $\varphi(x, y, z) = x + y + z - 1$; this affine form defines the plane given by the equation $x + y + z = 1$, which is the plane through the points $(0, 0, 1), (0, 1, 0)$, and $(1, 0, 0)$. Both of these hyperplanes are illustrated in Figure 8.2.

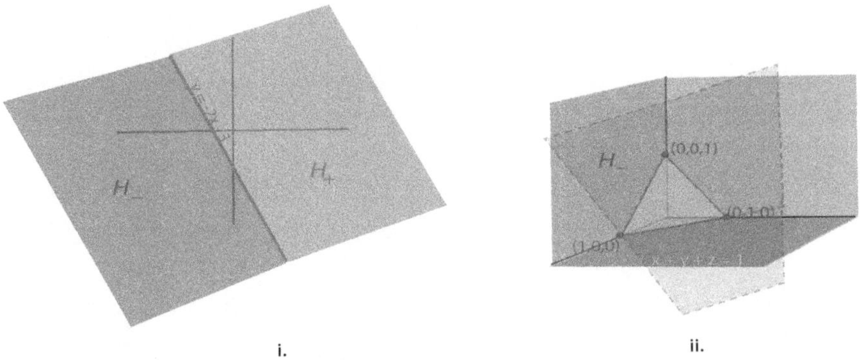

Fig. 8.2 Figure i. illustrates the hyperplane $H(\varphi)$ for $\varphi(x, y) = 2x + y + 3$, while Figure ii. illustrates the hyperplane $H(\varphi)$ for $\varphi(x, y, z) = x + y + z - 1$.

Definition 8.8. For any two vector $x, y \in \mathbb{R}^n$ with $x = (x_1, \ldots, x_n)$ and $y = (y_1, \ldots, y_n)$ we write $x \leq y$ iff $x_i \leq y_i$ for $i = 1, \ldots, n$, and $x \geq y$ iff $y \leq x$. In particular $x \geq 0$ iff $x_i \geq 0$ for $i = 1, \ldots, n$.

Certain special types of convex sets called cones and $\mathcal{H}$-polyhedra play an important role. The set of feasible solutions of a linear program is an $\mathcal{H}$-polyhedron, and cones play a crucial role in the proof of Proposition 9.1 and in the Farkas–Minkowski proposition (Proposition 11.2).

8.3 Cones, Polyhedral Cones, and $\mathcal{H}$-Polyhedra

Cones and polyhedral cones are defined as follows.

Definition 8.9. Given a nonempty subset $S \subseteq \mathbb{R}^n$, the *cone* $C = \text{cone}(S)$ spanned by S is the convex set

$$\text{cone}(S) = \left\{ \sum_{i=1}^{k} \lambda_i u_i,\ u_i \in S, \lambda_i \in \mathbb{R},\ \lambda_i \geq 0 \right\},$$

of positive combinations of vectors from S. If S consists of a finite set of vectors, the cone $C = \text{cone}(S)$ is called a *polyhedral cone*. Figure 8.3 illustrates a polyhedral cone.

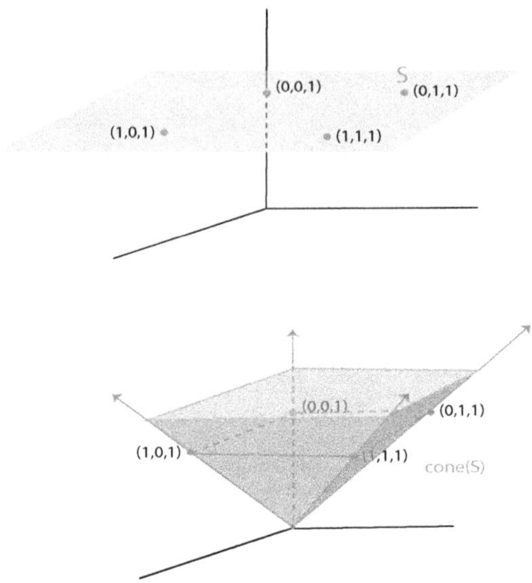

Fig. 8.3 Let $S = \{(0,0,1), (1,0,1), (1,1,1), (0,1,1)\}$. The polyhedral cone, cone(S), is the solid "pyramid" with apex at the origin and square cross sections.

Note that if some nonzero vector u belongs to a cone C, then $\lambda u \in C$ for all $\lambda \geq 0$, that is, the *ray* $\{\lambda u \mid \lambda \geq 0\}$ belongs to C.

Remark: The cones (and polyhedral cones) of Definition 8.9 are *always convex.* For this reason, we use the simpler terminology cone instead of convex cone. However, there are more general kinds of cones (see Definition 14.1) that are not convex (for example, a union of polyhedral cones or the linear cone generated by the curve in Figure 8.4), and if we were dealing with those we would refer to the cones of Definition 8.9 as convex cones.

Definition 8.10. An *$\mathcal{H}$-polyhedron*, for short a *polyhedron*, is any subset $\mathcal{P} = \bigcap_{i=1}^{s} C_i$ of $\mathbb{R}^n$ defined as the intersection of a finite number s of closed half-spaces C_i. An example of an $\mathcal{H}$-polyhedron is shown in Figure 8.6. An *$\mathcal{H}$-polytope* is a bounded $\mathcal{H}$-polyhedron, which means that there is a closed ball $B_r(x)$ of center x and radius $r > 0$ such that $\mathcal{P} \subseteq B_r(x)$. An example of a $\mathcal{H}$-polytope is shown in Figure 8.5.

By convention, we agree that $\mathbb{R}^n$ itself is an $\mathcal{H}$-polyhedron.

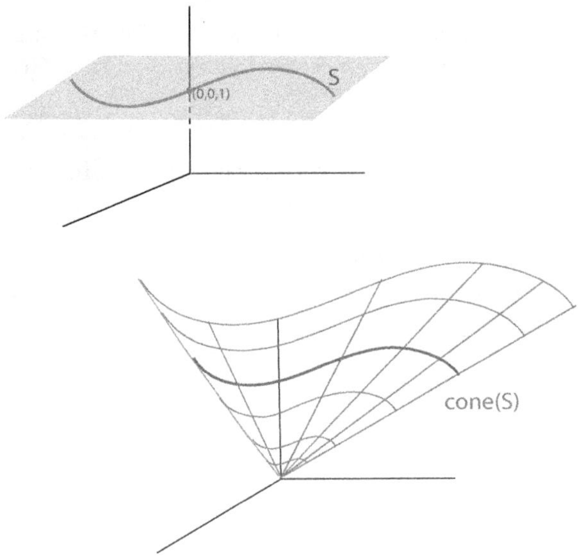

Fig. 8.4 Let S be a planar curve in $z = 1$. The linear cone of S, consisting of all half rays connecting S to the origin, is not convex.

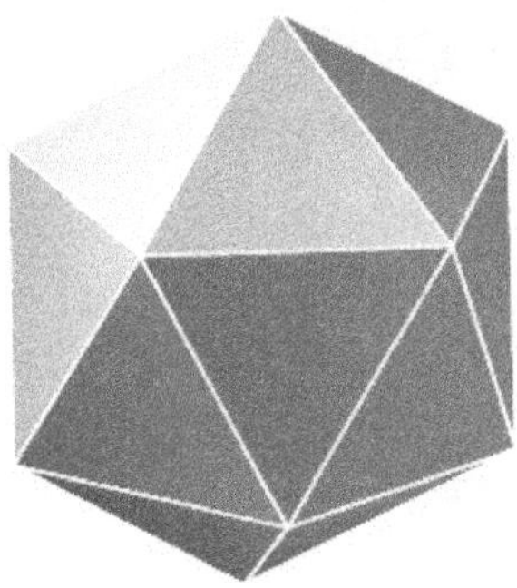

Fig. 8.5 An icosahedron is an example of an $\mathcal{H}$-polytope.

Remark: The $\mathcal{H}$-polyhedra of Definition 8.10 are always convex. For this reason, as in the case of cones we use the simpler terminology $\mathcal{H}$-polyhedron instead of convex $\mathcal{H}$-polyhedron. In algebraic topology, there are more general polyhedra that are not convex.

It can be shown that an $\mathcal{H}$-polytope $\mathcal{P}$ is equal to the convex hull of finitely many points (the extreme points of $\mathcal{P}$). This is a nontrivial result whose proof takes a significant amount of work; see Gallier [Gallier (2016)] and Ziegler [Ziegler (1997)].

An unbounded $\mathcal{H}$-polyhedron is not equal to the convex hull of finite set of points. To obtain an equivalent notion we introduce the notion of a $\mathcal{V}$-polyhedron.

Definition 8.11. A *$\mathcal{V}$-polyhedron* is any convex subset $A \subseteq \mathbb{R}^n$ of the form

$$A = \operatorname{conv}(Y) + \operatorname{cone}(V) = \{a + v \mid a \in \operatorname{conv}(Y),\ v \in \operatorname{cone}(V)\},$$

where $Y \subseteq \mathbb{R}^n$ and $V \subseteq \mathbb{R}^n$ are *finite* (possibly empty).

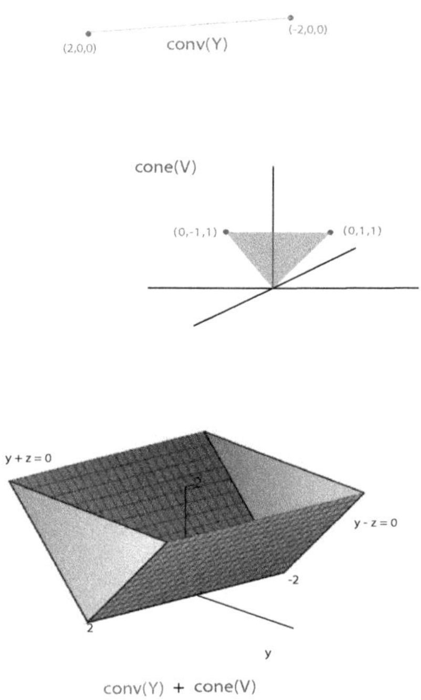

Fig. 8.6 The "triangular trough" determined by the inequalities $y - z \leq 0$, $y + z \geq 0$, and $-2 \leq x \leq 2$ is an $\mathcal{H}$-polyhedron and an $\mathcal{V}$-polyhedron, where $Y = \{(2,0,0), (-2,0,0)\}$ and $V = \{(0,1,1), (0,-1,1)\}$.

When $V = \emptyset$ we simply have a *polytope*, and when $Y = \emptyset$ or $Y = \{0\}$, we simply have a cone.

It can be shown that every $\mathcal{H}$-polyhedron is a $\mathcal{V}$-polyhedron and conversely. This is one of the major theorems in the theory of polyhedra, and its proof is nontrivial. For a complete proof, see Gallier [Gallier (2016)] and Ziegler [Ziegler (1997)].

Every polyhedral cone is closed. This is an important fact that is used in the proof of several other key results such as Proposition 9.1 and the Farkas–Minkowski proposition (Proposition 11.2).

Although it seems obvious that a polyhedral cone should be closed, a rigorous proof is not entirely trivial.

Indeed, the fact that a polyhedral cone is closed relies crucially on the fact that C is spanned by a *finite* number of vectors, because the cone generated by an infinite set may not be closed. For example, consider the closed disk $D \subseteq \mathbb{R}^2$ of center $(0,1)$ and radius 1, which is tangent to the x-axis at the origin. Then the cone(D) consists of the open upper half-plane *plus* the origin $(0,0)$, but this set is not closed.

Proposition 8.2. *Every polyhedral cone C is closed.*

Proof. This is proven by showing that

(1) Every primitive cone is closed, where a *primitive cone* is a polyhedral cone spanned by linearly independent vectors.
(2) A polyhedral cone C is the union of finitely many primitive cones.

Assume that $(a_1, \ldots, a_m)$ are linearly independent vectors in $\mathbb{R}^n$, and consider any sequence $(x^{(k)})_{k\geq 0}$

$$x^{(k)} = \sum_{i=1}^{m} \lambda_i^{(k)} a_i$$

of vectors in the primitive cone cone$(\{a_1, \ldots, a_m\})$, which means that $\lambda_j^{(k)} \geq 0$ for $i = 1, \ldots, m$ and all $k \geq 0$. The vectors $x^{(k)}$ belong to the subspace U spanned by $(a_1, \ldots, a_m)$, and U is closed. Assume that the sequence $(x^{(k)})_{k\geq 0}$ converges to a limit $x \in \mathbb{R}^n$. Since U is closed and $x^{(k)} \in U$ for all $k \geq 0$, we have $x \in U$. If we write $x = x_1 a_1 + \cdots + x_m a_m$, we would like to prove that $x_i \geq 0$ for $i = 1, \ldots, m$. The sequence the $(x^{(k)})_{k\geq 0}$ converges to x iff

$$\lim_{k \mapsto \infty} \left\| x^{(k)} - x \right\| = 0,$$

iff

$$\lim_{k \mapsto \infty} \Big(\sum_{i=1}^{m} |\lambda_i^{(k)} - x_i|^2 \Big)^{1/2} = 0$$

iff

$$\lim_{k \mapsto \infty} \lambda_i^{(k)} = x_i, \quad i = 1, \ldots, m.$$

Since $\lambda_i^{(k)} \geq 0$ for $i = 1, \ldots, m$ and all $k \geq 0$, we have $x_i \geq 0$ for $i = 1, \ldots, m$, so $x \in \text{cone}(\{a_1, \ldots, a_m\})$.

Next, assume that x belongs to the polyhedral cone C. Consider a positive combination

$$x = \lambda_1 a_1 + \cdots + \lambda_k a_k, \tag{$*_1$}$$

for some nonzero $a_1, \ldots, a_k \in C$, with $\lambda_i \geq 0$ and with k *minimal*. Since k is minimal, we must have $\lambda_i > 0$ for $i = 1, \ldots, k$. We claim that $(a_1, \ldots, a_k)$ are linearly independent.

If not, there is some nontrivial linear combination

$$\mu_1 a_1 + \cdots + \mu_k a_k = 0, \tag{$*_2$}$$

and since the a_i are nonzero, $\mu_j \neq 0$ for some at least some j. We may assume that $\mu_j < 0$ for some j (otherwise, we consider the family $(-\mu_i)_{1 \leq i \leq k}$), so let

$$J = \{j \in \{1, \ldots, k\} \mid \mu_j < 0\}.$$

For any $t \in \mathbb{R}$, since $x = \lambda_1 a_1 + \cdots + \lambda_k a_k$, using $(*_2)$ we get

$$x = (\lambda_1 + t\mu_1) a_1 + \cdots + (\lambda_k + t\mu_k) a_k, \tag{$*_3$}$$

and if we pick

$$t = \min_{j \in J} \left(-\frac{\lambda_j}{\mu_j} \right) \geq 0,$$

we have $(\lambda_i + t\mu_i) \geq 0$ for $i = 1, \ldots, k$, but $\lambda_j + t\mu_j = 0$ for some $j \in J$, so $(*_3)$ is an expression of x with less that k nonzero coefficients, contradicting the minimality of k in $(*_1)$. Therefore, $(a_1, \ldots, a_k)$ are linearly independent.

Since a polyhedral cone C is spanned by finitely many vectors, there are finitely many primitive cones (corresponding to linearly independent subfamilies), and since every $x \in C$, belongs to some primitive cone, C is the union of a finite number of primitive cones. Since every primitive cone is closed, as a union of finitely many closed sets, C itself is closed.

The above facts are also proven in Matousek and Gardner [Matousek and Gartner (2007)] (Chapter 6, Section 5, Lemmas 6.5.3, 6.5.4, and 6.5.5). □

Another way to prove that a polyhedral cone C is closed is to show that C is also a $\mathcal{H}$-polyhedron. This takes even more work; see Gallier [Gallier (2016)] (Chapter 4, Section 4, Proposition 4.16). Yet another proof is given in Lax [Lax (2007)] (Chapter 13, Theorem 1).

8.4 Summary

The main concepts and results of this chapter are listed below:

- Affine combination.
- Affine hull.
- Affine subspace; direction of an affine subspace, dimension of an affine subspace.
- Convex combination.
- Convex set, dimension of a convex set.
- Convex hull.
- Affine form.
- Affine hyperplane, half-spaces.
- Cone, polyhedral cone.
- $\mathcal{H}$-polyhedron, $\mathcal{H}$-polytope.
- $\mathcal{V}$-polyhedron, polytope.
- Primitive cone.

8.5 Problems

Problem 8.1. Prove Proposition 8.1.

Problem 8.2. Describe an icosahedron both as an $\mathcal{H}$-polytope and as a $\mathcal{V}$-polytope. Do the same thing for a dodecahedron. What do you observe?

Chapter 9

Linear Programs

In this chapter we introduce linear programs and the basic notions relating to this concept. We define the $\mathcal{H}$-polyhedron $\mathcal{P}(A, b)$ of feasible solutions. Then we define bounded and unbounded linear programs and the notion of optimal solution. We define slack variables and the important notion of *linear program in standard form.*

We show that if a linear program in standard form has a feasible solution and is bounded above, then it has an optimal solution. This is not an obvious result and the proof relies on the fact that a polyhedral cone is closed (this result was shown in the previous chapter).

Next we show that in order to find optimal solutions it suffices to consider solutions of a special form called *basic feasible solutions.* We prove that if a linear program in standard form has a feasible solution and is bounded above, then some basic feasible solution is an optimal solution (Theorem 9.1).

Geometrically, a basic feasible solution corresponds to a *vertex.* In Theorem 9.2 we prove that a basic feasible solution of a linear program in standard form is a vertex of the polyhedron $\mathcal{P}(A, b)$. Finally, we prove that if a linear program in standard form has some feasible solution, then it has a basic feasible solution (see Theorem 9.3). This fact allows the simplex algorithm described in the next chapter to get started.

9.1 Linear Programs, Feasible Solutions, Optimal Solutions

The purpose of linear programming is to solve the following type of optimization problem.

Definition 9.1. *A Linear Program* (P) is the following kind of optimization problem:

$$\begin{aligned}
&\text{maximize} \quad cx\\
&\text{subject to}\\
&\quad a_1x \le b_1\\
&\quad \cdots\\
&\quad a_mx \le b_m\\
&\quad x \ge 0,
\end{aligned}$$

where $x \in \mathbb{R}^n$, $c, a_1, \ldots, a_m \in (\mathbb{R}^n)^*$, $b_1, \ldots, b_m \in \mathbb{R}$.

The linear form c defines the *objective function* $x \mapsto cx$ of the Linear Program (P) (from $\mathbb{R}^n$ to $\mathbb{R}$), and the inequalities $a_ix \le b_i$ and $x_j \ge 0$ are called the *constraints* of the Linear Program (P).

If we define the $m \times n$ matrix

$$A = \begin{pmatrix} a_1 \\ \vdots \\ a_m \end{pmatrix}$$

whose rows are the row vectors $a_1, \ldots, a_m$ and b as the column vector

$$b = \begin{pmatrix} b_1 \\ \vdots \\ b_m \end{pmatrix},$$

the m inequality constraints $a_ix \le b_i$ can be written in matrix form as

$$Ax \le b.$$

Thus the Linear Program (P) can also be stated as the Linear Program (P):

$$\begin{aligned}
&\text{maximize} \quad cx\\
&\text{subject to} \quad Ax \le b \text{ and } x \ge 0.
\end{aligned}$$

We should note that in many applications, the natural primal optimization problem is actually the *minimization* of some objective function $cx = c_1x_1 + \cdots + c_nx_n$, rather its maximization. For example, many of the optimization problems considered in Papadimitriou and Steiglitz [Papadimitriou and Steiglitz (1998)] are minimization problems.

Of course, minimizing cx is equivalent to maximizing $-cx$, so our presentation covers minimization too.

Here is an explicit example of a linear program of Type (P):

Example 9.1.

$$\begin{aligned} \text{maximize} \quad & x_1 + x_2 \\ \text{subject to} \quad & \\ & x_2 - x_1 \le 1 \\ & x_1 + 6x_2 \le 15 \\ & 4x_1 - x_2 \le 10 \\ & x_1 \ge 0,\ x_2 \ge 0, \end{aligned}$$

and in matrix form

$$\begin{aligned} \text{maximize} \quad & (1\ 1) \begin{pmatrix} x_1 \\ x_2 \end{pmatrix} \\ \text{subject to} \quad & \\ & \begin{pmatrix} -1 & 1 \\ 1 & 6 \\ 4 & -1 \end{pmatrix} \begin{pmatrix} x_1 \\ x_2 \end{pmatrix} \le \begin{pmatrix} 1 \\ 15 \\ 10 \end{pmatrix} \\ & x_1 \ge 0,\ x_2 \ge 0. \end{aligned}$$

It turns out that $x_1 = 3, x_2 = 2$ yields the maximum of the objective function $x_1 + x_2$, which is 5. This is illustrated in Figure 9.1. Observe that the set of points that satisfy the above constraints is a convex region cut out by half planes determined by the lines of equations

$$\begin{aligned} x_2 - x_1 &= 1 \\ x_1 + 6x_2 &= 15 \\ 4x_1 - x_2 &= 10 \\ x_1 &= 0 \\ x_2 &= 0. \end{aligned}$$

In general, each constraint $a_i x \le b_i$ corresponds to the affine form φ_i given by $\varphi_i(x) = a_i x - b_i$ and defines the half-space $H_-(\varphi_i)$, and each inequality $x_j \ge 0$ defines the half-space $H_+(x_j)$. The intersection of these half-spaces is the set of solutions of all these constraints. It is a (possibly empty) $\mathcal{H}$-polyhedron denoted $\mathcal{P}(A, b)$.

Definition 9.2. If $\mathcal{P}(A, b) = \emptyset$, we say that the Linear Program (P) has *no feasible solution*, and otherwise any $x \in \mathcal{P}(A, b)$ is called a *feasible solution* of (P).

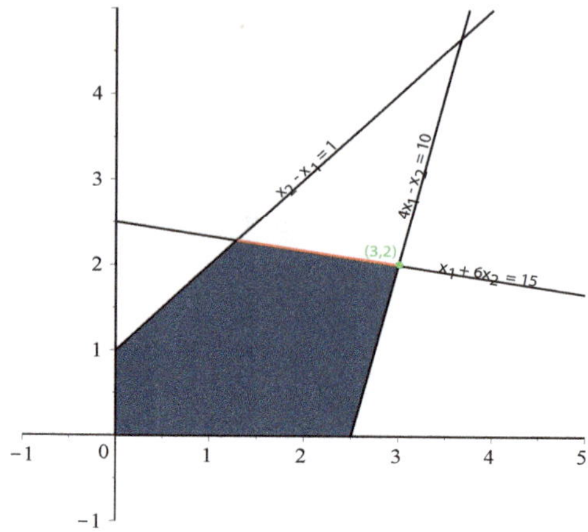

Fig. 9.1 The $\mathcal{H}$-polyhedron associated with Example 9.1. The green point $(3, 2)$ is the unique optimal solution.

The linear program shown in Example 9.2 obtained by reversing the direction of the inequalities $x_2 - x_1 \leq 1$ and $4x_1 - x_2 \leq 10$ in the linear program of Example 9.1 has no feasible solution; see Figure 9.2.

Example 9.2.

$$\begin{aligned} &\text{maximize} \quad x_1 + x_2 \\ &\text{subject to} \\ &\qquad x_1 - x_2 \leq -1 \\ &\qquad x_1 + 6x_2 \leq 15 \\ &\qquad x_2 - 4x_1 \leq -10 \\ &\qquad x_1 \geq 0,\ x_2 \geq 0. \end{aligned}$$

Assume $\mathcal{P}(A, b) \neq \emptyset$, so that the Linear Program (P) has a feasible solution. In this case, consider the image $\{cx \in \mathbb{R} \mid x \in \mathcal{P}(A, b)\}$ of $\mathcal{P}(A, b)$ under the objective function $x \mapsto cx$.

Definition 9.3. If the set $\{cx \in \mathbb{R} \mid x \in \mathcal{P}(A, b)\}$ is unbounded above, then we say that the Linear Program (P) is *unbounded*.

The linear program shown in Example 9.3 obtained from the linear

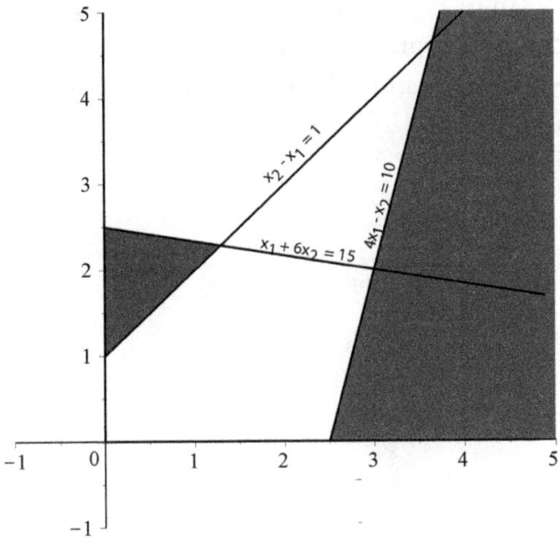

Fig. 9.2 There is no $\mathcal{H}$-polyhedron associated with Example 9.2 since the blue and purple regions do not overlap.

program of Example 9.1 by deleting the constraints $4x_1 - x_2 \leq 10$ and $x_1 + 6x_2 \leq 15$ is unbounded.

Example 9.3.

$$\begin{aligned} &\text{maximize} \quad x_1 + x_2 \\ &\text{subject to} \\ &\qquad\qquad x_2 - x_1 \leq 1 \\ &\qquad\qquad x_1 \geq 0,\ x_2 \geq 0. \end{aligned}$$

Otherwise, we will prove shortly that if μ is the least upper bound of the set $\{cx \in \mathbb{R} \mid x \in \mathcal{P}(A, b)\}$, then there is some $p \in \mathcal{P}(A, b)$ such that

$$cp = \mu,$$

that is, the objective function $x \mapsto cx$ has a maximum value μ on $\mathcal{P}(A, b)$ which is achieved by some $p \in \mathcal{P}(A, b)$.

Definition 9.4. If the set $\{cx \in \mathbb{R} \mid x \in \mathcal{P}(A, b)\}$ is nonempty and bounded above, any point $p \in \mathcal{P}(A, b)$ such that $cp = \max\{cx \in \mathbb{R} \mid x \in \mathcal{P}(A, b)\}$ is called an *optimal solution* (or *optimum*) of (P). Optimal solutions are often denoted by an upper $*$; for example, p^*.

The linear program of Example 9.1 has a unique optimal solution $(3, 2)$, but observe that the linear program of Example 9.4 in which the objective function is $(1/6)x_1 + x_2$ has infinitely many optimal solutions; the maximum of the objective function is $15/6$ which occurs along the points of orange boundary line in Figure 9.1.

Example 9.4.

$$\begin{aligned} &\text{maximize} \quad \frac{1}{6}x_1 + x_2 \\ &\text{subject to} \\ &\qquad x_2 - x_1 \leq 1 \\ &\qquad x_1 + 6x_2 \leq 15 \\ &\qquad 4x_1 - x_2 \leq 10 \\ &\qquad x_1 \geq 0, \ x_2 \geq 0. \end{aligned}$$

The proof that if the set $\{cx \in \mathbb{R} \mid x \in \mathcal{P}(A, b)\}$ is nonempty and bounded above, then there is an optimal solution $p \in \mathcal{P}(A, b)$, is not as trivial as it might seem. It relies on the fact that a polyhedral cone is closed, a fact that was shown in Section 8.3.

We also use a trick that makes the proof simpler, which is that a Linear Program (P) with inequality constraints $Ax \leq b$

$$\begin{aligned} &\text{maximize} \quad cx \\ &\text{subject to} \quad Ax \leq b \text{ and } x \geq 0, \end{aligned}$$

is equivalent to the Linear Program (P_2) with equality constraints

$$\begin{aligned} &\text{maximize} \quad \widehat{c}\,\widehat{x} \\ &\text{subject to} \quad \widehat{A}\widehat{x} = b \text{ and } \widehat{x} \geq 0, \end{aligned}$$

where $\widehat{A}$ is an $m \times (n + m)$ matrix, $\widehat{c}$ is a linear form in $(\mathbb{R}^{n+m})^*$, and $\widehat{x} \in \mathbb{R}^{n+m}$, given by

$$\widehat{A} = \begin{pmatrix} A & I_m \end{pmatrix}, \quad \widehat{c} = \begin{pmatrix} c & 0_m^\top \end{pmatrix}, \quad \text{and} \quad \widehat{x} = \begin{pmatrix} x \\ z \end{pmatrix},$$

with $x \in \mathbb{R}^n$ and $z \in \mathbb{R}^m$.

Indeed, $\widehat{A}\widehat{x} = b$ and $\widehat{x} \geq 0$ iff

$$Ax + z = b, \quad x \geq 0, \ z \geq 0,$$

iff

$$Ax \leq b, \quad x \geq 0,$$

and $\widehat{c}\,\widehat{x} = cx$.

Definition 9.5. The variables z are called *slack variables*, and a linear program of the form (P_2) is called a linear program in *standard form*.

The result of converting the linear program of Example 9.4 to standard form is the program shown in Example 9.5.

Example 9.5.

$$\begin{aligned}
&\text{maximize} \quad \frac{1}{6}x_1 + x_2 \\
&\text{subject to} \\
&\qquad x_2 - x_1 + z_1 = 1 \\
&\qquad x_1 + 6x_2 + z_2 = 15 \\
&\qquad 4x_1 - x_2 + z_3 = 10 \\
&\qquad x_1 \geq 0,\ x_2 \geq 0,\ z_1 \geq 0,\ z_2 \geq 0,\ z_3 \geq 0.
\end{aligned}$$

We can now prove that if a linear program has a feasible solution and is bounded, then it has an optimal solution.

Proposition 9.1. *Let (P_2) be a linear program in standard form, with equality constraint $Ax = b$. If $\mathcal{P}(A, b)$ is nonempty and bounded above, and if μ is the least upper bound of the set $\{cx \in \mathbb{R} \mid x \in \mathcal{P}(A, b)\}$, then there is some $p \in \mathcal{P}(A, b)$ such that*

$$cp = \mu,$$

that is, the objective function $x \mapsto cx$ has a maximum value μ on $\mathcal{P}(A, b)$ which is achieved by some optimum solution $p \in \mathcal{P}(A, b)$.

Proof. Since $\mu = \sup\{cx \in \mathbb{R} \mid x \in \mathcal{P}(A, b)\}$, there is a sequence $(x^{(k)})_{k \geq 0}$ of vectors $x^{(k)} \in \mathcal{P}(A, b)$ such that $\lim_{k \mapsto \infty} cx^{(k)} = \mu$. In particular, if we write $x^{(k)} = (x_1^{(k)}, \ldots, x_n^{(k)})$ we have $x_j^{(k)} \geq 0$ for $j = 1, \ldots, n$ and for all $k \geq 0$. Let $\widetilde{A}$ be the $(m + 1) \times n$ matrix

$$\widetilde{A} = \begin{pmatrix} c \\ A \end{pmatrix},$$

and consider the sequence $(\widetilde{A}x^{(k)})_{k \geq 0}$ of vectors $\widetilde{A}x^{(k)} \in \mathbb{R}^{m+1}$. We have

$$\widetilde{A}x^{(k)} = \begin{pmatrix} c \\ A \end{pmatrix} x^{(k)} = \begin{pmatrix} cx^{(k)} \\ Ax^{(k)} \end{pmatrix} = \begin{pmatrix} cx^{(k)} \\ b \end{pmatrix},$$

since by hypothesis $x^{(k)} \in \mathcal{P}(A,b)$, and the constraints are $Ax = b$ and $x \geq 0$. Since by hypothesis $\lim_{k \mapsto \infty} cx^{(k)} = \mu$, the sequence $(\widetilde{A}x^{(k)})_{k \geq 0}$ converges to the vector $\begin{pmatrix} \mu \\ b \end{pmatrix}$. Now, observe that each vector $\widetilde{A}x^{(k)}$ can be written as the convex combination

$$\widetilde{A}x^{(k)} = \sum_{j=1}^{n} x_j^{(k)} \widetilde{A}^j,$$

with $x_j^{(k)} \geq 0$ and where $\widetilde{A}^j \in \mathbb{R}^{m+1}$ is the jth column of $\widetilde{A}$. Therefore, $\widetilde{A}x^{(k)}$ belongs to the polyheral cone

$$C = \text{cone}(\widetilde{A}^1, \ldots, \widetilde{A}^n) = \{\widetilde{A}x \mid x \in \mathbb{R}^n,\ x \geq 0\},$$

and since by Proposition 8.2 this cone is closed, $\lim_{k \geq \infty} \widetilde{A}x^{(k)} \in C$, which means that there is some $u \in \mathbb{R}^n$ with $u \geq 0$ such that

$$\begin{pmatrix} \mu \\ b \end{pmatrix} = \lim_{k \geq \infty} \widetilde{A}x^{(k)} = \widetilde{A}u = \begin{pmatrix} cu \\ Au \end{pmatrix},$$

that is, $cu = \mu$ and $Au = b$. Hence, u is an optimal solution of (P_2). $\square$

The next question is, how do we find such an optimal solution? It turns out that for linear programs in standard form where the constraints are of the form $Ax = b$ and $x \geq 0$, there are always optimal solutions of a special type called *basic feasible solutions*.

9.2 Basic Feasible Solutions and Vertices

If the system $Ax = b$ has a solution and if some row of A is a linear combination of other rows, then the corresponding equation is redundant, so we may assume that the rows of A are linearly independent; that is, we may assume that A has rank m, so $m \leq n$.

Definition 9.6. If A is an $m \times n$ matrix, for any nonempty subset K of $\{1, \ldots, n\}$, let A_K be the submatrix of A consisting of the columns of A whose indices belong to K. We denote the jth column of the matrix A by A^j.

Definition 9.7. Given a Linear Program (P_2)

$$\begin{aligned} &\text{maximize} && cx \\ &\text{subject to} && Ax = b \text{ and } x \geq 0, \end{aligned}$$

where A has rank m, a vector $x \in \mathbb{R}^n$ is a *basic feasible solution* of (P) if $x \in \mathcal{P}(A,b) \neq \emptyset$, and if there is some subset K of $\{1, \ldots, n\}$ of size m such that

(1) The matrix A_K is invertible (that is, the columns of A_K are linearly independent).
(2) $x_j = 0$ for all $j \notin K$.

The subset K is called a *basis* of x. Every index $k \in K$ is called *basic*, and every index $j \notin K$ is called *nonbasic*. Similarly, the columns A^k corresponding to indices $k \in K$ are called *basic*, and the columns A^j corresponding to indices $j \notin K$ are called *nonbasic*. The variables corresponding to basic indices $k \in K$ are called *basic variables*, and the variables corresponding to indices $j \notin K$ are called *nonbasic*.

For example, the linear program

$$\begin{array}{ll} \text{maximize} & x_1 + x_2 \\ \text{subject to} & x_1 + x_2 + x_3 = 1 \text{ and } x_1 \geq 0,\ x_2 \geq 0,\ x_3 \geq 0, \end{array} \qquad (*)$$

has three basic feasible solutions; the basic feasible solution $K = \{1\}$ corresponds to the point $(1, 0, 0)$; the basic feasible solution $K = \{2\}$ corresponds to the point $(0, 1, 0)$; the basic feasible solution $K = \{3\}$ corresponds to the point $(0, 0, 1)$. Each of these points corresponds to the vertices of the slanted purple triangle illustrated in Figure 9.3. The vertices $(1, 0, 0)$ and $(0, 1, 0)$ optimize the objective function with a value of 1.

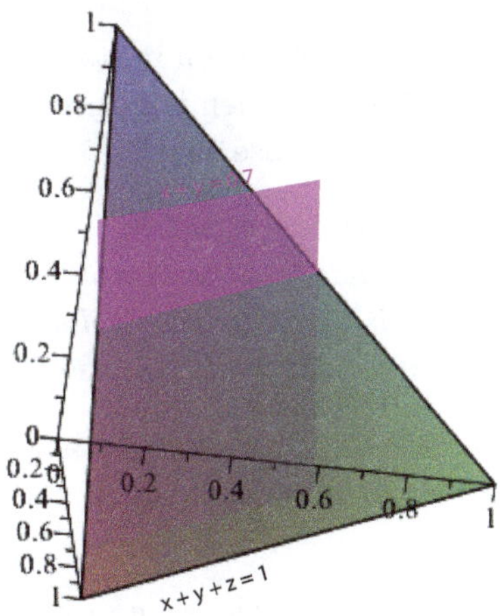

Fig. 9.3 The $\mathcal{H}$-polytope associated with Linear Program $(*)$. The objective function (with $x_1 \to x$ and $x_2 \to y$) is represented by vertical planes parallel to the purple plane $x + y = 0.7$, and reaches it maximal value when $x + y = 1$.

We now show that if the Standard Linear Program (P_2) as in Definition 9.7 has some feasible solution and is bounded above, then some basic feasible solution is an optimal solution. We follow Matousek and Gardner [Matousek and Gartner (2007)] (Chapter 4, Section 2, Theorem 4.2.3).

First we obtain a more convenient characterization of a basic feasible solution.

Proposition 9.2. *Given any Standard Linear Program (P_2) where $Ax = b$ and A is an $m \times n$ matrix of rank m, for any feasible solution x, if $J_> = \{j \in \{1, \ldots, n\} \mid x_j > 0\}$, then x is a basic feasible solution iff the columns of the matrix $A_{J_>}$ are linearly independent.*

Proof. If x is a basic feasible solution, then there is some subset $K \subseteq \{1, \ldots, n\}$ of size m such that the columns of A_K are linearly independent and $x_j = 0$ for all $j \notin K$, so by definition, $J_> \subseteq K$, which implies that the columns of the matrix $A_{J_>}$ are linearly independent.

Conversely, assume that x is a feasible solution such that the columns of the matrix $A_{J_>}$ are linearly independent. If $|J_>| = m$, we are done since we can pick $K = J_>$ and then x is a basic feasible solution. If $|J_>| < m$, we can extend $J_>$ to an m-element subset K by adding $m - |J_>|$ column indices so that the columns of A_K are linearly independent, which is possible since A has rank m. $\square$

Next we prove that if a linear program in standard form has any feasible solution x_0 and is bounded above, then is has some basic feasible solution $\widetilde{x}$ which is as good as x_0, in the sense that $c\widetilde{x} \geq cx_0$.

Proposition 9.3. *Let (P_2) be any standard linear program with objective function cx, where $Ax = b$ and A is an $m \times n$ matrix of rank m. If (P_2) is bounded above and if x_0 is some feasible solution of (P_2), then there is some basic feasible solution $\widetilde{x}$ such that $c\widetilde{x} \geq cx_0$.*

Proof. Among the feasible solutions x such that $cx \geq cx_0$ (x_0 is one of them) pick one with the *maximum* number of coordinates x_j equal to 0, say $\widetilde{x}$. Let

$$K = J_> = \{j \in \{1, \ldots, n\} \mid \widetilde{x}_j > 0\}$$

and let $s = |K|$. We claim that $\widetilde{x}$ is a basic feasible solution, and by construction $c\widetilde{x} \geq cx_0$.

If the columns of A_K are linearly independent, then by Proposition 9.2 we know that $\widetilde{x}$ is a basic feasible solution and we are done.

Otherwise, the columns of A_K are linearly dependent, so there is some nonzero vector $v = (v_1, \ldots, v_s)$ such that $A_K v = 0$. Let $w \in \mathbb{R}^n$ be the vector obtained by extending v by setting $w_j = 0$ for all $j \notin K$. By construction,

$$Aw = A_K v = 0.$$

We will derive a contradiction by exhibiting a feasible solution $x(t_0)$ such that $cx(t_0) \geq cx_0$ with more zero coordinates than $\widetilde{x}$.

For this we claim that we may assume that w satisfies the following two conditions:

(1) $cw \geq 0$.
(2) There is some $j \in K$ such that $w_j < 0$.

If $cw = 0$ and if Condition (2) fails, since $w \neq 0$, we have $w_j > 0$ for some $j \in K$, in which case we can use $-w$, for which $w_j < 0$.

If $cw < 0$, then $c(-w) > 0$, so we may assume that $cw > 0$. If $w_j > 0$ for all $j \in K$, since $\widetilde{x}$ is feasible, $\widetilde{x} \geq 0$, and so $x(t) = \widetilde{x} + tw \geq 0$ for all $t \geq 0$. Furthermore, since $Aw = 0$ and $\widetilde{x}$ is feasible, we have

$$Ax(t) = A\widetilde{x} + tAw = b,$$

and thus $x(t)$ is feasible for all $t \geq 0$. We also have

$$cx(t) = c\widetilde{x} + tcw.$$

Since $cw > 0$, as $t > 0$ goes to infinity the objective function $cx(t)$ also tends to infinity, contradicting the fact that is is bounded above. Therefore, some w satisfying Conditions (1) and (2) above must exist.

We show that there is some $t_0 > 0$ such that $cx(t_0) \geq cx_0$ and $x(t_0) = \widetilde{x} + t_0 w$ is feasible, yet $x(t_0)$ has more zero coordinates than $\widetilde{x}$, a contradiction.

Since $x(t) = \widetilde{x} + tw$, we have

$$x(t)_i = \widetilde{x}_i + tw_i,$$

so if we let $I = \{i \in \{1, \ldots, n\} \mid w_i < 0\} \subseteq K$, which is nonempty since w satisfies Condition (2) above, if we pick

$$t_0 = \min_{i \in I} \left\{ \frac{-\widetilde{x}_i}{w_i} \right\},$$

then $t_0 > 0$, because $w_i < 0$ for all $i \in I$, and by definition of K we have $\widetilde{x}_i > 0$ for all $i \in K$. By the definition of $t_0 > 0$ and since $\widetilde{x} \geq 0$, we have

$$x(t_0)_j = \widetilde{x}_j + t_0 w_j \geq 0 \quad \text{for all } j \in K,$$

so $x(t_0) \geq 0$, and $x(t_0)_i = 0$ for some $i \in I$. Since $Ax(t_0) = b$ (for any t), $x(t_0)$ is a feasible solution,

$$cx(t_0) = c\widetilde{x} + t_0 cw \geq cx_0 + t_0 cw \geq cx_0,$$

and $x(t_0)_i = 0$ for some $i \in I$, we see that $x(t_0)$ has more zero coordinates than $\widetilde{x}$, a contradiction. □

Proposition 9.3 implies the following important result.

Theorem 9.1. *Let (P_2) be any standard linear program with objective function cx, where $Ax = b$ and A is an $m \times n$ matrix of rank m. If (P_2) has some feasible solution and if it is bounded above, then some basic feasible solution $\widetilde{x}$ is an optimal solution of (P_2).*

Proof. By Proposition 9.3, for any feasible solution x there is some basic feasible solution $\widetilde{x}$ such that $cx \leq c\widetilde{x}$. But there are only finitely many basic feasible solutions, so one of them has to yield the maximum of the objective function. □

Geometrically, basic solutions are exactly the vertices of the polyhedron $\mathcal{P}(A, b)$, a notion that we now define.

Definition 9.8. Given an $\mathcal{H}$-polyhedron $\mathcal{P} \subseteq \mathbb{R}^n$, a *vertex* of $\mathcal{P}$ is a point $v \in \mathcal{P}$ with property that there is some nonzero linear form $c \in (\mathbb{R}^n)^*$ and some $\mu \in \mathbb{R}$, such that v is the unique point of $\mathcal{P}$ for which the map $x \mapsto cx$ has the maximum value μ; that is, $cy < cv = \mu$ for all $y \in \mathcal{P} - \{v\}$. Geometrically, this means that the hyperplane of equation $cy = \mu$ touches $\mathcal{P}$ exactly at v. More generally, a convex subset F of $\mathcal{P}$ is a *k-dimensional face* of $\mathcal{P}$ if F has dimension k and if there is some affine form $\varphi(x) = cx - \mu$ such that $cy = \mu$ for all $y \in F$, and $cy < \mu$ for all $y \in \mathcal{P} - F$. A 1-dimensional face is called an *edge*.

The concept of a vertex is illustrated in Figure 9.4, while the concept of an edge is illustrated in Figure 9.5.

Since a k-dimensional face F of $\mathcal{P}$ is equal to the intersection of the hyperplane $H(\varphi)$ of equation $cx = \mu$ with $\mathcal{P}$, it is indeed convex and the notion of dimension makes sense. Observe that a 0-dimensional face of $\mathcal{P}$ is a vertex. If $\mathcal{P}$ has dimension d, then the $(d-1)$-dimensional faces of $\mathcal{P}$ are called its *facets*.

Fig. 9.4 The cube centered at the origin with diagonal through $(-1, -1, -1)$ and $(1, 1, 1)$ has eight vertices. The vertex $(1, 1, 1)$ is associated with the linear form $x + y + z = 3$.

If (P) is a linear program in standard form, then its basic feasible solutions are exactly the vertices of the polyhedron $\mathcal{P}(A, b)$. To prove this fact we need the following simple proposition

Proposition 9.4. *Let $Ax = b$ be a linear system where A is an $m \times n$ matrix of rank m. For any subset $K \subseteq \{1, \ldots, n\}$ of size m, if A_K is invertible, then there is at most one basic feasible solution $x \in \mathbb{R}^n$ with $x_j = 0$ for all $j \notin K$ (of course, $x \geq 0$)*

Proof. In order for x to be feasible we must have $Ax = b$. Write $N = \{1, \ldots, n\} - K$, x_K for the vector consisting of the coordinates of x with indices in K, and x_N for the vector consisting of the coordinates of x with indices in N. Then

$$Ax = A_K x_K + A_N x_N = b.$$

In order for x to be a basic feasible solution we must have $x_N = 0$, so

$$A_K x_K = b.$$

Since by hypothesis A_K is invertible, $x_K = A_K^{-1} b$ is uniquely determined. If $x_K \geq 0$ then x is a basic feasible solution, otherwise it is not. This proves that there is at most one basic feasible solution $x \in \mathbb{R}^n$ with $x_j = 0$ for all $j \notin K$. $\square$

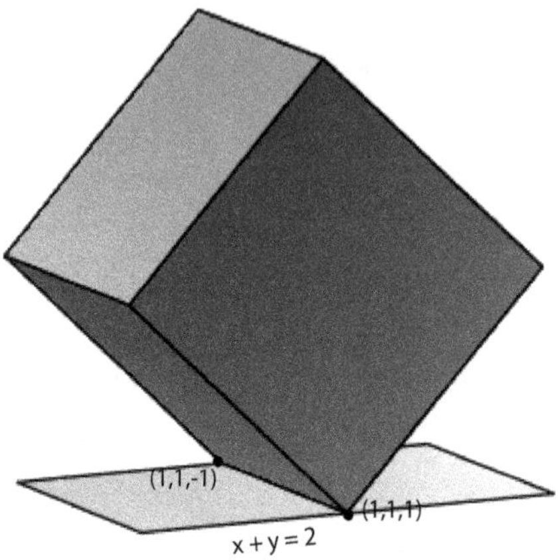

Fig. 9.5 The cube centered at the origin with diagonal through $(-1,-1,-1)$ and $(1,1,1)$ has twelve edges. The edge from $(1,1,-1)$ to $(1,1,1)$ is associated with the linear form $x+y=2$.

Theorem 9.2. *Let (P) be a linear program in standard form, where $Ax = b$ and A is an $m \times n$ matrix of rank m. For every $v \in \mathcal{P}(A,b)$, the following conditions are equivalent:*

(1) v is a vertex of the Polyhedron $\mathcal{P}(A,b)$.
(2) v is a basic feasible solution of the Linear Program (P).

Proof. First, assume that v is a vertex of $\mathcal{P}(A,b)$, and let $\varphi(x) = cx - \mu$ be a linear form such that $cy < \mu$ for all $y \in \mathcal{P}(A,b)$ and $cv = \mu$. This means that v is the unique point of $\mathcal{P}(A,b)$ for which the objective function $x \mapsto cx$ has the maximum value μ on $\mathcal{P}(A,b)$, so by Theorem 9.1, since this maximum is achieved by some basic feasible solution, by uniqueness v must be a basic feasible solution.

Conversely, suppose v is a basic feasible solution of (P) corresponding to a subset $K \subseteq \{1,\ldots,n\}$ of size m. Let $\widehat{c} \in (\mathbb{R}^n)^*$ be the linear form defined by

$$\widehat{c}_j = \begin{cases} 0 & \text{if } j \in K \\ -1 & \text{if } j \notin K. \end{cases}$$

By construction $\widehat{c}\,v = 0$ and $\widehat{c}\,x \leq 0$ for any $x \geq 0$, hence the function $x \mapsto \widehat{c}\,x$ on $\mathcal{P}(A,b)$ has a maximum at v. Furthermore, $\widehat{c}\,x < 0$ for any

$x \geq 0$ such that $x_j > 0$ for some $j \notin K$. However, by Proposition 9.4, the vector v is the only basic feasible solution such that $v_j = 0$ for all $j \notin K$, and therefore v is the only point of $\mathcal{P}(A, b)$ maximizing the function $x \mapsto \widehat{c}x$, so it is a vertex. $\square$

In theory, to find an optimal solution we try all $\binom{n}{m}$ possible m-elements subsets K of $\{1, \ldots, n\}$ and solve for the corresponding unique solution x_K of $A_K x = b$. Then we check whether such a solution satisfies $x_K \geq 0$, compute cx_K, and return some feasible x_K for which the objective function is maximum. This is a totally impractical algorithm.

A practical algorithm is the *simplex algorithm*. Basically, the simplex algorithm tries to "climb" in the polyhderon $\mathcal{P}(A, b)$ from vertex to vertex along edges (using basic feasible solutions), trying to maximize the objective function. We present the simplex algorithm in the next chapter. The reader may also consult texts on linear programming. In particular, we recommend Matousek and Gardner [Matousek and Gartner (2007)], Chvatal [Chvatal (1983)], Papadimitriou and Steiglitz [Papadimitriou and Steiglitz (1998)], Bertsimas and Tsitsiklis [Bertsimas and Tsitsiklis (1997)], Ciarlet [Ciarlet (1989)], Schrijver [Schrijver (1999)], and Vanderbei [Vanderbei (2014)].

Observe that Theorem 9.1 asserts that if a Linear Program (P) in standard form (where $Ax = b$ and A is an $m \times n$ matrix of rank m) has some feasible solution and is bounded above, then some basic feasible solution is an optimal solution. By Theorem 9.2, the polyhedron $\mathcal{P}(A, b)$ must have some vertex.

But suppose we only know that $\mathcal{P}(A, b)$ is nonempty; that is, we don't know that the objective function cx is bounded above. Does $\mathcal{P}(A, b)$ have some vertex?

The answer to the above question is *yes*, and this is important because the simplex algorithm needs an initial basic feasible solution to get started. Here we prove that if $\mathcal{P}(A, b)$ is nonempty, then it must contain a vertex. This proof still doesn't constructively yield a vertex, but we will see in the next chapter that the simplex algorithm always finds a vertex if there is one (provided that we use a pivot rule that prevents cycling).

Theorem 9.3. *Let (P) be a linear program in standard form, where $Ax = b$ and A is an $m \times n$ matrix of rank m. If $\mathcal{P}(A, b)$ is nonempty (there is a feasible solution), then $\mathcal{P}(A, b)$ has some vertex; equivalently, (P) has some basic feasible solution.*

Proof. The proof relies on a trick, which is to add slack variables

$x_{n+1}, \ldots, x_{n+m}$ and use the new objective function $-(x_{n+1} + \cdots + x_{n+m})$.

If we let $\widehat{A}$ be the $m \times (m+n)$-matrix, and x, $\overline{x}$, and $\widehat{x}$ be the vectors given by

$$\widehat{A} = \begin{pmatrix} A \; I_m \end{pmatrix}, \quad x = \begin{pmatrix} x_1 \\ \vdots \\ x_n \end{pmatrix} \in \mathbb{R}^n, \quad \overline{x} = \begin{pmatrix} x_{n+1} \\ \vdots \\ x_{n+m} \end{pmatrix} \in \mathbb{R}^m, \quad \widehat{x} = \begin{pmatrix} x \\ \overline{x} \end{pmatrix} \in \mathbb{R}^{n+m},$$

then consider the Linear Program $(\widehat{P})$ in standard form

$$\begin{aligned} &\text{maximize} && -(x_{n+1} + \cdots + x_{n+m}) \\ &\text{subject to} && \widehat{A}\,\widehat{x} = b \text{ and } \widehat{x} \geq 0. \end{aligned}$$

Since $x_i \geq 0$ for all i, the objective function $-(x_{n+1} + \cdots + x_{n+m})$ is bounded above by 0. The system $\widehat{A}\,\widehat{x} = b$ is equivalent to the system

$$Ax + \overline{x} = b,$$

so for every feasible solution $u \in \mathcal{P}(A, b)$, since $Au = b$, the vector $(u, 0_m)$ is also a feasible solution of $(\widehat{P})$, in fact an optimal solution since the value of the objective function $-(x_{n+1} + \cdots + x_{n+m})$ for $\overline{x} = 0$ is 0. By Proposition 9.3, the linear program $(\widehat{P})$ has some basic feasible solution (u^*, w^*) for which the value of the objective function is greater than or equal to the value of the objective function for $(u, 0_m)$, and since $(u, 0_m)$ is an optimal solution, (u^*, w^*) is also an optimal solution of $(\widehat{P})$. This implies that $w^* = 0$, since otherwise the objective function $-(x_{n+1} + \cdots + x_{n+m})$ would have a strictly negative value.

Therefore, $(u^*, 0_m)$ is a basic feasible solution of $(\widehat{P})$, and thus the columns corresponding to nonzero components of u^* are linearly independent. Some of the coordinates of u^* could be equal to 0, but since A has rank m we can add columns of A to obtain a basis K associated with u^*, and u^* is indeed a basic feasible solution of (P). $\square$

The definition of a basic feasible solution can be adapted to linear programs where the constraints are of the form $Ax \leq b$, $x \geq 0$; see Matousek and Gardner [Matousek and Gartner (2007)] (Chapter 4, Section 4, Definition 4.4.2).

The most general type of linear program allows constraints of the form $a_i x \geq b_i$ or $a_i x = b_i$ besides constraints of the form $a_i x \leq b_i$. The variables x_i may also take negative values. It is always possible to convert such programs to the type considered in Definition 9.1. We proceed as follows.

Every constraint $a_i x \geq b_i$ is replaced by the constraint $-a_i x \leq -b_i$. Every equality constraint $a_i x = b_i$ is replaced by the two constraints $a_i x \leq b_i$ and $-a_i x \leq -b_i$.

If there are n variables x_i, we create n new variables y_i and n new variables z_i and replace every variable x_i by $y_i - z_i$. We also add the $2n$ constraints $y_i \geq 0$ and $z_i \geq 0$. If the constraints are given by the inequalities $Ax \leq b$, we now have constraints given by

$$(A \; -A)\begin{pmatrix} y \\ z \end{pmatrix} \leq b, \quad y \geq 0, \, z \geq 0.$$

We replace the objective function cx by $cy - cz$.

Remark: We also showed that we can replace the inequality constraints $Ax \leq b$ by equality constraints $Ax = b$, by adding slack variables constrained to be nonnegative.

9.3 Summary

The main concepts and results of this chapter are listed below:

- Linear program.
- Objective function, constraints.
- Feasible solution.
- Bounded and unbounded linear programs.
- Optimal solution, optimum.
- Slack variables, linear program in standard form.
- Basic feasible solution.
- Basis of a variable.
- Basic, nonbasic index, basic, nonbasic variable.
- Vertex, face, edge, facet.

9.4 Problems

Problem 9.1. Convert the following program to standard form:

$$\begin{aligned} &\text{maximize} \quad x_1 + x_2 \\ &\text{subject to} \\ &\qquad\qquad x_2 - x_1 \leq 1 \\ &\qquad\qquad x_1 + 6x_2 \leq 15 \\ &\qquad\quad -4x_1 + x_2 \geq 10. \end{aligned}$$

Problem 9.2. Convert the following program to standard form:

$$\begin{aligned} &\text{maximize} \quad 3x_1 - 2x_2 \\ &\text{subject to} \\ &\qquad 2x_1 - x_2 \leq 4 \\ &\qquad x_1 + 3x_2 \geq 5 \\ &\qquad x_2 \geq 0. \end{aligned}$$

Problem 9.3. The notion of basic feasible solution for linear programs where the constraints are of the form $Ax \leq b$, $x \geq 0$ is defined as follows. A basic feasible solution of a (general) linear program with n variables is a feasible solution for which some n linearly independent constraints hold with equality.

Prove that the definition of a basic feasible solution for linear programs in standard form is a special case of the above definition.

Problem 9.4. Consider the linear program

$$\begin{aligned} &\text{maximize} \quad x_1 + x_2 \\ &\text{subject to} \\ &\qquad x_1 + x_2 \leq 1. \end{aligned}$$

Show that none of the optimal solutions are basic.

Problem 9.5. The *standard n-simplex* is the subset Δ^n of $\mathbb{R}^{n+1}$ given by

$$\Delta^n = \{(x_1, \ldots, x_{n+1}) \in \mathbb{R}^{n+1} \mid x_1 + \cdots + x_{n+1} = 1,\ x_i \geq 0,\ 1 \leq i \leq n+1\}.$$

(1) Prove that Δ^n is convex and that it is the convex hull of the $n+1$ vectors $e_1, \ldots e_{n+1}$, where e_i is the ith canonical unit basis vector, $i = 1, \ldots, n+1$.

(2) Prove that Δ^n is the intersection of $n+1$ half spaces and determine the hyperplanes defining these half-spaces.

Remark: The volume under the standard simplex Δ^n is $1/(n+1)!$.

Problem 9.6. The n-dimensional *cross-polytope* is the subset XP_n of $\mathbb{R}^n$ given by

$$XP_n = \{(x_1, \ldots, x_n) \in \mathbb{R}^n \mid |x_1| + \cdots + |x_n| \leq 1\}.$$

(1) Prove that XP_n is convex and that it is the convex hull of the $2n$ vectors $\pm e_i$, where e_i is the ith canonical unit basis vector, $i = 1, \ldots, n$.

(2) Prove that XP_n is the intersection of 2^n half spaces and determine the hyperplanes defining these half-spaces.

Remark: The volume of XP_n is $2^n/n!$.

Problem 9.7. The n-dimensional *hypercube* is the subset C_n of $\mathbb{R}^n$ given by

$$C_n = \{(x_1, \ldots, x_n) \in \mathbb{R}^n \mid |x_i| \leq 1, 1 \leq i \leq n\}.$$

(1) Prove that C_n is convex and that it is the convex hull of the 2^n vectors $(\pm 1, \ldots, \pm 1)$, $i = 1, \ldots, n$.

(2) Prove that C_n is the intersection of $2n$ half spaces and determine the hyperplanes defining these half-spaces.

Remark: The volume of C_n is 2^n.

Chapter 10

The Simplex Algorithm

10.1 The Idea Behind the Simplex Algorithm

The simplex algorithm, due to Dantzig, applies to a linear program (P) in standard form, where the constraints are given by $Ax = b$ and $x \geq 0$, with A a $m \times n$ matrix of rank m, and with an objective function $x \mapsto cx$. This algorithm either reports that (P) has no feasible solution, or that (P) is unbounded, or yields an optimal solution. Geometrically, the algorithm climbs from vertex to vertex in the polyhedron $\mathcal{P}(A, b)$, trying to improve the value of the objective function. Since vertices correspond to basic feasible solutions, the simplex algorithm actually works with basic feasible solutions.

Recall that a basic feasible solution x is a feasible solution for which there is a subset $K \subseteq \{1, \dots, n\}$ of size m such that the matrix A_K consisting of the columns of A whose indices belong to K are linearly independent, and that $x_j = 0$ for all $j \notin K$. We also let $J_>(x)$ be the set of indices

$$J_>(x) = \{j \in \{1, \dots, n\} \mid x_j > 0\},$$

so for a basic feasible solution x associated with K, we have $J_>(x) \subseteq K$. In fact, by Proposition 9.2, a feasible solution x is a basic feasible solution iff the columns of $A_{J_>(x)}$ are linearly independent.

If $J_>(x)$ had cardinality m for all basic feasible solutions x, then the simplex algorithm would make progress at every step, in the sense that it would strictly increase the value of the objective function. Unfortunately, it is possible that $|J_>(x)| < m$ for certain basic feasible solutions, and in this case a step of the simplex algorithm may not increase the value of the objective function. Worse, in rare cases, it is possible that the algorithm enters an infinite loop. This phenomenon called *cycling* can be detected, but in this case the algorithm fails to give a conclusive answer.

Fortunately, there are ways of preventing the simplex algorithm from cycling (for example, Bland's rule discussed later), although proving that these rules work correctly is quite involved.

The potential "bad" behavior of a basic feasible solution is recorded in the following definition.

Definition 10.1. Given a Linear Program (P) in standard form where the constraints are given by $Ax = b$ and $x \geq 0$, with A an $m \times n$ matrix of rank m, a basic feasible solution x is *degenerate* if $|J_>(x)| < m$, otherwise it is *nondegenerate*.

The origin 0_n, if it is a basic feasible solution, is degenerate. For a less trivial example, $x = (0, 0, 0, 2)$ is a degenerate basic feasible solution of the following linear program in which $m = 2$ and $n = 4$.

Example 10.1.

$$\begin{aligned} &\text{maximize} \quad x_2 \\ &\text{subject to} \\ &\qquad -x_1 + x_2 + x_3 = 0 \\ &\qquad x_1 + x_4 = 2 \\ &\qquad x_1 \geq 0,\ x_2 \geq 0,\ x_3 \geq 0,\ x_4 \geq 0. \end{aligned}$$

The matrix A and the vector b are given by

$$A = \begin{pmatrix} -1 & 1 & 1 & 0 \\ 1 & 0 & 0 & 1 \end{pmatrix}, \quad b = \begin{pmatrix} 0 \\ 2 \end{pmatrix},$$

and if $x = (0, 0, 0, 2)$, then $J_>(x) = \{4\}$. There are two ways of forming a set of two linearly independent columns of A containing the fourth column.

Given a basic feasible solution x associated with a subset K of size m, since the columns of the matrix A_K are linearly independent, by abuse of language we call the columns of A_K a *basis* of x.

If u is a vertex of (P), that is, a basic feasible solution of (P) associated with a basis K (of size m), in "normal mode," the simplex algorithm tries to move along an edge from the vertex u to an adjacent vertex v (with $u, v \in \mathcal{P}(A, b) \subseteq \mathbb{R}^n$) corresponding to a basic feasible solution whose basis is obtained by replacing one of the basic vectors A^k with $k \in K$ by another nonbasic vector A^j for some $j \notin K$, in such a way that the value of the objective function is increased.

Let us demonstrate this process on an example.

Example 10.2. Let (P) be the following linear program in standard form.

$$\begin{aligned}
&\text{maximize} \quad x_1 + x_2\\
&\text{subject to}\\
&\qquad\qquad -x_1 + x_2 + x_3 = 1\\
&\qquad\qquad\qquad x_1 + x_4 = 3\\
&\qquad\qquad\qquad x_2 + x_5 = 2\\
&\qquad x_1 \geq 0,\ x_2 \geq 0,\ x_3 \geq 0,\ x_4 \geq 0,\ x_5 \geq 0.
\end{aligned}$$

The matrix A and the vector b are given by

$$A = \begin{pmatrix} -1 & 1 & 1 & 0 & 0 \\ 1 & 0 & 0 & 1 & 0 \\ 0 & 1 & 0 & 0 & 1 \end{pmatrix}, \quad b = \begin{pmatrix} 1 \\ 3 \\ 2 \end{pmatrix}.$$

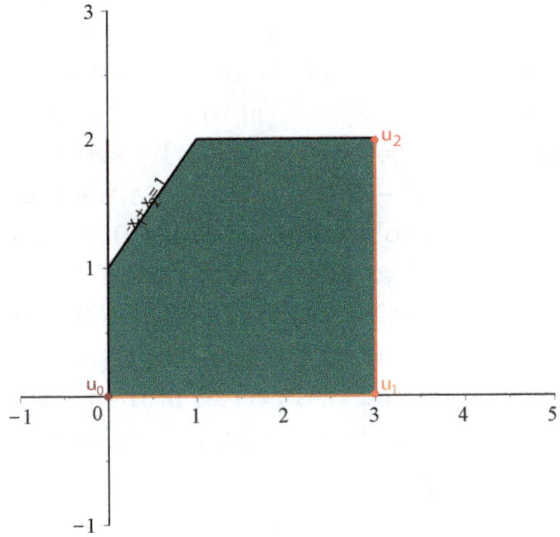

Fig. 10.1 The planar $\mathcal{H}$-polyhedron associated with Example 10.2. The initial basic feasible solution is the origin. The simplex algorithm first moves along the horizontal orange line to feasible solution at vertex u_1. It then moves along the vertical red line to obtain the optimal feasible solution u_2.

The vector $u_0 = (0, 0, 1, 3, 2)$ corresponding to the basis $K = \{3, 4, 5\}$ is a basic feasible solution, and the corresponding value of the objective function is $0 + 0 = 0$. Since the columns (A^3, A^4, A^5) corresponding to $K = \{3, 4, 5\}$ are linearly independent we can express A^1 and A^2 as

$$\begin{aligned} A^1 &= -A^3 + A^4 \\ A^2 &= A^3 + A^5. \end{aligned}$$

Since

$$1A^3 + 3A^4 + 2A^5 = Au_0 = b,$$

for any $\theta \in \mathbb{R}$, we have

$$\begin{aligned} b &= 1A^3 + 3A^4 + 2A^5 - \theta A^1 + \theta A^1 \\ &= 1A^3 + 3A^4 + 2A^5 - \theta(-A^3 + A^4) + \theta A^1 \\ &= \theta A^1 + (1 + \theta)A^3 + (3 - \theta)A^4 + 2A^5, \end{aligned}$$

and

$$\begin{aligned} b &= 1A^3 + 3A^4 + 2A^5 - \theta A^2 + \theta A^2 \\ &= 1A^3 + 3A^4 + 2A^5 - \theta(A^3 + A^5) + \theta A^2 \\ &= \theta A^2 + (1 - \theta)A^3 + 3A^4 + (2 - \theta)A^5. \end{aligned}$$

In the first case, the vector $(\theta, 0, 1 + \theta, 3 - \theta, 2)$ is a feasible solution iff $0 \leq \theta \leq 3$, and the new value of the objective function is θ.

In the second case, the vector $(0, \theta, 1 - \theta, 3, 2 - \theta, 1)$ is a feasible solution iff $0 \leq \theta \leq 1$, and the new value of the objective function is also θ.

Consider the first case. It is natural to ask whether we can get another vertex and increase the objective function by setting to zero one of the coordinates of $(\theta, 0, 1 + \theta, 3 - \theta, 2)$, in this case the fourth one, by picking $\theta = 3$. This yields the feasible solution $(3, 0, 4, 0, 2)$, which corresponds to the basis (A^1, A^3, A^5), and so is indeed a basic feasible solution, with an improved value of the objective function equal to 3. Note that A^4 *left* the basis (A^3, A^4, A^5) and A^1 *entered* the new basis (A^1, A^3, A^5).

We can now express A^2 and A^4 in terms of the basis (A^1, A^3, A^5), which is easy to do since we already have A^1 and A^2 in term of (A^3, A^4, A^5), and A^1 and A^4 are swapped. Such a step is called a *pivoting step*. We obtain

$$\begin{aligned} A^2 &= A^3 + A^5 \\ A^4 &= A^1 + A^3. \end{aligned}$$

Then we repeat the process with $u_1 = (3, 0, 4, 0, 2)$ and the basis (A^1, A^3, A^5). We have

$$\begin{aligned} b &= 3A^1 + 4A^3 + 2A^5 - \theta A^2 + \theta A^2 \\ &= 3A^1 + 4A^3 + 2A^5 - \theta(A^3 + A^5) + \theta A^2 \\ &= 3A^1 + \theta A^2 + (4 - \theta)A^3 + (2 - \theta)A^5, \end{aligned}$$

and

$$\begin{aligned} b &= 3A^1 + 4A^3 + 2A^5 - \theta A^4 + \theta A^4 \\ &= 3A^1 + 4A^3 + 2A^5 - \theta(A^1 + A^3) + \theta A^4 \\ &= (3 - \theta)A^1 + (4 - \theta)A^3 + \theta A^4 + 2A^5. \end{aligned}$$

In the first case, the point $(3, \theta, 4 - \theta, 0, 2 - \theta)$ is a feasible solution iff $0 \leq \theta \leq 2$, and the new value of the objective function is $3 + \theta$. In the second case, the point $(3 - \theta, 0, 4 - \theta, \theta, 2)$ is a feasible solution iff $0 \leq \theta \leq 3$, and the new value of the objective function is $3 - \theta$. To increase the objective function, we must choose the first case and we pick $\theta = 2$. Then we get the feasible solution $u_2 = (3, 2, 2, 0, 0)$, which corresponds to the basis (A^1, A^2, A^3), and thus is a basic feasible solution. The new value of the objective function is 5.

Next we express A^4 and A^5 in terms of the basis (A^1, A^2, A^3). Again this is easy to do since we just swapped A^5 and A^2 (a pivoting step), and we get

$$\begin{aligned} A^5 &= A^2 - A^3 \\ A^4 &= A^1 + A^3. \end{aligned}$$

We repeat the process with $u_2 = (3, 2, 2, 0, 0)$ and the basis (A^1, A^2, A^3). We have

$$\begin{aligned} b &= 3A^1 + 2A^2 + 2A^3 - \theta A^4 + \theta A^4 \\ &= 3A^1 + 2A^2 + 2A^3 - \theta(A^1 + A^3) + \theta A^4 \\ &= (3 - \theta)A^1 + 2A^2 + (2 - \theta)A^3 + \theta A^4, \end{aligned}$$

and

$$\begin{aligned} b &= 3A^1 + 2A^2 + 2A^3 - \theta A^5 + \theta A^5 \\ &= 3A^1 + 2A^2 + 2A^3 - \theta(A^2 - A^3) + \theta A^5 \\ &= 3A^1 + (2 - \theta)A^2 + (2 + \theta)A^3 + \theta A^5. \end{aligned}$$

In the first case, the point $(3 - \theta, 2, 2 - \theta, \theta, 0)$ is a feasible solution iff $0 \leq \theta \leq 2$, and the value of the objective function is $5 - \theta$. In the second case, the point $(3, 2 - \theta, 2 + \theta, 0, \theta)$ is a feasible solution iff $0 \leq \theta \leq 2$, and the value of the objective function is also $5 - \theta$. Since we must have $\theta \geq 0$ to have a feasible solution, there is no way to increase the objective function. In this situation, it turns out that we have reached an optimal solution, in our case $u_2 = (3, 2, 2, 0, 0)$, with the maximum of the objective function equal to 5.

We could also have applied the simplex algorithm to the vertex $u_0 = (0,0,1,3,2)$ and to the vector $(0,\theta,1-\theta,3,2-\theta,1)$, which is a feasible solution iff $0 \leq \theta \leq 1$, with new value of the objective function θ. By picking $\theta = 1$, we obtain the feasible solution $(0,1,0,3,1)$, corresponding to the basis (A^2, A^4, A^5), which is indeed a vertex. The new value of the objective function is 1. Then we express A^1 and A^3 in terms the basis (A^2, A^4, A^5) obtaining

$$\begin{aligned} A^1 &= A^4 - A^3 \\ A^3 &= A^2 - A^5, \end{aligned}$$

and repeat the process with $(0,1,0,3,1)$ and the basis (A^2, A^4, A^5). After three more steps we will reach the optimal solution $u_2 = (3,2,2,0,0)$.

Let us go back to the linear program of Example 10.1 with objective function x_2 and where the matrix A and the vector b are given by

$$A = \begin{pmatrix} -1 & 1 & 1 & 0 \\ 1 & 0 & 0 & 1 \end{pmatrix}, \quad b = \begin{pmatrix} 0 \\ 2 \end{pmatrix}.$$

Recall that $u_0 = (0,0,0,2)$ is a degenerate basic feasible solution, and the objective function has the value 0. See Figure 10.2 for a planar picture of the $\mathcal{H}$-polyhedron associated with Example 10.1.

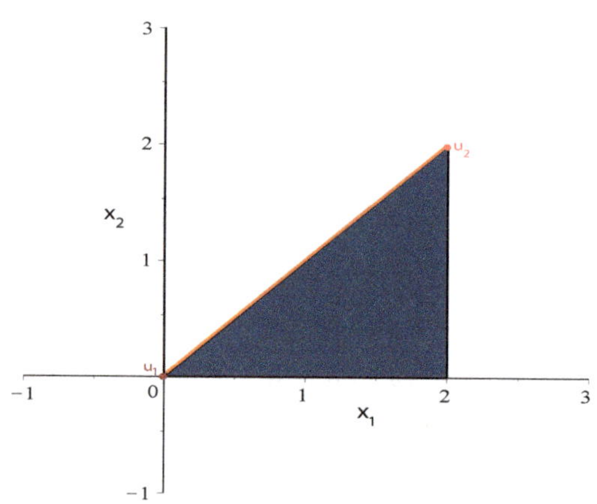

Fig. 10.2 The planar $\mathcal{H}$-polyhedron associated with Example 10.1. The initial basic feasible solution is the origin. The simplex algorithm moves along the slanted orange line to the apex of the triangle.

Pick the basis (A^3, A^4). Then we have

$$
\begin{aligned}
A^1 &= -A^3 + A^4 \\
A^2 &= A^3,
\end{aligned}
$$

and we get

$$
\begin{aligned}
b &= 2A^4 - \theta A^1 + \theta A^1 \\
&= 2A^4 - \theta(-A^3 + A^4) + \theta A^1 \\
&= \theta A^1 + \theta A^3 + (2 - \theta)A^4,
\end{aligned}
$$

and

$$
\begin{aligned}
b &= 2A^4 - \theta A^2 + \theta A^2 \\
&= 2A^4 - \theta A^3 + \theta A^2 \\
&= \theta A^2 - \theta A^3 + 2A^4.
\end{aligned}
$$

In the first case, the point $(\theta, 0, \theta, 2 - \theta)$ is a feasible solution iff $0 \leq \theta \leq 2$, and the value of the objective function is 0, and in the second case the point $(0, \theta, -\theta, 2)$ is a feasible solution iff $\theta = 0$, and the value of the objective function is θ. However, since we must have $\theta = 0$ in the second case, there is no way to increase the objective function either.

It turns out that in order to make the cases considered by the simplex algorithm as mutually exclusive as possible, since in the second case the coefficient of θ in the value of the objective function is nonzero, namely 1, we should choose the second case. We must pick $\theta = 0$, but we can swap the vectors A^3 and A^2 (because A^2 is coming in and A^3 has the coefficient $-\theta$, which is the reason why θ must be zero), and we obtain the basic feasible solution $u_1 = (0, 0, 0, 2)$ with the new basis (A^2, A^4). Note that this basic feasible solution corresponds to the same vertex $(0, 0, 0, 2)$ as before, but the basis has changed. The vectors A^1 and A^3 can be expressed in terms of the basis (A^2, A^4) as

$$
\begin{aligned}
A^1 &= -A^2 + A^4 \\
A^3 &= A^2.
\end{aligned}
$$

We now repeat the procedure with $u_1 = (0, 0, 0, 2)$ and the basis (A^2, A^4), and we get

$$
\begin{aligned}
b &= 2A^4 - \theta A^1 + \theta A^1 \\
&= 2A^4 - \theta(-A^2 + A^4) + \theta A^1 \\
&= \theta A^1 + \theta A^2 + (2 - \theta)A^4,
\end{aligned}
$$

and

$$\begin{aligned} b &= 2A^4 - \theta A^3 + \theta A^3 \\ &= 2A^4 - \theta A^2 + \theta A^3 \\ &= -\theta A^2 + \theta A^3 + 2A^4. \end{aligned}$$

In the first case, the point $(\theta, \theta, 0, 2 - \theta)$ is a feasible solution iff $0 \leq \theta \leq 2$ and the value of the objective function is θ, and in the second case the point $(0, -\theta, \theta, 2)$ is a feasible solution iff $\theta = 0$ and the value of the objective function is θ. In order to increase the objective function we must choose the first case and pick $\theta = 2$. We obtain the feasible solution $u_2 = (2, 2, 0, 0)$ whose corresponding basis is (A^1, A^2) and the value of the objective function is 2.

The vectors A^3 and A^4 are expressed in terms of the basis (A^1, A^2) as

$$\begin{aligned} A^3 &= A^2 \\ A^4 &= A^1 + A^3, \end{aligned}$$

and we repeat the procedure with $u_2 = (2, 2, 0, 0)$ and the basis (A^1, A^2). We get

$$\begin{aligned} b &= 2A^1 + 2A^2 - \theta A^3 + \theta A^3 \\ &= 2A^1 + 2A^2 - \theta A^2 + \theta A^3 \\ &= 2A^1 + (2 - \theta)A^2 + \theta A^3, \end{aligned}$$

and

$$\begin{aligned} b &= 2A^1 + 2A^2 - \theta A^4 + \theta A^4 \\ &= 2A^1 + 2A^2 - \theta(A^1 + A^3) + \theta A^4 \\ &= (2 - \theta)A^1 + 2A^2 - \theta A^3 + \theta A^4. \end{aligned}$$

In the first case, the point $(2, 2 - \theta, 0, \theta)$ is a feasible solution iff $0 \leq \theta \leq 2$ and the value of the objective function is $2 - \theta$, and in the second case, the point $(2 - \theta, 2, -\theta, \theta)$ is a feasible solution iff $\theta = 0$ and the value of the objective function is 2. This time there is no way to improve the objective function and we have reached an optimal solution $u_2 = (2, 2, 0, 0)$ with the maximum of the objective function equal to 2.

Let us now consider an example of an unbounded linear program.

Example 10.3. Let (P) be the following linear program in standard form.

$$\begin{aligned}
&\text{maximize} \quad x_1 \\
&\text{subject to} \\
&\qquad\qquad x_1 - x_2 + x_3 = 1 \\
&\qquad\qquad -x_1 + x_2 + x_4 = 2 \\
&\qquad\qquad x_1 \geq 0,\ x_2 \geq 0,\ x_3 \geq 0,\ x_4 \geq 0.
\end{aligned}$$

The matrix A and the vector b are given by

$$A = \begin{pmatrix} 1 & -1 & 1 & 0 \\ -1 & 1 & 0 & 1 \end{pmatrix}, \quad b = \begin{pmatrix} 1 \\ 2 \end{pmatrix}.$$

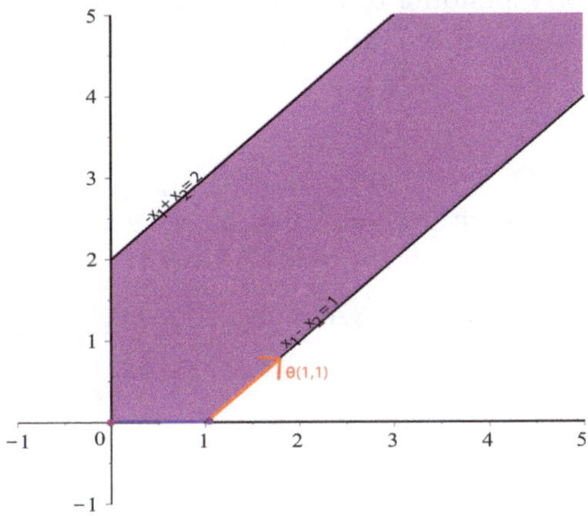

Fig. 10.3 The planar $\mathcal{H}$-polyhedron associated with Example 10.3. The initial basic feasible solution is the origin. The simplex algorithm first moves along the horizontal indigo line to basic feasible solution at vertex $(1,0)$. Any optimal feasible solution occurs by moving along the boundary line parameterized by the orange arrow $\theta(1,1)$.

The vector $u_0 = (0,0,1,2)$ corresponding to the basis $K = \{3,4\}$ is a basic feasible solution, and the corresponding value of the objective function is 0. The vectors A^1 and A^2 are expressed in terms of the basis (A^3, A^4) by

$$\begin{aligned}
A^1 &= A^3 - A^4 \\
A^2 &= -A^3 + A^4.
\end{aligned}$$

Starting with $u_0 = (0, 0, 1, 2)$, we get

$$\begin{aligned} b &= A^3 + 2A^4 - \theta A^1 + \theta A^1 \\ &= A^3 + 2A^4 - \theta(A^3 - A^4) + \theta A^1 \\ &= \theta A^1 + (1-\theta)A^3 + (2+\theta)A^4, \end{aligned}$$

and

$$\begin{aligned} b &= A^3 + 2A^4 - \theta A^2 + \theta A^2 \\ &= A^3 + 2A^4 - \theta(-A^3 + A^4) + \theta A^2 \\ &= \theta A^2 + (1+\theta)A^3 + (2-\theta)A^4. \end{aligned}$$

In the first case, the point $(\theta, 0, 1-\theta, 2+\theta)$ is a feasible solution iff $0 \leq \theta \leq 1$ and the value of the objective function is θ, and in the second case, the point $(0, \theta, 1+\theta, 2-\theta)$ is a feasible solution iff $0 \leq \theta \leq 2$ and the value of the objective function is 0. In order to increase the objective function we must choose the first case, and we pick $\theta = 1$. We get the feasible solution $u_1 = (1, 0, 0, 3)$ corresponding to the basis (A^1, A^4), so it is a basic feasible solution, and the value of the objective function is 1.

The vectors A^2 and A^3 are given in terms of the basis (A^1, A^4) by

$$\begin{aligned} A^2 &= -A^1 \\ A^3 &= A^1 + A^4. \end{aligned}$$

Repeating the process with $u_1 = (1, 0, 0, 3)$, we get

$$\begin{aligned} b &= A^1 + 3A^4 - \theta A^2 + \theta A^2 \\ &= A^1 + 3A^4 - \theta(-A^1) + \theta A^2 \\ &= (1+\theta)A^1 + \theta A^2 + 3A^4, \end{aligned}$$

and

$$\begin{aligned} b &= A^1 + 3A^4 - \theta A^3 + \theta A^3 \\ &= A^1 + 3A^4 - \theta(A^1 + A^4) + \theta A^3 \\ &= (1-\theta)A^1 + \theta A^3 + (3-\theta)A^4. \end{aligned}$$

In the first case, the point $(1+\theta, \theta, 0, 3)$ is a feasible solution for all $\theta \geq 0$ and the value of the objective function if $1+\theta$, and in the second case, the point $(1-\theta, 0, \theta, 3-\theta)$ is a feasible solution iff $0 \leq \theta \leq 1$ and the value of the objective function is $1-\theta$. This time, we are in the situation where the points

$$(1+\theta, \theta, 0, 3) = (1, 0, 0, 3) + \theta(1, 1, 0, 0), \quad \theta \geq 0$$

form an infinite ray in the set of feasible solutions, and the objective function $1+\theta$ is unbounded from above on this ray. This indicates that our linear program, although feasible, is unbounded.

Let us now describe a step of the simplex algorithm in general.

10.2 The Simplex Algorithm in General

We assume that we already have an initial vertex u_0 to start from. This vertex corresponds to a basic feasible solution with basis K_0. We will show later that it is always possible to find a basic feasible solution of a Linear Program (P) is standard form, or to detect that (P) has no feasible solution.

The idea behind the simplex algorithm is this: Given a pair (u, K) consisting of a basic feasible solution u and a basis K for u, find another pair (u^+, K^+) consisting of another basic feasible solution u^+ and a basis K^+ for u^+, such that K^+ is obtained from K by deleting some basic index $k^- \in K$ and adding some nonbasic index $j^+ \notin K$, in such a way that the value of the objective function increases (preferably strictly). The step which consists in swapping the vectors A^{k^-} and A^{j^+} is called a *pivoting step*.

Let u be a given vertex corresponds to a basic feasible solution with basis K. Since the m vectors A^k corresponding to indices $k \in K$ are linearly independent, they form a basis, so for every nonbasic $j \notin K$, we write

$$A^j = \sum_{k \in K} \gamma_k^j A^k. \tag{$*$}$$

We let $\gamma_K^j \in \mathbb{R}^m$ be the vector given by $\gamma_K^j = (\gamma_k^j)_{k \in K}$. Actually, since the vector γ_K^j depends on K, to be very precise we should denote its components by $(\gamma_K^j)_k$, but to simplify notation we usually write γ_k^j instead of $(\gamma_K^j)_k$ (unless confusion arises). We will explain later how the coefficients γ_k^j can be computed efficiently.

Since u is a feasible solution we have $u \geq 0$ and $Au = b$, that is,

$$\sum_{k \in K} u_k A^k = b. \tag{$**$}$$

For every nonbasic $j \notin K$, a candidate for entering the basis K, we try to find a new vertex $u(\theta)$ that improves the objective function, and for this we add $-\theta A^j + \theta A^j = 0$ to b in equation $(**)$ and then replace the occurrence of A^j in $-\theta A^j$ by the right hand side of equation $(*)$ to obtain

$$\begin{aligned} b &= \sum_{k \in K} u_k A^k - \theta A^j + \theta A^j \\ &= \sum_{k \in K} u_k A^k - \theta \Big(\sum_{k \in K} \gamma_k^j A^k \Big) + \theta A^j \\ &= \sum_{k \in K} \Big(u_k - \theta \gamma_k^j \Big) A^k + \theta A^j. \end{aligned}$$

Consequently, the vector $u(\theta)$ appearing on the right-hand side of the above equation given by

$$u(\theta)_i = \begin{cases} u_i - \theta\gamma_i^j & \text{if } i \in K \\ \theta & \text{if } i = j \\ 0 & \text{if } i \notin K \cup \{j\} \end{cases}$$

automatically satisfies the constraints $Au(\theta) = b$, and this vector is a feasible solution iff

$$\theta \geq 0 \quad \text{and} \quad u_k \geq \theta\gamma_k^j \quad \text{for all } k \in K.$$

Obviously $\theta = 0$ is a solution, and if

$$\theta^j = \min\left\{ \frac{u_k}{\gamma_k^j} \;\middle|\; \gamma_k^j > 0,\, k \in K \right\} > 0,$$

then we have a range of feasible solutions for $0 \leq \theta \leq \theta^j$. The value of the objective function for $u(\theta)$ is

$$cu(\theta) = \sum_{k \in K} c_k(u_k - \theta\gamma_k^j) + \theta c_j = cu + \theta\left(c_j - \sum_{k \in K} \gamma_k^j c_k\right).$$

Since the potential change in the objective function is

$$\theta\left(c_j - \sum_{k \in K} \gamma_k^j c_k\right)$$

and $\theta \geq 0$, if $c_j - \sum_{k \in K} \gamma_k^j c_k \leq 0$, then the objective function can't be increased.

However, if $c_{j^+} - \sum_{k \in K} \gamma_k^{j^+} c_k > 0$ for some $j^+ \notin K$, and if $\theta^{j^+} > 0$, then the objective function can be strictly increased by choosing any $\theta > 0$ such that $\theta \leq \theta^{j^+}$, so it is natural to zero at least one coefficient of $u(\theta)$ by picking $\theta = \theta^{j^+}$, which also maximizes the increase of the objective function. In this case (Case below (B2)), we obtain a new feasible solution $u^+ = u(\theta^{j^+})$.

Now, if $\theta^{j^+} > 0$, then there is some index $k \in K$ such $u_k > 0$, $\gamma_k^{j^+} > 0$, and $\theta^{j^+} = u_k/\gamma_k^{j^+}$, so we can pick such an index k^- for the vector A^{k^-} leaving the basis K. We claim that $K^+ = (K - \{k^-\}) \cup \{j^+\}$ is a basis. This is because the coefficient $\gamma_{k^-}^{j^+}$ associated with the column A^{k^-} is nonzero (in fact, $\gamma_{k^-}^{j^+} > 0$), so equation $(*)$, namely

$$A^{j^+} = \gamma_{k^-}^{j^+} A^{k^-} + \sum_{k \in K - \{k^-\}} \gamma_k^{j^+} A^k,$$

yields the equation

$$A^{k^-} = (\gamma_{k^-}^{j^+})^{-1} A^{j^+} - \sum_{k \in K - \{k^-\}} (\gamma_{k^-}^{j^+})^{-1} \gamma_k^{j^+} A^k,$$

and these equations imply that the subspaces spanned by the vectors $(A^k)_{k \in K}$ and the vectors $(A^k)_{k \in K^+}$ are identical. However, K is a basis of dimension m so this subspace has dimension m, and since K^+ also has m elements, it must be a basis. Therefore, $u^+ = u(\theta^{j^+})$ is a basic feasible solution.

The above case is the most common one, but other situations may arise. In what follows, we discuss all eventualities.

Case (A).

We have $c_j - \sum_{k \in K} \gamma_k^j c_k \leq 0$ for all $j \notin K$. Then it turns out that u is an *optimal solution*. Otherwise, we are in Case (B).

Case (B).

We have $c_j - \sum_{k \in K} \gamma_k^j c_k > 0$ for some $j \notin K$ (not necessarily unique). There are three subcases.

Case (B1).

If for some $j \notin K$ as above we also have $\gamma_k^j \leq 0$ for all $k \in K$, since $u_k \geq 0$ for all $k \in K$, this places no restriction on θ, and the objective function is *unbounded above*. This is demonstrated by Example 10.3 with $K = \{3, 4\}$ and $j = 2$ since $\gamma_{\{3,4\}}^2 = (-1, 0)$.

Case (B2).

There is some index $j^+ \notin K$ such that simultaneously

(1) $c_{j^+} - \sum_{k \in K} \gamma_k^{j^+} c_k > 0$, which means that the objective function can potentially be increased;
(2) There is some $k \in K$ such that $\gamma_k^{j^+} > 0$, and for every $k \in K$, if $\gamma_k^{j^+} > 0$ then $u_k > 0$, which implies that $\theta^{j^+} > 0$.

If we pick $\theta = \theta^{j^+}$ where

$$\theta^{j^+} = \min\left\{ \frac{u_k}{\gamma_k^{j^+}} \;\middle|\; \gamma_k^{j^+} > 0,\ k \in K \right\} > 0,$$

then the feasible solution u^+ given by

$$u_i^+ = \begin{cases} u_i - \theta^{j^+} \gamma_i^{j^+} & \text{if } i \in K \\ \theta^{j^+} & \text{if } i = j^+ \\ 0 & \text{if } i \notin K \cup \{j^+\} \end{cases}$$

is a vertex of $\mathcal{P}(A, b)$. If we pick any index $k^- \in K$ such that $\theta^{j^+} = u_{k^-}/\gamma_{k^-}^{j+}$, then $K^+ = (K - \{k^-\}) \cup \{j^+\}$ is a basis for u^+. The vector A^{j^+} enters the new basis K^+, and the vector A^{k^-} leaves the old basis K. This is a *pivoting step*. The objective function increases strictly. This is demonstrated by Example 10.2 with $K = \{3, 4, 5\}$, $j = 1$, and $k = 4$. Then $\gamma_{\{3,4,5\}}^1 = (-1, 1, 0)$, with $\gamma_4^1 = 1$. Since $u = (0, 0, 1, 3, 2)$, $\theta^1 = \frac{u_4}{\gamma_4^1} = 3$, and the new optimal solutions becomes $u^+ = (3, 0, 1-3(-1), 3-3(1), 2-3(0)) = (3, 0, 4, 0, 2)$.

Case (B3).

There is some index $j \notin K$ such that $c_j - \sum_{k \in K} \gamma_k^j c_k > 0$, and for each of the indices $j \notin K$ satisfying the above property we have simultaneously

(1) $c_j - \sum_{k \in K} \gamma_k^j c_k > 0$, which means that the objective function can potentially be increased;
(2) There is some $k \in K$ such that $\gamma_k^j > 0$, and $u_k = 0$, which implies that $\theta^j = 0$.

Consequently, the objective function *does not change*. In this case, u is a degenerate basic feasible solution.

We can associate to $u^+ = u$ a new basis K^+ as follows: Pick any index $j^+ \notin K$ such that

$$c_{j^+} - \sum_{k \in K} \gamma_k^{j^+} c_k > 0,$$

and any index $k^- \in K$ such that

$$\gamma_{k^-}^{j^+} > 0,$$

and let $K^+ = (K - \{k^-\}) \cup \{j^+\}$. As in Case (B2), the vector A^{j^+} enters the new basis K^+, and the vector A^{k^-} leaves the old basis K. This is a *pivoting step*. However, the objective function *does not change* since $\theta^{j+} = 0$. This is demonstrated by Example 10.1 with $K = \{3, 4\}$, $j = 2$, and $k = 3$.

It is easy to prove that in Case (A) the basic feasible solution u is an optimal solution, and that in Case (B1) the linear program is unbounded. We already proved that in Case (B2) the vector u^+ and its basis K^+ constitutes a basic feasible solution, and the proof in Case (B3) is similar. For details, see Ciarlet [Ciarlet (1989)] (Chapter 10).

It is convenient to reinterpret the various cases considered by introducing the following sets:

$$B_1 = \Big\{ j \notin K \mid c_j - \sum_{k \in K} \gamma_k^j c_k > 0,\ \max_{k \in K} \gamma_k^j \leq 0 \Big\}$$

$$B_2 = \Big\{ j \notin K \mid c_j - \sum_{k \in K} \gamma_k^j c_k > 0,$$
$$\max_{k \in K} \gamma_k^j > 0,\ \min\Big\{ \frac{u_k}{\gamma_k^j} \ \Big|\ k \in K, \gamma_k^j > 0 \Big\} > 0 \Big\}$$

$$B_3 = \Big\{ j \notin K \mid c_j - \sum_{k \in K} \gamma_k^j c_k > 0,\ \max_{k \in K} \gamma_k^j > 0,$$
$$\min\Big\{ \frac{u_k}{\gamma_k^j} \ \Big|\ k \in K, \gamma_k^j > 0 \Big\} = 0 \Big\},$$

and

$$B = B_1 \cup B_2 \cup B_3 = \Big\{ j \notin K \mid c_j - \sum_{k \in K} \gamma_k^j c_k > 0 \Big\}.$$

Then it is easy to see that the following equivalences hold:

$$\text{Case (A)} \Longleftrightarrow B = \emptyset, \qquad \text{Case (B)} \Longleftrightarrow B \neq \emptyset$$
$$\text{Case (B1)} \Longleftrightarrow B_1 \neq \emptyset$$
$$\text{Case (B2)} \Longleftrightarrow B_2 \neq \emptyset$$
$$\text{Case (B3)} \Longleftrightarrow B_3 \neq \emptyset.$$

Furthermore, Cases (A) and (B), Cases (B1) and (B3), and Cases (B2) and (B3) are mutually exclusive, while Cases (B1) and (B2) are not.

If Case (B1) and Case (B2) arise simultaneously, we opt for Case (B1) which says that the Linear Program (P) is unbounded and terminate the algorithm.

Here are a few remarks about the method.

In Case (B2), which is the path followed by the algorithm most frequently, various choices have to be made for the index $j^+ \notin K$ for which $\theta^{j^+} > 0$ (the new index in K^+). Similarly, various choices have to be made for the index $k^- \in K$ leaving K, but such choices are typically less important.

Similarly in Case (B3), various choices have to be made for the new index $j^+ \notin K$ going into K^+. In Cases (B2) and (B3), criteria for making such choices are called *pivot rules*.

Case (B3) only arises when u is a degenerate vertex. But even if u is degenerate, Case (B2) may arise if $u_k > 0$ whenever $\gamma_k^j > 0$. It may also happen that u is nondegenerate but as a result of Case (B2), the new vertex u^+ is degenerate because at least two components $u_{k_1} - \theta^{j^+}\gamma_{k_1}^{j^+}$ and $u_{k_2} - \theta^{j^+}\gamma_{k_2}^{j^+}$ vanish for some distinct $k_1, k_2 \in K$.

Cases (A) and (B1) correspond to situations where the algorithm terminates, and Case (B2) can only arise a finite number of times during execution of the simplex algorithm, since the objective function is strictly increased from vertex to vertex and there are only finitely many vertices. Therefore, if the simplex algorithm is started on any initial basic feasible solution u_0, then one of three mutually exclusive situations may arise:

(1) There is a finite sequence of occurrences of Case (B2) and/or Case (B3) ending with an occurrence of Case (A). Then the last vertex produced by the algorithm is an optimal solution. This is what occurred in Examples 10.1 and 10.2.
(2) There is a finite sequence of occurrences of Case (B2) and/or Case (B3) ending with an occurrence of Case (B1). We conclude that the problem is unbounded, and thus has no solution. This is what occurred in Example 10.3.
(3) There is a finite sequence of occurrences of Case (B2) and/or Case (B3), followed by an infinite sequence of Case (B3). If this occurs, the algorithm visits the some basis twice. This a phenomenon known as *cycling*. In this eventually the algorithm fails to come to a conclusion.

There are examples for which cycling occur, although this is rare in practice. Such an example is given in Chvatal [Chvatal (1983)]; see Chapter 3, pages 31–32, for an example with seven variables and three equations that cycles after six iterations under a certain pivot rule.

The third possibility can be avoided by the choice of a suitable pivot rule. Two of these rules are *Bland's rule* and the *lexicographic rule*; see Chvatal [Chvatal (1983)] (Chapter 3, pages 34–38).

Bland's rule says: choose the smallest of the eligible incoming indices $j^+ \notin K$, and similarly choose the smallest of the eligible outgoing indices $k^- \in K$.

It can be proven that cycling cannot occur if Bland's rule is chosen as the pivot rule. The proof is very technical; see Chvatal [Chvatal (1983)] (Chapter 3, pages 37–38), Matousek and Gardner [Matousek and Gartner (2007)] (Chapter 5, Theorem 5.8.1), and Papadimitriou and Steiglitz [Papadimitriou and Steiglitz (1998)] (Section 2.7). Therefore, assuming that some initial basic feasible solution is provided, and using a suitable pivot rule (such as Bland's rule), the simplex algorithm always terminates and either yields an optimal solution or reports that the linear program is unbounded. Unfortunately, Bland's rules is one of the slowest pivot rules.

The choice of a pivot rule affects greatly the number of pivoting steps that the simplex algorithms goes through. It is not our intention here to explain the various pivot rules. We simply mention the following rules, referring the reader to Matousek and Gardner [Matousek and Gartner (2007)] (Chapter 5, Section 5.7) or to the texts cited in Section 8.1.

(1) Largest coefficient, or Dantzig's rule.
(2) Largest increase.
(3) Steepest edge.
(4) Bland's Rule.
(5) Random edge.

The steepest edge rule is one of the most popular. The idea is to maximize the ratio

$$\frac{c(u^+ - u)}{\|u^+ - u\|}.$$

The random edge rule picks the index $j^+ \notin K$ of the entering basis vector uniformly at random among all eligible indices.

Let us now return to the issue of the initialization of the simplex algorithm. We use the Linear Program $(\widehat{P})$ introduced during the proof of Theorem 9.3.

Consider a Linear Program $(P2)$

$$\begin{aligned} &\text{maximize} \quad cx \\ &\text{subject to} \quad Ax = b \text{ and } x \geq 0, \end{aligned}$$

in standard form where A is an $m \times n$ matrix of rank m.

First, observe that since the constraints are equations, we can ensure that $b \geq 0$, because every equation $a_i x = b_i$ where $b_i < 0$ can be replaced by $-a_i x = -b_i$. The next step is to introduce the Linear Program $(\widehat{P})$ in standard form

$$\begin{aligned} &\text{maximize} \quad -(x_{n+1} + \cdots + x_{n+m}) \\ &\text{subject to} \quad \widehat{A}\widehat{x} = b \text{ and } \widehat{x} \geq 0, \end{aligned}$$

where $\widehat{A}$ and $\widehat{x}$ are given by

$$\widehat{A} = \begin{pmatrix} A & I_m \end{pmatrix}, \quad \widehat{x} = \begin{pmatrix} x_1 \\ \vdots \\ x_{n+m} \end{pmatrix}.$$

Since we assumed that $b \geq 0$, the vector $\widehat{x} = (0_n, b)$ is a feasible solution of $(\widehat{P})$, in fact a basic feasible solutions since the matrix associated with the indices $n+1, \ldots, n+m$ is the identity matrix I_m. Furthermore, since $x_i \geq 0$ for all i, the objective function $-(x_{n+1} + \cdots + x_{n+m})$ is bounded above by 0.

If we execute the simplex algorithm with a pivot rule that prevents cycling, starting with the basic feasible solution $(0_n, d)$, since the objective function is bounded by 0, the simplex algorithm terminates with an optimal solution given by some basic feasible solution, say (u^*, w^*), with $u^* \in \mathbb{R}^n$ and $w^* \in \mathbb{R}^m$.

As in the proof of Theorem 9.3, for every feasible solution $u \in \mathcal{P}(A, b)$, the vector $(u, 0_m)$ is an optimal solution of $(\widehat{P})$. Therefore, if $w^* \neq 0$, then $\mathcal{P}(A, b) = \emptyset$, since otherwise for every feasible solution $u \in \mathcal{P}(A, b)$ the vector $(u, 0_m)$ would yield a value of the objective function $-(x_{n+1} + \cdots + x_{n+m})$ equal to 0, but (u^*, w^*) yields a strictly negative value since $w^* \neq 0$.

Otherwise, $w^* = 0$, and u^* is a feasible solution of $(P2)$. Since $(u^*, 0_m)$ is a basic feasible solution of $(\widehat{P})$ the columns corresponding to nonzero components of u^* are linearly independent. Some of the coordinates of u^* could be equal to 0, but since A has rank m we can add columns of A to obtain a basis K^* associated with u^*, and u^* is indeed a basic feasible solution of $(P2)$.

Running the simplex algorithm on the Linear Program $\widehat{P}$ to obtain an initial feasible solution (u_0, K_0) of the linear program $(P2)$ is called *Phase I* of the simplex algorithm. Running the simplex algorithm on the Linear Program $(P2)$ with some initial feasible solution (u_0, K_0) is called *Phase II* of the simplex algorithm. If a feasible solution of the Linear Program $(P2)$ is readily available then Phase I is skipped. Sometimes, at the end of Phase I, an optimal solution of $(P2)$ is already obtained.

In summary, we proved the following fact worth recording.

Proposition 10.1. *For any Linear Program* $(P2)$

$$\begin{aligned} &\text{maximize} && cx \\ &\text{subject to} && Ax = b \text{ and } x \geq 0, \end{aligned}$$

in standard form, where A *is an* $m \times n$ *matrix of rank* m *and* $b \geq 0$, *consider the Linear Program* $(\widehat{P})$ *in standard form*

$$\begin{aligned} &\text{maximize} && -(x_{n+1} + \cdots + x_{n+m}) \\ &\text{subject to} && \widehat{A}\widehat{x} = b \text{ and } \widehat{x} \geq 0. \end{aligned}$$

The simplex algorithm with a pivot rule that prevents cycling started on the basic feasible solution $\widehat{x} = (0_n, b)$ *of* $(\widehat{P})$ *terminates with an optimal solution* (u^*, w^*).

(1) *If* $w^* \neq 0$, *then* $\mathcal{P}(A, b) = \emptyset$, *that is, the Linear Program* $(P2)$ *has no feasible solution.*

(2) *If* $w^* = 0$, *then* $\mathcal{P}(A, b) \neq \emptyset$, *and* u^* *is a basic feasible solution of* $(P2)$ *associated with some basis* K.

Proposition 10.1 shows that determining whether the polyhedron $\mathcal{P}(A, b)$ defined by a system of equations $Ax = b$ and inequalities $x \geq 0$ is nonempty is decidable. This decision procedure uses a fail-safe version of the simplex algorithm (that prevents cycling), and the proof that it always terminates and returns an answer is nontrivial.

10.3 How to Perform a Pivoting Step Efficiently

We now discuss briefly how to perform the computation of (u^+, K^+) from a basic feasible solution (u, K).

In order to avoid applying permutation matrices it is preferable to allow a basis K to be a sequence of indices, possibly out of order. Thus, for any $m \times n$ matrix A (with $m \leq n$) and any sequence $K = (k_1, k_2, \cdots, k_m)$ of m elements with $k_i \in \{1, \ldots, n\}$, the matrix A_K denotes the $m \times m$ matrix whose ith column is the k_ith column of A, and similarly for any vector $u \in \mathbb{R}^n$ (resp. any linear form $c \in (\mathbb{R}^n)^*$), the vector $u_K \in \mathbb{R}^m$ (the linear form $c_K \in (\mathbb{R}^m)^*$) is the vector whose ith entry is the k_ith entry in u (resp. the linear whose ith entry is the k_ith entry in c).

For each nonbasic $j \notin K$, we have

$$A^j = \gamma^j_{k_1} A^{k_1} + \cdots + \gamma^j_{k_m} A^{k_m} = A_K \gamma^j_K,$$

so the vector γ^j_K is given by $\gamma^j_K = A_K^{-1} A^j$, that is, by solving the system

$$A_K \gamma^j_K = A^j. \qquad (*_\gamma)$$

To be very precise, since the vector γ^j_K depends on K its components should be denoted by $(\gamma^j_K)_{k_i}$, but as we said before, to simplify notation we write $\gamma^j_{k_i}$ instead of $(\gamma^j_K)_{k_i}$.

In order to decide which case applies ((A), (B1), (B2), (B3)), we need to compute the numbers $c_j - \sum_{k \in K} \gamma^j_k c_k$ for all $j \notin K$. For this, observe that

$$c_j - \sum_{k \in K} \gamma^j_k c_k = c_j - c_K \gamma^j_K = c_j - c_K A_K^{-1} A^j.$$

If we write $\beta_K = c_K A_K^{-1}$, then

$$c_j - \sum_{k \in K} \gamma_k^j c_k = c_j - \beta_K A^j,$$

and we see that $\beta_K^\top \in \mathbb{R}^m$ is the solution of the system $\beta_K^\top = (A_K^{-1})^\top c_k^\top$, which means that $\beta_K^\top$ is the solution of the system

$$A_K^\top \beta_K^\top = c_K^\top. \qquad (*_\beta)$$

Remark: Observe that since u is a basis feasible solution of (P), we have $u_j = 0$ for all $j \notin K$, so u is the solution of the equation $A_K u_K = b$. As a consequence, the value of the objective function for u is $cu = c_K u_K = c_K A_K^{-1} b$. This fact will play a crucial role in Section 11.2 to show that when the simplex algorithm terminates with an optimal solution of the Linear Program (P), then it also produces an optimal solution of the Dual Linear Program (D).

Assume that we have a basic feasible solution u, a basis K for u, and that we also have the matrix A_K as well its inverse A_K^{-1} (perhaps implicitly) and also the inverse $(A_K^\top)^{-1}$ of $A_K^\top$ (perhaps implicitly). Here is a description of an iteration step of the simplex algorithm, following almost exactly Chvatal (Chvatal [Chvatal (1983)], Chapter 7, Box 7.1).

An Iteration Step of the (Revised) Simplex Method

Step 1. Compute the numbers $c_j - \sum_{k \in K} \gamma_k^j c_k = c_j - \beta_K A^j$ for all $j \notin K$, and for this, compute $\beta_K^\top$ as the solution of the system

$$A_K^\top \beta_K^\top = c_K^\top.$$

If $c_j - \beta_K A^j \leq 0$ for all $j \notin K$, stop and return the optimal solution u (Case (A)).

Step 2. If Case (B) arises, use a pivot rule to determine which index $j^+ \notin K$ should enter the new basis K^+ (the condition $c_{j^+} - \beta_K A^{j^+} > 0$ should hold).

Step 3. Compute $\max_{k \in K} \gamma_k^{j^+}$. For this, solve the linear system

$$A_K \gamma_K^{j^+} = A^{j^+}.$$

Step 4. If $\max_{k \in K} \gamma_k^{j^+} \leq 0$, then stop and report that Linear Program (P) is unbounded (Case (B1)).

Step 5. If $\max_{k \in K} \gamma_k^{j^+} > 0$, use the ratios $u_k / \gamma_k^{j^+}$ for all $k \in K$ such that $\gamma_k^{j^+} > 0$ to compute θ^{j^+}, and use a pivot rule to determine which index $k^- \in K$ such that $\theta^{j^+} = u_{k^-} / \gamma_{k^-}^{j+}$ should leave K (Case (B2)).

If $\max_{k\in K} \gamma_k^{j^+} = 0$, then use a pivot rule to determine which index k^- for which $\gamma_{k^-}^{j^+} > 0$ should leave the basis K (Case (B3)).

Step 6. Update u, K, and A_K, to u^+ and K^+, and A_{K^+}. During this step, given the basis K specified by the sequence $K = (k_1, \ldots, k_\ell, \ldots, k_m)$, with $k^- = k_\ell$, then K^+ is the sequence obtained by replacing k_ℓ by the incoming index j^+, so $K^+ = (k_1, \ldots, j^+, \ldots, k_m)$ with j^+ in the ℓth slot.

The vector u is easily updated. To compute A_{K^+} from A_K we take advantage of the fact that A_K and A_{K^+} only differ by a *single column*, namely the ℓth column A^{j^+}, which is given by the linear combination

$$A^{j^+} = A_K \gamma_K^{j^+}.$$

To simplify notation, denote $\gamma_K^{j^+}$ by γ, and recall that $k^- = k_\ell$. If $K = (k_1, \ldots, k_m)$, then $A_K = [A^{k_1} \cdots A^{k^-} \cdots A^{i_m}]$, and since A_{K^+} is the result of replacing the ℓth column A^{k^-} of A_K by the column A^{j^+}, we have

$$A_{K^+} = [A^{k_1} \cdots A^{j^+} \cdots A^{i_m}] = [A^{k_1} \cdots A_K\gamma \cdots A^{i_m}] = A_K E(\gamma),$$

where $E(\gamma)$ is the following invertible matrix obtained from the identity matrix I_m by replacing its ℓth column by γ:

$$E(\gamma) = \begin{pmatrix} 1 & & & \gamma_1 & & & \\ & \ddots & & \vdots & & & \\ & & 1 & \gamma_{\ell-1} & & & \\ & & & \gamma_\ell & & & \\ & & & \gamma_{\ell+1} & 1 & & \\ & & & \vdots & & \ddots & \\ & & & \gamma_m & & & 1 \end{pmatrix}.$$

Since $\gamma_\ell = \gamma_{k^-}^{j^+} > 0$, the matrix $E(\gamma)$ is invertible, and it is easy to check that its inverse is given by

$$E(\gamma)^{-1} = \begin{pmatrix} 1 & & & -\gamma_\ell^{-1}\gamma_1 & & & \\ & \ddots & & \vdots & & & \\ & & 1 & -\gamma_\ell^{-1}\gamma_{\ell-1} & & & \\ & & & \gamma_\ell^{-1} & & & \\ & & & -\gamma_\ell^{-1}\gamma_{\ell+1} & 1 & & \\ & & & \vdots & & \ddots & \\ & & & -\gamma_\ell^{-1}\gamma_m & & & 1 \end{pmatrix},$$

which is very cheap to compute. We also have

$$A_{K^+}^{-1} = E(\gamma)^{-1} A_K^{-1}.$$

Consequently, if A_K and A_K^{-1} are available, then A_{K^+} and $A_{K^+}^{-1}$ can be computed cheaply in terms of A_K and A_K^{-1} and matrices of the form $E(\gamma)$. Then the systems $(*_\gamma)$ to find the vectors γ_K^j can be solved cheaply.

Since

$$A_{K^+}^\top = E(\gamma)^\top A_K^\top$$

and

$$(A_{K^+}^\top)^{-1} = (A_K^\top)^{-1}(E(\gamma)^\top)^{-1},$$

the matrices $A_{K^+}^\top$ and $(A_{K^+}^\top)^{-1}$ can also be computed cheaply from $A_K^\top$, $(A_K^\top)^{-1}$, and matrices of the form $E(\gamma)^\top$. Thus the systems $(*_\beta)$ to find the linear forms β_K can also be solved cheaply.

A matrix of the form $E(\gamma)$ is called an *eta matrix*; see Chvatal [Chvatal (1983)] (Chapter 7). We showed that the matrix A_{K^s} obtained after s steps of the simplex algorithm can be written as

$$A_{K^s} = A_{K^{s-1}} E_s$$

for some eta matrix E_s, so A_{k^s} can be written as the product

$$A_{K^s} = E_1 E_2 \cdots E_s$$

of s eta matrices. Such a factorization is called an *eta factorization.* The eta factorization can be used to either invert A_{K^s} or to solve a system of the form $A_{K_s}\gamma = A^{j^+}$ iteratively. Which method is more efficient depends on the sparsity of the E_i.

In summary, there are cheap methods for finding the next basic feasible solution (u^+, K^+) from (u, K). We simply wanted to give the reader a flavor of these techniques. We refer the reader to texts on linear programming for detailed presentations of methods for implementing efficiently the simplex method. In particular, the *revised simplex method* is presented in Chvatal [Chvatal (1983)], Papadimitriou and Steiglitz [Papadimitriou and Steiglitz (1998)], Bertsimas and Tsitsiklis [Bertsimas and Tsitsiklis (1997)], and Vanderbei [Vanderbei (2014)].

10.4 The Simplex Algorithm Using Tableaux

We now describe a formalism for presenting the simplex algorithm, namely *(full) tableaux*. This is the traditional formalism used in all books, modulo minor variations. A particularly nice feature of the tableau formalism is that the update of a tableau can be performed using elementary row operations *identical* to the operations used during the reduction of a matrix to row reduced echelon form (rref). What differs is the criterion for the choice of the pivot.

Since the quantities $c_j - c_K\gamma_K^j$ play a crucial role in determining which column A^j should come into the basis, the notation $\overline{c}_j$ is used to denote $c_j - c_K\gamma_K^j$, which is called the *reduced cost* of the variable x_j. The reduced costs actually depend on K so to be very precise we should denote them by $(\overline{c}_K)_j$, but to simplify notation we write $\overline{c}_j$ instead of $(\overline{c}_K)_j$. We will see shortly how $(\overline{c}_{K^+})_i$ is computed in terms of $(\overline{c}_K)_i$.

Observe that the data needed to execute the next step of the simplex algorithm are

(1) The current basic solution u_K and its basis $K = (k_1, \ldots, k_m)$.
(2) The reduced costs $\overline{c}_j = c_j - c_K A_K^{-1} A^j = c_j - c_K\gamma_K^j$, for all $j \notin K$.
(3) The vectors $\gamma_K^j = (\gamma_{k_i}^j)_{i=1}^m$ for all $j \notin K$, that allow us to express each A^j as $A_K\gamma_K^j$.

All this information can be packed into a $(m+1) \times (n+1)$ matrix called a *(full) tableau* organized as follows:

$c_K u_K$	$\overline{c}_1$	$\cdots$	$\overline{c}_j$	$\cdots$	$\overline{c}_n$
u_{k_1}	γ_1^1	$\cdots$	γ_1^j	$\cdots$	γ_1^n
$\vdots$	$\vdots$		$\vdots$		$\vdots$
u_{k_m}	γ_m^1	$\cdots$	γ_m^j	$\cdots$	γ_m^n

It is convenient to think as the first row as Row 0, and of the first column as Column 0. Row 0 contains the current value of the objective function and the reduced costs. Column 0, except for its top entry, contains the components of the current basic solution u_K, and the remaining columns, except for their top entry, contain the vectors γ_K^j. Observe that the γ_K^j corresponding to indices j in K constitute a permutation of the identity matrix I_m. The entry $\gamma_{k^-}^{j^+}$ is called the *pivot* element. A tableau together with the new basis $K^+ = (K - \{k^-\}) \cup \{j^+\}$ contains all the data needed to compute the new u_{K^+}, the new $\gamma_{K^+}^j$, and the new reduced costs $(\overline{c}_{K^+})_j$.

If we define the $m \times n$ matrix Γ as the matrix $\Gamma = [\gamma_K^1 \; \cdots \; \gamma_K^n]$ whose jth column is γ_K^j, and $\overline{c}$ as the row vector $\overline{c} = (\overline{c}_1 \; \cdots \; \overline{c}_n)$, then the above tableau is denoted concisely by

$c_K u_K$	$\overline{c}$
u_K	Γ

We now show that the update of a tableau can be performed using elementary row operations identical to the operations used during the reduction of a matrix to row reduced echelon form (rref).

If $K = (k_1, \ldots, k_m)$, j^+ is the index of the incoming basis vector, $k^- = k_\ell$ is the index of the column leaving the basis, and if $K^+ = (k_1, \ldots, k_{\ell-1}, j^+, k_{\ell+1}, \ldots, k_m)$, since $A_{K^+} = A_K E(\gamma_K^{j^+})$, the new columns $\gamma_{K^+}^j$ are computed in terms of the old columns γ_K^j using $(*_\gamma)$ and the equations

$$\gamma_{K^+}^j = A_{K^+}^{-1} A^j = E(\gamma_K^{j^+})^{-1} A_K^{-1} A^j = E(\gamma_K^{j^+})^{-1} \gamma_K^j.$$

Consequently, the matrix Γ^+ is given in terms of Γ by

$$\Gamma^+ = E(\gamma_K^{j^+})^{-1} \Gamma.$$

But the matrix $E(\gamma_K^{j^+})^{-1}$ is of the form

$$E(\gamma_K^{j^+})^{-1} = \begin{pmatrix} 1 & & & -(\gamma_{k^-}^{j^+})^{-1}\gamma_{k_1}^{j^+} & & & \\ & \ddots & & \vdots & & & \\ & & 1 & -(\gamma_{k^-}^{j^+})^{-1}\gamma_{k_{\ell-1}}^{j^+} & & & \\ & & & (\gamma_{k^-}^{j^+})^{-1} & & & \\ & & & -(\gamma_{k^-}^{j^+})^{-1}\gamma_{k_{\ell+1}}^{j^+} & 1 & & \\ & & & \vdots & & \ddots & \\ & & & -(\gamma_{k^-}^{j^+})^{-1}\gamma_{k_m}^{j^+} & & & 1 \end{pmatrix},$$

with the column involving the γs in the ℓth column, and Γ^+ is obtained by applying the following elementary row operations to Γ:

(1) Multiply Row ℓ by $1/\gamma_{k^-}^{j^+}$ (the inverse of the pivot) to make the entry on Row ℓ and Column j^+ equal to 1.
(2) Subtract $\gamma_{k_i}^{j^+} \times$ (the normalized) Row ℓ from Row i, for $i = 1, \ldots, \ell - 1, \ell + 1, \ldots, m$.

These are *exactly* the elementary row operations that reduce the ℓth column $\gamma_K^{j^+}$ of Γ to the ℓth column of the identity matrix I_m. Thus, this step is identical to the sequence of steps that the procedure to convert a matrix to row reduced echelon from executes on the ℓth column of the matrix. The only difference is the criterion for the choice of the pivot.

Since the new basic solution u_{K^+} is given by $u_{K^+} = A_{K^+}^{-1} b$, we have

$$u_{K^+} = E(\gamma_K^{j^+})^{-1} A_K^{-1} b = E(\gamma_K^{j^+})^{-1} u_K.$$

This means that u_+ is obtained from u_K by applying exactly the *same* elementary row operations that were applied to Γ. Consequently, just as in the procedure for reducing a matrix to rref, we can apply elementary row operations to the matrix $[u_k\ \Gamma]$, which consists of rows $1, \ldots, m$ of the tableau.

Once the new matrix Γ^+ is obtained, the new reduced costs are given by the following proposition.

Proposition 10.2. *Given any Linear Program (P2) in standard form*

$$\begin{aligned}&\text{maximize} \quad cx\\&\text{subject to} \quad Ax = b \text{ and } x \geq 0,\end{aligned}$$

where A is an $m \times n$ matrix of rank m, if (u, K) is a basic (not necessarily feasible) solution of (P2) and if $K^+ = (K - \{k^-\}) \cup \{j^+\}$, with $K = (k_1, \ldots, k_m)$ and $k^- = k_\ell$, then for $i = 1, \ldots, n$ we have

$$c_i - c_{K^+}\gamma^i_{K^+} = c_i - c_K\gamma^i_K - \frac{\gamma^i_{k^-}}{\gamma^{j^+}_{k^-}}(c_{j^+} - c_K\gamma^{j^+}_K).$$

Using the reduced cost notation, the above equation is

$$(\overline{c}_{K^+})_i = (\overline{c}_K)_i - \frac{\gamma^i_{k^-}}{\gamma^{j^+}_{k^-}}(\overline{c}_K)_{j^+}.$$

Proof. Without any loss of generality and to simplify notation assume that $K = (1, \ldots, m)$ and write j for j^+ and ℓ for k_m. Since $\gamma^i_K = A_K^{-1}A^i$, $\gamma^i_{K^+} = A_{K^+}^{-1}A^i$, and $A_{K^+} = A_K E(\gamma^j_K)$, we have

$$\begin{aligned}c_i - c_{K^+}\gamma^i_{K^+} &= c_i - c_{K^+}A_{K^+}^{-1}A^i = c_i - c_{K^+}E(\gamma^j_K)^{-1}A_K^{-1}A^i\\&= c_i - c_{K^+}E(\gamma^j_K)^{-1}\gamma^i_K,\end{aligned}$$

where

$$E(\gamma^j_K)^{-1} = \begin{pmatrix}1 & & & -(\gamma^j_\ell)^{-1}\gamma^j_1 & & & \\ & \ddots & & \vdots & & & \\ & & 1 & -(\gamma^j_\ell)^{-1}\gamma^j_{\ell-1} & & & \\ & & & (\gamma^j_\ell)^{-1} & & & \\ & & & -(\gamma^j_\ell)^{-1}\gamma^j_{\ell+1} & 1 & & \\ & & & \vdots & & \ddots & \\ & & & -(\gamma^j_\ell)^{-1}\gamma^j_m & & & 1\end{pmatrix}$$

where the ℓth column contains the γs. Since $c_{K^+} = (c_1, \ldots, c_{\ell-1}, c_j, c_{\ell+1}, \ldots, c_m)$, we have

$$c_{K^+}E(\gamma_K^j)^{-1} = \left(c_1, \ldots, c_{\ell-1}, \frac{c_j}{\gamma_\ell^j} - \sum_{k=1,k\neq\ell}^{m} c_k\frac{\gamma_k^j}{\gamma_\ell^j}, c_{\ell+1}, \ldots, c_m\right),$$

and

$$\begin{aligned}
c_{K^+}E(\gamma_K^j)^{-1}\gamma_K^i &= \left(c_1 \;\cdots\; c_{\ell-1} \;\; \frac{c_j}{\gamma_\ell^j} - \sum_{k=1,k\neq\ell}^{m} c_k\frac{\gamma_k^j}{\gamma_\ell^j} \;\; c_{\ell+1} \;\cdots\; c_m\right)\begin{pmatrix}\gamma_1^i\\ \vdots\\ \gamma_{\ell-1}^i\\ \gamma_\ell^i\\ \gamma_{\ell+1}^i\\ \vdots\\ \gamma_m^i\end{pmatrix}\\
&= \sum_{k=1,k\neq\ell}^{m} c_k\gamma_k^i + \frac{\gamma_\ell^i}{\gamma_\ell^j}\left(c_j - \sum_{k=1,k\neq\ell}^{m} c_k\gamma_k^j\right)\\
&= \sum_{k=1,k\neq\ell}^{m} c_k\gamma_k^i + \frac{\gamma_\ell^i}{\gamma_\ell^j}\left(c_j + c_\ell\gamma_\ell^j - \sum_{k=1}^{m} c_k\gamma_k^j\right)\\
&= \sum_{k=1}^{m} c_k\gamma_k^i + \frac{\gamma_\ell^i}{\gamma_\ell^j}\left(c_j - \sum_{k=1}^{m} c_k\gamma_k^j\right)\\
&= c_K\gamma_K^i + \frac{\gamma_\ell^i}{\gamma_\ell^j}(c_j - c_K\gamma_K^j),
\end{aligned}$$

and thus

$$c_i - c_{K^+}\gamma_{K^+}^i = c_i - c_{K^+}E(\gamma_K^j)^{-1}\gamma_K^i = c_i - c_K\gamma_K^i - \frac{\gamma_\ell^i}{\gamma_\ell^j}(c_j - c_K\gamma_K^j),$$

as claimed. □

Since $(\gamma_{k^-}^1, \ldots, \gamma_{k^-}^n)$ is the ℓth row of Γ, we see that Proposition 10.2 shows that

$$\overline{c}_{K^+} = \overline{c}_K - \frac{(\overline{c}_K)_{j^+}}{\gamma_{k^-}^{j^+}}\Gamma_\ell, \tag{†}$$

where Γ_ℓ denotes the ℓ-th row of Γ and $\gamma_{k^-}^{j^+}$ is the pivot. This means that $\overline{c}_{K^+}$ is obtained by the elementary row operations which consist of first normalizing the ℓth row by dividing it by the pivot $\gamma_{k^-}^{j^+}$, and then subtracting $(\overline{c}_K)_{j^+} \times$ the normalized Row ℓ from $\overline{c}_K$. *These are exactly the row operations that make the reduced cost* $(\overline{c}_K)_{j^+}$ *zero.*

Remark: It easy easy to show that we also have

$$\overline{c}_{K^+} = c - c_{K^+}\Gamma^+.$$

We saw in Section 10.2 that the change in the objective function after a pivoting step during which column j^+ comes in and column k^- leaves is given by

$$\theta^{j^+}\left(c_{j^+} - \sum_{k\in K} \gamma_k^{j^+} c_k\right) = \theta^{j^+}(\overline{c}_K)_{j^+},$$

where

$$\theta^{j^+} = \frac{u_{k^-}}{\gamma_{k^-}^{j^+}}.$$

If we denote the value of the objective function $c_K u_K$ by z_K, then we see that

$$z_{K^+} = z_K + \frac{(\overline{c}_K)_{j^+}}{\gamma_{k^-}^{j^+}} u_{k^-}.$$

This means that the new value z_{K^+} of the objective function is obtained by first normalizing the ℓth row by dividing it by the pivot $\gamma_{k^-}^{j^+}$, and then adding $(\overline{c}_K)_{j^+}\times$ the zeroth entry of the normalized ℓth line by $(\overline{c}_K)_{j^+}$ to the zeroth entry of line 0.

In updating the reduced costs, we subtract rather than add $(\overline{c}_K)_{j^+}\times$ the normalized row ℓ from row 0. This suggests storing $-z_K$ as the zeroth entry on line 0 rather than z_K, because then all the entries row 0 are updated by the *same* elementary row operations. Therefore, from now on, we use tableau of the form

$-c_K u_K$	$\overline{c}_1$	$\cdots$	$\overline{c}_j$	$\cdots$	$\overline{c}_n$
u_{k_1}	γ_1^1	$\cdots$	γ_1^j	$\cdots$	γ_1^n
$\vdots$	$\vdots$		$\vdots$		$\vdots$
u_{k_m}	γ_m^1	$\cdots$	γ_m^j	$\cdots$	γ_m^n

The simplex algorithm first chooses the incoming column j^+ by picking some column for which $\overline{c}_j > 0$, and then chooses the outgoing column k^- by considering the ratios $u_k/\gamma_k^{j^+}$ for which $\gamma_k^{j^+} > 0$ (along column j^+), and picking k^- to achieve the minimum of these ratios.

Here is an illustration of the simplex algorithm using elementary row operations on an example from Papadimitriou and Steiglitz [Papadimitriou and Steiglitz (1998)] (Section 2.9).

Example 10.4. Consider the linear program

$$\begin{aligned}
&\text{maximize} \quad -2x_2 - x_4 - 5x_7\\
&\text{subject to}\\
&\qquad x_1 + x_2 + x_3 + x_4 = 4\\
&\qquad x_1 + x_5 = 2\\
&\qquad x_3 + x_6 = 3\\
&\qquad 3x_2 + x_3 + x_7 = 6\\
&\qquad x_1, x_2, x_3, x_4, x_5, x_6, x_7 \geq 0.
\end{aligned}$$

We have the basic feasible solution $u = (0, 0, 0, 4, 2, 3, 6)$, with $K = (4, 5, 6, 7)$. Since $c_K = (-1, 0, 0, -5)$ and $c = (0, -2, 0, -1, 0, 0 - 5)$ the first tableau is

34	1	14	6	0	0	0	0
$u_4 = 4$	1	1	1	1	0	0	0
$u_5 = 2$	(1)	0	0	0	1	0	0
$u_6 = 3$	0	0	1	0	0	1	0
$u_7 = 6$	0	3	1	0	0	0	1

Since $\overline{c}_j = c_j - c_K\gamma_K^j$, Row 0 is obtained by subtracting $-1\times$ Row 1 and $-5\times$ Row 4 from $c = (0, -2, 0, -1, 0, 0, -5)$. Let us pick Column $j^+ = 1$ as the incoming column. We have the ratios (for positive entries on Column 1)

$$4/1,\ 2/1,$$

and since the minimum is 2, we pick the outgoing column to be Column $k^- = 5$. The pivot 1 is indicated in red. The new basis is $K = (4, 1, 6, 7)$. Next we apply row operations to reduce Column 1 to the second vector of the identity matrix I_4. For this, we subtract Row 2 from Row 1. We get the tableau

34	1	14	6	0	0	0	0
$u_4 = 2$	0	1	1	1	-1	0	0
$u_1 = 2$	(1)	0	0	0	1	0	0
$u_6 = 3$	0	0	1	0	0	1	0
$u_7 = 6$	0	3	1	0	0	0	1

To compute the new reduced costs, we want to set $\bar{c}_1$ to 0, so we apply the identical row operations and subtract Row 2 from Row 0 to obtain the tableau

32	0	14	6	0	−1	0	0
$u_4 = 2$	0	1	⓵	1	−1	0	0
$u_1 = 2$	1	0	0	0	1	0	0
$u_6 = 3$	0	0	1	0	0	1	0
$u_7 = 6$	0	3	1	0	0	0	1

Next, pick Column $j^+ = 3$ as the incoming column. We have the ratios (for positive entries on Column 3)

$$2/1, \ 3/1, \ 6/1,$$

and since the minimum is 2, we pick the outgoing column to be Column $k^- = 4$. The pivot 1 is indicated in red and the new basis is $K = (3, 1, 6, 7)$. Next we apply row operations to reduce Column 3 to the first vector of the identity matrix I_4. For this, we subtract Row 1 from Row 3 and from Row 4 and obtain the tableau:

32	0	14	6	0	−1	0	0
$u_3 = 2$	0	1	⓵	1	−1	0	0
$u_1 = 2$	1	0	0	0	1	0	0
$u_6 = 1$	0	−1	0	−1	1	1	0
$u_7 = 4$	0	2	0	−1	1	0	1

To compute the new reduced costs, we want to set $\bar{c}_3$ to 0, so we subtract $6\times$ Row 1 from Row 0 to get the tableau

20	0	8	0	−6	5	0	0
$u_3 = 2$	0	⓵	1	1	−1	0	0
$u_1 = 2$	1	0	0	0	1	0	0
$u_6 = 1$	0	−1	0	−1	1	1	0
$u_7 = 4$	0	2	0	−1	1	0	1

Next we pick $j^+ = 2$ as the incoming column. We have the ratios (for positive entries on Column 2)

$$2/1, \ 4/2,$$

and since the minimum is 2, we pick the outgoing column to be Column $k^- = 3$. The pivot 1 is indicated in red and the new basis is $K = (2, 1, 6, 7)$. Next we apply row operations to reduce Column 2 to the first vector of the identity matrix I_4. For this, we add Row 1 to Row 3 and subtract 2× Row 1 from Row 4 to obtain the tableau:

20	0	8	0	−6	5	0	0
$u_2 = 2$	0	(1)	1	1	−1	0	0
$u_1 = 2$	1	0	0	0	1	0	0
$u_6 = 3$	0	0	1	0	0	1	0
$u_7 = 0$	0	0	−2	−3	3	0	1

To compute the new reduced costs, we want to set $\overline{c}_2$ to 0, so we subtract 8× Row 1 from Row 0 to get the tableau

4	0	0	−8	−14	13	0	0
$u_2 = 2$	0	1	1	1	−1	0	0
$u_1 = 2$	1	0	0	0	1	0	0
$u_6 = 3$	0	0	1	0	0	1	0
$u_7 = 0$	0	0	−2	−3	(3)	0	1

The only possible incoming column corresponds to $j^+ = 5$. We have the ratios (for positive entries on Column 5)

$$2/1,\ 0/3,$$

and since the minimum is 0, we pick the outgoing column to be Column $k^- = 7$. The pivot 3 is indicated in red and the new basis is $K = (2, 1, 6, 5)$. Since the minimum is 0, the basis $K = (2, 1, 6, 5)$ is degenerate (indeed, the component corresponding to the index 5 is 0). Next we apply row operations to reduce Column 5 to the fourth vector of the identity matrix I_4. For this, we multiply Row 4 by 1/3, and then add the normalized Row 4 to Row 1 and subtract the normalized Row 4 from Row 2 to obtain the tableau:

4	0	0	−8	−14	13	0	0
$u_2 = 2$	0	1	1/3	0	0	0	1/3
$u_1 = 2$	1	0	2/3	1	0	0	−1/3
$u_6 = 3$	0	0	1	0	0	1	0
$u_5 = 0$	0	0	−2/3	−1	(1)	0	1/3

To compute the new reduced costs, we want to set $\overline{c}_5$ to 0, so we subtract $13\times$ Row 4 from Row 0 to get the tableau

4	0	0	2/3	−1	0	0	−13/3
$u_2 = 2$	0	1	1/3	0	0	0	1/3
$u_1 = 2$	1	0	(2/3)	1	0	0	−1/3
$u_6 = 3$	0	0	1	0	0	1	0
$u_5 = 0$	0	0	−2/3	−1	1	0	1/3

The only possible incoming column corresponds to $j^+ = 3$. We have the ratios (for positive entries on Column 3)

$$2/(1/3) = 6,\ 2/(2/3) = 3,\ 3/1 = 3,$$

and since the minimum is 3, we pick the outgoing column to be Column $k^- = 1$. The pivot 2/3 is indicated in red and the new basis is $K = (2, 3, 6, 5)$. Next we apply row operations to reduce Column 3 to the second vector of the identity matrix I_4. For this, we multiply Row 2 by 3/2, subtract $(1/3)\times$ (normalized Row 2) from Row 1, and subtract normalized Row 2 from Row 3, and add $(2/3)\times$ (normalized Row 2) to Row 4 to obtain the tableau:

4	0	0	2/3	−1	0	0	−13/3
$u_2 = 1$	−1/2	1	0	−1/2	0	0	1/2
$u_3 = 3$	3/2	0	(1)	3/2	0	0	−1/2
$u_6 = 0$	−3/2	0	0	−3/2	0	1	1/2
$u_5 = 2$	1	0	0	0	1	0	0

To compute the new reduced costs, we want to set $\overline{c}_3$ to 0, so we subtract $(2/3)\times$ Row 2 from Row 0 to get the tableau

2	−1	0	0	−2	0	0	−4
$u_2 = 1$	−1/2	1	0	−1/2	0	0	1/2
$u_3 = 3$	3/2	0	1	3/2	0	0	−1/2
$u_6 = 0$	−3/2	0	0	−3/2	0	1	1/2
$u_5 = 2$	1	0	0	0	1	0	0

Since all the reduced cost are ≤ 0, we have reached an optimal solution, namely $(0, 1, 3, 0, 2, 0, 0, 0)$, with optimal value -2.

The progression of the simplex algorithm from one basic feasible solution to another corresponds to the visit of vertices of the polyhedron $\mathcal{P}$ associated with the constraints of the linear program illustrated in Figure 10.4.

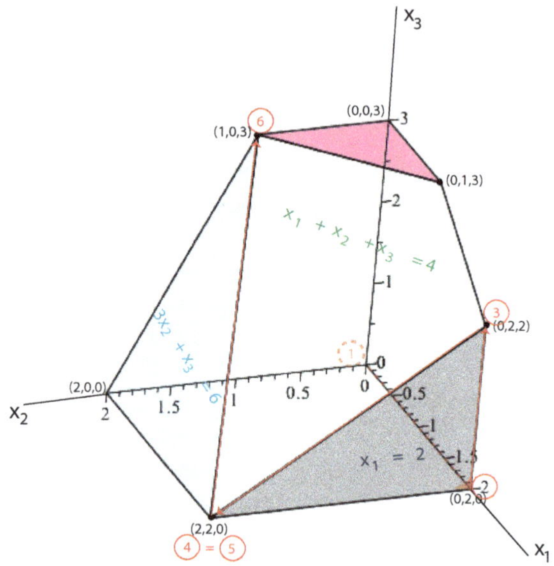

Fig. 10.4 The polytope $\mathcal{P}$ associated with the linear program optimized by the tableau method. The red arrowed path traces the progression of the simplex method from the origin to the vertex $(0, 1, 3)$.

As a final comment, if it is necessary to run Phase I of the simplex algorithm, in the event that the simplex algorithm terminates with an optimal solution $(u^*, 0_m)$ and a basis K^* such that some $u_i = 0$, then the basis K^* contains indices of basic columns A^j corresponding to slack variables that need to be *driven out* of the basis. This is easy to achieve by performing a pivoting step involving some other column j^+ corresponding to one of the original variables (not a slack variable) for which $(\gamma_{K^*})_i^{j^+} \neq 0$. In such a step, it doesn't matter whether $(\gamma_{K^*})_i^{j^+} < 0$ or $(\overline{c}_{K^*})_{j^+} \leq 0$. If the original matrix A has no redundant equations, such a step is always possible. Otherwise, $(\gamma_{K^*})_i^j = 0$ for all non-slack variables, so we detected that the ith equation is redundant and we can delete it.

Other presentations of the tableau method can be found in Bertsimas and Tsitsiklis [Bertsimas and Tsitsiklis (1997)] and Papadimitriou and Steiglitz [Papadimitriou and Steiglitz (1998)].

10.5 Computational Efficiency of the Simplex Method

Let us conclude with a few comments about the efficiency of the simplex algorithm. In *practice*, it was observed by Dantzig that for linear programs with $m < 50$ and $m + n < 200$, the simplex algorithms typically requires less than $3m/2$ iterations, but at most $3m$ iterations. This fact agrees with more recent empirical experiments with much larger programs that show that the number iterations is bounded by $3m$. Thus, it was somewhat of a shock in 1972 when Klee and Minty found a linear program with n variables and n equations for which the simplex algorithm with Dantzig's pivot rule requires requires $2^n - 1$ iterations. This program (taken from Chvatal [Chvatal (1983)], page 47) is reproduced below:

$$\begin{aligned}&\text{maximize} \quad \sum_{j=1}^{n} 10^{n-j} x_j\\ &\text{subject to}\\ &\quad \left(2\sum_{j=1}^{i-1} 10^{i-j} x_j\right) + x_i \leq 100^{i-1}\\ &\quad x_j \geq 0,\end{aligned}$$

for $i = 1, \ldots, n$ and $j = 1, \ldots, n$.

If $p = \max(m, n)$, then, in terms of worse case behavior, for all currently known pivot rules, the simplex algorithm has exponential complexity in p. However, as we said earlier, in practice, nasty examples such as the Klee–Minty example seem to be rare, and the number of iterations appears to be linear in m.

Whether or not a pivot rule (a clairvoyant rule) for which the simplex algorithms runs in polynomial time in terms of m is still an *open problem*.

The *Hirsch conjecture* claims that there is some pivot rule such that the simplex algorithm finds an optimal solution in $O(p)$ steps. The best bound known so far due to Kalai and Kleitman is $m^{1+\ln n} = (2n)^{\ln m}$. For more on this topic, see Matousek and Gardner [Matousek and Gartner (2007)] (Section 5.9) and Bertsimas and Tsitsiklis [Bertsimas and Tsitsiklis (1997)] (Section 3.7).

Researchers have investigated the problem of finding upper bounds on the expected number of pivoting steps if a randomized pivot rule is used. Bounds better than 2^m (but of course, not polynomial) have been found.

Understanding the complexity of linear programming, in particular of the simplex algorithm, is still ongoing. The interested reader is referred to Matousek and Gardner [Matousek and Gartner (2007)] (Chapter 5, Section 5.9) for some pointers.

In the next section we consider important theoretical criteria for determining whether a set of constraints $Ax \leq b$ and $x \geq 0$ has a solution or not.

10.6 Summary

The main concepts and results of this chapter are listed below:

- Degenerate and nondegenerate basic feasible solution.
- Pivoting step, pivot rule.
- Cycling.
- Bland's rule, Dantzig's rule, steepest edge rule, random edge rule, largest increase rule, lexicographic rule.
- Phase I and Phase II of the simplex algorithm.
- eta matrix, eta factorization.
- Revised simplex method.
- Reduced cost.
- Full tableaux.
- The Hirsch conjecture.

10.7 Problems

Problem 10.1. In Section 10.2 prove that if Case (A) arises, then the basic feasible solution u is an optimal solution. Prove that if Case (B1) arises, then the linear program is unbounded. Prove that if Case (B3) arises, then (u^+, K^+) is a basic feasible solution.

Problem 10.2. In Section 10.2 prove that the following equivalences hold:

$$\begin{aligned}
&\text{Case (A)} \Longleftrightarrow B = \emptyset, \qquad \text{Case (B)} \Longleftrightarrow B \neq \emptyset\\
&\text{Case (B1)} \Longleftrightarrow B_1 \neq \emptyset\\
&\text{Case (B2)} \Longleftrightarrow B_2 \neq \emptyset\\
&\text{Case (B3)} \Longleftrightarrow B_3 \neq \emptyset.
\end{aligned}$$

Furthermore, prove that Cases (A) and (B), Cases (B1) and (B3), and Cases (B2) and (B3) are mutually exclusive, while Cases (B1) and (B2) are not.

Problem 10.3. Consider the linear program (due to E.M.L. Beale):

$$\begin{aligned} \text{maximize} \quad & (3/4)x_1 - 150x_2 + (1/50)x_3 - 6x_4 \\ \text{subject to} \quad & \\ & (1/4)x_1 - 60x_2 - (1/25)x_3 + 9x_4 \leq 0 \\ & (1/4)x_1 - 90x_2 - (1/50)x_3 + 3x_4 \leq 0 \\ & x_3 \leq 1 \\ & x_1 \geq 0,\ x_2 \geq 0,\ x_3 \geq 0,\ x_4 \geq 0. \end{aligned}$$

(1) Convert the above program to standard form.

(2) Show that if we apply the simplex algorithm with the pivot rule which selects the column entering the basis as the column of smallest index, then the method cycles.

Problem 10.4. Read carefully the proof given by Chvatal that the lexicographic pivot rule and Bland's pivot rule prevent cycling; see Chvatal [Chvatal (1983)] (Chapter 3, pages 34–38).

Problem 10.5. Solve the following linear program (from Chvatal [Chvatal (1983)], Chapter 3, page 44) using the two-phase simplex algorithm:

$$\begin{aligned} \text{maximize} \quad & 3x_1 + x_2 \\ \text{subject to} \quad & \\ & x_1 - x_2 \leq -1 \\ & -x_1 - x_2 \leq -3 \\ & 2x_1 + x_2 \leq 4 \\ & x_1 \geq 0,\ x_2 \geq 0. \end{aligned}$$

Problem 10.6. Solve the following linear program (from Chvatal [Chvatal (1983)], Chapter 3, page 44) using the two-phase simplex algorithm:

$$\begin{aligned} \text{maximize} \quad & 3x_1 + x_2 \\ \text{subject to} \quad & \\ & x_1 - x_2 \leq -1 \\ & -x_1 - x_2 \leq -3 \\ & 2x_1 + x_2 \leq 2 \\ & x_1 \geq 0,\ x_2 \geq 0. \end{aligned}$$

Problem 10.7. Solve the following linear program (from Chvatal [Chvatal (1983)], Chapter 3, page 44) using the two-phase simplex algorithm:

$$\begin{aligned} \text{maximize} \quad & 3x_1 + x_2 \\ \text{subject to} \quad & \\ & x_1 - x_2 \le -1 \\ & -x_1 - x_2 \le -3 \\ & 2x_1 - x_2 \le 2 \\ & x_1 \ge 0,\ x_2 \ge 0. \end{aligned}$$

Problem 10.8. Show that the following linear program (from Chvatal [Chvatal (1983)], Chapter 3, page 43) is unbounded.

$$\begin{aligned} \text{maximize} \quad & x_1 + 3x_2 - x_3 \\ \text{subject to} \quad & \\ & 2x_1 + 2x_2 - x_3 \le 10 \\ & 3x_1 - 2x_2 + x_3 \le 10 \\ & x_1 - 3x_2 + x_3 \le 10 \\ & x_1 \ge 0,\ x_2 \ge 0,\ x_3 \ge 0. \end{aligned}$$

Hint. Try $x_1 = 0, x_3 = t$, and a suitable value for x_2.

Chapter 11

Linear Programming and Duality

11.1 Variants of the Farkas Lemma

If A is an $m \times n$ matrix and if $b \in \mathbb{R}^m$ is a vector, it is known from linear algebra that the linear system $Ax = b$ has no solution iff there is some linear form $y \in (\mathbb{R}^m)^*$ such that $yA = 0$ and $yb \neq 0$. This means that the linear from y vanishes on the columns $A^1, \ldots, A^n$ of A but does not vanish on b. Since the linear form y defines the linear hyperplane H of equation $yz = 0$ (with $z \in \mathbb{R}^m$), geometrically the equation $Ax = b$ has no solution iff there is a linear hyperplane H containing $A^1, \ldots, A^n$ and not containing b. This is a kind of separation theorem that says that the vectors $A^1, \ldots, A^n$ and b can be separated by some linear hyperplane H.

What we would like to do is to generalize this kind of criterion, first to a system $Ax = b$ subject to the constraints $x \geq 0$, and next to sets of inequality constraints $Ax \leq b$ and $x \geq 0$. There are indeed such criteria going under the name of *Farkas lemma*.

The key is a separation result involving polyhedral cones known as the Farkas–Minkowski proposition. We have the following fundamental separation lemma.

Proposition 11.1. *Let $C \subseteq \mathbb{R}^n$ be a closed nonempty (convex) cone. For any point $a \in \mathbb{R}^n$, if $a \notin C$, then there is a linear hyperplane H (through 0) such that*

(1) C lies in one of the two half-spaces determined by H.
(2) $a \notin H$.
(3) a lies in the other half-space determined by H.

We say that H strictly separates C and a.

Proposition 11.1, which is illustrated in Figure 11.1, is an easy consequence of another separation theorem that asserts that given any two nonempty closed convex sets A and B of $\mathbb{R}^n$ with A compact, there is a hyperplane H strictly separating A and B (which means that $A \cap H = \emptyset$, $B \cap H = \emptyset$, that A lies in one of the two half-spaces determined by H, and B lies in the other half-space determined by H); see Gallier [Gallier (2011)] (Chapter 7, Corollary 7.4 and Proposition 7.3). This proof is nontrivial and involves a geometric version of the Hahn–Banach theorem.

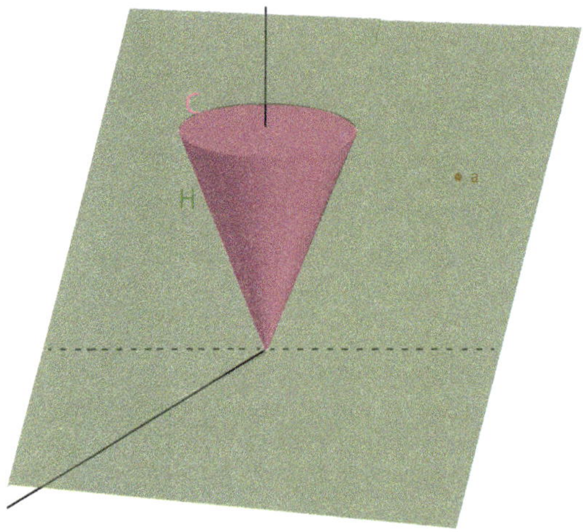

Fig. 11.1 In $\mathbb{R}^3$, the olive green hyperplane H separates the cone C from the orange point a.

The Farkas–Minkowski proposition is Proposition 11.1 applied to a polyhedral cone

$$C = \{\lambda_1 a_1 + \cdots + \lambda_n a_n \mid \lambda_i \geq 0,\ i = 1, \ldots, n\}$$

where $\{a_1, \ldots, a_n\}$ is a *finite* number of vectors $a_i \in \mathbb{R}^n$. By Proposition 8.2, any polyhedral cone is closed, so Proposition 11.1 applies and we obtain the following separation lemma.

Proposition 11.2. *(Farkas–Minkowski) Let $C \subseteq \mathbb{R}^n$ be a nonempty polyhedral cone $C = \text{cone}(\{a_1, \ldots, a_n\})$. For any point $b \in \mathbb{R}^n$, if $b \notin C$, then there is a linear hyperplane H (through 0) such that*

(1) C lies in one of the two half-spaces determined by H.

(2) $b \notin H$.
(3) b lies in the other half-space determined by H.

Equivalently, there is a nonzero linear form $y \in (\mathbb{R}^n)^*$ such that

(1) $ya_i \geq 0$ for $i = 1, \ldots, n$.
(2) $yb < 0$.

A direct proof of the Farkas–Minkowski proposition not involving Proposition 11.1 is given at the end of this section.

Remark: There is a generalization of the Farkas–Minkowski proposition applying to infinite dimensional real Hilbert spaces; see Theorem 12.2 (or Ciarlet [Ciarlet (1989)], Chapter 9).

Proposition 11.2 implies our first version of Farkas' lemma.

Proposition 11.3. *(Farkas Lemma, Version I) Let A be an $m \times n$ matrix and let $b \in \mathbb{R}^m$ be any vector. The linear system $Ax = b$ has no solution $x \geq 0$ iff there is some nonzero linear form $y \in (\mathbb{R}^m)^*$ such that $yA \geq 0_n^\top$ and $yb < 0$.*

Proof. First assume that there is some nonzero linear form $y \in (\mathbb{R}^m)^*$ such that $yA \geq 0$ and $yb < 0$. If $x \geq 0$ is a solution of $Ax = b$, then we get

$$yAx = yb,$$

but if $yA \geq 0$ and $x \geq 0$, then $yAx \geq 0$, and yet by hypothesis $yb < 0$, a contradiction.

Next assume that $Ax = b$ has no solution $x \geq 0$. This means that b does not belong to the polyhedral cone $C = \text{cone}(\{A^1, \ldots, A^n\})$ spanned by the columns of A. By Proposition 11.2, there is a nonzero linear form $y \in (\mathbb{R}^m)^*$ such that

(1) $yA^j \geq 0$ for $j = 1, \ldots, n$.
(2) $yb < 0$,

which says that $yA \geq 0_n^\top$ and $yb < 0$. □

Next consider the solvability of a system of inequalities of the form $Ax \leq b$ and $x \geq 0$.

Proposition 11.4. *(Farkas Lemma, Version II) Let A be an $m \times n$ matrix and let $b \in \mathbb{R}^m$ be any vector. The system of inequalities $Ax \leq b$ has no solution $x \geq 0$ iff there is some nonzero linear form $y \in (\mathbb{R}^m)^*$ such that $y \geq 0_m^\top$, $yA \geq 0_n^\top$ and $yb < 0$.*

Proof. We use the trick of linear programming which consists of adding "slack variables" z_i to convert inequalities $a_i x \leq b_i$ into equations $a_i x + z_i = b_i$ with $z_i \geq 0$ already discussed just before Definition 8.9. If we let $z = (z_1, \ldots, z_m)$, it is obvious that the system $Ax \leq b$ has a solution $x \geq 0$ iff the equation

$$(A \; I_m) \begin{pmatrix} x \\ z \end{pmatrix} = b$$

has a solution $\begin{pmatrix} x \\ z \end{pmatrix}$ with $x \geq 0$ and $z \geq 0$. Now by Farkas I, the above system has no solution with with $x \geq 0$ and $z \geq 0$ iff there is some nonzero linear form $y \in (\mathbb{R}^m)^*$ such that

$$y\,(A \; I_m) \geq 0_{n+m}^\top$$

and $yb < 0$, that is, $yA \geq 0_n^\top$, $y \geq 0_m^\top$ and $yb < 0$. □

In the next section we use Farkas II to prove the duality theorem in linear programming. Observe that by taking the negation of the equivalence in Farkas II we obtain a criterion of solvability, namely:

The system of inequalities $Ax \leq b$ has a solution $x \geq 0$ iff for every nonzero linear form $y \in (\mathbb{R}^m)^*$ such that $y \geq 0_m^\top$, if $yA \geq 0_n^\top$, then $yb \geq 0$.

We now prove the Farkas–Minkowski proposition without using Proposition 11.1. This approach uses a basic property of the distance function from a point to a closed set.

Definition 11.1. Let $X \subseteq \mathbb{R}^n$ be any nonempty set and let $a \in \mathbb{R}^n$ be any point. The *distance* $d(a, X)$ from a to X is defined as

$$d(a, X) = \inf_{x \in X} \|a - x\|.$$

Here, $\| \; \|$ denotes the Euclidean norm.

Proposition 11.5. *Let $X \subseteq \mathbb{R}^n$ be any nonempty set and let $a \in \mathbb{R}^n$ be any point. If X is closed, then there is some $z \in X$ such that $\|a - z\| = d(a, X)$.*

Proof. Since X is nonempty, pick any $x_0 \in X$, and let $r = \|a - x_0\|$. If $B_r(a)$ is the closed ball $B_r(a) = \{x \in \mathbb{R}^n \mid \|x - a\| \leq r\}$, then clearly

$$d(a, X) = \inf_{x \in X} \|a - x\| = \inf_{x \in X \cap B_r(a)} \|a - x\|.$$

Since $B_r(a)$ is compact and X is closed, $K = X \cap B_r(a)$ is also compact. But the function $x \mapsto \|a - x\|$ defined on the compact set K is continuous, and the image of a compact set by a continuous function is compact, so by Heine–Borel it has a minimum that is achieved by some $z \in K \subseteq X$. □

Remark: If U is a nonempty, closed and convex subset of a Hilbert space V, a standard result of Hilbert space theory (the projection lemma, see Proposition 12.4) asserts that for any $v \in V$ there is a *unique* $p \in U$ such that

$$\|v - p\| = \inf_{u \in U} \|v - u\| = d(v, U),$$

and

$$\langle p - v, u - p \rangle \geq 0 \quad \text{for all } u \in U.$$

Here $\|w\| = \sqrt{\langle w, w \rangle}$, where $\langle -, - \rangle$ is the inner product of the Hilbert space V.

We can now give a proof of the Farkas–Minkowski proposition (Proposition 11.2) that does not use Proposition 11.1. This proof is adapted from Matousek and Gardner [Matousek and Gartner (2007)] (Chapter 6, Sections 6.5).

Proof of the Farkas–Minkowski proposition. Let $C = \text{cone}(\{a_1, \ldots, a_m\})$ be a polyhedral cone (nonempty) and assume that $b \notin C$. By Proposition 8.2, the polyhedral cone is closed, and by Proposition 11.5 there is some $z \in C$ such that $d(b, C) = \|b - z\|$; that is, z is a point of C closest to b. Since $b \notin C$ and $z \in C$ we have $u = z - b \neq 0$, and we claim that the linear hyperplane H orthogonal to u does the job, as illustrated in Figure 11.2.

First let us show that

$$\langle u, z \rangle = \langle z - b, z \rangle = 0. \qquad (*_1)$$

This is trivial if $z = 0$, so assume $z \neq 0$. If $\langle u, z \rangle \neq 0$, then either $\langle u, z \rangle > 0$ or $\langle u, z \rangle < 0$. In either case we show that we can find some point $z' \in C$ closer to b than z is, a contradiction.

Case 1: $\langle u, z \rangle > 0$.

Let $z' = (1 - \alpha)z$ for any α such that $0 < \alpha < 1$. Then $z' \in C$ and since $u = z - b$,

$$z' - b = (1 - \alpha)z - (z - u) = u - \alpha z,$$

so

$$\|z' - b\|^2 = \|u - \alpha z\|^2 = \|u\|^2 - 2\alpha\langle u, z \rangle + \alpha^2 \|z\|^2 .$$

If we pick $\alpha > 0$ such that $\alpha < 2\langle u, z \rangle / \|z\|^2$, then $-2\alpha\langle u, z \rangle + \alpha^2 \|z\|^2 < 0$, so $\|z' - b\|^2 < \|u\|^2 = \|z - b\|^2$, contradicting the fact that z is a point of C closest to b.

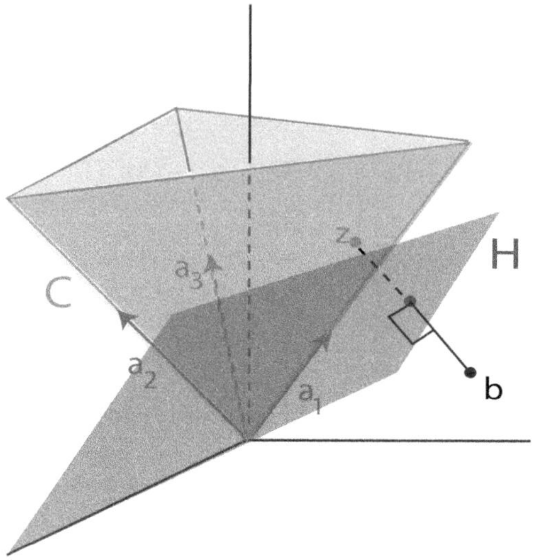

Fig. 11.2 The hyperplane H, perpendicular to $z - b$, separates the point b from $C = \mathrm{cone}(\{a_1, a_2, a_3\})$.

Case 2: $\langle u, z\rangle < 0$.

Let $z' = (1 + \alpha)z$ for any α such that $\alpha \geq -1$. Then $z' \in C$ and since $u = z - b$, we have $z' - b = (1 + \alpha)z - (z - u) = u + \alpha z$ so

$$\|z' - b\|^2 = \|u + \alpha z\|^2 = \|u\|^2 + 2\alpha\langle u, z\rangle + \alpha^2 \|z\|^2,$$

and if

$$0 < \alpha < -2\langle u, z\rangle / \|z\|^2,$$

then $2\alpha\langle u, z\rangle + \alpha^2 \|z\|^2 < 0$, so $\|z' - b\|^2 < \|u\|^2 = \|z - b\|^2$, a contradiction as above.

Therefore $\langle u, z\rangle = 0$. We have

$$\langle u, u\rangle = \langle u, z - b\rangle = \langle u, z\rangle - \langle u, b\rangle = -\langle u, b\rangle,$$

and since $u \neq 0$, we have $\langle u, u\rangle > 0$, so $\langle u, u\rangle = -\langle u, b\rangle$ implies that

$$\langle u, b\rangle < 0. \qquad (*_2)$$

It remains to prove that $\langle u, a_i\rangle \geq 0$ for $i = 1, \ldots, m$. Pick any $x \in C$ such that $x \neq z$. We claim that

$$\langle b - z, x - z\rangle \leq 0. \qquad (*_3)$$

Otherwise $\langle b - z, x - z\rangle > 0$, that is, $\langle z - b, x - z\rangle < 0$, and we show that we can find some point $z' \in C$ on the line segment $[z, x]$ closer to b than z is.

For any α such that $0 \leq \alpha \leq 1$, we have $z' = (1-\alpha)z + \alpha x = z + \alpha(x - z) \in C$, and since $z' - b = z - b + \alpha(x - z)$ we have

$$\|z' - b\|^2 = \|z - b + \alpha(x - z)\|^2 = \|z - b\|^2 + 2\alpha\langle z - b, x - z\rangle + \alpha^2 \|x - z\|^2,$$

so for any $\alpha > 0$ such that

$$\alpha < -2\langle z - b, x - z\rangle / \|x - z\|^2,$$

we have $2\alpha\langle z - b, x - z\rangle + \alpha^2 \|x - z\|^2 < 0$, which implies that $\|z' - b\|^2 < \|z - b\|^2$, contradicting that z is a point of C closest to b.

Since $\langle b - z, x - z\rangle \leq 0$, $u = z - b$, and by $(*_1)$, $\langle u, z\rangle = 0$, we have

$$0 \geq \langle b - z, x - z\rangle = \langle -u, x - z\rangle = -\langle u, x\rangle + \langle u, z\rangle = -\langle u, x\rangle,$$

which means that

$$\langle u, x\rangle \geq 0 \quad \text{for all } x \in C, \tag{$*_3$}$$

as claimed. In particular,

$$\langle u, a_i\rangle \geq 0 \quad \text{for } i = 1, \ldots, m. \tag{$*_4$}$$

Then by $(*_2)$ and $(*_4)$, the linear form defined by $y = u^\top$ satisfies the properties $yb < 0$ and $ya_i \geq 0$ for $i = 1, \ldots, m$, which proves the Farkas–Minkowski proposition. □

There are other ways of proving the Farkas–Minkowski proposition, for instance using minimally infeasible systems or Fourier–Motzkin elimination; see Matousek and Gardner [Matousek and Gartner (2007)] (Chapter 6, Sections 6.6 and 6.7).

11.2 The Duality Theorem in Linear Programming

Let (P) be the linear program

$$\begin{aligned} &\text{maximize} && cx \\ &\text{subject to} && Ax \leq b \text{ and } x \geq 0, \end{aligned}$$

with A an $m \times n$ matrix, and assume that (P) has a feasible solution and is bounded above. Since by hypothesis the objective function $x \mapsto cx$ is bounded on $\mathcal{P}(A, b)$, it might be useful to deduce an *upper bound* for cx

from the inequalities $Ax \leq b$, for any $x \in \mathcal{P}(A, b)$. We can do this as follows: for every inequality

$$a_i x \leq b_i \quad 1 \leq i \leq m,$$

pick a nonnegative scalar y_i, multiply both sides of the above inequality by y_i obtaining

$$y_i a_i x \leq y_i b_i \quad 1 \leq i \leq m,$$

(the direction of the inequality is preserved since $y_i \geq 0$), and then add up these m equations, which yields

$$(y_1 a_1 + \cdots + y_m a_m)x \leq y_1 b_1 + \cdots + y_m b_m.$$

If we can pick the $y_i \geq 0$ such that

$$c \leq y_1 a_1 + \cdots + y_m a_m,$$

then since $x_j \geq 0$, we have

$$cx \leq (y_1 a_1 + \cdots + y_m a_m)x \leq y_1 b_1 + \cdots + y_m b_m,$$

namely we found an upper bound of the value cx of the objective function of (P) for any feasible solution $x \in \mathcal{P}(A, b)$. If we let y be the linear form $y = (y_1, \ldots, y_m)$, then since

$$A = \begin{pmatrix} a_1 \\ \vdots \\ a_m \end{pmatrix}$$

$y_1 a_1 + \cdots + y_m a_m = yA$, and $y_1 b_1 + \cdots + y_m b_m = yb$, what we did was to look for some $y \in (\mathbb{R}^m)^*$ such that

$$c \leq yA, \quad y \geq 0,$$

so that we have

$$cx \leq yb \quad \text{for all } x \in \mathcal{P}(A, b). \tag{$*$}$$

Then it is natural to look for a "best" value of yb, namely a minimum value, which leads to the definition of the *dual* of the linear program (P), a notion due to John von Neumann.

Definition 11.2. Given any Linear Program (P)

$$\begin{aligned} &\text{maximize} \quad cx \\ &\text{subject to} \quad Ax \leq b \text{ and } x \geq 0, \end{aligned}$$

with A an $m \times n$ matrix, the *dual* (D) of (P) is the following optimization problem:

$$\begin{aligned} &\text{minimize} \quad yb \\ &\text{subject to} \quad yA \geq c \text{ and } y \geq 0, \end{aligned}$$

where $y \in (\mathbb{R}^m)^*$.

The variables $y_1, \ldots, y_m$ are called the *dual variables*. The original Linear Program (P) is called the *primal* linear program and the original variables $x_1, \ldots, x_n$ are the *primal variables*.

Here is an explicit example of a linear program and its dual.

Example 11.1. Consider the linear program illustrated by Figure 11.3

$$\begin{aligned} &\text{maximize} \quad 2x_1 + 3x_2 \\ &\text{subject to} \\ &\qquad 4x_1 + 8x_2 \leq 12 \\ &\qquad 2x_1 + x_2 \leq 3 \\ &\qquad 3x_1 + 2x_2 \leq 4 \\ &\qquad x_1 \geq 0,\ x_2 \geq 0. \end{aligned}$$

Its dual linear program is illustrated in Figure 11.4

$$\begin{aligned} &\text{minimize} \quad 12y_1 + 3y_2 + 4y_3 \\ &\text{subject to} \\ &\qquad 4y_1 + 2y_2 + 3y_3 \geq 2 \\ &\qquad 8y_1 + y_2 + 2y_3 \geq 3 \\ &\qquad y_1 \geq 0,\ y_2 \geq 0,\ y_3 \geq 0. \end{aligned}$$

It can be checked that $(x_1, x_2) = (1/2, 5/4)$ is an optimal solution of the primal linear program, with the maximum value of the objective function $2x_1 + 3x_2$ equal to $19/4$, and that $(y_1, y_2, y_3) = (5/16, 0, 1/4)$ is an optimal solution of the dual linear program, with the minimum value of the objective function $12y_1 + 3y_2 + 4y_3$ also equal to $19/4$.

Observe that in the Primal Linear Program (P), we are looking for a *vector* $x \in \mathbb{R}^n$ maximizing the form cx, and that the constraints are determined by the action of the *rows* of the matrix A on x. On the other hand, in the Dual Linear Program (D), we are looking for a *linear form*

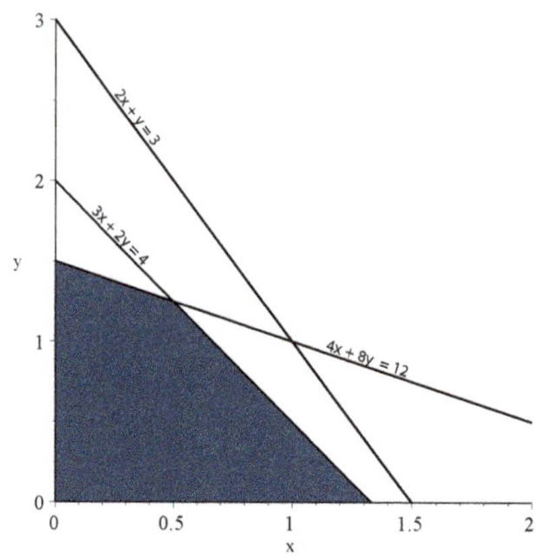

Fig. 11.3 The $\mathcal{H}$-polytope for the linear program of Example 11.1. Note $x_1 \to x$ and $x_2 \to y$.

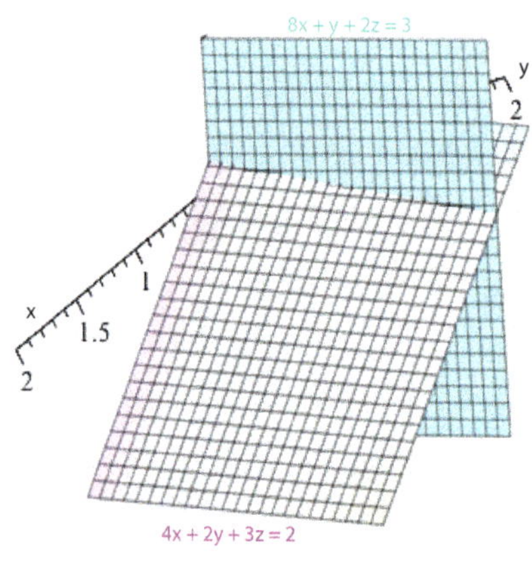

Fig. 11.4 The $\mathcal{H}$-polyhedron for the dual linear program of Example 11.1 is the spacial region "above" the pink plane and in "front" of the blue plane. Note $y_1 \to x$, $y_2 \to y$, and $y_3 \to z$.

$y \in (\mathbb{R}^*)^m$ minimizing the form yb, and the constraints are determined by the action of y on the *columns* of A. This is the sense in which (D) is the *dual* (P). In most presentations, the fact that (P) and (D) perform a search for a solution in spaces that are dual to each other is obscured by excessive use of transposition.

To convert the Dual Program (D) to a standard maximization problem we change the objective function yb to $-b^\top y^\top$ and the inequality $yA \geq c$ to $-A^\top y^\top \leq -c^\top$. The Dual Linear Program (D) is now stated as (D')

$$\begin{aligned} &\text{maximize} && -b^\top y^\top \\ &\text{subject to} && -A^\top y^\top \leq -c^\top \text{ and } y^\top \geq 0, \end{aligned}$$

where $y \in (\mathbb{R}^m)^*$. Observe that the dual in maximization form (D'') of the Dual Program (D') gives back the Primal Program (P).

The above discussion established the following inequality known as *weak duality*.

Proposition 11.6. *(Weak Duality) Given any Linear Program (P)*

$$\begin{aligned} &\text{maximize} && cx \\ &\text{subject to} && Ax \leq b \text{ and } x \geq 0, \end{aligned}$$

with A an $m \times n$ matrix, for any feasible solution $x \in \mathbb{R}^n$ of the Primal Problem (P) and every feasible solution $y \in (\mathbb{R}^m)^*$ of the Dual Problem (D), we have

$$cx \leq yb.$$

Definition 11.3. We say that the Dual Linear Program (D) is *bounded below* if
$\{yb \mid y^\top \in \mathcal{P}(-A^\top, -c^\top)\}$ is bounded below.

What happens if x^* is an optimal solution of (P) and if y^* is an optimal solution of (D)? We have $cx^* \leq y^*b$, but is there a "duality gap," that is, is it possible that $cx^* < y^*b$?

The answer is **no**, this is the *strong duality theorem*. Actually, the strong duality theorem asserts more than this.

Theorem 11.1. *(Strong Duality for Linear Programming) Let (P) be any linear program*

$$\begin{aligned} &\text{maximize} && cx \\ &\text{subject to} && Ax \leq b \text{ and } x \geq 0, \end{aligned}$$

with A an $m \times n$ matrix. The Primal Problem (P) has a feasible solution and is bounded above iff the Dual Problem (D) has a feasible solution and is bounded below. Furthermore, if (P) has a feasible solution and is bounded above, then for every optimal solution x^* of (P) and every optimal solution y^* of (D), we have

$$cx^* = y^*b.$$

Proof. If (P) has a feasible solution and is bounded above, then we know from Proposition 9.1 that (P) has some optimal solution. Let x^* be any optimal solution of (P). First we will show that (D) has a feasible solution v.

Let $\mu = cx^*$ be the maximum of the objective function $x \mapsto cx$. Then for any $\epsilon > 0$, the system of inequalities

$$Ax \leq b, \quad x \geq 0, \quad cx \geq \mu + \epsilon$$

has no solution, since otherwise μ would not be the maximum value of the objective function cx. We would like to apply Farkas II, so first we transform the above system of inequalities into the system

$$\begin{pmatrix} A \\ -c \end{pmatrix} x \leq \begin{pmatrix} b \\ -(\mu + \epsilon) \end{pmatrix}.$$

By Proposition 11.4 (Farkas II), there is some linear form $(\lambda, z) \in (\mathbb{R}^{m+1})^*$ such that $\lambda \geq 0$, $z \geq 0$,

$$(\lambda \; z) \begin{pmatrix} A \\ -c \end{pmatrix} \geq 0_m^\top,$$

and

$$(\lambda \; z) \begin{pmatrix} b \\ -(\mu + \epsilon) \end{pmatrix} < 0,$$

which means that

$$\lambda A - zc \geq 0_m^\top, \quad \lambda b - z(\mu + \epsilon) < 0,$$

that is,

$$\begin{aligned} \lambda A &\geq zc \\ \lambda b &< z(\mu + \epsilon) \\ \lambda &\geq 0, \;\; z \geq 0. \end{aligned}$$

On the other hand, since $x^* \geq 0$ is an optimal solution of the system $Ax \leq b$, by Farkas II again (by taking the negation of the equivalence), since $\lambda A \geq 0$ (for the same λ as before), we must have

$$\lambda b \geq 0. \tag{$*_1$}$$

We claim that $z > 0$. Otherwise, since $z \geq 0$, we must have $z = 0$, but then

$$\lambda b < z(\mu + \epsilon)$$

implies

$$\lambda b < 0, \tag{$*_2$}$$

and since $\lambda b \geq 0$ by $(*_1)$, we have a contradiction. Consequently, we can divide by $z > 0$ without changing the direction of inequalities, and we obtain

$$\begin{aligned}\frac{\lambda}{z}A &\geq c\\ \frac{\lambda}{z}b &< \mu + \epsilon\\ \frac{\lambda}{z} &\geq 0,\end{aligned}$$

which shows that $v = \lambda/z$ is a feasible solution of the Dual Problem (D). However, weak duality (Proposition 11.6) implies that $cx^* = \mu \leq yb$ for any feasible solution $y \geq 0$ of the Dual Program (D), so (D) is bounded below and by Proposition 9.1 applied to the version of (D) written as a maximization problem, we conclude that (D) has some optimal solution. For any optimal solution y^* of (D), since v is a feasible solution of (D) such that $vb < \mu + \epsilon$, we must have

$$\mu \leq y^* b < \mu + \epsilon,$$

and since our reasoning is valid for *any* $\epsilon > 0$, we conclude that $cx^* = \mu = y^* b$.

If we assume that the dual program (D) has a feasible solution and is bounded below, since the dual of (D) is (P), we conclude that (P) is also feasible and bounded above. □

The strong duality theorem can also be proven by the simplex method, because when it terminates with an optimal solution of (P), the final tableau also produces an optimal solution y of (D) that can be read off the reduced costs of columns $n+1, \ldots, n+m$ by flipping their signs. We follow the proof in Ciarlet [Ciarlet (1989)] (Chapter 10).

Theorem 11.2. *Consider the Linear Program (P),*

$$\begin{aligned}&\text{maximize} && cx\\ &\text{subject to} && Ax \leq b \text{ and } x \geq 0,\end{aligned}$$

its equivalent version $(P2)$ *in standard form,*

$$\begin{aligned}&\text{maximize} && \widehat{c}\,\widehat{x}\\ &\text{subject to} && \widehat{A}\widehat{x} = b \text{ and } \widehat{x} \geq 0,\end{aligned}$$

where $\widehat{A}$ is an $m \times (n+m)$ matrix, $\widehat{c}$ is a linear form in $(\mathbb{R}^{n+m})^*$, and $\widehat{x} \in \mathbb{R}^{n+m}$, given by

$$\widehat{A} = \begin{pmatrix} A \; I_m \end{pmatrix}, \quad \widehat{c} = \begin{pmatrix} c \; 0_m^\top \end{pmatrix}, \quad x = \begin{pmatrix} x_1 \\ \vdots \\ x_n \end{pmatrix}, \quad \overline{x} = \begin{pmatrix} x_{n+1} \\ \vdots \\ x_{n+m} \end{pmatrix}, \quad \widehat{x} = \begin{pmatrix} x \\ \overline{x} \end{pmatrix},$$

and the Dual (D) of (P) given by

$$\begin{aligned} &\text{minimize} \quad yb \\ &\text{subject to} \quad yA \geq c \ \textit{and} \ y \geq 0, \end{aligned}$$

where $y \in (\mathbb{R}^m)^*$. If the simplex algorithm applied to the Linear Program $(P2)$ terminates with an optimal solution $(\widehat{u}^*, K^*)$, where $\widehat{u}^*$ is a basic feasible solution and K^* is a basis for $\widehat{u}^*$, then $y^* = \widehat{c}_{K^*}\widehat{A}_{K^*}^{-1}$ is an optimal solution for (D) such that $\widehat{c}\,\widehat{u}^* = y^*b$. Furthermore, y^* is given in terms of the reduced costs by $y^* = -((\overline{c}_{K^*})_{n+1} \cdots (\overline{c}_{K^*})_{n+m})$.

Proof. We know that K^* is a subset of $\{1, \ldots, n+m\}$ consisting of m indices such that the corresponding columns of $\widehat{A}$ are linearly independent. Let $N^* = \{1, \ldots, n+m\} - K^*$. The simplex method terminates with an optimal solution in Case (A), namely when

$$\widehat{c}_j - \sum_{k \in k} \gamma_k^j \widehat{c}_k \leq 0 \quad \text{for all } j \in N^*,$$

where $\widehat{A^j} = \sum_{k \in K^*} \gamma_k^j \widehat{A^k}$, or using the notations of Section 10.3,

$$\widehat{c}_j - \widehat{c}_{K^*}\widehat{A}_{K^*}^{-1}\widehat{A}^j \leq 0 \quad \text{for all } j \in N^*.$$

The above inequalities can be written as

$$\widehat{c}_{N^*} - \widehat{c}_{K^*}\widehat{A}_{K^*}^{-1}\widehat{A}_{N^*} \leq 0_n^\top,$$

or equivalently as

$$\widehat{c}_{K^*}\widehat{A}_{K^*}^{-1}\widehat{A}_{N^*} \geq \widehat{c}_{N^*}. \tag{$*_1$}$$

The value of the objective function for the optimal solution $\widehat{u}^*$ is $\widehat{c}\,\widehat{u}^* = \widehat{c}_{K^*}\widehat{u}_{K^*}^*$, and since $\widehat{u}_{K^*}^*$ satisfies the equation $\widehat{A}_{K^*}\widehat{u}_{K^*}^* = b$, the value of the objective function is

$$\widehat{c}_{K^*}\,\widehat{u}_{K^*}^* = \widehat{c}_{K^*}\widehat{A}_{K^*}^{-1}b. \tag{$*_2$}$$

Then if we let $y^* = \widehat{c}_{K^*}\widehat{A}_{K^*}^{-1}$, obviously we have $y^*b = \widehat{c}_{K^*}\widehat{u}_{K^*}$, so if we can prove that y^* is a feasible solution of the Dual Linear program (D), by weak duality, y^* is an optimal solution of (D). We have

$$y^*\widehat{A}_{K^*} = \widehat{c}_{K^*}\widehat{A}_{K^*}^{-1}\widehat{A}_{K^*} = \widehat{c}_{K^*}, \tag{$*_3$}$$

and by $(*_1)$ we get

$$y^* \widehat{A}_{N^*} = \widehat{c}_{K^*} \widehat{A}_{K^*}^{-1} \widehat{A}_{N^*} \geq \widehat{c}_{N^*}. \tag{$*_4$}$$

Let P be the $(n+m) \times (n+m)$ permutation matrix defined so that

$$\widehat{A} P = (A \; I_m) \, P = \left(\widehat{A}_{K^*} \; \widehat{A}_{N^*} \right).$$

Then we also have

$$\widehat{c} P = (c \; 0_m^\top) \, P = \left(\widehat{c}_{K^*} \; \widehat{c}_{N^*} \right).$$

Using Equations $(*_3)$ and $(*_4)$ we obtain

$$y^* \left(\widehat{A}_{K^*} \; \widehat{A}_{N^*} \right) \geq \left(\widehat{c}_{K^*} \; \widehat{c}_{N^*} \right),$$

that is,

$$y^* \, (A \; I_m) \, P \geq (c \; 0_m^\top) \, P,$$

which is equivalent to

$$y^* \, (A \; I_m) \geq (c \; 0_m^\top),$$

that is

$$y^* A \geq c, \quad y \geq 0,$$

and these are exactly the conditions that say that y^* is a feasible solution of the Dual Program (D).

The reduced costs are given by $(\widehat{c}_{K^*})_i = \widehat{c}_i - \widehat{c}_{K^*} \widehat{A}_{K^*}^{-1} \widehat{A}^i$, for $i = 1, \ldots, n+m$. But for $i = n+j$ with $j = 1, \ldots, m$ each column $\widehat{A}^{n+j}$ is the jth vector of the identity matrix I_m and by definition $\widehat{c}_{n+j} = 0$, so

$$(\widehat{c}_{K^*})_{n+j} = -(\widehat{c}_{K^*} \widehat{A}_{K^*}^{-1})_j = -y_j^* \quad j = 1, \ldots, m,$$

as claimed. □

The fact that the above proof is fairly short is deceptive because this proof relies on the fact that there are versions of the simplex algorithm using pivot rules that prevent cycling, but the proof that such pivot rules work correctly is quite lengthy. Other proofs are given in Matousek and Gardner [Matousek and Gartner (2007)] (Chapter 6, Sections 6.3), Chvatal [Chvatal (1983)] (Chapter 5), and Papadimitriou and Steiglitz [Papadimitriou and Steiglitz (1998)] (Section 2.7).

Observe that since the last m rows of the final tableau are actually obtained by multiplying $[u \; \widehat{A}]$ by $\widehat{A}_{K^*}^{-1}$, the $m \times m$ matrix consisting of the last m columns and last m rows of the final tableau is $\widehat{A}_{K^*}^{-1}$ (basically, the simplex algorithm has performed the steps of a Gauss–Jordan reduction). This fact allows saving some steps in the primal dual method.

By combining weak duality and strong duality, we obtain the following theorem which shows that exactly four cases arise.

Theorem 11.3. *(Duality Theorem of Linear Programming) Let (P) be any linear program*

$$\begin{aligned} &\text{maximize} \quad cx \\ &\text{subject to} \quad Ax \leq b \text{ and } x \geq 0, \end{aligned}$$

and let (D) be its dual program

$$\begin{aligned} &\text{minimize} \quad yb \\ &\text{subject to} \quad yA \geq c \text{ and } y \geq 0, \end{aligned}$$

with A an $m \times n$ matrix. Then exactly one of the following possibilities occur:

(1) Neither (P) nor (D) has a feasible solution.
(2) (P) is unbounded and (D) has no feasible solution.
(3) (P) has no feasible solution and (D) is unbounded.
(4) Both (P) and (D) have a feasible solution. Then both have an optimal solution, and for every optimal solution x^* of (P) and every optimal solution y^* of (D), we have

$$cx^* = y^*b.$$

An interesting corollary of Theorem 11.3 is that there is a test to determine whether a Linear Program (P) has an optimal solution.

Corollary 11.1. *The Primal Program (P) has an optimal solution iff the following set of constraints is satisfiable:*

$$\begin{aligned} Ax &\leq b \\ yA &\geq c \\ cx &\geq yb \\ x \geq 0,\ y &\geq 0_m^\top. \end{aligned}$$

In fact, for any feasible solution (x^*, y^*) of the above system, x^* is an optimal solution of (P) and y^* is an optimal solution of (D)

11.3 Complementary Slackness Conditions

Another useful corollary of the strong duality theorem is the following result known as the *equilibrium theorem*.

Theorem 11.4. *(Equilibrium Theorem) For any Linear Program (P) and its Dual Linear Program (D) (with set of inequalities $Ax \leq b$ where A is an $m \times n$ matrix, and objective function $x \mapsto cx$), for any feasible solution x of (P) and any feasible solution y of (D), x and y are optimal solutions iff*

$$y_i = 0 \qquad \textit{for all } i \textit{ for which } \sum_{j=1}^{n} a_{ij}x_j < b_i \tag{$*_D$}$$

and

$$x_j = 0 \qquad \textit{for all } j \textit{ for which } \sum_{i=1}^{m} y_i a_{ij} > c_j. \tag{$*_P$}$$

Proof. First assume that $(*_D)$ and $(*_P)$ hold. The equations in $(*_D)$ say that $y_i = 0$ unless $\sum_{j=1}^{n} a_{ij}x_j = b_i$, hence

$$yb = \sum_{i=1}^{m} y_i b_i = \sum_{i=1}^{m} y_i \sum_{j=1}^{n} a_{ij}x_j = \sum_{i=1}^{m}\sum_{j=1}^{n} y_i a_{ij} x_j.$$

Similarly, the equations in $(*_P)$ say that $x_j = 0$ unless $\sum_{i=1}^{m} y_i a_{ij} = c_j$, hence

$$cx = \sum_{j=1}^{n} c_j x_j = \sum_{j=1}^{n}\sum_{i=1}^{m} y_i a_{ij} x_j.$$

Consequently, we obtain

$$cx = yb.$$

By weak duality (Proposition 11.6), we have

$$cx \leq yb = cx$$

for all feasible solutions x of (P), so x is an optimal solution of (P). Similarly,

$$yb = cx \leq yb$$

for all feasible solutions y of (D), so y is an optimal solution of (D).

Let us now assume that x is an optimal solution of (P) and that y is an optimal solution of (D). Then as in the proof of Proposition 11.6,

$$\sum_{j=1}^{n} c_j x_j \leq \sum_{i=1}^{m}\sum_{j=1}^{n} y_i a_{ij} x_j \leq \sum_{i=1}^{m} y_i b_i.$$

By strong duality, since x and y are optimal solutions the above inequalities are actually equalities, so in particular we have

$$\sum_{j=1}^{n}\left(c_j - \sum_{i=1}^{m} y_i a_{ij}\right)x_j = 0.$$

Since x and y are feasible, $x_i \geq 0$ and $y_j \geq 0$, so if $\sum_{i=1}^{m} y_i a_{ij} > c_j$, we must have $x_j = 0$. Similarly, we have

$$\sum_{i=1}^{m} y_i \left(\sum_{j=1}^{m} a_{ij}x_j - b_i\right) = 0,$$

so if $\sum_{j=1}^{m} a_{ij}x_j < b_i$, then $y_i = 0$. □

The equations in $(*_D)$ and $(*_P)$ are often called *complementary slackness conditions*. These conditions can be exploited to solve for an optimal solution of the primal problem with the help of the dual problem, and conversely. Indeed, if we guess a solution to one problem, then we may solve for a solution of the dual using the complementary slackness conditions, and then check that our guess was correct. This is the essence of the *primal-dual* method. To present this method, first we need to take a closer look at the dual of a linear program already in standard form.

11.4 Duality for Linear Programs in Standard Form

Let (P) be a linear program in standard form, where $Ax = b$ for some $m \times n$ matrix of rank m and some objective function $x \mapsto cx$ (of course, $x \geq 0$). To obtain the dual of (P) we convert the equations $Ax = b$ to the following system of inequalities involving a $(2m) \times n$ matrix:

$$\begin{pmatrix} A \\ -A \end{pmatrix} x \leq \begin{pmatrix} b \\ -b \end{pmatrix}.$$

Then if we denote the $2m$ dual variables by (y', y''), with $y', y'' \in (\mathbb{R}^m)^*$, the dual of the above program is

$$\begin{aligned} &\text{minimize} \quad y'b - y''b \\ &\text{subject to} \quad (y'\ y'') \begin{pmatrix} A \\ -A \end{pmatrix} \geq c \text{ and } y', y'' \geq 0, \end{aligned}$$

where $y', y'' \in (\mathbb{R}^m)^*$, which is equivalent to

$$\begin{aligned} &\text{minimize} \quad (y' - y'')b \\ &\text{subject to} \quad (y' - y'')A \geq c \text{ and } y', y'' \geq 0, \end{aligned}$$

where $y', y'' \in (\mathbb{R}^m)^*$. If we write $y = y' - y''$, we find that the above linear program is equivalent to the following Linear Program (D):

$$\begin{aligned} &\text{minimize} \quad yb \\ &\text{subject to} \quad yA \geq c, \end{aligned}$$

where $y \in (\mathbb{R}^m)^*$. Observe that y is *not required* to be nonnegative; it is arbitrary.

Next we would like to know what is the version of Theorem 11.2 for a linear program already in standard form. This is very simple.

Theorem 11.5. *Consider the Linear Program (P2) in standard form*

$$\begin{aligned} &\text{maximize} \quad cx \\ &\text{subject to} \quad Ax = b \text{ and } x \geq 0, \end{aligned}$$

and its Dual (D) given by

$$\begin{aligned} &\text{minimize} \quad yb \\ &\text{subject to} \quad yA \geq c, \end{aligned}$$

where $y \in (\mathbb{R}^m)^*$. If the simplex algorithm applied to the Linear Program $(P2)$ terminates with an optimal solution (u^*, K^*), where u^* is a basic feasible solution and K^* is a basis for u^*, then $y^* = c_{K^*}A_{K^*}^{-1}$ is an optimal solution for (D) such that $cu^* = y^*b$. Furthermore, if we assume that the simplex algorithm is started with a basic feasible solution (u_0, K_0) where $K_0 = (n - m + 1, \ldots, n)$ (the indices of the last m columns of A) and $A_{(n-m+1,\ldots,n)} = I_m$ (the last m columns of A constitute the identity matrix I_m), then the optimal solution $y^* = c_{K^*}A_{K^*}^{-1}$ for (D) is given in terms of the reduced costs by

$$y^* = c_{(n-m+1,\ldots,n)} - (\overline{c}_{K^*})_{(n-m+1,\ldots,n)},$$

and the $m \times m$ matrix consisting of last m columns and the last m rows of the final tableau is $A_{K^*}^{-1}$.

Proof. The proof of Theorem 11.2 applies with A instead of $\widehat{A}$, and we can show that

$$c_{K^*}A_{K^*}^{-1}A_{N^*} \geq c_{N^*},$$

and that $y^* = c_{K^*}A_{K^*}^{-1}$ satisfies, $cu^* = y^*b$, and

$$\begin{aligned} y^*A_{K^*} &= c_{K^*}A_{K^*}^{-1}A_{K^*} = c_{K^*}, \\ y^*A_{N^*} &= c_{K^*}A_{K^*}^{-1}A_{N^*} \geq c_{N^*}. \end{aligned}$$

Let P be the $n \times n$ permutation matrix defined so that

$$AP = \left(A_{K^*}\ A_{N^*}\right).$$

Then we also have

$$cP = \left(c_{K^*}\ c_{N^*}\right),$$

and using the above equations and inequalities we obtain

$$y^* \left(A_{K^*}\ A_{N^*}\right) \geq \left(c_{K^*}\ c_{N^*}\right),$$

that is, $y^* AP \geq cP$, which is equivalent to

$$y^* A \geq c,$$

which shows that y^* is a feasible solution of (D) (remember, y^* is arbitrary so there is no need for the constraint $y^* \geq 0$).

The reduced costs are given by

$$(\overline{c}_{K^*})_i = c_i - c_{K^*} A_{K^*}^{-1} A^i,$$

and since for $j = n - m + 1, \ldots, n$ the column A^j is the $(j + m - n)$th column of the identity matrix I_m, we have

$$(\overline{c}_{K^*})_j = c_j - (c_{K^*} A_{K^*}^{-1})_{j+m-n} \quad j = n - m + 1, \ldots, n,$$

that is,

$$y^* = c_{(n-m+1,\ldots,n)} - \left(\overline{c}_{K^*}\right)_{(n-m+1,\ldots,n)},$$

as claimed. Since the last m rows of the final tableau is obtained by multiplying $[u_0\ A]$ by $A_{K^*}^{-1}$, and the last m columns of A constitute I_m, the last m rows and the last m columns of the final tableau constitute $A_{K^*}^{-1}$. $\square$

Let us now take a look at the complementary slackness conditions of Theorem 11.4. If we go back to the version of (P) given by

$$\begin{aligned} &\text{maximize} \quad cx \\ &\text{subject to} \quad \begin{pmatrix} A \\ -A \end{pmatrix} x \leq \begin{pmatrix} b \\ -b \end{pmatrix} \ \text{and } x \geq 0, \end{aligned}$$

and to the version of (D) given by

$$\begin{aligned} &\text{minimize} \quad y'b - y''b \\ &\text{subject to} \quad (y'\ y'') \begin{pmatrix} A \\ -A \end{pmatrix} \geq c \ \text{and } y', y'' \geq 0, \end{aligned}$$

where $y', y'' \in (\mathbb{R}^m)^*$, since the inequalities $Ax \leq b$ and $-Ax \leq -b$ together imply that $Ax = b$, we have equality for all these inequality constraints, and so the Conditions $(*_D)$ place no constraints at all on y' and y'', while the Conditions $(*_P)$ assert that

$$x_j = 0 \qquad \text{for all } j \text{ for which } \textstyle\sum_{i=1}^m (y_i' - y_i'')a_{ij} > c_j.$$

If we write $y = y' - y''$, the above conditions are equivalent to

$$x_j = 0 \qquad \text{for all } j \text{ for which } \textstyle\sum_{i=1}^m y_i a_{ij} > c_j.$$

Thus we have the following version of Theorem 11.4.

Theorem 11.6. *(Equilibrium Theorem, Version 2) For any Linear Program $(P2)$ in standard form (with $Ax = b$ where A is an $m \times n$ matrix, $x \geq 0$, and objective function $x \mapsto cx$) and its Dual Linear Program (D), for any feasible solution x of (P) and any feasible solution y of (D), x and y are optimal solutions iff*

$$x_j = 0 \qquad \textit{for all } j \textit{ for which } \textstyle\sum_{i=1}^m y_i a_{ij} > c_j. \qquad (*_P)$$

Therefore, the slackness conditions applied to a Linear Program $(P2)$ in standard form and to its Dual (D) *only impose slackness conditions on the variables x_j of the primal problem.*

The above fact plays a crucial role in the primal-dual method.

11.5 The Dual Simplex Algorithm

Given a Linear Program $(P2)$ in standard form

$$\begin{aligned} &\text{maximize} && cx \\ &\text{subject to} && Ax = b \text{ and } x \geq 0, \end{aligned}$$

where A is an $m \times n$ matrix of rank m, if no obvious feasible solution is available *but if* $c \leq 0$, rather than using the method for finding a feasible solution described in Section 10.2 we may use a method known as the dual simplex algorithm. This method uses basic solutions (u, K) where $Au = b$ and $u_j = 0$ for all $u_j \notin K$, but does not require $u \geq 0$, so u may not be feasible. However, $y = c_K A_K^{-1}$ is required to be feasible for the dual program

$$\begin{aligned} &\text{minimize} && yb \\ &\text{subject to} && yA \geq c, \end{aligned}$$

where $y \in (\mathbb{R}^*)^m$. Since $c \leq 0$, observe that $y = 0_m^\top$ is a feasible solution of the dual.

If a basic solution u of $(P2)$ is found such that $u \geq 0$, then $cu = yb$ for $y = c_K A_K^{-1}$, and we have found an optimal solution u for $(P2)$ and y for (D). The dual simplex method makes progress by attempting to make negative components of u zero and by decreasing the objective function of the dual program.

The dual simplex method starts with a basic solution (u, K) of $Ax = b$ which is not feasible but for which $y = c_K A_K^{-1}$ is dual feasible. In many cases the original linear program is specified by a set of inequalities $Ax \leq b$ with some $b_i < 0$, so by adding slack variables it is easy to find such basic solution u, and if in addition $c \leq 0$, then because the cost associated with slack variables is 0, we see that $y = 0$ is a feasible solution of the dual.

Given a basic solution (u, K) of $Ax = b$ (feasible or not), $y = c_K A_K^{-1}$ is dual feasible iff $c_K A_K^{-1} A \geq c$, and since $c_K A_K^{-1} A_K = c_K$, the inequality $c_K A_K^{-1} A \geq c$ is equivalent to $c_K A_K^{-1} A_N \geq c_N$, that is,

$$c_N - c_K A_K^{-1} A_N \leq 0, \tag{$*_1$}$$

where $N = \{1, \ldots, n\} - K$. Equation $(*_1)$ is equivalent to

$$c_j - c_K \gamma_K^j \leq 0 \quad \text{for all } j \in N, \tag{$*_2$}$$

where $\gamma_K^j = A_K^{-1} A^j$. Recall that the notation $\overline{c}_j$ is used to denote $c_j - c_K \gamma_K^j$, which is called the *reduced cost* of the variable x_j.

As in the simplex algorithm we need to decide which column A^k leaves the basis K and which column A^j enters the new basis K^+, in such a way that $y^+ = c_{K^+} A_{K^+}^{-1}$ is a feasible solution of (D), that is, $c_{N^+} - c_{K^+} A_{K^+}^{-1} A_{N^+} \leq 0$, where $N^+ = \{1, \ldots, n\} - K^+$. We use Proposition 10.2 to decide which column k^- should leave the basis.

Suppose (u, K) is a solution of $Ax = b$ for which $y = c_K A_K^{-1}$ is dual feasible.

Case (A). If $u \geq 0$, then u is an optimal solution of $(P2)$.

Case (B). There is some $k \in K$ such that $u_k < 0$. In this case pick some $k^- \in K$ such that $u_{k^-} < 0$ (according to some pivot rule).

Case (B1). Suppose that $\gamma_{k^-}^j \geq 0$ for all $j \notin K$ (in fact, for all j, since $\gamma_{k^-}^j \in \{0, 1\}$ for all $j \in K$). If so, we we claim that $(P2)$ is not feasible.

Indeed, let v be some basic feasible solution. We have $v \geq 0$ and $Av = b$, that is,

$$\sum_{j=1}^{n} v_j A^j = b,$$

so by multiplying both sides by A_K^{-1} and using the fact that by definition $\gamma_K^j = A_K^{-1}A^j$, we obtain

$$\sum_{j=1}^{n} v_j \gamma_K^j = A_K^{-1}b = u_K.$$

But recall that by hypothesis $u_{k^-} < 0$, yet $v_j \geq 0$ and $\gamma_{k^-}^j \geq 0$ for all j, so the component of index k^- is zero or positive on the left, and negative on the right, a contradiction. Therefore, $(P2)$ is indeed not feasible.

Case (B2). We have $\gamma_{k^-}^j < 0$ for some j.

We pick the column A^j entering the basis among those for which $\gamma_{k^-}^j < 0$. Since we assumed that $c_j - c_K\gamma_K^j \leq 0$ for all $j \in N$ by $(*_2)$, consider

$$\begin{aligned}\mu^+ &= \max\left\{ -\frac{c_j - c_K\gamma_K^j}{\gamma_{k^-}^j} \;\middle|\; \gamma_{k^-}^j < 0,\; j \in N \right\} \\ &= \max\left\{ -\frac{\overline{c}_j}{\gamma_{k^-}^j} \;\middle|\; \gamma_{k^-}^j < 0,\; j \in N \right\} \leq 0,\end{aligned}$$

and the set

$$N(\mu^+) = \left\{ j \in N \;\middle|\; -\frac{\overline{c}_j}{\gamma_{k^-}^j} = \mu^+ \right\}.$$

We pick some index $j^+ \in N(\mu^+)$ as the index of the column entering the basis (using some pivot rule).

Recall that by hypothesis $c_i - c_K\gamma_K^i \leq 0$ for all $j \notin K$ and $c_i - c_K\gamma_K^i = 0$ for all $i \in K$. Since $\gamma_{k^-}^{j^+} < 0$, for any index i such that $\gamma_{k^-}^i \geq 0$, we have $-\gamma_{k^-}^i/\gamma_{k^-}^{j^+} \geq 0$, and since by Proposition 10.2

$$c_i - c_{K^+}\gamma_{K^+}^i = c_i - c_K\gamma_K^i - \frac{\gamma_{k^-}^i}{\gamma_{k^-}^{j^+}}(c_{j^+} - c_K\gamma_K^{j^+}),$$

we have $c_i - c_{K^+}\gamma_{K^+}^i \leq 0$. For any index i such that $\gamma_{k^-}^i < 0$, by the choice of $j^+ \in K^*$,

$$-\frac{c_i - c_K\gamma_K^i}{\gamma_{k^-}^i} \leq -\frac{c_{j^+} - c_K\gamma_K^{j^+}}{\gamma_{k^-}^{j^+}},$$

so

$$c_i - c_K\gamma_K^i - \frac{\gamma_{k^-}^i}{\gamma_{k^-}^{j^+}}(c_{j^+} - c_K\gamma_K^{j^+}) \leq 0,$$

and again, $c_i - c_{K^+}\gamma^i_{K^+} \leq 0$. Therefore, if we let $K^+ = (K - \{k^-\}) \cup \{j^+\}$, then $y^+ = c_{K^+}A^{-1}_{K^+}$ is dual feasible. As in the simplex algorithm, θ^+ is given by

$$\theta^+ = u_{k^-}/\gamma^{j^+}_{k^-} \geq 0,$$

and u^+ is also computed as in the simplex algorithm by

$$u^+_i = \begin{cases} u_i - \theta^{j^+}\gamma^{j^+}_i & \text{if } i \in K \\ \theta^{j^+} & \text{if } i = j^+ \\ 0 & \text{if } i \notin K \cup \{j^+\} \end{cases}.$$

The change in the objective function of the primal and dual program (which is the same, since $u_K = A^{-1}_K b$ and $y = c_K A^{-1}_K$ is chosen such that $cu = c_K u_K = yb$) is the same as in the simplex algorithm, namely

$$\theta^+\left(c^{j^+} - c_K\gamma^{j^+}_K\right).$$

We have $\theta^+ > 0$ and $c^{j^+} - c_K\gamma^{j^+}_K \leq 0$, so if $c^{j^+} - c_K\gamma^{j^+}_K < 0$, then the objective function of the dual program decreases strictly.

Case (B3). $\mu^+ = 0$.

The possibility that $\mu^+ = 0$, that is, $c^{j^+} - c_K\gamma^{j^+}_K = 0$, may arise. In this case, the objective function doesn't change. This is a case of degeneracy similar to the degeneracy that arises in the simplex algorithm. We still pick $j^+ \in N(\mu^+)$, but we need a pivot rule that prevents cycling. Such rules exist; see Bertsimas and Tsitsiklis [Bertsimas and Tsitsiklis (1997)] (Section 4.5) and Papadimitriou and Steiglitz [Papadimitriou and Steiglitz (1998)] (Section 3.6).

The reader surely noticed that the dual simplex algorithm is very similar to the simplex algorithm, except that the simplex algorithm preserves the property that (u, K) is (primal) feasible, whereas the dual simplex algorithm preserves the property that $y = c_K A^{-1}_K$ is dual feasible. One might then wonder whether the dual simplex algorithm is equivalent to the simplex algorithm applied to the dual problem. This is indeed the case, there is a one-to-one correspondence between the dual simplex algorithm and the simplex algorithm applied to the dual problem in maximization form. This correspondence is described in Papadimitriou and Steiglitz [Papadimitriou and Steiglitz (1998)] (Section 3.7).

The comparison between the simplex algorithm and the dual simplex algorithm is best illustrated if we use a description of these methods in terms of *(full) tableaux*.

Recall that a *(full) tableau* is an $(m+1) \times (n+1)$ matrix organized as follows:

$-c_K u_K$	$\overline{c}_1$	$\cdots$	$\overline{c}_j$	$\cdots$	$\overline{c}_n$
u_{k_1}	γ_1^1	$\cdots$	γ_1^j	$\cdots$	γ_1^n
$\vdots$	$\vdots$		$\vdots$		$\vdots$
u_{k_m}	γ_m^1	$\cdots$	γ_m^j	$\cdots$	γ_m^n

The top row contains the current value of the objective function and the reduced costs, the first column except for its top entry contain the components of the current basic solution u_K, and the remaining columns except for their top entry contain the vectors γ_K^j. Observe that the γ_K^j corresponding to indices j in K constitute a permutation of the identity matrix I_m. A tableau together with the new basis $K^+ = (K - \{k^-\}) \cup \{j^+\}$ contains all the data needed to compute the new u_{K^+}, the new $\gamma_{K^+}^j$, and the new reduced costs $\overline{c}_i - (\gamma_{k^-}^i / \gamma_{k^-}^{j^+})\overline{c}_{j^+}$.

When executing the simplex algorithm, we have $u_k \geq 0$ for all $k \in K$ (and $u_j = 0$ for all $j \notin K$), and the incoming column j^+ is determined by picking one of the column indices such that $\overline{c}_j > 0$. Then the index k^- of the leaving column is determined by looking at the minimum of the ratios $u_k / \gamma_k^{j^+}$ for which $\gamma_k^{j^+} > 0$ (along column j^+).

On the other hand, when executing the dual simplex algorithm, we have $\overline{c}_j \leq 0$ for all $j \notin K$ (and $\overline{c}_k = 0$ for all $k \in K$), and the outgoing column k^- is determined by picking one of the row indices such that $u_k < 0$. The index j^+ of the incoming column is determined by looking at the maximum of the ratios $-\overline{c}_j / \gamma_{k^-}^j$ for which $\gamma_{k^-}^j < 0$ (along row k^-).

More details about the comparison between the simplex algorithm and the dual simplex algorithm can be found in Bertsimas and Tsitsiklis [Bertsimas and Tsitsiklis (1997)] and Papadimitriou and Steiglitz [Papadimitriou and Steiglitz (1998)].

Here is an example of the the dual simplex method.

Example 11.2. Consider the following linear program in standard form:

Maximize $\quad -4x_1 - 2x_2 - x_3$

$$\text{subject to} \quad \begin{pmatrix} -1 & -1 & 2 & 1 & 0 & 0 \\ -4 & -2 & 1 & 0 & 1 & 0 \\ 1 & 1 & -4 & 0 & 0 & 1 \end{pmatrix} \begin{pmatrix} x_1 \\ x_2 \\ x_3 \\ x_4 \\ x_5 \\ x_6 \end{pmatrix} = \begin{pmatrix} -3 \\ -4 \\ 2 \end{pmatrix}, \; x_1, x_2, x_3, x_4, x_5, x_6 \geq 0.$$

We initialize the dual simplex procedure with (u, K) where $u = \begin{pmatrix} 0 \\ 0 \\ 0 \\ -3 \\ -4 \\ 2 \end{pmatrix}$ and $K = (4, 5, 6)$. The initial tableau, before explicitly calculating the reduced cost, is

0	$\overline{c}_1$	$\overline{c}_2$	$\overline{c}_3$	$\overline{c}_4$	$\overline{c}_5$	$\overline{c}_6$
$u_4 = -3$	-1	-1	2	1	0	0
$u_5 = -4$	-4	-2	1	0	1	0
$u_6 = 2$	1	1	-4	0	0	1

.

Since u has negative coordinates, Case (B) applies, and we will set $k^- = 4$. We must now determine whether Case (B1) or Case (B2) applies. This determination is accomplished by scanning the first three columns in the tableau and observing each column has a negative entry. Thus Case (B2) is applicable, and we need to determine the reduced costs. Observe that $c = (-4, -2, -1, 0, 0, 0)$, which in turn implies $c_{(4,5,6)} = (0, 0, 0)$. Equation $(*_2)$ implies that the nonzero reduced costs are

$$\overline{c}_1 = c_1 - c_{(4,5,6)} \begin{pmatrix} -1 \\ -4 \\ 1 \end{pmatrix} = -4$$

$$\overline{c}_2 = c_2 - c_{(4,5,6)} \begin{pmatrix} -1 \\ -2 \\ 1 \end{pmatrix} = -2$$

$$\overline{c}_3 = c_3 - c_{(4,5,6)} \begin{pmatrix} -2 \\ 1 \\ 4 \end{pmatrix} = -1,$$

and our tableau becomes

0	-4	-2	-1	0	0	0
$u_4 = -3$	-1	$\textcircled{-1}$	2	1	0	0
$u_5 = -4$	-4	-2	1	0	1	0
$u_6 = 2$	1	1	-4	0	0	1

.

Since $k^- = 4$, our pivot row is the first row of the tableau. To determine candidates for j^+, we scan this row, locate negative entries and compute

$$\mu^+ = \max\left\{ -\frac{\overline{c}_j}{\gamma_4^j} \;\middle|\; \gamma_4^j < 0,\, j \in \{1, 2, 3\} \right\} = \max\left\{ \frac{-2}{1}, \frac{-4}{1} \right\} = -2.$$

Since μ^+ occurs when $j = 2$, we set $j^+ = 2$. Our new basis is $K^+ = (2, 5, 6)$. We must normalize the first row of the tableau, namely multiply by -1, then add twice this normalized row to the second row, and subtract the normalized row from the third row to obtain the updated tableau.

0	-4	-2	-1	0	0	0
$u_2 = 3$	1	(1)	-2	-1	0	0
$u_5 = 2$	-2	0	-3	-2	1	0
$u_6 = -1$	0	0	-2	1	0	1

It remains to update the reduced costs and the value of the objective function by adding twice the normalized row to the top row.

6	-2	0	-5	-2	0	0
$u_2 = 3$	1	1	-2	-1	0	0
$u_5 = 2$	-2	0	-3	-2	1	0
$u_6 = -1$	0	0	(-2)	1	0	1

We now repeat the procedure of Case (B2) and set $k^- = 6$ (since this is the only negative entry of u^+). Our pivot row is now the third row of the updated tableau, and the new μ^+ becomes

$$\mu^+ = \max\left\{ -\frac{\overline{c}_j}{\gamma_6^j} \;\middle|\; \gamma_6^j < 0,\, j \in \{1, 3, 4\} \right\} = \max\left\{ \frac{-5}{2} \right\} = -\frac{5}{2},$$

which implies that $j^+ = 3$. Hence the new basis is $K^+ = (2, 5, 3)$, and we update the tableau by taking $-\frac{1}{2}$ of Row 3, adding twice the normalized Row 3 to Row 1, and adding three times the normalized Row 3 to Row 2.

6	-2	0	-5	-2	0	0
$u_2 = 4$	1	1	0	-2	0	-1
$u_5 = 7/2$	-2	0	0	$-7/2$	1	$-3/2$
$u_3 = 1/2$	0	0	(1)	$-1/2$	0	$-1/2$

It remains to update the objective function and the reduced costs by adding five times the normalized row to the top row.

$17/2$	-2	0	0	$-9/2$	0	$-5/2$
$u_2 = 4$	1	1	0	-2	0	-1
$u_5 = 7/2$	-2	0	0	$-\frac{7}{2}$	1	$-3/2$
$u_3 = 1/2$	0	0	(1)	$-1/2$	0	$-1/2$

Since u^+ has no negative entries, the dual simplex method terminates and objective function $-4x_1 - 2x_2 - x_3$ is maximized with $-\frac{17}{2}$ at $(0, 4, \frac{1}{2})$. See Figure 11.5.

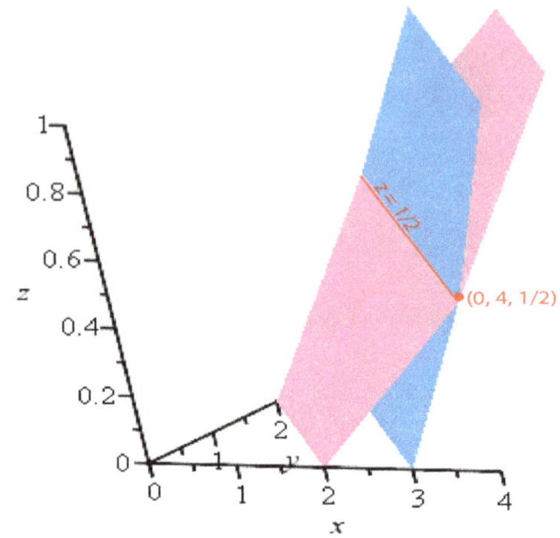

Fig. 11.5 The objective function $-4x_1 - 2x_2 - x_3$ is maximized at the intersection between the blue plane $-x_1 - x_2 + 2x_3 = -3$ and the pink plane $x_1 + x_2 - 4x_3 = 2$.

11.6 The Primal-Dual Algorithm

Let $(P2)$ be a linear program in standard form

$$\begin{aligned} &\text{maximize} \quad cx \\ &\text{subject to} \quad Ax = b \text{ and } x \geq 0, \end{aligned}$$

where A is an $m \times n$ matrix of rank m, and (D) be its dual given by

$$\begin{aligned} &\text{minimize} \quad yb \\ &\text{subject to} \quad yA \geq c, \end{aligned}$$

where $y \in (\mathbb{R}^m)^*$.

First we may assume that $b \geq 0$ by changing every equation $\sum_{j=1}^n a_{ij}x_j = b_i$ with $b_i < 0$ to $\sum_{j=1}^n -a_{ij}x_j = -b_i$. If we happen to have some feasible solution y of the dual program (D), we know from Theorem 11.6 that a feasible solution x of $(P2)$ is an optimal solution iff the equations in $(*_P)$ hold. If we denote by J the subset of $\{1, \ldots, n\}$ for which the equalities

$$yA^j = c_j$$

hold, then by Theorem 11.6 a feasible solution x of $(P2)$ is an optimal solution iff

$$x_j = 0 \quad \text{for all } j \notin J.$$

Let $|J| = p$ and $N = \{1, \ldots, n\} - J$. The above suggests looking for $x \in \mathbb{R}^n$ such that

$$\begin{aligned} \sum_{j \in J} x_j A^j &= b \\ x_j &\geq 0 \quad \text{for all } j \in J \\ x_j &= 0 \quad \text{for all } j \notin J, \end{aligned}$$

or equivalently

$$A_J x_J = b, \quad x_J \geq 0, \tag{$*_1$}$$

and

$$x_N = 0_{n-p}.$$

To search for such an x, we just need to look for a feasible x_J, and for this we can use the *Restricted Primal* linear program (RP) defined as follows:

$$\begin{aligned} &\text{maximize} && -(\xi_1 + \cdots + \xi_m) \\ &\text{subject to} && (A_J \; I_m) \begin{pmatrix} x_J \\ \xi \end{pmatrix} = b \text{ and } x, \xi \geq 0. \end{aligned}$$

Since by hypothesis $b \geq 0$ and the objective function is bounded above by 0, this linear program has an optimal solution (x_J^*, ξ^*).

If $\xi^* = 0$, then the vector $u^* \in \mathbb{R}^n$ given by $u_J^* = x_J^*$ and $u_N^* = 0_{n-p}$ is an optimal solution of (P).

Otherwise, $\xi^* > 0$ and we have failed to solve $(*_1)$. However we may try to use ξ^* to improve y. For this consider the Dual (DRP) of (RP):

$$\begin{aligned} &\text{minimize} && zb \\ &\text{subject to} && zA_J \geq 0 \\ & && z \geq -\mathbf{1}_m^\top. \end{aligned}$$

Observe that the Program (DRP) has the same objective function as the original Dual Program (D). We know by Theorem 11.5 that the optimal solution (x_J^*, ξ^*) of (RP) yields an optimal solution z^* of (DRP) such that

$$z^* b = -(\xi_1^* + \cdots + \xi_m^*) < 0.$$

In fact, if K^* is the basis associated with (x_J^*, ξ^*) and if we write

$$\widehat{A} = (A_J \; I_m)$$

and $\widehat{c} = [0_p^\top \;\; -\mathbf{1}^\top]$, then by Theorem 11.5 we have

$$z^* = \widehat{c}_{K^*} \widehat{A}_{K^*}^{-1} = -\mathbf{1}_m^\top - (\overline{c}_{K^*})_{(p+1,\ldots,p+m)},$$

where $(\overline{c}_{K^*})_{(p+1,\ldots,p+m)}$ denotes the row vector of reduced costs in the final tableau corresponding to the last m columns.

If we write

$$y(\theta) = y + \theta z^*,$$

then the new value of the objective function of (D) is

$$y(\theta)b = yb + \theta z^* b, \tag{$*_2$}$$

and since $z^*b < 0$, we have a chance of improving the objective function of (D), that is, decreasing its value for $\theta > 0$ small enough if $y(\theta)$ is feasible for (D). This will be the case iff $y(\theta)A \geq c$ iff

$$yA + \theta z^* A \geq c. \tag{$*_3$}$$

Now since y is a feasible solution of (D) we have $yA \geq c$, so if $z^*A \geq 0$, then $(*_3)$ is satisfied and $y(\theta)$ is a solution of (D) for all $\theta > 0$, which means that (D) is unbounded. But this implies that (P) is not feasible.

Let us take a closer look at the inequalities $z^*A \geq 0$. For $j \in J$, since z^* is an optimal solution of (DRP), we know that $z^*A_J \geq 0$, so if $z^*A^j \geq 0$ for all $j \in N$, then $(P2)$ is not feasible.

Otherwise, there is some $j \in N = \{1, \ldots, n\} - J$ such that

$$z^* A^j < 0,$$

and then since by the definition of N we have $yA^j > c_j$ for all $j \in N$, if we pick θ such that

$$0 < \theta \leq \frac{yA^j - c_j}{-z^* A^j} \quad j \in N,\; z^* A^j < 0,$$

then we decrease the objective function $y(\theta)b = yb + \theta z^* b$ of (D) (since $z^*b < 0$). Therefore we pick the best θ, namely

$$\theta^+ = \min\left\{ \frac{yA^j - c_j}{-z^* A^j} \;\middle|\; j \notin J,\; z^* A^j < 0 \right\} > 0. \tag{$*_4$}$$

Next we update y to $y^+ = y(\theta^+) = y + \theta^+ z^*$, we create the new restricted primal with the new subset

$$J^+ = \{ j \in \{1, \ldots, n\} \mid y^+ A^j = c_j \},$$

and repeat the process.

Here are the steps of the primal-dual algorithm.

Step 1. Find some feasible solution y of the Dual Program (D). We will show later that this is always possible.

Step 2. Compute

$$J^+ = \{j \in \{1, \ldots, n\} \mid yA^j = c_j\}.$$

Step 3. Set $J = J^+$ and solve the Problem (RP) using the simplex algorithm, starting from the optimal solution determined during the previous round, obtaining the optimal solution (x_J^*, ξ^*) with the basis K^*.

Step 4.

If $\xi^* = 0$, then stop with an optimal solution u^* for (P) such that $u_J^* = x_J^*$ and the other components of u^* are zero.

Else let

$$z^* = -\mathbf{1}_m^\top - (\overline{c}_{K^*})_{(p+1,\ldots,p+m)},$$

be the optimal solution of (DRP) corresponding to (x_J^*, ξ^*) and the basis K^*.

If $z^* A^j \geq 0$ for all $j \notin J$, then stop; the Program (P) has no feasible solution.

Else compute

$$\theta^+ = \min\left\{ -\frac{yA^j - c_j}{z^* A^j} \;\middle|\; j \notin J,\ z^* A^j < 0 \right\}, \qquad y^+ = y + \theta^+ z^*,$$

and

$$J^+ = \{j \in \{1, \ldots, n\} \mid y^+ A^j = c_j\}.$$

Go back to Step 3.

The following proposition shows that at each iteration we can start the Program (RP) with the optimal solution obtained at the previous iteration.

Proposition 11.7. *Every $j \in J$ such that A^j is in the basis of the optimal solution ξ^* belongs to the next index set J^+.*

Proof. Such an index $j \in J$ correspond to a variable ξ_j such that $\xi_j > 0$, so by complementary slackness, the constraint $z^* A^j \geq 0$ of the Dual Program (DRP) must be an equality, that is, $z^* A^j = 0$. But then we have

$$y^+ A^j = yA^j + \theta^+ z^* A^j = c_j,$$

which shows that $j \in J^+$. ☐

If (u^*, ξ^*) with the basis K^* is the optimal solution of the Program (RP), Proposition 11.7 together with the last property of Theorem 11.5 allows us to restart the (RP) in Step 3 with $(u^*, \xi^*)_{K^*}$ as initial solution (with basis K^*). For every $j \in J - J^+$, column j is deleted, and for every $j \in J^+ - J$, the new column A^j is computed by multiplying $\widehat{A}_{K^*}^{-1}$ and A^j, but $\widehat{A}_{K^*}^{-1}$ is the matrix $\Gamma^*[1{:}m; p+1{:}p+m]$ consisting of the last m columns of Γ^* in the final tableau, and the new reduced $\overline{c}_j$ is given by $c_j - z^* A^j$. Reusing the optimal solution of the previous (RP) may improve efficiency significantly.

Another crucial observation is that for any index $j_0 \in N$ such that $\theta^+ = (yA^{j_0} - c_{j_0})/(-z^* A^{j_0})$, we have

$$y^+ A_{j_0} = y A_{j_0} + \theta^+ z^* A^{j_0} = c_{j_0},$$

and so $j_0 \in J^+$. This fact that be used to ensure that the primal-dual algorithm terminates in a finite number of steps (using a pivot rule that prevents cycling); see Papadimitriou and Steiglitz [Papadimitriou and Steiglitz (1998)] (Theorem 5.4).

It remains to discuss how to pick some initial feasible solution y of the Dual Program (D). If $c_j \leq 0$ for $j = 1, \ldots, n$, then we can pick $y = 0$. If we are dealing with a minimization problem, the weight c_j are often nonnegative, so from the point of view of maximization we will have $-c_j \leq 0$ for all j, and we will be able to use $y = 0$ as a starting point.

Going back to our primal problem in maximization form and its dual in minimization form, we still need to deal with the situation where $c_j > 0$ for some j, in which case there may not be any obvious y feasible for (D). Preferably we would like to find such a y very cheaply.

There is a trick to deal with this situation. We pick some very large positive number M and add to the set of equations $Ax = b$ the new equation

$$x_1 + \cdots + x_n + x_{n+1} = M,$$

with the new variable x_{n+1} constrained to be nonnegative. If the Program (P) has a feasible solution, such an M exists. In fact it can shown that for any basic feasible solution $u = (u_1, \ldots, u_n)$, each $|u_i|$ is bounded by some expression depending only on A and b; see Papadimitriou and Steiglitz [Papadimitriou and Steiglitz (1998)] (Lemma 2.1). The proof is not difficult and relies on the fact that the inverse of a matrix can be expressed in terms of certain determinants (the adjugates). Unfortunately, this bound contains $m!$ as a factor, which makes it quite impractical.

Having added the new equation above, we obtain the new set of equations

$$\begin{pmatrix} A & 0_n \\ \mathbf{1}_n^\top & 1 \end{pmatrix} \begin{pmatrix} x \\ x_{n+1} \end{pmatrix} = \begin{pmatrix} b \\ M \end{pmatrix},$$

with $x \geq 0, x_{n+1} \geq 0$, and the new objective function given by

$$(c\; 0) \begin{pmatrix} x \\ x_{n+1} \end{pmatrix} = cx.$$

The dual of the above linear program is

$$\begin{aligned} &\text{minimize} \quad yb + y_{m+1}M \\ &\text{subject to} \quad yA^j + y_{m+1} \geq c_j \quad j = 1, \ldots, n \\ &\qquad\qquad\quad y_{m+1} \geq 0. \end{aligned}$$

If $c_j > 0$ for some j, observe that the linear form $\widetilde{y}$ given by

$$\widetilde{y}_i = \begin{cases} 0 & \text{if } 1 \leq i \leq m \\ \max_{1 \leq j \leq n}\{c_j\} > 0 \end{cases}$$

is a feasible solution of the new dual program. In practice, we can choose M to be a number close to the largest integer representable on the computer being used.

Here is an example of the primal-dual algorithm given in the Math 588 class notes of T. Molla [Molla (2015)].

Example 11.3. Consider the following linear program in standard form:

$$\text{Maximize} \quad -x_1 - 3x_2 - 3x_3 - x_4$$

$$\text{subject to} \quad \begin{pmatrix} 3 & 4 & -3 & 1 \\ 3 & -2 & 6 & -1 \\ 6 & 4 & 0 & 1 \end{pmatrix} \begin{pmatrix} x_1 \\ x_2 \\ x_3 \\ x_4 \end{pmatrix} = \begin{pmatrix} 2 \\ 1 \\ 4 \end{pmatrix} \quad \text{and } x_1, x_2, x_3, x_4 \geq 0.$$

The associated Dual Program (D) is

$$\text{Minimize} \quad 2y_1 + y_2 + 4y_3$$

$$\text{subject to} \quad (y_1\; y_2\; y_3) \begin{pmatrix} 3 & 4 & -3 & 1 \\ 3 & -2 & 6 & -1 \\ 6 & 4 & 0 & 1 \end{pmatrix} \geq (-1\; -3\; -3\; -1).$$

We initialize the primal-dual algorithm with the dual feasible point $y = (-1/3\;\; 0\;\; 0)$. Observe that only the first inequality of (D) is actually an equality, and hence $J = \{1\}$. We form the Restricted Primal Program $(RP1)$

$$\text{Maximize} \quad -(\xi_1 + \xi_2 + \xi_3)$$

$$\text{subject to} \quad \begin{pmatrix} 3\,1\,0\,0 \\ 3\,0\,1\,0 \\ 6\,0\,0\,1 \end{pmatrix} \begin{pmatrix} x_1 \\ \xi_1 \\ \xi_2 \\ \xi_3 \end{pmatrix} = \begin{pmatrix} 2 \\ 1 \\ 4 \end{pmatrix} \quad \text{and } x_1, \xi_1, \xi_2, \xi_3 \geq 0.$$

We now solve $(RP1)$ via the simplex algorithm. The initial tableau with $K = (2, 3, 4)$ and $J = \{1\}$ is

	x_1	ξ_1	ξ_2	ξ_3
7	12	0	0	0
$\xi_1 = 2$	3	1	0	0
$\xi_2 = 1$	(3)	0	1	0
$\xi_3 = 4$	6	0	0	1

.

For $(RP1)$, $\hat{c} = (0, -1, -1, -1)$, $(x_1, \xi_1, \xi_2, \xi_3) = (0, 2, 1, 4)$, and the nonzero reduced cost is given by

$$0 - (-1 \ \ -1 \ \ -1) \begin{pmatrix} 3 \\ 3 \\ 6 \end{pmatrix} = 12.$$

Since there is only one nonzero reduced cost, we must set $j^+ = 1$. Since $\min\{\xi_1/3, \xi_2/3, \xi_3/6\} = 1/3$, we see that $k^- = 3$ and $K = (2, 1, 4)$. Hence we pivot through the red circled 3 (namely we divide row 2 by 3, and then subtract $3\times$ (row 2) from row 1, $6\times$ (row 2) from row 3, and $12\times$ (row 2) from row 0), to obtain the tableau

	x_1	ξ_1	ξ_2	ξ_3
3	0	0	-4	0
$\xi_1 = 1$	0	1	-1	0
$x_1 = 1/3$	1	0	$1/3$	0
$\xi_3 = 2$	0	0	-2	1

.

At this stage the simplex algorithm for $(RP1)$ terminates since there are no positive reduced costs. Since the upper left corner of the final tableau is not zero, we proceed with Step 4 of the primal dual algorithm and compute

$$z^* = (-1 \ \ -1 \ \ -1) - (0 \ \ -4 \ \ 0) = (-1 \ \ 3 \ \ -1),$$

$$yA^2 - c_2 = (-1/300) \begin{pmatrix} 4 \\ -2 \\ 4 \end{pmatrix} + 3 = \frac{5}{3}, \qquad z^* A^2 = -(-1\,3\,-1) \begin{pmatrix} 4 \\ -2 \\ 4 \end{pmatrix} = 14,$$

$$yA^4 - c_4 = (-1/3\,0\,0)\begin{pmatrix}1\\-1\\1\end{pmatrix} + 1 = \frac{2}{3}, \qquad z^*A^4 = -(-1\,3\,-1)\begin{pmatrix}1\\-1\\1\end{pmatrix} = 5,$$

so

$$\theta^+ = \min\left\{\frac{5}{42}, \frac{2}{15}\right\} = \frac{5}{42},$$

and we conclude that the new feasible solution for (D) is

$$y^+ = (-1/3\ \ 0\ \ 0) + \frac{5}{42}(-1\ \ 3\ \ -1) = (-19/42\ \ 5/14\ \ -5/42).$$

When we substitute y^+ into (D), we discover that the first two constraints are equalities, and that the new J is $J = \{1, 2\}$. The new Reduced Primal $(RP2)$ is

$$\begin{aligned}
&\text{Maximize} \quad -(\xi_1 + \xi_2 + \xi_3)\\
&\text{subject to} \quad \begin{pmatrix}3 & 4 & 1 & 0 & 0\\ 3 & -2 & 0 & 1 & 0\\ 6 & 4 & 0 & 0 & 1\end{pmatrix}\begin{pmatrix}x_1\\x_2\\\xi_1\\\xi_2\\\xi_3\end{pmatrix} = \begin{pmatrix}2\\1\\4\end{pmatrix} \quad \text{and } x_1, x_2, \xi_1, \xi_2, \xi_3 \geq 0.
\end{aligned}$$

Once again, we solve $(RP2)$ via the simplex algorithm, where $\hat{c} = (0, 0, -1, -1, -1)$, $(x_1, x_2,\ \xi_1, \xi_2, \xi_3) = (1/3, 0, 1, 0, 2)$ and $K = (3, 1, 5)$. The initial tableau is obtained from the final tableau of the previous $(RP1)$ by adding a column corresponding the the variable x_2, namely

$$\widehat{A}_K^{-1}A^2 = \begin{pmatrix}1 & -1 & 0\\ 0 & 1/3 & 0\\ 0 & -2 & 1\end{pmatrix}\begin{pmatrix}4\\-2\\4\end{pmatrix} = \begin{pmatrix}6\\-2/3\\8\end{pmatrix},$$

with

$$\overline{c}_2 = c_2 - z^*A^2 = 0 - (-1\,3\,-1)\begin{pmatrix}4\\-2\\4\end{pmatrix} = 14,$$

and we get

	x_1	x_2	ξ_1	ξ_2	ξ_3
3	0	14	0	-4	0
$\xi_1 = 1$	0	(6)	1	-1	0
$x_1 = 1/3$	1	$-2/3$	0	$1/3$	0
$\xi_3 = 2$	0	8	0	-2	1

.

Note that $j^+ = 2$ since the only positive reduced cost occurs in column 2. Also observe that since $\min\{\xi_1/6, \xi_3/8\} = \xi_1/6 = 1/6$, we set $k^- = 3$, $K = (2, 1, 5)$ and pivot along the red 6 to obtain the tableau

	x_1	x_2	ξ_1	ξ_2	ξ_3
$2/3$	0	0	$-7/3$	$-5/3$	0
$x_2 = 1/6$	0	1	$1/6$	$-1/6$	0
$x_1 = 4/9$	1	0	$1/9$	$2/9$	0
$\xi_3 = 2/3$	0	0	$-4/3$	$-2/3$	1

.

Since the reduced costs are either zero or negative the simplex algorithm terminates, and we compute

$$z^* = (-1\ \ -1\ \ -1) - (-7/3\ \ -5/3\ \ 0) = (4/3\ \ 2/3\ \ -1),$$

$$y^+A^4 - c_4 = (-19/42\ \ 5/14\ \ -5/42)\begin{pmatrix}1\\-1\\1\end{pmatrix} + 1 = 1/14,$$

$$z^*A^4 = -(4/3\ \ 2/3\ \ -1)\begin{pmatrix}1\\-1\\1\end{pmatrix} = 1/3,$$

so

$$\theta^+ = \frac{3}{14},$$

$$y^+ = (-19/42\ \ 5/14\ \ -5/42) + \frac{5}{14}(4/3\ \ 2/3\ \ -1) = (-1/6\ \ 1/2\ \ -1/3).$$

When we plug y^+ into (D), we discover that the first, second, and fourth constraints are equalities, which implies $J = \{1, 2, 4\}$. Hence the new Restricted Primal $(RP3)$ is

$$\text{Maximize}\quad -(\xi_1 + \xi_2 + \xi_3)$$

$$\text{subject to}\quad \begin{pmatrix}3 & 4 & 1 & 1 & 0 & 0\\ 3 & -2 & -1 & 0 & 1 & 0\\ 6 & 4 & 1 & 0 & 0 & 1\end{pmatrix}\begin{pmatrix}x_1\\x_2\\x_4\\\xi_1\\\xi_2\\\xi_3\end{pmatrix} = \begin{pmatrix}2\\1\\4\end{pmatrix}\quad \text{and } x_1, x_2, x_4, \xi_1, \xi_2, \xi_3 \geq 0.$$

The initial tableau for $(RP3)$, with $\hat{c} = (0, 0, 0, -1, -1, -1)$, $(x_1, x_2, x_4, \xi_1, \xi_2, \xi_3) = (4/9, 1/6, 0, 0, 0, 2/3)$ and $K = (2, 1, 6)$, is obtained from the final tableau of the previous ($RP2$) by adding a column corresponding the the variable x_4, namely

$$\widehat{A}_K^{-1} A^4 = \begin{pmatrix} 1/6 & -1/6 & 0 \\ 1/9 & 2/9 & 0 \\ -4/3 & -2/3 & 1 \end{pmatrix} \begin{pmatrix} 1 \\ -1 \\ 1 \end{pmatrix} = \begin{pmatrix} 1/3 \\ -1/9 \\ 1/3 \end{pmatrix},$$

with

$$\overline{c}_4 = c_4 - z^* A^4 = 0 - (4/3 \; 2/3 \; -1) \begin{pmatrix} 1 \\ -1 \\ 1 \end{pmatrix} = 1/3,$$

and we get

	x_1	x_2	x_4	ξ_1	ξ_2	ξ_3
2/3	0	0	1/3	−7/3	−5/3	0
$x_2 = 1/6$	0	1	(1/3)	1/6	−1/6	0
$x_1 = 4/9$	1	0	−1/9	1/9	2/9	0
$\xi_3 = 2/3$	0	0	1/3	−4/3	−2/3	1

Since the only positive reduced cost occurs in column 3, we set $j^+ = 3$. Furthermore since $\min\{x_2/(1/3), \xi_3/(1/3)\} = x_2/(1/3) = 1/2$, we let $k^- = 2$, $K = (3, 1, 6)$, and pivot around the red circled 1/3 to obtain

	x_1	x_2	x_4	ξ_1	ξ_2	ξ_3
1/2	0	−1	0	−5/2	−3/2	0
$x_4 = 1/2$	0	3	1	1/2	−1/2	0
$x_1 = 1/2$	1	1/3	0	1/6	1/6	0
$\xi_3 = 1/2$	0	−1	0	−3/2	−1/2	1

At this stage there are no positive reduced costs, and we must compute

$$z^* = (-1 \; -1 \; -1) - (-5/2 \; -3/2 \; 0) = (3/2 \; 1/2 \; -1),$$

$$y^+ A^3 - c_3 = (-1/6 \; 1/2 \; -1/3) \begin{pmatrix} -3 \\ 6 \\ 0 \end{pmatrix} + 3 = 13/2,$$

$$z^* A^3 = -(3/2 \; 1/2 \; -1) \begin{pmatrix} -3 \\ 6 \\ 0 \end{pmatrix} = 3/2,$$

so

$$\theta^+ = \frac{13}{3},$$

$$y^+ = (-1/6\ 1/2\ \ -1/3) + \frac{13}{3}(3/2\ 1/2\ \ -1) = (19/3\ 8/3\ \ -14/3).$$

We plug y^+ into (D) and discover that the first, third, and fourth constraints are equalities. Thus, $J = \{1, 3, 4\}$ and the Restricted Primal $(RP4)$ is

$$\text{Maximize} \quad -(\xi_1 + \xi_2 + \xi_3)$$

$$\text{subject to} \quad \begin{pmatrix} 3 & -3 & 1 & 1 & 0 & 0 \\ 3 & 6 & -1 & 0 & 1 & 0 \\ 6 & 0 & 1 & 0 & 0 & 1 \end{pmatrix} \begin{pmatrix} x_1 \\ x_3 \\ x_4 \\ \xi_1 \\ \xi_2 \\ \xi_3 \end{pmatrix} = \begin{pmatrix} 2 \\ 1 \\ 4 \end{pmatrix} \quad \text{and } x_1, x_3, x_4, \xi_1, \xi_2, \xi_3 \geq 0.$$

The initial tableau for $(RP4)$, with $\hat{c} = (0, 0, 0, -1, -1, -1)$, $(x_1, x_3, x_4, \xi_1, \xi_2, \xi_3) = (1/2,\ 0, 1/2, 0, 0, 1/2)$ and $K = (3, 1, 6)$ is obtained from the final tableau of the previous $(RP3)$ by replacing the column corresponding to the variable x_2 by a column corresponding to the variable x_3, namely

$$\widehat{A}_K^{-1} A^3 = \begin{pmatrix} 1/2 & -1/2 & 0 \\ 1/6 & 1/6 & 0 \\ -3/2 & -1/2 & 1 \end{pmatrix} \begin{pmatrix} -3 \\ 6 \\ 0 \end{pmatrix} = \begin{pmatrix} -9/2 \\ 1/2 \\ 3/2 \end{pmatrix},$$

with

$$\overline{c}_3 = c_3 - z^* A^3 = 0 - (3/2\ 1/2\ -1) \begin{pmatrix} -3 \\ 6 \\ 0 \end{pmatrix} = 3/2,$$

and we get

	x_1	x_3	x_4	ξ_1	ξ_2	ξ_3
$1/2$	0	$3/2$	0	$-5/2$	$-3/2$	0
$x_4 = 1/2$	0	$-9/2$	1	$1/2$	$-1/2$	0
$x_1 = 1/2$	1	$1/2$	0	$1/6$	$1/6$	0
$\xi_3 = 1/2$	0	$3/2$ (circled)	0	$-3/2$	$-1/2$	1

By analyzing the top row of reduced cost, we see that $j^+ = 2$. Furthermore, since

$\min\{x_1/(1/2), \xi_3/(3/2)\} = \xi_3/(3/2) = 1/3$, we let $k^- = 6$, $K = (3, 1, 2)$, and pivot along the red circled 3/2 to obtain

	x_1	x_3	x_4	ξ_1	ξ_2	ξ_3
0	0	0	0	-1	-1	-1
$x_4 = 2$	0	0	1	-4	-2	3
$x_1 = 1/3$	1	0	0	2/3	1/3	$-1/3$
$x_3 = 1/3$	0	1	0	-1	$-1/3$	2/3

.

Since the upper left corner of the final tableau is zero and the reduced costs are all ≤ 0, we are finally finished. Then $y = (19/3 \;\; 8/3 \;\; -14/3)$ is an optimal solution of (D), but more importantly $(x_1, x_2, x_3, x_4) = (1/3, 0, 1/3, 2)$ is an optimal solution for our original linear program and provides an optimal value of $-10/3$.

The primal-dual algorithm for linear programming doesn't seem to be the favorite method to solve linear programs nowadays. But it is important because its basic principle, to use a restricted (simpler) primal problem involving an objective function with fixed weights, namely 1, and the dual problem to provide feedback to the primal by improving the objective function of the dual, has led to a whole class of combinatorial algorithms (often approximation algorithms) based on the primal-dual paradigm. The reader will get a taste of this kind of algorithm by consulting Papadimitriou and Steiglitz [Papadimitriou and Steiglitz (1998)], where it is explained how classical algorithms such as Dijkstra's algorithm for the shortest path problem, and Ford and Fulkerson's algorithm for max flow can be derived from the primal-dual paradigm.

11.7 Summary

The main concepts and results of this chapter are listed below:

- Strictly separating hyperplane.
- Farkas–Minkowski proposition.
- Farkas lemma, version I, Farkas lemma, version II.
- Distance of a point to a subset.
- Dual linear program, primal linear program.
- Dual variables, primal variables.
- Complementary slackness conditions.
- Dual simplex algorithm.
- Primal-dual algorithm.
- Restricted primal linear program.

11.8 Problems

Problem 11.1. Let $(v_1, \dots, v_n)$ be a sequence of n vectors in $\mathbb{R}^d$ and let V be the $d \times n$ matrix whose j-th column is v_j. Prove the equivalence of the following two statements:

(a) There is no nontrivial positive linear dependence among the v_j, which means that there is no nonzero vector, $y = (y_1, \dots, y_n) \in \mathbb{R}^n$, with $y_j \geq 0$ for $j = 1, \dots, n$, so that

$$y_1 v_1 + \cdots + y_n v_n = 0$$

or equivalently, $Vy = 0$.

(b) There is some vector, $c \in \mathbb{R}^d$, so that $c^\top V > 0$, which means that $c^\top v_j > 0$, for $j = 1, \dots, n$.

Problem 11.2. Check that the dual in maximization form (D'') of the Dual Program (D') (which is the dual of (P) in maximization form),

$$\begin{aligned} \text{maximize} \quad & -b^\top y^\top \\ \text{subject to} \quad & -A^\top y^\top \leq -c^\top \text{ and } y^\top \geq 0, \end{aligned}$$

where $y \in (\mathbb{R}^m)^*$, gives back the Primal Program (P).

Problem 11.3. In a General Linear Program (P) with n primal variables $x_1, \dots, x_n$ and objective function $\sum_{j=1}^n c_j x_j$ (to be maximized), the m constraints are of the form

$$\begin{aligned} \sum_{j=1}^n a_{ij} x_j &\leq b_i, \\ \sum_{j=1}^n a_{ij} x_j &\geq b_i, \\ \sum_{j=1}^n a_{ij} x_j &= b_i, \end{aligned}$$

for $i = 1, \dots, m$, and the variables x_j satisfy an inequality of the form

$$\begin{aligned} x_j &\geq 0, \\ x_j &\leq 0, \\ x_j &\in \mathbb{R}, \end{aligned}$$

for $j = 1, \ldots, n$. If $y_1, \ldots, y_m$ are the dual variables, show that the dual program of the linear program in standard form equivalent to (P) is equivalent to the linear program whose objective function is $\sum_{i=1}^m y_i b_i$ (to be minimized) and whose constraints are determined as follows:

$$\text{if} \quad \left\{\begin{matrix} x_j \geq 0 \\ x_j \leq 0 \\ x_j \in \mathbb{R} \end{matrix}\right\}, \quad \text{then} \quad \left\{\begin{matrix} \sum_{i=1}^m a_{ij} y_i \geq c_j \\ \sum_{i=1}^m a_{ij} y_i \leq c_j \\ \sum_{i=1}^m a_{ij} y_i = c_j \end{matrix}\right\},$$

and

$$\text{if} \quad \left\{\begin{matrix} \sum_{j=1}^n a_{ij} x_j \leq b_i \\ \sum_{j=1}^n a_{ij} x_j \geq b_i \\ \sum_{j=1}^n a_{ij} x_j = b_i \end{matrix}\right\}, \quad \text{then} \quad \left\{\begin{matrix} y_i \geq 0 \\ y_i \leq 0 \\ y_i \in \mathbb{R} \end{matrix}\right\}.$$

Problem 11.4. Apply the procedure of Problem 11.3 to show that the dual of the (general) linear program

$$\begin{aligned} &\text{maximize} \quad 3x_1 + 2x_2 + 5x_3 \\ &\text{subject to} \\ &\qquad 5x_1 + 3x_2 + x_3 = -8 \\ &\qquad 4x_1 + 2x_2 + 8x_3 \leq 23 \\ &\qquad 6x_1 + 7x_2 + 3x_3 \geq 1 \\ &\qquad x_1 \leq 4,\ x_3 \geq 0 \end{aligned}$$

is the (general) linear program:

$$\begin{aligned} &\text{minimize} \quad -8y_1 + 23y_2 - y_3 + 4y_4 \\ &\text{subject to} \\ &\qquad 5y_1 + 4y_2 - 6y_3 + y_4 = 3 \\ &\qquad 3y_1 + 2y_2 - 7y_3 = 2 \\ &\qquad y_1 + 8y_2 - 3y_3 \geq 5 \\ &\qquad y_2,\ y_3,\ y_4 \geq 0. \end{aligned}$$

Problem 11.5. (1) Prove that the dual of the (general) linear program

$$\begin{aligned}&\text{maximize} \quad cx\\&\text{subject to} \quad Ax = b \text{ and } x \in \mathbb{R}^n\end{aligned}$$

is

$$\begin{aligned}&\text{minimize} \quad yb\\&\text{subject to} \quad yA = c \text{ and } y \in \mathbb{R}^m.\end{aligned}$$

(2) Prove that the dual of the (general) linear program

$$\begin{aligned}&\text{maximize} \quad cx\\&\text{subject to} \quad Ax \geq b \text{ and } x \geq 0\end{aligned}$$

is

$$\begin{aligned}&\text{minimize} \quad yb\\&\text{subject to} \quad yA \geq c \text{ and } y \leq 0.\end{aligned}$$

Problem 11.6. Use the complementary slackness conditions to confirm that

$$x_1 = 2,\ x_2 = 4,\ x_3 = 0,\ x_4 = 0,\ x_5 = 7,\ x_6 = 0$$

is an optimal solution of the following linear program (from Chavatal [Chvatal (1983)], Chapter 5):

$$\begin{aligned}&\text{maximize} \quad 18x_1 - 7x_2 + 12x_3 + 5x_4 + 8x_6\\&\text{subject to}\end{aligned}$$

$$\begin{aligned}2x_1 - 6x_2 + 2x_3 + 7x_4 + 3x_5 + 8x_6 &\leq 1\\-3x_1 - x_2 + 4x_3 - 3x_4 + x_5 + 2x_6 &\leq -2\\8x_1 - 3x_2 + 5x_3 - 2x_4 + 2x_6 &\leq 4\\4x_1 + 8x_3 + 7x_4 - x_5 + 3x_6 &\leq 1\\5x_1 + 2x_2 - 3x_3 + 6x_4 - 2x_5 - x_6 &\leq 5\\x_1, x_2, x_3, x_4, x_5, x_6 &\geq 0.\end{aligned}$$

Problem 11.7. Check carefully that the dual simplex method is equivalent to the simplex method applied to the dual program in maximization form.

PART 3
NonLinear Optimization

Chapter 12

Basics of Hilbert Spaces

Most of the "deep" results about the existence of minima of real-valued functions proven in Chapter 13 rely on two fundamental results of Hilbert space theory:

(1) The projection lemma, which is a result about nonempty, closed, convex subsets of a Hilbert space V.
(2) The Riesz representation theorem, which allows us to express a continuous linear form on a Hilbert space V in terms of a vector in V and the inner product on V.

The correctness of the Karush–Kuhn–Tucker conditions appearing in Lagrangian duality follows from a version of the Farkas–Minkowski proposition, which also follows from the projection lemma.

Thus, we feel that it is indispensable to review some basic results of Hilbert space theory, although in most applications considered here the Hilbert space in question will be finite-dimensional. However, in optimization theory, there are many problems where we seek to find a *function* minimizing some type of energy functional (often given by a bilinear form), in which case we are dealing with an infinite dimensional Hilbert space, so it necessary to develop tools to deal with the more general situation of infinite-dimensional Hilbert spaces.

12.1 The Projection Lemma

Given a Hermitian space $\langle E, \varphi \rangle$, we showed in Section 13.1 (Vol. I) that the function $\| \, \| \colon E \to \mathbb{R}$ defined such that $\|u\| = \sqrt{\varphi(u,u)}$, is a norm on E. Thus, E is a normed vector space. If E is also complete, then it is a very interesting space.

Recall that completeness has to do with the convergence of Cauchy sequences. A normed vector space $\langle E, \| \, \| \rangle$ is automatically a metric space under the metric d defined such that $d(u, v) = \|v - u\|$ (see Chapter 2 for the definition of a normed vector space and of a metric space, or Lang [Lang (1996, 1997)], or Dixmier [Dixmier (1984)]). Given a metric space E with metric d, a sequence $(a_n)_{n\geq 1}$ of elements $a_n \in E$ is a *Cauchy sequence* iff for every $\epsilon > 0$, there is some $N \geq 1$ such that

$$d(a_m, a_n) < \epsilon \quad \text{for all} \quad m, n \geq N.$$

We say that E is *complete* iff every Cauchy sequence converges to a limit (which is unique, since a metric space is Hausdorff).

Every finite dimensional vector space over $\mathbb{R}$ or $\mathbb{C}$ is complete. For example, one can show by induction that given any basis $(e_1, \ldots, e_n)$ of E, the linear map $h\colon \mathbb{C}^n \to E$ defined such that

$$h((z_1, \ldots, z_n)) = z_1 e_1 + \cdots + z_n e_n$$

is a homeomorphism (using the *sup*-norm on $\mathbb{C}^n$). One can also use the fact that any two norms on a finite dimensional vector space over $\mathbb{R}$ or $\mathbb{C}$ are equivalent (see Chapter 8 (Vol. I), or Lang [Lang (1997)], Dixmier [Dixmier (1984)], Schwartz [Schwartz (1991)]).

However, if E has infinite dimension, it may not be complete. When a Hermitian space is complete, a number of the properties that hold for finite dimensional Hermitian spaces also hold for infinite dimensional spaces. For example, any closed subspace has an orthogonal complement, and in particular, a finite dimensional subspace has an orthogonal complement. Hermitian spaces that are also complete play an important role in analysis. Since they were first studied by Hilbert, they are called Hilbert spaces.

Definition 12.1. A (complex) Hermitian space $\langle E, \varphi \rangle$ which is a complete normed vector space under the norm $\| \, \|$ induced by φ is called a *Hilbert space*. A real Euclidean space $\langle E, \varphi \rangle$ which is complete under the norm $\| \, \|$ induced by φ is called a *real Hilbert space*.

All the results in this section hold for complex Hilbert spaces as well as for real Hilbert spaces. We state all results for the complex case only, since they also apply to the real case, and since the proofs in the complex case need a little more care.

Example 12.1. The space ℓ^2 of all countably infinite sequences $x = (x_i)_{i\in\mathbb{N}}$ of complex numbers such that $\sum_{i=0}^{\infty} |x_i|^2 < \infty$ is a Hilbert space. It will be shown later that the map $\varphi\colon \ell^2 \times \ell^2 \to \mathbb{C}$ defined such that

$$\varphi\left((x_i)_{i\in\mathbb{N}}, (y_i)_{i\in\mathbb{N}}\right) = \sum_{i=0}^{\infty} x_i\overline{y_i}$$

is well defined, and that ℓ^2 is a Hilbert space under φ. In fact, we will prove a more general result (Proposition A.3).

Example 12.2. The set $\mathcal{C}^\infty[a,b]$ of smooth functions $f\colon [a,b] \to \mathbb{C}$ is a Hermitian space under the Hermitian form

$$\langle f, g\rangle = \int_a^b f(x)\overline{g(x)}dx,$$

but it is not a Hilbert space because it is not complete. It is possible to construct its completion $\mathrm{L}^2([a,b])$, which turns out to be the space of Lebesgue integrable functions on $[a,b]$.

Theorem 2.4 yields a quick proof of the fact that any Hermitian space E (with Hermitian product $\langle -,-\rangle$) can be embedded in a Hilbert space E_h.

Theorem 12.1. *Given a Hermitian space $(E, \langle -,-\rangle)$ (resp. Euclidean space), there is a Hilbert space $(E_h, \langle -,-\rangle_h)$ and a linear map $\varphi\colon E \to E_h$, such that*

$$\langle u, v\rangle = \langle \varphi(u), \varphi(v)\rangle_h$$

for all $u, v \in E$, and $\varphi(E)$ is dense in E_h. Furthermore, E_h is unique up to isomorphism.

Proof. Let $(\widehat{E}, \|\,\|_{\widehat{E}})$ be the Banach space, and let $\varphi\colon E \to \widehat{E}$ be the linear isometry, given by Theorem 2.4. Let $\|u\| = \sqrt{\langle u, u\rangle}$ (with $u \in E$) and $E_h = \widehat{E}$. If E is a real vector space, we know from Section 11.1 (Vol. I) that the inner product $\langle -,-\rangle$ can be expressed in terms of the norm $\|u\|$ by the polarity equation

$$\langle u, v\rangle = \frac{1}{2}(\|u+v\|^2 - \|u\|^2 - \|v\|^2),$$

and if E is a complex vector space, we know from Section 13.1 (Vol. I) that we have the polarity equation

$$\langle u, v\rangle = \frac{1}{4}(\|u+v\|^2 - \|u-v\|^2 + i\|u+iv\|^2 - i\|u-iv\|^2).$$

By the Cauchy–Schwarz inequality, $|\langle u, v\rangle| \leq \|u\|\|v\|$, the map $\langle -,-\rangle\colon E\times E \to \mathbb{C}$ (resp. $\langle -,-\rangle\colon E\times E \to \mathbb{R}$) is continuous. However, it is not uniformly continuous, but we can get around this problem by using the polarity equations to extend it to a continuous map. By continuity, the polarity equations also hold in E_h, which shows that $\langle -, - \rangle$ extends to a positive definite Hermitian inner product (resp. Euclidean inner product) $\langle -, - \rangle_h$ on E_h induced by $\| \, \|_{\widehat{E}}$ extending $\langle -, - \rangle$. $\square$

Remark: We followed the approach in Schwartz [Schwartz (1980)] (Chapter XXIII, Section 42, Theorem 2). For other approaches, see Munkres [Munkres (2000)] (Chapter 7, Section 43), and Bourbaki [Bourbaki (1981)].

One of the most important facts about finite-dimensional Hermitian (and Euclidean) spaces is that they have orthonormal bases. This implies that, up to isomorphism, *every finite-dimensional Hermitian space is isomorphic to* $\mathbb{C}^n$ (for some $n \in \mathbb{N}$) and that the inner product is given by

$$\langle (x_1, \ldots, x_n), (y_1, \ldots, y_n) \rangle = \sum_{i=1}^{n} x_i \overline{y_i}.$$

Furthermore, every subspace W has an orthogonal complement $W^{\perp}$, and the inner product induces a natural duality between E and E^* (actually, between $\overline{E}$ and E^*) where E^* is the space of linear forms on E.

When E is a Hilbert space, E may be infinite dimensional, often of uncountable dimension. Thus, we can't expect that E always have an orthonormal basis. However, if we modify the notion of basis so that a "Hilbert basis" is an orthogonal family that is also dense in E, i.e., every $v \in E$ is the limit of a sequence of finite combinations of vectors from the Hilbert basis, then we can recover most of the "nice" properties of finite-dimensional Hermitian spaces. For instance, if $(u_k)_{k \in K}$ is a Hilbert basis, for every $v \in E$, we can define the Fourier coefficients $c_k = \langle v, u_k \rangle / \|u_k\|$, and then, v is the "sum" of its Fourier series $\sum_{k \in K} c_k u_k$. However, the cardinality of the index set K can be very large, and it is necessary to define what it means for a family of vectors indexed by K to be summable. We will do this in Section A.1. It turns out that every Hilbert space is isomorphic to a space of the form $\ell^2(K)$, where $\ell^2(K)$ is a generalization of the space of Example 12.1 (see Theorem A.1, usually called the Riesz–Fischer theorem).

Our first goal is to prove that a closed subspace of a Hilbert space has an orthogonal complement. We also show that duality holds if we redefine the dual E' of E to be the space of *continuous* linear maps on E. Our presentation closely follows Bourbaki [Bourbaki (1981)]. We also were inspired by Rudin [Rudin (1987)], Lang [Lang (1996, 1997)], Schwartz

[Schwartz (1991, 1980)], and Dixmier [Dixmier (1984)]. In fact, we highly recommend Dixmier [Dixmier (1984)] as a clear and simple text on the basics of topology and analysis. To achieve this goal, we must first prove the so-called projection lemma.

Recall that in a metric space E, a subset X of E is *closed* iff for every convergent sequence (x_n) of points $x_n \in X$, the limit $x = \lim_{n\to\infty} x_n$ also belongs to X. The *closure* $\overline{X}$ *of* X is the set of all limits of convergent sequences (x_n) of points $x_n \in X$. Obviously, $X \subseteq \overline{X}$. We say that the subset X of E is *dense in* E iff $E = \overline{X}$, the closure of X, which means that every $a \in E$ is the limit of some sequence (x_n) of points $x_n \in X$. Convex sets will again play a crucial role. In a complex vector space E, a subset $C \subseteq E$ is convex if $(1-\lambda)x + \lambda y \in C$ for all $x, y \in C$ and all *real* $\lambda \in [0,1]$. Observe that a subspace is convex.

First we state the following easy "parallelogram law," whose proof is left as an exercise.

Proposition 12.1. *If E is a Hermitian space, for any two vectors $u, v \in E$, we have*

$$\|u+v\|^2 + \|u-v\|^2 = 2(\|u\|^2 + \|v\|^2).$$

From the above, we get the following proposition:

Proposition 12.2. *If E is a Hermitian space, given any $d, \delta \in \mathbb{R}$ such that $0 \le \delta < d$, let*

$$B = \{u \in E \mid \|u\| < d\} \quad \text{and} \quad C = \{u \in E \mid \|u\| \le d + \delta\}.$$

For any convex set such A that $A \subseteq C - B$, we have

$$\|v - u\| \le \sqrt{12 d\delta},$$

for all $u, v \in A$ (see Figure 12.1).

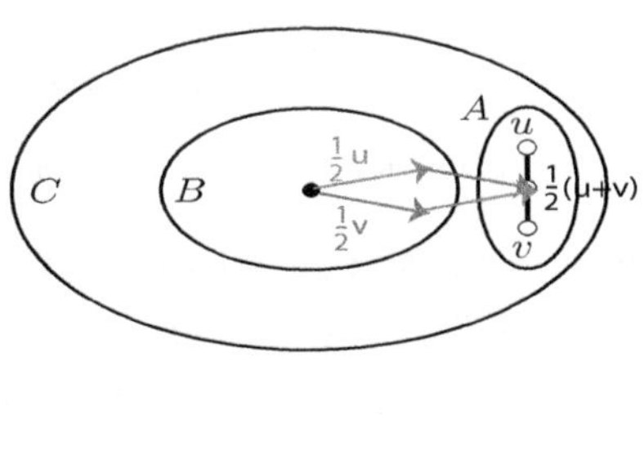

Fig. 12.1 Inequality of Proposition 12.2.

Proof. Since A is convex, $\frac{1}{2}(u+v) \in A$ if $u, v \in A$, and thus, $\|\frac{1}{2}(u+v)\| \geq d$. From the parallelogram equality written in the form

$$\left\|\frac{1}{2}(u+v)\right\|^2 + \left\|\frac{1}{2}(u-v)\right\|^2 = \frac{1}{2}\left(\|u\|^2 + \|v\|^2\right),$$

since $\delta < d$, we get

$$\left\|\frac{1}{2}(u-v)\right\|^2 = \frac{1}{2}\left(\|u\|^2 + \|v\|^2\right) - \left\|\frac{1}{2}(u+v)\right\|^2 \leq (d+\delta)^2 - d^2 = 2d\delta + \delta^2 \leq 3d\delta,$$

from which

$$\|v - u\| \leq \sqrt{12d\delta}.$$

$\square$

Definition 12.2. If X is a nonempty subset of a metric space (E, d), for any $a \in E$, recall that we define the *distance* $d(a, X)$ *of* a *to* X as

$$d(a, X) = \inf_{b \in X} d(a, b).$$

Also, the *diameter* $\delta(X)$ *of* X is defined by

$$\delta(X) = \sup\{d(a, b) \mid a, b \in X\}.$$

It is possible that $\delta(X) = \infty$.

We leave the following standard two facts as an exercise (see Dixmier [Dixmier (1984)]):

Proposition 12.3. *Let E be a metric space.*

(1) For every subset $X \subseteq E$, $\delta(X) = \delta(\overline{X})$.
(2) If E is a complete metric space, for every sequence (F_n) of closed nonempty subsets of E such that $F_{n+1} \subseteq F_n$, if $\lim_{n \to \infty} \delta(F_n) = 0$, then $\bigcap_{n=1}^{\infty} F_n$ consists of a single point.

We are now ready to prove the crucial projection lemma.

Proposition 12.4. *(Projection lemma) Let E be a Hilbert space.*

(1) For any nonempty convex and closed subset $X \subseteq E$, for any $u \in E$, there is a unique vector $p_X(u) \in X$ such that

$$\|u - p_X(u)\| = \inf_{v \in X} \|u - v\| = d(u, X).$$

See Figure 12.2.

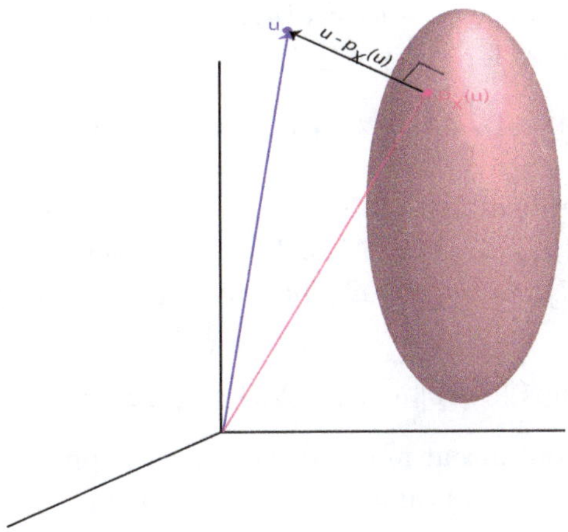

Fig. 12.2 Let X be the solid pink ellipsoid. The projection of the purple point u onto X is the magenta point $p_X(u)$.

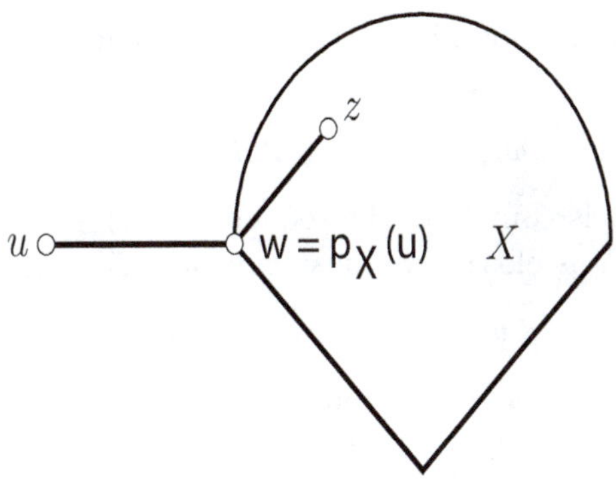

Fig. 12.3 Inequality of Proposition 12.4.

(2) *The vector $p_X(u)$ is the unique vector $w \in E$ satisfying the following property (see Figure 12.3):*

$$w \in X \quad \text{and} \quad \Re \langle u - w, z - w \rangle \leq 0 \quad \text{for all } z \in X. \qquad (*)$$

(3) *If X is a nonempty closed subspace of E, then the vector $p_X(u)$ is the unique vector $w \in E$ satisfying the following property:*

$$w \in X \quad \text{and} \quad \langle u - w, z \rangle = 0 \quad \text{for all } z \in X. \qquad (**)$$

Proof. (1) Let $d = \inf_{v \in X} \|u - v\| = d(u, X)$. We define a sequence X_n of subsets of X as follows: for every $n \geq 1$,

$$X_n = \left\{ v \in X \mid \|u - v\| \leq d + \frac{1}{n} \right\}.$$

It is immediately verified that each X_n is nonempty (by definition of d), convex, and that $X_{n+1} \subseteq X_n$. Also, by Proposition 12.2, (where $B = \{v \in E \mid \|u - v\| \leq d\}$, $C = \{v \in E \mid \|u - v\| \leq d + \frac{1}{n}\}$, and $A = X_n$), we have

$$\sup\{\|z - v\| \mid v, z \in X_n\} \leq \sqrt{12d/n},$$

and thus, $\bigcap_{n \geq 1} X_n$ contains at most one point; see Proposition 12.3(2). We will prove that $\bigcap_{n \geq 1} X_n$ contains exactly one point, namely, $p_X(u)$. For this, define a sequence $(w_n)_{n \geq 1}$ by picking some $w_n \in X_n$ for every $n \geq 1$. We claim that $(w_n)_{n \geq 1}$ is a Cauchy sequence. Given any $\epsilon > 0$, if we pick N such that

$$N > \frac{12d}{\epsilon^2},$$

since $(X_n)_{n \geq 1}$ is a monotonic decreasing sequence, which means that $X_{n+1} \subseteq X_n$ for all $n \geq 1$, for all $m, n \geq N$, we have

$$\|w_m - w_n\| \leq \sqrt{12d/N} < \epsilon,$$

as desired. Since E is complete, the sequence $(w_n)_{n \geq 1}$ has a limit w, and since $w_n \in X$ and X is closed, we must have $w \in X$. Also observe that

$$\|u - w\| \leq \|u - w_n\| + \|w_n - w\|,$$

and since w is the limit of $(w_n)_{n \geq 1}$ and

$$\|u - w_n\| \leq d + \frac{1}{n},$$

given any $\epsilon > 0$, there is some n large enough so that

$$\frac{1}{n} < \frac{\epsilon}{2} \quad \text{and} \quad \|w_n - w\| \leq \frac{\epsilon}{2},$$

and thus

$$\|u - w\| \leq d + \epsilon.$$

Since the above holds for every $\epsilon > 0$, we have $\|u - w\| = d$. Thus, $w \in X_n$ for all $n \geq 1$, which proves that $\bigcap_{n \geq 1} X_n = \{w\}$. Now any $z \in X$ such that $\|u - z\| = d(u, X) = d$ also belongs to every X_n, and thus $z = w$, proving the uniqueness of w, which we denote as $p_X(u)$. See Figure 12.4.

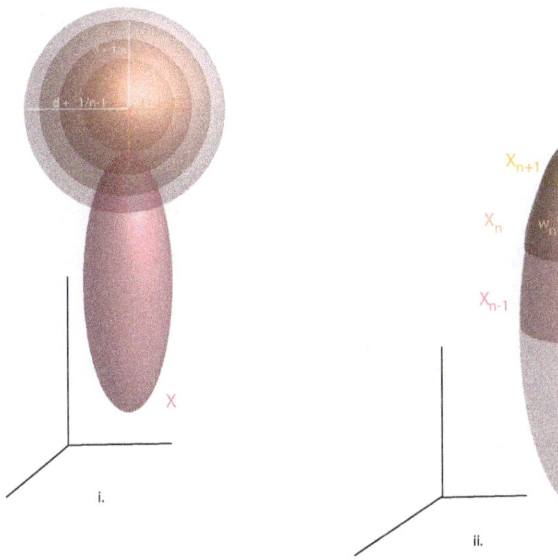

Fig. 12.4 Let X be the solid pink ellipsoid with $p_X(u) = w$ at its apex. Each X_n is the intersection of X and a solid sphere centered at u with radius $d + 1/n$. These intersections are the colored "caps" of Figure ii. The Cauchy sequence $(w_n)_{n\geq 1}$ is obtained by selecting a point in each colored X_n.

(2) Let $z \in X$. Since X is convex, $v = (1-\lambda)p_X(u) + \lambda z \in X$ for every λ, $0 \leq \lambda \leq 1$. Then by the definition of u, we have

$$\|u - v\| \geq \|u - p_X(u)\|$$

for all λ, $0 \leq \lambda \leq 1$, and since

$$\begin{aligned}\|u - v\|^2 &= \|u - p_X(u) - \lambda(z - p_X(u))\|^2 \\ &= \|u - p_X(u)\|^2 + \lambda^2\|z - p_X(u)\|^2 - 2\lambda \Re \langle u - p_X(u), z - p_X(u)\rangle,\end{aligned}$$

for all λ, $0 < \lambda \leq 1$, we get

$$\Re \langle u - p_X(u), z - p_X(u)\rangle = \frac{1}{2\lambda}\left(\|u - p_X(u)\|^2 - \|u - v\|^2\right) + \frac{\lambda}{2}\|z - p_X(u)\|^2. \quad (\dagger)$$

Since

$$\|u - v\| \geq \|u - p_X(u)\|,$$

we have

$$\|u - p_X(u)\|^2 - \|u - v\|^2 = (\|u - p_X(u)\| - \|u - v\|)(\|u - p_X(u)\| + \|u - v\|) \leq 0,$$

and since Equation (†) holds for all λ such that $0 < \lambda \leq 1$, if $\|u - p_X(u)\|^2 - \|u - v\|^2 < 0$, then for $\lambda > 0$ small enough we have

$$\frac{1}{2\lambda}\left(\|u - p_X(u)\|^2 - \|u - v\|^2\right) + \frac{\lambda}{2}\|z - p_X(u)\|^2 < 0,$$

and if $\|u - p_X(u)\|^2 - \|u - v\|^2 = 0$, then the limit of $\frac{\lambda}{2}\|z - p_X(u)\|^2$ as $\lambda > 0$ goes to zero is zero, so in all cases, by (†), we have

$$\Re\langle u - p_X(u), z - p_X(u)\rangle \leq 0.$$

Conversely, assume that $w \in X$ satisfies the condition

$$\Re\langle u - w, z - w\rangle \leq 0$$

for all $z \in X$. For all $z \in X$, we have

$$\|u - z\|^2 = \|u - w\|^2 + \|z - w\|^2 - 2\Re\langle u - w, z - w\rangle \geq \|u - w\|^2,$$

which implies that $\|u - w\| = d(u, X) = d$, and from (1), that $w = p_X(u)$.

(3) If X is a subspace of E and $w \in X$, when z ranges over X the vector $z - w$ also ranges over the whole of X so Condition $(*)$ is equivalent to

$$w \in X \quad \text{and} \quad \Re\langle u - w, z\rangle \leq 0 \quad \text{for all } z \in X. \tag{$*_1$}$$

Since X is a subspace, if $z \in X$, then $-z \in X$, which implies that $(*_1)$ is equivalent to

$$w \in X \quad \text{and} \quad \Re\langle u - w, z\rangle = 0 \quad \text{for all } z \in X. \tag{$*_2$}$$

Finally, since X is a subspace, if $z \in X$,then $iz \in X$, and this implies that

$$0 = \Re\langle u - w, iz\rangle = -i\Im\langle u - w, z\rangle,$$

so $\Im\langle u - w, z\rangle = 0$, but since we also have $\Re\langle u - w, z\rangle = 0$, we see that $(*_2)$ is equivalent to

$$w \in X \quad \text{and} \quad \langle u - w, z\rangle = 0 \quad \text{for all } z \in X, \tag{$**$}$$

as claimed. □

Definition 12.3. The vector $p_X(u)$ is called the *projection of* u *onto* X, and the map $p_X \colon E \to X$ is called the *projection of* E *onto* X.

In the case of a real Hilbert space, there is an intuitive geometric interpretation of the condition

$$\langle u - p_X(u), z - p_X(u)\rangle \leq 0$$

for all $z \in X$. If we restate the condition as

$$\langle u - p_X(u), p_X(u) - z\rangle \geq 0$$

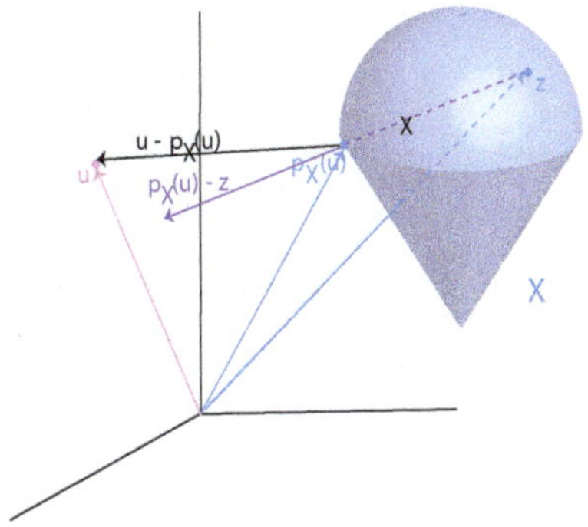

Fig. 12.5 Let X be the solid blue ice cream cone. The acute angle between the black vector $u - p_X(u)$ and the purple vector $p_X(u) - z$ is less than $\pi/2$.

for all $z \in X$, this says that the absolute value of the measure of the angle between the vectors $u - p_X(u)$ and $p_X(u) - z$ is at most $\pi/2$. See Figure 12.5. This makes sense, since X is convex, and points in X must be on the side opposite to the "tangent space" to X at $p_X(u)$, which is orthogonal to $u - p_X(u)$. Of course, this is only an intuitive description, since the notion of tangent space has not been defined!

If X is a closed subspace of E, then Condition $(**)$ says that the vector $u - p_X(u)$ is *orthogonal* to X, in the sense that $u - p_X(u)$ is orthogonal to every vector $z \in X$.

The map $p_X : E \to X$ is continuous as shown below.

Proposition 12.5. *Let E be a Hilbert space. For any nonempty convex and closed subset $X \subseteq E$, the map $p_X : E \to X$ is continuous. In fact, p_X satisfies the Lipschitz condition*

$$\|p_X(v) - p_X(u)\| \le \|v - u\| \quad \textit{for all } u, v \in E.$$

Proof. For any two vectors $u, v \in E$, let $x = p_X(u) - u$, $y = p_X(v) - p_X(u)$, and $z = v - p_X(v)$. Clearly, (as illustrated in Figure 12.6),

$$v - u = x + y + z,$$

and from Proposition 12.4(2), we also have

$$\Re\langle x, y\rangle \ge 0 \quad \text{and} \quad \Re\langle z, y\rangle \ge 0,$$

from which we get

$$\begin{aligned}\|v-u\|^2 &= \|x+y+z\|^2 = \|x+z+y\|^2\\ &= \|x+z\|^2 + \|y\|^2 + 2\Re\langle x, y\rangle + 2\Re\langle z, y\rangle\\ &\geq \|y\|^2 = \|p_X(v) - p_X(u)\|^2.\end{aligned}$$

However, $\|p_X(v) - p_X(u)\| \leq \|v-u\|$ obviously implies that p_X is continuous. □

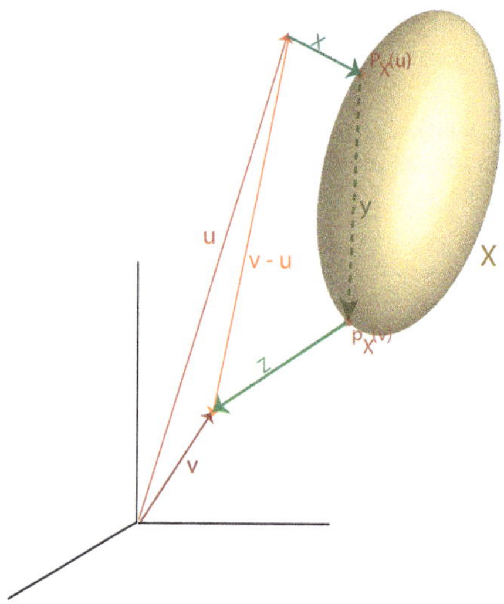

Fig. 12.6 Let X be the solid gold ellipsoid. The vector $v-u$ is the sum of the three green vectors, each of which is determined by the appropriate projections.

We can now prove the following important proposition.

Proposition 12.6. *Let E be a Hilbert space.*

(1) For any closed subspace $V \subseteq E$, we have $E = V \oplus V^{\perp}$, and the map $p_V \colon E \to V$ is linear and continuous.
(2) For any $u \in E$, the projection $p_V(u)$ is the unique vector $w \in E$ such that

$$w \in V \quad \textit{and} \quad \langle u-w, z\rangle = 0 \quad \textit{for all } z \in V.$$

Proof. (1) First, we prove that $u - p_V(u) \in V^\perp$ for all $u \in E$. For any $v \in V$, since V is a subspace, $z = p_V(u) + \lambda v \in V$ for all $\lambda \in \mathbb{C}$, and since V is convex and nonempty (since it is a subspace), and closed by hypothesis, by Proposition 12.4(2), we have

$$\Re(\overline{\lambda}\,\langle u - p_V(u), v\rangle) = \Re(\langle u - p_V(u), \lambda v\rangle = \Re\,\langle u - p_V(u), z - p_V(u)\rangle \leq 0$$

for all $\lambda \in \mathbb{C}$. In particular, the above holds for $\lambda = \langle u - p_V(u), v\rangle$, which yields

$$|\,\langle u - p_V(u), v\rangle\,| \leq 0,$$

and thus, $\langle u - p_V(u), v\rangle = 0$. See Figure 12.7. As a consequence, $u - p_V(u) \in V^\perp$ for all $u \in E$. Since $u = p_V(u) + u - p_V(u)$ for every $u \in E$, we have $E = V + V^\perp$. On the other hand, since $\langle -, -\rangle$ is positive definite, $V \cap V^\perp = \{0\}$, and thus $E = V \oplus V^\perp$.

We already proved in Proposition 12.5 that $p_V\colon E \to V$ is continuous. Also, since

$$\begin{aligned} &p_V(\lambda u + \mu v) - (\lambda p_V(u) + \mu p_V(v)) \\ &\qquad = p_V(\lambda u + \mu v) - (\lambda u + \mu v) + \lambda(u - p_V(u)) + \mu(v - p_V(v)), \end{aligned}$$

for all $u, v \in E$, and since the left-hand side term belongs to V, and from what we just showed, the right-hand side term belongs to $V^\perp$, we have

$$p_V(\lambda u + \mu v) - (\lambda p_V(u) + \mu p_V(v)) = 0,$$

showing that p_V is linear.

(2) This is basically obvious from (1). We proved in (1) that $u - p_V(u) \in V^\perp$, which is exactly the condition

$$\langle u - p_V(u), z\rangle = 0$$

for all $z \in V$. Conversely, if $w \in V$ satisfies the condition

$$\langle u - w, z\rangle = 0$$

for all $z \in V$, since $w \in V$, every vector $z \in V$ is of the form $y - w$, with $y = z + w \in V$, and thus, we have

$$\langle u - w, y - w\rangle = 0$$

for all $y \in V$, which implies the condition of Proposition 12.4(2):

$$\Re\,\langle u - w, y - w\rangle \leq 0$$

for all $y \in V$. By Proposition 12.4, $w = p_V(u)$ is the projection of u onto V. $\square$

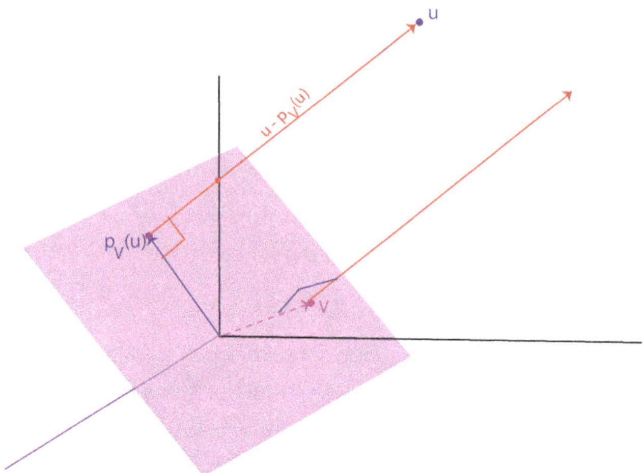

Fig. 12.7 Let V be the pink plane. The vector $u - p_V(u)$ is perpendicular to any $v \in V$.

Remark: If $p_V : E \to V$ is linear, then V is a subspace of E. It follows that if V is a closed convex subset of E, then $p_V : E \to V$ is linear iff V is a subspace of E.

Example 12.3. Let us illustrate the power of Proposition 12.6 on the following "least squares" problem. Given a real $m \times n$-matrix A and some vector $b \in \mathbb{R}^m$, we would like to solve the linear system

$$Ax = b$$

in the least-squares sense, which means that we would like to find some solution $x \in \mathbb{R}^n$ that minimizes the Euclidean norm $\|Ax - b\|$ of the error $Ax - b$. It is actually not clear that the problem has a solution, but it does! The problem can be restated as follows: Is there some $x \in \mathbb{R}^n$ such that

$$\|Ax - b\| = \inf_{y \in \mathbb{R}^n} \|Ay - b\|,$$

or equivalently, is there some $z \in \operatorname{Im}(A)$ such that

$$\|z - b\| = d(b, \operatorname{Im}(A)),$$

where $\operatorname{Im}(A) = \{Ay \in \mathbb{R}^m \mid y \in \mathbb{R}^n\}$, the image of the linear map induced by A. Since $\operatorname{Im}(A)$ is a closed subspace of $\mathbb{R}^m$, because we are in finite dimension, Proposition 12.6 tells us that there is a unique $z \in \operatorname{Im}(A)$ such that

$$\|z - b\| = \inf_{y \in \mathbb{R}^n} \|Ay - b\|,$$

and thus the problem always has a solution since $z \in \operatorname{Im}(A)$, and since there is at least some $x \in \mathbb{R}^n$ such that $Ax = z$ (by definition of $\operatorname{Im}(A)$). Note that such an x is not necessarily unique. Furthermore, Proposition 12.6 also tells us that $z \in \operatorname{Im}(A)$ is the solution of the equation

$$\langle z - b, w\rangle = 0 \quad \text{for all } w \in \operatorname{Im}(A),$$

or equivalently, that $x \in \mathbb{R}^n$ is the solution of

$$\langle Ax - b, Ay\rangle = 0 \quad \text{for all } y \in \mathbb{R}^n,$$

which is equivalent to

$$\langle A^\top(Ax - b), y\rangle = 0 \quad \text{for all } y \in \mathbb{R}^n,$$

and thus, since the inner product is positive definite, to $A^\top(Ax - b) = 0$, i.e.,

$$A^\top Ax = A^\top b.$$

Therefore, the solutions of the original least-squares problem are precisely the solutions of the the so-called *normal equations*

$$A^\top Ax = A^\top b,$$

discovered by Gauss and Legendre around 1800. We also proved that the normal equations always have a solution.

Computationally, it is best not to solve the normal equations directly, and instead, to use methods such as the QR-decomposition (applied to A) or the SVD-decomposition (in the form of the pseudo-inverse). We will come back to this point later on.

Here is another important corollary of Proposition 12.6.

Corollary 12.1. *For any continuous nonnull linear map $h\colon E \to \mathbb{C}$, the null space*

$$H = \operatorname{Ker} h = \{u \in E \mid h(u) = 0\} = h^{-1}(0)$$

is a closed hyperplane H, and thus, $H^\perp$ is a subspace of dimension one such that $E = H \oplus H^\perp$.

The above suggests defining the dual space of E as the set of all continuous maps $h\colon E \to \mathbb{C}$.

Remark: If $h\colon E \to \mathbb{C}$ is a linear map which is **not** continuous, then it can be shown that the hyperplane $H = \operatorname{Ker} h$ is dense in E! Thus, $H^\perp$

is reduced to the trivial subspace $\{0\}$. This goes against our intuition of what a hyperplane in $\mathbb{R}^n$ (or $\mathbb{C}^n$) is, and warns us not to trust our "physical" intuition too much when dealing with infinite dimensions. As a consequence, the map $\flat\colon E \to E^*$ introduced in Section 13.2 (Vol. I) (see just after Definition 12.4 below) is not surjective, since the linear forms of the form $u \mapsto \langle u, v\rangle$ (for some fixed vector $v \in E$) are continuous (the inner product is continuous).

12.2 Duality and the Riesz Representation Theorem

We now show that by redefining the dual space of a Hilbert space as the set of continuous linear forms on E we recover Theorem 13.6 (Vol. I).

Definition 12.4. Given a Hilbert space E, we define the *dual space E' of E* as the vector space of all continuous linear forms $h\colon E \to \mathbb{C}$. Maps in E' are also called *bounded linear operators, bounded linear functionals, or simply operators or functionals.*

As in Section 13.2 (Vol. I), for all $u, v \in E$, we define the maps $\varphi^l_u\colon E \to \mathbb{C}$ and $\varphi^r_v\colon E \to \mathbb{C}$ such that

$$\varphi^l_u(v) = \overline{\langle u, v\rangle},$$

and

$$\varphi^r_v(u) = \langle u, v\rangle\,.$$

In fact, $\varphi^l_u = \varphi^r_u$, and because the inner product $\langle -, -\rangle$ is continuous, it is obvious that φ^r_v is continuous and linear, so that $\varphi^r_v \in E'$. To simplify notation, we write φ_v instead of φ^r_v.

Theorem 13.6 (Vol. I) is generalized to Hilbert spaces as follows.

Proposition 12.7. *(Riesz representation theorem) Let E be a Hilbert space. Then the map $\flat\colon E \to E'$ defined such that*

$$\flat(v) = \varphi_v,$$

is semilinear, continuous, and bijective. Furthermore, for any continuous linear map $\psi \in E'$, if $u \in E$ is the unique vector such that

$$\psi(v) = \langle v, u\rangle \quad \textit{for all } v \in E,$$

then we have $\|\psi\| = \|u\|$, where

$$\|\psi\| = \sup\left\{ \frac{|\psi(v)|}{\|v\|} \;\middle|\; v \in E,\ v \neq 0\right\}.$$

Proof. The proof is basically identical to the proof of Theorem 13.6 (Vol. I), except that a different argument is required for the surjectivity of $\flat\colon E \to E'$, since E may not be finite dimensional. For any nonnull linear operator $h \in E'$, the hyperplane $H = \mathrm{Ker}\, h = h^{-1}(0)$ is a closed subspace of E, and by Proposition 12.6, $H^{\perp}$ is a subspace of dimension one such that $E = H \oplus H^{\perp}$. Then picking any nonnull vector $w \in H^{\perp}$, observe that H is also the kernel of the linear operator φ_w, with

$$\varphi_w(u) = \langle u, w \rangle ,$$

and thus, since any two nonzero linear forms defining the same hyperplane must be proportional, there is some nonzero scalar $\lambda \in \mathbb{C}$ such that $h = \lambda\varphi_w$. But then, $h = \varphi_{\overline{\lambda}w}$, proving that $\flat\colon E \to E'$ is surjective.

By the Cauchy–Schwarz inequality we have

$$|\psi(v)| = |\langle v, u \rangle| \leq \|v\| \, \|u\| ,$$

so by definition of $\|\psi\|$ we get

$$\|\psi\| \leq \|u\| .$$

Obviously $\psi = 0$ iff $u = 0$ so assume $u \neq 0$. We have

$$\|u\|^2 = \langle u, u \rangle = \psi(u) \leq \|\psi\| \, \|u\| ,$$

which yields $\|u\| \leq \|\psi\|$, and therefore $\|\psi\| = \|u\|$, as claimed. ☐

Proposition 12.7 is known as the *Riesz representation theorem or "Little Riesz Theorem."* It shows that the inner product on a Hilbert space induces a natural semilinear isomorphism between E and its dual E' (equivalently, a linear isomorphism between $\overline{E}$ and E'). This isomorphism is an isometry (it is preserves the norm).

Remark: Many books on quantum mechanics use the so-called Dirac notation to denote objects in the Hilbert space E and operators in its dual space E'. In the Dirac notation, an element of E is denoted as $|x\rangle$, and an element of E' is denoted as $\langle t|$. The scalar product is denoted as $\langle t| \cdot |x\rangle$. This uses the isomorphism between E and E', except that the inner product is assumed to be semi-linear on the left rather than on the right.

Proposition 12.7 allows us to define the adjoint of a linear map, as in the Hermitian case (see Proposition 13.8 (Vol. I)). Actually, we can prove a slightly more general result which is used in optimization theory.

If $\varphi\colon E \times E \to \mathbb{C}$ is a sesquilinear map on a normed vector space $(E, \| \|)$, then Proposition 2.26 is immediately adapted to prove that φ is continuous iff there is some constant $k \geq 0$ such that

$$|\varphi(u, v)| \leq k \|u\| \|v\| \quad \text{for all } u, v \in E.$$

Thus we define $\|\varphi\|$ as in Definition 2.25 by

$$\|\varphi\| = \sup \{|\varphi(x, y)| \mid \|x\| \leq 1, \ \|y\| \leq 1, \ x, y \in E\}.$$

Proposition 12.8. *Given a Hilbert space E, for every continuous sesquilinear map $\varphi\colon E \times E \to \mathbb{C}$, there is a unique continuous linear map $f_\varphi\colon E \to E$, such that*

$$\varphi(u, v) = \langle u, f_\varphi(v)\rangle \quad \textit{for all } u, v \in E.$$

We also have $\|f_\varphi\| = \|\varphi\|$. If φ is Hermitian, then f_φ is self-adjoint, that is

$$\langle u, f_\varphi(v)\rangle = \langle f_\varphi(u), v\rangle \quad \textit{for all } u, v \in E.$$

Proof. The proof is adapted from Rudin [Rudin (1991)] (Theorem 12.8). To define the function f_φ, we proceed as follows. For any fixed $v \in E$, define the linear map φ_v by

$$\varphi_v(u) = \varphi(u, v) \quad \text{for all } u \in E.$$

Since φ is continuous, φ_v is continuous. So by Proposition 12.7, there is a unique vector in E that we denote $f_\varphi(v)$ such that

$$\varphi_v(u) = \langle u, f_\varphi(v)\rangle \quad \text{for all } u \in E,$$

and $\|f_\varphi(v)\| = \|\varphi_v\|$. Let us check that the map $v \mapsto f_\varphi(v)$ is linear.

We have

$$\begin{aligned}
\varphi(u, v_1 + v_2) &= \varphi(u, v_1) + \varphi(u, v_2) && \varphi \text{ is additive} \\
&= \langle u, f_\varphi(v_1)\rangle + \langle u, f_\varphi(v_2)\rangle && \text{by definition of } f_\varphi \\
&= \langle u, f_\varphi(v_1) + f_\varphi(v_2)\rangle && \langle -, -\rangle \text{ is additive}
\end{aligned}$$

for all $u \in E$, and since $f_\varphi(v_1 + v_2)$ is the unique vector such that $\varphi(u, v_1 + v_2) = \langle u, f_\varphi(v_1 + v_2)\rangle$ for all $u \in E$, we must have

$$f_\varphi(v_1 + v_2) = f_\varphi(v_1) + f_\varphi(v_2).$$

For any $\lambda \in \mathbb{C}$ we have

$$\begin{aligned}
\varphi(u, \lambda v) &= \overline{\lambda}\varphi(u, v) && \varphi \text{ is sesquilinear} \\
&= \overline{\lambda}\langle u, f_\varphi(v)\rangle && \text{by definition of } f_\varphi \\
&= \langle u, \lambda f_\varphi(v)\rangle && \langle -, -\rangle \text{ is sesquilinear}
\end{aligned}$$

for all $u \in E$, and since $f_\varphi(\lambda v)$ is the unique vector such that $\varphi(u, \lambda v) = \langle u, f_\varphi(\lambda v)\rangle$ for all $u \in E$, we must have

$$f_\varphi(\lambda v) = \lambda f_\varphi(v).$$

Therefore f_φ is linear.

Then by definition of $\|\varphi\|$, we have

$$|\varphi_v(u)| = |\varphi(u, v)| \leq \|\varphi\| \, \|u\| \, \|v\| ,$$

which shows that $\|\varphi_v\| \leq \|\varphi\| \, \|v\|$. Since $\|f_\varphi(v)\| = \|\varphi_v\|$, we have

$$\|f_\varphi(v)\| \leq \|\varphi\| \, \|v\| ,$$

which shows that f_φ is continuous and that $\|f_\varphi\| \leq \|\varphi\|$. But by the Cauchy–Schwarz inequality we also have

$$|\varphi(u, v)| = |\langle u, f_\varphi(v)\rangle| \leq \|u\| \, \|f_\varphi(v)\| \leq \|u\| \, \|f_\varphi\| \, \|v\| ,$$

so $\|\varphi\| \leq \|f_\varphi\|$, and thus

$$\|f_\varphi\| = \|\varphi\| .$$

If φ is Hermitian, $\varphi(v, u) = \overline{\varphi(u, v)}$, so

$$\langle f_\varphi(u), v\rangle = \overline{\langle v, f_\varphi(u)\rangle} = \overline{\varphi(v, u)} = \varphi(u, v) = \langle u, f_\varphi(v)\rangle,$$

which shows that f_φ is self-adjoint. ☐

Proposition 12.9. *Given a Hilbert space E, for every continuous linear map $f\colon E \to E$, there is a unique continuous linear map $f^*\colon E \to E$, such that*

$$\langle f(u), v\rangle = \langle u, f^*(v)\rangle \quad \textit{for all } u, v \in E,$$

and we have $\|f^*\| = \|f\|$. The map f^* is called the adjoint of f.

Proof. The proof is adapted from Rudin [Rudin (1991)] (Section 12.9). By the Cauchy–Schwarz inequality, since

$$|\langle x, y\rangle| \leq \|x\| \, \|y\| ,$$

we see that the sesquilinear map $(x, y) \mapsto \langle x, y\rangle$ on $E \times E$ is continuous. Let $\varphi\colon E \times E \to \mathbb{C}$ be the sesquilinear map given by

$$\varphi(u, v) = \langle f(u), v\rangle \quad \text{for all } u, v \in E.$$

Since f is continuous and the inner product $\langle -, -\rangle$ is continuous, this is a continuous map. By Proposition 12.8, there is a unique linear map $f^*\colon E \to E$ such that

$$\langle f(u), v\rangle = \varphi(u, v) = \langle u, f^*(v)\rangle \quad \text{for all } u, v \in E,$$

with $\|f^*\| = \|\varphi\|$.

We can also prove that $\|\varphi\| = \|f\|$. First, by definition of $\|\varphi\|$ we have

$$\begin{aligned}\|\varphi\| &= \sup\{|\varphi(x,y)| \mid \|x\| \le 1,\ \|y\| \le 1\} \\ &= \sup\{|\langle f(x), y\rangle| \mid \|x\| \le 1,\ \|y\| \le 1\} \\ &\le \sup\{\|f(x)\|\,\|y\| \mid \|x\| \le 1, \|y\| \le 1\} \\ &\le \sup\{\|f(x)\| \mid \|x\| \le 1\} \\ &= \|f\|.\end{aligned}$$

In the other direction we have

$$\|f(x)\|^2 = \langle f(x), f(x)\rangle = \varphi(x, f(x)) \le \|\varphi\|\,\|x\|\,\|f(x)\|,$$

and if $f(x) \neq 0$ we get $\|f(x)\| \le \|\varphi\|\,\|x\|$. This inequality holds trivially if $f(x) = 0$, so we conclude that $\|f\| \le \|\varphi\|$. Therefore we have

$$\|\varphi\| = \|f\|,$$

as claimed, and consequently $\|f^*\| = \|\varphi\| = \|f\|$. □

It is easy to show that the adjoint satisfies the following properties:

$$\begin{aligned}(f+g)^* &= f^* + g^* \\ (\lambda f)^* &= \overline{\lambda} f^* \\ (f \circ g)^* &= g^* \circ f^* \\ f^{**} &= f.\end{aligned}$$

One can also show that $\|f^* \circ f\| = \|f\|^2$ (see Rudin [Rudin (1991)], Section 12.9).

As in the Hermitian case, given two Hilbert spaces E and F, the above results can be adapted to show that for any linear map $f\colon E \to F$, there is a unique linear map $f^*\colon F \to E$ such that

$$\langle f(u), v\rangle_2 = \langle u, f^*(v)\rangle_1$$

for all $u \in E$ and all $v \in F$. The linear map f^* is also called the adjoint of f.

12.3 Farkas–Minkowski Lemma in Hilbert Spaces

In this section $(V, \langle -, -\rangle)$ is assumed to be a *real* Hilbert space. The projection lemma can be used to show an interesting version of the Farkas–Minkowski lemma in a Hilbert space.

Given a finite sequence of vectors $(a_1, \ldots, a_m)$ with $a_i \in V$, let C be the polyhedral cone

$$C = \text{cone}(a_1, \ldots, a_m) = \left\{ \sum_{i=1}^{m} \lambda_i a_i \mid \lambda_i \geq 0,\ i = 1, \ldots, m \right\}.$$

For any vector $b \in V$, the Farkas–Minkowski lemma gives a criterion for checking whether $b \in C$.

In Proposition 8.2 we proved that every polyhedral cone cone$(a_1, \ldots, a_m)$ with $a_i \in \mathbb{R}^n$ is closed. Close examination of the proof shows that it goes through if $a_i \in V$ where V is any vector space possibly of infinite dimension, because the important fact is that the number m of these vectors is finite, not their dimension.

Theorem 12.2. *(Farkas–Minkowski Lemma in Hilbert Spaces) Let $(V, \langle -, - \rangle)$ be a real Hilbert space. For any finite sequence of vectors $(a_1, \ldots, a_m)$ with $a_i \in V$, if C is the polyhedral cone $C = \text{cone}(a_1, \ldots, a_m)$, for any vector $b \in V$, we have $b \notin C$ iff there is a vector $u \in V$ such that*

$$\langle a_i, u \rangle \geq 0 \quad i = 1, \ldots, m, \quad \text{and} \quad \langle b, u \rangle < 0.$$

Equivalently, $b \in C$ iff for all $u \in V$,

$$\text{if } \langle a_i, u \rangle \geq 0 \quad i = 1, \ldots, m, \quad \text{then} \quad \langle b, u \rangle \geq 0.$$

Proof. We follow Ciarlet [Ciarlet (1989)] (Chapter 9, Theorem 9.1.1). We already established in Proposition 8.2 that the polyhedral cone $C = \text{cone}(a_1, \ldots, a_m)$ is closed. Next we claim the following:

Claim: If C is a nonempty, closed, convex subset of a Hilbert space V, and $b \in V$ is any vector such that $b \notin C$, then there exist some $u \in V$ and infinitely many scalars $\alpha \in \mathbb{R}$ such that

$$\begin{aligned} \langle v, u \rangle &> \alpha \quad \text{for every } v \in C \\ \langle b, u \rangle &< \alpha. \end{aligned}$$

We use the projection lemma (Proposition 12.4) which says that since $b \notin C$ there is some unique $c = p_C(b) \in C$ such that

$$\begin{aligned} \|b - c\| &= \inf_{v \in C} \|b - v\| > 0 \\ \langle b - c, v - c \rangle &\leq 0 \quad \text{for all } v \in C, \end{aligned}$$

or equivalently

$$\begin{aligned} \|b - c\| &= \inf_{v \in C} \|b - v\| > 0 \\ \langle v - c, c - b \rangle &\geq 0 \quad \text{for all } v \in C. \end{aligned}$$

As a consequence, since $b \notin C$ and $c \in C$, we have $c - b \neq 0$, so

$$\langle v, c - b\rangle \geq \langle c, c - b\rangle > \langle b, c - b\rangle$$

because $\langle c, c - b\rangle - \langle b, c - b\rangle = \langle c - b, c - b\rangle > 0$, and if we pick $u = c - b$ and any α such that

$$\langle c, c - b\rangle > \alpha > \langle b, c - b\rangle,$$

the claim is satisfied.

We now prove the Farkas–Minkowski lemma. Assume that $b \notin C$. Since C is nonempty, convex, and closed, by the claim there is some $u \in V$ and some $\alpha \in \mathbb{R}$ such that

$$\begin{aligned} \langle v, u\rangle &> \alpha \quad \text{for every } v \in C \\ \langle b, u\rangle &< \alpha. \end{aligned}$$

But C is a polyhedral cone containing 0, so we must have $\alpha < 0$. Then for every $v \in C$, since C a polyhedral cone if $v \in C$ then $\lambda v \in C$ for all $\lambda > 0$, so by the above

$$\langle v, u\rangle > \frac{\alpha}{\lambda} \quad \text{for every } \lambda > 0,$$

which implies that

$$\langle v, u\rangle \geq 0.$$

Since $a_i \in C$ for $i = 1, \ldots, m$, we proved that

$$\langle a_i, u\rangle \geq 0 \quad i = 1, \ldots, m \quad \text{and} \quad \langle b, u\rangle < \alpha < 0,$$

which proves Farkas lemma. □

Remark: Observe that the claim established during the proof of Theorem 12.2 shows that the affine hyperplane $H_{u,\alpha}$ of equation $\langle v, u\rangle = \alpha$ for all $v \in V$ separates strictly C and $\{b\}$.

12.4 Summary

The main concepts and results of this chapter are listed below:

- Hilbert space.
- Projection lemma.
- Distance of a point to a subset, diameter.
- Projection onto a closed and convex subset.
- Orthogonal complement of a closed subspace.

- Dual of a Hilbert space.
- Bounded linear operator (or functional).
- Riesz representation theorem.
- Adjoint of a continuous linear map.
- Farkas–Minkowski lemma.

12.5 Problems

Problem 12.1. Let V be a Hilbert space. Prove that a subspace W of V is dense in V if and only if there is no nonzero vector orthogonal to W.

Problem 12.2. Prove that the adjoint satisfies the following properties:

$$\begin{aligned} (f+g)^* &= f^* + g^* \\ (\lambda f)^* &= \overline{\lambda} f^* \\ (f \circ g)^* &= g^* \circ f^* \\ f^{**} &= f. \end{aligned}$$

Problem 12.3. Prove that $\|f^* \circ f\| = \|f\|^2$.

Problem 12.4. Let V be a (real) Hilbert space and let C be a nonempty closed convex subset of V. Define the map $h\colon V \to \mathbb{R} \cup \{+\infty\}$ by

$$h(u) = \sup_{v \in C} \langle u, v \rangle.$$

Prove that

$$C = \bigcap_{u \in V} \{v \in V \mid (u, v) \leq h(u)\} = \bigcap_{u \in \Lambda_C} \{v \in V \mid (u, v) \leq h(u)\},$$

where $\Lambda_C = \{u \in C \mid h(u) \neq +\infty\}$.

Describe Λ_C when C is also a subspace of V.

Problem 12.5. Let A be a real $m \times n$ matrix, and let (u_k) be a sequence of vectors $u_k \in \mathbb{R}^n$ such that $u_k \geq 0$. Prove that if the sequence (Au_k) converges, then there is some $u \in \mathbb{R}^n$ such that

$$Au = \lim_{k \mapsto \infty} Au_k \quad \text{and} \quad u \geq 0.$$

Problem 12.6. Let V be a real Hilbert space, $(a_1, \ldots, a_m)$ a sequence of m vectors in V, b some vector in V, $(\alpha_1, \ldots, \alpha_m)$ a sequence of m real numbers, and β some real number. Prove that the inclusion

$$\{w \in V \mid \langle a_i, w \rangle \geq \alpha_i,\ 1 \leq i \leq m\} \subseteq \{w \in V \mid \langle b, w \rangle \geq \beta\}$$

holds if and only if there exist $\lambda_1, \ldots, \lambda_m \in \mathbb{R}$ such that $\lambda_i \geq 0$ for $i = 1, \ldots, m$ and

$$b = \sum_{i=1}^{m} \lambda_i a_i$$

$$\beta \leq \sum_{i=1}^{m} \lambda_i \alpha_i.$$

Chapter 13

General Results of Optimization Theory

This chapter is devoted to some general results of optimization theory. A main theme is to find sufficient conditions that ensure that an objective function has a minimum which is achieved. We define the notion of a coercive function. The most general result is Theorem 13.1, which applies to a coercive convex function on a convex subset of a separable Hilbert space. In the special case of a coercive quadratic functional, we obtain the Lions–Stampacchia theorem (Theorem 13.4), and the Lax–Milgram theorem (Theorem 13.5). We define elliptic functionals, which generalize quadratic functions defined by symmetric positive definite matrices. We define gradient descent methods, and discuss their convergence. A gradient descent method looks for a descent direction and a stepsize parameter, which is obtained either using an exact line search or a backtracking line search. A popular technique to find the search direction is steepest descent. In addition to steepest descent for the Euclidean norm, we discuss steepest descent for an arbitrary norm. We also consider a special case of steepest descent, Newton's method. This method converges faster than the other gradient descent methods, but it is quite expensive since it requires computing and storing Hessians. We also present the method of conjugate gradients and prove its correctness. We briefly discuss the method of gradient projection and the penalty method in the case of constrained optima.

13.1 Optimization Problems; Basic Terminology

The main goal of *optimization theory* is to construct *algorithms* to find solutions (often approximate) of problems of the form

$$\begin{aligned}&\text{find} \quad u\\ &\text{such that} \quad u \in U \ \text{ and } \ J(u) = \inf_{v \in U} J(v),\end{aligned}$$

where U is a given subset of a (real) vector space V (possibly infinite dimensional) and $J\colon \Omega \to \mathbb{R}$ is a function defined on some open subset Ω of V such that $U \subseteq \Omega$.

To be very clear, $\inf_{v\in U} J(v)$ denotes the *greatest lower bound* of the set of real numbers $\{J(u) \mid u \in U\}$. To make sure that we are on firm grounds, let us review the notions of greatest lower bound and least upper bound of a set of real numbers.

Let X be any nonempty subset of $\mathbb{R}$. The set $LB(X)$ of *lower bounds* of X is defined as

$$LB(X) = \{b \in \mathbb{R} \mid b \leq x \ \text{ for all } x \in X\}.$$

If the set X is not bounded below, which means that for every $r \in \mathbb{R}$ there is some $x \in X$ such that $x < r$, then $LB(X)$ is empty. Otherwise, if $LB(X)$ is nonempty, since it is bounded above by every element of X, by a fundamental property of the real numbers, the set $LB(X)$ has a greatest element denoted $\inf X$. The real number $\inf X$ is thus the *greatest lower bound* of X. In general, $\inf X$ does not belong to X, but if it does, then it is the least element of X.

If $LB(X) = \emptyset$, then X is *unbounded below* and $\inf X$ is undefined. In this case (with an abuse of notation), we write

$$\inf X = -\infty.$$

By convention, when $X = \emptyset$ we set

$$\inf \emptyset = +\infty.$$

For example, if $X = \{x \in \mathbb{R} \mid x \leq 0\}$, then $LB(X) = \emptyset$. On the other hand, if $X = \{1/n \mid n \in \mathbb{N} - \{0\}\}$, then $LB(X) = \{x \in \mathbb{R} \mid x \leq 0\}$ and $\inf X = 0$, which is not in X.

Similarly, the set $UB(X)$ of *upper bounds* of X is given by

$$UB(X) = \{u \in \mathbb{R} \mid x \leq u \ \text{ for all } x \in X\}.$$

If X is not bounded above, then $UB(X) = \emptyset$. Otherwise, if $UB(X) \neq \emptyset$, then it has least element denoted $\sup X$. Thus $\sup X$ is the *least upper bound* of X. If $\sup X \in X$, then it is the greatest element of X. If $UB(X) = \emptyset$, then

$$\sup X = +\infty.$$

By convention, when $X = \emptyset$ we set

$$\sup \emptyset = -\infty.$$

For example, if $X = \{x \in \mathbb{R} \mid x \geq 0\}$, then $LB(X) = \emptyset$. On the other hand, if $X = \{1 - 1/n \mid n \in \mathbb{N} - \{0\}\}$, then $UB(X) = \{x \in \mathbb{R} \mid x \geq 1\}$ and $\sup X = 1$, which is not in X.

The element $\inf_{v \in U} J(v)$ is just $\inf\{J(v) \mid v \in U\}$. The notation J^* is often used to denote $\inf_{v \in U} J(v)$. If the function J is not bounded below, which means that for every $r \in \mathbb{R}$, there is some $u \in U$ such that $J(u) < r$, then

$$\inf_{v \in U} J(v) = -\infty,$$

and we say that our minimization problem has no solution, or that it is unbounded (below). For example, if $V = \Omega = \mathbb{R}$, $U = \{x \in \mathbb{R} \mid x \leq 0\}$, and $J(x) = -x$, then the function $J(x)$ is not bounded below and $\inf_{v \in U} J(v) = -\infty$.

The issue is that J^* may not belong to $\{J(u) \mid u \in U\}$, that is, it may not be achieved by some element $u \in U$, and solving the above problem consists in finding some $u \in U$ that achieves the value J^* in the sense that $J(u) = J^*$. If no such $u \in U$ exists, again we say that our minimization problem has no solution.

The minimization problem

$$\begin{aligned} &\text{find} \quad u \\ &\text{such that} \quad u \in U \text{ and } J(u) = \inf_{v \in U} J(v) \end{aligned}$$

is often presented in the following more informal way:

$$\begin{aligned} &\text{minimize} \quad J(v) \\ &\text{subject to} \quad v \in U. \end{aligned} \qquad \textbf{(Problem M)}$$

A vector $u \in U$ such that $J(u) = \inf_{v \in U} J(v)$ is often called a *minimizer* of J over U. Some authors denote the set of minimizers of J over U by $\arg\min_{v \in U} J(v)$ and write

$$u \in \arg\min_{v \in U} J(v)$$

to express that u is such a minimizer. When such a minimizer is unique, by abuse of notation, this unique minimizer u is denoted by

$$u = \arg\min_{v \in U} J(v).$$

We prefer not to use this notation, although it seems to have invaded the literature.

If we need to maximize rather than minimize a function, then we try to find some $u \in U$ such that

$$J(u) = \sup_{v \in U} J(v).$$

Here $\sup_{v \in U} J(v)$ is the least upper bound of the set $\{J(u) \mid u \in U\}$. Some authors denote the set of *maximizers* of J over U by $\arg\max_{v \in U} J(v)$.

Remark: Some authors define an *extended real-valued function* as a function $f\colon \Omega \to \mathbb{R}$ which is allowed to take the value $-\infty$ or even $+\infty$ for some of its arguments. Although this may be convenient to deal with situations where we need to consider $\inf_{v \in U} J(v)$ or $\sup_{v \in U} J(v)$, such "functions" are really partial functions and we prefer not to use the notion of extended real-valued function.

In most cases, U is defined as the set of solutions of a finite sets of *constraints*, either equality constraints $\varphi_i(v) = 0$, or inequality constraints $\varphi_i(v) \leq 0$, where the $\varphi_i\colon \Omega \to \mathbb{R}$ are some given functions. The function J is often called the *functional* of the optimization problem. This is a slightly odd terminology, but it is justified if V is a function space.

The following questions arise naturally:

(1) Results concerning the *existence and uniqueness* of a solution for Problem (M). In the next section we state sufficient conditions either on the domain U or on the function J that ensure the existence of a solution.
(2) The *characterization* of the possible solutions of Problem M. These are conditions for any element $u \in U$ to be a solution of the problem. Such conditions usually involve the derivative dJ_u of J, and possibly the derivatives of the functions φ_i defining U. Some of these conditions become sufficient when the functions φ_i are convex.
(3) The effective construction of *algorithms*, typically iterative algorithms that construct a sequence $(u_k)_{k \geq 1}$ of elements of U whose limit is a solution $u \in U$ of our problem. It is then necessary to understand when and how quickly such sequences converge. Gradient descent methods fall under this category. As a general rule, unconstrained problems (for

which $U = \Omega = V$) are (much) easier to deal with than constrained problems (where $U \neq V$).

The material of this chapter is heavily inspired by Ciarlet [Ciarlet (1989)]. *In this chapter it is assumed that V is a real vector space with an inner product* $\langle -, - \rangle$. If V is infinite dimensional, then we assume that it is a real Hilbert space (it is complete). As usual, we write $\|u\| = \langle u, u \rangle^{1/2}$ for the norm associated with the inner product $\langle -, - \rangle$. The reader may want to review Section 12.1, especially the projection lemma and the Riesz representation theorem.

As a matter of terminology, if U is defined by inequality and equality constraints as

$$U = \{v \in \Omega \mid \varphi_i(v) \leq 0,\ i = 1, \ldots, m,\ \psi_j(v) = 0,\ j = 1, \ldots, p\},$$

if J and all the functions φ_i and ψ_j are affine, the problem is said to be *linear* (or a *linear program*), and otherwise *nonlinear*. If J is of the form

$$J(v) = \langle Av, v \rangle - \langle b, v \rangle$$

where A is a nonzero symmetric positive semidefinite matrix and the constraints are affine, the problem is called a *quadratic programming problem*. If the inner product $\langle -, - \rangle$ is the standard Euclidean inner product, J is also expressed as

$$J(v) = v^\top Av - b^\top v.$$

13.2 Existence of Solutions of an Optimization Problem

We begin with the case where U is a closed but possibly unbounded subset of $\mathbb{R}^n$. In this case the following type of functions arise.

Definition 13.1. A real-valued function $J \colon V \to \mathbb{R}$ defined on a normed vector space V is *coercive* iff for any sequence $(v_k)_{k \geq 1}$ of vectors $v_k \in V$, if $\lim_{k \mapsto \infty} \|v_k\| = \infty$, then

$$\lim_{k \mapsto \infty} J(v_k) = +\infty.$$

For example, the function $f(x) = x^2 + 2x$ is coercive, but an affine function $f(x) = ax + b$ is not.

Proposition 13.1. *Let U be a nonempty, closed subset of $\mathbb{R}^n$, and let $J \colon \mathbb{R}^n \to \mathbb{R}$ be a continuous function which is coercive if U is unbounded. Then there is a least one element $u \in \mathbb{R}^n$ such that*

$$u \in U \quad \text{and} \quad J(u) = \inf_{v \in U} J(v).$$

Proof. Since $U \neq \emptyset$, pick any $u_0 \in U$. Since J is coercive, there is some $r > 0$ such that for all $v \in \mathbb{R}^n$, if $\|v\| > r$ then $J(u_0) < J(v)$. It follows that J is minimized over the set

$$U_0 = U \cap \{v \in \mathbb{R}^n \mid \|v\| \leq r\}.$$

Since U is closed and since the closed ball $\{v \in \mathbb{R}^n \mid \|v\| \leq r\}$ is compact, U_0 is compact, but we know that any continuous function on a compact set has a minimum which is achieved. $\square$

The key point in the above proof is the fact that U_0 is compact. In order to generalize Proposition 13.1 to the case of an infinite dimensional vector space, we need some additional assumptions, and it turns out that the convexity of U and of the function J is sufficient. The key is that convex, closed and bounded subsets of a Hilbert space are "weakly compact."

Definition 13.2. Let V be a Hilbert space. A sequence $(u_k)_{k\geq 1}$ of vectors $u_k \in V$ *converges weakly* if there is some $u \in V$ such that

$$\lim_{k\mapsto\infty} \langle v, u_k\rangle = \langle v, u\rangle \quad \text{for every } v \in V.$$

Recall that a Hibert space is separable if it has a countable Hilbert basis (see Definition A.4). Also, in a Euclidean space (of finite dimension) V, the inner product induces an isomorphism between V and its dual V^*. In our case, we need the isomorphism $\sharp$ from V^* to V defined such that for every linear form $\omega \in V^*$, the vector $\omega^\sharp \in V$ is uniquely defined by the equation

$$\omega(v) = \langle v, \omega^\sharp\rangle \quad \text{for all } v \in V.$$

In a Hilbert space, the dual space V' is the set of all continuous linear forms $\omega\colon V \to \mathbb{R}$, and the existence of the isomorphism $\sharp$ between V' and V is given by the Riesz representation theorem; see Proposition 12.7. This theorem allows a generalization of the notion of gradient. Indeed, if $f\colon V \to \mathbb{R}$ is a function defined on the Hilbert space V and if f is differentiable at some point $u \in V$, then by definition, the derivative $df_u\colon V \to \mathbb{R}$ is a continuous linear form, so by the Riesz representation theorem (Proposition 12.7) there is a unique vector, denoted $\nabla f_u \in V$, such that

$$df_u(v) = \langle v, \nabla f_u\rangle \quad \text{for all } v \in V.$$

Definition 13.3. The unique vector ∇f_u such that

$$df_u(v) = \langle v, \nabla f_u\rangle \quad \text{for all } v \in V$$

is called the *gradient* of f at u.

Similarly, since the second derivative $\mathrm{D}^2 f_u \colon V \to V'$ of f induces a continuous symmetric billinear form from $V \times V$ to $\mathbb{R}$, by Proposition 12.8, there is a unique continuous self-adjoint linear map $\nabla^2 f_u \colon V \to V$ such that

$$\mathrm{D}^2 f_u(v, w) = \langle \nabla^2 f_u(v), w \rangle \quad \text{for all } v, w \in V.$$

The map $\nabla^2 f_u$ is a generalization of the *Hessian*.

The next theorem is a rather general result about the existence of minima of convex functions defined on convex domains. The proof is quite involved and can be omitted upon first reading.

Theorem 13.1. *Let U be a nonempty, convex, closed subset of a separable Hilbert space V, and let $J \colon V \to \mathbb{R}$ be a convex, differentiable function which is coercive if U is unbounded. Then there is a least one element $u \in V$ such that*

$$u \in U \quad \textit{and} \quad J(u) = \inf_{v \in U} J(v).$$

Proof. As in the proof of Proposition 13.1, since the function J is coercive, we may assume that U is bounded and convex (however, if V infinite dimensional, then U is not compact in general). The proof proceeds in four steps.

Step 1. Consider a *minimizing sequence* $(u_k)_{k \geq 0}$, namely a sequence of elements $u_k \in V$ such that

$$u_k \in U \quad \text{for all } k \geq 0, \qquad \lim_{k \mapsto \infty} J(u_k) = \inf_{v \in U} J(v).$$

At this stage, it is possible that $\inf_{v \in U} J(v) = -\infty$, but we will see that this is actually impossible. However, since U is bounded, the sequence $(u_k)_{k \geq 0}$ is bounded. Our goal is to prove that there is some subsequence of $(w_\ell)_{\ell \geq 0}$ of $(u_k)_{k \geq 0}$ that converges weakly.

Since the sequence $(u_k)_{k \geq 0}$ is bounded there is some constant $C > 0$ such that $\|u_k\| \leq C$ for all $k \geq 0$. Then by the Cauchy–Schwarz inequality, for every $v \in V$ we have

$$|\langle v, u_k \rangle| \leq \|v\| \, \|u_k\| \leq C \|v\|,$$

which shows that the sequence $(\langle v, u_k \rangle)_{k \geq 0}$ is bounded. Since V is a separable Hilbert space, there is a countable family $(v_k)_{k \geq 0}$ of vectors $v_k \in V$ which is dense in V. Since the sequence $(\langle v_1, u_k \rangle)_{k \geq 0}$ is bounded (in $\mathbb{R}$), we can find a convergent subsequence $((v_1, u_{i_1(j)}))_{j \geq 0}$. Similarly, since the sequence $(\langle v_2, u_{i_1(j)} \rangle)_{j \geq 0}$ is bounded, we can find a convergent subsequence

$(\langle v_2, u_{i_2(j)}\rangle)_{j\geq 0}$, and in general, since the sequence $(\langle v_k, u_{i_{k-1}(j)}\rangle)_{j\geq 0}$ is bounded, we can find a convergent subsequence $(\langle v_k, u_{i_k(j)}\rangle)_{j\geq 0}$.

We obtain the following infinite array:

$$\begin{pmatrix} \langle v_1, u_{i_1(1)}\rangle & \langle v_2, u_{i_2(1)}\rangle & \cdots & \langle v_k, u_{i_k(1)}\rangle & \cdots \\ \langle v_1, u_{i_1(2)}\rangle & \langle v_2, u_{i_2(2)}\rangle & \cdots & \langle v_k, u_{i_k(2)}\rangle & \cdots \\ \vdots & \vdots & \vdots & \vdots & \vdots \\ \langle v_1, u_{i_1(k)}\rangle & \langle v_2, u_{i_2(k)}\rangle & \cdots & \langle v_k, u_{i_k(k)}\rangle & \cdots \\ \vdots & \vdots & \vdots & \vdots & \vdots \end{pmatrix}$$

Consider the "diagonal" sequence $(w_\ell)_{\ell\geq 0}$ defined by

$$w_\ell = u_{i_\ell(\ell)}, \quad \ell \geq 0.$$

We are going to prove that for every $v \in V$, the sequence $(\langle v, w_\ell\rangle)_{\ell\geq 0}$ has a limit.

By construction, for every $k \geq 0$, the sequence $(\langle v_k, w_\ell\rangle)_{\ell\geq 0}$ has a limit, which is the limit of the sequence $(\langle v_k, u_{i_k(j)}\rangle)_{j\geq 0}$, since the sequence $(i_\ell(\ell))_{\ell\geq 0}$ is a subsequence of every sequence $(i_\ell(j))_{j\geq 0}$ for every $\ell \geq 0$.

Pick any $v \in V$ and any $\epsilon > 0$. Since $(v_k)_{k\geq 0}$ is dense in V, there is some v_k such that

$$\|v - v_k\| \leq \epsilon/(4C).$$

Then we have

$$\begin{aligned} |\langle v, w_\ell\rangle - \langle v, w_m\rangle| &= |\langle v, w_\ell - w_m\rangle| \\ &= |\langle v_k + v - v_k, w_\ell - w_m\rangle| \\ &= |\langle v_k, w_\ell - w_m\rangle + \langle v - v_k, w_\ell - w_m\rangle| \\ &\leq |\langle v_k, w_\ell\rangle - \langle v_k, w_m\rangle| + |\langle v - v_k, w_\ell - w_m\rangle|. \end{aligned}$$

By Cauchy–Schwarz and since $\|w_\ell - w_m\| \leq \|w_\ell\| + \|w_m\| \leq C + C = 2C$,

$$|\langle v - v_k, w_\ell - w_m\rangle| \leq \|v - v_k\|\,\|w_\ell - w_m\| \leq (\epsilon/(4C))2C = \epsilon/2,$$

so

$$|\langle v, w_\ell\rangle - \langle v, w_m\rangle| \leq |\langle v_k, w_\ell - w_m\rangle| + \epsilon/2.$$

With the element v_k held fixed, by a previous argument the sequence $(\langle v_k, w_\ell\rangle)_{\ell\geq 0}$ converges, so it is a Cauchy sequence. Consequently there is some ℓ_0 (depending on ϵ and v_k) such that

$$|\langle v_k, w_\ell\rangle - \langle v_k, w_m\rangle| \leq \epsilon/2 \quad \text{for all } \ell, m \geq \ell_0,$$

so we get

$$|\langle v, w_\ell \rangle - \langle v, w_m \rangle| \leq \epsilon/2 + \epsilon/2 = \epsilon \quad \text{for all } \ell, m \geq \ell_0.$$

This proves that the sequence $(\langle v, w_\ell \rangle)_{\ell \geq 0}$ is a Cauchy sequence, and thus it converges.

Define the function $g\colon V \to \mathbb{R}$ by

$$g(v) = \lim_{\ell \mapsto \infty} \langle v, w_\ell \rangle, \quad \text{for all } v \in V.$$

Since

$$|\langle v, w_\ell \rangle| \leq \|v\| \, \|w_\ell\| \leq C \, \|v\| \quad \text{for all } \ell \geq 0,$$

we have

$$|g(v)| \leq C \, \|v\| \, ,$$

so g is a continuous linear map. By the Riesz representation theorem (Proposition 12.7), there is a unique $u \in V$ such that

$$g(v) = \langle v, u \rangle \quad \text{for all } v \in V,$$

which shows that

$$\lim_{\ell \mapsto \infty} \langle v, w_\ell \rangle = \langle v, u \rangle \quad \text{for all } v \in V,$$

namely the subsequence $(w_\ell)_{\ell \geq 0}$ of the sequence $(u_k)_{k \geq 0}$ converges weakly to $u \in V$.

Step 2. We prove that the "weak limit" u of the sequence $(w_\ell)_{\ell \geq 0}$ belongs to U.

Consider the projection $p_U(u)$ of $u \in V$ onto the closed convex set U. Since $w_\ell \in U$, by Proposition 12.4(2) and the fact that U is convex and closed, we have

$$\langle p_U(u) - u, w_\ell - p_U(u) \rangle \geq 0 \quad \text{for all } \ell \geq 0.$$

The weak convergence of the sequence $(w_\ell)_{\ell \geq 0}$ to u implies that

$$\begin{aligned} 0 \leq \lim_{\ell \mapsto \infty} \langle p_U(u) - u, w_\ell - p_U(u) \rangle &= \langle p_U(u) - u, u - p_U(u) \rangle \\ &= - \|p_U(u) - u\| \leq 0, \end{aligned}$$

so $\|p_U(u) - u\| = 0$, which means that $p_U(u) = u$, and so $u \in U$.

Step 3. We prove that

$$J(v) \leq \liminf_{\ell \mapsto \infty} J(z_\ell)$$

for every sequence $(z_\ell)_{\ell \geq 0}$ converging weakly to some element $v \in V$.

Since J is assumed to be differentiable and convex, by Proposition 4.6(1) we have

$$J(v) + \langle \nabla J_v, z_\ell - v\rangle \le J(z_\ell) \quad \text{for all } \ell \ge 0,$$

and by definition of weak convergence

$$\lim_{\ell \mapsto \infty} \langle \nabla J_v, z_\ell \rangle = \langle \nabla J_v, v\rangle,$$

so $\lim_{\ell\mapsto\infty}\langle \nabla J_v, z_\ell - v\rangle = 0$, and by definition of lim inf we get

$$J(v) \le \liminf_{\ell\mapsto\infty} J(z_\ell)$$

for every sequence $(z_\ell)_{\ell\ge 0}$ converging weakly to some element $v \in V$.

Step 4. The weak limit $u \in U$ of the subsequence $(w_\ell)_{\ell \ge 0}$ extracted from the minimizing sequence $(u_k)_{k\ge 0}$ satisfies the equation

$$J(u) = \inf_{v\in U} J(v).$$

By Step (1) and Step (2) the subsequence $(w_\ell)_{\ell\ge 0}$ of the sequence $(u_k)_{k\ge 0}$ converges weakly to some element $u \in U$, so by Step (3) we have

$$J(u) \le \liminf_{\ell\mapsto\infty} J(w_\ell).$$

On the other hand, by definition of $(w_\ell)_{\ell\ge 0}$ as a subsequence of $(u_k)_{k\ge 0}$, since the sequence $(J(u_k))_{k\ge 0}$ converges to $J(v)$, we have

$$J(u) \le \liminf_{\ell\mapsto\infty} J(w_\ell) = \lim_{k\mapsto\infty} J(u_k) = \inf_{v\in U} J(v),$$

which proves that $u \in U$ achieves the minimum of J on U. □

Remark: Theorem 13.1 still holds if we only assume that J is convex and continuous. It also holds in a reflexive Banach space, of which Hilbert spaces are a special case; see Brezis [Brezis (2011)], Corollary 3.23.

Theorem 13.1 is a rather general theorem whose proof is quite involved. For functions J of a certain type, we can obtain existence and uniqueness results that are easier to prove. This is true in particular for quadratic functionals.

13.3 Minima of Quadratic Functionals

Definition 13.4. Let V be a real Hilbert space. A function $J\colon V \to \mathbb{R}$ is called a *quadratic functional* if it is of the form

$$J(v) = \frac{1}{2}a(v, v) - h(v),$$

where $a\colon V \times V \to \mathbb{R}$ is a bilinear form which is symmetric and continuous, and $h\colon V \to \mathbb{R}$ is a continuous linear form.

Definition 13.4 is a natural extension of the notion of a quadratic functional on $\mathbb{R}^n$. Indeed, by Proposition 12.8, there is a unique continuous self-adjoint linear map $A\colon V \to V$ such that

$$a(u,v) = \langle Au, v\rangle \quad \text{for all } u, v \in V,$$

and by the Riesz representation theorem (Proposition 12.7), there is a unique $b \in V$ such that

$$h(v) = \langle b, v\rangle \quad \text{for all } v \in V.$$

Consequently, J can be written as

$$J(v) = \frac{1}{2}\langle Av, v\rangle - \langle b, v\rangle \quad \text{for all } v \in V. \tag{1}$$

Since a is bilinear and h is linear, by Propositions 3.3 and 3.5, observe that the derivative of J is given by

$$dJ_u(v) = a(u,v) - h(v) \quad \text{for all } v \in V,$$

or equivalently by

$$dJ_u(v) = \langle Au, v\rangle - \langle b, v\rangle = \langle Au - b, v\rangle, \quad \text{for all } v \in V.$$

Thus the gradient of J is given by

$$\nabla J_u = Au - b, \tag{2}$$

just as in the case of a quadratic function of the form $J(v) = (1/2)v^\top Av - b^\top v$, where A is a symmetric $n \times n$ matrix and $b \in \mathbb{R}^n$. To find the second derivative $\mathrm{D}^2 J_u$ of J at u we compute

$$dJ_{u+v}(w) - dJ_u(w) = a(u+v, w) - h(w) - (a(u,w) - h(w)) = a(v,w),$$

so

$$\mathrm{D}^2 J_u(v,w) = a(v,w) = \langle Av, w\rangle,$$

which yields

$$\nabla^2 J_u = A. \tag{3}$$

We will also make use of the following formula.

Proposition 13.2. *If J is a quadratic functional, then*

$$J(u + \rho v) = \frac{\rho^2}{2} a(v,v) + \rho(a(u,v) - h(v)) + J(u).$$

Proof. Since a is symmetric bilinear and h is linear, we have

$$\begin{aligned} J(u + \rho v) &= \frac{1}{2}a(u + \rho v, u + \rho v) - h(u + \rho v) \\ & \frac{\rho^2}{2}a(v, v) + \rho a(u, v) + \frac{1}{2}a(u, u) - h(u) - \rho h(v) \\ &= \frac{\rho^2}{2}a(v, v) + \rho(a(u, v) - h(v)) + J(u). \end{aligned}$$

Since $dJ_u(v) = a(u, v) - h(v) = \langle Au - b, v\rangle$ and $\nabla J_u = Au - b$, we can also write

$$J(u + \rho v) = \frac{\rho^2}{2}a(v, v) + \rho\langle \nabla J_u, v\rangle + J(u),$$

as claimed. □

We have the following theorem about the existence and uniqueness of minima of quadratic functionals.

Theorem 13.2. *Given any Hilbert space V, let $J\colon V \to \mathbb{R}$ be a quadratic functional of the form*

$$J(v) = \frac{1}{2}a(v, v) - h(v).$$

Assume that there is some real number $\alpha > 0$ such that

$$a(v, v) \geq \alpha \left\|v\right\|^2 \quad \textit{for all } v \in V. \tag{$*_\alpha$}$$

If U is any nonempty, closed, convex subset of V, then there is a unique $u \in U$ such that

$$J(u) = \inf_{v \in U} J(v).$$

The element $u \in U$ satisfies the condition

$$a(u, v - u) \geq h(v - u) \quad \textit{for all } v \in U. \tag{$*$}$$

Conversely (with the same assumptions on U as above), if an element $u \in U$ satisfies $(*)$, then

$$J(u) = \inf_{v \in U} J(v).$$

If U is a subspace of V, then the above inequalities are replaced by the equations

$$a(u, v) = h(v) \quad \textit{for all } v \in U. \tag{$**$}$$

Proof. The key point is that the bilinear form a is actually an inner product in V. This is because it is positive definite, since $(*_\alpha)$ implies that

$$\sqrt{\alpha}\,\|v\| \leq (a(v,v))^{1/2},$$

and on the other hand the continuity of a implies that

$$a(v,v) \leq \|a\|\,\|v\|^2,$$

so we get

$$\sqrt{\alpha}\,\|v\| \leq (a(v,v))^{1/2} \leq \sqrt{\|a\|}\,\|v\|.$$

The above also shows that the norm $v \mapsto (a(v,v))^{1/2}$ induced by the inner product a is equivalent to the norm induced by the inner product $\langle -, - \rangle$ on V. Thus h is still continuous with respect to the norm $v \mapsto (a(v,v))^{1/2}$. Then by the Riesz representation theorem (Proposition 12.7), there is some unique $c \in V$ such that

$$h(v) = a(c,v) \quad \text{for all } v \in V.$$

Consequently, we can express $J(v)$ as

$$J(v) = \frac{1}{2}a(v,v) - a(c,v) = \frac{1}{2}a(v-c,v-c) - \frac{1}{2}a(c,c).$$

But then minimizing $J(v)$ over U is equivalent to minimizing $(a(v-c,v-c))^{1/2}$ over $v \in U$, and by the projection lemma (Proposition 12.4(1)) this is equivalent to finding the projection $p_U(c)$ of c on the closed convex set U with respect to the inner product a. Therefore, there is a unique $u = p_U(c) \in U$ such that

$$J(u) = \inf_{v \in U} J(v).$$

Also by Proposition 12.4(2), this unique element $u \in U$ is characterized by the condition

$$a(u-c, v-u) \geq 0 \quad \text{for all } v \in U.$$

Since

$$a(u-c,v-u) = a(u,v-u) - a(c,v-u) = a(u,v-u) - h(v-u),$$

the above inequality is equivalent to

$$a(u,v-u) \geq h(v-u) \quad \text{for all } v \in U. \tag{$*$}$$

If U is a subspace of V, then by Proposition 12.4(3) we have the condition

$$a(u-c,v) = 0 \quad \text{for all } v \in U,$$

which is equivalent to

$$a(u,v) = a(c,v) = h(v) \quad \text{for all } v \in U, \tag{$**$}$$

a claimed. $\square$

Note that the symmetry of the bilinear form a played a crucial role. Also, the inequalities

$$a(u, v - u) \geq h(v - u) \quad \text{for all } v \in U$$

are sometimes called *variational inequalities*.

Definition 13.5. A bilinear form $a\colon V \times V \to \mathbb{R}$ such that there is some real $\alpha > 0$ such that

$$a(v, v) \geq \alpha \left\|v\right\|^2 \quad \text{for all } v \in V$$

is said to be *coercive*.

Theorem 13.2 is the special case of Stampacchia's theorem and the Lax–Milgram theorem when $U = V$, and where a is a symmetric bilinear form. To prove Stampacchia's theorem in general, we need to recall the *contraction mapping theorem*.

Definition 13.6. Let (E, d) be a metric space. A map $f\colon E \to E$ is a *contraction* (or a *contraction mapping*) if there is some real number k such that $0 \leq k < 1$ and

$$d(f(u), f(v)) \leq kd(u, v) \quad \text{for all } u, v \in E.$$

The number k is often called a *Lipschitz constant*.

The following theorem is proven in Section 2.10; see Theorem 2.5. A proof can be also found in Apostol [Apostol (1974)], Dixmier [Dixmier (1984)], or Schwartz [Schwartz (1991)], among many sources. For the reader's convenience we restate this theorem.

Theorem 13.3. *(Contraction Mapping Theorem) Let (E, d) be a complete metric space. Every contraction $f\colon E \to E$ has a unique fixed point (that is, an element $u \in E$ such that $f(u) = u$).*

The contraction mapping theorem is also known as the *Banach fixed point theorem*.

Theorem 13.4. *(Lions–Stampacchia) Given a Hilbert space V, let $a\colon V \times V \to \mathbb{R}$ be a continuous bilinear form (not necessarily symmetric), let $h \in V'$ be a continuous linear form, and let J be given by*

$$J(v) = \frac{1}{2} a(v, v) - h(v), \quad v \in V.$$

If a is coercive, then for every nonempty, closed, convex subset U of V, there is a unique $u \in U$ such that

$$a(u, v-u) \geq h(v-u) \quad \text{for all } v \in U. \tag{$*$}$$

If a is symmetric, then $u \in U$ is the unique element of U such that

$$J(u) = \inf_{v \in U} J(v).$$

Proof. As discussed just after Definition 13.4, by Proposition 12.8, there is a unique continuous linear map $A\colon V \to V$ such that

$$a(u, v) = \langle Au, v \rangle \quad \text{for all } u, v \in V,$$

with $\|A\| = \|a\| = C$, and by the Riesz representation theorem (Proposition 12.7), there is a unique $b \in V$ such that

$$h(v) = \langle b, v \rangle \quad \text{for all } v \in V.$$

Consequently, J can be written as

$$J(v) = \frac{1}{2}\langle Av, v \rangle - \langle b, v \rangle \quad \text{for all } v \in V. \tag{$*_1$}$$

Since $\|A\| = \|a\| = C$, we have $\|Av\| \leq \|A\| \, \|v\| = C\,\|v\|$ for all $v \in V$. Using ($*_1$), the inequality ($*$) is equivalent to finding u such that

$$\langle Au, v-u \rangle \geq \langle b, v-u \rangle \quad \text{for all } v \in V. \tag{$*_2$}$$

Let $\rho > 0$ be a constant to be determined later. Then ($*_2$) is equivalent to

$$\langle \rho b - \rho Au + u - u, v - u \rangle \leq 0 \quad \text{for all } v \in V. \tag{$*_3$}$$

By the projection lemma (Proposition 12.4 (1) and (2)), ($*_3$) is equivalent to finding $u \in U$ such that

$$u = p_U(\rho b - \rho Au + u). \tag{$*_4$}$$

We are led to finding a fixed point of the function $F\colon V \to V$ given by

$$F(v) = p_U(\rho b - \rho Av + v).$$

By Proposition 12.5, the projection map p_U does not increase distance, so

$$\|F(v_1) - F(v_2)\| \leq \|v_1 - v_2 - \rho(Av_1 - Av_2)\|.$$

Since a is coercive we have

$$a(v, v) \geq \alpha \|v\|^2,$$

since $a(v, v) = \langle Av, v \rangle$ we have

$$\langle Av, v \rangle \geq \alpha \|v\|^2 \quad \text{for all } v \in V, \tag{$*_5$}$$

and since

$$\|Av\| \leq C\|v\| \quad \text{for all } v \in V, \tag{$*_6$}$$

we get

$$\begin{aligned}\|F(v_1) - F(v_2)\|^2 &\leq \|v_1 - v_2\|^2 - 2\rho\langle Av_1 - Av_2, v_1 - v_2\rangle \\ &\quad + \rho^2 \|Av_1 - Av_2\|^2 \\ &\leq \left(1 - 2\rho\alpha + \rho^2 C^2\right) \|v_1 - v_2\|^2.\end{aligned}$$

If we pick $\rho > 0$ such that $\rho < 2\alpha/C^2$, then

$$k^2 = 1 - 2\rho\alpha + \rho^2 C^2 < 1,$$

and then

$$\|F(v_1) - F(v_2)\| \leq k\|v_1 - v_2\|, \tag{$*_7$}$$

with $0 \leq k < 1$, which shows that F is a contraction. By Theorem 13.3, the map F has a unique fixed point $u \in U$, which concludes the proof of the first statement. If a is also symmetric, then the second statement is just the first part of Theorem 13.2. □

Remark: Many physical problems can be expressed in terms of an unknown function u that satisfies some inequality

$$a(u, v - u) \geq h(v - u) \quad \text{for all } v \in U,$$

for some set U of "admissible" functions which is closed and convex. The bilinear form a and the linear form h are often given in terms of integrals. The above inequality is called a *variational inequality*.

In the special case where $U = V$ we obtain the Lax–Milgram theorem.

Theorem 13.5. *(Lax–Milgram's Theorem) Given a Hilbert space V, let $a\colon V \times V \to \mathbb{R}$ be a continuous bilinear form (not necessarily symmetric), let $h \in V'$ be a continuous linear form, and let J be given by*

$$J(v) = \frac{1}{2}a(v, v) - h(v), \quad v \in V.$$

If a is coercive, which means that there is some $\alpha > 0$ such that

$$a(v, v) \geq \alpha\|v\|^2 \quad \textit{for all } v \in V,$$

then there is a unique $u \in V$ such that

$$a(u, v) = h(v) \quad \textit{for all } v \in V.$$

If a is symmetric, then $u \in V$ is the unique element of V such that

$$J(u) = \inf_{v \in V} J(v).$$

The Lax–Milgram theorem plays an important role in solving linear elliptic partial differential equations; see Brezis [Brezis (2011)].

We now consider various methods, known as gradient descents, to find minima of certain types of functionals.

13.4 Elliptic Functionals

We begin by defining the notion of an elliptic functional which generalizes the notion of a quadratic function defined by a symmetric positive definite matrix. Elliptic functionals are well adapted to the types of iterative methods described in this section and lend themselves well to an analysis of the convergence of these methods.

Definition 13.7. Given a Hilbert space V, a functional $J\colon V \to \mathbb{R}$ is said to be *elliptic* if it is continuously differentiable on V, and if there is some constant $\alpha > 0$ such that

$$\langle \nabla J_v - \nabla J_u, v - u\rangle \geq \alpha \left\| v - u \right\|^2 \quad \text{for all } u, v \in V.$$

The following proposition gathers properties of elliptic functionals that will be used later to analyze the convergence of various gradient descent methods.

Theorem 13.6. *Let V be a Hilbert space.*

(1) An elliptic functional $J\colon V \to \mathbb{R}$ is strictly convex and coercice. Furthermore, it satisfies the identity

$$J(v) - J(u) \geq \langle \nabla J_u, v - u\rangle + \frac{\alpha}{2} \left\| v - u \right\|^2 \quad \textit{for all } u, v \in V.$$

(2) If U is a nonempty, convex, closed subset of the Hilbert space V and if J is an elliptic functional, then Problem (P),

$$\begin{aligned} &\text{find} \quad u \\ &\text{such that} \quad u \in U \ \textit{and}\ J(u) = \inf_{v \in U} J(v) \end{aligned}$$

has a unique solution.

(3) Suppose the set U is convex and that the functional J is elliptic. Then an element $u \in U$ is a solution of Problem (P) if and only if it satisfies the condition

$$\langle \nabla J_u, v - u\rangle \geq 0 \quad \textit{for every } v \in U$$

in the general case, or

$$\nabla J_u = 0 \ \textit{if}\ U = V.$$

(4) A functional J which is twice differentiable in V is elliptic if and only if

$$\langle \nabla^2 J_u(w), w\rangle \geq \alpha \left\| w \right\|^2 \quad \textit{for all } u, w \in V.$$

Proof. (1) Since J is a C^1-function, by Taylor's formula with integral remainder in the case $m = 0$ (Theorem 3.9), we get

$$\begin{aligned}
J(v) - J(u) &= \int_0^1 dJ_{u+t(v-u)}(v-u)dt \\
&= \int_0^1 \langle \nabla J_{u+t(v-u)}, v-u\rangle dt \\
&= \langle \nabla J_u, v-u\rangle + \int_0^1 \langle \nabla J_{u+t(v-u)} - \nabla J_u, v-u\rangle dt \\
&= \langle \nabla J_u, v-u\rangle + \int_0^1 \frac{\langle \nabla J_{u+t(v-u)} - \nabla J_u, t(v-u)\rangle}{t} dt \\
&\geq \langle \nabla J_u, v-u\rangle + \int_0^1 \alpha t \left\|v-u\right\|^2 dt \qquad \text{since } J \text{ is elliptic} \\
&= \langle \nabla J_u, v-u\rangle + \frac{\alpha}{2}\left\|v-u\right\|^2.
\end{aligned}$$

Using the inequality

$$J(v) - J(u) \geq \langle \nabla J_u, v-u\rangle + \frac{\alpha}{2}\left\|v-u\right\|^2 \quad \text{for all } u, v \in V,$$

by Proposition 4.6(2), since

$$J(v) > J(u) + \langle \nabla J_u, v-u\rangle \quad \text{for all } u, v \in V,\ v \neq u,$$

the function J is strictly convex. It is coercive because (using Cauchy–Schwarz)

$$\begin{aligned}
J(v) &\geq J(0) + \langle \nabla J_0, v\rangle + \frac{\alpha}{2}\left\|v\right\|^2 \\
&\geq J(0) - \left\|\nabla J_0\right\| \left\|v\right\| + \frac{\alpha}{2}\left\|v\right\|^2,
\end{aligned}$$

and the term $(-\left\|\nabla J_0\right\| + \frac{\alpha}{2}\left\|v\right\|)\left\|v\right\|$ goes to $+\infty$ when $\left\|v\right\|$ tends to $+\infty$.

(2) Since by (1) the functional J is coercive, by Theorem 13.1, Problem (P) has a solution. Since J is strictly convex, by Theorem 4.5(2), it has a unique minimum.

(3) These are just the conditions of Theorem 4.5(3, 4).

(4) If J is twice differentiable, we showed in Section 3.5 that we have

$$D^2 J_u(w, w) = D_w(DJ)(u) = \lim_{\theta \mapsto 0} \frac{DJ_{u+\theta w}(w) - DJ_u(w)}{\theta},$$

and since

$$\begin{aligned}
D^2 J_u(w, w) &= \langle \nabla^2 J_u(w), w\rangle \\
DJ_{u+\theta w}(w) &= \langle \nabla J_{u+\theta w}, w\rangle \\
DJ_u(w) &= \langle \nabla J_u, w\rangle,
\end{aligned}$$

and since J is elliptic, for all $u, w \in V$ we can write

$$\begin{aligned}\langle \nabla^2 J_u(w), w\rangle &= \lim_{\theta \mapsto 0} \frac{\langle \nabla J_{u+\theta w} - \nabla J_u, w\rangle}{\theta} \\ &= \lim_{\theta \mapsto 0} \frac{\langle \nabla J_{u+\theta w} - \nabla J_u, \theta w\rangle}{\theta^2} \\ &\geq \theta \left\|w\right\|^2.\end{aligned}$$

Conversely, assume that the condition

$$\langle \nabla^2 J_u(w), w\rangle \geq \alpha \left\|w\right\|^2 \quad \text{for all } u, w \in V$$

holds. If we define the function $g\colon V \to \mathbb{R}$ by

$$g(w) = \langle \nabla J_w, v - u\rangle = dJ_w(v - u) = \mathrm{D}_{v-u} J(w),$$

where u and v are fixed vectors in V, then we have

$$\begin{aligned}dg_{u+\theta(v-u)}(v - u) &= \mathrm{D}_{v-u} g(u + \theta(v - u)) = \mathrm{D}_{v-u}\mathrm{D}_{v-u} J(u + \theta(v - u)) \\ &= \mathrm{D}^2 J_{u+\theta(v-u)}(v - u, v - u)\end{aligned}$$

and we can apply the Taylor–MacLaurin formula (Theorem 3.8 with $m = 0$) to g, and we get

$$\begin{aligned}\langle \nabla J_v - \nabla J_u, v - u\rangle &= g(v) - g(u) \\ &= dg_{u+\theta(v-u)}(v - u) \qquad (0 < \theta < 1) \\ &= \mathrm{D}^2 J_{u+\theta(v-u)}(v - u, v - u) \\ &= \langle \nabla^2 J_{u+\theta(v-u)}(v - u), v - u\rangle \\ &\geq \alpha \left\|v - u\right\|^2,\end{aligned}$$

which shows that J is elliptic. □

Corollary 13.1. *If $J\colon \mathbb{R}^n \to \mathbb{R}$ is a quadratic function given by*

$$J(v) = \frac{1}{2}\langle Av, v\rangle - \langle b, v\rangle$$

(where A is a symmetric $n \times n$ matrix and $\langle -, -\rangle$ is the standard Eucidean inner product), then J is elliptic iff A is positive definite.

This a consequence of Theorem 13.6 because

$$\langle \nabla^2 J_u(w), w\rangle = \langle Aw, w\rangle \geq \lambda_1 \left\|w\right\|^2$$

where λ_1 is the smallest eigenvalue of A; see Proposition 16.24 (Rayleigh–Ritz, Vol. I). Note that by Proposition 16.24 (Rayleigh–Ritz, Vol. I), we also have the following corollary.

Corollary 13.2. *If $J\colon \mathbb{R}^n \to \mathbb{R}$ is a quadratic function given by*

$$J(v) = \frac{1}{2}\langle Av, v\rangle - \langle b, v\rangle$$

then

$$\langle \nabla^2 J_u(w), w \rangle \leq \lambda_n \|w\|^2$$

where λ_n *is the largest eigenvalue of* A.

The above fact will be useful later on.

Similarly, given a quadratic functional J defined on a Hilbert space V, where

$$J(v) = \frac{1}{2}a(v, v) - h(v),$$

by Theorem 13.6 (4), the functional J is elliptic iff there is some $\alpha > 0$ such that

$$\langle \nabla^2 J_u(v), v \rangle = a(v, v) \geq \alpha \|v\|^2 \quad \text{for all } v \in V.$$

This is precisely the hypothesis $(*_\alpha)$ used in Theorem 13.2.

13.5 Iterative Methods for Unconstrained Problems

We will now describe methods for solving unconstrained minimization problems, that is, finding the minimum (or minima) of a functions J over the whole space V. These methods are *iterative*, which means that given some *initial* vector u_0, we construct a sequence $(u_k)_{k \geq 0}$ that converges to a minimum u of the function J.

The key step is define u_{k+1} from u_k, and a first idea is to reduce the problem to a simpler problem, namely the minimization of a function of a *single (real) variable*. For this, we need two perform two steps:

(1) Find a *descent direction* at u_k, which is a some nonzero vector d_k which is usually determined from the gradient of J at various points. The descent direction d_k must satisfy the inequality $\langle \nabla J_{u_k}, d_k \rangle < 0$.
(2) *Exact line search*: Find the minimum of the restriction of the function J along the line through u_k and parallel to the direction d_k. This means finding a real $\rho_k \in \mathbb{R}$ (depending on u_k and d_k) such that

$$J(u_k + \rho_k d_k) = \inf_{\rho \in \mathbb{R}} J(u_k + \rho d_k).$$

This problem only succeeds if ρ_k is *unique*, in which case we set

$$u_{k+1} = u_k + \rho_k d_k.$$

This step is often called a *line search* or *line minimization*, and ρ_k is called the *stepsize* parameter. See Figure 13.1.

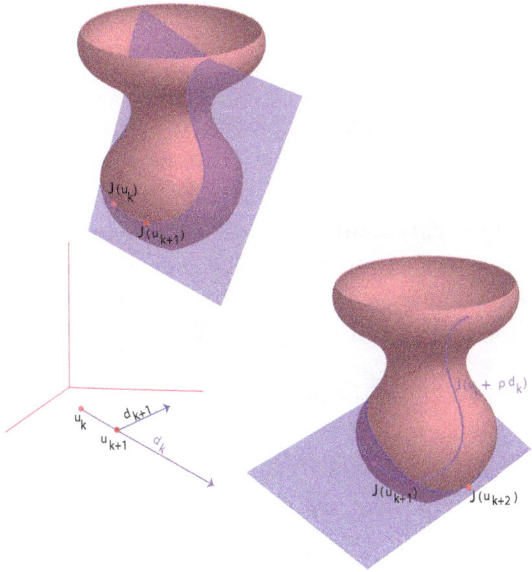

Fig. 13.1 Let $J\colon \mathbb{R}^2 \to \mathbb{R}$ be the function whose graph is represented by the pink surface. Given a point u_k in the xy-plane, and a direction d_k, we calculate first u_{k+1} and then u_{k+2}.

Proposition 13.3. *If J is a quadratic elliptic functional of the form*

$$J(v) = \frac{1}{2}a(v, v) - h(v),$$

then given d_k, there is a unique ρ_k solving the line search in Step (2).

Proof. This is because, by Proposition 13.2, we have

$$J(u_k + \rho d_k) = \frac{\rho^2}{2}a(d_k, d_k) + \rho\langle \nabla J_{u_k}, d_k\rangle + J(u_k),$$

and since $a(d_k, d_k) > 0$ (because J is elliptic), the above function of ρ has a unique minimum when its derivative is zero, namely

$$\rho\, a(d_k, d_k) + \langle \nabla J_{u_k}, d_k\rangle = 0. \qquad \square$$

Since Step (2) is often too costly, an alternative is

(3) *Backtracking line search*: Pick two constants α and β such that $0 < \alpha < 1/2$ and $0 < \beta < 1$, and set $t = 1$. Given a descent direction d_k at $u_k \in \mathrm{dom}(J)$,

while $J(u_k + td_k) > J(u_k) + \alpha t\langle \nabla J_{u_k}, d_k\rangle$ **do** $t := \beta t$;
$\rho_k = t$; $u_{k+1} = u_k + \rho_k d_k$.

Since d_k is a descent direction, we must have $\langle \nabla J_{u_k}, d_k\rangle < 0$, so for t small enough the condition $J(u_k + td_k) \leq J(u_k) + \alpha t\langle \nabla J_{u_k}, d_k\rangle$ will hold and the search will stop. It can be shown that the exit inequality $J(u_k + td_k) \leq J(u_k) + \alpha t\langle \nabla J_{u_k}, d_k\rangle$ holds for all $t \in (0, t_0]$, for some $t_0 > 0$. Thus the backtracking line search stops with a step length ρ_k that satisfies $\rho_k = 1$ or $\rho_k \in (\beta t_0, t_0]$. Care has to be exercised so that $u_k + \rho_k d_k \in \mathrm{dom}(J)$. For more details, see Boyd and Vandenberghe [Boyd and Vandenberghe (2004)] (Section 9.2).

We now consider one of the simplest methods for choosing the directions of descent in the case where $V = \mathbb{R}^n$, which is to pick the directions of the coordinate axes in a cyclic fashion. Such a method is called the *method of relaxation*.

If we write

$$u_k = (u_1^k, u_2^k, \dots, u_n^k),$$

then the components u_i^{k+1} of u_{k+1} are computed in terms of u_k by solving from top down the following system of equations:

$$\begin{aligned}
J(\mathbf{u_1^{k+1}}, u_2^k, u_3^k, \dots, u_n^k) &= \inf_{\lambda\in\mathbb{R}} J(\lambda, u_2^k, u_3^k, \dots, u_n^k)\\
J(u_1^{k+1}, \mathbf{u_2^{k+1}}, u_3^k, \dots, u_n^k) &= \inf_{\lambda\in\mathbb{R}} J(u_1^{k+1}, \lambda, u_3^k, \dots, u_n^k)\\
&\ \ \vdots\\
J(u_1^{k+1}, \dots, u_{n-1}^{k+1}, \mathbf{u_n^{k+1}}) &= \inf_{\lambda\in\mathbb{R}} J(u_1^{k+1}, \dots, u_{n-1}^{k+1}, \lambda).
\end{aligned}$$

Another and more informative way to write the above system is to define the vectors $u_{k;i}$ by

$$\begin{aligned}
u_{k;0} &= (u_1^k, u_2^k, \dots, u_n^k)\\
u_{k;1} &= (u_1^{k+1}, u_2^k, \dots, u_n^k)\\
&\ \ \vdots\\
u_{k;i} &= (u_1^{k+1}, \dots, u_i^{k+1}, u_{i+1}^k, \dots, u_n^k)\\
&\ \ \vdots\\
u_{k;n} &= (u_1^{k+1}, u_2^{k+1}, \dots, u_n^{k+1}).
\end{aligned}$$

Note that $u_{k;0} = u_k$ and $u_{k;n} = u_{k+1}$. Then our minimization problem can be written as

$$\begin{aligned} J(u_{k;1}) &= \inf_{\lambda \in \mathbb{R}} J(u_{k;0} + \lambda e_1) \\ &\vdots \\ J(u_{k;i}) &= \inf_{\lambda \in \mathbb{R}} J(u_{k;i-1} + \lambda e_i) \\ &\vdots \\ J(u_{k;n}) &= \inf_{\lambda \in \mathbb{R}} J(u_{k;n-1} + \lambda e_n), \end{aligned}$$

where e_i denotes the ith canonical basis vector in $\mathbb{R}^n$. If J is differentiable, necessary conditions for a minimum, which are also sufficient if J is convex, is that the directional derivatives $dJ_v(e_i)$ be all zero, that is,

$$\langle \nabla J_v, e_i \rangle = 0 \quad i = 0, \ldots, n.$$

The following result regarding the convergence of the method of relaxation is proven in Ciarlet [Ciarlet (1989)] (Chapter 8, Theorem 8.4.2).

Proposition 13.4. *If the functional $J\colon \mathbb{R}^n \to \mathbb{R}$ is elliptic, then the relaxation method converges.*

Remarks: The proof of Proposition 13.4 uses Theorem 13.6. The finite dimensionality of $\mathbb{R}^n$ also plays a crucial role. The differentiability of the function J is also crucial. Examples where the method loops forever if J is not differentiable can be given; see Ciarlet [Ciarlet (1989)] (Chapter 8, Section 8.4). The proof of Proposition 13.4 yields an *a priori* bound on the error $\|u - u_k\|$. If J is a quadratic functional

$$J(v) = \frac{1}{2} v^\top A v - b^\top v,$$

where A is a symmetric positive definite matrix, then $\nabla J_v = Av - b$, so the above method for solving for u_{k+1} in terms of u_k becomes the *Gauss–Seidel method* for solving a linear system; see Section 9.3 (Vol. I).

We now discuss gradient methods.

13.6 Gradient Descent Methods for Unconstrained Problems

The intuition behind these methods is that the convergence of an iterative method ought to be better if the difference $J(u_k) - J(u_{k+1})$ is as large as possible during every iteration step. To achieve this, it is natural to pick the descent direction to be the one *in the opposite direction of the gradient vector* ∇J_{u_k}. This choice is justified by the fact that we can write

$$J(u_k + w) = J(u_k) + \langle \nabla J_{u_k}, w \rangle + \epsilon(w) \|w\|, \quad \text{with } \lim_{w \mapsto 0} \epsilon(w) = 0.$$

If $\nabla J_{u_k} \neq 0$, the first-order part of the variation of the function J is bounded in absolute value by $\|\nabla J_{u_k}\| \|w\|$ (by the Cauchy–Schwarz inequality), with equality if ∇J_{u_k} and w are collinear.

Gradient descent methods pick the direction of descent to be $d_k = -\nabla J_{u_k}$, so that we have

$$u_{k+1} = u_k - \rho_k \nabla J_{u_k},$$

where we put a negative sign in front of of the variable ρ_k as a reminder that the descent direction is *opposite* to that of the gradient; a positive value is expected for the scalar ρ_k.

There are four standard methods to pick ρ_k:

(1) *Gradient method with fixed stepsize parameter*. This is the simplest and cheapest method which consists of using the *same* constant $\rho_k = \rho$ for all iterations.
(2) *Gradient method with variable stepsize parameter*. In this method, the parameter ρ_k is adjusted in the course of iterations according to various criteria.
(3) *Gradient method with optimal stepsize parameter*, also called *steepest descent method for the Euclidean norm*. This is a version of Method 2 in which ρ_k is determined by the following line search:

$$J(u_k - \rho_k \nabla J_{u_k}) = \inf_{\rho \in \mathbb{R}} J(u_k - \rho \nabla J_{u_k}).$$

This optimization problem only succeeds if the above minimization problem has a *unique* solution.
(4) *Gradient descent method with backtracking line search*. In this method, the step parameter is obtained by performing a backtracking line search.

We have the following useful result about the convergence of the gradient method with optimal parameter.

Proposition 13.5. *Let* $J\colon \mathbb{R}^n \to \mathbb{R}$ *be an elliptic functional. Then the gradient method with optimal stepsize parameter converges.*

Proof. Since J is elliptic, by Theorem 13.6(3), the functional J has a unique minimum u characterized by $\nabla J_u = 0$. Our goal is to prove that the sequence $(u_k)_{k\geq 0}$ constructed using the gradient method with optimal parameter converges to u, starting from any initial vector u_0. Without loss of generality we may assume that $u_{k+1} \neq u_k$ and $\nabla J_{u_k} \neq 0$ for all k, since otherwise the method converges in a finite number of steps.

Step 1. Show that any two consecutive descent directions are *orthogonal* and

$$J(u_k) - J(u_{k+1}) \geq \frac{\alpha}{2} \left\| u_k - u_{k+1} \right\|^2 .$$

Let $\varphi_k\colon \mathbb{R} \to \mathbb{R}$ be the function given by

$$\varphi_k(\rho) = J(u_k - \rho \nabla J_{u_k}).$$

Since the function φ_k is strictly convex and coercive, by Theorem 13.6(2), it has a unique minimum ρ_k which is the unique solution of the equation $\varphi_k'(\rho) = 0$. By the chain rule

$$\begin{aligned}\varphi_k'(\rho) &= dJ_{u_k - \rho\nabla J_{u_k}}(-\nabla J_{u_k}) \\ &= -\langle \nabla J_{u_k - \rho\nabla J_{u_k}}, \nabla J_{u_k}\rangle,\end{aligned}$$

and since $u_{k+1} = u_k - \rho_k \nabla J_{u_k}$ we get

$$\langle \nabla J_{u_{k+1}}, \nabla J_{u_k}\rangle = 0,$$

which shows that two consecutive descent directions are orthogonal.

Since $u_{k+1} = u_k - \rho_k \nabla J_{u_k}$ and we assumed that that $u_{k+1} \neq u_k$, we have $\rho_k \neq 0$, and we also get

$$\langle \nabla J_{u_{k+1}}, u_{k+1} - u_k\rangle = 0.$$

By the inequality of Theorem 13.6(1) we have

$$J(u_k) - J(u_{k+1}) \geq \frac{\alpha}{2} \left\| u_k - u_{k+1} \right\|^2 .$$

Step 2. Show that $\lim_{k\mapsto\infty} \left\| u_k - u_{k+1} \right\| = 0$.

It follows from the inequality proven in Step 1 that the sequence $(J(u_k))_{k\geq 0}$ is decreasing and bounded below (by $J(u)$, where u is the minimum of J), so it converges and we conclude that

$$\lim_{k\mapsto\infty} (J(u_k) - J(u_{k+1})) = 0,$$

which combined with the preceding inequality shows that

$$\lim_{k\mapsto\infty} \left\| u_k - u_{k+1} \right\| = 0.$$

Step 3. Show that $\left\| \nabla J_{u_k} \right\| \leq \left\| \nabla J_{u_k} - \nabla J_{u_{k+1}} \right\|$.

Using the orthogonality of consecutive descent directions, by Cauchy–Schwarz we have

$$\begin{aligned}\|\nabla J_{u_k}\|^2 &= \langle \nabla J_{u_k}, \nabla J_{u_k} - \nabla J_{u_{k+1}}\rangle \\ &\leq \|\nabla J_{u_k}\| \left\|\nabla J_{u_k} - \nabla J_{u_{k+1}}\right\|,\end{aligned}$$

so that

$$\|\nabla J_{u_k}\| \leq \left\|\nabla J_{u_k} - \nabla J_{u_{k+1}}\right\|.$$

Step 4. Show that $\lim_{k\mapsto\infty} \|\nabla J_{u_k}\| = 0$.

Since the sequence $(J(u_k))_{k\geq 0}$ is decreasing and the functional J is coercive, the sequence $(u_k)_{k\geq 0}$ must be bounded. By hypothesis, the derivative dJ is of J is continuous, so it is uniformly continuous over compact subsets of $\mathbb{R}^n$; here we are using the fact that $\mathbb{R}^n$ is finite dimensional. Hence, we deduce that for every $\epsilon > 0$, if $\|u_k - u_{k+1}\| < \epsilon$ then

$$\left\|dJ_{u_k} - dJ_{u_{k+1}}\right\|_2 < \epsilon.$$

But by definition of the operator norm and using the Cauchy–Schwarz inequality

$$\begin{aligned}\left\|dJ_{u_k} - dJ_{u_{k+1}}\right\|_2 &= \sup_{\|w\|\leq 1} |dJ_{u_k}(w) - dJ_{u_{k+1}}(w)| \\ &= \sup_{\|w\|\leq 1} |\langle \nabla J_{u_k} - \nabla J_{u_{k+1}}, w\rangle| \\ &\leq \left\|\nabla J_{u_k} - \nabla J_{u_{k+1}}\right\|.\end{aligned}$$

But we also have

$$\begin{aligned}\left\|\nabla J_{u_k} - \nabla J_{u_{k+1}}\right\|^2 &= \langle \nabla J_{u_k} - \nabla J_{u_{k+1}}, \nabla J_{u_k} - \nabla J_{u_{k+1}}\rangle \\ &= dJ_{u_k}(\nabla J_{u_k} - \nabla J_{u_{k+1}}) - dJ_{u_{k+1}}(\nabla J_{u_k} - \nabla J_{u_{k+1}}) \\ &\leq \left\|dJ_{u_k} - dJ_{u_{k+1}}\right\|_2^2,\end{aligned}$$

and so

$$\left\|dJ_{u_k} - dJ_{u_{k+1}}\right\|_2 = \left\|\nabla J_{u_k} - \nabla J_{u_{k+1}}\right\|.$$

It follows that since

$$\lim_{k\mapsto\infty} \|u_k - u_{k+1}\| = 0$$

then

$$\lim_{k\mapsto\infty} \left\|\nabla J_{u_k} - \nabla J_{u_{k+1}}\right\| = \lim_{k\mapsto\infty} \left\|dJ_{u_k} - dJ_{u_{k+1}}\right\|_2 = 0,$$

and using the fact that

$$\|\nabla J_{u_k}\| \leq \left\|\nabla J_{u_k} - \nabla J_{u_{k+1}}\right\|,$$

we obtain

$$\lim_{k \mapsto \infty} \|\nabla J_{u_k}\| = 0.$$

Step 5. Finally we can prove the convergence of the sequence $(u_k)_{k \geq 0}$.

Since J is elliptic and since $\nabla J_u = 0$ (since u is the minimum of J over $\mathbb{R}^n$), we have

$$\begin{aligned}\alpha \|u_k - u\|^2 &\leq \langle \nabla J_{u_k} - \nabla J_u, u_k - u\rangle \\ &= \langle \nabla J_{u_k}, u_k - u\rangle \\ &\leq \|\nabla J_{u_k}\| \, \|u_k - u\| .\end{aligned}$$

Hence, we obtain

$$\|u_k - u\| \leq \frac{1}{\alpha} \|\nabla J_{u_k}\| , \tag{b}$$

and since we showed that

$$\lim_{k \mapsto \infty} \|\nabla J_{u_k}\| = 0,$$

we see that the sequence $(u_k)_{k \geq 0}$ converges to the minimum u. □

Remarks: As with the previous proposition, the assumption of finite dimensionality is crucial. The proof provides an *a priori* bound on the error $\|u_k - u\|$.

If J is a an elliptic quadratic functional

$$J(v) = \frac{1}{2}\langle Av, v\rangle - \langle b, v\rangle,$$

we can use the orthogonality of the descent directions ∇J_{u_k} and $\nabla J_{u_{k+1}}$ to compute ρ_k. Indeed, we have $\nabla J_v = Av - b$, so

$$0 = \langle \nabla J_{u_{k+1}}, \nabla J_{u_k}\rangle = \langle A(u_k - \rho_k(Au_k - b)) - b, Au_k - b\rangle,$$

which yields

$$\rho_k = \frac{\|w_k\|^2}{\langle Aw_k, w_k\rangle}, \quad \text{with} \quad w_k = Au_k - b = \nabla J_{u_k}.$$

Consequently, a step of the iteration method takes the following form:

(1) Compute the vector

$$w_k = Au_k - b.$$

(2) Compute the scalar

$$\rho_k = \frac{\|w_k\|^2}{\langle Aw_k, w_k\rangle}.$$

(3) Compute the next vector u_{k+1} by

$$u_{k+1} = u_k - \rho_k w_k.$$

This method is of particular interest when the computation of Aw for a given vector w is cheap, which is the case if A is sparse.

Example 13.1. For a particular illustration of this method, we turn to the example provided by Shewchuk, with $A = \begin{pmatrix} 3 & 2 \\ 2 & 6 \end{pmatrix}$ and $b = \begin{pmatrix} 2 \\ -8 \end{pmatrix}$, namely

$$\begin{aligned} J(x,y) &= \frac{1}{2} \begin{pmatrix} x & y \end{pmatrix} \begin{pmatrix} 3 & 2 \\ 2 & 6 \end{pmatrix} \begin{pmatrix} x \\ y \end{pmatrix} - \begin{pmatrix} 2 & -8 \end{pmatrix} \begin{pmatrix} x \\ y \end{pmatrix} \\ &= \frac{3}{2}x^2 + 2xy + 3y^2 - 2x + 8y. \end{aligned}$$

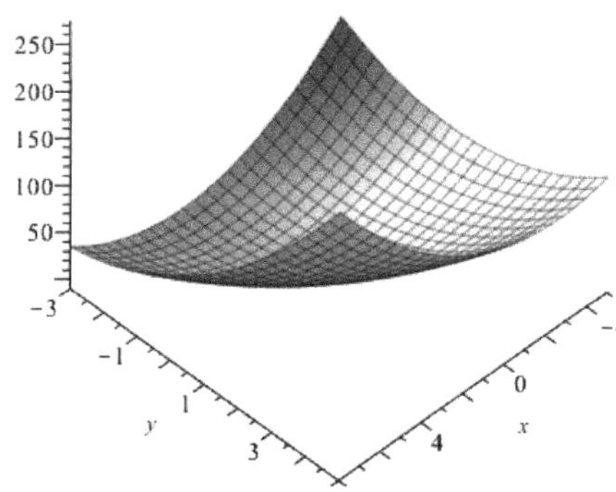

Fig. 13.2 The ellipsoid $J(x,y) = \frac{3}{2}x^2 + 2xy + 3y^2 - 2x + 8y$.

This quadratic ellipsoid, which is illustrated in Figure 13.2, has a unique minimum at $(2,-2)$. In order to find this minimum via the gradient descent with optimal step size parameter, we pick a starting point, say $u_k = (-2,-2)$, and calculate the search direction $w_k = \nabla J(-2,-2) = (-12,-8)$. Note that

$$\nabla J(x,y) = (3x + 2y - 2, 2x + 6y + 8) = \begin{pmatrix} 3 & 2 \\ 2 & 6 \end{pmatrix} \begin{pmatrix} x \\ y \end{pmatrix} - \begin{pmatrix} 2 \\ -8 \end{pmatrix}$$

is perpendicular to the appropriate elliptical level curve; see Figure 13.3. We next perform the line search along the line given by the equation $-8x + 12y = -8$ and determine ρ_k. See Figures 13.4 and 13.5. In particular, we find that

$$\rho_k = \frac{\|w_k\|^2}{\langle Aw_k, w_k \rangle} = \frac{13}{75}.$$

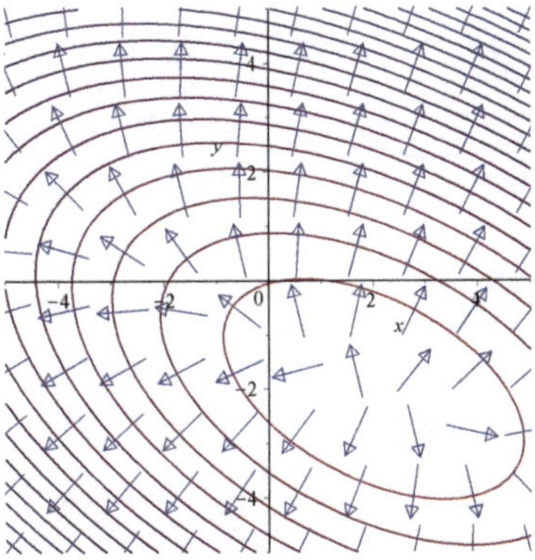

Fig. 13.3 The level curves of $J(x, y) = \frac{3}{2}x^2 + 2xy + 3y^2 - 2x + 8y$ and the associated gradient vector field $\nabla J(x, y) = (3x + 2y - 2, 2x + 6y + 8)$.

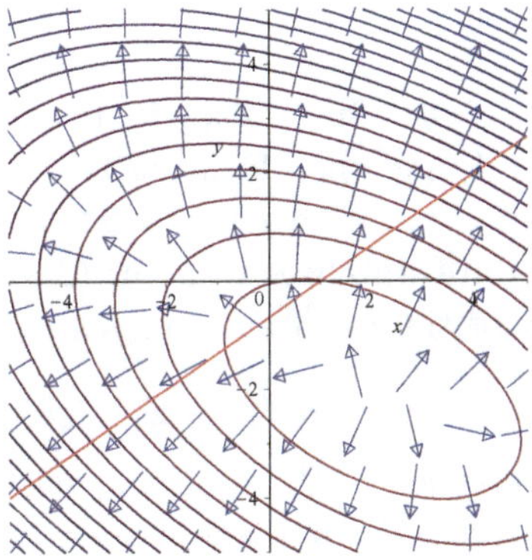

Fig. 13.4 The level curves of $J(x, y) = \frac{3}{2}x^2 + 2xy + 3y^2 - 2x + 8y$ and the red search line with direction $\nabla J(-2, -2) = (-12, -8)$.

This in turn gives us the new point

$$u_{k+1} = u_k - \frac{13}{75} w_k = (-2, -2) - \frac{13}{75}(-12, -8) = \left(\frac{2}{25}, -\frac{46}{75}\right),$$

and we continue the procedure by searching along the gradient direction $\nabla J(2/25, -46/75) = (-224/75, 112/25)$. Observe that $u_{k+1} = (\frac{2}{25}, -\frac{46}{75})$ has a gradient vector which is perpendicular to the search line with direction vector $w_k = \nabla J(-2, -2) = (-12 \;\, -8)$; see Figure 13.5. Geometrically this procedure corresponds to intersecting the plane $-8x + 12y = -8$ with the

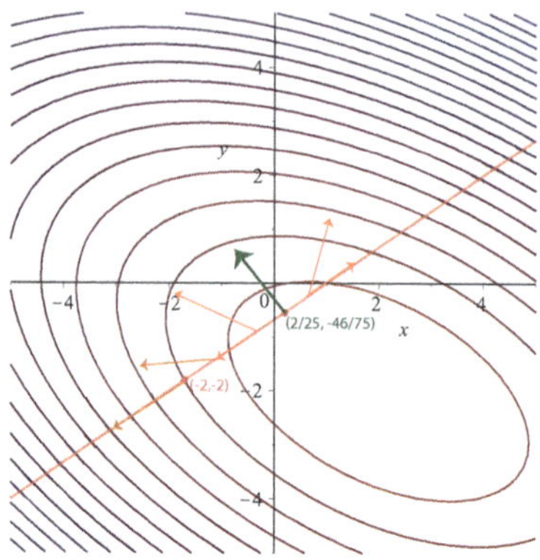

Fig. 13.5 Let $u_k = (-2, -2)$. When traversing along the red search line, we look for the green perpendicular gradient vector. This gradient vector, which occurs at $u_{k+1} = (2/25, -46/75)$, provides a minimal ρ_k, since it has no nonzero projection on the search line.

ellipsoid $J(x, y) = \frac{3}{2}x^2 + 2xy + 3y^2 - 2x + 8y$ to form the parabolic curve $f(x) = 25/6x^2 - 2/3x - 4$, and then locating the x-coordinate of its apex which occurs when $f'(x) = 0$, i.e., when $x = 2/25$; see Figure 13.6. After 31 iterations, this procedure stabilizes to point $(2, -2)$, which as we know, is the unique minimum of the quadratic ellipsoid $J(x, y) = \frac{3}{2}x^2 + 2xy + 3y^2 - 2x + 8y$.

A proof of the convergence of the gradient method with backtracking line search, under the hypothesis that J is strictly convex, is given in Boyd and Vandenberghe[Boyd and Vandenberghe (2004)] (Section 9.3.1). More details on this method and the steepest descent method for the Euclidean norm can also be found in Boyd and Vandenberghe [Boyd and Vandenberghe (2004)] (Section 9.3).

13.7 Convergence of Gradient Descent with Variable Stepsize

We now give a sufficient condition for the gradient method with variable stepsize parameter to converge. In addition to requiring J to be an elliptic functional, we add a Lipschitz condition on the gradient of J. This time the space V can be infinite dimensional.

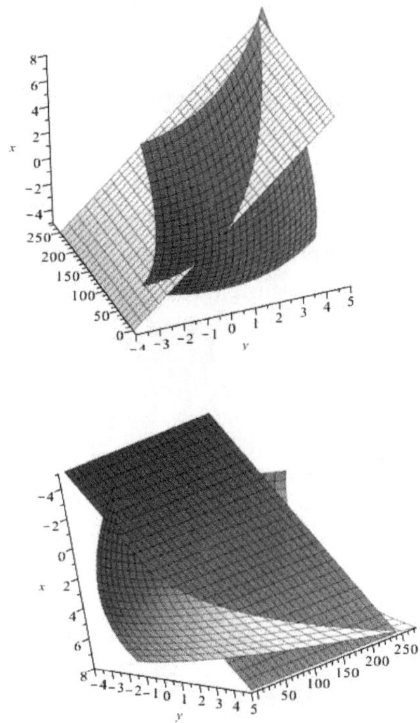

Fig. 13.6 Two views of the intersection between the plane $-8x + 12y = -8$ and the ellipsoid $J(x, y) = \frac{3}{2}x^2 + 2xy + 3y^2 - 2x + 8y$. The point u_{k+1} is the minimum of the parabolic intersection.

Proposition 13.6. *Let $J\colon V \to \mathbb{R}$ be a continuously differentiable functional defined on a Hilbert space V. Suppose there exists two constants $\alpha > 0$ and $M > 0$ such that*

$$\langle \nabla J_v - \nabla J_u, v - u \rangle \geq \alpha \|v - u\|^2 \quad \text{for all } u, v \in V,$$

and the Lipschitz condition

$$\|\nabla J_v - \nabla J_u\| \leq M \|v - u\| \quad \text{for all } u, v \in V.$$

If there exists two real numbers $a, b \in \mathbb{R}$ such that

$$0 < a \leq \rho_k \leq b \leq \frac{2\alpha}{M^2} \quad \text{for all } k \geq 0,$$

then the gradient method with variable stepsize parameter converges. Furthermore, there is some constant $\beta > 0$ (depending on α, M, a, b) such that

$$\beta < 1 \quad \textit{and} \quad \|u_k - u\| \leq \beta^k \|u_0 - u\|,$$

where $u \in M$ is the unique minimum of J.

Proof. By hypothesis the functional J is elliptic, so by Theorem 13.6(2) it has a unique minimum u characterized by the fact that $\nabla J_u = 0$. Then since $u_{k+1} = u_k - \rho_k \nabla J_{u_k}$, we can write

$$u_{k+1} - u = (u_k - u) - \rho_k \langle \nabla J_{u_k} - \nabla J_u \rangle. \qquad (*)$$

Using the inequalities

$$\langle \nabla J_{u_k} - \nabla J_u, u_k - u \rangle \geq \alpha \|u_k - u\|^2$$

and

$$\|\nabla J_{u_k} - \nabla J_u\| \leq M \|u_k - u\|,$$

and assuming that $\rho_k > 0$, it follows that

$$\begin{aligned}\|u_{k+1} - u\|^2 &= \|u_k - u\|^2 - 2\rho_k \langle \nabla J_{u_k} - \nabla J_u, u_k - u \rangle + \rho_k^2 \|\nabla J_{u_k} - \nabla J_u\|^2 \\ &\leq \left(1 - 2\alpha\rho_k + M^2\rho_k^2\right) \|u_k - u\|^2.\end{aligned}$$

Consider the function

$$T(\rho) = M^2\rho^2 - 2\alpha\rho + 1.$$

Its graph is a parabola intersecting the y-axis at $y = 1$ for $\rho = 0$, it has a minimum for $\rho = \alpha/M^2$, and it also has the value $y = 1$ for $\rho = 2\alpha/M^2$; see Figure 13.7. Therefore if we pick a, b and ρ_k such that

$$0 < a \leq \rho_k \leq b < \frac{2\alpha}{M^2},$$

we ensure that for $\rho \in [a, b]$ we have

$$T(\rho)^{1/2} = (M^2\rho^2 - 2\alpha\rho + 1)^{1/2} \leq (\max\{T(a), T(b)\})^{1/2} = \beta < 1.$$

Then by induction we get

$$\|u_{k+1} - u\| \leq \beta^{k+1} \|u_0 - u\|,$$

which proves convergence. ∎

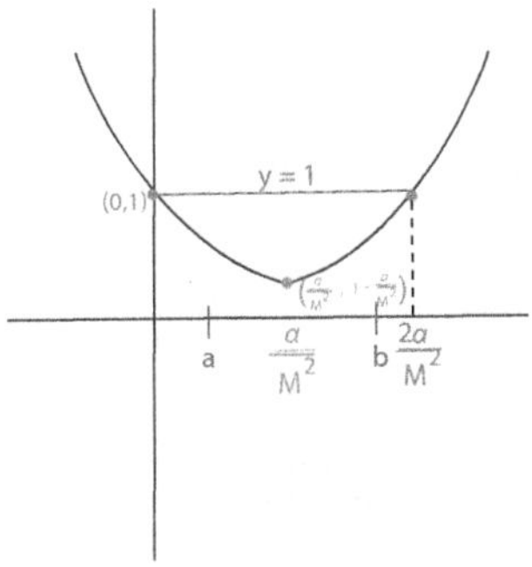

Fig. 13.7 The parabola $T(\rho)$ used in the proof of Proposition 13.6.

Remarks: In the proof of Proposition 13.6, it is the fact that V is complete which plays a crucial role. If J is twice differentiable, the hypothesis

$$\|\nabla J_v - \nabla J_u\| \leq M \|v - u\| \quad \text{for all } u, v \in V$$

can be expressed as

$$\sup_{v \in V} \left\|\nabla^2 J_v\right\| \leq M.$$

In the case of a quadratic elliptic functional defined over $\mathbb{R}^n$,

$$J(v) = \langle Av, v\rangle - \langle b, v\rangle,$$

the upper bound $2\alpha/M^2$ can be improved. In this case we have

$$\nabla J_v = Av - b,$$

and we know that we $\alpha = \lambda_1$ and $M = \lambda_n$ do the job, where λ_1 is the eigenvalue of A and λ_n is the largest eigenvalue of A. Hence we can pick a, b such that

$$0 < a \leq \rho_k \leq b < \frac{2\lambda_1}{\lambda_n^2}.$$

Since $u_{k+1} = u_k - \rho_k \nabla J_{u_k}$ and $\nabla J_{u_k} = Au_k - b$, we have

$$u_{k+1} - u = (u_k - u) - \rho_k(Au_k - Au) = (I - \rho_k A)(u_k - u),$$

so we get

$$\|u_{k+1} - u\| \leq \|I - \rho_k A\|_2 \|u_k - u\|.$$

However, since $I - \rho_k A$ is a symmetric matrix, $\|I - \rho_k A\|_2$ is the largest absolute value of its eigenvalues, so

$$\|I - \rho_k A\|_2 \leq \max\{|1 - \rho_k \lambda_1|, |1 - \rho_k \lambda_n|\}.$$

The function

$$\mu(\rho) = \max\{|1 - \rho\lambda_1|, |1 - \rho\lambda_n|\}$$

is a piecewise affine function, and it is easy to see that if we pick a, b such that

$$0 < a \leq \rho_k \leq b \leq \frac{2}{\lambda_n},$$

then

$$\max_{\rho \in [a,b]} \mu(\rho) \leq \max\{\mu(a), \mu(b)\} < 1.$$

Therefore, the upper bound $2\lambda_1/\lambda_n^2$ can be replaced by $2/\lambda_n$, which is typically much larger. A "good" pick for ρ_k is $2/(\lambda_1 + \lambda_n)$ (as opposed to λ_1/λ_n^2 for the first version). In this case

$$|1 - \rho_k\lambda_1| = |1 - \rho_k\lambda_n| = \frac{\lambda_n - \lambda_1}{\lambda_n + \lambda_1},$$

so we get

$$\beta = \frac{\lambda_n - \lambda_1}{\lambda_n + \lambda_1} = \frac{\frac{\lambda_n}{\lambda_1} - 1}{\frac{\lambda_n}{\lambda_1} + 1} = \frac{\text{cond}_2(A) - 1}{\text{cond}_2(A) + 1},$$

where $\text{cond}_2(A) = \lambda_n/\lambda_1$ is the condition number of the matrix A with respect to the spectral norm. Thus we see that the larger the condition number of A is, the slower the convergence of the method will be. This is not surprising since we already know that linear systems involving ill-conditioned matrices (matrices with a large condition number) are problematic and prone to numerical instability. One way to deal with this problem is to use a method known as preconditioning.

We only described the most basic gradient descent methods. There are numerous variants, and we only mention a few of these methods.

The method of *scaling* consists in using $-\rho_k D_k \nabla J_{u_k}$ as descent direction, where D_k is some suitably chosen symmetric positive definite matrix.

In the *gradient method with extrapolation*, u_{k+1} is determined by

$$u_{k+1} = u_k - \rho_k \nabla J_{u_k} + \beta_k(u_k - u_{k-1}).$$

Another rule for choosing the stepsize is *Armijo's rule*.

These methods, and others, are discussed in detail in Berstekas [Bertsekas (2015)].

Boyd and Vandenberghe discuss steepest descent methods for various types of norms besides the Euclidean norm; see Boyd and Vandenberghe [Boyd and Vandenberghe (2004)] (Section 9.4). Here is brief summary.

13.8 Steepest Descent for an Arbitrary Norm

The idea is to make $\langle \nabla J_{u_k}, d_k \rangle$ as negative as possible. To make the question sensible, we have to limit the size of d_k or normalize by the length of d_k.

Let $\| \, \|$ be any norm on $\mathbb{R}^n$. Recall from Section 13.7 in Volume I that the *dual norm* is defined by

$$\|y\|^D = \sup_{\substack{x \in \mathbb{R}^n \\ \|x\|=1}} |\langle x, y \rangle|.$$

Definition 13.8. A *normalized steepest descent direction* (with respect to the norm $\| \, \|$) is any unit vector $d_{\mathrm{nsd},k}$ which achieves the minimum of the set of reals

$$\{\langle \nabla J_{u_k}, d \rangle \mid \|d\| = 1\}.$$

By definition, $\|d_{\mathrm{nsd},k}\| = 1$.

A *unnormalized steepest descent direction* $d_{\mathrm{sd},k}$ is defined as

$$d_{\mathrm{sd},k} = \|\nabla J_{u_k}\|^D \, d_{\mathrm{nsd},k}.$$

It can be shown that

$$\langle \nabla J_{u_k}, d_{\mathrm{sd},k} \rangle = -(\|\nabla J_{u_k}\|^D)^2;$$

see Boyd and Vandenberghe [Boyd and Vandenberghe (2004)] (Section 9.4).

The *steepest descent method (with respect to the norm* $\| \, \|$*)* consists of the following steps: Given a starting point $u_0 \in \mathrm{dom}(J)$ do:

repeat

(1) Compute the steepest descent direction $d_{\mathrm{sd},k}$.
(2) Line search. Perform an exact or backtracking line search to find ρ_k.
(3) Update. $u_{k+1} = u_k + \rho_k d_{\mathrm{sd},k}$.

until stopping criterion is satisfied.

If $\| \, \|$ is the ℓ^2-norm, then we see immediately that $d_{\mathrm{sd},k} = -\nabla J_{u_k}$, so in this case the method *coincides* with the steepest descent method for the Euclidean norm as defined at the beginning of Section 13.6 in (3) and (4).

If P is a symmetric positive definite matrix, it is easy to see that $\|z\|_P = (z^\top P z)^{1/2} = \left\|P^{1/2} z\right\|_2$ is a norm. Then it can be shown that the normalized steepest descent direction is

$$d_{\mathrm{nsd,k}} = -(\nabla J_{u_k}^\top P^{-1} \nabla J_{u_k})^{-1/2} P^{-1} \nabla J_{u_k},$$

the dual norm is $\|z\|^D = \left\|P^{-1/2}z\right\|_2$, and the steepest descent direction with respect to $\| \|_P$ is given by

$$d_{\mathrm{sd},k} = -P^{-1}\nabla J_{u_k}.$$

A judicious choice for P can speed up the rate of convergence of the gradient descent method; see see Boyd and Vandenberghe [Boyd and Vandenberghe (2004)] (Section 9.4.1 and Section 9.4.4).

If $\| \|$ is the ℓ^1-norm, then it can be shown that $d_{\mathrm{nsd,k}}$ is determined as follows: let i be any index for which $\|\nabla J_{u_k}\|_\infty = |(\nabla J_{u_k})_i|$. Then

$$d_{\mathrm{nsd,k}} = -\mathrm{sign}\left(\frac{\partial J}{\partial x_i}(u_k)\right) e_i,$$

where e_i is the ith canonical basis vector, and

$$d_{\mathrm{sd,k}} = -\frac{\partial J}{\partial x_i}(u_k)e_i.$$

For more details, see Boyd and Vandenberghe [Boyd and Vandenberghe (2004)] (Section 9.4.2 and Section 9.4.4). It is also shown in Boyd and Vandenberghe [Boyd and Vandenberghe (2004)] (Section 9.4.3) that the steepest descent method converges for any norm $\| \|$ and any strictly convex function J.

One of the main goals in designing a gradient descent method is to ensure that the convergence factor is as small as possible, which means that the method converges as quickly as possible. Machine learning has been a catalyst for finding such methods. A method discussed in Strang [Strang (2019)] (Chapter VI, Section 4) consists in adding a *momentum term* to the gradient. In this method, u_{k+1} and d_{k+1} are determined by the following system of equations:

$$\begin{aligned} u_{k+1} &= u_k - \rho d_k \\ d_{k+1} - \nabla J_{u_{k+1}} &= \beta d_k. \end{aligned}$$

Of course the trick is to choose ρ and β in such a way that the convergence factor is as small as possible. If J is given by a quadratic functional, say $(1/2)u^\top Au - b^\top u$, then $\nabla J_{u_{k+1}} = Au_{k+1} - b$ so we obtain a linear system. It turns out that the rate of convergence of the method is determined by the largest and the smallest eigenvalues of A. Strang discusses this issue in the case of a 2×2 matrix. Convergence is significantly accelerated.

Another method is known as *Nesterov acceleration*. In this method,

$$u_{k+1} = u_k + \beta(u_k - u_{k-1}) - \rho\nabla J_{u_k + \gamma(u_k - u_{k-1})},$$

where β, ρ, γ are parameters. For details, see Strang [Strang (2019)] (Chapter VI, Section 4).

Lax also discusses other methods in which the step ρ_k is chosen using roots of Chebyshev polynomials; see Lax [Lax (2007)], Chapter 17, Sections 2–4.

A variant of Newton's method described in Section 5.2 can be used to find the minimum of a function belonging to a certain class of strictly convex functions. This method is the special case of the case where the norm is induced by a symmetric positive definite matrix P, namely $P = \nabla^2 J(x)$, the Hessian of J at x.

13.9 Newton's Method for Finding a Minimum

If $J\colon \Omega \to \mathbb{R}$ is a convex function defined on some open subset Ω of $\mathbb{R}^n$ which is twice differentiable and if its Hessian $\nabla^2 J(x)$ is symmetric positive definite for all $x \in \Omega$, then by Proposition 4.7(2), the function J is strictly convex. In this case, for any $x \in \Omega$, we have the quadratic norm induced by $P = \nabla^2 J(x)$ as defined in the previous section, given by

$$\|u\|_{\nabla^2 J(x)} = (u^\top \nabla^2 J(x)\, u)^{1/2}.$$

The steepest descent direction for this quadratic norm is given by

$$d_{\rm nt} = -(\nabla^2 J(x))^{-1} \nabla J_x.$$

The norm of $d_{\rm nt}$ for the the quadratic norm defined by $\nabla^2 J(x)$ is given by

$$\begin{aligned}(d_{\rm nt}^\top \nabla^2 J(x)\, d_{\rm nt})^{1/2} &= \left(-(\nabla J_x)^\top (\nabla^2 J(x))^{-1} \nabla^2 J(x)(-(\nabla^2 J(x))^{-1} \nabla J_x)\right)^{1/2} \\ &= \left((\nabla J_x)^\top (\nabla^2 J(x))^{-1} \nabla J_x\right)^{1/2}.\end{aligned}$$

Definition 13.9. Given a function $J\colon \Omega \to \mathbb{R}$ as above, for any $x \in \Omega$, the *Newton step* $d_{\rm nt}$ is defined by

$$d_{\rm nt} = -(\nabla^2 J(x))^{-1} \nabla J_x,$$

and the *Newton decrement* $\lambda(x)$ is defined by

$$\lambda(x) = \left((\nabla J_x)^\top (\nabla^2 J(x))^{-1} \nabla J_x\right)^{1/2}.$$

Observe that

$$\langle \nabla J_x, d_{\rm nt}\rangle = (\nabla J_x)^\top (-(\nabla^2 J(x))^{-1} \nabla J_x) = -\lambda(x)^2.$$

If $\nabla J_x \neq 0$, we have $\lambda(x) \neq 0$, so $\langle \nabla J_x, d_{\rm nt}\rangle < 0$, and $d_{\rm nt}$ is indeed a descent direction. The number $\langle \nabla J_x, d_{\rm nt}\rangle$ is the constant that shows up during a backtracking line search.

A nice feature of the Newton step and of the Newton decrement is that they are affine invariant. This means that if T is an invertible matrix and if we define g by $g(y) = J(Ty)$, if the Newton step associated with J is denoted by $d_{J,\mathrm{nt}}$ and similarly the Newton step associated with g is denoted by $d_{g,\mathrm{nt}}$, then it is shown in Boyd and Vandenberghe [Boyd and Vandenberghe (2004)] (Section 9.5.1) that

$$d_{g,\mathrm{nt}} = T^{-1} d_{J,\mathrm{nt}},$$

and so

$$x + d_{J,\mathrm{nt}} = T(y + d_{g,\mathrm{nt}}).$$

A similar properties applies to the Newton decrement.

Newton's method consists of the following steps: Given a starting point $u_0 \in \mathrm{dom}(J)$ and a tolerance $\epsilon > 0$ do:

repeat

(1) Compute the Newton step and decrement
$d_{\mathrm{nt},k} = -(\nabla^2 J(u_k))^{-1} \nabla J_{u_k}$ and $\lambda(u_k)^2 = (\nabla J_{u_k})^\top (\nabla^2 J(u_k))^{-1} \nabla J_{u_k}$.
(2) Stopping criterion. **quit** if $\lambda(u_k)^2/2 \le \epsilon$.
(3) Line Search. Perform an exact or backtracking line search to find ρ_k.
(4) Update. $u_{k+1} = u_k + \rho_k d_{\mathrm{nt},k}$.

Observe that this is essentially the descent procedure of Section 13.8 using the Newton step as search direction, except that the stopping criterion is checked just after computing the search direction, rather than after the update (a very minor difference).

The convergence of Newton's method is thoroughly analyzed in Boyd and Vandenberghe [Boyd and Vandenberghe (2004)] (Section 9.5.3). This analysis is made under the following assumptions:

(1) The function $J\colon \Omega \to \mathbb{R}$ is a convex function defined on some open subset Ω of $\mathbb{R}^n$ which is twice differentiable and its Hessian $\nabla^2 J(x)$ is symmetric positive definite for all $x \in \Omega$. This implies that there are two constants $m > 0$ and $M > 0$ such that $mI \preceq \nabla^2 J(x) \preceq MI$ for all $x \in \Omega$, which means that the eigenvalues of $\nabla^2 J(x)$ belong to $[m, M]$.
(2) The Hessian is Lipschitzian, which means that there is some $L \ge 0$ such that

$$\left\|\nabla^2 J(x) - \nabla^2 J(y)\right\|_2 \le L \left\|x, y\right\|_2 \quad \text{for all } x, y \in \Omega.$$

It turns out that the iterations of Newton's method fall into two phases, depending whether $\|\nabla J_{u_k}\|_2 \ge \eta$ or $\|\nabla J_{u_k}\|_2 < \eta$, where η is a number which depends on m, L, and the constant α used in the backtracking line search, and $\eta \leq m^2/L$.

(1) The first phase, called the *damped Newton phase*, occurs while $\|\nabla J_{u_k}\|_2 \geq \eta$. During this phase, the procedure can choose a step size $\rho_k = t < 1$, and there is some constant $\gamma > 0$ such that
$$J(u_{k+1}) - J(u_k) \leq -\gamma.$$

(2) The second phase, called the *quadratically convergent phase* or *pure Newton phase*, occurs while $\|\nabla J_{u_k}\|_2 < \eta$. During this phase, the step size $\rho_k = t = 1$ is always chosen, and we have
$$\frac{L}{2m^2} \left\|\nabla J_{u_{k+1}}\right\|_2 \leq \left(\frac{L}{2m^2} \|\nabla J_{u_k}\|_2\right)^2. \tag{$*_1$}$$

If we denote the minimal value of f by p^*, then the number of damped Newton steps is at most
$$\frac{J(u_0) - p^*}{\gamma}.$$

Equation ($*_1$) and the fact that $\eta \leq m^2/L$ shows that if $\|\nabla J_{u_k}\|_2 < \eta$, then $\left\|\nabla J_{u_{k+1}}\right\|_2 < \eta$. It follows by induction that for all $\ell \geq k$, we have
$$\frac{L}{2m^2} \left\|\nabla J_{u_{\ell+1}}\right\|_2 \leq \left(\frac{L}{2m^2} \|\nabla J_{u_\ell}\|_2\right)^2, \tag{$*_2$}$$
and thus (since $\eta \leq m^2/L$ and $\|\nabla J_{u_k}\|_2 < \eta$, we have $(L/m^2)\|\nabla J_{u_k}\|_2 < (L/m^2)\eta \leq 1$), so
$$\frac{L}{2m^2} \|\nabla J_{u_\ell}\|_2 \leq \left(\frac{L}{2m^2} \|\nabla J_{u_k}\|_2\right)^{2^{\ell-k}} \leq \left(\frac{1}{2}\right)^{2^{\ell-k}}, \quad \ell \geq k. \tag{$*_3$}$$
It is shown in Boyd and Vandenberghe [Boyd and Vandenberghe (2004)] (Section 9.1.2) that the hypothesis $mI \preceq \nabla^2 J(x)$ implies that
$$J(x) - p^* \leq \frac{1}{2m} \|\nabla J_x\|_2^2 \quad x \in \Omega.$$
As a consequence, by ($*_3$), we have
$$J(u_\ell) - p^* \leq \frac{1}{2m} \|\nabla J_{u_\ell}\|_2^2 \leq \frac{2m^3}{L^2} \left(\frac{1}{2}\right)^{2^{\ell-k+1}}. \tag{$*_4$}$$

Equation ($*_4$) shows that the convergence during the quadratically convergence phase is very fast. If we let
$$\epsilon_0 = \frac{2m^3}{L^2},$$

then Equation $(*_4)$ implies that we must have $J(u_\ell) - p^* \leq \epsilon$ after no more than

$$\log_2 \log_2(\epsilon_0/\epsilon)$$

iterations. The term $\log_2 \log_2(\epsilon_0/\epsilon)$ grows *extremely slowly* as ϵ goes to zero, and for practical purposes it can be considered constant, say five or six (six iterations gives an accuracy of about $\epsilon \approx 5 \cdot 10^{-20}\epsilon_0$).

In summary, the number of Newton iterations required to find a minimum of J is approximately bounded by

$$\frac{J(u_0) - p^*}{\gamma} + 6.$$

Examples of the application of Newton's method and further discussion of its efficiency are given in Boyd and Vandenberghe [Boyd and Vandenberghe (2004)] (Section 9.5.4). Basically, Newton's method has a faster convergence rate than gradient or steepest descent. Its main disadvantage is the cost for forming and storing the Hessian, and of computing the Newton step, which requires solving a linear system.

There are two major shortcomings of the convergence analysis of Newton's method as sketched above. The first is a practical one. The complexity estimates involve the constants m, M, and L, which are almost never known in practice. As a result, the bound on the number of steps required is almost never known specifically.

The second shortcoming is that although Newton's method itself is affine invariant, the analysis of convergence is very much dependent on the choice of coordinate system. If the coordinate system is changed, the constants m, M, L also change. This can be viewed as an aesthetic problem, but it would be nice if an analysis of convergence independent of an affine change of coordinates could be given.

Nesterov and Nemirovski discovered a condition on functions that allows an affine-invariant convergence analysis. This property, called *self-concordance*, is unfortunately not very intuitive.

Definition 13.10. A (partial) convex function f defined on $\mathbb{R}$ is *self-concordant* if

$$|f'''(x)| \leq 2(f''(x))^{3/2} \quad \text{for all } x \in \mathbb{R}.$$

A (partial) convex function f defined on $\mathbb{R}^n$ is *self-concordant* if for every nonzero $v \in \mathbb{R}^n$ and all $x \in \mathbb{R}^n$, the function $t \mapsto J(x+tv)$ is self-concordant.

Affine and convex quadratic functions are obviously self-concordant, since $f''' = 0$. There are many more interesting self-concordant functions, for example, the function
$X \mapsto -\log\det(X)$, where X is a symmetric positive definite matrix.

Self-concordance is discussed extensively in Boyd and Vandenberghe [Boyd and Vandenberghe (2004)] (Section 9.6). The main point of self-concordance is that a coordinate system-invariant proof of convergence can be given for a certain class of strictly convex self-concordant functions. This proof is given in Boyd and Vandenberghe [Boyd and Vandenberghe (2004)] (Section 9.6.4). Given a starting value u_0, we assume that the sublevel set $\{x \in \mathbb{R}^n \mid J(x) \leq J(u_0)\}$ is closed and that J is bounded below. Then there are two parameters η and γ as before, but *depending only on the parameters* α, β *involved in the line search*, such that:

(1) If $\lambda(u_k) > \eta$, then
$$J(u_{k+1}) - J(u_k) \leq -\gamma.$$

(2) If $\lambda(u_k) \leq \eta$, then the backtracking line search selects $t = 1$ and we have
$$2\lambda(u_{k+1}) \leq (2\lambda(u_k))^2.$$

As a consequence, for all $\ell \geq k$, we have
$$J(u_\ell) - p^* \leq \lambda(u_\ell)^2 \leq \left(\frac{1}{2}\right)^{2^{\ell-k+1}}.$$
In the end, accuracy $\epsilon > 0$ is achieved in at most
$$\frac{20 - 8\alpha}{\alpha\beta(1-2\alpha)^2}(J(u_0) - p^*) + \log_2\log_2(1/\epsilon)$$
iterations, where α and β are the constants involved in the line search. This bound is obviously independent of the chosen coordinate system.

Contrary to intuition, the descent direction $d_k = -\nabla J_{u_k}$ given by the opposite of the gradient is *not* always optimal. In the next section we will see how a better direction can be picked; this is the method of *conjugate gradients*.

13.10 Conjugate Gradient Methods for Unconstrained Problems

The conjugate gradient method due to Hestenes and Stiefel (1952) is a gradient descent method that applies to an elliptic quadratic functional $J\colon \mathbb{R}^n \to \mathbb{R}$ given by
$$J(v) = \frac{1}{2}\langle Av, v\rangle - \langle b, v\rangle,$$

where A is an $n \times n$ symmetric positive definite matrix. Although it is presented as an iterative method, it terminates in at most n steps.

As usual, the conjugate gradient method starts with some arbitrary initial vector u_0 and proceeds through a sequence of iteration steps generating (better and better) approximations u_k of the optimal vector u minimizing J. During an iteration step, two vectors need to be determined:

(1) The descent direction d_k.
(2) The next approximation u_{k+1}. To find u_{k+1}, we need to find the stepsize $\rho_k > 0$ and then

$$u_{k+1} = u_k - \rho_k d_k.$$

Typically, ρ_k is found by performing a line search along the direction d_k, namely we find ρ_k as the real number such that the function $\rho \mapsto J(u_k - \rho d_k)$ is minimized.

We saw in Proposition 13.5 that during execution of the gradient method with optimal stepsize parameter that any two consecutive descent directions are orthogonal. The new twist with the conjugate gradient method is that given $u_0, u_1, \ldots, u_k$, the next approximation u_{k+1} is obtained as the solution of the problem which consists in minimizing J over the affine subspace $u_k + \mathcal{G}_k$, where $\mathcal{G}_k$ is the subspace of $\mathbb{R}^n$ spanned by the gradients

$$\nabla J_{u_0}, \nabla J_{u_1}, \ldots, \nabla J_{u_k}.$$

We may assume that $\nabla J_{u_\ell} \neq 0$ for $\ell = 0, \ldots, k$, since the method terminates as soon as $\nabla J_{u_k} = 0$. A priori the subspace $\mathcal{G}_k$ has dimension $\leq k+1$, but we will see that in fact it has dimension $k+1$. Then we have

$$u_k + \mathcal{G}_k = \left\{ u_k + \sum_{i=0}^{k} \alpha_i \nabla J_{u_i} \;\middle|\; \alpha_i \in \mathbb{R},\ 0 \leq i \leq k \right\},$$

and our minimization problem is to find u_{k+1} such that

$$u_{k+1} \in u_k + \mathcal{G}_k \quad \text{and} \quad J(u_{k+1}) = \in J_{v \in u_k + \mathcal{G}_k} J(v).$$

In the gradient method with optimal stepsize parameter the descent direction d_k is proportional to the gradient ∇J_{u_k}, but in the conjugate gradient method, d_k is equal to ∇J_{u_k} corrected by some multiple of d_{k-1}.

The conjugate gradient method is superior to the gradient method with optimal stepsize parameter for the following reasons proved correct later:

(a) The gradients ∇J_{u_i} and ∇J_{u_j} are orthogonal for all i, j with $0 \leq i < j \leq k$. This implies that if $\nabla J_{u_i} \neq 0$ for $i = 0, \ldots, k$, then the vectors ∇J_{u_i} are linearly independent, so the method stops in at most n steps.

(b) If we write $\Delta_\ell = u_{\ell+1} - u_\ell = -\rho_\ell d_\ell$, the second remarkable fact about the conjugate gradient method is that the vectors Δ_ℓ satisfy the following conditions:

$$\langle A\Delta_\ell, \Delta_i \rangle = 0 \quad 0 \leq i < \ell \leq k.$$

The vectors Δ_ℓ and Δ_i are said to be *conjugate* with respect to the matrix A (or A-*conjugate*). As a consequence, if $\Delta_\ell \neq 0$ for $\ell = 0, \ldots, k$, then the vectors Δ_ℓ are linearly independent.

(c) There is a simple formula to compute d_{k+1} from d_k, and to compute ρ_k.

We now prove the above facts. We begin with (a).

Proposition 13.7. *Assume that* $\nabla J_{u_i} \neq 0$ *for* $i = 0, \ldots, k$. *Then the minimization problem, find* u_{k+1} *such that*

$$u_{k+1} \in u_k + \mathcal{G}_k \quad \textit{and} \quad J(u_{k+1}) = \inf_{v \in u_k + \mathcal{G}_k} J(v),$$

has a unique solution, and the gradients ∇J_{u_i} *and* ∇J_{u_j} *are orthogonal for all* i, j *with* $0 \leq i < j \leq k$.

Proof. The affine space $u_\ell + \mathcal{G}_\ell$ is closed and convex, and since J is a quadratic elliptic functional it is coercise and strictly convex, so by Theorem 13.6(2) it has a unique minimum in $u_\ell + \mathcal{G}_\ell$. This minimum $u_{\ell+1}$ is also the minimum of the problem, find $u_{\ell+1}$ such that

$$u_{\ell+1} \in u_\ell + \mathcal{G}_\ell \quad \text{and} \quad J(u_{\ell+1}) = \inf_{v \in \mathcal{G}_\ell} J(u_\ell + v),$$

and since $\mathcal{G}_\ell$ is a vector space, by Theorem 4.4 we must have

$$dJ_{u_\ell}(w) = 0 \quad \text{for all } w \in \mathcal{G}_\ell,$$

that is

$$\langle \nabla J_{u_\ell}, w \rangle = 0 \quad \text{for all } w \in \mathcal{G}_\ell.$$

Since $\mathcal{G}_\ell$ is spanned by $(\nabla J_{u_0}, \nabla J_{u_1}, \ldots, \nabla J_{u_\ell})$, we obtain

$$\langle \nabla J_{u_\ell}, \nabla J_{u_j} \rangle = 0, \quad 0 \leq j < \ell,$$

and since this holds for $\ell = 0, \ldots, k$, we get

$$\langle \nabla J_{u_i}, \nabla J_{u_j} \rangle = 0, \quad 0 \leq i < j \leq k,$$

which shows the second part of the proposition. $\square$

As a corollary of Proposition 13.7, if $\nabla J_{u_i} \neq 0$ for $i = 0, \ldots, k$, then the vectors ∇J_{u_i} are linearly independent and $\mathcal{G}_k$ has dimension $k+1$. Therefore, the conjugate gradient method terminates in at most n steps. Here is an example of a problem for which the gradient descent with optimal stepsize parameter does not converge in a finite number of steps.

Example 13.2. Let $J\colon \mathbb{R}^2 \to \mathbb{R}$ be the function given by

$$J(v_1, v_2) = \frac{1}{2}(\alpha_1 v_1^2 + \alpha_2 v_2^2),$$

where $0 < \alpha_1 < \alpha_2$. The minimum of J is attained at $(0,0)$. Unless the initial vector $u_0 = (u_1^0, u_2^0)$ has the property that either $u_1^0 = 0$ or $u_2^0 = 0$, we claim that the gradient descent with optimal stepsize parameter does not converge in a finite number of steps. Observe that

$$\nabla J_{(v_1,v_2)} = \begin{pmatrix} \alpha_1 v_1 \\ \alpha_2 v_2 \end{pmatrix}.$$

As a consequence, given u_k, the line search for finding ρ_k and u_{k+1} yields $u_{k+1} = (0,0)$ iff there is some $\rho \in \mathbb{R}$ such that

$$u_1^k = \rho\alpha_1 u_1^k \quad \text{and} \quad u_2^k = \rho\alpha_2 u_2^k.$$

Since $\alpha_1 \neq \alpha_2$, this is only possible if either $u_1^k = 0$ or $u_2^k = 0$. The formulae given just before Proposition 13.6 yield

$$u_1^{k+1} = \frac{\alpha_2^2(\alpha_2 - \alpha_1)u_1^k(u_2^k)^2}{\alpha_1^3(u_1^k)^2 + \alpha_2^3(u_2^k)^2}, \quad u_2^{k+1} = \frac{\alpha_1^2(\alpha_1 - \alpha_2)u_2^k(u_1^k)^2}{\alpha_1^3(u_1^k)^2 + \alpha_2^3(u_2^k)^2},$$

which implies that if $u_1^k \neq 0$ and $u_2^k \neq 0$, then $u_1^{k+1} \neq 0$ and $u_2^{k+1} \neq 0$, so the method runs forever from any initial vector $u_0 = (u_1^0, u_2^0)$ such that $u_1^0 \neq 0$ and, $u_2^0 \neq 0$.

We now prove (b).

Proposition 13.8. *Assume that $\nabla J_{u_i} \neq 0$ for $i = 0, \ldots, k$, and let $\Delta_\ell = u_{\ell+1} - u_\ell$, for $\ell = 0, \ldots, k$. Then $\Delta_\ell \neq 0$ for $\ell = 0, \ldots, k$, and*

$$\langle A\Delta_\ell, \Delta_i \rangle = 0, \quad 0 \leq i < \ell \leq k.$$

The vectors $\Delta_0, \ldots, \Delta_k$ are linearly independent.

Proof. Since J is a quadratic functional we have

$$\nabla J_{v+w} = A(v+w) - b = Av - b + Aw = \nabla J_v + Aw.$$

It follows that

$$\nabla J_{u_{\ell+1}} = \nabla J_{u_\ell + \Delta_\ell} = \nabla J_{u_\ell} + A\Delta_\ell, \quad 0 \leq \ell \leq k. \qquad (*_1)$$

By Proposition 13.7, since

$$\langle \nabla J_{u_i}, \nabla J_{u_j} \rangle = 0, \quad 0 \leq i < j \leq k,$$

we get

$$0 = \langle \nabla J_{u_\ell+1}, \nabla J_{u_\ell} \rangle = \|\nabla J_{u_\ell}\|^2 + \langle A\Delta_\ell, \nabla J_{u_\ell} \rangle, \quad \ell = 0, \ldots, k,$$

and since by hypothesis $\nabla J_{u_i} \neq 0$ for $i = 0, \ldots, k$, we deduce that

$$\Delta_\ell \neq 0, \quad \ell = 0, \ldots, k.$$

If $k \geq 1$, for $i = 0, \ldots, \ell - 1$ and $\ell \leq k$ we also have

$$\begin{aligned} 0 = \langle \nabla J_{u_{\ell+1}}, \nabla J_{u_i} \rangle &= \langle \nabla J_{u_\ell}, \nabla J_{u_i} \rangle + \langle A\Delta_\ell, \nabla J_{u_i} \rangle \\ &= \langle A\Delta_\ell, \nabla J_{u_i} \rangle. \end{aligned}$$

Since $\Delta_j = u_{j+1} - u_j \in \mathcal{G}_j$ and $\mathcal{G}_j$ is spanned by $(\nabla J_{u_0}, \nabla J_{u_1}, \ldots, \nabla J_{u_j})$, we obtain

$$\langle A\Delta_\ell, \Delta_j \rangle = 0, \quad 0 \leq j < \ell \leq k.$$

For the last statement of the proposition, let $w_0, w_1, \ldots, w_k$ be any $k+1$ nonzero vectors such that

$$\langle Aw_i, w_j \rangle = 0, \quad 0 \leq i < j \leq k.$$

We claim that $w_0, w_1, \ldots, w_k$ are linearly independent.

If we have a linear dependence $\sum_{i=0}^{k} \lambda_i w_i = 0$, then we have

$$0 = \left\langle A\left(\sum_{i=0}^{k} \lambda_i w_i\right), w_j \right\rangle = \sum_{i=0}^{k} \lambda_i \langle Aw_i, w_j \rangle = \lambda_j \langle Aw_j, w_j \rangle.$$

Since A is symmetric positive definite (because J is a quadratic elliptic functional) and $w_j \neq 0$, we must have $\lambda_j = 0$ for $j = 0, \ldots, k$. Therefore the vectors $w_0, w_1, \ldots, w_k$ are linearly independent. □

Remarks:

(1) Since A is symmetric positive definite, the bilinear map $(u, v) \mapsto \langle Au, v \rangle$ is an inner product $\langle -, - \rangle_A$ on $\mathbb{R}^n$. Consequently, two vectors u, v are *conjugate* with respect to the matrix A (or A-*conjugate*), which means that $\langle Au, v \rangle = 0$, iff u and v are orthogonal with respect to the inner product $\langle -, - \rangle_A$.

(2) By picking the descent direction to be $-\nabla J_{u_k}$, the gradient descent method with optimal stepsize parameter treats the level sets $\{u \mid J(u) = J(u_k)\}$ as if they were spheres. The conjugate gradient method is more subtle, and takes the "geometry" of the level set $\{u \mid J(u) = J(u_k)\}$ into account, through the notion of conjugate directions.
(3) The notion of conjugate direction has its origins in the theory of projective conics and quadrics where A is a 2×2 or a 3×3 matrix and where u and v are conjugate iff $u^\top A v = 0$.
(4) The terminology conjugate gradient is somewhat misleading. It is not the gradients who are conjugate directions, but the descent directions.

By definition of the vectors $\Delta_\ell = u_{\ell+1} - u_\ell$, we can write

$$\Delta_\ell = \sum_{i=0}^{\ell} \delta_i^\ell \, \nabla J_{u_i}, \quad 0 \leq \ell \leq k. \tag{$*_2$}$$

In matrix form, we can write

$$(\Delta_0 \; \Delta_1 \; \cdots \; \Delta_k) = (\nabla J_{u_0} \; \nabla J_{u_1} \; \cdots \; \nabla J_{u_k}) \begin{pmatrix} \delta_0^0 & \delta_0^1 & \cdots & \delta_0^{k-1} & \delta_0^k \\ 0 & \delta_1^1 & \cdots & \delta_1^{k-1} & \delta_1^k \\ 0 & 0 & \cdots & \delta_2^{k-1} & \delta_2^k \\ \vdots & \vdots & \ddots & \vdots & \vdots \\ 0 & 0 & \cdots & 0 & \delta_k^k \end{pmatrix},$$

which implies that $\delta_\ell^\ell \neq 0$ for $\ell = 0, \ldots, k$.

In view of the above fact, since Δ_ℓ and d_ℓ are collinear, it is convenient to write the descent direction d_ℓ as

$$d_\ell = \sum_{i=0}^{\ell-1} \lambda_i^\ell \, \nabla J_{u_i} + \nabla J_{u_\ell}, \quad 0 \leq \ell \leq k. \tag{$*_3$}$$

Our next goal is to compute u_{k+1}, assuming that the coefficients λ_i^k are known for $i = 0, \ldots, k$, and then to find simple formulae for the λ_i^k.

The problem reduces to finding ρ_k such that

$$J(u_k - \rho_k d_k) = \inf_{\rho \in \mathbb{R}} J(u_k - \rho d_k),$$

and then $u_{k+1} = u_k - \rho_k d_k$. In fact, by $(*_2)$, since

$$\Delta_k = \sum_{i=0}^{k} \delta_i^k \, \nabla J_{u_i} = \delta_k^k \left(\sum_{i=0}^{k-1} \frac{\delta_i^k}{\delta_k^k} \, \nabla J_{u_i} + \nabla J_{u_k} \right),$$

we must have

$$\Delta_k = \delta_k^k d_k \quad \text{and} \quad \rho_k = -\delta_k^k. \tag{$*_4$}$$

Remarkably, the coefficients λ_i^k and the descent directions d_k can be computed easily using the following formulae.

Proposition 13.9. *Assume that $\nabla J_{u_i} \neq 0$ for $i = 0, \dots, k$. If we write*

$$d_\ell = \sum_{i=0}^{\ell-1} \lambda_i^\ell \nabla J_{u_i} + \nabla J_{u_\ell}, \quad 0 \leq \ell \leq k,$$

then we have

$$(\dagger) \quad \begin{cases} \lambda_i^k = \dfrac{\|\nabla J_{u_k}\|^2}{\|\nabla J_{u_i}\|^2}, & 0 \leq i \leq k-1, \\ d_0 = \nabla J_{u_0} & \\ d_\ell = \nabla J_{u_\ell} + \dfrac{\|\nabla J_{u_\ell}\|^2}{\|\nabla J_{u_{\ell-1}}\|^2} d_{\ell-1}, & 1 \leq \ell \leq k. \end{cases}$$

Proof. Since by $(*_4)$ we have $\Delta_k = \delta_k^k d_k$, $\delta_k^k \neq 0$, (by Proposition 13.8) we have

$$\langle A\Delta_\ell, \Delta_i \rangle = 0, \quad 0 \leq i < \ell \leq k.$$

By $(*_1)$ we have $\nabla J_{u_{\ell+1}} = \nabla J_{u_\ell} + A\Delta_\ell$, and since A is a symmetric matrix, we have

$$0 = \langle Ad_k, \Delta_\ell \rangle = \langle d_k, A\Delta_\ell \rangle = \langle d_k, \nabla J_{u_{\ell+1}} - \nabla J_{u_\ell} \rangle,$$

for $\ell = 0, \dots, k-1$. Since

$$d_k = \sum_{i=0}^{k-1} \lambda_i^k \nabla J_{u_i} + \nabla J_{u_k},$$

we have

$$\left\langle \sum_{i=0}^{k-1} \lambda_i^k \nabla J_{u_i} + \nabla J_{u_k}, \nabla J_{u_{\ell+1}} - \nabla J_{u_\ell} \right\rangle = 0, \quad 0 \leq \ell \leq k-1.$$

Since by Proposition 13.7 the gradients ∇J_{u_i} are pairwise orthogonal, the above equations yield

$$\begin{aligned} -\lambda_{k-1}^k \|\nabla J_{u_{k-1}}\|^2 + \|\nabla J_k\|^2 &= 0 \\ -\lambda_\ell^k \|\nabla J_{u_\ell}\|^2 + \lambda_{\ell+1}^k \|\nabla J_{\ell+1}\|^2 &= 0, \quad 0 \leq \ell \leq k-2, \quad k \geq 2, \end{aligned}$$

and an easy induction yields

$$\lambda_i^k = \frac{\|\nabla J_{u_k}\|^2}{\|\nabla J_{u_i}\|^2}, \quad 0 \leq i \leq k-1.$$

Consequently, using $(*_3)$ we have

$$\begin{aligned} d_k &= \sum_{i=0}^{k-1} \frac{\|\nabla J_{u_k}\|^2}{\|\nabla J_{u_i}\|^2} \nabla J_{u_i} + \nabla J_{u_k} \\ &= \nabla J_{u_k} + \frac{\|\nabla J_{u_k}\|^2}{\|\nabla J_{u_{k-1}}\|^2} \left(\sum_{i=0}^{k-2} \frac{\|\nabla J_{u_{k-1}}\|^2}{\|\nabla J_{u_i}\|^2} \nabla J_{u_i} + \nabla J_{u_{k-1}} \right) \\ &= \nabla J_{u_k} + \frac{\|\nabla J_{u_k}\|^2}{\|\nabla J_{u_{k-1}}\|^2} d_{k-1}, \end{aligned}$$

which concludes the proof. □

It remains to compute ρ_k, which is the solution of the line search

$$J(u_k - \rho_k d_k) = \inf_{\rho \in \mathbb{R}} J(u_k - \rho d_k).$$

Since J is a quadratic functional, a basic computation left to the reader shows that the function to be minimized is

$$\rho \mapsto \frac{\rho^2}{2} \langle A d_k, d_k \rangle - \rho \langle \nabla J_{u_k}, d_k \rangle + J(u_k),$$

whose minimum is obtained when its derivative is zero, that is,

$$\rho_k = \frac{\langle \nabla J_{u_k}, d_k \rangle}{\langle A d_k, d_k \rangle}. \qquad (*_5)$$

In summary, the conjugate gradient method finds the minimum u of the elliptic quadratic functional

$$J(v) = \frac{1}{2} \langle Av, a \rangle - \langle b, v \rangle$$

by computing the sequence of vectors $u_1, d_1, \ldots, u_{k-1}, d_{k-1}, u_k$, starting from any vector u_0, with

$$d_0 = \nabla J_{u_0}.$$

If $\nabla J_{u_0} = 0$, then the algorithm terminates with $u = u_0$. Otherwise, for $k \geq 0$, assuming that $\nabla J_{u_i} \neq 0$ for $i = 1, \ldots, k$, compute

$$(*_6) \qquad \begin{cases} \rho_k = \dfrac{\langle \nabla J_{u_k}, d_k \rangle}{\langle A d_k, d_k \rangle} \\ u_{k+1} = u_k - \rho_k d_k \\ d_{k+1} = \nabla J_{u_{k+1}} + \dfrac{\|\nabla J_{u_{k+1}}\|^2}{\|\nabla J_{u_k}\|^2} d_k. \end{cases}$$

If $\nabla J_{u_{k+1}} = 0$, then the algorithm terminates with $u = u_{k+1}$.

As we showed before, the algorithm terminates in at most n iterations.

Example 13.3. Let us take the example of Section 13.6 and apply the conjugate gradient procedure. Recall that

$$\begin{aligned} J(x,y) &= \frac{1}{2}\begin{pmatrix} x & y \end{pmatrix}\begin{pmatrix} 3 & 2 \\ 2 & 6 \end{pmatrix}\begin{pmatrix} x \\ y \end{pmatrix} - \begin{pmatrix} x & y \end{pmatrix}\begin{pmatrix} x \\ y \end{pmatrix} \\ &= \frac{3}{2}x^2 + 2xy + 3y^2 - 2x + 8y. \end{aligned}$$

Note that $\nabla J_v = (3x + 2y - 2, 2x + 6y + 8)$,

Initialize the procedure by setting

$$u_0 = (-2, -2), \qquad d_0 = \nabla J_{u_0} = (-12, -8)$$

Step 1 involves calculating

$$\begin{aligned} \rho_0 &= \frac{\langle \nabla J_{u_0}, d_0 \rangle}{\langle A d_0, d_0 \rangle} = \frac{13}{75} \\ u_1 &= u_0 - \rho_0 d_0 = (-2, -2) - \frac{13}{75}(-12, -8) = \left(\frac{2}{25}, -\frac{46}{75}\right) \\ d_1 &= \nabla J_{u_1} + \frac{||\nabla J_{u_1}||^2}{||\nabla J_{u_0}||^2} d_0 = \left(-\frac{2912}{625}, \frac{18928}{5625}\right). \end{aligned}$$

Observe that ρ_0 and u_1 are precisely the *same* as in the case the case of gradient descent with optimal step size parameter. The difference lies in the calculation of d_1. As we will see, this change will make a *huge* difference in the convergence to the unique minimum $u = (2, -2)$.

We continue with the conjugate gradient procedure and calculate Step 2 as

$$\begin{aligned} \rho_1 &= \frac{\langle \nabla J_{u_1}, d_1 \rangle}{\langle A d_1, d_1 \rangle} = \frac{75}{82} \\ u_2 &= u_1 - \rho_1 d_1 = \left(\frac{2}{25}, -\frac{46}{75}\right) - \frac{75}{82}\left(-\frac{2912}{625}, \frac{18928}{5625}\right) = (2, -2) \\ d_2 &= \nabla J_{u_2} + \frac{||\nabla J_{u_2}||^2}{||\nabla J_{u_1}||^2} d_1 = (0, 0). \end{aligned}$$

Since $\nabla J_{u_2} = 0$, the procedure terminates in *two* steps, as opposed to the 31 steps needed for gradient descent with optimal step size parameter.

Hestenes and Stiefel realized that Equations $(*_6)$ can be modified to make the computations more efficient, by having only one evaluation of the matrix A on a vector, namely d_k. The idea is to compute ∇_{u_k} inductively.

Since by $(*_1)$ and $(*_4)$ we have $\nabla J_{u_{\ell+1}} = \nabla J_{u_\ell} + A\Delta_\ell = \nabla J_{u_\ell} - \rho_k A d_k$, the gradient $\nabla J_{u_{\ell+1}}$ can be computed iteratively:

$$\begin{aligned} \nabla J_0 &= Au_0 - b \\ \nabla J_{u_{\ell+1}} &= \nabla J_{u_\ell} - \rho_k A d_k. \end{aligned}$$

Since by Proposition 13.9 we have

$$d_k = \nabla J_{u_k} + \frac{\left\|\nabla J_{u_k}\right\|^2}{\left\|\nabla J_{u_{k-1}}\right\|^2} d_{k-1}$$

and since d_{k-1} is a linear combination of the gradients ∇J_{u_i} for $i = 0, \ldots, k-1$, which are all orthogonal to ∇J_{u_k}, we have

$$\rho_k = \frac{\langle \nabla J_{u_k}, d_k \rangle}{\langle A d_k, d_k \rangle} = \frac{\left\|\nabla J_{u_k}\right\|^2}{\langle A d_k, d_k \rangle}.$$

It is customary to introduce the term r_k defined as

$$\nabla J_{u_k} = Au_k - b \tag{$*_7$}$$

and to call it the *residual*. Then the conjugate gradient method consists of the following steps. We initialize the method starting from any vector u_0 and set

$$d_0 = r_0 = Au_0 - b.$$

The main iteration step is $(k \geq 0)$:

$$(*_8) \quad \begin{cases} \rho_k = \dfrac{\left\|r_k\right\|^2}{\langle A d_k, d_k \rangle} \\ u_{k+1} = u_k - \rho_k d_k \\ r_{k+1} = r_k + \rho_k A d_k \\ \beta_{k+1} = \dfrac{\left\|r_{k+1}\right\|^2}{\left\|r_k\right\|^2} \\ d_{k+1} = r_{k+1} + \beta_{k+1} d_k. \end{cases}$$

Beware that some authors define the residual r_k as $r_k = b - Au_k$ and the descent direction d_k as $-d_k$. In this case, the second equation becomes

$$u_{k+1} = u_k + \rho_k d_k.$$

Since $d_0 = r_0$, the equations

$$\begin{aligned} r_{k+1} &= r_k - \rho_k A d_k \\ d_{k+1} &= r_{k+1} - \beta_{k+1} d_k \end{aligned}$$

imply by induction that the subspace $\mathcal{G}_k$ is spanned by $(r_0, r_1, \ldots, r_k)$ and $(d_0, d_1, \ldots, d_k)$ is the subspace spanned by

$$(r_0, Ar_0, A^2 r_0, \ldots, A^k r_0).$$

Such a subspace is called a *Krylov subspace*.

If we define the *error* e_k as $e_k = u_k - u$, then $e_0 = u_0 - u$ and $Ae_0 = Au_0 - Au = Au_0 - b = d_0 = r_0$, and then because

$$u_{k+1} = u_k - \rho_k d_k$$

we see that

$$e_{k+1} = e_k - \rho_k d_k.$$

Since d_k belongs to the subspace spanned by $(r_0, Ar_0, A^2 r_0, \ldots, A^k r_0)$ and $r_0 = Ae_0$, we see that d_k belongs to the subspace spanned by $(Ae_0, A^2 e_0, A^3 e_0, \ldots, A^{k+1} e_0)$, and then by induction we see that e_{k+1} belongs to the subspace spanned by $(e_0, Ae_0, A^2 e_0, A^3 e_0, \ldots, A^{k+1} e_0)$. This means that there is a polynomial P_k of degree $\leq k$ such that $P_k(0) = 1$ and

$$e_k = P_k(A) e_0.$$

This is an important fact because it allows for an analysis of the convergence of the conjugate gradient method; see Trefethen and Bau [Trefethen and Bau III (1997)] (Lecture 38). For this, since A is symmetric positive definite, we know that $\langle u, v\rangle_A = \langle Av, u\rangle$ is an inner product on $\mathbb{R}^n$ whose associated norm is denoted by $\|v\|_A$. Then observe that if $e(v) = v - u$, then

$$\begin{aligned}
\|e(v)\|_A^2 &= \langle Av - Au, v - u\rangle \\
&= \langle Av, v\rangle - 2\langle Au, v\rangle + \langle Au, u\rangle \\
&= \langle Av, v\rangle - 2\langle b, v\rangle + \langle b, u\rangle \\
&= 2J(v) + \langle b, u\rangle.
\end{aligned}$$

It follows that $v = u_k$ minimizes $\|e(v)\|_A$ on $u_{k-1} + \mathcal{G}_{k-1}$ since u_k minimizes J on $u_{k-1} + \mathcal{G}_{k-1}$. Since $e_k = P_k(A)e_0$ for some polynomial P_k of degree $\leq k$ such that $P_k(0) = 1$, if we let $\mathcal{P}_k$ be the set of polynomials $P(t)$ of degree $\leq k$ such that $P(0) = 1$, then we have

$$\|e_k\|_A = \inf_{P \in \mathcal{P}_k} \|P(A)e_0\|_A.$$

Since A is a symmetric positive definite matrix it has real positive eigenvalues $\lambda_1, \ldots, \lambda_n$ and there is an orthonormal basis of eigenvectors $h_1, \ldots, h_n$ so that if we write $e_0 = \sum_{j=1}^n a_j h_j$, then we have

$$\|e_0\|_A^2 = \langle Ae_0, e_0\rangle = \left\langle \sum_{i=1}^n a_i \lambda_i h_i, \sum_{j=1}^n a_j h_j \right\rangle = \sum_{j=1}^n a_j^2 \lambda_j$$

and

$$\|P(A)e_0\|_A^2 = \langle AP(A)e_0, P(A)e_0\rangle = \left\langle \sum_{i=1}^{n} a_i\lambda_i P(\lambda_i)h_i, \sum_{j=1}^{n} a_j P(\lambda_j)h_j \right\rangle$$
$$= \sum_{j=1}^{n} a_j^2 \lambda_j (P(\lambda_j))^2.$$

These equations imply that

$$\|e_k\|_A \leq \left(\inf_{P\in\mathcal{P}_k} \max_{1\leq i\leq n} |P(\lambda_i)|\right) \|e_0\|_A \,.$$

It can be shown that the conjugate gradient method requires of the order of

n^3 additions,

n^3 multiplications,

$2n$ divisions.

In theory, this is worse than the number of elementary operations required by the Cholesky method. Even though the conjugate gradient method does not seem to be the best method for *full* matrices, it usually outperforms other methods for *sparse* matrices. The reason is that the matrix A only appears in the computation of the vector Ad_k. If the matrix A is banded (for example, tridiagonal), computing Ad_k is very cheap and there is no need to store the entire matrix A, in which case the conjugate gradient method is fast. Also, although in theory, up to n iterations may be required, in practice, convergence may occur after a much smaller number of iterations.

Using the inequality

$$\|e_k\|_A \leq \left(\inf_{P\in\mathcal{P}_k} \max_{1\leq i\leq n} |P(\lambda_i)|\right) \|e_0\|_A \,,$$

by choosing P to be a shifted Chebyshev polynomial, it can be shown that

$$\|e_k\|_A \leq 2\left(\frac{\sqrt{\kappa}-1}{\sqrt{\kappa}+1}\right)^k \|e_0\|_A \,,$$

where $\kappa = \mathrm{cond}_2(A)$; see Trefethen and Bau [Trefethen and Bau III (1997)] (Lecture 38, Theorem 38.5). Thus the rate of convergence of the conjugate gradient method is governed by the ratio

$$\frac{\sqrt{\mathrm{cond}_2(A)}-1}{\sqrt{\mathrm{cond}_2(A)}+1},$$

where $\text{cond}_2(A) = \lambda_n/\lambda_1$ is the condition number of the matrix A. Since A is positive definite, λ_1 is its smallest eigenvalue and λ_n is its largest eigenvalue.

The above fact leads to the process of *preconditioning*, a method which consists in replacing the matrix of a linear system $Ax = b$ by an "equivalent" one for example $M^{-1}A$ (since M is invertible, the system $Ax = b$ is equivalent to the system $M^{-1}Ax = M^{-1}b$), where M is chosen so that $M^{-1}A$ is still symmetric positive definite and has a smaller condition number than A; see Trefethen and Bau [Trefethen and Bau III (1997)] (Lecture 40) and Demmel [Demmel (1997)] (Section 6.6.5).

The method of conjugate gradients can be generalized to functionals that are not necessarily quadratic. The stepsize parameter ρ_k is still determined by a line search which consists in finding ρ_k such that

$$J(u_k - \rho_k d_k) = \inf_{\rho \in \mathbb{R}} J(u_k - \rho d_k).$$

This is more difficult than in the quadratic case and in general there is no guarantee that ρ_k is unique, so some criterion to pick ρ_k is needed. Then

$$u_{k+1} = u_k - \rho_k d_k,$$

and the next descent direction can be chosen in two ways:

(1) (*Polak–Ribière*)

$$d_k = \nabla J_{u_k} + \frac{\langle \nabla J_{u_k}, \nabla J_{u_k} - \nabla J_{u_{k-1}} \rangle}{\left\|\nabla J_{u_{k-1}}\right\|^2} d_{k-1},$$

(2) (*Fletcher–Reeves*)

$$d_k = \nabla J_{u_k} + \frac{\left\|\nabla J_{u_k}\right\|^2}{\left\|\nabla J_{u_{k-1}}\right\|^2} d_{k-1}.$$

Consecutive gradients are no longer orthogonal so these methods may run forever. There are various sufficient criteria for convergence. In practice, the Polak–Ribière method converges faster. There is no longer any guarantee that these methods converge to a global minimum.

13.11 Gradient Projection Methods for Constrained Optimization

We now consider the problem of finding the minimum of a convex functional $J \colon V \to \mathbb{R}$ over a nonempty, convex, closed subset U of a Hilbert space V. By Theorem 4.5(3), the functional J has a minimum at $u \in U$ iff

$$dJ_u(v - u) \geq 0 \quad \text{for all } v \in U,$$

which can be expressed as

$$\langle \nabla J_u, v - u\rangle \geq 0 \quad \text{for all } v \in U.$$

On the other hand, by the projection lemma (Proposition 12.4), the condition for a vector $u \in U$ to be the projection of an element $w \in V$ onto U is

$$\langle u - w, v - u\rangle \geq 0 \quad \text{for all } v \in U.$$

These conditions are obviously analogous, and we can make this analogy more precise as follows. If $p_U : V \to U$ is the projection map onto U, we have the following chain of equivalences:

$u \in U$ and $J(u) = \inf_{v \in U} J(v)$ iff

$u \in U$ and $\langle \nabla J_u, v - u\rangle \geq 0$ for every $v \in U$, iff

$u \in U$ and $\langle u - (u - \rho \nabla J_u), v - u\rangle \geq 0$ for every $v \in U$ and every $\rho > 0$,

iff $u = p_U(u - \rho \nabla J_u)$ for every $\rho > 0$.

In other words, for every $\rho > 0$, $u \in V$ is a *fixed-point* of the function $g : V \to U$ given by

$$g(v) = p_U(v - \rho \nabla J_v).$$

The above suggests finding u by the method of successive approximations for finding the fixed-point of a contracting mapping, namely given any initial $u_0 \in V$, to define the sequence $(u_k)_{k \geq 0}$ such that

$$u_{k+1} = p_U(u_k - \rho_k \nabla J_{u_k}),$$

where the parameter $\rho_k > 0$ is chosen at each step. This method is called the *projected-gradient method with variable stepsize parameter*. Observe that if $U = V$, then this is just the gradient method with variable stepsize. We have the following result about the convergence of this method.

Proposition 13.10. *Let $J : V \to \mathbb{R}$ be a continuously differentiable functional defined on a Hilbert space V, and let U be nonempty, convex, closed subset of V. Suppose there exists two constants $\alpha > 0$ and $M > 0$ such that*

$$\langle \nabla J_v - \nabla J_u, v - u\rangle \geq \alpha \left\|v - u\right\|^2 \quad \textit{for all } u, v \in V,$$

and

$$\left\|\nabla J_v - \nabla J_u\right\| \leq M \left\|v - u\right\| \quad \textit{for all } u, v \in V.$$

If there exists two real numbers $a, b \in \mathbb{R}$ such that

$$0 < a \leq \rho_k \leq b \leq \frac{2\alpha}{M^2} \quad \textit{for all } k \geq 0,$$

then the projected-gradient method with variable stepsize parameter converges. Furthermore, there is some constant $\beta > 0$ (depending on α, M, a, b) such that

$$\beta < 1 \quad \textit{and} \quad \|u_k - u\| \leq \beta^k \|u_0 - u\|,$$

where $u \in M$ is the unique minimum of J.

Proof. For every $\rho_k \geq 0$, define the function $g_k \colon V \to U$ by

$$g_k(v) = p_U(v - \rho_k \nabla J_v).$$

By Proposition 12.5, the projection map p_U has Lipschitz constant 1, so using the inequalities assumed to hold in the proposition, we have

$$\begin{aligned}
\|g_k(v_1) - g_k(v_2)\|^2 &= \|p_U(v_1 - \rho_k \nabla J_{v_1}) - p_U(v_2 - \rho_k \nabla J_{v_2})\|^2 \\
&\leq \|(v_1 - v_2) - \rho_k(\nabla J_{v_1} - \nabla J_{v_2})\|^2 \\
&= \|v_1 - v_2\|^2 - 2\rho_k \langle \nabla J_{v_1} - \nabla J_{v_2}, v_1 - v_2 \rangle \\
&\quad + \rho_k^2 \|\nabla J_{v_1} - \nabla J_{v_2}\|^2 \\
&\leq \left(1 - 2\alpha\rho_k + M^2\rho_k^2\right) \|v_1 - v_2\|^2.
\end{aligned}$$

As in the proof of Proposition 13.6, we know that if a and b satisfy the conditions $0 < a \leq \rho_k \leq b \leq \frac{2\alpha}{M^2}$, then there is some β such that

$$\left(1 - 2\alpha\rho_k + M^2\rho_k^2\right)^{1/2} \leq \beta < 1 \quad \text{for all } k \geq 0.$$

Since the minimizing point $u \in U$ is a fixed point of g_k for all k, by letting $v_1 = u_k$ and $v_2 = u$, we get

$$\|u_{k+1} - u\| = \|g_k(u_k) - g_k(u)\| \leq \beta \|u_k - u\|,$$

which proves the convergence of the sequence $(u_k)_{k\geq 0}$. ☐

In the case of an elliptic quadratic functional

$$J(v) = \frac{1}{2}\langle Av, a\rangle - \langle b, v\rangle$$

defined on $\mathbb{R}^n$, the reasoning just after the proof of Proposition 13.6 can be immediately adapted to show that convergence takes place as long as a, b and ρ_k are chosen such that

$$0 < a \leq \rho_k \leq b \leq \frac{2}{\lambda_n}.$$

In theory, Proposition 13.10 gives a guarantee of the convergence of the projected-gradient method. Unfortunately, because computing the projection $p_U(v)$ effectively is generally impossible, the range of practical applications of Proposition 13.10 is rather limited. One exception is the case where U is a product $\prod_{i=1}^m [a_i, b_i]$ of closed intervals (where $a_i = -\infty$ or $b_i = +\infty$ is possible). In this case, it is not hard to show that

$$p_U(w)_i = \begin{cases} a_i & \text{if } w_i < a_i \\ w_i & \text{if } a_i \leq w_i \leq b_i \\ b_i & \text{if } b_i < w_i. \end{cases}$$

In particular, this is the case if

$$U = \mathbb{R}^n_+ = \{v \in \mathbb{R}^n \mid v \geq 0\}$$

and if

$$J(v) = \frac{1}{2}\langle Av, a\rangle - \langle b, v\rangle$$

is an elliptic quadratic functional on $\mathbb{R}^n$. Then the vector $u_{k+1} = (u_1^{k+1}, \ldots, u_n^{k+1})$ is given in terms of $u_k = (u_1^k, \ldots, u_n^k)$ by

$$u_i^{k+1} = \max\{u_i^k - \rho_k(Au_k - b)_i, 0\}, \quad 1 \leq i \leq n.$$

13.12 Penalty Methods for Constrained Optimization

In the case where $V = \mathbb{R}^n$, another method to deal with constrained optimization is to incorporate the domain U into the objective function J by adding a penalty function.

Definition 13.11. Given a nonempty closed convex subset U of $\mathbb{R}^n$, a function $\psi\colon \mathbb{R}^n \to \mathbb{R}$ is called a *penalty function* for U if ψ is convex and continuous and if the following conditions hold:

$$\psi(v) \geq 0 \quad \text{for all} \quad v \in \mathbb{R}^n, \quad \text{and} \quad \psi(v) = 0 \quad \text{iff} \quad v \in U.$$

The following proposition shows that the use of penalty functions reduces a constrained optimization problem to a sequence of unconstrained optimization problems.

Proposition 13.11. *Let $J\colon \mathbb{R}^n \to \mathbb{R}$ be a continuous, coercive, strictly convex function, U be a nonempty, convex, closed subset of $\mathbb{R}^n$, $\psi\colon \mathbb{R}^n \to \mathbb{R}$ be a penalty function for U, and let $J_\epsilon\colon \mathbb{R}^n \to \mathbb{R}$ be the penalized objective function given by*

$$J_\epsilon(v) = J(v) + \frac{1}{\epsilon}\psi(v) \quad \textit{for all } v \in \mathbb{R}^n.$$

Then for every $\epsilon > 0$, there exists a unique element $u_\epsilon \in \mathbb{R}^n$ such that

$$J_\epsilon(u_\epsilon) = \inf_{v \in \mathbb{R}^n} J_\epsilon(v).$$

Furthermore, if $u \in U$ is the unique minimizer of J over U, so that $J(u) = \inf_{v \in U} J(v)$, then

$$\lim_{\epsilon \mapsto 0} u_\epsilon = u.$$

Proof. Observe that since J is coercive, since $\psi(v) \geq 0$ for all $v \in \mathbb{R}^n$, and $J_\epsilon = J + (1/\epsilon)\psi$, we have $J_\epsilon(v) \geq J(v)$ for all $v \in \mathbb{R}^n$, so J_ϵ is also coercive. Since J is strictly convex and $(1/\epsilon)\psi$ is convex, it is immediately checked that $J_\epsilon = J + (1/\epsilon)\psi$ is also strictly convex. Then by Proposition 13.1 (and the fact that J and J_ϵ are strictly convex), J has a unique minimizer $u \in U$, and J_ϵ has a unique minimizer $u_\epsilon \in \mathbb{R}^n$.

Since $\psi(u) = 0$ iff $u \in U$, and $\psi(v) \geq 0$ for all $v \in \mathbb{R}^n$, we have $J_\epsilon(u) = J(u)$, and since u_ϵ is the minimizer of J_ϵ we have $J_\epsilon(u_\epsilon) \leq J_\epsilon(u)$, so we obtain

$$J(u_\epsilon) \leq J(u_\epsilon) + \frac{1}{\epsilon}\,\psi(u_\epsilon) = J_\epsilon(u_\epsilon) \leq J_\epsilon(u) = J(u),$$

that is,

$$J_\epsilon(u_\epsilon) \leq J(u). \qquad (*_1)$$

Since J is coercive, the family $(u_\epsilon)_{\epsilon>0}$ is bounded. By compactness (since we are in $\mathbb{R}^n$), there exists a subsequence $(u_{\epsilon(i)})_{i\geq 0}$ with $\lim_{i\mapsto\infty} \epsilon(i) = 0$ and some element $u' \in \mathbb{R}^n$ such that

$$\lim_{i\mapsto\infty} u_{\epsilon(i)} = u'.$$

From the inequality $J(u_\epsilon) \leq J(u)$ proven in $(*_1)$ and the continuity of J, we deduce that

$$J(u') = \lim_{i\mapsto\infty} J(u_{\epsilon(i)}) \leq J(u). \qquad (*_2)$$

By definition of $J_\epsilon(u_\epsilon)$ and $(*_1)$, we have

$$0 \leq \psi(u_{\epsilon(i)}) \leq \epsilon(i)(J(u) - J(u_{\epsilon(i)})),$$

and since the sequence $(u_{\epsilon(i)})_{i\geq 0}$ converges, the numbers $J(u) - J(u_{\epsilon(i)})$ are bounded independently of i. Consequently, since $\lim_{i\mapsto\infty} \epsilon(i) = 0$ and since the function ψ is continuous, we have

$$0 = \lim_{i\mapsto\infty} \psi(u_{\epsilon(i)}) = \psi(u'),$$

which shows that $u' \in U$. Since by $(*_2)$ we have $J(u') \leq J(u)$, and since both $u, u' \in U$ and u is the unique minimizer of J over U we must have $u' = u$. Therefore u' is the unique minimizer of J over U. But then the whole family $(u_\epsilon)_{\epsilon>0}$ converges to u since we can use the same argument as above for *every* subsequence of $(u_\epsilon)_{\epsilon>0}$. ☐

Note that a convex function $\psi\colon \mathbb{R}^n \to \mathbb{R}$ is automatically continuous, so the assumption of continuity is redundant.

As an application of Proposition 13.11, if U is given by

$$U = \{v \in \mathbb{R}^n \mid \varphi_i(v) \leq 0,\ i = 1, \ldots, m\},$$

where the functions $\varphi_i\colon \mathbb{R}^n \to \mathbb{R}$ are convex, we can take ψ to be the function given by

$$\psi(v) = \sum_{i=1}^{m} \max\{\varphi_i(v), 0\}.$$

In practice, the applicability of the penalty-function method is limited by the difficulty to construct effectively "good" functions ψ, for example, differentiable ones. Note that in the above example the function ψ is not differentiable. A better penalty function is

$$\psi(v) = \sum_{i=1}^{m} (\max\{\varphi_i(v), 0\})^2.$$

Another way to deal with constrained optimization problems is to use *duality*. This approach is investigated in Chapter 14.

13.13 Summary

The main concepts and results of this chapter are listed below:

- Minimization, minimizer.
- Coercive functions.
- Minima of quadratic functionals.
- The theorem of Lions and Stampacchia.
- Lax–Milgram's theorem.
- Elliptic functionals.
- Descent direction, exact line search, backtracking line search.
- Method of relaxation.
- Gradient descent.
- Gradient descent method with fixed stepsize parameter.
- Gradient descent method with variable stepsize parameter.
- Steepest descent method for the Euclidean norm.
- Gradient descent method with backtracking line search.
- Normalized steepest descent direction.
- Unnormalized steepest descent direction.
- Steepest descent method (with respect to the norm $\| \ \|$).

- Momentum term.
- Newton's method.
- Newton step.
- Newton decrement.
- Damped Newton phase.
- Quadratically convergent phase.
- Self-concordant functions.
- Conjugate gradient method.
- Projected gradient methods.
- Penalty methods.

13.14 Problems

Problem 13.1. Consider the function $J\colon \mathbb{R}^n \to \mathbb{R}$ given by

$$J(v) = \frac{1}{2}\langle Av, v\rangle - \langle b, v\rangle + g(v),$$

where A is a real $n \times n$ symmetric positive definite matrix, $b \in \mathbb{R}^n$, and $g\colon \mathbb{R}^n \to \mathbb{R}$ is a continuous (not necessarily differentiable) convex function such that $g(v) \geq 0$ for all $v \in \mathbb{R}^n$, and let U be a nonempty, bounded, closed, convex subset of $\mathbb{R}^n$.

(1) Prove that there is a unique element $u \in U$ such that

$$J(u) = \inf_{v \in U} J(v).$$

Hint. Prove that J is strictly convex on $\mathbb{R}^n$.

(2) Check that

$$J(v) - J(u) = \langle Au - b, v - u\rangle + g(v) - g(u) + \frac{1}{2}\langle A(v-u), v-u\rangle.$$

Prove that an element $u \in U$ minimizes J in U iff

$$\langle Au - b, v - u\rangle + g(v) - g(u) \geq 0 \quad \text{for all } v \in U.$$

Problem 13.2. Consider n piecewise C^1 functions $\varphi_i\colon [0,1] \to \mathbb{R}$ and assume that these functions are linearly independent and that

$$\sum_{i=1}^{n} \varphi_i(x) = 1 \quad \text{for all } x \in [0,1].$$

Let $J\colon \mathbb{R}^n \to \mathbb{R}$ be the function given by

$$J(v) = \sum_{i,j=1}^{n} a_{iij} v_i v_j + \sum_{i=1}^{n} b_i v_i,$$

where $v = (v_1, \dots, v_n)$ and

$$a_{ij} = \int_0^1 \varphi_i'(t)\varphi_j'(t)dt, \quad b_i = \int_0^1 \varphi_i(t)dt.$$

(1) Let U_1 be the subset of $\mathbb{R}^n$ given by

$$U_1 = \left\{ v \in \mathbb{R}^n \mid \sum_{i=1}^n b_i v_i = 0 \right\}.$$

Consider the problem of finding a minimum of J over U_1. Prove that the Lagrange multiplier λ for which the Lagrangian has a critical point is $\lambda = -1$.

(2) Prove that the map defined on U_1 by

$$\|v\| = \left(\int_0^1 \left(\sum_{i=1}^n v_i \varphi_i'(x) \right)^2 dx \right)^{1/2}$$

is a norm. Prove that J is elliptic on U_1 with this norm. Prove that J has a unique minimum on U_1.

(3) Consider the the subset of $\mathbb{R}^n$ given by

$$U_2 = \left\{ v \in \mathbb{R}^n \mid \sum_{i=1}^n (\varphi_i(1) + \varphi_i(0)) v_i = 0 \right\}.$$

Consider the problem of finding a minimum of J over U_2. Prove that the Lagrange multiplier λ for which the Lagrangian has a critical point is $\lambda = -1/2$. Prove that J is elliptic on U_2 with the same norm as in (2). Prove that J has a unique minimum on U_2.

(4) Consider the the subset of $\mathbb{R}^n$ given by

$$U_3 = \left\{ v \in \mathbb{R}^n \mid \sum_{i=1}^n (\varphi_i(1) - \varphi_i(0)) v_i = 0 \right\}.$$

This time show that the necessary condition for having a minimum on U_3 yields the equation $1 + \lambda(1 - 1) = 0$. Conclude that J does not have a minimum on U_3.

Problem 13.3. Let A be a real $n \times n$ symmetric positive definite matrix and let $b \in \mathbb{R}^n$.

(1) Prove that if we apply the steepest descent method (for the Euclidean norm) to

$$J(v) = \frac{1}{2}\langle Av, v \rangle - \langle b, v \rangle,$$

and if we define the norm $\|v\|_A$ by

$$\|v\|_A = \langle Av, v\rangle^{1/2},$$

we get the inequality

$$\|u_{k+1} - u\|_A^2 \leq \|u_k - u\|_A^2 \left(1 - \frac{\|A(u_k - u)\|_2^4}{\|u_k - u\|_A^2 \|A(u_k - u)\|_A^2}\right).$$

(2) Using a diagonalization of A, where the eigenvalues of A are denoted $0 < \lambda_1 \leq \lambda_2 \leq \cdots \leq \lambda_n$, prove that

$$\|u_{k+1} - u\|_A \leq \frac{\mathrm{cond}_2(A) - 1}{\mathrm{cond}_2(A) + 1} \|u_k - u\|_A,$$

where $\mathrm{cond}_2(A) = \lambda_n/\lambda_1$, and thus

$$\|u_k - u\|_A \leq \left(\frac{\mathrm{cond}_2(A) - 1}{\mathrm{cond}_2(A) + 1}\right)^k \|u_0 - u\|_A.$$

(3) Prove that when $\mathrm{cond}_2(A) = 1$, then $A = I$ and the method converges in one step.

Problem 13.4. Prove that the method of Polak–Ribière converges if $J\colon \mathbb{R}^n \to \mathbb{R}$ is elliptic and a C^2 function.

Problem 13.5. Prove that the backtracking line search method described in Section 13.5 has the property that for t small enough the condition $J(u_k + td_k) \leq J(u_k) + \alpha t\langle \nabla J_{u_k}, d_k\rangle$ will hold and the search will stop. Prove that the exit inequality $J(u_k + td_k) \leq J(u_k) + \alpha t\langle \nabla J_{u_k}, d_k\rangle$ holds for all $t \in (0, t_0]$, for some $t_0 > 0$, so the backtracking line search stops with a step length ρ_k that satisfies $\rho_k = 1$ or $\rho_k \in (\beta t_0, t_0]$.

Problem 13.6. Let $d_{\mathrm{nsd},k}$ and $d_{\mathrm{sd},k}$ be the normalized and unnormalized descent directions of the steepest descent method for an arbitrary norm (see Section 13.8). Prove that

$$\begin{aligned}
\langle \nabla J_{u_k}, d_{\mathrm{nsd},k}\rangle &= -\left\|\nabla J_{u_k}\right\|^D \\
\langle \nabla J_{u_k}, d_{\mathrm{sd},k}\rangle &= -(\left\|\nabla J_{u_k}\right\|^D)^2 \\
d_{\mathrm{sd},k} &= \arg\min_v \left(\langle \nabla J_{u_k}, v\rangle + \frac{1}{2}\|v\|^2\right).
\end{aligned}$$

Problem 13.7. If P is a symmetric positive definite matrix, prove that $\|z\|_P = (z^\top P z)^{1/2} = \left\|P^{1/2} z\right\|_2$ is a norm. Prove that the normalized steepest descent direction is

$$d_{\mathrm{nsd,k}} = -(\nabla J_{u_k}^\top P^{-1} \nabla J_{u_k})^{-1/2} P^{-1} \nabla J_{u_k},$$

the dual norm is $\|z\|^D = \left\|P^{-1/2}z\right\|_2$, and the steepest descent direction with respect to $\| \|_P$ is given by

$$d_{\mathrm{sd},k} = -P^{-1}\nabla J_{u_k}.$$

Problem 13.8. If $\| \|$ is the ℓ^1-norm, then show that $d_{\mathrm{nsd,k}}$ is determined as follows: let i be any index for which $\|\nabla J_{u_k}\|_\infty = |(\nabla J_{u_k})_i|$. Then

$$d_{\mathrm{nsd,k}} = -\mathrm{sign}\left(\frac{\partial J}{\partial x_i}(u_k)\right) e_i,$$

where e_i is the ith canonical basis vector, and

$$d_{\mathrm{sd,k}} = -\frac{\partial J}{\partial x_i}(u_k)e_i.$$

Problem 13.9. (From Boyd and Vandenberghe [Boyd and Vandenberghe (2004)], Problem 9.12). If $\nabla^2 f(x)$ is singular (or very ill-conditioned), the Newton step $d_{\mathrm{nt}} = -(\nabla^2 J(x))^{-1}\nabla J_x$ is not well defined. Instead we can define a search direction d_{tr} as the solution of the problem

$$\begin{aligned} &\text{minimize} && (1/2)\langle Hv, v\rangle + \langle g, v\rangle \\ &\text{subject to} && \|v\|_2 \leq \gamma, \end{aligned}$$

where $H = \nabla^2 f_x$, $g = \nabla f_x$, and γ is some positive constant. The idea is to use a *trust region*, which is the closed ball $\{v \mid \|v\|_2 \leq \gamma\}$. The point $x + d_{\mathrm{tr}}$ minimizes the second-order approximation of f at x, subject to the constraint that

$$\|x + d_{\mathrm{tr}} - x\|_2 \leq \gamma.$$

The parameter γ, called the *trust parameter*, reflects our confidence in the second-order approximation.

Prove that d_{tr} minimizes

$$\frac{1}{2}\langle Hv, v\rangle + \langle g, v\rangle + \widehat{\beta}\,\|v\|_2^2,$$

for some $\widehat{\beta}$.

Problem 13.10. (From Boyd and Vandenberghe [Boyd and Vandenberghe (2004)], Problem 9.9). Prove that the Newton decrement $\lambda(x)$ is given by

$$\lambda(x) = \sup_{v \neq 0} -\frac{\langle \nabla J_x, v\rangle}{(\langle \nabla^2 J_x v, v\rangle)^{1/2}}.$$

Problem 13.11. Show that the function f given by $f(x) = \log(e^x + e^{-x})$ has a unique minimum for $x^* = 0$. Run Newton's method with fixed step size $t = 1$, starting with $x_0 = 1$, and then $x_0 = 1.1$. What do you observe?

Problem 13.12. Write a `Matlab` program implementing the conjugate gradient method. Test your program with the $n \times n$ matrix

$$A_n = \begin{pmatrix} 2 & -1 & 0 & \cdots & 0 \\ -1 & 2 & -1 & \dots & 0 \\ \vdots & \ddots & \ddots & \ddots & \vdots \\ 0 & \cdots & -1 & 2 & -1 \\ 0 & \cdots & 0 & -1 & 2 \end{pmatrix}$$

and various right-hand sides, for various values of n. Verify that the running time is $O(n^{3/2})$.

Chapter 14

Introduction to Nonlinear Optimization

This chapter contains the most important results of nonlinear optimization theory.

In Chapter 4 we investigated the problem of determining when a function $J\colon \Omega \to \mathbb{R}$ defined on some open subset Ω of a normed vector space E has a local extremum in a subset U of Ω defined by equational constraints, namely

$$U = \{x \in \Omega \mid \varphi_i(x) = 0,\ 1 \leq i \leq m\},$$

where the functions $\varphi_i\colon \Omega \to \mathbb{R}$ are continuous (and usually differentiable). Theorem 4.1 gave a necessary condition in terms of the Lagrange multipliers. In Section 4.3 we assumed that U was a convex subset of Ω; then Theorem 4.4 gave us a necessary condition for the function $J\colon \Omega \to \mathbb{R}$ to have a local minimum at u with respect to U if dJ_u exists, namely

$$dJ_u(v - u) \geq 0 \quad \text{for all } v \in U.$$

Our first goal is to find a necessary criterion for a function $J\colon \Omega \to \mathbb{R}$ to have a minimum on a subset U, even is this subset is *not* convex. This can be done by introducing a notion of "tangent cone" at a point $u \in U$. We define the cone of feasible directions and then state a necessary condition for a function to have local minimum on a set U that is not necessarily convex in terms of the cone of feasible directions. The cone of feasible directions is not always convex, but it is if the constraints are inequality constraints. An inequality constraint $\varphi(u) \leq 0$ is said to be *active* is $\varphi(u) = 0$. One can also define the notion of *qualified constraint*. Theorem 14.1 gives necessary conditions for a function J to have a minimum on a subset U defined by qualified inequality constraints in terms of the Karush–Kuhn–Tucker conditions (for short KKT conditions), which involve nonnegative Lagrange multipliers. The proof relies on a version of the Farkas–Minkowski lemma.

Some of the KTT conditions assert that $\lambda_i \varphi_i(u) = 0$, where $\lambda_i \geq 0$ is the Lagrange multiplier associated with the constraint $\varphi_i \leq 0$. To some extent, this implies that active constraints are more important than inactive constraints, since if $\varphi_i(u) < 0$ is an inactive constraint, then $\lambda_i = 0$. In general, the KKT conditions are useless unless the constraints are convex. In this case, there is a manageable notion of qualified constraint given by Slater's conditions. Theorem 14.2 gives necessary conditions for a function J to have a minimum on a subset U defined by convex inequality constraints in terms of the Karush–Kuhn–Tucker conditions. Furthermore, if J is also convex and if the KKT conditions hold, then J has a global minimum.

In Section 14.4, we apply Theorem 14.2 to the special case where the constraints are equality constraints, which can be expressed as $Ax = b$. In the special case where the convex objective function J is a convex quadratic functional of the form

$$J(x) = \frac{1}{2} x^\top P x + q^\top x + r,$$

where P is a $n \times n$ symmetric positive semidefinite matrix, the necessary and sufficient conditions for having a minimum are expressed by a linear system involving a matrix called the KKT matrix. We discuss conditions that guarantee that the KKT matrix is invertible, and how to solve the KKT system. We also briefly discuss variants of Newton's method dealing with equality constraints.

We illustrate the KKT conditions on an interesting example, the so-called hard margin support vector machine; see Sections 14.5 and 14.6. The problem is a classification problem, or more accurately a separation problem. Suppose we have two nonempty disjoint finite sets of p *blue* points $\{u_i\}_{i=1}^p$ and q *red* points $\{v_j\}_{j=1}^q$ in $\mathbb{R}^n$. Our goal is to find a hyperplane H of equation $w^\top x - b = 0$ (where $w \in \mathbb{R}^n$ is a nonzero vector and $b \in \mathbb{R}$), such that all the blue points u_i are in one of the two open half-spaces determined by H, and all the red points v_j are in the other open half-space determined by H.

If the two sets are indeed separable, then in general there are infinitely many hyperplanes separating them. Vapnik had the idea to find a hyperplane that maximizes the smallest distance between the points and the hyperplane. Such a hyperplane is indeed unique and is called a maximal hard margin hyperplane, or hard margin support vector machine. The support vectors are those for which the constraints are active.

Section 14.7 contains the most important results of the chapter. The notion of Lagrangian duality is presented. Given a primal optimization problem (P) consisting in minimizing an objective function $J(v)$ with respect to some inequality constraints $\varphi_i(v) \leq 0$, $i = 1, \ldots, m$, we define the *dual function* $G(\mu)$ as the result of minimizing the Lagrangian

$$L(v, \mu) = J(v) + \sum_{i=1}^{m} \mu_i \varphi_i(v)$$

with respect to v, with $\mu \in \mathbb{R}^m_+$. The dual program (D) is then to maximize $G(\mu)$ with respect to $\mu \in \mathbb{R}^m_+$. It turns out that G is a concave function, and the dual program is an unconstrained maximization. This is actually a misleading statement because G is generally a partial function, so maximizing $G(\mu)$ is equivalent to a constrained maximization problem in which the constraints specify the domain of G, but in many cases, we obtain a dual program simpler than the primal program. If d^* is the optimal value of the dual program and if p^* is the optimal value of the primal program, we always have

$$d^* \leq p^*,$$

which is known as *weak duality*. Under certain conditions, $d^* = p^*$, that is, the duality gap is zero, in which case we say that *strong duality* holds. Also, under certain conditions, a solution of the dual yields a solution of the primal, and if the primal has an optimal solution, then the dual has an optimal solution, but beware that the converse is generally false (see Theorem 14.5). We also show how to deal with equality constraints, and discuss the use of conjugate functions to find the dual function. Our coverage of Lagrangian duality is quite thorough, but we do not discuss more general orderings such as the semidefinite ordering. For these topics which belong to convex optimization, the reader is referred to Boyd and Vandenberghe [Boyd and Vandenberghe (2004)].

Our approach in this chapter is very much inspired by Ciarlet [Ciarlet (1989)] because we find it one of the more direct, and it is general enough to accommodate Hilbert spaces. The field of nonlinear optimization and convex optimization is vast and there are many books on the subject. Among those we recommend (in alphabetic order) Bertsekas [Bertsekas (2009, 2015, 2016)], Bertsekas, Nedić, and Ozdaglar [Bertsekas *et al.* (2003)], Boyd and Vandenberghe [Boyd and Vandenberghe (2004)], Luenberger [Luenberger (1997)], and Luenberger and Ye [Luenberger and Ye (2016)].

14.1 The Cone of Feasible Directions

Let V be a normed vector space and let U be a nonempty subset of V. For any point $u \in U$, consider any converging sequence $(u_k)_{k \geq 0}$ of vectors

$u_k \in U$ having u as their limit, with $u_k \neq u$ for all $k \geq 0$, and look at the sequence of "unit chords,"

$$\frac{u_k - u}{\|u_k - u\|}.$$

This sequence could oscillate forever, or it could have a limit, some unit vector $\widehat{w} \in V$. In the second case, all nonzero vectors $\lambda\widehat{w}$ for all $\lambda > 0$, belong to an object called the cone of feasible directions at u. First, we need to define the notion of cone.

Definition 14.1. Given a (real) vector space V, a nonempty subset $C \subseteq V$ is a *cone with apex* 0 (for short, a *cone*), if for any $v \in V$, if $v \in C$, then $\lambda v \in C$ for all $\lambda > 0$ ($\lambda \in \mathbb{R}$). For any $u \in V$, a *cone with apex* u is any nonempty subset of the form $u + C = \{u + v \mid v \in C\}$, where C is a cone with apex 0; see Figure 14.1.

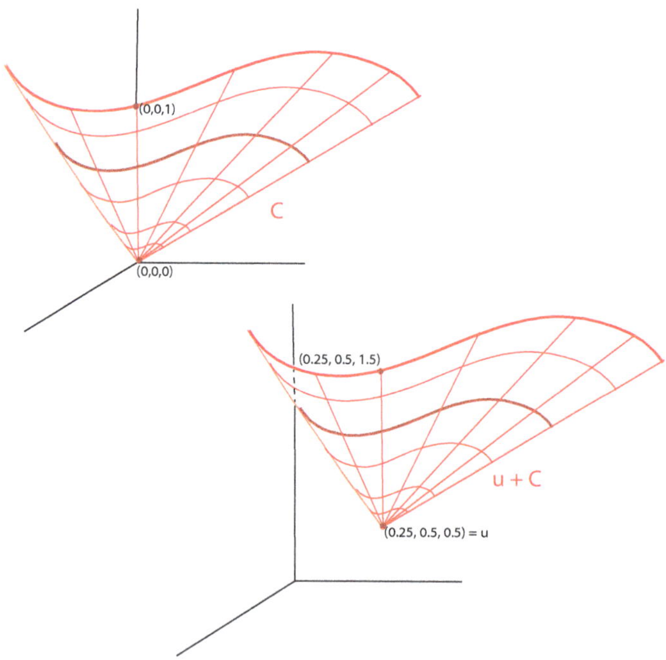

Fig. 14.1 Let C be the cone determined by the bold orange curve through $(0, 0, 1)$ in the plane $z = 1$. Then $u + C$, where $u = (0.25, 0.5, 0.5)$, is the affine translate of C via the vector u.

Observe that a cone with apex 0 (or u) is not necessarily convex, and that 0 does not necessarily belong to C (resp. u does not necessarily belong to $u + C$) (although in the case of the cone of feasible directions $C(u)$ we have $0 \in C(u)$). The condition for being a cone only asserts that if a nonzero vector v belongs to C, then the open ray $\{\lambda v \mid \lambda > 0\}$ (resp. the affine open ray $u + \{\lambda v \mid \lambda > 0\}$) also belongs to C.

Definition 14.2. Let V be a normed vector space and let U be a nonempty subset of V. For any point $u \in U$, the *cone $C(u)$ of feasible directions at u* is the union of $\{0\}$ and the set of all nonzero vectors $w \in V$ for which there exists some convergent sequence $(u_k)_{k\geq 0}$ of vectors such that

(1) $u_k \in U$ and $u_k \neq u$ for all $k \geq 0$, and $\lim_{k\mapsto\infty} u_k = u$.

(2) $\lim_{k\mapsto\infty} \dfrac{u_k - u}{\|u_k - u\|} = \dfrac{w}{\|w\|}$, with $w \neq 0$.

Condition (2) can also be expressed as follows: there is a sequence $(\delta_k)_{k\geq 0}$ of vectors $\delta_k \in V$ such that

$$u_k = u + \|u_k - u\| \frac{w}{\|w\|} + \|u_k - u\| \delta_k, \quad \lim_{k\mapsto\infty} \delta_k = 0, \;\; w \neq 0.$$

Figure 14.2 illustrates the construction of w in $C(u)$.

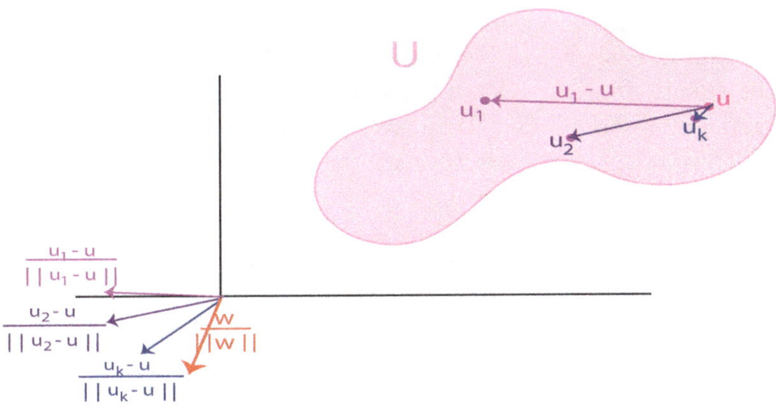

Fig. 14.2 Let U be the pink region in $\mathbb{R}^2$ with fuchsia point $u \in U$. For any sequence $(u_k)_{k\geq 0}$ of points in U which converges to u, form the chords $u_k - u$ and take the limit to construct the red vector w.

Clearly, the cone $C(u)$ of feasible directions at u is a cone with apex 0, and $u+C(u)$ is a cone with apex u. Obviously, it would be desirable to have conditions on U that imply that $C(u)$ is a convex cone. Such conditions will be given later on.

Observe that the cone $C(u)$ of feasible directions at u contains the velocity vectors at u of all curves γ in U through u. If $\gamma\colon (-1,1) \to U$ is such a curve with $\gamma(0) = u$, and if $\gamma'(u) \neq 0$ exists, then there is a sequence $(u_k)_{k\geq 0}$ of vectors in U converging to u as in Definition 14.2, with $u_k = \gamma(t_k)$ for some sequence $(t_k)_{k\geq 0}$ of reals $t_k > 0$ such that $\lim_{k\mapsto\infty} t_k = 0$, so that

$$u_k - u = t_k\gamma'(0) + t_k\epsilon_k, \quad \lim_{k\mapsto\infty} \epsilon_k = 0,$$

and we get

$$\lim_{k\mapsto\infty} \frac{u_k - u}{\|u_k - u\|} = \frac{\gamma'(0)}{\|\gamma'(0)\|}.$$

For an illustration of this paragraph in $\mathbb{R}^2$, see Figure 14.3.

Example 14.1. In $V = \mathbb{R}^2$, let φ_1 and φ_2 be given by

$$\begin{aligned}\varphi_1(u_1, u_2) &= -u_1 - u_2 \\ \varphi_2(u_1, u_2) &= u_1(u_1^2 + u_2^2) - (u_1^2 - u_2^2),\end{aligned}$$

and let

$$U = \{(u_1, u_2) \in \mathbb{R}^2 \mid \varphi_1(u_1, u_2) \leq 0,\ \varphi_2(u_1, u_2) \leq 0\}.$$

The region U is shown in Figure 14.4 and is bounded by the curve given by the equation $\varphi_1(u_1, u_2) = 0$, that is, $-u_1 - u_2 = 0$, the line of slope -1 through the origin, and the curve given by the equation $u_1(u_1^2 + u_2^2) - (u_1^2 - u_2^2) = 0$, a nodal cubic through the origin. We obtain a parametric definition of this curve by letting $u_2 = tu_1$, and we find that

$$u_1(t) = \frac{u_1^2(t) - u_2^2(t)}{u_1^2(t) + u_2^2(t)} = \frac{1 - t^2}{1 + t^2}, \quad u_2(t) = \frac{t(1 - t^2)}{1 + t^2}.$$

The tangent vector at t is given by $(u_1'(t), u_2'(t))$ with

$$u_1'(t) = \frac{-2t(1 + t^2) - (1 - t^2)2t}{(1 + t^2)^2} = \frac{-4t}{(1 + t^2)^2}$$

and

$$\begin{aligned}u_2'(t) &= \frac{(1 - 3t^2)(1 + t^2) - (t - t^3)2t}{(1 + t^2)^2} = \frac{1 - 2t^2 - 3t^4 - 2t^2 + 2t^4}{(1 + t^2)^2} \\ &= \frac{1 - 4t^2 - t^4}{(1 + t^2)^2}.\end{aligned}$$

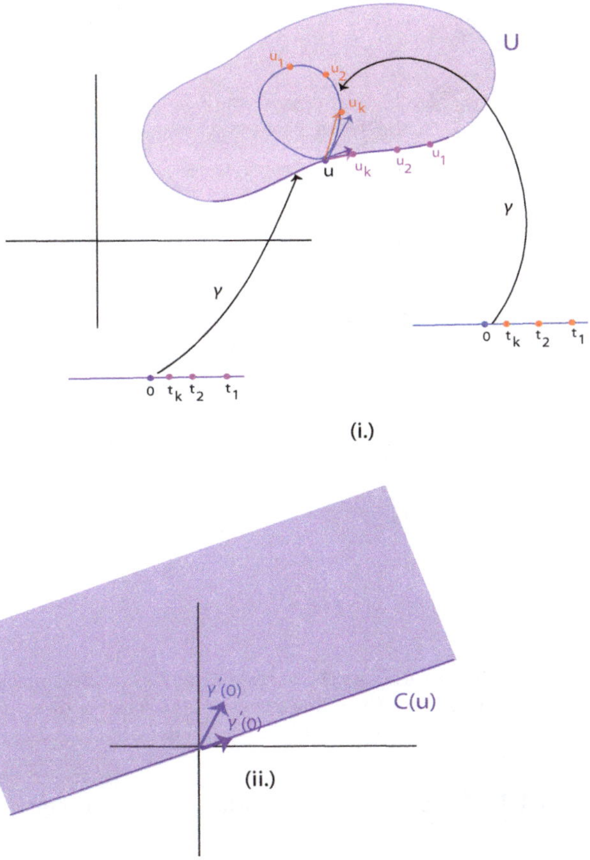

Fig. 14.3 Let U be purple region in $\mathbb{R}^2$ and u be the designated point on the boundary of U. Figure (i.) illustrates two curves through u and two sequences $(u_k)_{k\geq 0}$ converging to u. The limit of the chords $u_k - u$ corresponds to the tangent vectors for the appropriate curve. Figure (ii.) illustrates the half plane $C(u)$ of feasible directions.

The nodal cubic passes through the origin for $t = \pm 1$, and for $t = -1$ the tangent vector is $(1, -1)$, and for $t = 1$ the tangent vector is $(-1, -1)$. The cone of feasible directions $C(0)$ at the origin is given by

$$C(0) = \{(u_1, u_2) \in \mathbb{R}^2 \mid u_1 + u_2 \geq 0,\ |u_1| \geq |u_2|\}.$$

This is not a convex cone since it contains the sector delineated by the lines $u_2 = u_1$ and $u_2 = -u_1$, but also the ray supported by the vector $(-1, 1)$.

The two crucial properties of the cone of feasible directions are shown in the following proposition.

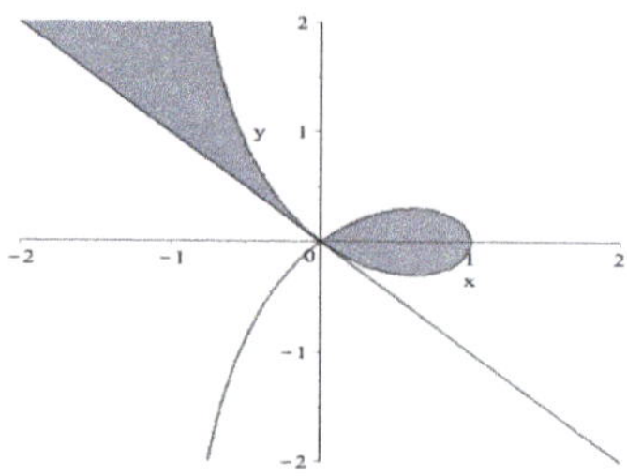

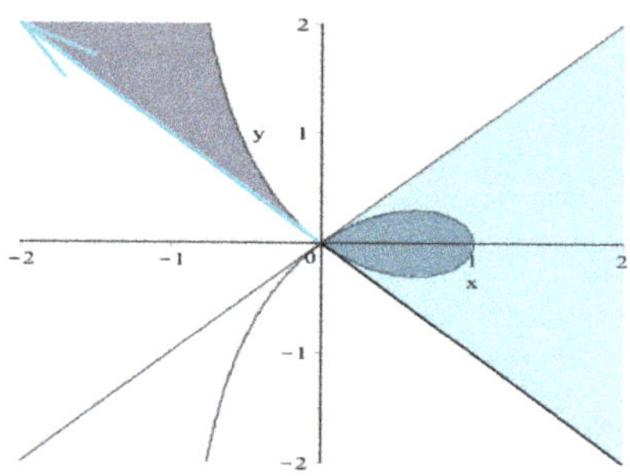

Fig. 14.4 Figure (i.) illustrates U as the shaded gray region which lies between the line $y = -x$ and nodal cubic. Figure (ii.) shows the cone of feasible directions, $C(0)$, as the union of the turquoise triangular cone and the turquoise directional ray $(-1, 1)$.

Proposition 14.1. *Let U be any nonempty subset of a normed vector space V.*

(1) For any $u \in U$, the cone $C(u)$ of feasible directions at u is closed.
(2) Let $J \colon \Omega \to \mathbb{R}$ be a function defined on an open subset Ω containing U. If J has a local minimum with respect to the set U at a point $u \in U$, and if J'_u exists at u, then

$$J'_u(v - u) \geq 0 \quad \text{for all } v \in u + C(u).$$

Proof. (1) Let $(w_n)_{n \geq 0}$ be a sequence of vectors $w_n \in C(u)$ converging to a limit $w \in V$. We may assume that $w \neq 0$, since $0 \in C(u)$ by definition, and thus we may also assume that $w_n \neq 0$ for all $n \geq 0$. By definition, for every $n \geq 0$, there is a sequence $(u^n_k)_{k \geq 0}$ of vectors in V and some $w_n \neq 0$ such that

(1) $u^n_k \in U$ and $u^n_k \neq u$ for all $k \geq 0$, and $\lim_{k \mapsto \infty} u^n_k = u$.

(2) There is a sequence $(\delta_k^n)_{k\geq 0}$ of vectors $\delta_k^n \in V$ such that

$$u_k^n = u + \|u_k^n - u\| \frac{w_n}{\|w_n\|} + \|u_k^n - u\| \, \delta_k^n, \quad \lim_{k\mapsto\infty} \delta_k^n = 0, \;\; w_n \neq 0.$$

Let $(\epsilon_n)_{n\geq 0}$ be a sequence of real numbers $\epsilon_n > 0$ such that $\lim_{n\mapsto\infty} \epsilon_n = 0$ (for example, $\epsilon_n = 1/(n+1)$). Due to the convergence of the sequences (u_k^n) and (δ_k^n) for every fixed n, there exist an integer $k(n)$ such that

$$\left\| u_{k(n)}^n - u \right\| \leq \epsilon_n, \qquad \left\| \delta_{k(n)}^n \right\| \leq \epsilon_n.$$

Consider the sequence $(u_{k(n)}^n)_{n\geq 0}$. We have

$$u_{k(n)}^n \in U, \;\; u_{k(n)}^n \neq 0, \;\; \text{for all } n \geq 0, \;\; \lim_{n\mapsto\infty} u_{k(n)}^n = u,$$

and we can write

$$u_{k(n)}^n = u + \left\| u_{k(n)}^n - u \right\| \frac{w}{\|w\|} + \left\| u_{k(n)}^n - u \right\| \left(\delta_{k(n)}^n + \left(\frac{w_n}{\|w_n\|} - \frac{w}{\|w\|} \right) \right).$$

Since $\lim_{k\mapsto\infty}(w_n/\|w_n\|) = w/\|w\|$, we conclude that $w \in C(u)$. See Figure 14.5.

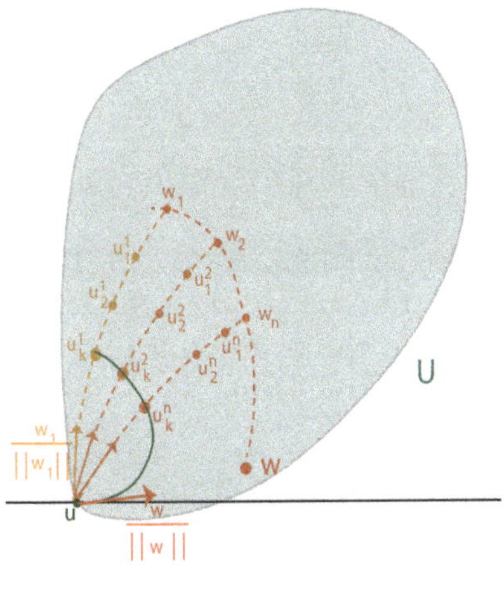

Fig. 14.5 Let U be the mint green region in $\mathbb{R}^2$ with $u = (0,0)$. Let $(w_n)_{n\geq 0}$ be a sequence of vectors (points) along the upper dashed curve which converge to w. By following the dashed orange longitudinal curves, and selecting an appropriate vector(point), we construct the dark green curve in U, which passes through u, and at u has tangent vector proportional to w.

(2) Let $w = v - u$ be any nonzero vector in the cone $C(u)$, and let $(u_k)_{k\,\geq 0}$ be a sequence of vectors in $U - \{u\}$ such that

(1) $\lim_{k\mapsto\infty} u_k = u$.
(2) There is a sequence $(\delta_k)_{k\geq 0}$ of vectors $\delta_k \in V$ such that

$$u_k - u = \|u_k - u\| \frac{w}{\|w\|} + \|u_k - u\| \, \delta_k, \quad \lim_{k\mapsto\infty} \delta_k = 0, \;\; w \neq 0,$$

(3) $J(u) \leq J(u_k)$ for all $k \geq 0$.

Since J is differentiable at u, we have

$$0 \leq J(u_k) - J(u) = J'_u(u_k - u) + \|u_k - u\| \, \epsilon_k, \tag{$*$}$$

for some sequence $(\epsilon_k)_{k\geq 0}$ such that $\lim_{k\mapsto\infty} \epsilon_k = 0$. Since J'_u is linear and continuous, and since

$$u_k - u = \|u_k - u\| \frac{w}{\|w\|} + \|u_k - u\| \, \delta_k, \quad \lim_{k\mapsto\infty} \delta_k = 0, \;\; w \neq 0,$$

$(*)$ implies that

$$0 \leq \frac{\|u_k - u\|}{\|w\|} \, (J'_u(w) + \eta_k),$$

with

$$\eta_k = \|w\| \, (J'_u(\delta_k) + \epsilon_k).$$

Since J'_u is continuous, we have $\lim_{k\mapsto\infty} \eta_k = 0$. But then $J'_u(w) \geq 0$, since if $J'_u(w) < 0$, then for k large enough the expression $J'_u(w) + \eta_k$ would be negative, and since $u_k \neq u$, the expression $(\|u_k - u\| \,/\, \|w\|)(J'_u(w) + \eta_k)$ would also be negative, a contradiction. □

From now on we assume that U is defined by a set of inequalities, that is

$$U = \{x \in \Omega \mid \varphi_i(x) \leq 0, \;\; 1 \leq i \leq m\},$$

where the functions $\varphi_i \colon \Omega \to \mathbb{R}$ are continuous (and usually differentiable). As we explained earlier, an equality constraint $\varphi_i(x) = 0$ is treated as the conjunction of the two inequalities $\varphi_i(x) \leq 0$ and $-\varphi_i(x) \leq 0$. Later on we will see that when the functions φ_i are convex, since $-\varphi_i$ is not necessarily convex, it is desirable to treat equality constraints separately, but for the time being we won't.

14.2 Active Constraints and Qualified Constraints

Our next goal is find sufficient conditions for the cone $C(u)$ to be convex, for any $u \in U$. For this we assume that the functions φ_i are differentiable at u. It turns out that the constraints φ_i that matter are those for which $\varphi_i(u) = 0$, namely the constraints that are tight, or as we say, active.

Definition 14.3. Given m functions $\varphi_i \colon \Omega \to \mathbb{R}$ defined on some open subset Ω of some vector space V, let U be the set defined by

$$U = \{x \in \Omega \mid \varphi_i(x) \leq 0, \ 1 \leq i \leq m\}.$$

For any $u \in U$, a constraint φ_i is said to be *active* at u if $\varphi_i(u) = 0$, else *inactive* at u if $\varphi_i(u) < 0$.

If a constraint φ_i is active at u, this corresponds to u being on a piece of the boundary of U determined by some of the equations $\varphi_i(u) = 0$; see Figure 14.6.

Definition 14.4. For any $u \in U$, with

$$U = \{x \in \Omega \mid \varphi_i(x) \leq 0, \ 1 \leq i \leq m\},$$

we define $I(u)$ as the set of indices

$$I(u) = \{i \in \{1, \ldots, m\} \mid \varphi_i(u) = 0\}$$

where the constraints are active. We define the set $C^*(u)$ as

$$C^*(u) = \{v \in V \mid (\varphi'_i)_u(v) \leq 0, \ i \in I(u)\}.$$

Since each $(\varphi'_i)_u$ is a linear form, the subset

$$C^*(u) = \{v \in V \mid (\varphi'_i)_u(v) \leq 0, \ i \in I(u)\}$$

is the intersection of half spaces passing through the origin, so it is a convex set, and obviously it is a cone. If $I(u) = \emptyset$, then $C^*(u) = V$.

The special kinds of $\mathcal{H}$-polyhedra of the form $C^*(u)$ cut out by hyperplanes through the origin are called *$\mathcal{H}$-cones*. It can be shown that every $\mathcal{H}$-cone is a polyhedral cone (also called a *$\mathcal{V}$-cone*), and conversely. The proof is nontrivial; see Gallier [Gallier (2016)] and Ziegler [Ziegler (1997)].

We will prove shortly that we always have the inclusion

$$C(u) \subseteq C^*(u).$$

However, the inclusion can be strict, as in Example 14.1. Indeed for $u = (0, 0)$ we have $I(0, 0) = \{1, 2\}$ and since

$$(\varphi'_1)_{(u_1,u_2)} = (-1 \ \ -1), \quad (\varphi'_2)_{(u_1,u_2)} = (3u_1^2 + u_2^2 - 2u_1 \ \ 2u_1u_2 + 2u_2),$$

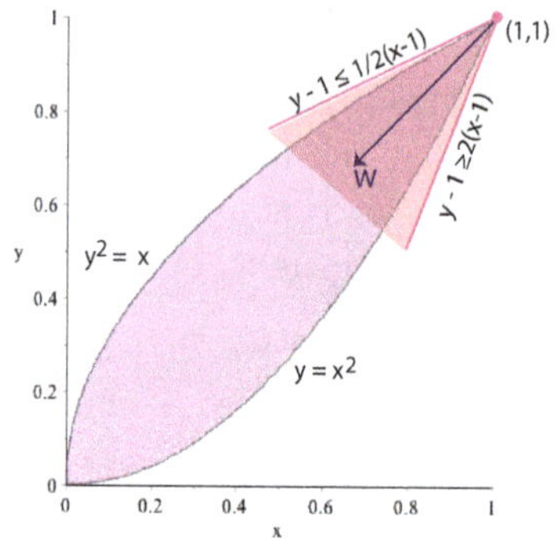

(i.)

(ii.)

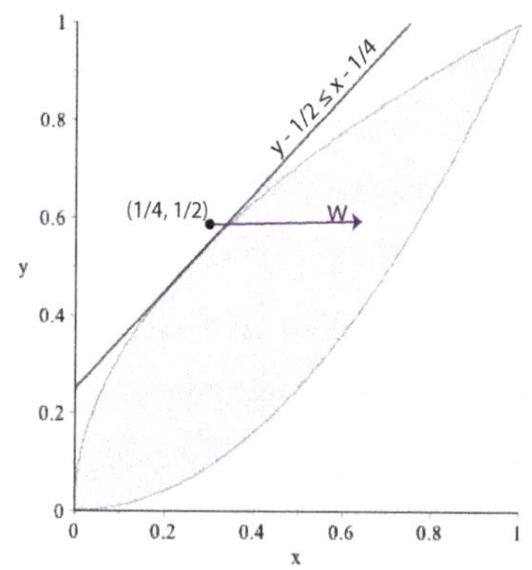

Fig. 14.6 Let U be the light purple planar region which lies between the curves $y = x^2$ and $y^2 = x$. Figure (i.) illustrates the boundary point $(1, 1)$ given by the equalities $y - x^2 = 0$ and $y^2 - x = 0$. The affine translate of cone of feasible directions, $C(1, 1)$, is illustrated by the pink triangle whose sides are the tangent lines to the boundary curves. Figure (ii.) illustrates the boundary point $(1/4, 1/2)$ given by the equality $y^2 - x = 0$. The affine translate of $C(1/4, 1/2)$ is the lilac half space bounded by the tangent line to $y^2 = x$ through $(1/4, 1/2)$.

we have $(\varphi_2')_{(0,0)} = (0\ 0)$, and thus $C^*(0) = \{(u_1, u_2) \in \mathbb{R}^2 \mid u_1 + u_2 \geq 0\}$ as illustrated in Figure 14.7.

The conditions stated in the following definition are sufficient conditions that imply that $C(u) = C^*(u)$, as we will prove next.

Definition 14.5. For any $u \in U$, with

$$U = \{x \in \Omega \mid \varphi_i(x) \leq 0,\ 1 \leq i \leq m\},$$

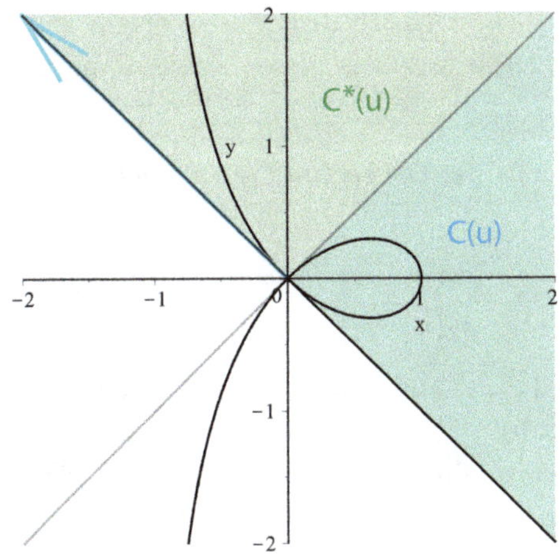

Fig. 14.7 For $u = (0, 0)$, $C^*(u)$ is the sea green half space given by $u_1 + u_2 \geq 0$. This half space strictly contains $C(u)$, namely the union of the turquoise triangular cone and the directional ray $(-1, 1)$.

if the functions φ_i are differentiable at u (in fact, we only this for $i \in I(u)$), we say that the constraints are *qualified* at u if the following conditions hold:

(a) Either the constraints φ_i are affine for all $i \in I(u)$, or
(b) There is some nonzero vector $w \in V$ such that the following conditions hold for all $i \in I(u)$:
   (i) $(\varphi_i')_u(w) \leq 0$.
   (ii) If φ_i is not affine, then $(\varphi_i')_u(w) < 0$.

Condition (b)(ii) implies that u is not a critical point of φ_i for every $i \in I(u)$, so there is no singularity at u in the zero locus of φ_i. Intuitively, if the constraints are qualified at u then the boundary of U near u behaves "nicely".

The boundary points illustrated in Figure 14.6 are qualified. Observe that $U = \{x \in \mathbb{R}^2 \mid \varphi_1(x, y) = y^2 - x \leq 0,\ \varphi_2(x, y) = x^2 - y \leq 0\}$. For $u = (1, 1)$, $I(u) = \{1, 2\}$, $(\varphi_1')_{(1,1)} = (-1\ \ 2)$, $(\varphi_2')_{(1,1)} = (2\ \ -1)$, and $w = (-1, -1)$ ensures that $(\varphi_1')_{(1,1)}$ and $(\varphi_1')_{(1,1)}$ satisfy Condition (b) of Definition 14.5. For $u = (1/4, 1/2)$, $I(u) = \{1\}$, $(\varphi_1')_{(1,1)} = (-1\ \ 1)$, and $w = (1, 0)$ will satisfy Condition (b).

In Example 14.1, the constraint $\varphi_2(u_1, u_2) = 0$ is not qualified at the origin because $(\varphi_2')_{(0,0)} = (0, 0)$; in fact, the origin is a self-intersection. In the example below, the origin is also a singular point, but for a different reason.

Example 14.2. Consider the region $U \subseteq \mathbb{R}^2$ determined by the two curves given by

$$\begin{aligned}\varphi_1(u_1, u_2) &= u_2 - \max(0, u_1^3)\\ \varphi_2(u_1, u_2) &= u_1^4 - u_2.\end{aligned}$$

We have $I(0,0) = \{1, 2\}$, and since $(\varphi_1)'_{(0,0)}(w_1, w_2) = (0\ 1)\binom{w_1}{w_2} = w_2$ and $(\varphi_2')_{(0,0)}(w_1, w_2) = (0\ \ -1)\binom{w_1}{w_2} = -w_2$, we have $C^*(0) = \{(u_1, u_2) \in \mathbb{R}^2 \mid u_2 = 0\}$, but the constraints are not qualified at $(0,0)$ since it is impossible to have simultaneously $(\varphi_1')_{(0,0)}(w_1, w_2) < 0$ and $(\varphi_2')_{(0,0)}(w_1, w_2) < 0$, so in fact $C(0) = \{(u_1, u_2) \in \mathbb{R}^2 \mid u_1 \geq 0,\ u_2 = 0\}$ is strictly contained in $C^*(0)$; see Figure 14.8.

Proposition 14.2. *Let u be any point of the set*

$$U = \{x \in \Omega \mid \varphi_i(x) \leq 0,\ 1 \leq i \leq m\},$$

where Ω is an open subset of the normed vector space V, and assume that the functions φ_i are differentiable at u (in fact, we only this for $i \in I(u)$). Then the following facts hold:

(1) The cone $C(u)$ of feasible directions at u is contained in the convex cone $C^*(u)$; that is,

$$C(u) \subseteq C^*(u) = \{v \in V \mid (\varphi_i')_u(v) \leq 0,\ i \in I(u)\}.$$

(2) If the constraints are qualified at u (and the functions φ_i are continuous at u for all $i \notin I(u)$ if we only assume φ_i differentiable at u for all $i \in I(u)$), then

$$C(u) = C^*(u).$$

Proof. (1) For every $i \in I(u)$, since $\varphi_i(v) \leq 0$ for all $v \in U$ and $\varphi_i(u) = 0$, the function $-\varphi_i$ has a local minimum at u with respect to U, so by Proposition 14.1(2), we have

$$(-\varphi_i')_u(v) \geq 0 \quad \text{for all } v \in C(u),$$

which is equivalent to $(\varphi_i')_u(v) \leq 0$ for all $v \in C(u)$ and for all $i \in I(u)$, that is, $u \in C^*(u)$.

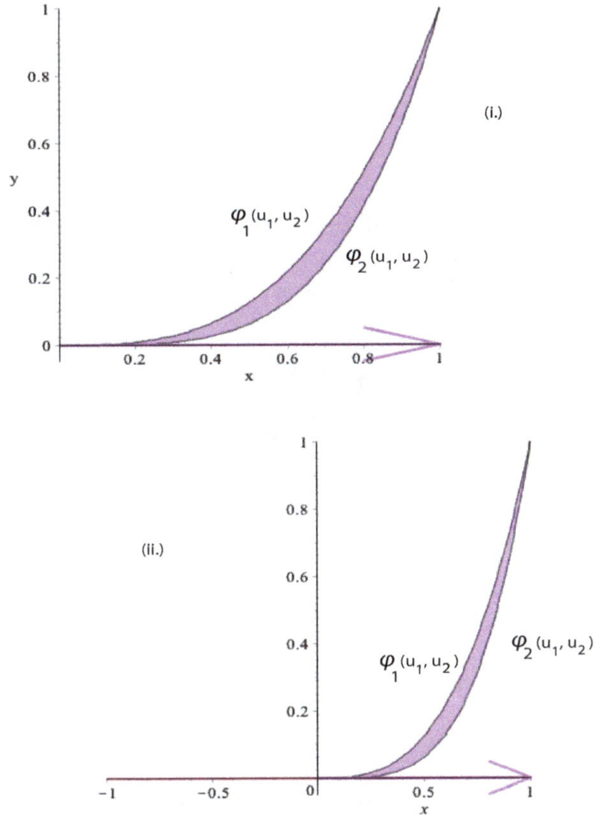

Fig. 14.8 Figures (i.) and (ii.) illustrate the purple moon shaped region associated with Example 14.2. Figure (i.) also illustrates $C(0)$, the cone of feasible directions, while Figure (ii.) illustrates the strict containment of $C(0)$ in $C^*(0)$.

(2)(a) First, let us assume that φ_i is affine for every $i \in I(u)$. Recall that φ_i must be given by $\varphi_i(v) = h_i(v) + c_i$ for all $v \in V$, where h_i is a linear form and $c_i \in \mathbb{R}$. Since the derivative of a linear map at any point is itself,

$$(\varphi_i')_u(v) = h_i(v) \quad \text{for all } v \in V.$$

Pick any nonzero $w \in C^*(u)$, which means that $(\varphi_i')_u(w) \leq 0$ for all $i \in I(u)$. For any sequence $(\epsilon_k)_{k\geq 0}$ of reals $\epsilon_k > 0$ such that $\lim_{k\mapsto\infty} \epsilon_k = 0$, let $(u_k)_{k\geq 0}$ be the sequence of vectors in V given by

$$u_k = u + \epsilon_k w.$$

We have $u_k - u = \epsilon_k w \neq 0$ for all $k \geq 0$ and $\lim_{k \mapsto \infty} u_k = u$. Furthermore, since the functions φ_i are continuous for all $i \notin I$, we have

$$0 > \varphi_i(u) = \lim_{k \mapsto \infty} \varphi_i(u_k),$$

and since φ_i is affine and $\varphi_i(u) = 0$ for all $i \in I$, we have $\varphi_i(u) = h_i(u) + c_i = 0$, so

$$\begin{aligned} \varphi_i(u_k) &= h_i(u_k) + c_i = h_i(u_k) - h_i(u) = h_i(u_k - u) = (\varphi'_i)_u(u_k - u) \\ &= \epsilon_k (\varphi'_i)_u(w) \leq 0, \end{aligned}$$

that is,

$$\varphi_i(u_k) \leq 0, \tag{$*_0$}$$

which implies that $u_k \in U$ for all k large enough. Since

$$\frac{u_k - u}{\|u_k - u\|} = \frac{w}{\|w\|} \quad \text{for all } k \geq 0,$$

we conclude that $w \in C(u)$. See Figure 14.9.

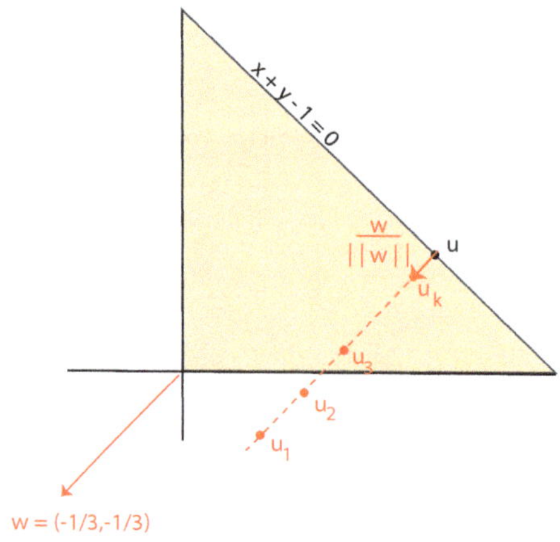

Fig. 14.9 Let U be the peach triangle bounded by the lines $y = 0$, $x = 0$, and $y = -x + 1$. Let u satisfy the affine constraint $\varphi(x, y) = y + x - 1$. Since $\varphi'_{(x,y)} = (1\ 1)$, set $w = (-1, -1)$ and approach u along the line $u + tw$.

(2)(b) Let us now consider the case where some function φ_i is not affine for some $i \in I(u)$. Let $w \neq 0$ be some vector in V such that Condition (b) of Definition 14.5 holds, namely: for all $i \in I(u)$, we have

(i) $(\varphi_i')_u(w) \leq 0$.
(ii) If φ_i is not affine, then $(\varphi_i')_u(w) < 0$.

Pick any nonzero vector $v \in C^*(u)$, which means that $(\varphi_i')_u(v) \leq 0$ for all $i \in I(u)$, and let $\delta > 0$ be any positive real number such that $v + \delta w \neq 0$. For any sequence $(\epsilon_k)_{k\geq 0}$ of reals $\epsilon_k > 0$ such that $\lim_{k\mapsto\infty} \epsilon_k = 0$, let $(u_k)_{k\geq 0}$ be the sequence of vectors in V given by

$$u_k = u + \epsilon_k(v + \delta w).$$

We have $u_k - u = \epsilon_k(v + \delta w) \neq 0$ for all $k \geq 0$ and $\lim_{k\mapsto\infty} u_k = u$. Furthermore, since the functions φ_i are continuous for all $i \notin I(u)$, we have

$$0 > \varphi_i(u) = \lim_{k\mapsto\infty} \varphi_i(u_k) \quad \text{for all } i \notin I(u). \tag{$*_1$}$$

Equation $(*_0)$ of the previous case shows that for all $i \in I(u)$ such that φ_i is affine, since $(\varphi_i')_u(v) \leq 0$, $(\varphi_i')_u(w) \leq 0$, and $\epsilon_k, \delta > 0$, we have

$$\varphi_i(u_k) = \epsilon_k((\varphi_i')_u(v) + \delta(\varphi_i')_u(w)) \leq 0 \quad \text{for all } i \in I(u) \text{ and } \varphi_i \text{ affine}. \tag{$*_2$}$$

Furthermore, since φ_i is differentiable and $\varphi_i(u) = 0$ for all $i \in I(u)$, if φ_i is not affine we have

$$\varphi_i(u_k) = \epsilon_k((\varphi_i')_u(v) + \delta(\varphi_i')_u(w)) + \epsilon_k \left\|u_k - u\right\| \eta_k(u_k - u)$$

with $\lim_{\|u_k - u\|\mapsto 0} \eta_k(u_k - u) = 0$, so if we write $\alpha_k = \|u_k - u\| \eta_k(u_k - u)$, we have

$$\varphi_i(u_k) = \epsilon_k((\varphi_i')_u(v) + \delta(\varphi_i')_u(w) + \alpha_k)$$

with $\lim_{k\mapsto\infty} \alpha_k = 0$, and since $(\varphi_i')_u(v) \leq 0$, we obtain

$$\varphi_i(u_k) \leq \epsilon_k(\delta(\varphi_i')_u(w) + \alpha_k) \quad \text{for all } i \in I(u) \text{ and } \varphi_i \text{ not affine}. \tag{$*_3$}$$

Equations $(*_1), (*_2), (*_3)$ show that $u_k \in U$ for k sufficiently large, where in $(*_3)$, since $(\varphi_i')_u(w) < 0$ and $\delta > 0$, even if $\alpha_k > 0$, when $\lim_{k\mapsto\infty} \alpha_k = 0$, we will have $\delta(\varphi_i')_u(w) + \alpha_k < 0$ for k large enough, and thus $\epsilon_k(\delta(\varphi_i')_u(w) + \alpha_k) < 0$ for k large enough.

Since

$$\frac{u_k - u}{\|u_k - u\|} = \frac{v + \delta w}{\|v + \delta w\|}$$

for all $k \geq 0$, we conclude that $v + \delta w \in C(u)$ for $\delta > 0$ small enough. But now the sequence $(v_n)_{n\geq 0}$ given by

$$v_n = v + \epsilon_n w$$

converges to v, and for n large enough, $v_n \in C(u)$. Since by Proposition 14.1(1), the cone $C(u)$ is closed, we conclude that $v \in C(u)$. See Figure 14.10.

In all cases, we proved that $C^*(u) \subseteq C(u)$, as claimed. ☐

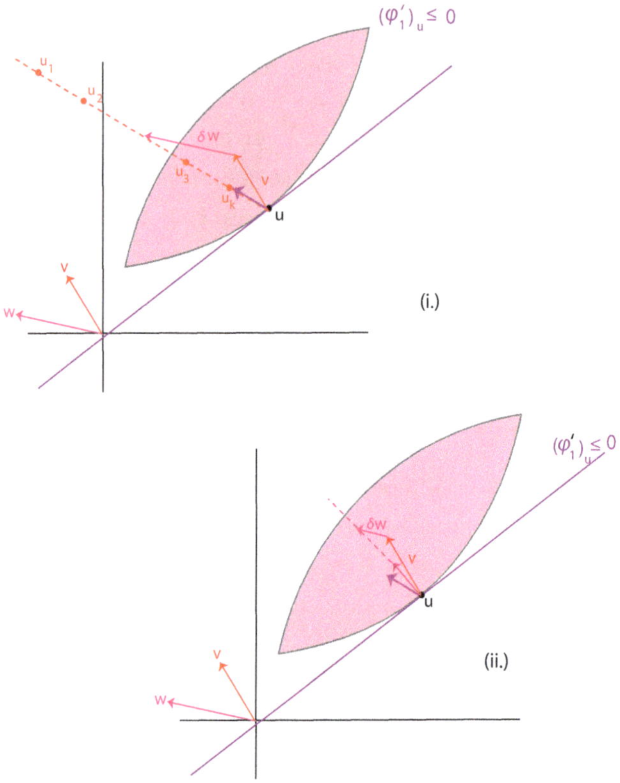

Fig. 14.10 Let U be the pink lounge in $\mathbb{R}^2$. Let u satisfy the non-affine constraint $\varphi_1(u)$. Choose vectors v and w in the half space $(\varphi_1')_u \leq 0$. Figure (i.) approaches u along the line $u+t(\delta w+v)$ and shows that $v+\delta w \in C(u)$ for fixed δ. Figure (ii.) varies δ in order that the purple vectors approach v as $\delta \to \infty$.

In the case of m affine constraints $a_i x \leq b_i$, for some linear forms a_i and some $b_i \in \mathbb{R}$, for any point $u \in \mathbb{R}^n$ such that $a_i u = b_i$ for all $i \in I(u)$, the cone $C(u)$ consists of all $v \in \mathbb{R}^n$ such that $a_i v \leq 0$, so $u + C(u)$ consists of all points $u + v$ such that

$$a_i(u+v) \leq b_i \quad \text{for all } i \in I(u),$$

which is the cone cut out by the hyperplanes determining some face of the polyhedron defined by the m constraints $a_i x \leq b_i$.

We are now ready to prove one of the most important results of nonlinear optimization.

14.3 The Karush–Kuhn–Tucker Conditions

If the domain U is defined by inequality constraints satisfying mild differentiability conditions and if the constraints at u are qualified, then there is a necessary condition for the function J to have a local minimum at $u \in U$ involving generalized Lagrange multipliers. The proof uses a version of Farkas lemma. In fact, the necessary condition stated next holds for infinite-dimensional vector spaces because there a version of Farkas lemma holding for *real* Hilbert spaces, but we will content ourselves with the version holding for finite dimensional normed vector spaces. For the more general version, see Theorem 12.2 (or Ciarlet [Ciarlet (1989)], Chapter 9).

We will be using the following version of Farkas lemma.

Proposition 14.3. *(Farkas Lemma, Version I) Let A be a real $m \times n$ matrix and let $b \in \mathbb{R}^m$ be any vector. The linear system $Ax = b$ has no solution $x \geq 0$ iff there is some nonzero linear form $y \in (\mathbb{R}^m)^*$ such that $yA \geq 0_n^\top$ and $yb < 0$.*

We will use the version of Farkas lemma obtained by taking a contrapositive, namely: *if $yA \geq 0_n^\top$ implies $yb \geq 0$ for all linear forms $y \in (\mathbb{R}^m)^*$, then linear system $Ax = b$ some solution $x \geq 0$.*

Actually, it is more convenient to use a version of Farkas lemma applying to a Euclidean vector space (with an inner product denoted $\langle -, - \rangle$). This version also applies to an infinite dimensional real Hilbert space; see Theorem 12.2. Recall that in a Euclidean space V the inner product induces an isomorphism between V and V', the space of continuous linear forms on V. In our case, we need the isomorphism $\sharp$ from V' to V defined such that for every linear form $\omega \in V'$, the vector $\omega^\sharp \in V$ is uniquely defined by the equation

$$\omega(v) = \langle v, \omega^\sharp \rangle \quad \text{for all } v \in V.$$

In $\mathbb{R}^n$, the isomorphism between $\mathbb{R}^n$ and $(\mathbb{R}^n)^*$ amounts to *transposition*: if $y \in (\mathbb{R}^n)^*$ is a linear form and $v \in \mathbb{R}^n$ is a vector, then

$$yv = v^\top y^\top.$$

The version of the Farkas–Minskowski lemma in term of an inner product is as follows.

Proposition 14.4. *(Farkas–Minkowski) Let V be a Euclidean space of* *finite dimension with inner product $\langle -, - \rangle$ (more generally, a Hilbert space). For any finite family $(a_1, \ldots, a_m)$ of m vectors $a_i \in V$ and any vector $b \in V$, for any $v \in V$,*

$$\text{if } \langle a_i, v\rangle \geq 0 \text{ for } i = 1, \ldots, m \text{ implies that } \langle b, v\rangle \geq 0,$$

then there exist $\lambda_1, \ldots, \lambda_m \in \mathbb{R}$ such that

$$\lambda_i \geq 0 \text{ for } i = 1, \ldots, m, \text{ and } b = \sum_{i=1}^{m} \lambda_i a_i,$$

that is, b belong to the polyhedral cone $\mathrm{cone}(a_1, \ldots, a_m)$.

Proposition 14.4 is the special case of Theorem 12.2 which holds for real Hilbert spaces.

We can now prove the following theorem.

Theorem 14.1. *Let $\varphi_i \colon \Omega \to \mathbb{R}$ be m constraints defined on some open subset Ω of a finite-dimensional Euclidean vector space V (more generally, a real Hilbert space V), let $J \colon \Omega \to \mathbb{R}$ be some function, and let U be given by*

$$U = \{x \in \Omega \mid \varphi_i(x) \leq 0, \ 1 \leq i \leq m\}.$$

For any $u \in U$, let

$$I(u) = \{i \in \{1, \ldots, m\} \mid \varphi_i(u) = 0\},$$

and assume that the functions φ_i are differentiable at u for all $i \in I(u)$ and continuous at u for all $i \notin I(u)$. If J is differentiable at u, has a local minimum at u with respect to U, and if the constraints are qualified at u, then there exist some scalars $\lambda_i(u) \in \mathbb{R}$ for all $i \in I(u)$, such that

$$J'_u + \sum_{i \in I(u)} \lambda_i(u)(\varphi'_i)_u = 0, \quad \text{and} \quad \lambda_i(u) \geq 0 \text{ for all } i \in I(u).$$

The above conditions are called the Karush–Kuhn–Tucker optimality conditions. Equivalently, in terms of gradients, the above conditions are expressed as

$$\nabla J_u + \sum_{i \in I(u)} \lambda_i(u) \nabla(\varphi_i)_u = 0, \quad \text{and} \quad \lambda_i(u) \geq 0 \text{ for all } i \in I(u).$$

Proof. By Proposition 14.1(2), we have

$$J'_u(w) \geq 0 \quad \text{for all } w \in C(u), \tag{$*_1$}$$

and by Proposition 14.2(2), we have $C(u) = C^*(u)$, where

$$C^*(u) = \{v \in V \mid (\varphi'_i)_u(v) \leq 0,\ \ i \in I(u)\}, \tag{$*_2$}$$

so $(*_1)$ can be expressed as: for all $w \in V$,

$$\text{if } w \in C^*(u) \text{ then } J'_u(w) \geq 0,$$

or

$$\text{if } -(\varphi'_i)_u(w) \geq 0 \text{ for all } i \in I(u), \text{ then } J'_u(w) \geq 0. \tag{$*_3$}$$

Under the isomorphism $\sharp$, the vector $(J'_u)^\sharp$ is the gradient ∇J_u, so that

$$J'_u(w) = \langle w, \nabla J_u \rangle, \tag{$*_4$}$$

and the vector $((\varphi'_i)_u)^\sharp$ is the gradient $\nabla(\varphi_i)_u$, so that

$$(\varphi'_i)_u(w) = \langle w, \nabla(\varphi_i)_u \rangle. \tag{$*_5$}$$

Using equations $(*_4)$ and $(*_5)$, equation $(*_3)$ can be written as: for all $w \in V$,

$$\text{if } \langle w, -\nabla(\varphi_i)_u \rangle \geq 0 \text{ for all } i \in I(u), \text{ then } \langle w, \nabla J_u \rangle \geq 0. \tag{$*_6$}$$

By the Farkas–Minkowski proposition (Proposition 14.4), there exist some scalars $\lambda_i(u)$ for all $i \in I(u)$, such that $\lambda_i(u) \geq 0$ and

$$\nabla J_u = \sum_{i \in I(u)} \lambda_i(u)(-\nabla(\varphi_i)_u),$$

that is

$$\nabla J_u + \sum_{i \in I(u)} \lambda_i(u)\nabla(\varphi_i)_u = 0,$$

and using the inverse of the isomorphism $\sharp$ (which is linear), we get

$$J'_u + \sum_{i \in I(u)} \lambda_i(u)(\varphi'_i)_u = 0,$$

as claimed. ☐

Since the constraints are inequalities of the form $\varphi_i(x) \leq 0$, there is a way of expressing the Karush–Kuhn–Tucker optimality conditions, often abbreviated as *KKT conditions*, in a way that does not refer explicitly to the index set $I(u)$:

$$J'_u + \sum_{i=1}^{m} \lambda_i(u)(\varphi'_i)_u = 0, \tag{KKT$_1$}$$

and

$$\sum_{i=1}^{m} \lambda_i(u)\varphi_i(u) = 0, \quad \lambda_i(u) \geq 0, \quad i = 1, \ldots, m. \tag{KKT$_2$}$$

Indeed, if we have the strict inequality $\varphi_i(u) < 0$ (the constraint φ_i is inactive at u), since all the terms $\lambda_i(u)\varphi_i(u)$ are nonpositive, we must have $\lambda_i(u) = 0$; that is, we only need to consider the $\lambda_i(u)$ for all $i \in I(u)$. Yet another way to express the conditions in (KKT$_2$) is

$$\lambda_i(u)\varphi_i(u) = 0, \quad \lambda_i(u) \geq 0, \quad i = 1, \ldots, m. \tag{KKT$'_2$}$$

In other words, for any $i \in \{1, \ldots, m\}$, if $\varphi_i(u) < 0$, then $\lambda_i(u) = 0$; that is,

- *if the constraint φ_i is inactive at u, then $\lambda_i(u) = 0$.*

By contrapositive, if $\lambda_i(u) \neq 0$, then $\varphi_i(u) = 0$; that is,

- *if $\lambda_i(u) \neq 0$, then the constraint φ_i is active at u.*

The conditions in (KKT$'_2$) are referred to as *complementary slackness* conditions.

The scalars $\lambda_i(u)$ are often called *generalized Lagrange multipliers*. If $V = \mathbb{R}^n$, the necessary conditions of Theorem 14.1 are expressed as the following system of equations and inequalities in the unknowns $(u_1, \ldots, u_n) \in \mathbb{R}^n$ and $(\lambda_1, \ldots, \lambda_m) \in \mathbb{R}^m_+$:

$$\begin{aligned}
\frac{\partial J}{\partial x_1}(u) + \lambda_1 \frac{\partial \varphi_1}{\partial x_1}(u) + \cdots + \lambda_m \frac{\partial \varphi_m}{\partial x_1}(u) &= 0 \\
\vdots \qquad\qquad &\ \ \vdots \\
\frac{\partial J}{\partial x_n}(u) + \lambda_1 \frac{\partial \varphi_n}{\partial x_1}(u) + \cdots + \lambda_m \frac{\partial \varphi_m}{\partial x_n}(u) &= 0 \\
\lambda_1\varphi_1(u) + \cdots + \lambda_m\varphi_m(u) &= 0 \\
\varphi_1(u) &\leq 0 \\
\vdots \quad &\ \ \vdots \\
\varphi_m(u) &\leq 0 \\
\lambda_1, \ldots, \lambda_m &\geq 0.
\end{aligned}$$

Example 14.3. Let J, φ_1 and φ_2 be the functions defined on $\mathbb{R}$ by

$$\begin{aligned}
J(x) &= x \\
\varphi_1(x) &= -x \\
\varphi_2(x) &= x - 1.
\end{aligned}$$

In this case

$$U = \{x \in \mathbb{R} \mid -x \leq 0,\ x - 1 \leq 0\} = [0, 1].$$

Since the constraints are affine, they are automatically qualified for any $u \in [0, 1]$. The system of equations and inequalities shown above becomes

$$\begin{aligned} 1 - \lambda_1 + \lambda_2 &= 0 \\ -\lambda_1 x + \lambda_2(x - 1) &= 0 \\ -x &\leq 0 \\ x - 1 &\leq 0 \\ \lambda_1, \lambda_2 &\geq 0. \end{aligned}$$

The first equality implies that $\lambda_1 = 1 + \lambda_2$. The second equality then becomes

$$-(1 + \lambda_2)x + \lambda_2(x - 1) = 0,$$

which implies that $\lambda_2 = -x$. Since $0 \leq x \leq 1$, or equivalently $-1 \leq -x \leq 0$, and $\lambda_2 \geq 0$, we conclude that $\lambda_2 = 0$ and $\lambda_1 = 1$ is the solution associated with $x = 0$, the minimum of $J(x) = x$ over $[0, 1]$. Observe that the case $x = 1$ corresponds to the maximum and not a minimum of $J(x) = x$ over $[0, 1]$.

Remark. Unless the linear forms $(\varphi'_i)_u$ for $i \in I(u)$ are linearly independent, the $\lambda_i(u)$ are generally not unique. Also, if $I(u) = \emptyset$, then the KKT conditions reduce to $J'_u = 0$. This is not surprising because in this case u belongs to the relative interior of U.

If the constraints are all affine equality constraints, then the KKT conditions are a bit simpler. We will consider this case shortly.

The conditions for the qualification of nonaffine constraints are hard (if not impossible) to use in practice, because they depend on $u \in U$ and on the derivatives $(\varphi'_i)_u$. Thus it is desirable to find simpler conditions. Fortunately, this is possible if the nonaffine functions φ_i are *convex*.

Definition 14.6. Let $U \subseteq \Omega \subseteq V$ be given by

$$U = \{x \in \Omega \mid \varphi_i(x) \leq 0,\ 1 \leq i \leq m\},$$

where Ω is an open subset of the Euclidean vector space V. If the functions $\varphi_i \colon \Omega \to \mathbb{R}$ are convex, we say that the constraints are *qualified* if the following conditions hold:

(a) Either the constraints φ_i are affine for all $i = 1, \ldots, m$ and $U \neq \emptyset$, or

(b) There is some vector $v \in \Omega$ such that the following conditions hold for $i = 1, \ldots, m$:

(i) $\varphi_i(v) \leq 0$.

(ii) If φ_i is not affine, then $\varphi_i(v) < 0$.

The above qualification conditions are known as *Slater's conditions*.

Condition (b)(i) also implies that U has nonempty relative interior. If Ω is convex, then U is also convex. This is because for all $u, v \in \Omega$, if $u \in U$ and $v \in U$, that is $\varphi_i(u) \leq 0$ and $\varphi_i(v) \leq 0$ for $i = 1, \ldots, m$, since the functions φ_i are convex, for all $\theta \in [0, 1]$ we have

$$\begin{aligned} \varphi_i((1-\theta)u + \theta v) &\leq (1-\theta)\varphi_i(u) + \theta\varphi_i(v) && \text{since } \varphi_i \text{ is convex} \\ &\leq 0 && \text{since } 1-\theta \geq 0, \theta \geq 0, \\ & && \varphi_i(u) \leq 0, \varphi_i(v) \leq 0, \end{aligned}$$

and any intersection of convex sets is convex.

It is important to observe that a *nonaffine equality constraint* $\varphi_i(u) = 0$ is *never* qualified.

Indeed, $\varphi_i(u) = 0$ is equivalent to $\varphi_i(u) \leq 0$ and $-\varphi_i(u) \leq 0$, so if these constraints are qualified and if φ_i is not affine then there is some nonzero vector $v \in \Omega$ such that both $\varphi_i(v) < 0$ and $-\varphi_i(v) < 0$, which is impossible. For this reason, *equality constraints are often assumed to be affine.*

The following theorem yields a more flexible version of Theorem 14.1 for constraints given by convex functions. If in addition, the function J is also *convex*, then the KKT conditions are also a *sufficient* condition for a local minimum.

Theorem 14.2. *Let $\varphi_i \colon \Omega \to \mathbb{R}$ be m convex constraints defined on some open convex subset Ω of a finite-dimensional Euclidean vector space V (more generally, a real Hilbert space V), let $J \colon \Omega \to \mathbb{R}$ be some function, let U be given by*

$$U = \{x \in \Omega \mid \varphi_i(x) \leq 0,\ 1 \leq i \leq m\},$$

and let $u \in U$ be any point such that the functions φ_i and J are differentiable at u.

(1) If J has a local minimum at u with respect to U, and if the constraints are qualified, then there exist some scalars $\lambda_i(u) \in \mathbb{R}$, such that the KKT condition hold:

$$J'_u + \sum_{i=1}^{m} \lambda_i(u)(\varphi'_i)_u = 0$$

and

$$\sum_{i=1}^{m} \lambda_i(u)\varphi_i(u) = 0, \quad \lambda_i(u) \geq 0, \quad i = 1, \ldots, m.$$

Equivalently, in terms of gradients, the above conditions are expressed as

$$\nabla J_u + \sum_{i=1}^{m} \lambda_i(u)\nabla(\varphi_i)_u = 0,$$

and

$$\sum_{i=1}^{m} \lambda_i(u)\varphi_i(u) = 0, \quad \lambda_i(u) \geq 0, \quad i = 1, \ldots, m.$$

(2) *Conversely, if the restriction of J to U is convex and if there exist scalars $(\lambda_1, \ldots, \lambda_m) \in \mathbb{R}^m_+$ such that the KKT conditions hold, then the function J has a (global) minimum at u with respect to U.*

Proof. (1) It suffices to prove that if the convex constraints are qualified according to Definition 14.6, then they are qualified according to Definition 14.5, since in this case we can apply Theorem 14.1.

If $v \in \Omega$ is a vector such that Condition (b) of Definition 14.6 holds and if $v \neq u$, for any $i \in I(u)$, since $\varphi_i(u) = 0$ and since φ_i is convex, by Proposition 4.6(1),

$$\varphi_i(v) \geq \varphi_i(u) + (\varphi_i')_u(v - u) = (\varphi_i')_u(v - u),$$

so if we let $w = v - u$ then

$$(\varphi_i')_u(w) \leq \varphi_i(v),$$

which shows that the nonaffine constraints φ_i for $i \in I(u)$ are qualified according to Definition 14.5, by Condition (b) of Definition 14.6.

If $v = u$, then the constraints φ_i for which $\varphi_i(u) = 0$ must be affine (otherwise, Condition (b)(ii) of Definition 14.6 would be false), and in this case we can pick $w = 0$.

(2) Let v be any arbitrary point in the convex subset U. Since $\varphi_i(v) \leq 0$ and $\lambda_i \geq 0$ for $i = 1, \ldots, m$, we have $\sum_{i=1}^{m} \lambda_i\varphi_i(v) \leq 0$, and using the fact that

$$\sum_{i=1}^{m} \lambda_i(u)\varphi_i(u) = 0, \quad \lambda_i(u) \geq 0, \quad i = 1, \ldots, m,$$

we have $\lambda_i = 0$ if $i \notin I(u)$ and $\varphi_i(u) = 0$ if $i \in I(u)$, so we have

$$\begin{aligned}
J(u) &\leq J(u) - \sum_{i=1}^{m} \lambda_i \varphi_i(v) \\
&\leq J(u) - \sum_{i \in I(u)} \lambda_i (\varphi_i(v) - \varphi_i(u)) && \lambda_i = 0 \text{ if } i \notin I(u), \\
& && \varphi_i(u) = 0 \text{ if } i \in I(u) \\
&\leq J(u) - \sum_{i \in I(u)} \lambda_i (\varphi_i')_u (v - u) && \text{(by Proposition 4.6)(1)} \\
&\leq J(u) + J_u'(v - u) && \text{(by the KKT conditions)} \\
&\leq J(v) && \text{(by Proposition 4.6)(1)},
\end{aligned}$$

and this shows that u is indeed a (global) minimum of J over U. □

It is important to note that when *both* the constraints, the domain of definition Ω, *and* the objective function J are *convex*, if the KKT conditions hold for some $u \in U$ and some $\lambda \in \mathbb{R}^m_+$, then Theorem 14.2 implies that J has a (global) minimum at u with respect to U, *independently* of any assumption on the qualification of the constraints.

The above theorem suggests introducing the function $L\colon \Omega \times \mathbb{R}^m_+ \to \mathbb{R}$ given by

$$L(v, \lambda) = J(v) + \sum_{i=1}^{m} \lambda_i \varphi_i(v),$$

with $\lambda = (\lambda_1, \ldots, \lambda_m)$. The function L is called the *Lagrangian* of the *Minimization Problem* (P):

$$\begin{aligned}
&\text{minimize} && J(v) \\
&\text{subject to} && \varphi_i(v) \leq 0, \quad i = 1, \ldots, m.
\end{aligned}$$

The KKT conditions of Theorem 14.2 imply that for any $u \in U$, if the vector $\lambda = (\lambda_1, \ldots, \lambda_m)$ is known and if u is a minimum of J on U, then

$$\begin{aligned}
\frac{\partial L}{\partial u}(u) &= 0 \\
J(u) &= L(u, \lambda).
\end{aligned}$$

The Lagrangian technique "absorbs" the constraints into the new objective function L and reduces the problem of finding a constrained minimum of the function J, to the problem of finding an unconstrained minimum of the function $L(v, \lambda)$. This is the main point of Lagrangian duality which will be treated in the next section.

A case that arises often in practice is the case where the constraints φ_i are affine. If so, the m constraints $a_i x \leq b_i$ can be expressed in matrix form as $Ax \leq b$, where A is an $m \times n$ matrix whose ith row is the row vector a_i. The KKT conditions of Theorem 14.2 yield the following corollary.

Proposition 14.5. *If U is given by*

$$U = \{x \in \Omega \mid Ax \leq b\},$$

where Ω is an open convex subset of $\mathbb{R}^n$ and A is an $m \times n$ matrix, and if J is differentiable at u and J has a local minimum at u, then there exist some vector $\lambda \in \mathbb{R}^m$, such that

$$\begin{aligned} &\nabla J_u + A^\top \lambda = 0 \\ &\lambda_i \geq 0 \ \text{and} \ \ \text{if } a_i u < b_i, \text{ then } \lambda_i = 0, \ i = 1, \ldots, m. \end{aligned}$$

If the function J is convex, then the above conditions are also sufficient for J to have a minimum at $u \in U$.

Another case of interest is the generalization of the minimization problem involving the affine constraints of a linear program in standard form, that is, equality constraints $Ax = b$ with $x \geq 0$, where A is an $m \times n$ matrix. In our formalism, this corresponds to the $2m + n$ constraints

$$\begin{aligned} a_i x - b_i &\leq 0, & i &= 1, \ldots, m \\ -a_i x + b_i &\leq 0, & i &= 1, \ldots, m \\ -x_j &\leq 0, & i &= 1, \ldots, n. \end{aligned}$$

In matrix form, they can be expressed as

$$\begin{pmatrix} A \\ -A \\ -I_n \end{pmatrix} \begin{pmatrix} x_1 \\ \vdots \\ x_n \end{pmatrix} \leq \begin{pmatrix} b \\ -b \\ 0_n \end{pmatrix}.$$

If we introduce the generalized Lagrange multipliers λ_i^+ and λ_i^- for $i = 1, \ldots, m$ and μ_j for $j = 1, \ldots, n$, then the KKT conditions are

$$\nabla J_u + \begin{pmatrix} A^\top & -A^\top & -I_n \end{pmatrix} \begin{pmatrix} \lambda^+ \\ \lambda^- \\ \mu \end{pmatrix} = 0_n,$$

that is,

$$\nabla J_u + A^\top \lambda^+ - A^\top \lambda^- - \mu = 0,$$

and $\lambda^+, \lambda^-, \mu \geq 0$, and if $a_i u < b_i$, then $\lambda_i^+ = 0$, if $-a_i u < -b_i$, then $\lambda_i^- = 0$, and if $-u_j < 0$, then $\mu_j = 0$. But the constraints $a_i u = b_i$ hold for $i = 1, \ldots, m$, so this places no restriction on the λ_i^+ and λ_i^-, and if we write $\lambda_i = \lambda_i^+ - \lambda_i^-$, then we have

$$\nabla J_u + A^\top \lambda = \mu,$$

with $\mu_j \geq 0$, and if $u_j > 0$ then $\mu_j = 0$, for $j = 1, \ldots, n$.

Thus we proved the following proposition (which is slight generalization of Proposition 8.7.2 in Matousek and Gardner [Matousek and Gartner (2007)]).

Proposition 14.6. *If U is given by*

$$U = \{x \in \Omega \mid Ax = b,\ x \geq 0\},$$

where where Ω is an open convex subset of $\mathbb{R}^n$ and A is an $m \times n$ matrix, and if J is differentiable at u and J has a local minimum at u, then there exist two vectors $\lambda \in \mathbb{R}^m$ and $\mu \in \mathbb{R}^n$, such that

$$\nabla J_u + A^\top \lambda = \mu,$$

with $\mu_j \geq 0$, and if $u_j > 0$ then $\mu_j = 0$, for $j = 1, \ldots, n$. Equivalently, there exists a vector $\lambda \in \mathbb{R}^m$ such that

$$(\nabla J_u)_j + (A^j)^\top \lambda \quad \begin{cases} = 0 & \text{if } u_j > 0 \\ \geq 0 & \text{if } u_j = 0, \end{cases}$$

where A^j is the jth column of A. If the function J is convex, then the above conditions are also sufficient for J to have a minimum at $u \in U$.

Yet another special case that arises frequently in practice is the minimization problem involving the affine equality constraints $Ax = b$, where A is an $m \times n$ matrix, with no restriction on x. Reviewing the proof of Proposition 14.6, we obtain the following proposition.

Proposition 14.7. *If U is given by*

$$U = \{x \in \Omega \mid Ax = b\},$$

where Ω is an open convex subset of $\mathbb{R}^n$ and A is an $m \times n$ matrix, and if J is differentiable at u and J has a local minimum at u, then there exist some vector $\lambda \in \mathbb{R}^m$ such that

$$\nabla J_u + A^\top \lambda = 0.$$

Equivalently, there exists a vector $\lambda \in \mathbb{R}^m$ such that

$$(\nabla J_u)_j + (A^j)^\top \lambda = 0,$$

where A^j is the jth column of A. If the function J is convex, then the above conditions are also sufficient for J to have a minimum at $u \in U$.

Observe that in Proposition 14.7, the λ_i are just standard Lagrange multipliers, with no restriction of positivity. Thus, Proposition 14.7 is a slight generalization of Theorem 4.1 that requires A to have rank m, but in the case of equational affine constraints, this assumption is unnecessary.

Here is an application of Proposition 14.7 to the *interior point method* in linear programming.

Example 14.4. In linear programming, the interior point method using a central path uses a logarithmic barrier function to keep the solutions $x \in \mathbb{R}^n$ of the equation $Ax = b$ away from boundaries by forcing $x > 0$, which means that $x_i > 0$ for all i; see Matousek and Gardner [Matousek and Gartner (2007)] (Section 7.2). Write

$$\mathbb{R}^n_{++} = \{x \in \mathbb{R}^n \mid x_i > 0,\ i = 1, \ldots, n\}.$$

Observe that $\mathbb{R}^n_{++}$ is open and convex. For any $\mu > 0$, we define the function f_μ defined on $\mathbb{R}^n_{++}$ by

$$f_\mu(x) = c^\top x + \mu \sum_{i=1}^{n} \ln x_i,$$

where $c \in \mathbb{R}^n$.

We would like to find necessary conditions for f_μ to have a maximum on

$$U = \{x \in \mathbb{R}^n_{++} \mid Ax = b\},$$

or equivalently to solve the following problem:

$$\begin{aligned}
&\text{maximize} \quad f_\mu(x) \\
&\text{subject to} \\
&\qquad\qquad\quad Ax = b \\
&\qquad\qquad\quad x > 0.
\end{aligned}$$

Since maximizing f_μ is equivalent to minimizing $-f_\mu$, by Proposition 14.7, if x is an optimal of the above problem then there is some $y \in \mathbb{R}^m$ such that

$$-\nabla f_\mu(x) + A^\top y = 0.$$

Since

$$\nabla f_\mu(x) = \begin{pmatrix} c_1 + \frac{\mu}{x_1} \\ \vdots \\ c_n + \frac{\mu}{x_n} \end{pmatrix},$$

we obtain the equation

$$c + \mu \begin{pmatrix} \frac{1}{x_1} \\ \vdots \\ \frac{1}{x_n} \end{pmatrix} = A^\top y.$$

To obtain a more convenient formulation, we define $s \in \mathbb{R}^n_{++}$ such that

$$s = \mu \begin{pmatrix} \frac{1}{x_1} \\ \vdots \\ \frac{1}{x_n} \end{pmatrix}$$

which implies that

$$\begin{pmatrix} s_1 x_1 & \cdots & s_n x_n \end{pmatrix} = \mu \mathbf{1}_n^\top,$$

and we obtain the following necessary conditions for f_μ to have a maximum:

$$\begin{aligned} Ax &= b \\ A^\top y - s &= c \\ \begin{pmatrix} s_1 x_1 & \cdots & s_n x_n \end{pmatrix} &= \mu \mathbf{1}_n^\top \\ s, x &> 0. \end{aligned}$$

It is not hard to show that if the primal linear program with objective function $c^\top x$ and equational constraints $Ax = b$ and the dual program with objective function $b^\top y$ and inequality constraints $A^\top y \geq c$ have interior feasible points x and y, which means that $x > 0$ and $s > 0$ (where $s = A^\top y - c$), then the above system of equations has a unique solution such that x is the unique maximizer of f_μ on U; see Matousek and Gardner [Matousek and Gartner (2007)] (Section 7.2, Lemma 7.2.1).

A particularly important application of Proposition 14.7 is the situation where $\Omega = \mathbb{R}^n$.

14.4 Equality Constrained Minimization

In this section we consider the following Program (P):

$$\begin{aligned} &\text{minimize} \quad J(v) \\ &\text{subject to} \quad Av = b, \ v \in \mathbb{R}^n, \end{aligned}$$

where J is a convex differentiable function and A is an $m \times n$ matrix of rank $m < n$ (the number of equality constraints is less than the number of variables, and these constraints are independent), and $b \in \mathbb{R}^m$.

According to Proposition 14.7 (with $\Omega = \mathbb{R}^n$), Program (P) has a minimum at $x \in \mathbb{R}^n$ if and only if there exist some Lagrange multipliers $\lambda \in \mathbb{R}^m$ such that the following equations hold:

$$\begin{aligned} Ax &= b && \text{(pfeasibilty)} \\ \nabla J_x + A^\top \lambda &= 0. && \text{(dfeasibility)} \end{aligned}$$

The set of linear equations $Ax = b$ is called the *primal feasibility equations* and the set of (generally nonlinear) equations $\nabla J_x + A^\top \lambda = 0$ is called the set of *dual feasibility equations*.

In general, it is impossible to solve these equations analytically, so we have to use numerical approximation procedures, most of which are variants of Newton's method. In special cases, for example if J is a quadratic functional, the dual feasibility equations are also linear, a case that we consider in more detail.

Suppose J is a convex quadratic functional of the form

$$J(x) = \frac{1}{2} x^\top P x + q^\top x + r,$$

where P is a $n \times n$ symmetric positive semidefinite matrix, $q \in \mathbb{R}^n$ and $r \in \mathbb{R}$. In this case

$$\nabla J_x = Px + q,$$

so the feasibility equations become

$$\begin{aligned} Ax &= b \\ Px + q + A^\top \lambda &= 0, \end{aligned}$$

which in matrix form become

$$\begin{pmatrix} P & A^\top \\ A & 0 \end{pmatrix} \begin{pmatrix} x \\ \lambda \end{pmatrix} = \begin{pmatrix} -q \\ b \end{pmatrix}. \qquad \text{(KKT-eq)}$$

The matrix of the linear system is usually called the *KKT-matrix*. Observe that the KKT matrix was already encountered in Proposition 6.3 with a different notation; there we had $P = A^{-1}$, $A = B^\top$, $q = b$, and $b = f$.

If the KKT matrix is invertible, then its unique solution (x^*, λ^*) yields a unique minimum x^* of Problem (P). If the KKT matrix is singular but the System (KKT-eq) is solvable, then *any solution* (x^*, λ^*) yields a minimum x^* of Problem (P).

Proposition 14.8. *If the System (KKT-eq) is not solvable, then Program (P) is unbounded below.*

Proof. We use the fact shown in Section 10.8 of Volume I, that a linear system $Bx = c$ has no solution iff there is some y that $B^\top y = 0$ and $y^\top c \neq 0$. By changing y to $-y$ if necessary, we may assume that $y^\top c > 0$. We apply this fact to the linear system (KKT-eq), so B is the KKT-matrix, which is symmetric, and we obtain the condition that there exist $v \in \mathbb{R}^n$ and $\lambda \in \mathbb{R}^m$ such that

$$Pv + A^\top \lambda = 0, \quad Av = 0, \quad -q^\top v + b^\top \lambda > 0.$$

Since the $m \times n$ matrix A has rank m and $b \in \mathbb{R}^m$, the system $Ax = b$, is solvable, so for any feasible x_0 (which means that $Ax_0 = b$), since $Av = 0$, the vector $x = x_0 + tv$ is also a feasible solution for all $t \in \mathbb{R}$. Using the fact that $Pv = -A^\top \lambda$, $v^\top P = -\lambda^\top A$, $Av = 0$, $x_0^\top A^\top = b^\top$, and P is symmetric, we have

$$\begin{aligned}
J(x_0 + tv) &= J(x_0) + (v^\top P x_0 + q^\top v)t + (1/2)(v^\top P v)t^2 \\
&= J(x_0) + (x_0^\top P v + q^\top v)t - (1/2)(\lambda^\top A v)t^2 \\
&= J(x_0) + (-x_0^\top A^\top \lambda + q^\top v)t \\
&= J(x_0) - (b^\top \lambda - q^\top v)t,
\end{aligned}$$

and since $-q^\top v + b^\top \lambda > 0$, the above expression goes to $-\infty$ when t goes to $+\infty$. $\square$

It is obviously important to have criteria to decide whether the KKT-matrix is invertible. There are indeed such criteria, as pointed in Boyd and Vandenberghe [Boyd and Vandenberghe (2004)] (Chapter 10, Exercise 10.1).

Proposition 14.9. *The invertibility of the KKT-matrix*

$$\begin{pmatrix} P & A^\top \\ A & 0 \end{pmatrix}$$

is equivalent to the following conditions:

(1) For all $x \in \mathbb{R}^n$, if $Ax = 0$ with $x \neq 0$, then $x^\top P x > 0$; that is, P is positive definite on the kernel of A.
(2) The kernels of A and P only have 0 in common ($(\mathrm{Ker}\, A) \cap (\mathrm{Ker}\, P) = \{0\}$).
(3) There is some $n \times (n - m)$ matrix F such that $\mathrm{Im}(F) = \mathrm{Ker}\,(A)$ and $F^\top P F$ is symmetric positive definite.
(4) There is some symmetric positive semidefinite matrix Q such that $P + A^\top Q A$ is symmetric positive definite. In fact, $Q = I$ works.

Proof sketch. Recall from Proposition 5.11in Volume I that a square matrix B is invertible iff its kernel is reduced to $\{0\}$; equivalently, for all x, if $Bx = 0$, then $x = 0$. Assume that Condition (1) holds. We have

$$\begin{pmatrix} P & A^\top \\ A & 0 \end{pmatrix} \begin{pmatrix} v \\ w \end{pmatrix} = \begin{pmatrix} 0 \\ 0 \end{pmatrix}$$

iff

$$Pv + A^\top w = 0, \quad Av = 0. \tag{$*$}$$

We deduce that

$$v^\top Pv + v^\top A^\top w = 0,$$

and since

$$v^\top A^\top w = (Av)^\top w = 0w = 0,$$

we obtain $v^\top Pv = 0$. Since Condition (1) holds, because $v \in \operatorname{Ker} A$, we deduce that $v = 0$. Then $A^\top w = 0$, but since the $m \times n$ matrix A has rank m, the $n \times m$ matrix $A^\top$ also has rank m, so its columns are linearly independent, and so $w = 0$. Therefore the KKT-matrix is invertible.

Conversely, assume that the KKT-matrix is invertible, yet the assumptions of Condition (1) fail. This means there is some $v \neq 0$ such that $Av = 0$ and $v^\top Pv = 0$. We claim that $Pv = 0$. This is because if P is a symmetric positive semidefinite matrix, then for any v, we have $v^\top Pv = 0$ iff $Pv = 0$.

If $Pv = 0$, then obviously $v^\top Pv = 0$, so assume the converse, namely $v^\top Pv = 0$. Since P is a symmetric positive semidefinite matrix, it can be diagonalized as

$$P = R^\top \Sigma R,$$

where R is an orthogonal matrix and Σ is a diagonal matrix

$$\Sigma = \operatorname{diag}(\lambda_1, \ldots, \lambda_s, 0, \ldots, 0),$$

where s is the rank of P and $\lambda_1 \geq \cdots \geq \lambda_s > 0$. Then $v^\top Pv = 0$ is equivalent to

$$v^\top R^\top \Sigma Rv = 0,$$

equivalently

$$(Rv)^\top \Sigma Rv = 0.$$

If we write $Rv = y$, then we have

$$0 = (Rv)^\top \Sigma Rv = y^\top \Sigma y = \sum_{i=1}^{s} \lambda_i y_i^2,$$

and since $\lambda_i > 0$ for $i = 1, \ldots, s$, this implies that $y_i = 0$ for $i = 1, \ldots, s$. Consequently, $\Sigma y = \Sigma R v = 0$, and so $Pv = R^\top \Sigma R v = 0$, as claimed. Since $v \neq 0$, the vector $(v, 0)$ is a nontrivial solution of equations $(*)$, a contradiction of the invertibility assumption of the KKT-matrix.

Observe that we proved that $Av = 0$ and $Pv = 0$ iff $Av = 0$ and $v^\top P v = 0$, so we easily obtain the fact that Condition (2) is equivalent to the invertibility of the KKT-matrix. Parts (3) and (4) are left as an exercise. □

In particular, if P is positive definite, then Proposition 14.9(4) applies, as we already know from Proposition 6.3. In this case, we can solve for x by elimination. We get

$$x = -P^{-1}(A^\top \lambda + q), \quad \text{where} \quad \lambda = -(AP^{-1}A^\top)^{-1}(b + AP^{-1}q).$$

In practice, we do not invert P and $AP^{-1}A^\top$. Instead, we solve the linear systems

$$\begin{aligned} Pz &= q \\ PE &= A^\top \\ (AE)\lambda &= -(b + Az) \\ Px &= -(A^\top \lambda + q). \end{aligned}$$

Observe that $(AP^{-1}A^\top)^{-1}$ is the Schur complement of P in the KKT matrix.

Since the KKT-matrix is symmetric, if it is invertible, we can convert it to $LDL^\top$ form using Proposition 7.6of Volume I. This method is only practical when the problem is small or when A and P are sparse.

If the KKT-matrix is invertible but P is not, then we can use a trick involving Proposition 14.9. We find a symmetric positive semidefinite matrix Q such that $P + A^\top Q A$ is symmetric positive definite, and since a solution (v, w) of the KKT-system should have $Av = b$, we also have $A^\top Q A v = A^\top Q b$, so the KKT-system is equivalent to

$$\begin{pmatrix} P + A^\top Q A & A^\top \\ A & 0 \end{pmatrix} \begin{pmatrix} v \\ w \end{pmatrix} = \begin{pmatrix} -q + A^\top Q b \\ b \end{pmatrix},$$

and since $P + A^\top Q A$ is symmetric positive definite, we can solve this system by elimination.

Another way to solve Problem (P) is to use variants of Newton's method as described in Section 13.9 dealing with equality constraints. Such methods are discussed extensively in Boyd and Vandenberghe [Boyd and Vandenberghe (2004)] (Chapter 10, Sections 10.2–10.4).

There are two variants of this method:

(1) The first method, called *feasible start Newton method*, assumes that the starting point u_0 is feasible, which means that $Au_0 = b$. The Newton step $d_{\rm nt}$ is a feasible direction, which means that $Ad_{\rm nt} = 0$.
(2) The second method, called *infeasible start Newton method*, does *not* assume that the starting point u_0 is feasible, which means that $Au_0 = b$ may not hold. This method is a little more complicated than the other method.

We only briefly discuss the feasible start Newton method, leaving it to the reader to consult Boyd and Vandenberghe [Boyd and Vandenberghe (2004)] (Chapter 10, Section 10.3) for a discussion of the infeasible start Newton method.

The Newton step $d_{\rm nt}$ is the solution of the linear system

$$\begin{pmatrix} \nabla^2 J(x) & A^\top \\ A & 0 \end{pmatrix} \begin{pmatrix} d_{\rm nt} \\ w \end{pmatrix} = \begin{pmatrix} -\nabla J_x \\ 0 \end{pmatrix}.$$

The Newton decrement $\lambda(x)$ is defined as in Section 13.9 as

$$\lambda(x) = (d_{\rm nt}^\top \nabla^2 J(x)\, d_{\rm nt})^{1/2} = \left((\nabla J_x)^\top (\nabla^2 J(x))^{-1} \nabla J_x\right)^{1/2}.$$

Newton's method with equality constraints (with feasible start) consists of the following steps: Given a starting point $u_0 \in \mathrm{dom}(J)$ with $Au_0 = b$, and a tolerance $\epsilon > 0$ do:

repeat

(1) Compute the Newton step and decrement
$d_{{\rm nt},k} = -(\nabla^2 J(u_k))^{-1} \nabla J_{u_k}$ and $\lambda(u_k)^2 = (\nabla J_{u_k})^\top (\nabla^2 J(u_k))^{-1} \nabla J_{u_k}$.
(2) Stopping criterion. **quit** if $\lambda(u_k)^2/2 \leq \epsilon$.
(3) Line Search. Perform an exact or backtracking line search to find ρ_k.
(4) Update. $u_{k+1} = u_k + \rho_k d_{{\rm nt},k}$.

Newton's method requires that the KKT-matrix be invertible. Under some mild assumptions, Newton's method (with feasible start) converges; see Boyd and Vandenberghe [Boyd and Vandenberghe (2004)] (Chapter 10, Section 10.2.4).

We now give an example illustrating Proposition 14.5, the *Support Vector Machine* (abbreviated as *SVM*).

14.5 Hard Margin Support Vector Machine; Version I

In this section we describe the following *classification problem*, or perhaps more accurately, *separation problem* (into two classes). Suppose we have two nonempty disjoint finite sets of p *blue* points $\{u_i\}_{i=1}^p$ and q *red* points $\{v_j\}_{j=1}^q$ in $\mathbb{R}^n$ (for simplicity, you may assume that these points are in the plane, that is, $n = 2$). Our goal is to find a hyperplane H of equation $w^\top x - b = 0$ (where $w \in \mathbb{R}^n$ is a nonzero vector and $b \in \mathbb{R}$), such that all the blue points u_i are in one of the two open half-spaces determined by H, and all the red points v_j are in the other open half-space determined by H; see Figure 14.11.

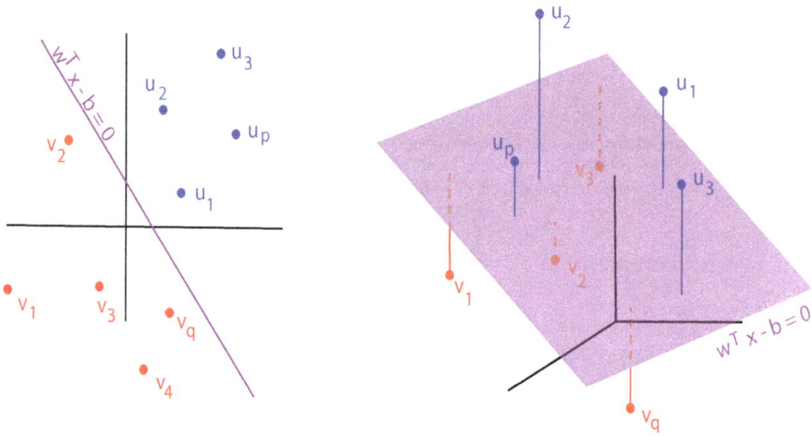

Fig. 14.11 Two examples of the SVM separation problem. The left figure is SVM in $\mathbb{R}^2$, while the right figure is SVM in $\mathbb{R}^3$.

Without loss of generality, we may assume that

$$\begin{aligned} w^\top u_i - b &> 0 && \text{for } i = 1, \ldots, p \\ w^\top v_j - b &< 0 && \text{for } j = 1, \ldots, q. \end{aligned}$$

Of course, separating the blue and the red points may be impossible, as we see in Figure 14.12 for four points where the line segments (u_1, u_2) and (v_1, v_2) intersect. If a hyperplane separating the two subsets of blue and red points exists, we say that they are *linearly separable*.

Remark: Write $m = p + q$. The reader should be aware that in machine learning the classification problem is usually defined as follows. We assign m so-called *class labels* $y_k = \pm 1$ to the data points in such a way that $y_i = +1$ for each blue point u_i, and $y_{p+j} = -1$ for each red point v_j, and we denote the m points by x_k, where $x_k = u_k$ for $k = 1, \ldots, p$ and

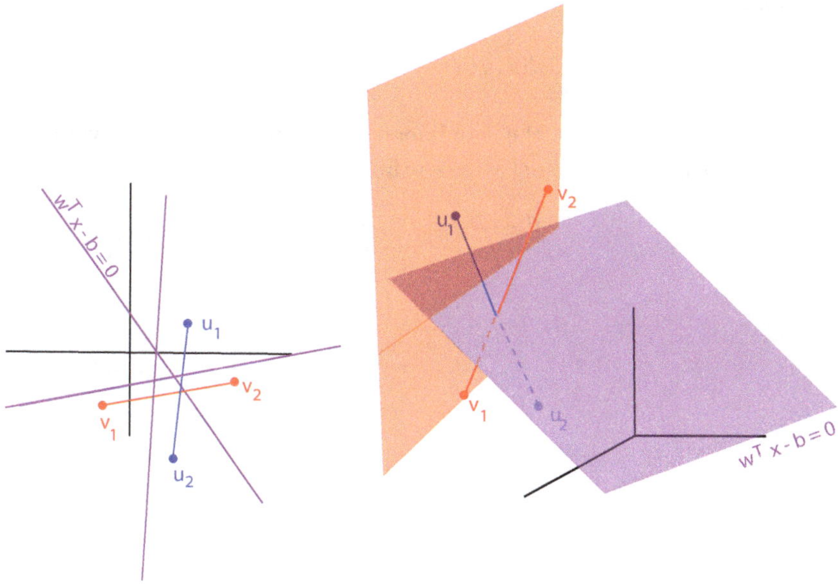

Fig. 14.12 Two examples in which it is impossible to find purple hyperplanes which separate the red and blue points.

$x_k = v_{k-p}$ for $k = p+1, \ldots, p+q$. Then the classification constraints can be written as

$$y_k(w^\top x_k - b) > 0 \quad \text{for } k = 1, \ldots, m.$$

The set of pairs $\{(x_1, y_1), \ldots, (x_m, y_m)\}$ is called a set of *training data* (or *training set*).

In the sequel, we will not use the above method, and we will stick to our two subsets of p *blue* points $\{u_i\}_{i=1}^p$ and q *red* points $\{v_j\}_{j=1}^q$.

Since there are infinitely many hyperplanes separating the two subsets (if indeed the two subsets are linearly separable), we would like to come up with a "good" criterion for choosing such a hyperplane.

The idea that was advocated by Vapnik (see Vapnik [Vapnik (1998)]) is to consider the distances $d(u_i, H)$ and $d(v_j, H)$ from *all* the points to the hyperplane H, and to pick a hyperplane H that maximizes the smallest of these distances. In machine learning this strategy is called finding a *maximal margin hyperplane*, or *hard margin support vector machine*, which definitely sounds more impressive.

Since the distance from a point x to the hyperplane H of equation

$w^\top x - b = 0$ is

$$d(x, H) = \frac{|w^\top x - b|}{\|w\|},$$

(where $\|w\| = \sqrt{w^\top w}$ is the Euclidean norm of w), it is convenient to temporarily assume that $\|w\| = 1$, so that

$$d(x, H) = |w^\top x - b|.$$

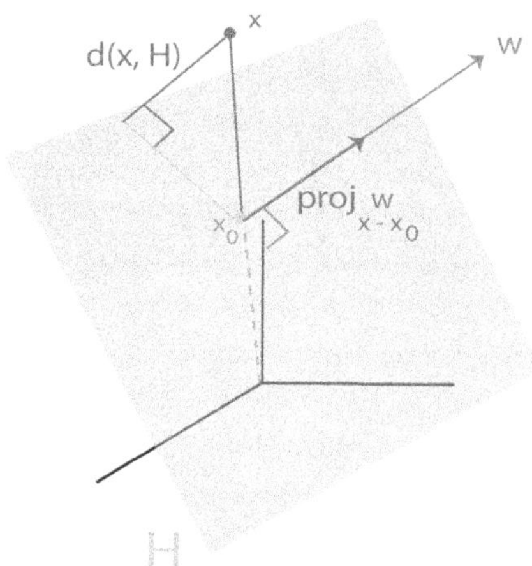

Fig. 14.13 In $\mathbb{R}^3$, the distance from a point to the plane $w^\top x - b = 0$ is given by the projection onto the normal w.

See Figure 14.13. Then with our sign convention, we have

$$\begin{aligned} d(u_i, H) &= w^\top u_i - b & i &= 1, \ldots, p \\ d(v_j, H) &= -w^\top v_j + b & j &= 1, \ldots, q. \end{aligned}$$

If we let

$$\delta = \min\{d(u_i, H),\ d(v_j, H) \mid 1 \leq i \leq p,\ 1 \leq j \leq q\},$$

then the hyperplane H should chosen so that

$$\begin{aligned} w^\top u_i - b &\geq \delta & i &= 1, \ldots, p \\ -w^\top v_j + b &\geq \delta & j &= 1, \ldots, q, \end{aligned}$$

and such that $\delta > 0$ is *maximal*. The distance δ is called the *margin* associated with the hyperplane H. This is indeed one way of formulating the two-class separation problem as an optimization problem with a linear objective function $J(\delta, w, b) = \delta$, and affine and quadratic constraints (SVM_{h1}):

$$\begin{aligned} &\text{maximize} \quad \delta \\ &\text{subject to} \\ &\qquad w^\top u_i - b \geq \delta \qquad i = 1, \ldots, p \\ &\qquad -w^\top v_j + b \geq \delta \qquad j = 1, \ldots, q \\ &\qquad \|w\| \leq 1. \end{aligned}$$

Observe that the Problem (SVM_{h1}) has an optimal solution $\delta > 0$ iff the two subsets are linearly separable. We used the constraint $\|w\| \leq 1$ rather than $\|w\| = 1$ because the former is qualified, whereas the latter is not. But if (w, b, δ) is an optimal solution, then $\|w\| = 1$, as shown in the following proposition.

Proposition 14.10. *If (w, b, δ) is an optimal solution of Problem (SVM_{h1}), so in particular $\delta > 0$, then we must have $\|w\| = 1$.*

Proof. First, if $w = 0$, then we get the two inequalities

$$-b \geq \delta, \quad b \geq \delta,$$

which imply that $b \leq -\delta$ and $b \geq \delta$ for some positive δ, which is impossible. But then, if $w \neq 0$ and $\|w\| < 1$, by dividing both sides of the inequalities by $\|w\| < 1$ we would obtain the better solution $(w/\|w\|, b/\|w\|, \delta/\|w\|)$, since $\|w\| < 1$ implies that $\delta/\|w\| > \delta$. ☐

We now prove that if the two subsets are linearly separable, then Problem (SVM_{h1}) has a unique optimal solution.

Theorem 14.3. *If two disjoint subsets of p blue points $\{u_i\}_{i=1}^p$ and q red points $\{v_j\}_{j=1}^q$ are linearly separable, then Problem (SVM_{h1}) has a unique optimal solution consisting of a hyperplane of equation $w^\top x - b = 0$ separating the two subsets with maximum margin δ. Furthermore, if we define $c_1(w)$ and $c_2(w)$ by*

$$\begin{aligned} c_1(w) &= \min_{1 \leq i \leq p} w^\top u_i \\ c_2(w) &= \max_{1 \leq j \leq q} w^\top v_j, \end{aligned}$$

then w is the unique maximum of the function

$$\rho(w) = \frac{c_1(w) - c_2(w)}{2}$$

over the convex subset U of $\mathbb{R}^n$ given by the inequalities

$$\begin{aligned} w^\top u_i - b &\geq \delta \qquad i = 1, \ldots, p \\ -w^\top v_j + b &\geq \delta \qquad j = 1, \ldots, q \\ \|w\| &\leq 1, \end{aligned}$$

and

$$b = \frac{c_1(w) + c_2(w)}{2}.$$

Proof. Our proof is adapted from Vapnik [Vapnik (1998)] (Chapter 10, Theorem 10.1). For any separating hyperplane H, since

$$\begin{aligned} d(u_i, H) &= w^\top u_i - b \qquad & i = 1, \ldots, p \\ d(v_j, H) &= -w^\top v_j + b \qquad & j = 1, \ldots, q, \end{aligned}$$

and since the smallest distance to H is

$$\begin{aligned} \delta &= \min\{d(u_i, H),\ d(v_j, H) \mid 1 \leq i \leq p,\ 1 \leq j \leq q\} \\ &= \min\{w^\top u_i - b,\ -w^\top v_j + b \mid 1 \leq i \leq p,\ 1 \leq j \leq q\} \\ &= \min\{\min\{w^\top u_i - b \mid 1 \leq i \leq p\}, \min\{-w^\top v_j + b \mid 1 \leq j \leq q\}\} \\ &= \min\{\min\{w^\top u_i \mid 1 \leq i \leq p\} - b\}, \min\{-w^\top v_j \mid 1 \leq j \leq q\} + b\} \\ &= \min\{\min\{w^\top u_i \mid 1 \leq i \leq p\} - b\}, -\max\{w^\top v_j \mid 1 \leq j \leq q\} + b\} \\ &= \min\{c_1(w) - b, -c_2(w) + b\}, \end{aligned}$$

in order for δ to be maximal we must have

$$c_1(w) - b = -c_2(w) + b,$$

which yields

$$b = \frac{c_1(w) + c_2(w)}{2}.$$

In this case,

$$c_1(w) - b = \frac{c_1(w) - c_2(w)}{2} = -c_2(w) + b,$$

so the maximum margin δ is indeed obtained when $\rho(w) = (c_1(w) - c_2(w))/2$ is maximal over U. Conversely, it is easy to see that any hyperplane of equation $w^\top x - b = 0$ associated with a w maximizing ρ over U and $b = (c_1(w) + c_2(w))/2$ is an optimal solution.

It remains to show that an optimal separating hyperplane exists and is unique. Since the unit ball is compact, U (as defined in Theorem 14.3) is compact, and since the function $w \mapsto \rho(w)$ is continuous, it achieves its maximum for some w_0 such that $\|w_0\| \leq 1$. Actually, we must have $\|w_0\| = 1$, since otherwise, by the reasoning used in Proposition 14.10, $w_0/\|w_0\|$ would be an even better solution. Therefore, w_0 is on the boundary of U. But ρ is a concave function (as an infimum of affine functions), so if it had two distinct maxima w_0 and w_0' with $\|w_0\| = \|w_0'\| = 1$, these would be global maxima since U is also convex, so we would have $\rho(w_0) = \rho(w_0')$ and then ρ would also have the same value along the segment (w_0, w_0') and in particular at $(w_0 + w_0')/2$, an interior point of U, a contradiction. $\square$

We can proceed with the above formulation (SVM_{h1}) but there is a way to reformulate the problem so that the constraints are all *affine*, which might be preferable since they will be *automatically qualified.*

14.6 Hard Margin Support Vector Machine; Version II

Since $\delta > 0$ (otherwise the data would not be separable into two disjoint sets), we can divide the affine constraints by δ to obtain

$$\begin{aligned} w'^\top u_i - b' &\geq 1 & i &= 1, \ldots, p \\ -w'^\top v_j + b' &\geq 1 & j &= 1, \ldots, q, \end{aligned}$$

except that now, w' is not necessarily a unit vector. To obtain the distances to the hyperplane H, we need to divide by $\|w'\|$ and then we have

$$\begin{aligned} \frac{w'^\top u_i - b'}{\|w'\|} &\geq \frac{1}{\|w'\|} & i &= 1, \ldots, p \\ \frac{-w'^\top v_j + b'}{\|w'\|} &\geq \frac{1}{\|w'\|} & j &= 1, \ldots, q, \end{aligned}$$

which means that the shortest distance from the data points to the hyperplane is $1/\|w'\|$. Therefore, we wish to maximize $1/\|w'\|$, that is, to minimize $\|w'\|$, so we obtain the following optimization Problem (SVM_{h2}):

Hard margin SVM (SVM_{h2}):

$$\text{minimize} \quad \frac{1}{2}\|w\|^2$$

subject to

$$\begin{aligned} w^\top u_i - b &\geq 1 & i &= 1, \ldots, p \\ -w^\top v_j + b &\geq 1 & j &= 1, \ldots, q. \end{aligned}$$

The objective function $J(w) = 1/2 \left\|w\right\|^2$ is convex, so Proposition 14.5 applies and gives us a necessary and sufficient condition for having a minimum in terms of the KKT conditions. First observe that the trivial solution $w = 0$ is impossible, because the blue constraints would be

$$-b \geq 1,$$

that is $b \leq -1$, and the red constraints would be

$$b \geq 1,$$

but these are contradictory. **Our goal is to find w and b, and optionally, δ.** We proceed in four steps first demonstrated on the following example.

Suppose that $p = q = n = 2$, so that we have two blue points

$$u_1^\top = (u_{11}, u_{12}) \qquad u_2^\top = (u_{21}, u_{22}),$$

two red points

$$v_1^\top = (v_{11}, v_{12}) \qquad v_2^\top = (v_{21}, v_{22}),$$

and

$$w^\top = (w_1, w_2).$$

Step 1: Write the constraints in matrix form. Let

$$C = \begin{pmatrix} -u_{11} & -u_{12} & 1 \\ -u_{21} & -u_{22} & 1 \\ v_{11} & v_{12} & -1 \\ v_{21} & v_{22} & -1 \end{pmatrix} \qquad d = \begin{pmatrix} -1 \\ -1 \\ -1 \\ -1 \end{pmatrix}. \tag{M}$$

The constraints become

$$C\begin{pmatrix} w \\ b \end{pmatrix} = \begin{pmatrix} -u_{11} & -u_{12} & 1 \\ -u_{21} & -u_{22} & 1 \\ v_{11} & v_{12} & -1 \\ v_{21} & v_{22} & -1 \end{pmatrix} \begin{pmatrix} w_1 \\ w_2 \\ b \end{pmatrix} \leq \begin{pmatrix} -1 \\ -1 \\ -1 \\ -1 \end{pmatrix}. \tag{C}$$

Step 2: Write the objective function in matrix form.

$$J(w_1, w_2, b) = \frac{1}{2} \begin{pmatrix} w_1 & w_2 & b \end{pmatrix} \begin{pmatrix} 1 & 0 & 0 \\ 0 & 1 & 0 \\ 0 & 0 & 0 \end{pmatrix} \begin{pmatrix} w_1 \\ w_2 \\ b \end{pmatrix}. \tag{O}$$

Step 3: Apply Proposition 14.5 to solve for w in terms of λ and μ. We obtain

$$\begin{pmatrix} w_1 \\ w_2 \\ 0 \end{pmatrix} + \begin{pmatrix} -u_{11} & -u_{21} & v_{11} & v_{21} \\ -u_{12} & -u_{22} & v_{12} & v_{22} \\ 1 & 1 & -1 & -1 \end{pmatrix} \begin{pmatrix} \lambda_1 \\ \lambda_2 \\ \mu_1 \\ \mu_2 \end{pmatrix} = \begin{pmatrix} 0 \\ 0 \\ 0 \end{pmatrix},$$

i.e.,

$$\nabla J_{(w,b)} + C^\top \begin{pmatrix} \lambda \\ \mu \end{pmatrix} = 0_3, \qquad \lambda^\top = (\lambda_1, \lambda_2),\ \ \mu^\top = (\mu_1, \mu_2).$$

Then

$$\begin{pmatrix} w_1 \\ w_2 \\ 0 \end{pmatrix} = \begin{pmatrix} u_{11} & u_{21} & -v_{11} & -v_{21} \\ u_{12} & u_{22} & -v_{12} & -v_{22} \\ -1 & -1 & 1 & 1 \end{pmatrix} \begin{pmatrix} \lambda_1 \\ \lambda_2 \\ \mu_1 \\ \mu_2 \end{pmatrix},$$

which implies

$$w = \begin{pmatrix} w_1 \\ w_2 \end{pmatrix} = \lambda_1 \begin{pmatrix} u_{11} \\ u_{12} \end{pmatrix} + \lambda_2 \begin{pmatrix} u_{21} \\ u_{22} \end{pmatrix} - \mu_1 \begin{pmatrix} v_{11} \\ v_{12} \end{pmatrix} - \mu_2 \begin{pmatrix} v_{21} \\ v_{22} \end{pmatrix} \tag{$*_1$}$$

with respect to

$$\mu_1 + \mu_2 - \lambda_1 - \lambda_2 = 0. \tag{$*_2$}$$

Step 4: Rewrite the constraints at (C) using $(*_1)$. In particular $C\begin{pmatrix} w \\ b \end{pmatrix} \leq d$ becomes

$$\begin{pmatrix} -u_{11} & -u_{12} & 1 \\ -u_{21} & -u_{22} & 1 \\ v_{11} & v_{12} & -1 \\ v_{21} & v_{22} & -1 \end{pmatrix} \begin{pmatrix} u_{11} & u_{21} & -v_{11} & -v_{21} & 0 \\ u_{12} & u_{22} & -v_{21} & -v_{22} & 0 \\ 0 & 0 & 0 & 0 & 1 \end{pmatrix} \begin{pmatrix} \lambda_1 \\ \lambda_2 \\ \mu_1 \\ \mu_2 \\ b \end{pmatrix} \leq \begin{pmatrix} -1 \\ -1 \\ -1 \\ -1 \end{pmatrix}.$$

Rewriting the previous equation in "block" format gives us

$$-\begin{pmatrix} -u_{11} & -u_{12} \\ -u_{21} & -u_{22} \\ v_{11} & v_{12} \\ v_{21} & v_{22} \end{pmatrix} \begin{pmatrix} -u_{11} & -u_{21} & v_{11} & v_{21} \\ -u_{12} & -u_{22} & v_{21} & v_{22} \end{pmatrix} \begin{pmatrix} \lambda_1 \\ \lambda_2 \\ \mu_1 \\ \mu_2 \end{pmatrix} + b \begin{pmatrix} 1 \\ 1 \\ -1 \\ -1 \end{pmatrix} + \begin{pmatrix} 1 \\ 1 \\ 1 \\ 1 \end{pmatrix} \leq \begin{pmatrix} 0 \\ 0 \\ 0 \\ 0 \end{pmatrix},$$

which with the definition

$$X = \begin{pmatrix} -u_{11} & -u_{21} & v_{11} & v_{21} \\ -u_{12} & -u_{22} & v_{21} & v_{22} \end{pmatrix}$$

yields

$$-X^\top X\begin{pmatrix}\lambda\\ \mu\end{pmatrix}+b\begin{pmatrix}\mathbf{1}_2\\ -\mathbf{1}_2\end{pmatrix}+\mathbf{1}_4\leq 0_4. \qquad (*_3)$$

Let us now consider the general case.

Step 1: Write the constraints in matrix form. First we rewrite the constraints as

$$\begin{aligned}
-u_i^\top w+b &\leq -1 & i &= 1,\ldots,p\\
v_j^\top w-b &\leq -1 & j &= 1,\ldots,q,
\end{aligned}$$

and we get the $(p+q)\times(n+1)$ matrix C and the vector $d\in\mathbb{R}^{p+q}$ given by

$$C=\begin{pmatrix}-u_1^\top & 1\\ \vdots & \vdots\\ -u_p^\top & 1\\ v_1^\top & -1\\ \vdots & \vdots\\ v_q^\top & -1\end{pmatrix},\quad d=\begin{pmatrix}-1\\ \vdots\\ -1\end{pmatrix},$$

so the set of inequality constraints is

$$C\begin{pmatrix}w\\ b\end{pmatrix}\leq d.$$

Step 2: The objective function in matrix form is given by

$$J(w,b)=\frac{1}{2}\begin{pmatrix}w^\top & b\end{pmatrix}\begin{pmatrix}I_n & 0_n\\ 0_n^\top & 0\end{pmatrix}\begin{pmatrix}w\\ b\end{pmatrix}.$$

Note that the corresponding matrix is symmetric positive semidefinite, but it is *not* invertible. Thus, the function J is convex but not strictly convex. This will cause some minor trouble in finding the dual function of the problem.

Step 3: If we introduce the generalized Lagrange multipliers $\lambda\in\mathbb{R}^p$ and $\mu\in\mathbb{R}^q$, according to Proposition 14.5, the first KKT condition is

$$\nabla J_{(w,b)}+C^\top\begin{pmatrix}\lambda\\ \mu\end{pmatrix}=0_{n+1},$$

with $\lambda\geq 0,\mu\geq 0$. By the result of Example 3.5,

$$\nabla J_{(w,b)}=\begin{pmatrix}I_n & 0_n\\ 0_n^\top & 0\end{pmatrix}\begin{pmatrix}w\\ b\end{pmatrix}=\begin{pmatrix}w\\ 0\end{pmatrix},$$

so we get

$$\begin{pmatrix} w \\ 0 \end{pmatrix} = -C^\top \begin{pmatrix} \lambda \\ \mu \end{pmatrix},$$

that is,

$$\begin{pmatrix} w \\ 0 \end{pmatrix} = \begin{pmatrix} u_1 & \cdots & u_p & -v_1 & \cdots & -v_q \\ -1 & \cdots & -1 & 1 & \cdots & 1 \end{pmatrix} \begin{pmatrix} \lambda \\ \mu \end{pmatrix}.$$

Consequently,

$$w = \sum_{i=1}^{p} \lambda_i u_i - \sum_{j=1}^{q} \mu_j v_j, \qquad (*_1)$$

and

$$\sum_{j=1}^{q} \mu_j - \sum_{i=1}^{p} \lambda_i = 0. \qquad (*_2)$$

Step 4: Rewrite the constraint using $(*_1)$. Plugging the above expression for w into the constraints $C \begin{pmatrix} w \\ b \end{pmatrix} \leq d$ we get

$$\begin{pmatrix} -u_1^\top & 1 \\ \vdots & \vdots \\ -u_p^\top & 1 \\ v_1^\top & -1 \\ \vdots & \vdots \\ v_q^\top & -1 \end{pmatrix} \begin{pmatrix} u_1 & \cdots & u_p & -v_1 & \cdots & -v_q & 0_n \\ 0 & \cdots & 0 & 0 & \cdots & 0 & 1 \end{pmatrix} \begin{pmatrix} \lambda \\ \mu \\ b \end{pmatrix} \leq \begin{pmatrix} -1 \\ \vdots \\ -1 \end{pmatrix},$$

so if let X be the $n \times (p+q)$ matrix given by

$$X = \begin{pmatrix} -u_1 & \cdots & -u_p & v_1 & \cdots & v_q \end{pmatrix},$$

we obtain

$$w = -X \begin{pmatrix} \lambda \\ \mu \end{pmatrix}, \qquad (*_1')$$

and the above inequalities are written in matrix form as

$$\begin{pmatrix} X^\top & \begin{matrix} \mathbf{1}_p \\ -\mathbf{1}_q \end{matrix} \end{pmatrix} \begin{pmatrix} -X & 0_n \\ 0_{p+q}^\top & 1 \end{pmatrix} \begin{pmatrix} \lambda \\ \mu \\ b \end{pmatrix} \leq -\mathbf{1}_{p+q};$$

that is,

$$-X^\top X \begin{pmatrix} \lambda \\ \mu \end{pmatrix} + b \begin{pmatrix} \mathbf{1}_p \\ -\mathbf{1}_q \end{pmatrix} + \mathbf{1}_{p+q} \leq 0_{p+q}. \qquad (*_3)$$

Equivalently, the ith inequality is

$$-\sum_{j=1}^{p} u_i^\top u_j \lambda_j + \sum_{k=1}^{q} u_i^\top v_k \mu_k + b + 1 \leq 0 \qquad i = 1, \ldots, p,$$

and the $(p+j)$th inequality is

$$\sum_{i=1}^{p} v_j^\top u_i \lambda_i - \sum_{k=1}^{q} v_j^\top v_k \mu_k - b + 1 \leq 0 \qquad j = 1, \ldots, q.$$

We also have $\lambda \geq 0, \mu \geq 0$. Furthermore, if the ith inequality is inactive, then $\lambda_i = 0$, and if the $(p+j)$th inequality is inactive, then $\mu_j = 0$. Since the constraints are affine and since J is convex, if we can find $\lambda \geq 0$, $\mu \geq 0$, and b such that the inequalities in $(*_3)$ are satisfied, and $\lambda_i = 0$ and $\mu_j = 0$ when the corresponding constraint is inactive, then by Proposition 14.5 we have an optimum solution.

Remark: The second KKT condition can be written as

$$\begin{pmatrix}\lambda^\top & \mu^\top\end{pmatrix} \left(-X^\top X \begin{pmatrix}\lambda \\ \mu\end{pmatrix} + b \begin{pmatrix}\mathbf{1}_p \\ -\mathbf{1}_q\end{pmatrix} + \mathbf{1}_{p+q} \right) = 0;$$

that is,

$$-\begin{pmatrix}\lambda^\top & \mu^\top\end{pmatrix} X^\top X \begin{pmatrix}\lambda \\ \mu\end{pmatrix} + b \begin{pmatrix}\lambda^\top & \mu^\top\end{pmatrix} \begin{pmatrix}\mathbf{1}_p \\ -\mathbf{1}_q\end{pmatrix} + \begin{pmatrix}\lambda^\top & \mu^\top\end{pmatrix} \mathbf{1}_{p+q} = 0.$$

Since $(*_2)$ says that $\sum_{i=1}^{p} \lambda_i = \sum_{j=1}^{q} \mu_j$, the second term is zero, and by $(*_1')$ we get

$$w^\top w = \begin{pmatrix}\lambda^\top & \mu^\top\end{pmatrix} X^\top X \begin{pmatrix}\lambda \\ \mu\end{pmatrix} = \sum_{i=1}^{p} \lambda_i + \sum_{j=1}^{q} \mu_j.$$

Thus, we obtain a simple expression for $\|w\|^2$ in terms of λ and μ.

The vectors u_i and v_j for which the i-th inequality is active and the $(p+j)$th inequality is active are called *support vectors*. For every vector u_i or v_j that is not a support vector, the corresponding inequality is inactive, so $\lambda_i = 0$ and $\mu_j = 0$. Thus we see that *only the support vectors contribute to a solution.* If we can *guess* which vectors u_i and v_j are support vectors, namely, those for which $\lambda_i \neq 0$ and $\mu_j \neq 0$, then for each support vector u_i we have an equation

$$-\sum_{j=1}^{p} u_i^\top u_j \lambda_j + \sum_{k=1}^{q} u_i^\top v_k \mu_k + b + 1 = 0,$$

and for each support vector v_j we have an equation

$$\sum_{i=1}^{p} v_j^\top u_i\lambda_i - \sum_{k=1}^{q} v_j^\top v_k\mu_k - b + 1 = 0,$$

with $\lambda_i = 0$ and $\mu_j = 0$ for all non-support vectors, so together with the equation $(*_2)$ we have a linear system with an equal number of equations and variables, which is solvable if our separation problem has a solution. Thus, in principle we can find λ, μ, and b by solving a linear system.

Remark: We can first solve for λ and μ (by eliminating b), and by $(*_1)$ and since $w \neq 0$, there is a least some nonzero λ_{i_0} and thus some nonzero μ_{j_0}, so the corresponding inequalities are equations

$$-\sum_{j=1}^{p} u_{i_0}^\top u_j\lambda_j + \sum_{k=1}^{q} u_{i_0}^\top v_k\mu_k + b + 1 = 0$$

$$\sum_{i=1}^{p} v_{j_0}^\top u_i\lambda_i - \sum_{k=1}^{q} v_{j_0}^\top v_k\mu_k - b + 1 = 0,$$

so b is given in terms of λ and μ by

$$b = \frac{1}{2}(u_{i_0}^\top + v_{j_0}^\top)\left(\sum_{i=1}^{p} \lambda_i u_i - \sum_{j=1}^{p} \mu_j v_j\right).$$

Using the dual of the Lagrangian, we can solve for λ and μ, but typically b is not determined, so we use the above method to find b.

The above nondeterministic procedure in which we guess which vectors are support vectors is not practical. We will see later that a practical method for solving for λ and μ consists in maximizing the dual of the Lagrangian.

If w is an optimal solution, then $\delta = 1/\|w\|$ is the shortest distance from the support vectors to the separating hyperplane $H_{w,b}$ of equation $w^\top x - b = 0$. If we consider the two hyperplanes $H_{w,b+1}$ and $H_{w,b-1}$ of equations

$$w^\top x - b - 1 = 0 \quad \text{and} \quad w^\top x - b + 1 = 0,$$

then $H_{w,b+1}$ and $H_{w,b-1}$ are two hyperplanes parallel to the hyperplane $H_{w,b}$ and the distance between them is 2δ. Furthermore, $H_{w,b+1}$ contains the support vectors u_i, $H_{w,b-1}$ contains the support vectors v_j, and there are no data points u_i or v_j in the open region between these two hyperplanes containing the separating hyperplane $H_{w,b}$ (called a "slab" by Boyd and

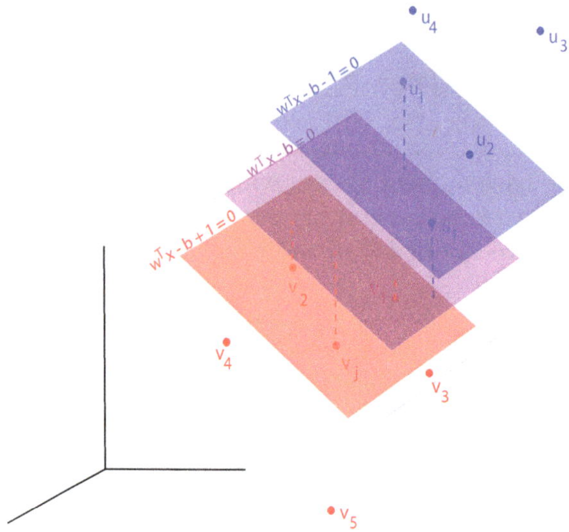

Fig. 14.14 In $\mathbb{R}^3$, the solution to Hard Margin SVM$_{h2}$ is the purple plane sandwiched between the red plane $w^\top x - b + 1 = 0$ and the blue plane $w^\top x - b - 1 = 0$, each of which contains the appropriate support vectors u_i and v_j.

Vandenberghe; see [Boyd and Vandenberghe (2004)], Section 8.6). This situation is illustrated in Figure 14.14.

Even if $p = 1$ and $q = 2$, a solution is not obvious. In the plane, there are four possibilities:

(1) If u_1 is on the segment (v_1, v_2), there is no solution.
(2) If the projection h of u_1 onto the line determined by v_1 and v_2 is between v_1 and v_2, that is $h = (1 - \alpha)v_1 + \alpha_2 v_2$ with $0 \leq \alpha \leq 1$, then it is the line parallel to $v_2 - v_1$ and equidistant to u and both v_1 and v_2, as illustrated in Figure 14.15.
(3) If the projection h of u_1 onto the line determined by v_1 and v_2 is to the right of v_2, that is $h = (1 - \alpha)v_1 + \alpha_2 v_2$ with $\alpha > 1$, then it is the bisector of the line segment (u_1, v_2).
(4) If the projection h of u_1 onto the line determined by v_1 and v_2 is to the left of v_1, that is $h = (1 - \alpha)v_1 + \alpha_2 v_2$ with $\alpha < 0$, then it is the bisector of the line segment (u_1, v_1).

If $p = q = 1$, we can find a solution explicitly. Then $(*_2)$ yields

$$\lambda = \mu,$$

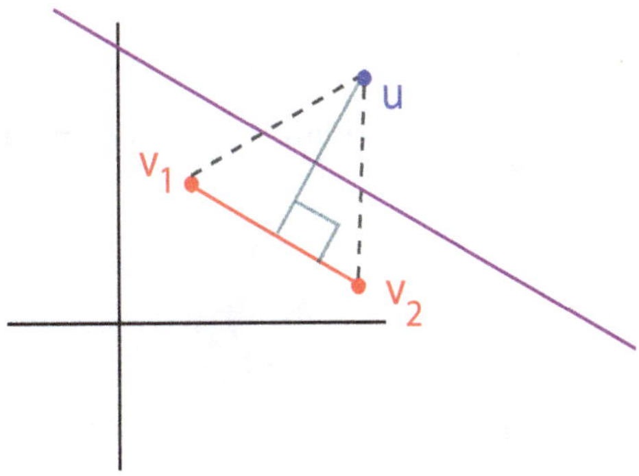

Fig. 14.15 The purple line, which is the bisector of the altitude of the isosceles triangle, separates the two red points from the blue point in a manner which satisfies Hard Margin SVM_{h2}.

and if we guess that the constraints are active, the corresponding equality constraints are

$$\begin{aligned} -u^\top u\lambda + u^\top v\mu + b + 1 &= 0 \\ u^\top v\lambda - v^\top v\mu - b + 1 &= 0, \end{aligned}$$

so we get

$$\begin{aligned} (-u^\top u + u^\top v)\lambda + b + 1 &= 0 \\ (u^\top v - v^\top v)\lambda - b + 1 &= 0, \end{aligned}$$

Adding up the two equations we find

$$(2u^\top v - u^\top u - v^\top v)\lambda + 2 = 0,$$

that is

$$\lambda = \frac{2}{(u-v)^\top (u-v)}.$$

By subtracting the first equation from the second, we find

$$(u^\top u - v^\top v)\lambda - 2b = 0,$$

which yields

$$b = \lambda \frac{(u^\top u - v^\top v)}{2} = \frac{u^\top u - v^\top v}{(u-v)^\top (u-v)}.$$

Then by $(*_1)$ we obtain

$$w = \frac{2(u-v)}{(u-v)^\top (u-v)}.$$

We verify easily that

$$2(u_1 - v_1)x_1 + \cdots + 2(u_n - v_n)x_n = (u_1^2 + \cdots + u_n^2) - (v_1^2 + \cdots + v_n^2)$$

is the equation of the bisector hyperplane between u and v; see Figure 14.16.

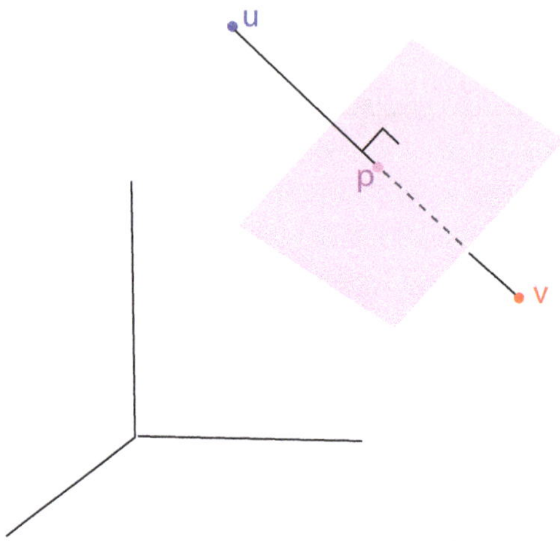

Fig. 14.16 In $\mathbb{R}^3$, the solution to Hard Margin SVM_{h2} for the points u and v is the purple perpendicular planar bisector of $u - v$.

In the next section we will derive the dual of the optimization problem discussed in this section. We will also consider a more flexible solution involving a *soft margin*.

14.7 Lagrangian Duality and Saddle Points

In this section we investigate methods to solve the *Minimization Problem (P)*:

$$\begin{aligned} &\text{minimize} && J(v) \\ &\text{subject to} && \varphi_i(v) \leq 0, \quad i = 1, \ldots, m. \end{aligned}$$

It turns out that under certain conditions the original Problem (P), called *primal problem*, can be solved in two stages with the help another Problem (D), called the *dual problem*. The Dual Problem (D) is a *maximization problem* involving a function G, called the *Lagrangian dual*, and it is obtained by *minimizing* the *Lagrangian* $L(v, \mu)$ of Problem (P) over the variable $v \in \mathbb{R}^n$, holding μ fixed, where $L\colon \Omega \times \mathbb{R}^m_+ \to \mathbb{R}$ is given by

$$L(v, \mu) = J(v) + \sum_{i=1}^{m} \mu_i \varphi_i(v),$$

with $\mu \in \mathbb{R}^m_+$.

The two steps of the method are:

(1) Find the dual function $\mu \mapsto G(\mu)$ explicitly by solving the minimization problem of finding the minimum of $L(v, \mu)$ with respect to $v \in \Omega$, holding μ fixed. This is an unconstrained minimization problem (with $v \in \Omega$). If we are lucky, a unique minimizer u_μ such that $G(\mu) = L(u_\mu, \mu)$ can be found. We will address the issue of uniqueness later on.

(2) Solve the maximization problem of finding the maximum of the function $\mu \mapsto G(\mu)$ over all $\mu \in \mathbb{R}^m_+$. This is basically an unconstrained problem, except for the fact that $\mu \in \mathbb{R}^m_+$.

If Steps (1) and (2) are successful, under some suitable conditions on the function J and the constraints φ_i (for example, if they are convex), for any solution $\lambda \in \mathbb{R}^m_+$ obtained in Step (2), the vector u_λ obtained in Step (1) is an optimal solution of Problem (P). This is proven in Theorem 14.5.

In order to prove Theorem 14.5, which is our main result, we need two intermediate technical results of independent interest involving the notion of saddle point.

The local minima of a function $J\colon \Omega \to \mathbb{R}$ over a domain U defined by inequality constraints are saddle points of the Lagrangian $L(v, \mu)$ associated with J and the constraints φ_i. Then, under some mild hypotheses, the set of solutions of the *Minimization Problem (P)*

$$\begin{aligned} &\text{minimize} && J(v) \\ &\text{subject to} && \varphi_i(v) \leq 0, \quad i = 1, \dots, m \end{aligned}$$

coincides with the set of first arguments of the saddle points of the Lagrangian

$$L(v, \mu) = J(v) + \sum_{i=1}^{m} \mu_i \varphi_i(v).$$

This is proved in Theorem 14.4. To prove Theorem 14.5, we also need Proposition 14.11, a basic property of saddle points.

Definition 14.7. Let $L\colon \Omega \times M \to \mathbb{R}$ be a function defined on a set of the form $\Omega \times M$, where Ω and M are open subsets of two normed vector spaces. A point $(u, \lambda) \in \Omega \times M$ is a *saddle point* of L if u is a minimum of the function $L(-, \lambda)\colon \Omega \to \mathbb{R}$ given by $v \mapsto L(v, \lambda)$ for all $v \in \Omega$ and λ fixed, and λ is a maximum of the function $L(u, -)\colon M \to \mathbb{R}$ given by $\mu \mapsto L(u, \mu)$ for all $\mu \in M$ and u fixed; equivalently,

$$\sup_{\mu \in M} L(u, \mu) = L(u, \lambda) = \inf_{v \in \Omega} L(v, \lambda).$$

Note that the order of the arguments u and λ is important. The second set M will be the set of generalized multipliers, and this is why we use the symbol M. Typically, $M = \mathbb{R}^m_+$.

A saddle point is often depicted as a mountain pass, which explains the terminology; see Figure 14.17. However, this is a bit misleading since other situations are possible; see Figure 14.18.

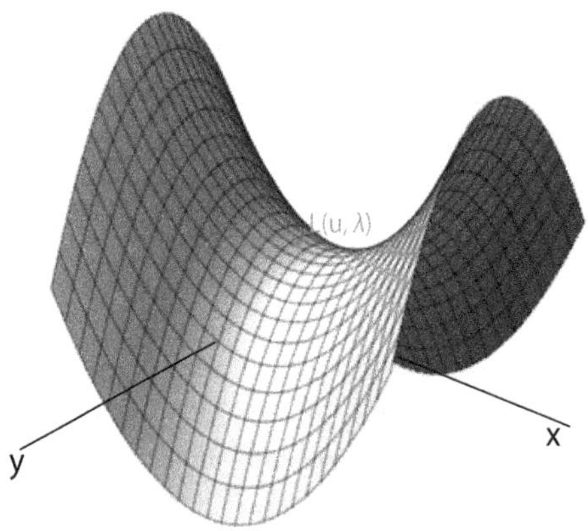

Fig. 14.17 A three-dimensional rendition of a saddle point $L(u, \lambda)$ for the function $L(u, \lambda) = u^2 - \lambda^2$. The plane $x = u$ provides a maximum as the apex of a downward opening parabola, while the plane $y = \lambda$ provides a minimum as the apex of an upward opening parabola.

Proposition 14.11. *If (u, λ) is a saddle point of a function $L\colon \Omega \times M \to \mathbb{R}$,*

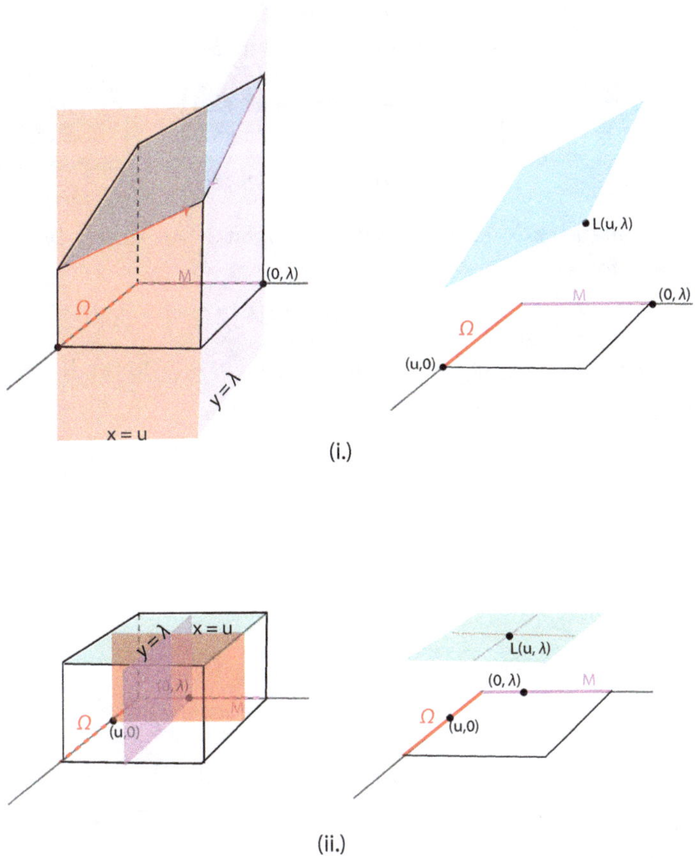

Fig. 14.18 Let $\Omega = \{[t, 0, 0] \mid 0 \leq t \leq 1\}$ and $M = \{[0, t, 0] \mid 0 \leq t \leq 1\}$. In Figure (*i.*), $L(u, \lambda)$ is the blue slanted quadrilateral whose forward vertex is a saddle point. In Figure (*ii.*), $L(u, \lambda)$ is the planar green rectangle composed entirely of saddle points.

then

$$\sup_{\mu \in M} \inf_{v \in \Omega} L(v, \mu) = L(u, \lambda) = \inf_{v \in \Omega} \sup_{\mu \in M} L(v, \mu).$$

Proof. First we prove that the following inequality always holds:

$$\sup_{\mu \in M} \inf_{v \in \Omega} L(v, \mu) \leq \inf_{v \in \Omega} \sup_{\mu \in M} L(v, \mu). \tag{$*_1$}$$

Pick any $w \in \Omega$ and any $\rho \in M$. By definition of inf (the greatest lower bound) and sup (the least upper bound), we have

$$\inf_{v\in\Omega} L(v,\rho) \le L(w,\rho) \le \sup_{\mu\in M} L(w,\mu).$$

The cases where $\inf_{v\in\Omega} L(v,\rho) = -\infty$ or where $\sup_{\mu\in M} L(w,\mu) = +\infty$ may arise, but this is not a problem. Since

$$\inf_{v\in\Omega} L(v,\rho) \le \sup_{\mu\in M} L(w,\mu)$$

and the right-hand side is independent of ρ, it is an upper bound of the left-hand side for all ρ, so

$$\sup_{\mu\in M}\inf_{v\in\Omega} L(v,\mu) \le \sup_{\mu\in M} L(w,\mu).$$

Since the left-hand side is independent of w, it is a lower bound for the right-hand side for all w, so we obtain $(*_1)$:

$$\sup_{\mu\in M}\inf_{v\in\Omega} L(v,\mu) \le \inf_{v\in\Omega}\sup_{\mu\in M} L(v,\mu).$$

To obtain the reverse inequality, we use the fact that (u,λ) is a saddle point, so

$$\inf_{v\in\Omega}\sup_{\mu\in M} L(v,\mu) \le \sup_{\mu\in M} L(u,\mu) = L(u,\lambda)$$

and

$$L(u,\lambda) = \inf_{v\in\Omega} L(v,\lambda) \le \sup_{\mu\in M}\inf_{v\in\Omega} L(v,\mu),$$

and these imply that

$$\inf_{v\in\Omega}\sup_{\mu\in M} L(v,\mu) \le \sup_{\mu\in M}\inf_{v\in\Omega} L(v,\mu), \tag{$*_2$}$$

as desired. □

We now return to our main Minimization Problem (P):

$$\begin{aligned} &\text{minimize} \quad J(v) \\ &\text{subject to} \quad \varphi_i(v) \le 0, \quad i = 1,\ldots,m, \end{aligned}$$

where $J\colon \Omega \to \mathbb{R}$ and the constraints $\varphi_i\colon \Omega \to \mathbb{R}$ are some functions defined on some open subset Ω of some finite-dimensional Euclidean vector space V (more generally, a real Hilbert space V).

Definition 14.8. The *Lagrangian* of the Minimization Problem (P) defined above is the function $L\colon \Omega \times \mathbb{R}^m_+ \to \mathbb{R}$ given by

$$L(v,\mu) = J(v) + \sum_{i=1}^m \mu_i\varphi_i(v),$$

with $\mu = (\mu_1,\ldots,\mu_m)$. The numbers μ_i are called *generalized Lagrange multipliers*.

The following theorem shows that under some suitable conditions, every solution u of the Problem (P) is the first argument of a saddle point (u, λ) of the Lagrangian L, and conversely, if (u, λ) is a saddle point of the Lagrangian L, then u is a solution of the Problem (P).

Theorem 14.4. *Consider Problem (P) defined above where $J\colon \Omega \to \mathbb{R}$ and the constraints $\varphi_i\colon \Omega \to \mathbb{R}$ are some functions defined on some open subset Ω of some finite-dimensional Euclidean vector space V (more generally, a real Hilbert space V). The following facts hold.*

(1) If $(u, \lambda) \in \Omega \times \mathbb{R}^m_+$ is a saddle point of the Lagrangian L associated with Problem (P), then $u \in U$, u is a solution of Problem (P), and $J(u) = L(u, \lambda)$.
(2) If Ω is convex (open), if the functions φ_i $(1 \leq i \leq m)$ and J are convex and differentiable at the point $u \in U$, if the constraints are qualified, and if $u \in U$ is a minimum of Problem (P), then there exists some vector $\lambda \in \mathbb{R}^m_+$ such that the pair $(u, \lambda) \in \Omega \times \mathbb{R}^m_+$ is a saddle point of the Lagrangian L.

Proof. (1) Since (u, λ) is a saddle point of L we have $\sup_{\mu \in \mathbb{R}^m_+} L(u, \mu) = L(u, \lambda)$ which implies that $L(u, \mu) \leq L(u, \lambda)$ for all $\mu \in \mathbb{R}^m_+$, which means that

$$J(u) + \sum_{i=1}^m \mu_i \varphi_i(u) \leq J(u) + \sum_{i=1}^m \lambda_i \varphi_i(u),$$

that is,

$$\sum_{i=1}^m (\mu_i - \lambda_i)\varphi_i(u) \leq 0 \quad \text{for all } \mu \in \mathbb{R}^m_+.$$

If we let each μ_i be large enough, then $\mu_i - \lambda_i > 0$, and if we had $\varphi_i(u) > 0$, then the term $(\mu_i - \lambda_i)\varphi_i(u)$ could be made arbitrarily large and positive, so we conclude that $\varphi_i(u) \leq 0$ for $i = 1, \ldots, m$, and consequently, $u \in U$. For $\mu = 0$, we conclude that $\sum_{i=1}^m \lambda_i \varphi_i(u) \geq 0$. However, since $\lambda_i \geq 0$ and $\varphi_i(u) \leq 0$, (since $u \in U$), we have $\sum_{i=1}^m \lambda_i \varphi_i(u) \leq 0$. Combining these two inequalities shows that

$$\sum_{i=1}^m \lambda_i \varphi_i(u) = 0. \tag{$*_1$}$$

This shows that $J(u) = L(u, \lambda)$. Since the inequality $L(u, \lambda) \leq L(v, \lambda)$ is

$$J(u) + \sum_{i=1}^m \lambda_i \varphi_i(u) \leq J(v) + \sum_{i=1}^m \lambda_i \varphi_i(v),$$

by $(*_1)$ we obtain

$$\begin{aligned} J(u) &\le J(v) + \sum_{i=1}^{m} \lambda_i \varphi_i(v) && \text{for all } v \in \Omega \\ &\le J(v) && \text{for all } v \in U \text{ (since } \varphi_i(v) \le 0 \text{ and } \lambda_i \ge 0), \end{aligned}$$

which shows that u is a minimum of J on U.

(2) The hypotheses required to apply Theorem 14.2(1) are satisfied. Consequently if $u \in U$ is a solution of Problem (P), then there exists some vector $\lambda \in \mathbb{R}^m_+$ such that the KKT conditions hold:

$$J'(u) + \sum_{i=1}^{m} \lambda_i (\varphi'_i)_u = 0 \quad \text{and} \quad \sum_{i=1}^{m} \lambda_i \varphi_i(u) = 0.$$

The second equation yields

$$L(u, \mu) = J(u) + \sum_{i=1}^{m} \mu_i \varphi_i(u) \le J(u) = J(u) + \sum_{i=1}^{m} \lambda_i \varphi_i(u) = L(u, \lambda),$$

that is,

$$L(u, \mu) \le L(u, \lambda) \quad \text{for all } \mu \in \mathbb{R}^m_+ \tag{$*_2$}$$

(since $\varphi_i(u) \le 0$ as $u \in U$), and since the function $v \mapsto J(v) + \sum_{i=1} \lambda_i \varphi_i(v) = L(v, \lambda)$ is convex as a sum of convex functions, by Theorem 4.5(4), the first equation is a sufficient condition for the existence of minimum. Consequently,

$$L(u, \lambda) \le L(v, \lambda) \quad \text{for all } v \in \Omega, \tag{$*_3$}$$

and $(*_2)$ and $(*_3)$ show that (u, λ) is a saddle point of L. □

To recap what we just proved, under some mild hypotheses, the set of solutions of the Minimization Problem (P)

$$\begin{aligned} &\text{minimize} && J(v) \\ &\text{subject to} && \varphi_i(v) \le 0, \quad i = 1, \ldots, m \end{aligned}$$

coincides with the set of first arguments of the saddle points of the Lagrangian

$$L(v, \mu) = J(v) + \sum_{i=1}^{m} \mu_i \varphi_i(v),$$

and for any optimum $u \in U$ of Problem (P), we have $J(u) = L(u, \lambda)$.

Therefore, if we knew some particular second argument λ of these saddle points, then the *constrained* Problem (P) would be replaced by the *unconstrained* Problem (P_λ):

$$\begin{aligned} &\text{find } u_\lambda \in \Omega \text{ such that} \\ &L(u_\lambda, \lambda) = \inf_{v \in \Omega} L(v, \lambda). \end{aligned}$$

How do we find such an element $\lambda \in \mathbb{R}^m_+$*?*

For this, remember that for a saddle point (u_λ, λ), by Proposition 14.11, we have

$$L(u_\lambda, \lambda) = \inf_{v \in \Omega} L(v, \lambda) = \sup_{\mu \in \mathbb{R}^m_+} \inf_{v \in \Omega} L(v, \mu),$$

so we are naturally led to introduce the function $G \colon \mathbb{R}^m_+ \to \mathbb{R}$ given by

$$G(\mu) = \inf_{v \in \Omega} L(v, \mu) \quad \mu \in \mathbb{R}^m_+,$$

and then λ will be a solution of the problem

$$\begin{aligned} &\text{find } \lambda \in \mathbb{R}^m_+ \text{ such that} \\ &G(\lambda) = \sup_{\mu \in \mathbb{R}^m_+} G(\mu), \end{aligned}$$

which is equivalent to the *Maximization Problem* (D):

$$\begin{aligned} &\text{maximize} \quad G(\mu) \\ &\text{subject to} \quad \mu \in \mathbb{R}^m_+. \end{aligned}$$

Definition 14.9. Given the Minimization Problem (P)

$$\begin{aligned} &\text{minimize} \quad J(v) \\ &\text{subject to} \quad \varphi_i(v) \leq 0, \quad i = 1, \ldots, m, \end{aligned}$$

where $J \colon \Omega \to \mathbb{R}$ and the constraints $\varphi_i \colon \Omega \to \mathbb{R}$ are some functions defined on some open subset Ω of some finite-dimensional Euclidean vector space V (more generally, a real Hilbert space V), the function $G \colon \mathbb{R}^m_+ \to \mathbb{R}$ given by

$$G(\mu) = \inf_{v \in \Omega} L(v, \mu) \quad \mu \in \mathbb{R}^m_+,$$

is called the *Lagrange dual function* (or simply *dual function*). The *Problem* (D)

$$\begin{aligned} &\text{maximize} \quad G(\mu) \\ &\text{subject to} \quad \mu \in \mathbb{R}^m_+ \end{aligned}$$

is called the *Lagrange dual problem*. The Problem (P) is often called the *primal problem*, and (D) is the *dual problem*. The variable μ is called the *dual variable*. The variable $\mu \in \mathbb{R}^m_+$ is said to be *dual feasible* if $G(\mu)$ is defined (not $-\infty$). If $\lambda \in \mathbb{R}^m_+$ is a maximum of G, then we call it a *dual optimal* or an *optimal Lagrange multiplier*.

Since

$$L(v, \mu) = J(v) + \sum_{i=1}^{m} \mu_i \varphi_i(v),$$

the function $G(\mu) = \inf_{v \in \Omega} L(v, \mu)$ is the pointwise infimum of some affine functions of μ, so it is *concave*, even if the φ_i are not convex. One of the main advantages of the dual problem over the primal problem is that it is a *convex optimization problem*, since we wish to maximize a concave objective function G (thus minimize $-G$, a convex function), and the constraints $\mu \geq 0$ are convex. In a number of practical situations, the dual function G can indeed be computed.

To be perfectly rigorous, we should mention that the dual function G is actually a *partial function*, because it takes the value $-\infty$ when the map $v \mapsto L(v, \mu)$ is unbounded below.

Example 14.5. Consider the Linear Program (P)

$$\begin{aligned} &\text{minimize} \quad c^\top v \\ &\text{subject to} \quad Av \leq b, \ \ v \geq 0, \end{aligned}$$

where A is an $m \times n$ matrix. The constraints $v \geq 0$ are rewritten as $-v_i \leq 0$, so we introduce Lagrange multipliers $\mu \in \mathbb{R}^m_+$ and $\nu \in \mathbb{R}^n_+$, and we have the Lagrangian

$$\begin{aligned} L(v, \mu, \nu) &= c^\top v + \mu^\top (Av - b) - \nu^\top v \\ &= -b^\top \mu + (c + A^\top \mu - \nu)^\top v. \end{aligned}$$

The linear function $v \mapsto (c + A^\top \mu - \nu)^\top v$ is unbounded below unless $c + A^\top \mu - \nu = 0$, so the dual function $G(\mu, \nu) = \inf_{v \in \mathbb{R}^n} L(v, \mu, \nu)$ is given for all $\mu \geq 0$ and $\nu \geq 0$ by

$$G(\mu, \nu) = \begin{cases} -b^\top \mu & \text{if } A^\top \mu - \nu + c = 0, \\ -\infty & \text{otherwise.} \end{cases}$$

The domain of G is a proper subset of $\mathbb{R}^m_+ \times \mathbb{R}^n_+$.

Observe that the value $G(\mu,\nu)$ of the function G, when it is defined, is independent of the second argument ν. Since we are interested in maximizing G, this suggests introducing the function $\widehat{G}$ of the single argument μ given by

$$\widehat{G}(\mu) = -b^\top \mu,$$

which is defined for all $\mu \in \mathbb{R}^m_+$.

Of course, $\sup_{\mu\in\mathbb{R}^m_+} \widehat{G}(\mu)$ and $\sup_{(\mu,\nu)\in\mathbb{R}^m_+\times\mathbb{R}^n_+} G(\mu,\, \nu)$ are generally different, but note that $\widehat{G}(\mu) = G(\mu,\nu)$ iff there is some $\nu \in \mathbb{R}^n_+$ such that $A^\top\mu - \nu + c = 0$ iff $A^\top\mu + c \geq 0$. Therefore, finding $\sup_{(\mu,\nu)\in\mathbb{R}^m_+\times\mathbb{R}^n_+} G(\mu,\nu)$ is equivalent to the constrained Problem (D_1)

$$\begin{aligned} &\text{maximize} && -b^\top\mu \\ &\text{subject to} && A^\top\mu \geq -c, \;\; \mu \geq 0. \end{aligned}$$

The above problem is the dual of the Linear Program (P).

In summary, the dual function G of a primary Problem (P) often contains hidden inequality constraints that define its domain, and sometimes it is possible to make these domain constraints $\psi_1(\mu) \leq 0, \ldots, \psi_p(\mu) \leq 0$ explicit, to define a new function $\widehat{G}$ that depends only on $q < m$ of the variables μ_i and is defined for all values $\mu_i \geq 0$ of these variables, and to replace the Maximization Problem (D), find $\sup_{\mu\in\mathbb{R}^m_+} G(\mu)$, by the constrained Problem (D_1)

$$\begin{aligned} &\text{maximize} && \widehat{G}(\mu) \\ &\text{subject to} && \psi_i(\mu) \leq 0, \quad i = 1, \ldots, p. \end{aligned}$$

Problem (D_1) is different from the Dual Program (D), but it is equivalent to (D) as a maximization problem.

14.8 Weak and Strong Duality

Another important property of the dual function G is that it provides a *lower bound* on the value of the objective function J. Indeed, we have

$$G(\mu) \leq L(u,\mu) \leq J(u) \quad \text{for all } u \in U \text{ and all } \mu \in \mathbb{R}^m_+, \tag{$\dagger$}$$

since $\mu \geq 0$ and $\varphi_i(u) \leq 0$ for $i = 1, \ldots, m$, so

$$G(\mu) = \inf_{v\in\Omega} L(v,\mu) \leq L(u,\mu) = J(u) + \sum_{i=1}^m \mu_i\varphi_i(u) \leq J(u).$$

If the Primal Problem (P) has a minimum denoted p^* and the Dual Problem (D) has a maximum denoted d^*, then the above inequality implies that

$$d^* \leq p^* \tag{$\dagger_w$}$$

known as *weak duality*. Equivalently, for every optimal solution λ^* of the dual problem and every optimal solution u^* of the primal problem, we have

$$G(\lambda^*) \leq J(u^*). \tag{$\dagger_{w'}$}$$

In particular, if $p^* = -\infty$, which means that the primal problem is unbounded below, then the dual problem is unfeasible. Conversely, if $d^* = +\infty$, which means that the dual problem is unbounded above, then the primal problem is unfeasible.

Definition 14.10. The difference $p^* - d^* \geq 0$ is called the *optimal duality gap*. If the duality gap is zero, that is, $p^* = d^*$, then we say that *strong duality* holds.

Even when the duality gap is strictly positive, the inequality $(\dagger_w)$ can be helpful to find a lower bound on the optimal value of a primal problem that is difficult to solve, since the dual problem is *always* convex.

If the primal problem and the dual problem are feasible and if the optimal values p^* and d^* are finite and $p^* = d^*$ (no duality gap), then the complementary slackness conditions hold for the inequality constraints.

Proposition 14.12. *(Complementary Slackness) Given the Minimization Problem (P)*

$$\begin{aligned} &\text{minimize} && J(v) \\ &\text{subject to} && \varphi_i(v) \leq 0, \quad i = 1, \ldots, m, \end{aligned}$$

and its Dual Problem (D)

$$\begin{aligned} &\text{maximize} && G(\mu) \\ &\text{subject to} && \mu \in \mathbb{R}^m_+, \end{aligned}$$

if both (P) and (D) are feasible, $u \in U$ is an optimal solution of (P), $\lambda \in \mathbb{R}^m_+$ is an optimal solution of (D), and $J(u) = G(\lambda)$, then

$$\sum_{i=1}^m \lambda_i \varphi_i(u) = 0.$$

In other words, if the constraint φ_i is inactive at u, then $\lambda_i = 0$.

Proof. Since $J(u) = G(\lambda)$ we have

$$
\begin{aligned}
J(u) &= G(\lambda) \\
&= \inf_{v\in\Omega}\left(J(v) + \sum_{i=1}^{m} \lambda_i \varphi_i(v)\right) && \text{by definition of } G \\
&\leq J(u) + \sum_{i=1}^{m} \lambda_i \varphi_i(u) && \text{the greatest lower bound is a lower bound} \\
&\leq J(u) && \text{since } \lambda_i \geq 0, \varphi_i(u) \leq 0.
\end{aligned}
$$

which implies that $\sum_{i=1}^{m} \lambda_i \varphi_i(u) = 0$. ☐

Going back to Example 14.5, we see that weak duality says that for any feasible solution u of the Primal Problem (P), that is, some $u \in \mathbb{R}^n$ such that

$$Au \leq b, \ \ u \geq 0,$$

and for any feasible solution $\mu \in \mathbb{R}^m$ of the Dual Problem (D_1), that is,

$$A^\top \mu \geq -c, \ \ \mu \geq 0,$$

we have

$$-b^\top \mu \leq c^\top u.$$

Actually, if u and λ are optimal, then we know from Theorem 11.1 that strong duality holds, namely $-b^\top \mu = c^\top u$, but the proof of this fact is nontrivial.

The following theorem establishes a link between the solutions of the Primal Problem (P) and those of the Dual Problem (D). It also gives sufficient conditions for the duality gap to be zero.

Theorem 14.5. *Consider the Minimization Problem (P):*

$$
\begin{aligned}
&\text{minimize} \quad J(v) \\
&\text{subject to} \quad \varphi_i(v) \leq 0, \quad i = 1, \ldots, m,
\end{aligned}
$$

where the functions J and φ_i are defined on some open subset Ω of a finite-dimensional Euclidean vector space V (more generally, a real Hilbert space V).

(1) Suppose the functions $\varphi_i \colon \Omega \to \mathbb{R}$ are continuous, and that for every $\mu \in \mathbb{R}^m_+$, the Problem (P_μ):

$$
\begin{aligned}
&\text{minimize} \quad L(v, \mu) \\
&\text{subject to} \quad v \in \Omega,
\end{aligned}
$$

has a unique solution u_μ, *so that*

$$L(u_\mu, \mu) = \inf_{v \in \Omega} L(v, \mu) = G(\mu),$$

and the function $\mu \mapsto u_\mu$ *is continuous (on* $\mathbb{R}^m_+$*). Then the function* G *is differentiable for all* $\mu \in \mathbb{R}^m_+$, *and*

$$G'_\mu(\xi) = \sum_{i=1}^{m} \xi_i \varphi_i(u_\mu) \quad \text{for all } \xi \in \mathbb{R}^m.$$

If λ *is any solution of Problem* (D):

$$\begin{aligned} &\text{maximize} \quad G(\mu) \\ &\text{subject to} \quad \mu \in \mathbb{R}^m_+, \end{aligned}$$

then the solution u_λ *of the corresponding problem* (P_λ) *is a solution of Problem* (P).

(2) *Assume Problem* (P) *has some solution* $u \in U$, *and that* Ω *is convex (open), the functions* φ_i $(1 \leq i \leq m)$ *and* J *are convex and differentiable at* u, *and that the constraints are qualified. Then Problem* (D) *has a solution* $\lambda \in \mathbb{R}^m_+$, *and* $J(u) = G(\lambda)$; *that is, the duality gap is zero.*

Proof. (1) Our goal is to prove that for any solution λ of Problem (D), the pair (u_λ, λ) is a saddle point of L. By Theorem 14.4(1), the point $u_\lambda \in U$ is a solution of Problem (P).

Since $\lambda \in \mathbb{R}^m_+$ is a solution of Problem (D), by definition of $G(\lambda)$ and since u_λ satisfies Problem (P_λ), we have

$$G(\lambda) = \inf_{v \in \Omega} L(v, \lambda) = L(u_\lambda, \lambda),$$

which is one of the two equations characterizing a saddle point. In order to prove the second equation characterizing a saddle point,

$$\sup_{\mu \in \mathbb{R}^m_+} L(u_\mu, \mu) = L(u_\lambda, \lambda),$$

we will begin by proving that the function G is differentiable for all $\mu \in \mathbb{R}^m_+$, in order to be able to apply Theorem 4.4 to conclude that since G has a maximum at λ, that is, $-G$ has minimum at λ, then $-G'_\lambda(\mu - \lambda) \geq 0$ for all $\mu \in \mathbb{R}^m_+$. In fact, we prove that

$$G'_\mu(\xi) = \sum_{i=1}^{m} \xi_i \varphi_i(u_\mu) \quad \text{for all } \xi \in \mathbb{R}^m. \qquad (*_{\text{deriv}})$$

Consider any two points μ and $\mu+\xi$ in $\mathbb{R}^m_+$. By definition of u_μ we have

$$L(u_\mu, \mu) \leq L(u_{\mu+\xi}, \mu),$$

which means that

$$J(u_\mu) + \sum_{i=1}^m \mu_i \varphi_i(u_\mu) \leq J(u_{\mu+\xi}) + \sum_{i=1}^m \mu_i \varphi_i(u_{\mu+\xi}), \qquad (*_1)$$

and since $G(\mu) = L(u_\mu, \mu) = J(u_\mu) + \sum_{i=1}^m \mu_i \varphi_i(u_\mu)$ and $G(\mu+\xi) = L(u_{\mu+\xi}, \mu+\xi) = J(u_{\mu+\xi}) + \sum_{i=1}^m (\mu_i + \xi_i)\varphi_i(u_{\mu+\xi})$, we have

$$G(\mu+\xi) - G(\mu) = J(u_{\mu+\xi}) - J(u_\mu) + \sum_{i=1}^m (\mu_i + \xi_i)\varphi_i(u_{\mu+\xi}) - \sum_{i=1}^m \mu_i \varphi_i(u_\mu). \qquad (*_2)$$

Since $(*_1)$ can be written as

$$0 \leq J(u_{\mu+\xi}) - J(u_\mu) + \sum_{i=1}^m \mu_i \varphi_i(u_{\mu+\xi}) - \sum_{i=1}^m \mu_i \varphi_i(u_\mu),$$

by adding $\sum_{i=1}^m \xi_i \varphi_i(u_{\mu+\xi})$ to both sides of the above inequality and using $(*_2)$ we get

$$\sum_{i=1}^m \xi_i \varphi_i(u_{\mu+\xi}) \leq G(\mu+\xi) - G(\mu). \qquad (*_3)$$

By definition of $u_{\mu+\xi}$ we have

$$L(u_{\mu+\xi}, \mu+\xi) \leq L(u_\mu, \mu+\xi),$$

which means that

$$J(u_{\mu+\xi}) + \sum_{i=1}^m (\mu_i + \xi_i)\varphi_i(u_{\mu+\xi}) \leq J(u_\mu) + \sum_{i=1}^m (\mu_i + \xi_i)\varphi_i(u_\mu). \qquad (*_4)$$

This can be written as

$$J(u_{\mu+\xi}) - J(u_\mu) + \sum_{i=1}^m (\mu_i + \xi_i)\varphi_i(u_{\mu+\xi}) - \sum_{i=1}^m (\mu_i + \xi_i)\varphi_i(u_\mu) \leq 0,$$

and by adding $\sum_{i=1}^m \xi_i \varphi_i(u_\mu)$ to both sides of the above inequality and using $(*_2)$ we get

$$G(\mu+\xi) - G(\mu) \leq \sum_{i=1}^m \xi_i \varphi_i(u_\mu). \qquad (*_5)$$

By putting $(*_3)$ and $(*_5)$ together we obtain

$$\sum_{i=1}^m \xi_i \varphi_i(u_{\mu+\xi}) \leq G(\mu+\xi) - G(\mu) \leq \sum_{i=1}^m \xi_i \varphi_i(u_\mu). \qquad (*_6)$$

Consequently there is some $\theta \in [0,1]$ such that

$$G(\mu+\xi) - G(\mu) = (1-\theta)\Big(\sum_{i=1}^m \xi_i\varphi_i(u_\mu)\Big) + \theta\Big(\sum_{i=1}^m \xi_i\varphi_i(u_{\mu+\xi})\Big)$$

$$= \sum_{i=1}^m \xi_i\varphi_i(u_\mu) + \theta\Big(\sum_{i=1}^m \xi_i(\varphi_i(u_{\mu+\xi}) - \varphi_i(u_\mu))\Big).$$

Since by hypothesis the functions $\mu \mapsto u_\mu$ (from $\mathbb{R}^m_+$ to Ω) and $\varphi_i \colon \Omega \to \mathbb{R}$ are continuous, for any $\mu \in \mathbb{R}^m_+$ we can write

$$G(\mu+\xi) - G(\mu) = \sum_{i=1}^m \xi_i\varphi_i(u_\mu) + \|\xi\|\,\epsilon(\xi), \quad \text{with } \lim_{\xi\mapsto 0}\epsilon(\xi) = 0, \quad (*_7)$$

for any $\|\ \|$ norm on $\mathbb{R}^m$. Equation $(*_7)$ show that G is differentiable for any $\mu \in \mathbb{R}^m_+$, and that

$$G'_\mu(\xi) = \sum_{i=1}^m \xi_i\varphi_i(u_\mu) \quad \text{for all } \xi \in \mathbb{R}^m. \quad (*_8)$$

Actually there is a small problem, namely that the notion of derivative was defined for a function defined on an *open* set, but $\mathbb{R}^m_+$ is not open. The difficulty only arises to ensure that the derivative is unique, but in our case we have a unique expression for the derivative so there is no problem as far as defining the derivative. There is still a potential problem, which is that we would like to apply Theorem 4.4 to conclude that since G has a maximum at λ, that is, $-G$ has a minimum at λ, then

$$-G'_\lambda(\mu - \lambda) \geq 0 \quad \text{for all } \mu \in \mathbb{R}^m_+, \quad (*_9)$$

but the hypotheses of Theorem 4.4 require the domain of the function to be open. Fortunately, close examination of the proof of Theorem 4.4 shows that the proof still holds with $U = \mathbb{R}^m_+$. Therefore, $(*_8)$ holds, Theorem 4.4 is valid, which in turn implies

$$G'_\lambda(\mu - \lambda) \leq 0 \quad \text{for all } \mu \in \mathbb{R}^m_+, \quad (*_{10})$$

which, using the expression for G'_λ given in $(*_8)$ gives

$$\sum_{i=1}^m \mu_i\varphi_i(u_\lambda) \leq \sum_{i=1}^m \lambda_i\varphi_i(u_\lambda), \quad \text{for all } \mu \in \mathbb{R}^m_+. \quad (*_{11})$$

As a consequence of $(*_{11})$, we obtain

$$L(u_\lambda, \mu) = J(u_\lambda) + \sum_{i=1}^m \mu_i\varphi_i(u_\lambda)$$

$$\leq J(u_\lambda) + \sum_{i=1}^m \lambda_i\varphi_i(u_\lambda) = L(u_\lambda, \lambda),$$

for all $\mu \in \mathbb{R}^m_+$, that is,

$$L(u_\lambda, \mu) \leq L(u_\lambda, \lambda), \quad \text{for all } \mu \in \mathbb{R}^m_+, \qquad (*_{12})$$

which implies the second inequality

$$\sup_{\mu \in \mathbb{R}^m_+} L(u_\mu, \mu) = L(u_\lambda, \lambda)$$

stating that (u_λ, λ) is a saddle point. Therefore, (u_λ, λ) is a saddle point of L, as claimed.

(2) The hypotheses are exactly those required by Theorem 14.4(2), thus there is some $\lambda \in \mathbb{R}^m_+$ such that (u, λ) is a saddle point of the Lagrangian L, and by Theorem 14.4(1) we have $J(u) = L(u, \lambda)$. By Proposition 14.11, we have

$$J(u) = L(u, \lambda) = \inf_{v \in \Omega} L(v, \lambda) = \sup_{\mu \in \mathbb{R}^m_+} \inf_{v \in \Omega} L(v, \mu),$$

which can be rewritten as

$$J(u) = G(\lambda) = \sup_{\mu \in \mathbb{R}^m_+} G(\mu).$$

In other words, Problem (D) has a solution, and $J(u) = G(\lambda)$. ☐

Remark: Note that Theorem 14.5(2) could have already be obtained as a consequence of Theorem 14.4(2), but the dual function G was not yet defined. If (u, λ) is a saddle point of the Lagrangian L (defined on $\Omega \times \mathbb{R}^m_+$), then by Proposition 14.11, the vector λ is a solution of Problem (D). Conversely, under the hypotheses of Part (1) of Theorem 14.5, if λ is a solution of Problem (D), then (u_λ, λ) is a saddle point of L. *Consequently, under the above hypotheses, the set of solutions of the Dual Problem (D) coincide with the set of second arguments λ of the saddle points (u, λ) of L.* In some sense, this result is the "dual" of the result stated in Theorem 14.4, namely that the set of solutions of Problem (P) coincides with set of first arguments u of the saddle points (u, λ) of L.

Informally, in Theorem 14.5(1), the hypotheses say that if $G(\mu)$ can be "computed nicely," in the sense that there is a unique minimizer u_μ of $L(v, \mu)$ (with $v \in \Omega$) such that $G(\mu) = L(u_\mu, \mu)$, and if a maximizer λ of $G(\mu)$ (with $\mu \in \mathbb{R}^m_+$) can be determined, then u_λ yields the minimum value of J, that is, $p^* = J(u_\lambda)$. If the constraints are qualified and if the functions J and φ_i are convex and differentiable, then since the KKT conditions hold, the duality gap is zero; that is,

$$G(\lambda) = L(u_\lambda, \lambda) = J(u_\lambda).$$

Example 14.6. Going back to Example 14.5 where we considered the linear program (P)

$$\begin{aligned}&\text{minimize} \quad c^\top v\\ &\text{subject to} \quad Av \leq b, \ v \geq 0,\end{aligned}$$

with A an $m \times n$ matrix, the Lagrangian $L(\mu, \nu)$ is given by

$$L(v, \mu, \nu) = -b^\top \mu + (c + A^\top \mu - \nu)^\top v,$$

and we found that the dual function $G(\mu, \nu) = \inf_{v \in \mathbb{R}^n} L(v, \mu, \nu)$ is given for all $\mu \geq 0$ and $\nu \geq 0$ by

$$G(\mu, \nu) = \begin{cases} -b^\top \mu & \text{if } A^\top \mu - \nu + c = 0, \\ -\infty & \text{otherwise.} \end{cases}$$

The hypotheses of Theorem 14.5(1) certainly fail since there are infinitely $u_{\mu,\nu} \in \mathbb{R}^n$ such that $G(\mu, \nu) = \inf_{v \in \mathbb{R}^n} L(v, \mu, \nu) = L(u_{\mu,\nu}, \mu, \nu)$. Therefore, the dual function G is no help in finding a solution of the Primal Problem (P). As we saw earlier, if we consider the modified dual Problem (D_1) then strong duality holds, but this *does not* follow from Theorem 14.5, and a different proof is required.

Thus, we have the somewhat counter-intuitive situation that the *general* theory of Lagrange duality does not apply, at least directly, to linear programming, a fact that is not sufficiently emphasized in many expositions. A separate treatment of duality is required.

Unlike the case of linear programming, which needs a separate treatment, Theorem 14.5 applies to the optimization problem involving a convex quadratic objective function and a set of affine inequality constraints. So in some sense, convex quadratic programming is simpler than linear programming!

Example 14.7. Consider the quadratic objective function

$$J(v) = \frac{1}{2} v^\top A v - v^\top b,$$

where A is an $n \times n$ matrix which is symmetric positive definite, $b \in \mathbb{R}^n$, and the constraints are affine inequality constraints of the form

$$Cv \leq d,$$

where C is an $m \times n$ matrix and $d \in \mathbb{R}^m$. For the time being, we do not assume that C has rank m. Since A is symmetric positive definite, J

is strictly convex, as implied by Proposition 4.6 (see Example 4.6). The Lagrangian of this quadratic optimization problem is given by

$$\begin{aligned} L(v,\mu) &= \frac{1}{2}v^\top Av - v^\top b + (Cv-d)^\top \mu \\ &= \frac{1}{2}v^\top Av - v^\top(b - C^\top \mu) - \mu^\top d. \end{aligned}$$

Since A is symmetric positive definite, by Proposition 6.2, the function $v \mapsto L(v,\mu)$ has a unique minimum obtained for the solution u_μ of the linear system

$$Av = b - C^\top \mu;$$

that is,

$$u_\mu = A^{-1}(b - C^\top \mu).$$

This shows that the Problem (P_μ) has a unique solution which depends continuously on μ. Then any solution λ of the dual problem, $u_\lambda = A^{-1}(b - C^\top \lambda)$ is an optimal solution of the primal problem.

We compute $G(\mu)$ as follows:

$$\begin{aligned} G(\mu) = L(u_\mu,\mu) &= \frac{1}{2}u_\mu^\top Au_\mu - u_\mu^\top(b - C^\top\mu) - \mu^\top d \\ &= \frac{1}{2}u_\mu^\top(b - C^\top\mu) - u_\mu^\top(b - C^\top\mu) - \mu^\top d \\ &= -\frac{1}{2}u_\mu^\top(b - C^\top\mu) - \mu^\top d \\ &= -\frac{1}{2}(b - C^\top\mu)^\top A^{-1}(b - C^\top\mu) - \mu^\top d \\ &= -\frac{1}{2}\mu^\top CA^{-1}C^\top\mu + \mu^\top(CA^{-1}b - d) - \frac{1}{2}b^\top A^{-1}b. \end{aligned}$$

Since A is symmetric positive definite, the matrix $CA^{-1}C^\top$ is symmetric positive semidefinite. Since A^{-1} is also symmetric positive definite, $\mu^\top CA^{-1}C^\top\mu = 0$ iff $(C^\top\mu)^\top A^{-1}(C^\top\mu) = 0$ iff $C^\top\mu = 0$ implies $\mu = 0$, that is, $\operatorname{Ker} C^\top = (0)$, which is equivalent to $\operatorname{Im}(C) = \mathbb{R}^m$, namely if C has rank m (in which case, $m \leq n$). Thus $CA^{-1}C^\top$ is symmetric positive definite iff C has rank m.

We showed just after Theorem 13.6 that the functional $v \mapsto (1/2)v^\top Av$ is elliptic iff A is symmetric positive definite, and Theorem 13.6 shows that an elliptic functional is coercive, which is the hypothesis used in Theorem 13.2. Therefore, by Theorem 13.2, if the inequalities $Cx \leq d$ have a solution, the primal problem has a unique solution. In this case, as a consequence, by Theorem 14.5(2), the function $-G(\mu)$ always has a minimum, which is unique if C has rank m. The fact that $-G(\mu)$ has a minimum is not obvious when C has rank $< m$, since in this case $CA^{-1}C^\top$ is not invertible.

We also verify easily that the gradient of G is given by

$$\nabla G_\mu = Cu_\mu - d = -CA^{-1}C^\top \mu + CA^{-1}b - d.$$

Observe that since $CA^{-1}C^\top$ is symmetric positive semidefinite, $-G(\mu)$ is convex.

Therefore, if C has rank m, a solution of Problem (P) is obtained by finding the unique solution λ of the equation

$$-CA^{-1}C^\top \mu + CA^{-1}b - d = 0,$$

and then the minimum u_λ of Problem (P) is given by

$$u_\lambda = A^{-1}(b - C^\top \lambda).$$

If C has rank $< m$, then we can find $\lambda \geq 0$ by finding a feasible solution of the linear program whose set of constraints is given by

$$-CA^{-1}C^\top \mu + CA^{-1}b - d = 0,$$

using the standard method of adding nonnegative slack variables $\xi_1, \ldots, \xi_m$ and maximizing $-(\xi_1 + \cdots + \xi_m)$.

14.9 Handling Equality Constraints Explicitly

Sometimes it is desirable to handle equality constraints explicitly (for instance, this is what Boyd and Vandenberghe do, see [Boyd and Vandenberghe (2004)]). The only difference is that the Lagrange multipliers associated with *equality constraints* are *not required* to be nonnegative, as we now show.

Consider the *Optimization Problem* (P')

$$\begin{aligned} &\text{minimize} && J(v) \\ &\text{subject to} && \varphi_i(v) \leq 0, \quad i = 1, \ldots, m \\ & && \psi_j(v) = 0, \quad j = 1, \ldots, p. \end{aligned}$$

We treat each equality constraint $\psi_j(u) = 0$ as the conjunction of the inequalities $\psi_j(u) \leq 0$ and $-\psi_j(u) \leq 0$, and we associate Lagrange multipliers $\lambda \in \mathbb{R}^m_+$, and $\nu^+, \nu^- \in \mathbb{R}^p_+$. Assuming that the constraints are qualified, by Theorem 14.1, the KKT conditions are

$$J'_u + \sum_{i=1}^m \lambda_i (\varphi'_i)_u + \sum_{j=1}^p \nu_j^+ (\psi'_j)_u - \sum_{j=1}^p \nu_j^- (\psi'_j)_u = 0,$$

and

$$\sum_{i=1}^{m} \lambda_i \varphi_i(u) + \sum_{j=1}^{p} \nu_j^+ \psi_j(u) - \sum_{j=1}^{p} \nu_j^- \psi_j(u) = 0,$$

with $\lambda \geq 0, \nu^+ \geq 0, \nu^- \geq 0$. Since $\psi_j(u) = 0$ for $j = 1, \ldots, p$, these equations can be rewritten as

$$J'_u + \sum_{i=1}^{m} \lambda_i (\varphi'_i)_u + \sum_{j=1}^{p} (\nu_j^+ - \nu_j^-)(\psi'_j)_u = 0,$$

and

$$\sum_{i=1}^{m} \lambda_i \varphi_i(u) = 0$$

with $\lambda \geq 0, \nu^+ \geq 0, \nu^- \geq 0$, and if we introduce $\nu_j = \nu_j^+ - \nu_j^-$ we obtain the following KKT conditions for programs with explicit equality constraints:

$$J'_u + \sum_{i=1}^{m} \lambda_i (\varphi'_i)_u + \sum_{j=1}^{p} \nu_j (\psi'_j)_u = 0,$$

and

$$\sum_{i=1}^{m} \lambda_i \varphi_i(u) = 0$$

with $\lambda \geq 0$ and $\nu \in \mathbb{R}^p$ arbitrary.

Let us now assume that the functions φ_i and ψ_j are *convex*. As we explained just after Definition 14.6, nonaffine equality constraints are never qualified. Thus, in order to generalize Theorem 14.2 to explicit equality constraints, we assume that the *equality constraints* ψ_j are *affine*.

Theorem 14.6. *Let $\varphi_i \colon \Omega \to \mathbb{R}$ be m convex inequality constraints and $\psi_j \colon \Omega \to \mathbb{R}$ be p affine equality constraints defined on some open convex subset Ω of a finite-dimensional Euclidean vector space V (more generally, a real Hilbert space V), let $J \colon \Omega \to \mathbb{R}$ be some function, let U be given by*

$$U = \{x \in \Omega \mid \varphi_i(x) \leq 0, \;\; \psi_j(x) = 0, \;\; 1 \leq i \leq m, 1 \leq j \leq p\},$$

and let $u \in U$ be any point such that the functions φ_i and J are differentiable at u, and the functions ψ_j are affine.

(1) *If J has a local minimum at u with respect to U, and if the constraints are qualified, then there exist some vectors $\lambda \in \mathbb{R}^m_+$ and $\nu \in \mathbb{R}^p$, such that the KKT condition hold:*

$$J'_u + \sum_{i=1}^{m} \lambda_i(u)(\varphi'_i)_u + \sum_{j=1}^{p} \nu_j(\psi'_j)_u = 0,$$

and

$$\sum_{i=1}^{m} \lambda_i(u)\varphi_i(u) = 0, \quad \lambda_i \geq 0, \quad i = 1, \ldots, m.$$

Equivalently, in terms of gradients, the above conditions are expressed as

$$\nabla J_u + \sum_{i=1}^{m} \lambda_i \nabla(\varphi_i)_u + \sum_{j=1}^{p} \nu_j \nabla(\psi_j)_u = 0$$

and

$$\sum_{i=1}^{m} \lambda_i(u)\varphi_i(u) = 0, \quad \lambda_i \geq 0, \quad i = 1, \ldots, m.$$

(2) *Conversely, if the restriction of J to U is convex and if there exist vectors $\lambda \in \mathbb{R}^m_+$ and $\nu \in \mathbb{R}^p$ such that the KKT conditions hold, then the function J has a (global) minimum at u with respect to U.*

The Lagrangian $L(v, \lambda, \nu)$ of Problem (P') is defined as

$$L(v, \mu, \nu) = J(v) + \sum_{i=1}^{m} \mu_i \varphi_i(v) + \sum_{j=1}^{p} \nu_i \psi_j(v),$$

where $v \in \Omega$, $\mu \in \mathbb{R}^m_+$, and $\nu \in \mathbb{R}^p$.

The function $G \colon \mathbb{R}^m_+ \times \mathbb{R}^p \to \mathbb{R}$ given by

$$G(\mu, \nu) = \inf_{v \in \Omega} L(v, \mu, \nu) \quad \mu \in \mathbb{R}^m_+,\ \nu \in \mathbb{R}^p$$

is called the *Lagrange dual function* (or *dual function*), and the *Dual Problem (D') is*

$$\begin{aligned} &\text{maximize} \quad G(\mu, \nu) \\ &\text{subject to} \quad \mu \in \mathbb{R}^m_+,\ \nu \in \mathbb{R}^p. \end{aligned}$$

Observe that the Lagrange multipliers ν *are not restricted* to be nonnegative.

Theorem 14.4 and Theorem 14.5 are immediately generalized to Problem (P'). We only state the new version of 14.5, leaving the new version of Theorem 14.4 as an exercise.

Theorem 14.7. *Consider the minimization problem (P'):*

$$\begin{aligned} &\text{minimize} \quad J(v) \\ &\text{subject to} \quad \varphi_i(v) \le 0, \quad i = 1, \ldots, m \\ &\qquad\qquad\quad \psi_j(v) = 0, \quad j = 1, \ldots, p. \end{aligned}$$

where the functions J, φ_i are defined on some open subset Ω of a finite-dimensional Euclidean vector space V (more generally, a real Hilbert space V), and the functions ψ_j are affine.

(1) Suppose the functions $\varphi_i \colon \Omega \to \mathbb{R}$ are continuous, and that for every $\mu \in \mathbb{R}^m_+$ and every $\nu \in \mathbb{R}^p$, the Problem $(P_{\mu,\nu})$:

$$\begin{aligned} &\text{minimize} \quad L(v, \mu, \nu) \\ &\text{subject to} \quad v \in \Omega, \end{aligned}$$

has a unique solution $u_{\mu,\nu}$, so that

$$L(u_{\mu,\nu}, \mu, \nu) = \inf_{v \in \Omega} L(v, \mu, \nu) = G(\mu, \nu),$$

and the function $(\mu, \nu) \mapsto u_{\mu,\nu}$ is continuous (on $\mathbb{R}^m_+ \times \mathbb{R}^p$). Then the function G is differentiable for all $\mu \in \mathbb{R}^m_+$ and all $\nu \in \mathbb{R}^p$, and

$$G'_{\mu,\nu}(\xi, \zeta) = \sum_{i=1}^m \xi_i \varphi_i(u_{\mu,\nu}) + \sum_{j=1}^p \zeta_j \psi_j(u_{\mu,\nu})$$

for all $\xi \in \mathbb{R}^m$ and all $\zeta \in \mathbb{R}^p$.

If (λ, η) is any solution of Problem (D):

$$\begin{aligned} &\text{maximize} \quad G(\mu, \nu) \\ &\text{subject to} \quad \mu \in \mathbb{R}^m_+, \ \nu \in \mathbb{R}^p, \end{aligned}$$

then the solution $u_{\lambda,\eta}$ of the corresponding Problem $(P_{\lambda,\eta})$ is a solution of Problem (P').

(2) Assume Problem (P') has some solution $u \in U$, and that Ω is convex (open), the functions φ_i $(1 \le i \le m)$ and J are convex, differentiable at u, and that the constraints are qualified. Then Problem (D') has a solution $(\lambda, \eta) \in \mathbb{R}^m_+ \times \mathbb{R}^p$, and $J(u) = G(\lambda, \eta)$; that is, the duality gap is zero.

In the next section we derive the dual function and the dual program of the optimization problem of Section 14.6 (Hard margin SVM), which involves both inequality and equality constraints. We also derive the KKT conditions associated with the dual program.

14.10 Dual of the Hard Margin Support Vector Machine

Recall the **Hard margin SVM** problem (SVM_{h2}):

$$\begin{aligned}
&\text{minimize} \quad \frac{1}{2}\left\|w\right\|^2, \qquad w \in \mathbb{R}^n \\
&\text{subject to} \\
&\qquad w^\top u_i - b \geq 1 \qquad i = 1, \ldots, p \\
&\qquad - w^\top v_j + b \geq 1 \qquad j = 1, \ldots, q.
\end{aligned}$$

We proceed in six steps.
Step 1: Write the constraints in matrix form.
The inequality constraints are written as

$$C\begin{pmatrix} w \\ b \end{pmatrix} \leq d,$$

where C is a $(p+q) \times (n+1)$ matrix C and $d \in \mathbb{R}^{p+q}$ is the vector given by

$$C = \begin{pmatrix} -u_1^\top & 1 \\ \vdots & \vdots \\ -u_p^\top & 1 \\ v_1^\top & -1 \\ \vdots & \vdots \\ v_q^\top & -1 \end{pmatrix}, \qquad d = \begin{pmatrix} -1 \\ \vdots \\ -1 \end{pmatrix} = -\mathbf{1}_{p+q}.$$

If we let X be the $n \times (p+q)$ matrix given by

$$X = \begin{pmatrix} -u_1 & \cdots & -u_p & v_1 & \cdots & v_q \end{pmatrix},$$

then

$$C = \begin{pmatrix} X^\top & \begin{matrix} \mathbf{1}_p \\ -\mathbf{1}_q \end{matrix} \end{pmatrix}$$

and so

$$C^\top = \begin{pmatrix} X \\ \mathbf{1}_p^\top \; -\mathbf{1}_q^\top \end{pmatrix}.$$

Step 2: Write the objective function in matrix form.
The objective function is given by

$$J(w, b) = \frac{1}{2}\begin{pmatrix} w^\top & b \end{pmatrix}\begin{pmatrix} I_n & 0_n \\ 0_n^\top & 0 \end{pmatrix}\begin{pmatrix} w \\ b \end{pmatrix}.$$

Note that the corresponding matrix is symmetric positive semidefinite, but it is *not* invertible. Thus the function J is convex but not strictly convex.

Step 3: Write the Lagrangian in matrix form.

As in Example 14.7, we obtain the Lagrangian

$$L(w,b,\lambda,\mu) = \frac{1}{2}\begin{pmatrix} w^\top & b \end{pmatrix}\begin{pmatrix} I_n & 0_n \\ 0_n^\top & 0 \end{pmatrix}\begin{pmatrix} w \\ b \end{pmatrix} - \begin{pmatrix} w^\top & b \end{pmatrix}\left(0_{n+1} - C^\top\begin{pmatrix} \lambda \\ \mu \end{pmatrix}\right) + \begin{pmatrix} \lambda^\top & \mu^\top \end{pmatrix}\mathbf{1}_{p+q},$$

that is,

$$L(w,b,\lambda,\mu) = \frac{1}{2}\begin{pmatrix} w^\top & b \end{pmatrix}\begin{pmatrix} I_n & 0_n \\ 0_n^\top & 0 \end{pmatrix}\begin{pmatrix} w \\ b \end{pmatrix} + \begin{pmatrix} w^\top & b \end{pmatrix}\begin{pmatrix} X\begin{pmatrix} \lambda \\ \mu \end{pmatrix} \\ \mathbf{1}_p^\top\lambda - \mathbf{1}_q^\top\mu \end{pmatrix} + \begin{pmatrix} \lambda^\top & \mu^\top \end{pmatrix}\mathbf{1}_{p+q}.$$

Step 4: Find the dual function $G(\lambda,\mu)$.

In order to find the dual function $G(\lambda,\mu)$, we need to minimize $L(w,b,\lambda,\mu)$ with respect to w and b and for this, since the objective function J is convex and since $\mathbb{R}^{n+1}$ is convex and open, we can apply Theorem 4.5, which gives a necessary and sufficient condition for a minimum. The gradient of $L(w,b,\lambda,\mu)$ with respect to w and b is

$$\begin{aligned}\nabla L_{w,b} &= \begin{pmatrix} I_n & 0_n \\ 0_n^\top & 0 \end{pmatrix}\begin{pmatrix} w \\ b \end{pmatrix} + \begin{pmatrix} X\begin{pmatrix} \lambda \\ \mu \end{pmatrix} \\ \mathbf{1}_p^\top\lambda - \mathbf{1}_q^\top\mu \end{pmatrix} \\ &= \begin{pmatrix} w \\ 0 \end{pmatrix} + \begin{pmatrix} X\begin{pmatrix} \lambda \\ \mu \end{pmatrix} \\ \mathbf{1}_p^\top\lambda - \mathbf{1}_q^\top\mu \end{pmatrix}.\end{aligned}$$

The necessary and sufficient condition for a minimum is

$$\nabla L_{w,b} = 0,$$

which yields

$$w = -X\begin{pmatrix} \lambda \\ \mu \end{pmatrix} \tag{$*_1$}$$

and

$$\mathbf{1}_p^\top\lambda - \mathbf{1}_q^\top\mu = 0. \tag{$*_2$}$$

The second equation can be written as

$$\sum_{i=1}^{p} \lambda_i = \sum_{j=1}^{q} \mu_j. \qquad (*_3)$$

Plugging back w from $(*_1)$ into the Lagrangian and using $(*_2)$ we get

$$G(\lambda, \mu) = -\frac{1}{2} \begin{pmatrix} \lambda^\top & \mu^\top \end{pmatrix} X^\top X \begin{pmatrix} \lambda \\ \mu \end{pmatrix} + \begin{pmatrix} \lambda^\top & \mu^\top \end{pmatrix} \mathbf{1}_{p+q}; \qquad (*_4)$$

of course, $\begin{pmatrix} \lambda^\top & \mu^\top \end{pmatrix} \mathbf{1}_{p+q} = \sum_{i=1}^{p} \lambda_i + \sum_{j=1}^{q} \mu_j$. Actually, to be perfectly rigorous $G(\lambda, \mu)$ is only defined on the intersection of the hyperplane of equation $\sum_{i=1}^{p} \lambda_i = \sum_{j=1}^{q} \mu_j$ with the convex octant in $\mathbb{R}^{p+q}$ given by $\lambda \geq 0, \mu \geq 0$, so for all $\lambda \in \mathbb{R}^p_+$ and all $\mu \in \mathbb{R}^q_+$, we have

$$G(\lambda, \mu) = \begin{cases} -\frac{1}{2} \begin{pmatrix} \lambda^\top & \mu^\top \end{pmatrix} X^\top X \begin{pmatrix} \lambda \\ \mu \end{pmatrix} + \begin{pmatrix} \lambda^\top & \mu^\top \end{pmatrix} \mathbf{1}_{p+q} & \text{if } \sum_{i=1}^{p} \lambda_i = \sum_{j=1}^{q} \mu_j \\ -\infty & \text{otherwise.} \end{cases}$$

Note that the condition

$$\sum_{i=1}^{p} \lambda_i = \sum_{j=1}^{q} \mu_j$$

is Condition $(*_2)$ of Example 14.6, which is not surprising.

Step 5: Write the dual program in matrix form.

Maximizing the dual function $G(\lambda, \mu)$ over its domain of definition is equivalent to maximizing

$$\widehat{G}(\lambda, \mu) = -\frac{1}{2} \begin{pmatrix} \lambda^\top & \mu^\top \end{pmatrix} X^\top X \begin{pmatrix} \lambda \\ \mu \end{pmatrix} + \begin{pmatrix} \lambda^\top & \mu^\top \end{pmatrix} \mathbf{1}_{p+q}$$

subject to the constraint

$$\sum_{i=1}^{p} \lambda_i = \sum_{j=1}^{q} \mu_j,$$

so we formulate the dual program as,

$$\begin{aligned} &\text{maximize} \quad -\frac{1}{2} \begin{pmatrix} \lambda^\top & \mu^\top \end{pmatrix} X^\top X \begin{pmatrix} \lambda \\ \mu \end{pmatrix} + \begin{pmatrix} \lambda^\top & \mu^\top \end{pmatrix} \mathbf{1}_{p+q} \\ &\text{subject to} \\ &\qquad \sum_{i=1}^{p} \lambda_i = \sum_{j=1}^{q} \mu_j \\ &\qquad \lambda \geq 0, \; \mu \geq 0, \end{aligned}$$

or equivalently,

$$\text{minimize} \quad \frac{1}{2}\begin{pmatrix}\lambda^\top & \mu^\top\end{pmatrix} X^\top X \begin{pmatrix}\lambda \\ \mu\end{pmatrix} - \begin{pmatrix}\lambda^\top & \mu^\top\end{pmatrix} \mathbf{1}_{p+q}$$

subject to

$$\sum_{i=1}^{p} \lambda_i = \sum_{j=1}^{q} \mu_j$$

$$\lambda \geq 0, \ \mu \geq 0.$$

The constraints of the dual program are a lot simpler than the constraints

$$\begin{pmatrix} X^\top & \begin{matrix}\mathbf{1}_p \\ -\mathbf{1}_q\end{matrix}\end{pmatrix} \begin{pmatrix} w \\ b \end{pmatrix} \leq -\mathbf{1}_{p+q}$$

of the primal program because these constraints have been "absorbed" by the objective function $\widehat{G}(\lambda, \nu)$ of the dual program which involves the matrix $X^\top X$. The matrix $X^\top X$ is symmetric positive semidefinite, but not invertible in general.

Step 6: Solve the dual program.

This step involves using numerical procedures typically based on gradient descent to find λ and μ, for example, ADMM from Section 16.6. Once λ and μ are determined, w is determined by $(*_1)$ and b is determined as in Section 14.6 using the fact that there is at least some i_0 such that $\lambda_{i_0} > 0$ and some j_0 such that $\mu_{j_0} > 0$.

Remarks:

(1) Since the constraints are affine and the objective function is convex, by Theorem 14.7(2) the duality gap is zero, so for any minimum w of $J(w, b) = (1/2)w^\top w$ and any maximum (λ, μ) of G, we have

$$J(w, b) = \frac{1}{2} w^\top w = G(\lambda, \mu).$$

But by $(*_1)$

$$w = -X \begin{pmatrix}\lambda \\ \mu\end{pmatrix} = \sum_{i=1}^{p} \lambda_i u_i - \sum_{j=1}^{q} \mu_j v_j,$$

so

$$\begin{pmatrix}\lambda^\top & \mu^\top\end{pmatrix} X^\top X \begin{pmatrix}\lambda \\ \mu\end{pmatrix} = w^\top w,$$

and we get

$$\frac{1}{2}w^\top w = -\frac{1}{2}\left(\lambda^\top\ \mu^\top\right) X^\top X \begin{pmatrix}\lambda\\ \mu\end{pmatrix} + \left(\lambda^\top\ \mu^\top\right)\mathbf{1}_{p+q}$$

$$= -\frac{1}{2}w^\top w + \left(\lambda^\top\ \mu^\top\right)\mathbf{1}_{p+q},$$

so

$$w^\top w = \left(\lambda^\top\ \mu^\top\right)\mathbf{1}_{p+q} = \sum_{i=1}^{p}\lambda_i + \sum_{j=1}^{q}\mu_j,$$

which yields

$$G(\lambda,\mu) = \frac{1}{2}\left(\sum_{i=1}^{p}\lambda_i + \sum_{j=1}^{q}\mu_j\right).$$

The above formulae are stated in Vapnik [Vapnik (1998)] (Chapter 10, Section 1).

(2) It is instructive to compute the Lagrangian of the dual program and to derive the KKT conditions for this Lagrangian.
The conditions $\lambda \geq 0$ being equivalent to $-\lambda \leq 0$, and the conditions $\mu \geq 0$ being equivalent to $-\mu \leq 0$, we introduce Lagrange multipliers $\alpha \in \mathbb{R}^p_+$ and $\beta \in \mathbb{R}^q_+$ as well as a multiplier $\rho \in \mathbb{R}$ for the equational constraint, and we form the Lagrangian

$$L(\lambda,\mu,\alpha,\beta,\rho) = \frac{1}{2}\left(\lambda^\top\ \mu^\top\right) X^\top X \begin{pmatrix}\lambda\\ \mu\end{pmatrix} - \left(\lambda^\top\ \mu^\top\right)\mathbf{1}_{p+q}$$

$$- \sum_{i=1}^{p}\alpha_i\lambda_i - \sum_{j=1}^{q}\beta_j\mu_j + \rho\Big(\sum_{j=1}^{q}\mu_j - \sum_{i=1}^{p}\lambda_i\Big).$$

It follows that the KKT conditions are

$$X^\top X \begin{pmatrix}\lambda\\ \mu\end{pmatrix} - \mathbf{1}_{p+q} - \begin{pmatrix}\alpha\\ \beta\end{pmatrix} + \rho\begin{pmatrix}-\mathbf{1}_p\\ \mathbf{1}_q\end{pmatrix} = 0_{p+q}, \qquad (*_4)$$

and $\alpha_i\lambda_i = 0$ for $i = 1,\ldots,p$ and $\beta_j\mu_j = 0$ for $j = 1,\ldots,q$.
But $(*_4)$ is equivalent to

$$-X^\top X \begin{pmatrix}\lambda\\ \mu\end{pmatrix} + \rho\begin{pmatrix}\mathbf{1}_p\\ -\mathbf{1}_q\end{pmatrix} + \mathbf{1}_{p+q} + \begin{pmatrix}\alpha\\ \beta\end{pmatrix} = 0_{p+q},$$

which is precisely the result of adding $\alpha \geq 0$ and $\beta \geq 0$ as slack variables to the inequalities $(*_3)$ of Example 14.6, namely

$$-X^\top X \begin{pmatrix}\lambda\\ \mu\end{pmatrix} + b\begin{pmatrix}\mathbf{1}_p\\ -\mathbf{1}_q\end{pmatrix} + \mathbf{1}_{p+q} \leq 0_{p+q},$$

to make them equalities, where ρ plays the role of b.

When the constraints are *affine*, the dual function $G(\lambda,\nu)$ can be expressed in terms of the conjugate of the objective function J.

14.11 Conjugate Function and Legendre Dual Function

The notion of conjugate function goes back to Legendre and plays an important role in classical mechanics for converting a Lagrangian to a Hamiltonian; see Arnold [Arnold (1989)] (Chapter 3, Sections 14 and 15).

Definition 14.11. Let $f\colon A \to \mathbb{R}$ be a function defined on some subset A of $\mathbb{R}^n$. The *conjugate* f^* of the function f is the partial function $f^*\colon \mathbb{R}^n \to \mathbb{R}$ defined by

$$f^*(y) = \sup_{x\in A}(\langle y, x\rangle - f(x)) = \sup_{x\in A}(y^\top x - f(x)), \quad y \in \mathbb{R}^n.$$

The conjugate of a function is also called the *Fenchel conjugate*, or *Legendre transform* when f is differentiable.

As the pointwise supremum of a family of affine functions in y, the conjugate function f^* is *convex*, even if the original function f is not convex.

By definition of f^* we have

$$f(x) + f^*(y) \geq \langle x, y\rangle = x^\top y,$$

whenever the left-hand side is defined. The above is known as *Fenchel's inequality* (or *Young's inequality* if f is differentiable).

If $f\colon A \to \mathbb{R}$ is convex (so A is convex) and if $\mathbf{epi}(f)$ is closed, then it can be shown that $f^{**} = f$. In particular, this is true if $A = \mathbb{R}^n$.

The domain of f^* can be very small, even if the domain of f is big. For example, if $f\colon \mathbb{R} \to \mathbb{R}$ is the affine function given by $f(x) = ax + b$ (with $a, b \in \mathbb{R}$), then the function $x \mapsto yx - ax - b$ is unbounded above unless $y = a$, so

$$f^*(y) = \begin{cases} -b & \text{if } y = a \\ +\infty & \text{otherwise.} \end{cases}$$

The domain of f^* can also be bigger than the domain of f; see Example 14.8(3).

The conjugates of many functions that come up in optimization are derived in Boyd and Vandenberghe; see [Boyd and Vandenberghe (2004)], Section 3.3. We mention a few that will be used in this chapter.

Example 14.8.

(1) *Negative logarithm*: $f(x) = -\log x$, with $\mathrm{dom}(f) = \{x \in \mathbb{R} \mid x > 0\}$. The function $x \mapsto yx + \log x$ is unbounded above if $y \geq 0$, and when $y < 0$, its maximum is obtained iff its derivative is zero, namely

$$y + \frac{1}{x} = 0.$$

Substituting for $x = -1/y$ in $yx + \log x$, we obtain $-1 + \log(-1/y) = -1 - \log(-y)$, so we have

$$f^*(y) = -\log(-y) - 1,$$

with $\mathrm{dom}(f^*) = \{y \in \mathbb{R} \mid y < 0\}$.

(2) *Exponential*: $f(x) = e^x$, with $\mathrm{dom}(f) = \mathbb{R}$. The function $x \mapsto yx - e^x$ is unbounded if $y < 0$. When $y > 0$, it reaches a maximum iff its derivative is zero, namely

$$y - e^x = 0.$$

Substituting for $x = \log y$ in $yx - e^x$, we obtain $y \log y - y$, so we have

$$f^*(y) = y \log y - y,$$

with $\mathrm{dom}(f^*) = \{y \in \mathbb{R} \mid y \geq 0\}$, with the convention that $0 \log 0 = 0$.

(3) *Negative Entropy*: $f(x) = x \log x$, with $\mathrm{dom}(f) = \{x \in \mathbb{R} \mid x \geq 0\}$, with the convention that $0 \log 0 = 0$. The function $x \mapsto yx - x \log x$ is bounded above for all $y > 0$, and it attains its maximum when its derivative is zero, namely

$$y - \log x - 1 = 0.$$

Substituting for $x = e^{y-1}$ in $yx - x \log x$, we obtain $ye^{y-1} - e^{y-1}(y-1) = e^{y-1}$, which yields

$$f^*(y) = e^{y-1},$$

with $\mathrm{dom}(f^*) = \mathbb{R}$.

(4) *Strictly convex quadratic function*: $f(x) = \frac{1}{2}x^\top Ax$, where A is an $n \times n$ symmetric positive definite matrix, with $\mathrm{dom}(f) = \mathbb{R}^n$. The function $x \mapsto y^\top x - \frac{1}{2}x^\top Ax$ has a unique maximum when is gradient is zero, namely

$$y = Ax.$$

Substituting for $x = A^{-1}y$ in $y^\top x - \frac{1}{2}x^\top Ax$, we obtain

$$y^\top A^{-1}y - \frac{1}{2}y^\top A^{-1}y = -\frac{1}{2}y^\top A^{-1}y,$$

so

$$f^*(y) = -\frac{1}{2}y^\top A^{-1}y$$

with $\mathrm{dom}(f^*) = \mathbb{R}^n$.

(5) *Log-determinant*: $f(X) = \log\det(X^{-1})$, where X is an $n \times n$ symmetric positive definite matrix. Then

$$f(Y) = \log\det((-Y)^{-1}) - n,$$

where Y is an $n \times n$ symmetric negative definite matrix; see Boyd and Vandenberghe; see [Boyd and Vandenberghe (2004)], Section 3.3.1, Example 3.23.

(6) *Norm on* $\mathbb{R}^n$: $f(x) = \|x\|$ for any norm $\|\;\|$ on $\mathbb{R}^n$, with $\mathrm{dom}(f) = \mathbb{R}^n$. Recall from Section 13.7 in Vol. I that the dual norm $\|\;\|^D$ of the norm $\|\;\|$ (with respect to the canonical inner product $x \cdot y = y^\top x$ on $\mathbb{R}^n$ is given by

$$\|y\|^D = \sup_{\|x\|=1} |y^\top x|,$$

and that

$$|y^\top x| \leq \|x\| \, \|y\|^D .$$

We have

$$\begin{aligned} f^*(y) &= \sup_{x \in \mathbb{R}^n} (y^\top x - \|x\|) \\ &= \sup_{x \in \mathbb{R}^n, x \neq 0} \left(y^\top \frac{x}{\|x\|} - 1 \right) \|x\| \\ &\leq \sup_{x \in \mathbb{R}^n, x \neq 0} (\|y\|^D - 1) \|x\| , \end{aligned}$$

so if $\|y\|^D > 1$ this last term goes to $+\infty$, but if $\|y\|^D \leq 1$, then its maximum is 0. Therefore,

$$f^*(y) = \|y\|^* = \begin{cases} 0 & \text{if } \|y\|^D \leq 1 \\ +\infty & \text{otherwise.} \end{cases}$$

(7) *Norm squared*: $f(x) = \frac{1}{2}\|x\|^2$ for any norm $\|\;\|$ on $\mathbb{R}^n$, with $\mathrm{dom}(f) = \mathbb{R}^n$. Since $|y^\top x| \leq \|x\| \, \|y\|^D$, we have

$$y^\top x - (1/2)\|x\|^2 \leq \|y\|^D \|x\| - (1/2)\|x\|^2 .$$

The right-hand side is a quadratic function of $\|x\|$ which achieves its maximum at $\|x\| = \|y\|^D$, with maximum value $(1/2)(\|y\|^D)^2$. Therefore

$$y^\top x - (1/2)\|x\|^2 \leq (1/2)\left(\|y\|^D\right)^2$$

for all x, which shows that

$$f^*(y) \leq (1/2)\left(\|y\|^D\right)^2.$$

By definition of the dual norm and because the unit sphere is compact, for any $y \in \mathbb{R}^n$, there is some $x \in \mathbb{R}^n$ such that $\|x\| = 1$ and $y^\top x = \|y\|^D$, so multiplying both sides by $\|y\|^D$ we obtain

$$y^\top \|y\|^D x = \left(\|y\|^D\right)^2$$

and for $z = \|y\|^D x$, since $\|x\| = 1$ we have $\|z\| = \|y\|^D \|x\| = \|y\|^D$, so we get

$$y^\top z - (1/2)(\|z\|)^2 = \left(\|y\|^D\right)^2 - (1/2)\left(\|y\|^D\right)^2 = (1/2)\left(\|y\|^D\right)^2,$$

which shows that the upper bound $(1/2)\left(\|y\|^D\right)^2$ is achieved. Therefore,

$$f^*(y) = \frac{1}{2}\left(\|y\|^D\right)^2,$$

and $\mathrm{dom}(f^*) = \mathbb{R}^n$.

(8) *Log-sum-exp function:* $f(x) = \log\left(\sum_{i=1}^n e^{x_i}\right)$, where $x = (x_1, \ldots, x_n) \in \mathbb{R}^n$. To determine the values of $y \in \mathbb{R}^n$ for which the maximum of $g(x) = y^\top x - f(x)$ over $x \in \mathbb{R}^n$ is attained, we compute its gradient and we find

$$\nabla f_x = \begin{pmatrix} y_1 - \dfrac{e^{x_1}}{\sum_{i=1}^n e^{x_i}} \\ \vdots \\ y_n - \dfrac{e^{x_n}}{\sum_{i=1}^n e^{x_i}} \end{pmatrix}.$$

Therefore, $(y_1, \ldots, y_n)$ must satisfy the system of equations

$$y_j = \frac{e^{x_j}}{\sum_{i=1}^n e^{x_i}}, \quad j = 1, \ldots, n. \tag{$*$}$$

The condition $\sum_{i=1}^n y_i = 1$ is obviously necessary, as well as the conditions $y_i > 0$, for $i = 1, \ldots, n$. Conversely, if $\mathbf{1}^\top y = 1$ and $y > 0$, then $x_j = \log y_i$ for $i = 1, \ldots, n$ is a solution. Since $(*)$ implies that

$$x_i = \log y_i + \log\left(\sum_{i=1}^n e^{x_i}\right), \tag{$**$}$$

we get

$$\begin{aligned} y^\top x - f(x) &= \sum_{i=1}^n y_i x_i - \log\left(\sum_{i=1}^n e^{x_i}\right) \\ &= \sum_{i=1}^n y_i \log y_i + \sum_{i=1}^n y_i \log\left(\sum_{i=1}^n e^{x_i}\right) - \log\left(\sum_{i=1}^n e^{x_i}\right) \quad (**) \\ &= \sum_{i=1}^n y_i \log y_i + \left(\sum_{i=1}^n y_i - 1\right) \log\left(\sum_{i=1}^n e^{x_i}\right) \\ &= \sum_{i=1}^n y_i \log y_i, \qquad \text{since } \textstyle\sum_{i=1}^n y_i = 1. \end{aligned}$$

Consequently, if $f^*(y)$ is defined, then $f^*(y) = \sum_{i=1}^n y_i \log y_i$. If we agree that $0 \log 0 = 0$, then it is an easy exercise (or, see Boyd and Vandenberghe [Boyd and Vandenberghe (2004)], Section 3.3, Example 3.25) to show that

$$f^*(y) = \begin{cases} \sum_{i=1}^n y_i \log y_i & \text{if } \mathbf{1}^\top y = 1 \text{ and } y \geq 0 \\ \infty & \text{otherwise.} \end{cases}$$

Thus we obtain the negative entropy restricted to the domain $\mathbf{1}^\top y = 1$ and $y \geq 0$.

If $f \colon \mathbb{R}^n \to \mathbb{R}$ is convex and differentiable, then x^* maximizes $x^\top y - f(x)$ iff x^* minimizes $-x^\top y + f(x)$ iff

$$\nabla f_{x^*} = y,$$

and so

$$f^*(y) = (x^*)^\top \nabla f_{x^*} - f(x^*).$$

Consequently, if we can solve the equation

$$\nabla f_z = y$$

for z given y, then we obtain $f^*(y)$.

It can be shown that if f is twice differentiable, strictly convex, and surlinear, which means that

$$\lim_{\|y\| \mapsto +\infty} \frac{f(y)}{\|y\|} = +\infty,$$

then there is a unique x_y such that $\nabla f_{x_y} = y$, so that

$$f^*(y) = x_y^\top \nabla f_{x_y} - f(x_y),$$

and f^* is differentiable with

$$\nabla f_y^* = x_y.$$

We now return to our optimization problem.

Proposition 14.13. *Consider Problem* (P),

$$\begin{aligned} &\text{minimize} && J(v) \\ &\text{subject to} && Av \leq b \\ &&& Cv = d, \end{aligned}$$

with affine inequality and equality constraints (with A an $m \times n$ matrix, C an $p \times n$ matrix, $b \in \mathbb{R}^m$, $d \in \mathbb{R}^p$). The dual function $G(\lambda, \nu)$ is given by

$$G(\lambda, \nu) = \begin{cases} -b^\top \lambda - d^\top \nu - J^*(-A^\top \lambda - C^\top \nu) & \text{if } -A^\top \lambda - C^\top \nu \in \text{dom}(J^*), \\ -\infty & \text{otherwise}, \end{cases}$$

for all $\lambda \in \mathbb{R}^m_+$ and all $\nu \in \mathbb{R}^p$, where J^* is the conjugate of J.

Proof. The Lagrangian associated with the above program is

$$\begin{aligned} L(v, \lambda, \nu) &= J(v) + (Av - b)^\top \lambda + (Cv - d)^\top \nu \\ &= -b^\top \lambda - d^\top \nu + J(v) + (A^\top \lambda + C^\top \nu)^\top v, \end{aligned}$$

with $\lambda \in \mathbb{R}^m_+$ and $\nu \in \mathbb{R}^p$. By definition

$$\begin{aligned} G(\lambda, \nu) &= -b^\top \lambda - d^\top \nu + \inf_{v \in \mathbb{R}^n} (J(v) + (A^\top \lambda + C^\top \nu)^\top v) \\ &= -b^\top \lambda - d^\top \nu - \sup_{v \in \mathbb{R}^n} (-(A^\top \lambda + C^\top \nu)^\top v - J(v)) \\ &= -b^\top \lambda - d^\top \nu - J^*(-A^\top \lambda - C^\top \nu). \end{aligned}$$

Therefore, for all $\lambda \in \mathbb{R}^m_+$ and all $\nu \in \mathbb{R}^p$, we have

$$G(\lambda, \nu) = \begin{cases} -b^\top \lambda - d^\top \nu - J^*(-A^\top \lambda - C^\top \nu) & \text{if } -A^\top \lambda - C^\top \nu \in \text{dom}(J^*), \\ -\infty & \text{otherwise}, \end{cases}$$

as claimed. ∎

As application of Proposition 14.13, consider the following example.

Example 14.9. Consider the following problem:

$$\begin{aligned} &\text{minimize} && \|v\| \\ &\text{subject to} && Av = b, \end{aligned}$$

where $\| \;\|$ is any norm on $\mathbb{R}^n$. Using the result of Example 14.8(6), we obtain

$$G(\nu) = -b^\top \nu - \left\| -A^\top \nu \right\|^*,$$

that is,

$$G(\nu) = \begin{cases} -b^\top \nu & \text{if } \left\| A^\top \nu \right\|^D \leq 1 \\ -\infty & \text{otherwise.} \end{cases}$$

In the special case where $\| \;\| = \| \;\|_2$, we also have $\| \;\|^D = \| \;\|_2$.

Another interesting application is to the entropy minimization problem.

Example 14.10. Consider the following problem known as *entropy minimization*:

$$\begin{aligned} \text{minimize} \quad & f(x) = \sum_{i=1}^n x_i \log x_i \\ \text{subject to} \quad & Ax \leq b \\ & \mathbf{1}^\top x = 1, \end{aligned}$$

where $\operatorname{dom}(f) = \{x \in \mathbb{R}^n \mid x \geq 0\}$. By Example 14.8(3), the conjugate of the negative entropy function $u \log u$ is e^{v-1}, so we easily see that

$$f^*(y) = \sum_{i=1}^n e^{y_i - 1},$$

which is defined on $\mathbb{R}^n$. Proposition 14.13 implies that the dual function $G(\lambda, \mu)$ of the entropy minimization problem is given by

$$G(\lambda, \mu) = -b^\top \lambda - \mu - e^{-\mu-1} \sum_{i=1}^n e^{-(A^i)^\top \lambda},$$

for all $\lambda \in \mathbb{R}^n_+$ and all $\mu \in \mathbb{R}$, where A^i is the ith column of A. It follows that the dual program is:

$$\begin{aligned} \text{maximize} \quad & -b^\top \lambda - \mu - e^{-\mu-1} \sum_{i=1}^n e^{-(A^i)^\top \lambda} \\ \text{subject to} \quad & \lambda \geq 0. \end{aligned}$$

We can simplify this problem by maximizing over the variable $\mu \in \mathbb{R}$. For fixed λ, the objective function is maximized when the derivative is zero, that is,

$$-1 + e^{-\mu-1} \sum_{i=1}^n e^{-(A^i)^\top \lambda} = 0,$$

which yields

$$\mu = \log\left(\sum_{i=1}^{n} e^{-(A^i)^\top \lambda}\right) - 1.$$

By plugging the above value back into the objective function of the dual, we obtain the following program:

$$\begin{aligned} &\text{maximize} \quad -b^\top \lambda - \log\left(\sum_{i=1}^{n} e^{-(A^i)^\top \lambda}\right) \\ &\text{subject to} \;\; \lambda \geq 0. \end{aligned}$$

The entropy minimization problem is another problem for which Theorem 14.6 applies, and thus can be solved using the dual program. Indeed, the Lagrangian of the primal program is given by

$$L(x, \lambda, \mu) = \sum_{i-1}^{n} x_i \log x_i + \lambda^\top (Ax - b) + \mu(\mathbf{1}^\top x - 1).$$

Using the second derivative criterion for convexity, we see that $L(x, \lambda, \mu)$ is strictly convex for $x \in \mathbb{R}^n_+$ and is bounded below, so it has a unique minimum which is obtain by setting the gradient ∇L_x to zero. We have

$$\nabla L_x = \begin{pmatrix} \log x_1 + 1 + (A^1)^\top \lambda + \mu \\ \vdots \\ \log x_n + 1 + (A^n)^\top \lambda + \mu. \end{pmatrix}$$

so by setting ∇L_x to 0 we obtain

$$x_i = e^{-((A^n)^\top \lambda + \mu + 1)}, \quad i = 1, \ldots, n. \tag{$*$}$$

By Theorem 14.6, since the objective function is convex and the constraints are affine, if the primal has a solution then so does the dual, and λ and μ constitute an optimal solution of the dual, then $x = (x_1, \ldots, x_n)$ given by the equations in $(*)$ is an optimal solution of the primal.

Other examples are given in Boyd and Vandenberghe; see [Boyd and Vandenberghe (2004)], Section 5.1.6.

The derivation of the dual function of Problem (SVM$_{h1}$) from Section 14.5 involves a similar type of reasoning.

Example 14.11. Consider the Hard Margin Problem (SVM$_{h1}$):

$$\begin{aligned} &\text{maximize} \quad \delta \\ &\text{subject to} \\ &\qquad w^\top u_i - b \geq \delta \qquad i = 1, \ldots, p \\ &\qquad -w^\top v_j + b \geq \delta \qquad j = 1, \ldots, q \\ &\qquad \|w\|_2 \leq 1, \end{aligned}$$

which is converted to the following minimization problem:

$$
\begin{aligned}
&\text{minimize} \quad -2\delta\\
&\text{subject to}\\
&\qquad w^\top u_i - b \geq \delta \qquad i = 1, \ldots, p\\
&\qquad -w^\top v_j + b \geq \delta \qquad j = 1, \ldots, q\\
&\qquad \|w\|_2 \leq 1.
\end{aligned}
$$

We replaced δ by 2δ because this will make it easier to find a nice geometric interpretation. Recall from Section 14.5 that Problem (SVM_{h1}) has a an optimal solution iff $\delta > 0$, in which case $\|w\| = 1$.

The corresponding Lagrangian with $\lambda \in \mathbb{R}^p_+, \mu \in \mathbb{R}^q_+, \gamma \in \mathbb{R}^+$, is

$$
\begin{aligned}
L(w, b, \delta, \lambda, \mu, \gamma) &= -2\delta + \sum_{i=1}^p \lambda_i(\delta + b - w^\top u_i) + \sum_{j=1}^q \mu_j(\delta - b + w^\top v_j)\\
&\quad + \gamma(\|w\|_2 - 1)\\
&= w^\top\left(-\sum_{i=1}^p \lambda_i u_i + \sum_{j=1}^q \mu_j v_j\right) + \gamma\,\|w\|_2\\
&\quad + \left(\sum_{i=1}^p \lambda_i - \sum_{j=1}^q \mu_j\right) b + \left(-2 + \sum_{i=1}^p \lambda_i + \sum_{j=1}^q \mu_j\right)\delta - \gamma.
\end{aligned}
$$

Next to find the dual function $G(\lambda, \mu, \gamma)$ we need to minimize $L(w, b, \delta, \lambda, \mu, \gamma)$ with respect to w, b and δ, so its gradient with respect to w, b and δ must be zero. This implies that

$$
\begin{aligned}
\sum_{i=1}^p \lambda_i - \sum_{j=1}^q \mu_j &= 0\\
-2 + \sum_{i=1}^p \lambda_i + \sum_{j=1}^q \mu_j &= 0,
\end{aligned}
$$

which yields

$$
\sum_{i=1}^p \lambda_i = \sum_{j=1}^q \mu_j = 1.
$$

Observe that we did not compute the partial derivative with respect to w because it does not yield any useful information due to the presence of the term $\|w\|_2$ (as opposed to $\|w\|_2^2$). Our minimization problem is reduced to:

find

$$
\begin{aligned}
&\inf_{w, \|w\| \leq 1} \left(w^\top \left(\sum_{j=1}^{q} \mu_j v_j - \sum_{i=1}^{p} \lambda_i u_i \right) + \gamma \|w\|_2 - \gamma \right) \\
&= -\gamma - \gamma \inf_{w, \|w\| \leq 1} \left(-w^\top \frac{1}{\gamma} \left(\sum_{j=1}^{q} \mu_j v_j - \sum_{i=1}^{p} \lambda_i u_i \right) + \|-w\|_2 \right) \\
&= \begin{cases} -\gamma & \text{if } \left\| \frac{1}{\gamma} \left(\sum_{j=1}^{q} \mu_j v_j - \sum_{i=1}^{p} \lambda_i u_i \right) \right\|_2^D \leq 1 \\ -\infty & \text{otherwise} \end{cases} \quad \text{by Example 14.8(6)} \\
&= \begin{cases} -\gamma & \text{if } \left\| \sum_{j=1}^{q} \mu_j v_j - \sum_{i=1}^{p} \lambda_i u_i \right\|_2 \leq \gamma \\ -\infty & \text{otherwise.} \end{cases} \quad \text{since } \| \ \|_2^D = \| \ \|_2 \text{ and } \gamma > 0
\end{aligned}
$$

It is immediately verified that the above formula is still correct if $\gamma = 0$. Therefore

$$
G(\lambda, \mu, \gamma) = \begin{cases} -\gamma & \text{if } \left\| \sum_{j=1}^{q} \mu_j v_j - \sum_{i=1}^{p} \lambda_i u_i \right\|_2 \leq \gamma \\ -\infty & \text{otherwise.} \end{cases}
$$

Since $\left\| \sum_{j=1}^{q} \mu_j v_j - \sum_{i=1}^{p} \lambda_i u_i \right\|_2 \leq \gamma$ iff $-\gamma \leq - \left\| \sum_{j=1}^{q} \mu_j v_j - \sum_{i=1}^{p} \lambda_i u_i \right\|_2$, the dual program, maximizing $G(\lambda, \mu, \gamma)$, is equivalent to

$$
\begin{aligned}
&\text{maximize} \quad - \left\| \sum_{j=1}^{q} \mu_j v_j - \sum_{i=1}^{p} \lambda_i u_i \right\|_2 \\
&\text{subject to} \\
&\qquad \sum_{i=1}^{p} \lambda_i = 1, \ \ \lambda \geq 0 \\
&\qquad \sum_{j=1}^{q} \mu_j = 1, \ \ \mu \geq 0,
\end{aligned}
$$

equivalently

$$\begin{aligned}
&\text{minimize} \quad \left\| \sum_{j=1}^{q} \mu_j v_j - \sum_{i=1}^{p} \lambda_i u_i \right\|_2 \\
&\text{subject to} \\
&\qquad \sum_{i=1}^{p} \lambda_i = 1, \ \ \lambda \geq 0 \\
&\qquad \sum_{j=1}^{q} \mu_j = 1, \ \ \mu \geq 0.
\end{aligned}$$

Geometrically, $\sum_{i=1}^{p} \lambda_i u_i$ with $\sum_{i=1}^{p} \lambda_i = 1$ and $\lambda \geq 0$ is a convex combination of the u_is, and $\sum_{j=1}^{q} \mu_j v_j$ with $\sum_{j=1}^{q} \mu_j = 1$ and $\mu \geq 0$ is a convex combination of the v_js, so the dual program is to minimize the distance between the polyhedron $\text{conv}(u_1, \ldots, u_p)$ (the convex hull of the u_is) and the polyhedron $\text{conv}(v_1, \ldots, v_q)$ (the convex hull of the v_js). Since both polyhedra are compact, the shortest distance between then is achieved. In fact, there is some vertex u_i such that if $P(u_i)$ is its projection onto $\text{conv}(v_1, \ldots, v_q)$ (which exists by Hilbert space theory), then the length of the line segment $(u_i, P(u_i))$ is the shortest distance between the two polyhedra (and similarly there is some vertex v_j such that if $P(v_j)$ is its projection onto $\text{conv}(u_1, \ldots, u_p)$ then the length of the line segment $(v_j, P(v_j))$ is the shortest distance between the two polyhedra).

If the two subsets are separable, in which case Problem (SVM_{h1}) has an optimal solution $\delta > 0$, because the objective function is convex and the convex constraint $\|w\|_2 \leq 1$ is qualified since δ may be negative, by Theorem 14.5(2) the duality gap is zero, so δ is half of the minimum distance between the two convex polyhedra $\text{conv}(u_1, \ldots, u_p)$ and $\text{conv}(v_1, \ldots, v_q)$; see Figure 14.19.

It should be noted that the constraint $\|w\| \leq 1$ yields a formulation of the dual problem which has the advantage of having a nice geometric interpretation: finding the minimal distance between the convex polyhedra $\text{conv}(u_1, \ldots, u_p)$ and $\text{conv}(v_1, \ldots, v_q)$. Unfortunately this formulation is not useful for actually solving the problem. However, if the equivalent constraint $\|w\|^2$ $(= w^\top w) \leq 1$ is used, then the dual problem is much more useful as a solving tool.

In Chapter 18 we consider the case where the sets of points $\{u_1, \ldots, u_p\}$ and $\{v_1, \ldots, v_q\}$ are not linearly separable.

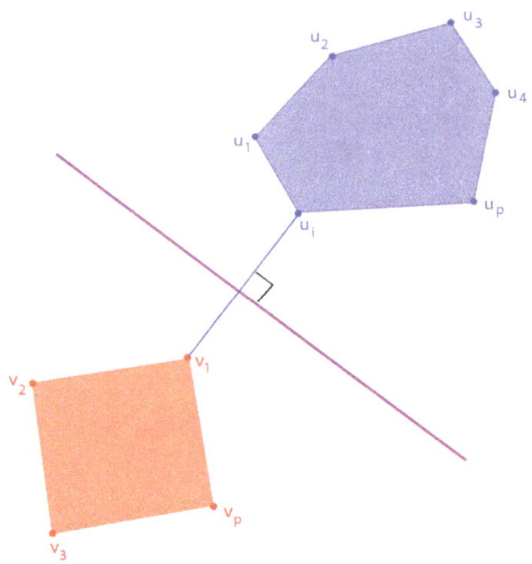

Fig. 14.19 In $\mathbb{R}^2$ the convex hull of the u_is, namely the blue hexagon, is separated from the convex hull of the v_js, i.e., the red square, by the purple hyperplane (line) which is the perpendicular bisector to the blue line segment between u_i and v_1, where this blue line segment is the shortest distance between the two convex polygons.

14.12 Some Techniques to Obtain a More Useful Dual Program

In some cases, it is advantageous to reformulate a primal optimization problem to obtain a more useful dual problem. Three different reformulations are proposed in Boyd and Vandenberghe; see [Boyd and Vandenberghe (2004)], Section 5.7:

(1) Introducing new variables and associated equality constraints.
(2) Replacing the objective function with an increasing function of the original function.
(3) Making explicit constraints implicit, that is, incorporating them into the domain of the objective function.

We only give illustrations of (1) and (2) and refer the reader to Boyd and Vandenberghe [Boyd and Vandenberghe (2004)] (Section 5.7) for more examples of these techniques.

Consider the unconstrained program:

$$\text{minimize} \quad f(Ax + b),$$

where A is an $m \times n$ matrix and $b \in \mathbb{R}^m$. While the conditions for a zero duality gap are satisfied, the Lagrangian is

$$L(x) = f(Ax + b),$$

so the dual function G is the constant function whose value is

$$G = \inf_{x \in \mathbb{R}^n} f(Ax + b),$$

which is not useful at all.

Let us reformulate the problem as

$$\begin{aligned} &\text{minimize} \quad f(y) \\ &\text{subject to} \\ &\qquad\qquad Ax + b = y, \end{aligned}$$

where we introduced the new variable $y \in \mathbb{R}^m$ and the equality constraint $Ax + b = y$. The two problems are obviously equivalent. The Lagrangian of the reformulated problem is

$$L(x, y, \mu) = f(y) + \mu^\top (Ax + b - y)$$

where $\mu \in \mathbb{R}^m$. To find the dual function $G(\mu)$ we minimize $L(x, y, \mu)$ over x and y. Minimizing over x we see that $G(\mu) = -\infty$ unless $A^\top \mu = 0$, in which case we are left with

$$G(\mu) = b^\top \mu + \inf_y (f(y) - \mu^\top y) = b^\top \mu - \inf_y (\mu^\top y - f(y)) = b^\top \mu - f^*(\mu),$$

where f^* is the conjugate of f. It follows that the dual program can be expressed as

$$\begin{aligned} &\text{maximize} \quad b^\top \mu - f^*(\mu) \\ &\text{subject to} \\ &\qquad\qquad A^\top \mu = 0. \end{aligned}$$

This formulation of the dual is much more useful than the dual of the original program.

Example 14.12. As a concrete example, consider the following unconstrained program:

$$\text{minimize} \quad f(x) = \log\left(\sum_{i=1}^n e^{(A^i)^\top x + b_i} \right)$$

where A^i is a column vector in $\mathbb{R}^n$. We reformulate the problem by introducing new variables and equality constraints as follows:

$$\text{minimize} \quad f(y) = \log\left(\sum_{i=1}^{n} e^{y_i}\right)$$

$$\text{subject to}$$

$$Ax + b = y,$$

where A is the $n \times n$ matrix whose columns are the vectors A^i and $b = (b_1, \ldots, b_n)$. Since by Example 14.8(8), the conjugate of the log-sum-exp function $f(y) = \log\left(\sum_{i=1}^{n} e^{y_i}\right)$ is

$$f^*(\mu) = \begin{cases} \sum_{i=1}^{n} \mu_i \log \mu_i & \text{if } \mathbf{1}^\top \mu = 1 \text{ and } \mu \geq 0 \\ \infty & \text{otherwise,} \end{cases}$$

the dual of the reformulated problem can be expressed as

$$\text{maximize} \quad b^\top \mu - \log\left(\sum_{i=1}^{n} \mu_i \log \mu_i\right)$$

$$\text{subject to}$$

$$\begin{aligned} \mathbf{1}^\top \mu &= 1 \\ A^\top \mu &= 0 \\ \mu &\geq 0, \end{aligned}$$

an entropy maximization problem.

Example 14.13. Similarly the unconstrained norm minimization problem

$$\text{minimize} \quad \|Ax - b\|,$$

where $\| \ \|$ is any norm on $\mathbb{R}^m$, has a dual function which is a constant, and is not useful. This problem can be reformulated as

$$\text{minimize} \quad \|y\|$$

$$\text{subject to}$$

$$Ax - b = y.$$

By Example 14.8(6), the conjugate of the norm is given by

$$\|y\|^* = \begin{cases} 0 & \text{if } \|y\|^D \leq 1 \\ +\infty & \text{otherwise,} \end{cases}$$

so the dual of the reformulated program is:

$$\begin{aligned} &\text{maximize} \quad b^\top \mu \\ &\text{subject to} \\ &\qquad \|\mu\|^D \leq 1 \\ &\qquad A^\top \mu = 0. \end{aligned}$$

Here is now an example of (2), replacing the objective function with an increasing function of the the original function.

Example 14.14. The norm minimization of Example 14.13 can be reformulated as

$$\begin{aligned} &\text{minimize} \quad \frac{1}{2}\|y\|^2 \\ &\text{subject to} \\ &\qquad Ax - b = y. \end{aligned}$$

This program is obviously equivalent to the original one. By Example 14.8(7), the conjugate of the square norm is given by

$$\frac{1}{2}\left(\|y\|^D\right)^2,$$

so the dual of the reformulated program is

$$\begin{aligned} &\text{maximize} \quad -\frac{1}{2}\left(\|\mu\|^D\right)^2 + b^\top \mu \\ &\text{subject to} \\ &\qquad A^\top \mu = 0. \end{aligned}$$

Note that this dual is different from the dual obtained in Example 14.13.

The objective function of the dual program in Example 14.13 is linear, but we have the nonlinear constraint $\|\mu\|^D \leq 1$. On the other hand, the objective function of the dual program of Example 14.14 is quadratic, whereas its constraints are affine. We have other examples of this trade-off with the Programs (SVM$_{h2}$) (quadratic objective function, affine constraints), and (SVM$_{h1}$) (linear objective function, one nonlinear constraint).

Sometimes, it is also helpful to replace a constraint by an increasing function of this constraint; for example, to use the constraint $\|w\|_2^2$ $(= w^\top w) \leq 1$ instead of $\|w\|_2 \leq 1$.

In Chapter 19 we revisit the problem of solving an overdetermined or underdetermined linear system $Ax = b$ considered in Volume I, Section 21.1 from a different point of view.

14.13 Uzawa's Method

Let us go back to our Minimization Problem

$$\begin{aligned} &\text{minimize} \quad J(v) \\ &\text{subject to} \quad \varphi_i(v) \leq 0, \quad i = 1, \ldots, m, \end{aligned}$$

where the functions J and φ_i are defined on some open subset Ω of a finite-dimensional Euclidean vector space V (more generally, a real Hilbert space V). As usual, let

$$U = \{v \in V \mid \varphi_i(v) \leq 0, \ 1 \leq i \leq m\}.$$

If the functional J satisfies the inequalities of Proposition 13.10 and if the functions φ_i are convex, in theory, the projected-gradient method converges to the unique minimizer of J over U. Unfortunately, it is usually impossible to compute the projection map $p_U \colon V \to U$.

On the other hand, the domain of the Lagrange dual function $G \colon \mathbb{R}^m_+ \to \mathbb{R}$ given by

$$G(\mu) = \inf_{v \in \Omega} L(v, \mu) \quad \mu \in \mathbb{R}^m_+,$$

is $\mathbb{R}^m_+$, where

$$L(v, \mu) = J(v) + \sum_{i=1}^{m} \mu_i \varphi_i(v)$$

is the Lagrangian of our problem. Now the projection p_+ from $\mathbb{R}^m$ to $\mathbb{R}^m_+$ is very simple, namely

$$(p_+(\lambda))_i = \max\{\lambda_i, 0\}, \quad 1 \leq i \leq m.$$

It follows that the projection-gradient method should be applicable to the *Dual Problem (D)*:

$$\begin{aligned} &\text{maximize} \quad G(\mu) \\ &\text{subject to} \quad \mu \in \mathbb{R}^m_+. \end{aligned}$$

If the hypotheses of Theorem 14.5 hold, then a solution λ of the Dual Program (D) yields a solution u_λ of the primal problem.

Uzawa's method is essentially the gradient method with fixed stepsize applied to the Dual Problem (D). However, it is designed to yield a solution of the primal problem.

Uzawa's method:

Given an arbitrary initial vector $\lambda^0 \in \mathbb{R}^m_+$, two sequences $(\lambda^k)_{k \geq 0}$ and $(u^k)_{k \geq 0}$ are constructed, with $\lambda^k \in \mathbb{R}^m_+$ and $u^k \in V$.

Assuming that $\lambda^0, \lambda^1, \ldots, \lambda^k$ are known, u^k and λ^{k+1} are determined as follows:

u^k is the unique solution of the minimization problem, find $u^k \in V$ such that

$$\text{(UZ)} \quad \begin{cases} J(u^k) + \sum_{i=1}^m \lambda_i^k \varphi_i(u^k) = \inf_{v \in V} \left(J(v) + \sum_{i=1}^m \lambda_i^k \varphi_i(v) \right); \text{ and} \\ \qquad \lambda_i^{k+1} = \max\{\lambda_i^k + \rho\varphi_i(u^k), 0\}, \quad 1 \le i \le m, \end{cases}$$

where $\rho > 0$ is a suitably chosen parameter.

Recall that in the proof of Theorem 14.5 we showed ($*_{\text{deriv}}$), namely

$$G'_{\lambda^k}(\xi) = \langle \nabla G_{\lambda^k}, \xi \rangle = \sum_{i=1}^m \xi_i \varphi_i(u^k),$$

which means that $(\nabla G_{\lambda^k})_i = \varphi_i(u^k)$. Then the second equation in (UZ) corresponds to the gradient-projection step

$$\lambda^{k+1} = p_+(\lambda^k + \rho \nabla G_{\lambda^k}).$$

Note that because the problem is a maximization problem we use a positive sign instead of a negative sign. Uzawa's method is indeed a gradient method.

Basically, Uzawa's method replaces a constrained optimization problem by a sequence of unconstrained optimization problems involving the Lagrangian of the (primal) problem.

Interestingly, under certain hypotheses, it is possible to prove that the sequence of approximate solutions $(u^k)_{k \ge 0}$ converges to the minimizer u of J over U, even if the sequence $(\lambda^k)_{k \ge 0}$ does not converge. We prove such a result when the constraints φ_i are *affine*.

Theorem 14.8. *Suppose $J\colon \mathbb{R}^n \to \mathbb{R}$ is an elliptic functional, which means that J is continuously differentiable on $\mathbb{R}^n$, and there is some constant $\alpha > 0$ such that*

$$\langle \nabla J_v - \nabla J_u, v - u \rangle \ge \alpha \|v - u\|^2 \quad \textit{for all } u, v \in \mathbb{R}^n,$$

and that U is a nonempty closed convex subset given by

$$U = \{v \in \mathbb{R}^n \mid Cv \le d\},$$

where C is a real $m \times n$ matrix and $d \in \mathbb{R}^m$. If the scalar ρ satisfies the condition

$$0 < \rho < \frac{2\alpha}{\|C\|_2^2},$$

where $\|C\|_2$ *is the spectral norm of* C*, then the sequence* $(u^k)_{k\geq 0}$ *computed by Uzawa's method converges to the unique minimizer* $u \in U$ *of* J*.*

Furthermore, if C *has rank* m*, then the sequence* $(\lambda^k)_{k\geq 0}$ *converges to the unique maximizer of the Dual Problem* (D)*.*

Proof. *Step 1.* We establish algebraic conditions relating the unique minimizer $u \in U$ of J over U and some $\lambda \in \mathbb{R}^m_+$ such that (u, λ) is a saddle point.

Since J is elliptic and U is nonempty closed and convex, by Theorem 13.6, the functional J is strictly convex, so it has a unique minimizer $u \in U$. Since J is convex and the constraints are affine, by Theorem 14.5(2) the Dual Problem (D) has at least one solution. By Theorem 14.4(2), there is some $\lambda \in \mathbb{R}^m_+$ such that (u, λ) is a saddle point of the Lagrangian L.

If we define the affine function φ by

$$\varphi(v) = (\varphi_1(v), \ldots, \varphi_m(v)) = Cv - d,$$

then the Lagrangian $L(v, \mu)$ can be written as

$$L(v, \mu) = J(v) + \sum_{i=1}^{m} \mu_i \varphi_i(v) = J(v) + \langle C^\top \mu, v\rangle - \langle \mu, d\rangle.$$

Since

$$L(u, \lambda) = \inf_{v\in\mathbb{R}^n} L(v, \lambda),$$

by Theorem 4.5(4) we must have

$$\nabla J_u + C^\top \lambda = 0, \tag{$*_1$}$$

and since

$$G(\lambda) = L(u, \lambda) = \sup_{\mu\in\mathbb{R}^m_+} L(u, \mu),$$

by Theorem 4.5(3) (and since maximizing a function g is equivalent to minimizing $-g$), we must have

$$G'_\lambda(\mu - \lambda) \leq 0 \quad \text{for all } \mu \in \mathbb{R}^m_+,$$

and since as noted earlier $\nabla G_\lambda = \varphi(u)$, we get

$$\langle \varphi(u), \mu - \lambda\rangle \leq 0 \quad \text{for all } \mu \in \mathbb{R}^m_+. \tag{$*_2$}$$

As in the proof of Proposition 13.10, $(*_2)$ can be expressed as follows for every $\rho > 0$:

$$\langle \lambda - (\lambda + \rho\varphi(u)), \mu - \lambda\rangle \geq 0 \quad \text{for all } \mu \in \mathbb{R}^m_+, \tag{$**_2$}$$

which shows that λ can be viewed as the projection onto $\mathbb{R}^m_+$ of the vector $\lambda + \rho\varphi(u)$. In summary we obtain the equations

$$(\dagger_1) \quad \begin{cases} \nabla J_u + C^\top \lambda = 0 \\ \lambda = p_+(\lambda + \rho\varphi(u)). \end{cases}$$

Step 2. We establish algebraic conditions relating the unique solution u_k of the minimization problem arising during an iteration of Uzawa's method in (UZ) and λ^k.

Observe that the Lagrangian $L(v, \mu)$ is strictly convex as a function of v (as the sum of a strictly convex function and an affine function). As in the proof of Theorem 13.6(1) and using Cauchy–Schwarz, we have

$$\begin{aligned} J(v) + \langle C^\top \mu, v\rangle &\geq J(0) + \langle \nabla J_0, v\rangle + \frac{\alpha}{2}\left\|v\right\|^2 + \langle C^\top \mu, v\rangle \\ &\geq J(0) - \left\|\nabla J_0\right\| \left\|v\right\| - \left\|C^\top \mu\right\| \left\|v\right\| + \frac{\alpha}{2}\left\|v\right\|^2, \end{aligned}$$

and the term $(-\left\|\nabla J_0\right\| - \left\|C^\top \mu\right\| \left\|v\right\| + \frac{\alpha}{2}\left\|v\right\|)\left\|v\right\|$ goes to $+\infty$ when $\left\|v\right\|$ tends to $+\infty$, so $L(v, \mu)$ is coercive as a function of v. Therefore, the minimization problem find u^k such that

$$J(u^k) + \sum_{i=1}^m \lambda_i^k \varphi_i(u^k) = \inf_{v\in\mathbb{R}^n}\left(J(v) + \sum_{i=1}^m \lambda_i^k \varphi_i(v)\right)$$

has a unique solution $u^k \in \mathbb{R}^n$. It follows from Theorem 4.5(4) that the vector u^k must satisfy the equation

$$\nabla J_{u^k} + C^\top \lambda^k = 0, \tag{$*_3$}$$

and since by definition of Uzawa's method

$$\lambda^{k+1} = p_+(\lambda^k + \rho\varphi(u^k)), \tag{$*_4$}$$

we obtain the equations

$$(\dagger_2) \quad \begin{cases} \nabla J_{u^k} + C^\top \lambda^k = 0 \\ \lambda^{k+1} = p_+(\lambda^k + \rho\varphi(u^k)). \end{cases}$$

Step 3. By subtracting the first of the two equations of $(\dagger_1)$ and $(\dagger_2)$ we obtain

$$\nabla J_{u^k} - \nabla J_u + C^\top(\lambda^k - \lambda) = 0,$$

and by subtracting the second of the two equations of $(\dagger_1)$ and $(\dagger_2)$ and using Proposition 12.5, we obtain

$$\left\|\lambda^{k+1} - \lambda\right\| \leq \left\|\lambda^k - \lambda + \rho C(u^k - u)\right\|.$$

In summary, we proved

$$(\dagger) \qquad \begin{cases} \nabla J_{u^k} - \nabla J_u + C^\top(\lambda^k - \lambda) = 0 \\ \left\|\lambda^{k+1} - \lambda\right\| \leq \left\|\lambda^k - \lambda + \rho C(u^k - u)\right\|. \end{cases}$$

Step 4. Convergence of the sequence $(u^k)_{k\geq 0}$ to u.

Squaring both sides of the inequality in $(\dagger)$ we obtain

$$\left\|\lambda^{k+1} - \lambda\right\|^2 \leq \left\|\lambda^k - \lambda\right\|^2 + 2\rho\langle C^\top(\lambda^k - \lambda), u_k - u\rangle + \rho^2 \left\|C(u^k - u)\right\|^2.$$

Using the equation in $(\dagger)$ and the inequality

$$\langle \nabla J_{u^k} - \nabla J_u, u^k - u\rangle \geq \alpha \left\|u^k - u\right\|^2,$$

we get

$$\begin{aligned} \left\|\lambda^{k+1} - \lambda\right\|^2 &\leq \left\|\lambda^k - \lambda\right\|^2 - 2\rho\langle \nabla J_{u^k} - \nabla J_u, u^k - u\rangle + \rho^2 \left\|C(u^k - u)\right\|^2 \\ &\leq \left\|\lambda^k - \lambda\right\|^2 - \rho(2\alpha - \rho \left\|C\right\|_2^2) \left\|u^k - u\right\|^2. \end{aligned}$$

Consequently, if

$$0 \leq \rho \leq \frac{2\alpha}{\left\|C\right\|_2^2},$$

we have

$$\left\|\lambda^{k+1} - \lambda\right\| \leq \left\|\lambda^k - \lambda\right\|, \quad \text{for all } k \geq 0. \tag{$*_5$}$$

By $(*_5)$, the sequence $(\left\|\lambda^k - \lambda\right\|)_{k\geq 0}$ is nonincreasing and bounded below by 0, so it converges, which implies that

$$\lim_{k\mapsto\infty} \left(\left\|\lambda^{k+1} - \lambda\right\| - \left\|\lambda^k - \lambda\right\|\right) = 0,$$

and since

$$\left\|\lambda^{k+1} - \lambda\right\|^2 \leq \left\|\lambda^k - \lambda\right\|^2 - \rho(2\alpha - \rho \left\|C\right\|_2^2) \left\|u^k - u\right\|^2,$$

we also have

$$\rho(2\alpha - \rho \left\|C\right\|_2^2) \left\|u^k - u\right\|^2 \leq \left\|\lambda^k - \lambda\right\|^2 - \left\|\lambda^{k+1} - \lambda\right\|^2.$$

So if

$$0 < \rho < \frac{2\alpha}{\left\|C\right\|_2^2},$$

then $\rho(2\alpha - \rho \left\|C\right\|_2^2) > 0$, and we conclude that

$$\lim_{k\mapsto\infty} \left\|u^k - u\right\| = 0,$$

that is, the sequence $(u^k)_{k\geq 0}$ converges to u.

Step 5. Convergence of the sequence $(\lambda^k)_{k\geq 0}$ to λ if C has rank m.

Since the sequence $(\|\lambda^k - \lambda\|)_{k\geq 0}$ is nonincreasing, the sequence $(\lambda^k)_{k\geq 0}$ is bounded, and thus it has a convergent subsequence $(\lambda^{i(k)})_{i\geq 0}$ whose limit is some $\lambda' \in \mathbb{R}^m_+$. Since J' is continuous, by $(\dagger_2)$ we have

$$\nabla J_u + C^\top \lambda' = \lim_{i\mapsto\infty} (\nabla J_{u^{i(k)}} + C^\top \lambda^{i(k)}) = 0. \qquad (*_6)$$

If C has rank m, then $\mathrm{Im}(C) = \mathbb{R}^m$, which is equivalent to $\mathrm{Ker}\,(C^\top) = (0)$, so $C^\top$ is injective and since by $(\dagger_1)$ we also have $\nabla J_u + C^\top \lambda = 0$, we conclude that $\lambda' = \lambda$. The above reasoning applies to any subsequence of $(\lambda^k)_{k\geq 0}$, so $(\lambda^k)_{k\geq 0}$ converges to λ. $\square$

In the special case where J is an elliptic quadratic functional

$$J(v) = \frac{1}{2}\langle Av, v\rangle - \langle b, v\rangle,$$

where A is symmetric positive definite, by $(\dagger_2)$ an iteration of Uzawa's method gives

$$\begin{aligned} &Au^k - b + C^\top \lambda^k = 0 \\ &\lambda_i^{k+1} = \max\{(\lambda^k + \rho(Cu^k - d))_i, 0\}, \quad 1 \leq i \leq m. \end{aligned}$$

Theorem 14.8 implies that Uzawa's method converges if

$$0 < \rho < \frac{2\lambda_1}{\|C\|_2^2},$$

where λ_1 is the smallest eigenvalue of A.

If we solve for u^k using the first equation, we get

$$\lambda^{k+1} = p_+(\lambda^k + \rho(-CA^{-1}C^\top \lambda^k + CA^{-1}b - d)). \qquad (*_7)$$

In Example 14.7 we showed that the gradient of the dual function G is given by

$$\nabla G_\mu = Cu_\mu - d = -CA^{-1}C^\top \mu + CA^{-1}b - d,$$

so $(*_7)$ can be written as

$$\lambda^{k+1} = p_+(\lambda^k + \rho\nabla G_{\lambda^k});$$

this shows that Uzawa's method is indeed the gradient method with fixed stepsize applied to the dual program.

14.14 Summary

The main concepts and results of this chapter are listed below:

- The cone of feasible directions.
- Cone with apex.
- Active and inactive constraints.
- Qualified constraint at u.
- Farkas lemma.
- Farkas–Minkowski lemma.
- Karush–Kuhn–Tucker optimality conditions (or *KKT*-conditions).
- Complementary slackness conditions.
- Generalized Lagrange multipliers.
- Qualified convex constraint.
- Lagrangian of a minimization problem.
- Equality constrained minimization.
- KKT matrix.
- Newton's method with equality constraints (feasible start and infeasible start).
- Hard margin support vector machine
- Training data
- Linearly separable sets of points.
- Maximal margin hyperplane.
- Support vectors
- Saddle points.
- Lagrange dual function.
- Lagrange dual program.
- Duality gap.
- Weak duality.
- Strong duality.
- Handling equality constraints in the Lagrangian.
- Dual of the Hard Margin SVM (SVM_{h2}).
- Conjugate functions and Legendre dual functions.
- Dual of the Hard Margin SVM (SVM_{h1}).
- Uzawa's Method.

14.15 Problems

Problem 14.1. Prove (3) and (4) of Proposition 14.9.

Problem 14.2. Assume that in Theorem 14.5, $V = \mathbb{R}^n$, J is elliptic and the constraints φ_i are of the form

$$\sum_{j=1}^{n} c_{ij} v_j \leq d_i,$$

that is, affine. Prove that the Problem (P_μ) has a unique solution which is continuous in μ.

Problem 14.3. (1) Prove that the set of saddle points of a function $L\colon \Omega \times M \to \mathbb{R}$ is of the form $V_0 \times M_0$, for some $V_0 \subseteq \Omega$ and some $M_0 \subseteq M$.

(2) Assume that Ω and M are convex subsets of some normed vector spaces, assume that for any fixed $v \in \Omega$ the map

$$\mu \mapsto L(v, \mu) \quad \text{is concave,}$$

and for any fixed $\mu \in M$ the map

$$v \mapsto L(v, \mu) \quad \text{is convex.}$$

Prove that $V_0 \times M_0$ is convex.

(3) Prove that if for every fixed $\mu \in M$ the map

$$v \mapsto L(v, \mu) \quad \text{is strictly convex,}$$

then V_0 has a most one element.

Problem 14.4. Prove that the conjugate of the function f given by $f(X) = \log\det(X^{-1})$, where X is an $n \times n$ symmetric positive definite matrix, is

$$f^*(Y) = \log\det((-Y)^{-1}) - n,$$

where Y is an $n \times n$ symmetric negative definite matrix.

Problem 14.5. (From Boyd and Vandenberghe [Boyd and Vandenberghe (2004)], Problem 5.12) Given an $m \times n$ matrix A and any vector $b \in \mathbb{R}^n$, consider the problem

$$\begin{aligned} &\text{minimize} \quad -\sum_{i=1}^{m} \log(b_i - a_i x) \\ &\text{subject to} \quad Ax < b, \end{aligned}$$

where a_i is the ith row of A. This is called the *analytic centering problem*. It can be shown that the problem has a unique solution iff the open polyhedron $\{x \in \mathbb{R}^n \mid Ax < b\}$ is nonempty and bounded.

(1) Prove that necessary and sufficient conditions for the problem to have an optimal solution are

$$Ax < b, \quad \sum_{i=1}^{m} \frac{a_i^\top}{b_i - a_x x} = 0.$$

(2) Derive a dual program for the above program.
Hint. First introduce new variables y_i and equations $y_i = b_i - a_i x$.

Problem 14.6. (From Boyd and Vandenberghe [Boyd and Vandenberghe (2004)], Problem 5.13) A *Boolean linear program* is the following optimization problem:

$$\begin{aligned} &\text{minimize} \quad c^\top x \\ &\text{subject to} \quad Ax \leq b \\ &\qquad\qquad\quad x_i \in \{0, 1\}, \ i = 1, \ldots, n, \end{aligned}$$

where A is an $m \times n$ matrix, $c \in \mathbb{R}^n$ and $b \in \mathbb{R}^m$. The fact that the solutions $x \in \mathbb{R}^n$ are constrained to have coordinates x_i taking the values 0 or 1 makes it a hard problem. The above problem can be stated as a program with quadratic constraints:

$$\begin{aligned} &\text{minimize} \quad c^\top x \\ &\text{subject to} \quad Ax \leq b \\ &\qquad\qquad\quad x_i(1 - x_i) = 0, \ i = 1, \ldots, n. \end{aligned}$$

(1) Derive the Lagrangian dual of the above program.

(2) A way to approximate a solution of the Boolean linear program is to consider its linear relaxation where the constraints $x_i \in \{0, 1\}$ are replaced by the *linear constraints* $0 \leq x_i \leq 1$:

$$\begin{aligned} &\text{minimize} \quad c^\top x \\ &\text{subject to} \quad Ax \leq b \\ &\qquad\qquad\quad 0 \leq x_i \leq 1, \ i = 1, \ldots, n. \end{aligned}$$

Find the dual linear program of the above linear program. Show that the maxima of the dual programs in (1) and (2) are the same.

Chapter 15

Subgradients and Subdifferentials of Convex Functions ⊛

In this chapter we consider some deeper aspects of the theory of convex functions that are not necessarily differentiable at every point of their domain. Some substitute for the gradient is needed. Fortunately, for convex functions, there is such a notion, namely subgradients. Geometrically, given a (proper) convex function f, the subgradients at x are vectors normal to supporting hyperplanes to the epigraph of the function at $(x, f(x))$. The subdifferential $\partial f(x)$ to f at x is the set of all subgradients at x. A crucial property is that f is differentiable at x iff $\partial f(x) = \{\nabla f_x\}$, where ∇f_x is the gradient of f at x. Another important property is that a (proper) convex function f attains its minimum at x iff $0 \in \partial f(x)$. A major motivation for developing this more sophisticated theory of "differentiation" of convex functions is to extend the Lagrangian framework to convex functions that are not necessarily differentiable.

Experience shows that the applicability of convex optimization is significantly increased by considering extended real-valued functions, namely functions $f\colon S \to \mathbb{R} \cup \{-\infty, +\infty\}$, where S is some subset of $\mathbb{R}^n$ (usually convex). This is reminiscent of what happens in measure theory, where it is natural to consider functions that take the value $+\infty$. We already encountered functions that take the value $-\infty$ as a result of a minimization that does not converge. For example, if $J(u, v) = u$, and we have the affine constraint $v = 0$, for any fixed λ, the minimization problem

$$\begin{aligned} &\text{minimize} \quad u + \lambda v \\ &\text{subject to} \quad v = 0, \end{aligned}$$

yields the solution $u = -\infty$ and $v = 0$.

Until now, we chose not to consider functions taking the value $-\infty$, and instead we considered partial functions, but it turns out to be convenient to admit functions taking the value $-\infty$.

Allowing functions to take the value $+\infty$ is also a convenient alternative to dealing with partial functions. This situation is well illustrated by the indicator function of a convex set.

Definition 15.1. Let $C \subseteq \mathbb{R}^n$ be any subset of $\mathbb{R}^n$. The *indicator function* I_C of C is the function given by

$$I_C(u) = \begin{cases} 0 & \text{if } u \in C \\ +\infty & \text{if } u \notin C. \end{cases}$$

The indicator function I_C is a variant of the characteristic function χ_C of the set C (defined such that $\chi_C(u) = 1$ if $u \in C$ and $\chi_C(u) = 0$ if $u \notin C$). Rockafellar denotes the indicator function I_C by $\delta(-|C)$; that is, $\delta(u|C) = I_C(u)$; see Rockafellar [Rockafellar (1970)], page 28.

Given a partial function $f\colon \mathbb{R}^n \to \mathbb{R} \cup \{-\infty\}$, by setting $f(u) = +\infty$ if $u \notin \text{dom}(f)$, we convert the partial function f into a total function with values in $\mathbb{R} \cup \{-\infty, +\infty\}$. Still, one has to remember that such functions are really partial functions, but $-\infty$ and $+\infty$ play different roles. The value $f(x) = -\infty$ indicates that computing $f(x)$ using a *minimization procedure did not terminate*, but $f(x) = +\infty$ means that the *function f is really undefined at x*.

The definition of a convex function $f\colon S \to \mathbb{R} \cup \{-\infty, +\infty\}$ needs to be slightly modified to accommodate the infinite values $\pm\infty$. The cleanest definition uses the notion of epigraph.

A remarkable and very useful fact is that the optimization problem

$$\begin{aligned} &\text{minimize} \quad J(u) \\ &\text{subject to} \quad u \in C, \end{aligned}$$

where C is a closed convex set in $\mathbb{R}^n$ and J is a convex function can be rewritten in term of the indicator function I_C of C, as

$$\begin{aligned} &\text{minimize} \quad J(u) + I_C(z) \\ &\text{subject to} \quad u - z = 0. \end{aligned}$$

But $J(u) + I_C(z)$ is not differentiable, even if J is, which forces us to deal with convex functions which are not differentiable

Convex functions are not necessarily differentiable, but if a convex function f has a finite value $f(u)$ at u (which means that $f(u) \in \mathbb{R}$), then it has a one-sided directional derivative at u. Another crucial notion is the notion of subgradient, which is a substitute for the notion of gradient when the function f is not differentiable at u.

In Section 15.1, we introduce extended real-valued functions, which are functions that may also take the values $\pm\infty$. In particular, we define proper convex functions, and the closure of a convex function. Subgradients and subdifferentials are defined in Section 15.2. We discuss some properties of subgradients in Section 15.3 and Section 15.4. In particular, we relate subgradients to one-sided directional derivatives. In Section 15.5, we discuss the problem of finding the minimum of a proper convex function and give some criteria in terms of subdifferentials. In Section 15.6, we sketch the generalization of the results presented in Chapter 14 about the Lagrangian framework to programs allowing an objective function and inequality constraints which are convex but not necessarily differentiable. In fact, it is fair to say that the theory of extended real-valued convex functions and the notions of subgradient and subdifferential developed in this chapter constitute the machinery needed to extend the Lagrangian framework to convex functions that are not necessarily differentiable.

This chapter relies heavily on Rockafellar [Rockafellar (1970)]. Some of the results in this chapter are also discussed in Bertsekas [Bertsekas (2009); Bertsekas *et al.* (2003); Bertsekas (2015)]. It should be noted that Bertsekas has developed a framework to discuss duality that he refers to as the *min common/max crossing* framework, for short MC/MC. Although this framework is elegant and interesting in its own right, the fact that Bertsekas relies on it to prove properties of subdifferentials makes it little harder for a reader to "jump in."

15.1 Extended Real-Valued Convex Functions

We extend the ordering on $\mathbb{R}$ by setting

$$-\infty < x < +\infty, \quad \text{for all } x \in \mathbb{R}.$$

Definition 15.2. A (total) function $f\colon \mathbb{R}^n \to \mathbb{R} \cup \{-\infty, +\infty\}$ is called an *extended real-valued function*. For any $x \in \mathbb{R}^n$, we say that $f(x)$ is *finite* if $f(x) \in \mathbb{R}$ (equivalently, $f(x) \neq \pm\infty$). The function f is *finite* if $f(x)$ is finite for all $x \in \mathbb{R}^n$.

Adapting slightly Definition 4.8, given a function $f\colon \mathbb{R}^n \to \mathbb{R} \cup \{-\infty, +\infty\}$, the *epigraph* of f is the subset of $\mathbb{R}^{n+1}$ given by

$$\mathbf{epi}(f) = \{(x, y) \in \mathbb{R}^{n+1} \mid f(x) \leq y\}.$$

See Figure 15.1.

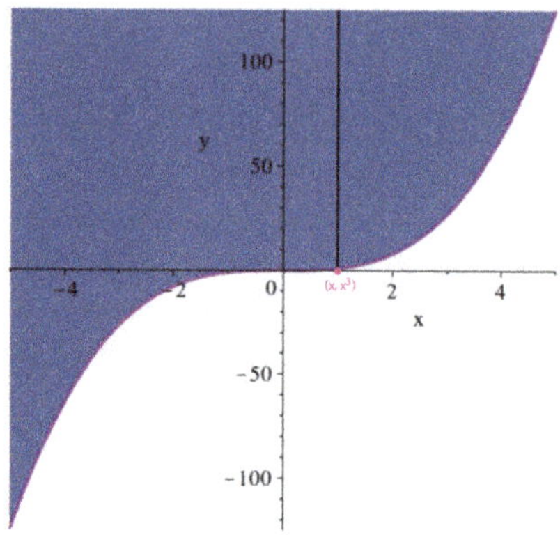

Fig. 15.1 Let $f\colon \mathbb{R} \to \mathbb{R} \cup \{-\infty, +\infty\}$ be given by $f(x) = x^3$ for $x \in \mathbb{R}$. Its graph in $\mathbb{R}^2$ is the magenta curve, and its epigraph is the union of the magenta curve and blue region "above" this curve. For any point $x \in \mathbb{R}$, $\mathbf{epi}(f)$ contains the ray which starts at (x, x^3) and extends upward.

If S is a nonempty subset of $\mathbb{R}^n$, the epigraph of the restriction of f to S is defined as

$$\mathbf{epi}(f|S) = \{(x, y) \in \mathbb{R}^{n+1} \mid f(x) \leq y,\ x \in S\}.$$

Observe the following facts:

(1) For any $x \in S$, if $f(x) = -\infty$, then $\mathbf{epi}(f)$ contains the "vertical line" $\{(x, y) \mid y \in \mathbb{R}\}$ in $\mathbb{R}^{n+1}$.
(2) For any $x \in S$, if $f(x) \in \mathbb{R}$, then $\mathbf{epi}(f)$ contains the ray $\{(x, y)\} \mid f(x) \leq y\}$ in $\mathbb{R}^{n+1}$.
(3) For any $x \in S$, if $f(x) = +\infty$, then $\mathbf{epi}(f)$ does not contain any point (x, y), with $y \in \mathbb{R}$.
(4) We have $\mathbf{epi}(f) = \emptyset$ iff f corresponds to the partial function undefined everywhere; that is, $f(x) = +\infty$ for all $x \in \mathbb{R}^n$.

Definition 15.3. Given a nonempty subset S of $\mathbb{R}^n$, a (total) function $f\colon \mathbb{R}^n \to \mathbb{R} \cup \{-\infty, +\infty\}$ is *convex on* S if its epigraph $\mathbf{epi}(f|S)$ is convex as a subset of $\mathbb{R}^{n+1}$. See Figure 15.2. The function f is *concave* on S if $-f$ is convex on S. The function f is *affine* on S if it is finite, convex, and concave. If $S = \mathbb{R}^n$, we simply that f is *convex* (resp. *concave*, resp. *affine*).

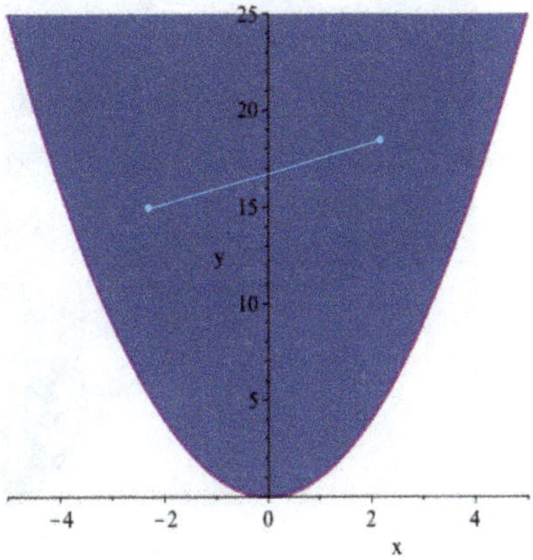

Fig. 15.2 Let $f\colon \mathbb{R} \to \mathbb{R} \cup \{-\infty, +\infty\}$ be given by $f(x) = x^2$ for $x \in \mathbb{R}$. Its graph in $\mathbb{R}^2$ is the magenta curve, and its epigraph is the union of the magenta curve and blue region "above" this curve. Observe that $\mathbf{epi}(f)$ is a convex set of $\mathbb{R}^2$ since the aqua line segment connecting any two points is contained within the epigraph.

Definition 15.4. Given any function $f\colon \mathbb{R}^n \to \mathbb{R} \cup \{-\infty, +\infty\}$, the *effective domain* $\mathrm{dom}(f)$ of f is given by

$$\mathrm{dom}(f) = \{x \in \mathbb{R}^n \mid (\exists y \in \mathbb{R})((x, y) \in \mathbf{epi}(f))\} = \{x \in \mathbb{R}^n \mid f(x) < +\infty\}.$$

Observe that the effective domain of f contains the vectors $x \in \mathbb{R}^n$ such that $f(x) = -\infty$, but excludes the vectors $x \in \mathbb{R}^n$ such that $f(x) = +\infty$.

Example 15.1. The above fact is illustrated by the function $f\colon \mathbb{R} \to \mathbb{R} \cup \{-\infty, +\infty\}$ where

$$f(x) = \begin{cases} -x^2 & \text{if } x \geq 0 \\ +\infty & \text{if } x < 0. \end{cases}$$

The epigraph of this function is illustrated Figure 15.3. By definition $\mathrm{dom}(f) = [0, \infty)$.

If f is a convex function, since $\mathrm{dom}(f)$ is the image of $\mathbf{epi}(f)$ by a linear map (a projection), *it is convex*.

By definition, $\mathbf{epi}(f|S)$ is convex iff for any (x_1, y_1) and (x_2, y_2) with $x_1, x_2 \in S$ and $y_1, y_2 \in \mathbb{R}$ such that $f(x_1) \leq y_1$ and $f(x_2) \leq y_2$, for every λ such that $0 \leq \lambda \leq 1$, we have

$$(1 - \lambda)(x_1, y_1) + \lambda(x_2, y_2) = ((1 - \lambda)x_1 + \lambda x_2, (1 - \lambda)y_1 + \lambda y_2) \in \mathbf{epi}(f|S),$$

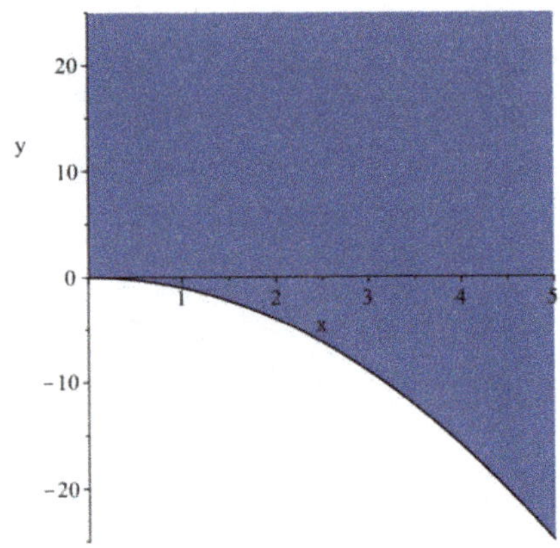

Fig. 15.3 The epigraph of the concave function $f(x) = -x^2$ if $x \geq 0$ and $+\infty$ otherwise.

which means that $(1-\lambda)x_1 + \lambda x_2 \in S$ and

$$f((1-\lambda)x_1 + \lambda x_2) \leq (1-\lambda)y_1 + \lambda y_2. \qquad (*)$$

Thus S *must be convex* and $f((1-\lambda)x_1 + \lambda x_2) < +\infty$. Condition $(*)$ is a little awkward, since it does not refer explicitly to $f(x_1)$ and $f(x_2)$, as these values may be $-\infty$, in which case it is not clear what the expression $(1-\lambda)f(x_1) + \lambda f(x_2)$ means.

In order to perform arithmetic operations involving $-\infty$ and $+\infty$, we adopt the following conventions:

$$\begin{aligned}
\alpha + (+\infty) &= +\infty + \alpha = +\infty & -\infty < \alpha \leq +\infty \\
\alpha + -\infty &= -\infty + \alpha = -\infty & -\infty \leq \alpha < +\infty \\
\alpha(+\infty) &= (+\infty)\alpha = +\infty & 0 < \alpha \leq +\infty \\
\alpha(-\infty) &= (-\infty)\alpha = -\infty & 0 < \alpha \leq +\infty \\
\alpha(+\infty) &= (+\infty)\alpha = -\infty & -\infty \leq \alpha \leq 0 \\
\alpha(-\infty) &= (-\infty)\alpha = +\infty & -\infty \leq \alpha < 0 \\
0(+\infty) &= (+\infty)0 = 0 & 0(-\infty) = (-\infty)0 = 0 \\
-(-\infty) &= +\infty & \\
\inf \emptyset &= +\infty & \sup \emptyset = -\infty.
\end{aligned}$$

The expressions $+\infty + (-\infty)$ and $-\infty + (+\infty)$ are *meaningless*.

The following characterizations of convex functions are easy to show.

Proposition 15.1. *Let C be a nonempty convex subset of $\mathbb{R}^n$.*

(1) A function $f: C \to \mathbb{R}^n \cup \{+\infty\}$ *is convex on* C *iff*

$$f((1-\lambda)x + \lambda y) \leq (1-\lambda)f(x) + \lambda f(y)$$

for all $x, y \in C$ *and all* λ *such that* $0 < \lambda < 1$.

(2) A function $f: \mathbb{R}^n \to \mathbb{R}^n \cup \{-\infty, +\infty\}$ *is convex iff*

$$f((1-\lambda)x + \lambda y) < (1-\lambda)\alpha + \lambda\beta$$

for all $\alpha, \beta \in \mathbb{R}$, *all* $x, y \in \mathbb{R}^n$ *such that* $f(x) < \alpha$ *and* $f(y) < \beta$, *and all* λ *such that* $0 < \lambda < 1$.

The "good" convex functions that we would like to deal with are defined below.

Definition 15.5. A convex function $f: \mathbb{R}^n \to \mathbb{R} \cup \{-\infty, +\infty\}$ is *proper*[1] if its epigraph is nonempty and does not contain any vertical line. Equivalently, a convex function f is proper if $f(x) > -\infty$ for all $x \in \mathbb{R}^n$ and $f(x) < +\infty$ for some $x \in \mathbb{R}^n$. A convex function which is not proper is called an *improper function*.

Observe that a convex function f is proper iff $\text{dom}(f) \neq \emptyset$ and if the restriction of f to $\text{dom}(f)$ is a finite function.

It is immediately verified that a set C is convex iff its indicator function I_C is convex, and clearly, the indicator function of a convex set is proper.

The important object of study is the set of proper functions, but improper functions can't be avoided.

Example 15.2. Here is an example of an improper convex function $f: \mathbb{R} \to \mathbb{R} \cup \{-\infty, +\infty\}$:

$$f(x) = \begin{cases} -\infty & \text{if } |x| < 1 \\ 0 & \text{if } |x| = 1 \\ +\infty & \text{if } |x| > 1 \end{cases}$$

Observe that $\text{dom}(f) = [-1, 1]$, and that $\mathbf{epi}(f)$ is not closed. See Figure 15.4.

[1]This terminology is unfortunate because it clashes with the notion of a proper function from topology, which has to do with the preservation of compact subsets under inverse images.

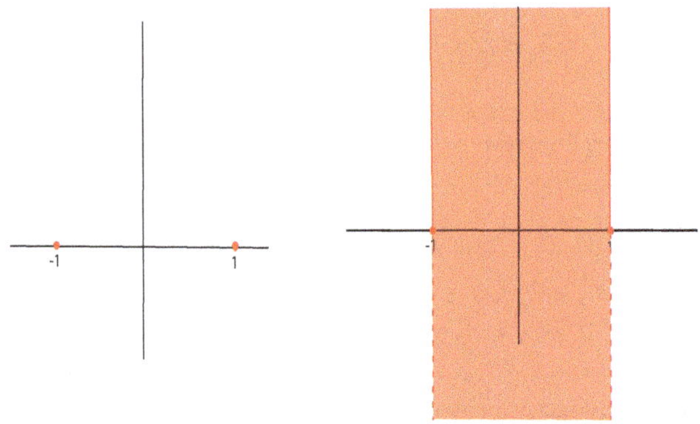

Fig. 15.4 The improper convex function of Example 15.2 and its epigraph depicted as a rose colored region of $\mathbb{R}^2$.

Functions whose epigraph are closed tend to have better properties. To characterize such functions we introduce sublevel sets.

Definition 15.6. Given a function $f\colon \mathbb{R}^n \to \mathbb{R} \cup \{-\infty, +\infty\}$, for any $\alpha \in \mathbb{R} \cup \{-\infty, +\infty\}$, the *sublevel sets* $\mathrm{sublev}_\alpha(f)$ and $\mathrm{sublev}_{<\alpha}(f)$ are the sets

$$\mathrm{sublev}_\alpha(f) = \{x \in \mathbb{R}^n \mid f(x) \le \alpha\}$$
$$\mathrm{sublev}_{<\alpha}(f) = \{x \in \mathbb{R}^n \mid f(x) < \alpha\}.$$

For the improper convex function of Example 15.2, we have

$\mathrm{sublev}_{-\infty}(f) = (-1, 1)$ while $\mathrm{sublev}_{<-\infty}(f) = \emptyset$.
$\mathrm{sublev}_\alpha(f) = (-1, 1) = \mathrm{sublev}_{<\alpha}(f)$ whenever $-\infty < \alpha < 0$.
$\mathrm{sublev}_0(f) = [-1, 1]$ while $\mathrm{sublev}_{<0}(f) = (-1, 1)$.
$\mathrm{sublev}_\alpha(f) = [-1, 1] = \mathrm{sublev}_{<\alpha}(f)$ whenever $0 < \alpha < +\infty$.
$\mathrm{sublev}_{+\infty}(f) = \mathbb{R}$ while $\mathrm{sublev}_{<+\infty}(f) = [-1, 1]$.

A useful corollary of Proposition 15.1 is the following result whose (easy) proof can be found in Rockafellar [Rockafellar (1970)] (Theorem 4.6).

Proposition 15.2. *If f is any convex function on $\mathbb{R}^n$, then for every $\alpha \in \mathbb{R} \cup \{-\infty, +\infty\}$, the sublevel sets* $\mathrm{sublev}_\alpha(f)$ *and* $\mathrm{sublev}_{<\alpha}(f)$ *are convex.*

Definition 15.7. A function $f\colon \mathbb{R}^n \to \mathbb{R} \cup \{-\infty, +\infty\}$ is *lower semi-continuous* if the sublevel sets $\mathrm{sublev}_\alpha(f) = \{x \in \mathbb{R}^n \mid f(x) \le \alpha\}$ are closed for all $\alpha \in \mathbb{R}$.

Observe that the improper convex function of Example 15.2 is not lower semi-continuous since $\text{sublev}_\alpha(f) = (-1, 1)$ whenever $-\infty < \alpha < 0$. This result reflects the fact that epigraph is not closed as shown in the following proposition; see Rockafellar [Rockafellar (1970)] (Theorem 7.1).

Proposition 15.3. *Let $f\colon \mathbb{R}^n \to \mathbb{R} \cup \{-\infty, +\infty\}$ be any function. The following properties are equivalent:*

(1) The function f is lower semi-continuous.
(2) The epigraph of f is a closed set in $\mathbb{R}^{n+1}$.

The notion of the closure of convex function plays an important role. It is a bit subtle because a convex function may be improper.

Definition 15.8. Let $f\colon \mathbb{R}^n \to \mathbb{R} \cup \{-\infty, +\infty\}$ be any function. The function whose epigraph is the closure of the epigraph $\mathbf{epi}(f)$ of f (in $\mathbb{R}^{n+1}$) is called the *lower semi-continuous hull* of f. If f is a convex function and if $f(x) > -\infty$ for all $x \in \mathbb{R}^n$, then the *closure* $\text{cl}(f)$ *of* f is equal to its lower semi-continuous hull, else if $f(x) = -\infty$ for some $x \in \mathbb{R}^n$, then the *closure* $\text{cl}(f)$ *of* f is the constant function with value $-\infty$. A convex function f is *closed* if $f = \text{cl}(f)$.

Definition 15.8 implies that there are *only two closed improper convex functions*: the constant function with value $-\infty$ and the constant function with value $+\infty$. Also, by Proposition 15.3, *a proper convex function is closed iff it is equal to its lower semi-continuous hull iff its epigraph is nonempty and closed.*

Given a convex set C in $\mathbb{R}^n$, the interior $\text{int}(C)$ of C (the largest open subset of $\mathbb{R}^n$ contained in C) is often not interesting because C may have dimension smaller than n. For example, a (closed) triangle in $\mathbb{R}^3$ has empty interior.

The remedy is to consider the affine hull $\text{aff}(C)$ of C, which is the smallest affine set containing C; see Section 8.2. The dimension of C is the dimension of $\text{aff}(C)$. Then the relative interior of C is the interior of C in $\text{aff}(C)$ endowed with the subspace topology induced on $\text{aff}(C)$. More explicitly, we can make the following definition.

Definition 15.9. Let C be a subset of $\mathbb{R}^n$. The *relative interior* of C is the set

$$\mathbf{relint}(C) = \{x \in C \mid B_\epsilon(x) \cap \text{aff}(C) \subseteq C \quad \text{for some } \epsilon > 0\},$$

where $B_\epsilon(x) = \{y \in \mathbb{R}^n \mid \|x - y\|_2 < \epsilon\}$, the open ball of center x and radius ϵ. The *relative boundary* of C is defined as $\overline{C} - \textbf{relint}(C)$, where $\overline{C}$ is the closure of C in $\mathbb{R}^n$ (the smallest closed subset of $\mathbb{R}^n$ containing C).

Remark. Observe that $\text{int}(C) \subseteq \textbf{relint}(C)$. Rockafellar denotes the relative interior of a set C by $\textbf{ri}(C)$.

The following result from Rockafellar [Rockafellar (1970)] (Theorem 7.2) tells us that an improper convex function mostly takes infinite values, except perhaps at relative boundary points of its effective domain.

Proposition 15.4. *If f is an improper convex function, then $f(x) = -\infty$ for every $x \in \textbf{relint}(\text{dom}(f))$. Thus an improper convex function takes infinite values, except at relative boundary points of its effective domain.*

Example 15.2 illustrates Proposition 15.4.

The following result also holds; see Rockafellar [Rockafellar (1970)] (Corollary 7.2.3).

Proposition 15.5. *If f is a convex function whose effective domain is relatively open, which means that $\textbf{relint}(\text{dom}(f)) = \text{dom}(f)$, then either $f(x) > -\infty$ for all $x \in \mathbb{R}^n$, or $f(x) = \pm\infty$ for all $x \in \mathbb{R}^n$.*

We also have the following result showing that the closure of a proper convex function does not differ much from the original function; see Rockafellar [Rockafellar (1970)] (Theorem 7.4).

Proposition 15.6. *Let $f \colon \mathbb{R}^n \to \mathbb{R} \cup \{+\infty\}$ be a proper convex function. Then $\text{cl}(f)$ is a closed proper convex function, and $\text{cl}(f)$ agrees with f on $\text{dom}(f)$ except possibly at relative boundary points.*

Example 15.3. For an example of Propositions 15.6 and 15.5, let $f \colon \mathbb{R} \to \mathbb{R} \cup \{+\infty\}$ be the proper convex function

$$f(x) = \begin{cases} x^2 & \text{if } x < 1 \\ +\infty & \text{if } |x| \geq 1. \end{cases}$$

Then $\text{cl}(f)$ is

$$\text{cl}f(x) = \begin{cases} x^2 & \text{if } x \leq 1 \\ +\infty & \text{if } |x| > 1, \end{cases}$$

and $\text{cl}f(x) = f(x)$ whenever $x \in (-\infty, 1) = \textbf{relint}(\text{dom}(f)) = \text{dom}(f)$. Furthermore, since $\textbf{relint}(\text{dom}(f)) = \text{dom}(f)$, $f(x) > -\infty$ for all $x \in \mathbb{R}$. See Figure 15.5.

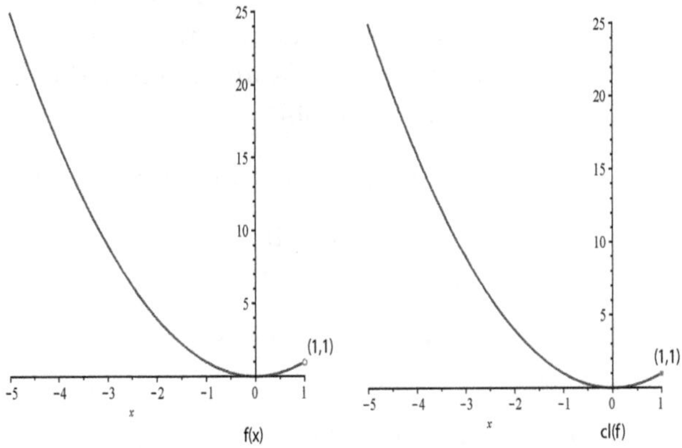

Fig. 15.5 The proper convex function of Example 15.3 and its closure. These two functions only differ at the relative boundary point of dom(f), namely $x = 1$.

Small miracle: *the indicator function* I_C *of any closed convex set is proper and closed*. Indeed, for any $\alpha \in \mathbb{R}$ the sublevel set $\{x \in \mathbb{R}^n \mid I_C(x) \leq \alpha\}$ is either empty if $\alpha < 0$, or equal to C if $\alpha \geq 0$, and C is closed.

We now discuss briefly continuity properties of convex functions. The fact that a convex function f can take the values $\pm\infty$ causes a difficulty, so we consider the restriction of f to its effective domain. There is still a problem because an improper function may take the value $-\infty$. However, if we consider any subset C of dom(f) which is relatively open, which means that $\mathbf{relint}(C) = C$, then $C \subseteq \mathbf{relint}(\text{dom}(f))$, so by Proposition 15.4, the function f has the constant value $-\infty$ on C, and so it can be considered to be continuous on C. Thus we are led to consider proper functions.

Definition 15.10. Given a proper convex function f, for any subset $S \subseteq \text{dom}(f)$, we say that f is *continuous relative to* S if the restriction of f to S is continuous, with S endowed with the subspace topology.

The following result is proven in Rockafellar [Rockafellar (1970)] (Theorem 10.1).

Proposition 15.7. *If f is a proper convex function, then f is continuous on any convex relatively open subset C ($\mathbf{relint}(C) = C$) contained in its effective domain* dom(f)*, in particular relative to* $\mathbf{relint}(\text{dom}(f))$.

As a corollary, any convex function f which is finite on $\mathbb{R}^n$ is continuous.

The behavior of a convex function at relative boundary points of the effective domain can be tricky. Here is an example due to Rockafellar [Rockafellar (1970)] illustrating the problems.

Example 15.4. Consider the proper convex function (on $\mathbb{R}^2$) given by

$$f(x,y) = \begin{cases} y^2/(2x) & \text{if } x > 0 \\ 0 & \text{if } x = 0, y = 0 \\ +\infty & \text{otherwise.} \end{cases}$$

We have

$$\mathrm{dom}(f) = \{(x,y) \in \mathbb{R}^2 \mid x > 0\} \cup \{(0,0)\}.$$

See Figure 15.6.

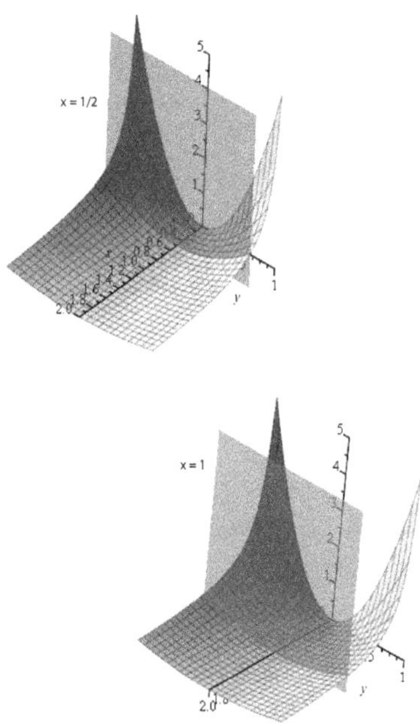

Fig. 15.6 The proper convex function of Example 15.4. When intersected by vertical planes of the form $x = \alpha$, for $\alpha > 0$, the trace is an upward parabola. When α is close to zero, this parabola approximates the positive z axis.

The function f is continuous on the open right half-plane $\{(x,y) \in \mathbb{R}^2 \mid x > 0\}$, but not at $(0,0)$. The limit of $f(x,y)$ when (x,y) approaches $(0,0)$

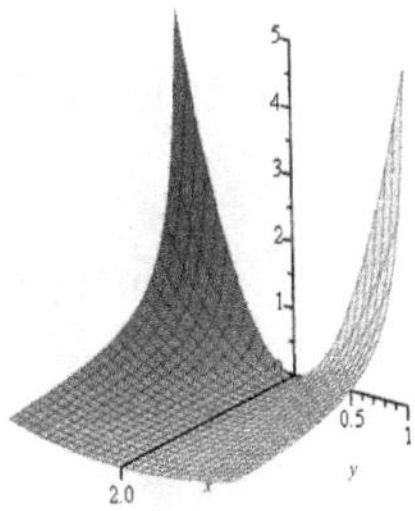

Figure a

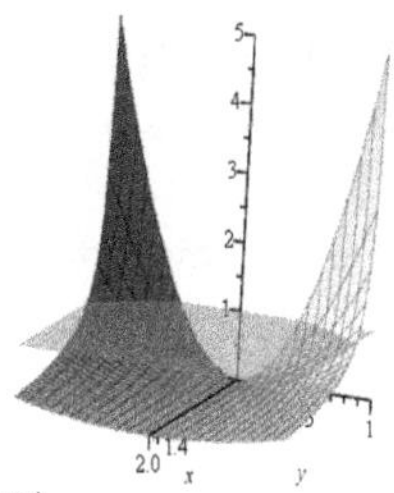

Figure b

Fig. 15.7 Figure (a) illustrates the proper convex function of Example 15.4. Figure (b) illustrates the approach to $(0,0)$ along the planar parabolic curve $(y^2/2, y)$. Then $f(y^2/2, y) = 1$ and Figure (b) shows the intersection of the surface with the plane $z = 1$.

on the parabola of equation $x = y^2/(2\alpha)$ is α for any $\alpha > 0$. See Figure 15.7 However, it is easy to see that the limit along any line segment from $(0,0)$ to a point in the open right half-plane is 0.

We conclude this quick tour of the basic properties of convex functions with a result involving the Lipschitz condition.

Definition 15.11. Let $f\colon E \to F$ be a function between normed vector spaces E and F, and let U be a nonempty subset of E. We say that f *Lipschitzian on* U (or *has the Lipschitz condition on* U) if there is some $c \geq 0$ such that

$$\|f(x) - f(y)\|_F \leq c\,\|x - y\|_E \quad \text{for all } x, y \in U.$$

Obviously, if f is Lipschitzian on U it is uniformly continuous on U. The following result is proven in Rockafellar [Rockafellar (1970)] (Theorem 10.4).

Proposition 15.8. *Let f be a proper convex function, and let S be any (nonempty) closed bounded subset of* **relint**(dom(f)). *Then f is Lipschitzian on S.*

In particular, a finite convex function on $\mathbb{R}^n$ is Lipschitzian on every compact subset of $\mathbb{R}^n$. However, such a function may not be Lipschitzian on $\mathbb{R}^n$ as a whole.

15.2 Subgradients and Subdifferentials

We saw in the previous section that proper convex functions have "good" continuity properties. Remarkably, if f is a convex function, for any $x \in \mathbb{R}^n$ such that $f(x)$ is finite, the one-sided derivative $f'(x; u)$ exists for all $u \in \mathbb{R}^n$; This result has been shown at least since 1893, as noted by Stoltz (see Rockafellar [Rockafellar (1970)], page 428). Directional derivatives will be discussed in Section 15.3. If f is differentiable at x, then of course

$$df_x(u) = \langle \nabla f_x, u \rangle \quad \text{for all } u \in \mathbb{R}^n,$$

where ∇f_x is the gradient of f at x.

But even if f is not differentiable at x, it turns out that for "most" $x \in \text{dom}(f)$, in particular if $x \in \mathbf{relint}(\text{dom}(f))$, there is a nonempty closed convex set $\partial f(x)$ which may be viewed as a generalization of the gradient ∇f_x. This convex set of $\mathbb{R}^n$, $\partial f(x)$, called the *subdifferential of f at x*, has some of the properties of the gradient ∇f_x. The vectors in $\partial f(x)$ are called *subgradients* at x. For example, if f is a proper convex function, then f achieves its minimum at $x \in \mathbb{R}^n$ iff $0 \in \partial f(x)$. Some of the theorems of Chapter 14 can be generalized to convex functions that are not necessarily differentiable by replacing conditions involving gradients by conditions involving subdifferentials. These generalizations are crucial for the justification that various iterative methods for solving optimization programs converge. For example, they are used to prove the convergence of the ADMM method discussed in Chapter 16.

One should note that the notion of subdifferential is not just a gratuitous mathematical generalization. The remarkable fact that the optimization problem

$$\begin{aligned} &\text{minimize} \quad J(u) \\ &\text{subject to} \quad u \in C, \end{aligned}$$

where C is a closed convex set in $\mathbb{R}^n$ can be rewritten as

$$\begin{aligned} &\text{minimize} \quad J(u) + I_C(z) \\ &\text{subject to} \quad u - z = 0, \end{aligned}$$

where I_C is the indicator function of C, forces us to deal with functions such as $J(u) + I_C(z)$ which are not differentiable, even if J is. ADMM can cope with this situation (under certain conditions), and subdifferentials cannot be avoided in justifying its convergence. However, it should be said that the subdifferential $\partial f(x)$ is a theoretical tool that is never computed in practice (except in very special simple cases).

To define subgradients we need to review (affine) hyperplanes.

Recall that an *affine form* $\varphi\colon \mathbb{R}^n \to \mathbb{R}$ is a function of the form

$$\varphi(x) = h(x) + c, \quad x \in \mathbb{R}^n,$$

where $h\colon \mathbb{R}^n \to \mathbb{R}$ is a linear form and $c \in \mathbb{R}$ is some constant. An *affine hyperplane* $H \subseteq \mathbb{R}^n$ is the kernel of any nonconstant affine form $\varphi\colon \mathbb{R}^n \to \mathbb{R}$ (which means that the linear form h defining φ is not the zero linear form),

$$H = \varphi^{-1}(0) = \{x \in \mathbb{R}^n \mid \varphi(x) = 0\}.$$

Any two nonconstant affine forms φ and ψ defining the same (affine) hyperplane H, in the sense that $H = \varphi^{-1}(0) = \psi^{-1}(0)$, must be proportional, which means that there is some nonzero $\alpha \in \mathbb{R}$ such that $\psi = \alpha\varphi$.

A nonconstant affine form φ also defines the two *half spaces* H_+ and H_- given by

$$H_+ = \{x \in \mathbb{R}^n \mid \varphi(x) \geq 0\}, \quad H_- = \{x \in \mathbb{R}^n \mid \varphi(x) \leq 0\}.$$

Clearly, $H_+ \cap H_- = H$, their common boundary. See Figure 15.8. The choice of sign is somewhat arbitrary, since the affine form $\alpha\varphi$ with $\alpha < 0$ defines the half spaces with H_- and H_+ (the half spaces are swapped).

By the duality induced by the Euclidean inner product on $\mathbb{R}^n$, a linear form $h\colon \mathbb{R}^n \to \mathbb{R}$ corresponds to a *unique* vector $u \in \mathbb{R}^n$ such that

$$h(x) = \langle x, u\rangle \quad \text{for all } x \in \mathbb{R}^n.$$

Then if φ is the affine form given by $\varphi(x) = \langle x, u\rangle + c$, this affine form is nonconstant iff $u \neq 0$, and u is normal to the hyperplane H, in the sense that if $x_0 \in H$ is any fixed vector in H, and x is any vector in H, then $\langle x - x_0, u\rangle = 0$.

Indeed, $x_0 \in H$ means that $\langle x_0, u\rangle + c = 0$, and $x \in H$ means that $\langle x, u\rangle + c = 0$, so we get $\langle x_0, u\rangle = \langle x, u\rangle$, which implies $\langle x - x_0, u\rangle = 0$.

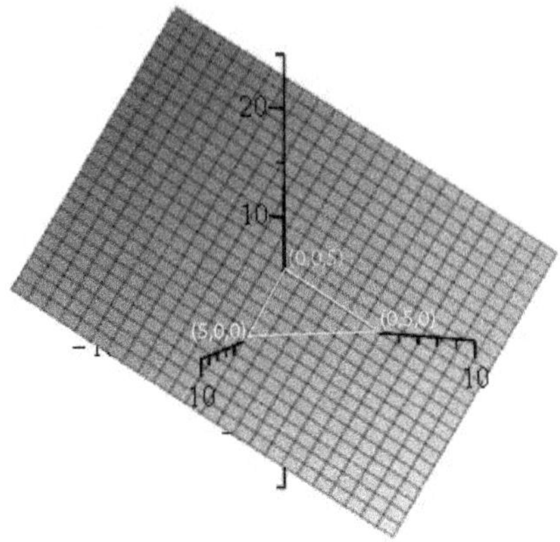

Fig. 15.8 The affine hyperplane $H = \{x \in \mathbb{R}^3 \mid x + y + z - 2 = 0\}$. The half space H_+ faces the viewer and contains the point $(0, 0, 10)$, while the half space H_- is behind H and contains the point $(0, 0, 0)$.

Here is an observation which plays a key role in defining the notion of subgradient. An illustration of the following proposition is provided by Figure 15.9.

Proposition 15.9. *Let $\varphi\colon \mathbb{R}^n \to \mathbb{R}$ be a nonconstant affine form. Then the map $\omega\colon \mathbb{R}^{n+1} \to \mathbb{R}$ given by*

$$\omega(x, \alpha) = \varphi(x) - \alpha, \quad x \in \mathbb{R}^n,\ \alpha \in \mathbb{R},$$

is a nonconstant affine form defining a hyperplane $\mathcal{H} = \omega^{-1}(0)$ which is the graph of the affine form φ. Furthermore, this hyperplane is nonvertical in $\mathbb{R}^{n+1}$, in the sense that $\mathcal{H}$ cannot be defined by a nonconstant affine form $(x, \alpha) \mapsto \psi(x)$ which does not depend on α.

Proof. Indeed, φ is of the form $\varphi(x) = h(x) + c$ for some nonzero linear form h, so

$$\omega(x, \alpha) = h(x) - \alpha + c.$$

Since h is linear, the map $(x, \alpha) = h(x) - \alpha$ is obviously linear and nonzero, so ω is a nonconstant affine form defining a hyperplane $\mathcal{H}$ in $\mathbb{R}^{n+1}$. By definition,

$$\mathcal{H} = \{(x, \alpha) \in \mathbb{R}^{n+1} \mid \omega(x, \alpha) = 0\} = \{(x, \alpha) \in \mathbb{R}^{n+1} \mid \varphi(x) - \alpha = 0\},$$

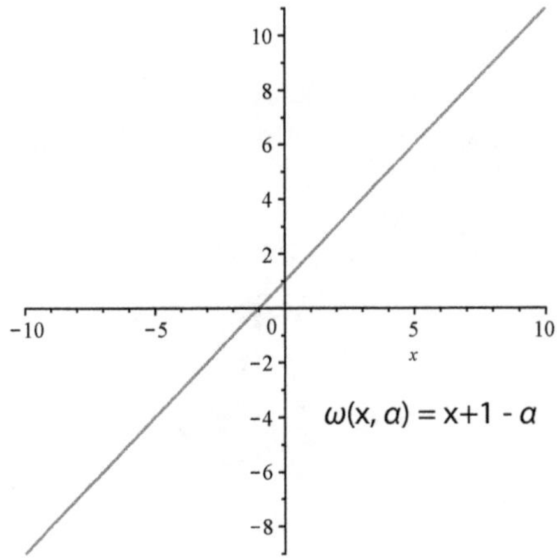

Fig. 15.9 Let $\varphi\colon \mathbb{R} \to \mathbb{R}$ be the affine form $\varphi(x) = x + 1$. Let $\omega\colon \mathbb{R}^2 \to \mathbb{R}$ be the affine form $\omega(x, \alpha) = x + 1 - \alpha$. The hyperplane $\mathcal{H} = \omega^{-1}(0)$ is the red line with equation $x - \alpha + 1 = 0$.

which is the graph of φ. If $\mathcal{H}$ was a vertical hyperplane, then $\mathcal{H}$ would be defined by a nonconstant affine form ψ independent of α, but the affine form ω given by $\omega(x, \alpha) = \varphi(x) - \alpha$ and the affine form $\psi(x)$ can't be proportional, a contradiction. ☐

We say that $\mathcal{H}$ is the *hyperplane (in $\mathbb{R}^{n+1}$) induced by the affine form $\varphi\colon \mathbb{R}^n \to \mathbb{R}$*. Also recall the notion of supporting hyperplane to a convex set.

Definition 15.12. If C is a nonempty convex set in $\mathbb{R}^n$ and x is a vector in C, an affine hyperplane H is a *supporting hyperplane to C at x* if

(1) $x \in H$.
(2) Either $C \subseteq H_+$ or $C \subseteq H_-$.

See Figure 15.10. Equivalently, there is some nonconstant affine form φ such that $\varphi(z) = \langle z, u\rangle - c$ for all $z \in \mathbb{R}^n$, for some nonzero $u \in \mathbb{R}^n$ and some $c \in \mathbb{R}$, such that

(1) $\langle x, u\rangle = c$.
(2) $\langle z, u\rangle \leq c$ for all $z \in C$

The notion of vector normal to a cone is defined as follows.

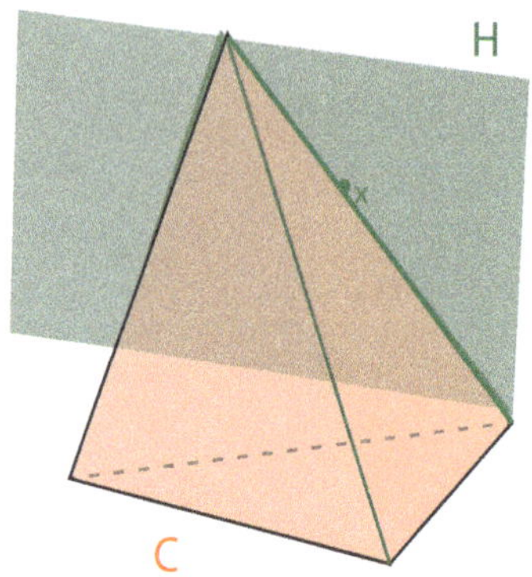

Fig. 15.10 Let C be the solid peach tetrahedron in $\mathbb{R}^3$. The green plane H is a supporting hyperplane to the point x since $x \in H$ and $C \subseteq H_+$, i.e., H only intersects C on the edge containing x and so the tetrahedron lies in "front" of H.

Definition 15.13. Given a nonempty convex set C in $\mathbb{R}^n$, for any $a \in C$, a vector $u \in \mathbb{R}^n$ is *normal to C at a* if

$$\langle z - a, u\rangle \leq 0 \quad \text{for all } z \in C.$$

In other words, u does not make an acute angle with any line segment in C with a as endpoint. The set of all vectors u normal to C is called the *normal cone* to C at a and is denoted by $N_C(a)$. See Figure 15.11.

It is easy to check that the normal cone to C at a is a convex cone. Also, if the hyperplane H defined by an affine form $\varphi(z) = \langle z, u\rangle - c$ with $u \neq 0$ is a supporting hyperplane to C at x, since $\langle z, u\rangle \leq c$ for all $z \in C$ and $\langle x, u\rangle = c$, we have $\langle z - x, u\rangle \leq 0$ for all $z \in C$, *which means that u is normal to C at x*. This concept is illustrated by Figure 15.12.

The notion of subgradient can be motived as follows. A function $f\colon \mathbb{R}^n \to \mathbb{R}$ is differentiable at $x \in \mathbb{R}^n$ if

$$f(x + y) = f(x) + df_x(y) + \epsilon(y)\, \|y\|_2\,,$$

for all $y \in \mathbb{R}^n$ in some nonempty subset containing x, where $df_x\colon \mathbb{R}^n \to \mathbb{R}$ is a linear form, and ϵ is some function such that $\lim_{\|y\| \mapsto 0} \epsilon(y) = 0$. Furthermore,

$$df_x(y) = \langle y, \nabla f_x\rangle \quad \text{for all } y \in \mathbb{R}^n,$$

where ∇f_x is the *gradient* of f at x, so

$$f(x + y) = f(x) + \langle y, \nabla f_x\rangle + \epsilon(y)\, \|y\|_2\,.$$

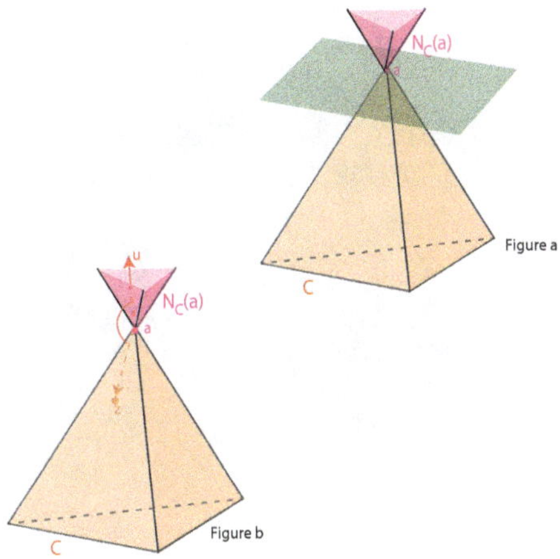

Fig. 15.11 Let C be the solid peach tetrahedron in $\mathbb{R}^3$. The small upside-down magenta tetrahedron is the translate of $N_C(a)$. Figure (a) shows that the normal cone is separated from C by the horizontal green supporting hyperplane. Figure (b) shows that any vector $u \in N_C(a)$ does not make an acute angle with a line segment in C emanating from a.

If we assume that f is convex, it makes sense to replace the equality sign by the inequality sign $\geq$ in the above equation and to drop the "error term" $\epsilon(y)\left\|y\right\|_2$, so a vector u is a subgradient of f at x if

$$f(x+y) \geq f(x) + \langle y, u\rangle \quad \text{for all } y \in \mathbb{R}^n.$$

Thus we are led to the following definition.

Definition 15.14. Let $f\colon \mathbb{R}^n \to \mathbb{R} \cup \{-\infty, +\infty\}$ be a convex function. For any $x \in \mathbb{R}^n$, a *subgradient of f at x* is any vector $u \in \mathbb{R}^n$ such that

$$f(z) \geq f(x) + \langle z - x, u\rangle, \quad \text{for all } z \in \mathbb{R}^n. \qquad (*_{\text{subgrad}})$$

The above inequality is called the *subgradient inequality*. The set of all subgradients of f at x is denoted $\partial f(x)$ and is called the *subdifferential* of f at x. If $\partial f(x) \neq \emptyset$, then we say that *$f$ is subdifferentiable at x*.

Assume that $f(x)$ is finite. Observe that the subgradient inequality says that 0 is a subgradient at x iff f has a global minimum at x. In this case, the hyperplane $\mathcal{H}$ (in $\mathbb{R}^{n+1}$) defined by the affine form $\omega(x, \alpha) = f(x) - \alpha$ is a horizontal supporting hyperplane to the epigraph $\mathbf{epi}(f)$ at $(x, f(x))$. If $u \in \partial f(x)$ and $u \neq 0$, then $(*_{\text{subgrad}})$ says that the hyperplane induced by the affine form $z \mapsto \langle z - x, u\rangle + f(x)$ as in Proposition 15.9 is a nonvertical

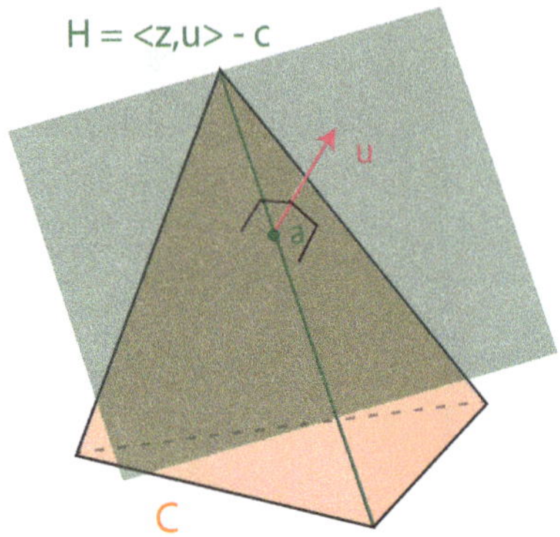

Fig. 15.12 Let C be the solid peach tetrahedron in $\mathbb{R}^3$. The green plane H defined by $\varphi(z) = \langle z, u\rangle - c$ is a supporting hyperplane to C at a. The pink normal to H, namely the vector u, is also normal to C at a.

supporting hyperplane $\mathcal{H}$ (in $\mathbb{R}^{n+1}$) to the epigraph $\mathbf{epi}(f)$ at $(x, f(x))$. The vector $(u, -1) \in \mathbb{R}^{n+1}$ is normal to the hyperplane $\mathcal{H}$. See Figure 15.13.

Indeed, if $u \neq 0$, the hyperplane $\mathcal{H}$ is given by

$$\mathcal{H} = \{(y, \alpha) \in \mathbb{R}^{n+1} \mid \omega(y, \alpha) = 0\}$$

with $\omega(y, \alpha) = \langle y - x, u\rangle + f(x) - \alpha$, so $\omega(x, f(x)) = 0$, which means that $(x, f(x)) \in \mathcal{H}$. Also, for any $(z, \beta) \in \mathbf{epi}(f)$, by $(*_{\text{subgrad}})$, we have

$$\omega(z, \beta) = \langle z - x, u\rangle + f(x) - \beta \leq f(z) - \beta \leq 0,$$

since $(z, \beta) \in \mathbf{epi}(f)$, so $\mathbf{epi}(f) \subseteq \mathcal{H}_-$, and $\mathcal{H}$ is a nonvertical supporting hyperplane (in $\mathbb{R}^{n+1}$) to the epigraph $\mathbf{epi}(f)$ at $(x, f(x))$. Since

$$\omega(y, \alpha) = \langle y - x, u\rangle + f(x) - \alpha = \langle (y - x, \alpha), (u, -1)\rangle + f(x),$$

the vector $(u, -1)$ is indeed normal to the hyperplane $\mathcal{H}$.

The above facts are important and recorded as the following proposition.

Proposition 15.10. *If $f(x)$ is finite, then f is subdifferentiable at x if and only if there is a nonvertical supporting hyperplane (in $\mathbb{R}^{n+1}$) to the epigraph* $\mathbf{epi}(f)$ *at $(x, f(x))$. In this case, there is a linear form φ (over $\mathbb{R}^n$) such that $f(x) \geq \varphi(x)$ for all $x \in \mathbb{R}^n$. We can pick φ given by $\varphi(y) = \langle y - x, u\rangle + f(x)$ for all $y \in \mathbb{R}^n$.*

It is easy to see that $\partial f(x)$ is closed and convex. The set $\partial f(x)$ may be empty, or reduced to a single element. In $\partial f(x)$ consists of a single

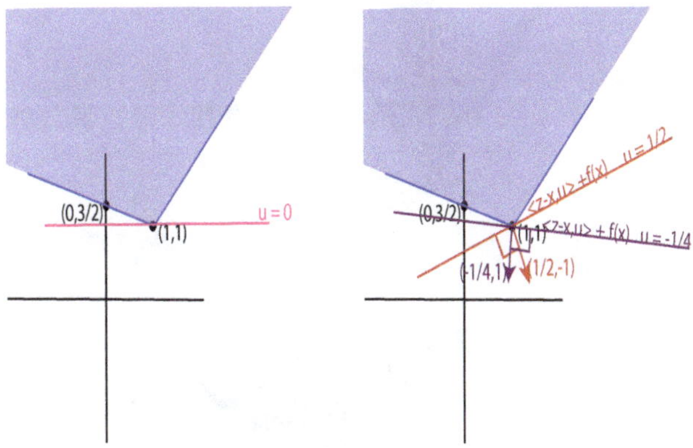

Fig. 15.13 Let $f\colon \mathbb{R} \to \mathbb{R} \cup \{-\infty, +\infty\}$ be the piecewise function defined by $f(x) = x+1$ for $x \geq 1$ and $f(x) = -\frac{1}{2}x + \frac{3}{2}$ for $x < 1$. Its epigraph is the shaded blue region in $\mathbb{R}^2$. Since f has minimum at $x = 1$, $0 \in \partial f(1)$, and the graph of $f(x)$ has a horizontal supporting hyperplane at $(1, 1)$. Since $\{\frac{1}{2}, -\frac{1}{4}\} \subset \partial f(1)$, the maroon line $\frac{1}{2}(x-1)+1$ (with normal $(\frac{1}{2}, -1)$) and the violet line $-\frac{1}{4}(x-1)+1$ (with normal $(-\frac{1}{4}, -1)$) are supporting hyperplanes to the graph of $f(x)$ at $(1, 1)$.

element it can be shown that f is finite near x, differentiable at x, and that $\partial f(x) = \{\nabla f_x\}$, the gradient of f at x.

Example 15.5. The ℓ^2 norm $f(x) = \|x\|_2$ is subdifferentiable for all $x \in \mathbb{R}^n$, in fact differentiable for all $x \neq 0$. For $x = 0$, the set $\partial f(0)$ consists of all $u \in \mathbb{R}^n$ such that

$$\|z\|_2 \geq \langle z, u\rangle \quad \text{for all } z \in \mathbb{R}^n,$$

namely (by Cauchy–Schwarz), the Euclidean unit ball $\{u \in \mathbb{R}^n \mid \|u\|_2 \leq 1\}$. See Figure 15.14.

Example 15.6. For the ℓ^∞ norm if $f(x) = \|x\|_\infty$, we leave it as an exercise to show that $\partial f(0)$ is the polyhedron

$$\partial f(0) = \text{conv}\{\pm e_1, \ldots, \pm e_n\}.$$

See Figure 15.15. One can also work out what is $\partial f(x)$ if $x \neq 0$, but this is more complicated; see Rockafellar [Rockafellar (1970)], page 215.

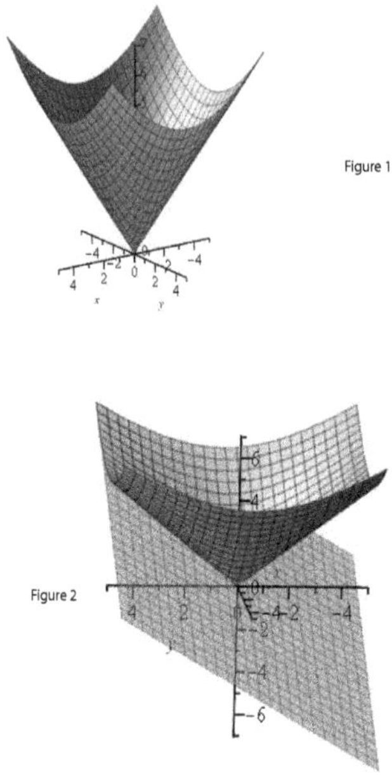

Fig. 15.14 Figure (1) shows the graph in $\mathbb{R}^3$ of $f(x, y) = \|(x, y)\|_2 = \sqrt{x^2 + y^2}$. Figure (2) shows the supporting hyperplane with normal $(\frac{1}{\sqrt{2}}, \frac{1}{\sqrt{2}}, -1)$, where $(\frac{1}{\sqrt{2}}, \frac{1}{\sqrt{2}}) \in \partial f(0)$.

Example 15.7. The following function is an example of a proper convex function which is not subdifferentiable everywhere:

$$f(x) = \begin{cases} -(1 - |x|^2)^{1/2} & \text{if } |x| \le 1 \\ +\infty & \text{otherwise.} \end{cases}$$

See Figure 15.16. We leave it as an exercise to show that f is subdifferentiable (in fact differentiable) at x when $|x| < 1$, but $\partial f(x) = \emptyset$ when $|x| \ge 1$, even though $x \in \mathrm{dom}(f)$ for $|x| = 1$.

Example 15.8. The subdifferential of an indicator function is interesting. Let C be a nonempty convex set. By definition, $u \in \partial I_C(x)$ iff

$$I_C(z) \ge I_C(x) + \langle z - x, u \rangle \quad \text{for all } z \in \mathbb{R}^n.$$

Since C is nonempty, there is some $z \in C$ such that $I_C(z) = 0$, so the above condition implies that $x \in C$ (otherwise $I_C(x) = +\infty$ but $0 \ge$

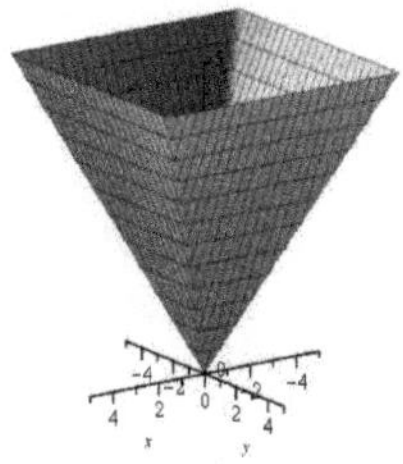

Figure 1

Figure 2

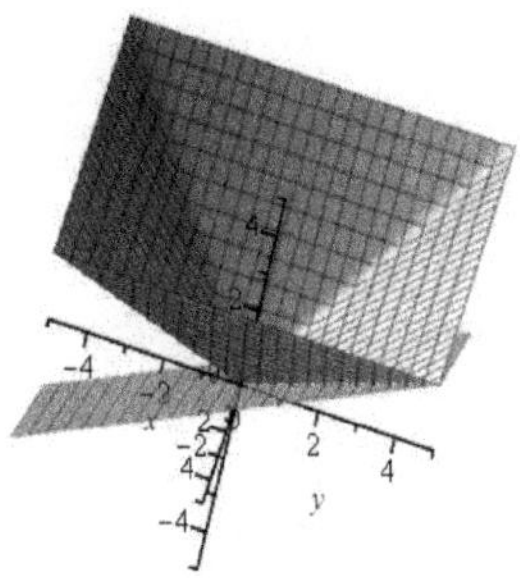

Fig. 15.15 Figure (1) shows the graph in $\mathbb{R}^3$ of $f(x,y) = \|(x,y)\|_\infty = \sup\{|x|,|y|\}$. Figure (2) shows the supporting hyperplane with normal $(\frac{1}{2},\frac{1}{2},-1)$, where $(\frac{1}{2},\frac{1}{2}) \in \partial f(0)$.

$+\infty + \langle z-u,u\rangle$ is impossible), so $0 \geq \langle z-x,u\rangle$ for all $z \in C$, which means that z is normal to C at x. Therefore, *$\partial I_C(x)$ is the normal cone $N_C(x)$ to C at x.*

Example 15.9. The subdifferentials of the indicator function f of the nonnegative orthant of $\mathbb{R}^n$ reveal a connection to complementary slackness conditions. Recall that this indicator function is given by

$$f(x_1,\ldots,x_n) = \begin{cases} 0 & \text{if } x_i \geq 0,\ 1 \leq i \leq n, \\ +\infty & \text{otherwise.} \end{cases}$$

By Example 15.8, the subgradients y of f at $x \geq 0$ form the normal cone to the nonnegative orthant at x. This means that $y \in N_C(x)$ iff

$$\langle z-x,y\rangle \leq 0 \quad \text{for all } z \geq 0$$

iff

$$\langle z,y\rangle \leq \langle x,y\rangle \quad \text{for all } z \geq 0.$$

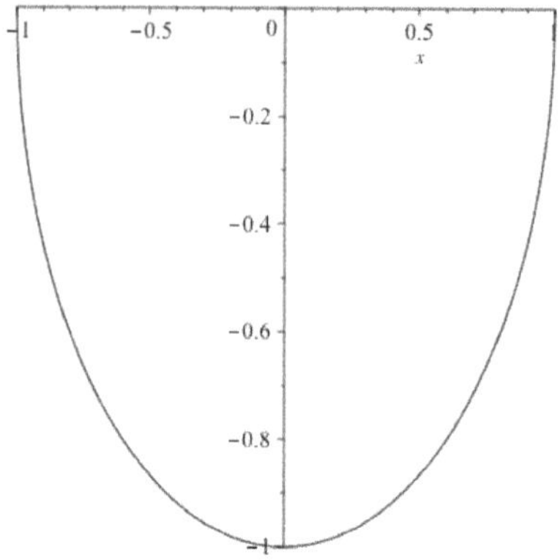

Fig. 15.16 The graph of the function in Example 15.7.

In particular, for $z = 0$ we get $\langle x, y\rangle \geq 0$, and for $z = 2x \geq 0$, we have $\langle x, y\rangle \leq 0$, so $\langle x, y\rangle = 0$. As a consequence, $y \in N_C(x)$ iff $\langle x, y\rangle = 0$ and

$$\langle z, y\rangle \leq 0 \quad \text{for all } z \geq 0.$$

For $z = e_j \geq 0$, we get $y_j \leq 0$. Conversely, if $y \leq 0$ and $\langle x, y\rangle = 0$, since $x \geq 0$, we get $\langle z, y\rangle \leq 0$ for all $z \geq 0$, and so

$$\partial f(x) = \{y = (y_1, \ldots, y_n) \in \mathbb{R}^n \mid y \leq 0,\ \langle x, y\rangle = 0\}.$$

But for $x \geq 0$ and $y \leq 0$ we have $\langle x, y\rangle = \sum_{j=1}^n x_j y_j = 0$ iff $x_j y_j = 0$ for $j = 1, \ldots, n$, thus we see that $y \in \partial f(x)$ iff we have

$$x_j \geq 0,\ y_j \leq 0,\ x_j y_j = 0, \quad 1 \leq j \leq n,$$

which are complementary slackness conditions.

Supporting hyperplanes to the epigraph of a proper convex function f can be used to prove a property which plays a key role in optimization theory. The proof uses a classical result of convex geometry, namely the Minkowski supporting hyperplane theorem.

Theorem 15.1. *(Minkowski) Let C be a nonempty convex set in $\mathbb{R}^n$. For any point $a \in C - \mathbf{relint}(C)$, there is a supporting hyperplane H to C at a.*

Theorem 15.1 is proven in Rockafellar [Rockafellar (1970)] (Theorem 11.6). See also Berger [Berger (1990a)] (Proposition 11.5.2). The proof is not as simple as one might expect, and is based on a geometric version of the Hahn–Banach theorem.

In order to prove Theorem 15.2 below we need two technical propositions.

Proposition 15.11. *Let C be any nonempty convex set in $\mathbb{R}^n$. For any $x \in \mathbf{relint}(C)$ and any $y \in \overline{C}$, we have $(1-\lambda)x + \lambda y \in \mathbf{relint}(C)$ for all λ such that $0 \leq \lambda < 1$. In other words, the line segment from x to y including x and excluding y lies entirely within $\mathbf{relint}(C)$.*

Proposition 15.11 is proven in Rockafellar [Rockafellar (1970)] (Theorem 6.1). The proof is not difficult but quite technical.

Proposition 15.12. *For any proper convex function f on $\mathbb{R}^n$, we have*

$$\mathbf{relint}(\mathbf{epi}(f)) = \{(x,\mu) \in \mathbb{R}^{n+1} \mid x \in \mathbf{relint}(\mathrm{dom}(f)),\ f(x) < \mu\}.$$

Proof. Proposition 15.12 is proven in Rockafellar [Rockafellar (1970)] (Lemma 7.3). By working in the affine hull of $\mathbf{epi}(f)$, the statement of Proposition 15.12 is equivalent to

$$\mathrm{int}(\mathbf{epi}(f)) = \{(x,\mu) \in \mathbb{R}^{m+1} \mid x \in \mathrm{int}(\mathrm{dom}(f)),\ f(x) < \mu\},$$

assuming that the affine hull of $\mathbf{epi}(f)$ has dimension $m+1$. See Figure (1) of Figure 15.17. The inclusion $\subseteq$ is obvious, so we only need to prove the reverse inclusion. Then for any $z \in \mathrm{int}(\mathrm{dom}(f))$, we can find a convex polyhedral subset $P = \mathrm{conv}(a_1, \ldots, a_{m+1})$ with $a_1, \ldots, a_{m+1} \in \mathrm{dom}(f)$ such that $z \in \mathrm{int}(P)$. Let

$$\alpha = \max\{f(a_1), \ldots, f(a_{m+1})\}.$$

Since any $x \in P$ can be expressed as

$$x = \lambda_1 a_1 + \cdots + \lambda_{m+1} a_{m+1}, \quad \lambda_1 + \cdots + \lambda_{m+1} = 1,\ \lambda_i \geq 0,$$

and since f is convex we have

$$f(x) \leq \lambda_1 f(a_1) + \cdots + \lambda_{m+1} f(a_{m+1}) \leq (\lambda_1 + \cdots + \lambda_{m+1})\alpha = \alpha \quad \text{for all } x \in P.$$

The above shows that the open subset

$$\{(x,\mu) \in \mathbb{R}^{m+1} \mid x \in \mathrm{int}(P),\ \alpha < \mu\}$$

is contained in $\mathbf{epi}(f)$. See Figure (2) of Figure 15.17. In particular, for every $\mu > \alpha$, we have

$$(z,\mu) \in \mathrm{int}(\mathbf{epi}(f)).$$

Thus for any $\beta \in \mathbb{R}$ such that $\beta > f(z)$, we see that (z,β) belongs to the relative interior of the vertical line segment $\{(z,\mu) \in \mathbb{R}^{m+1} \mid f(z) \leq \mu \leq \alpha + \beta + 1\}$ which meets $\mathrm{int}(\mathbf{epi}(f))$. See Figure (3) of Figure 15.17. By Proposition 15.11, $(z,\beta) \in \mathrm{int}(\mathbf{epi}(f))$. ☐

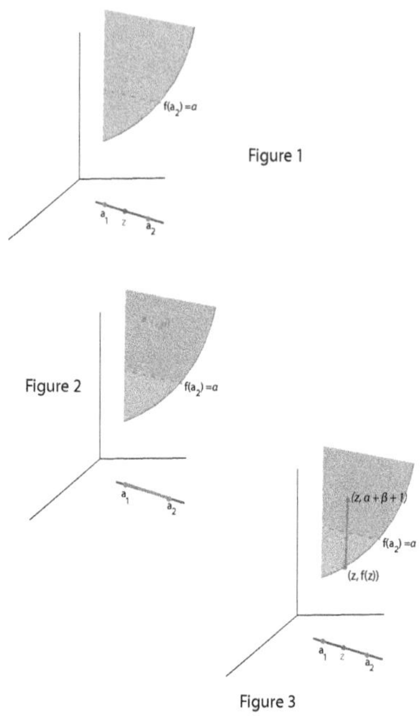

Fig. 15.17 Figure (1) illustrates $\mathbf{epi}(f)$, where $\mathbf{epi}(f)$ is contained in a vertical plane of affine dimension 2. Figure (2) illustrates the magenta open subset $\{(x, \mu) \in \mathbb{R}^2 \mid x \in \text{int}(P),\ \alpha < \mu\}$ of $\mathbf{epi}(f)$. Figure (3) illustrates the vertical line segment $\{(z, \mu) \in \mathbb{R}^2 \mid f(z) \leq \mu \leq \alpha + \beta + 1\}$.

We can now prove the following important theorem.

Theorem 15.2. *Let f be a proper convex function on $\mathbb{R}^n$. For any $x \in$ $\mathbf{relint}(\text{dom}(f))$, there is a nonvertical supporting hyperplane $\mathcal{H}$ to $\mathbf{epi}(f)$ at $(x, f(x))$. Consequently f is subdifferentiable for all $x \in \mathbf{relint}(\text{dom}(f))$, and there is some affine form $\varphi\colon \mathbb{R}^n \to \mathbb{R}$ such that $f(x) \geq \varphi(x)$ for all $x \in \mathbb{R}^n$.*

Proof. By Proposition 15.2, for any $x \in \mathbf{relint}(\text{dom}(f))$, we have $(x, \mu) \in \mathbf{relint}(\mathbf{epi}(f))$ for all $\mu \in \mathbb{R}$ such that $f(x) < \mu$. Since by definition of $\mathbf{epi}(f)$ we have $(x, f(x)) \in \mathbf{epi}(f) - \mathbf{relint}(\mathbf{epi}(f))$, by Minkowski's theorem (Theorem 15.1), there is a supporting hyperplane $\mathcal{H}$ to $\mathbf{epi}(f)$ through $(x, f(x))$. Since $x \in \mathbf{relint}(\text{dom}(f))$ and f is proper, the hyperplane $\mathcal{H}$

is not a vertical hyperplane. By Definition 15.14, the function f is subdifferentiable at x, and the subgradient inequality shows that if we let $\varphi(z) = f(x) + \langle z - x, u\rangle$, then φ is an affine form such that $f(x) \geq \varphi(x)$ for all $x \in \mathbb{R}^n$. □

Intuitively, a proper convex function can't decrease faster than an affine function. It is surprising how much work it takes to prove such an "obvious" fact.

Remark: Consider the proper convex function $f\colon \mathbb{R} \to \mathbb{R} \cup \{+\infty\}$ given by

$$f(x) = \begin{cases} -\sqrt{x} & \text{if } x \geq 0 \\ +\infty & \text{if } x < 0. \end{cases}$$

We have $\text{dom}(f) = [0, +\infty)$, f is differentiable for all $x > 0$, but it is not subdifferentiable at $x = 0$. The only supporting hyperplane to $\mathbf{epi}(f)$ at $(0, 0)$ is the vertical line of equation $x = 0$ (the y-axis) as illustrated by Figure 15.18.

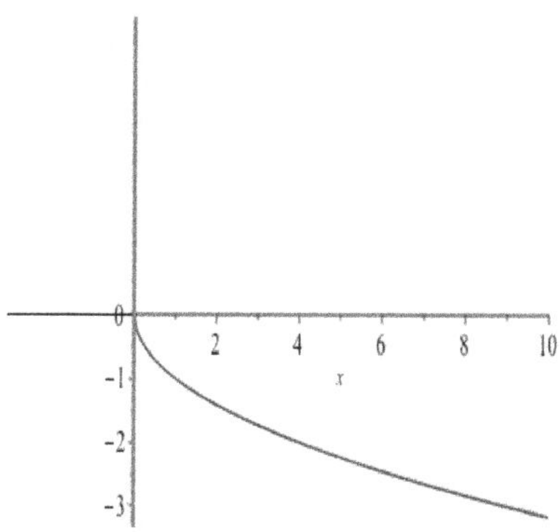

Fig. 15.18 The graph of the partial function $f(x) = -\sqrt{x}$ and its red vertical supporting hyperplane at $x = 0$.

15.3 Basic Properties of Subgradients and Subdifferentials

A major tool to prove properties of subgradients is a variant of the notion of directional derivative.

Definition 15.15. Let $f\colon \mathbb{R}^n \to \mathbb{R} \cup \{-\infty, +\infty\}$ be any function. For any $x \in \mathbb{R}^n$ such that $f(x)$ is finite ($f(x) \in \mathbb{R}$), for any $u \in \mathbb{R}^n$, the *one-sided directional derivative* $f'(x; u)$ is defined to be the limit

$$f'(x; u) = \lim_{\lambda \downarrow 0} \frac{f(x + \lambda u) - f(x)}{\lambda}$$

if it exists ($-\infty$ and $+\infty$ being allowed as limits). See Figure 15.19. The above notation for the limit means that we consider the limit when $\lambda > 0$ tends to 0.

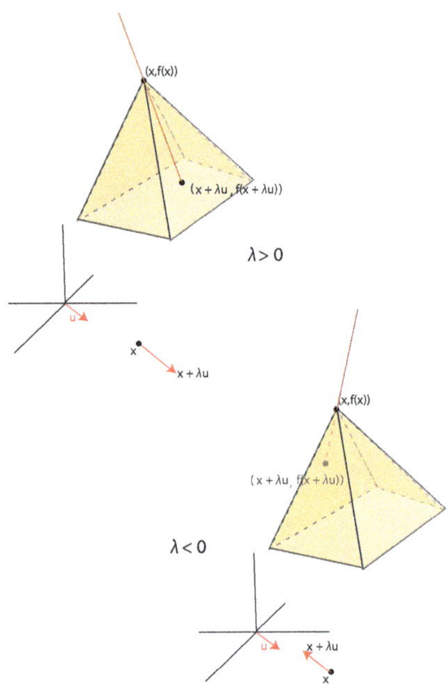

Fig. 15.19 Let $f\colon \mathbb{R}^2 \to \mathbb{R} \cup \{-\infty, +\infty\}$ be the function whose graph (in $\mathbb{R}^3$) is the surface of the peach pyramid. The top figure illustrates that $f'(x; u)$ is the slope of the slanted burnt orange line, while the bottom figure depicts the line associated with $\lim_{\lambda \uparrow 0} \frac{f(x+\lambda u) - f(x)}{\lambda}$.

Note that

$$\lim_{\lambda \uparrow 0} \frac{f(x + \lambda u) - f(x)}{\lambda}$$

denotes the one-sided limit when $\lambda < 0$ tends to zero, and that

$$-f'(x; -u) = \lim_{\lambda \uparrow 0} \frac{f(x + \lambda u) - f(x)}{\lambda},$$

so *the (two-sided) directional derivative* $D_u f(x)$ *exists iff* $-f'(x;-u) = f'(x;u)$. Also, if f is differentiable at x, then

$$f'(x;u) = \langle \nabla f_x, u \rangle, \quad \text{for all } u \in \mathbb{R}^n,$$

where ∇f_x is the gradient of f at x. Here is the first remarkable result.

Proposition 15.13. *Let* $f\colon \mathbb{R}^n \to \mathbb{R} \cup \{-\infty, +\infty\}$ *be a convex function. For any* $x \in \mathbb{R}^n$*, if* $f(x)$ *is finite, then the function*

$$\lambda \mapsto \frac{f(x+\lambda u) - f(x)}{\lambda}$$

is a nondecreasing function of $\lambda > 0$*, so that* $f'(x;u)$ *exists for any* $u \in \mathbb{R}^n$*, and*

$$f'(x;u) = \inf_{\lambda > 0} \frac{f(x+\lambda u) - f(x)}{\lambda}.$$

Furthermore, $f'(x;u)$ *is a positively homogeneous convex function of* u *(which means that* $f'(x;\alpha u) = \alpha f'(x;u)$ *for all* $\alpha \in \mathbb{R}$ *with* $\alpha > 0$ *and all* $u \in \mathbb{R}^n$*),* $f'(x;0) = 0$*, and*

$$-f'(x;-u) \leq f'(x;u) \quad \textit{for all } u \in \mathbb{R}^n$$

Proposition 15.13 is proven in Rockafellar [Rockafellar (1970)] (Theorem 23.1). The proof is not difficult but not very informative.

Remark: As a convex function of u, it can be shown that the effective domain of the function $u \mapsto f'(x;u)$ is the convex cone generated by $\mathrm{dom}(f) - x$.

We will now state without proof some of the most important properties of subgradients and subdifferentials. Complete details can be found in Rockafellar [Rockafellar (1970)] (Part V, Section 23).

In order to state the next proposition, we need the following definition.

Definition 15.16. For any convex set C in $\mathbb{R}^n$, the *support function* $\delta^*(-|C)$ of C is defined by

$$\delta^*(x|C) = \sup_{y \in C} \langle x, y \rangle, \quad x \in \mathbb{R}^n.$$

According to Definition 14.11, the conjugate of the indicator function I_C of a convex set C is given by

$$I_C^*(x) = \sup_{y \in \mathbb{R}^n} (\langle x, y \rangle - I_C(y)) = \sup_{y \in C} \langle x, y \rangle = \delta^*(x|C).$$

Thus $\delta^*(-|C) = I_C^*$*, the conjugate of the indicator function* I_C.

The following proposition relates directional derivatives at x and the subdifferential at x.

Proposition 15.14. *Let $f\colon \mathbb{R}^n \to \mathbb{R} \cup \{-\infty, +\infty\}$ be a convex function. For any $x \in \mathbb{R}^n$, if $f(x)$ is finite, then a vector $u \in \mathbb{R}^n$ is a subgradient to f at x if and only if*

$$f'(x; y) \geq \langle y, u \rangle \quad \text{for all } y \in \mathbb{R}^n.$$

Furthermore, the closure of the convex function $y \mapsto f'(x; y)$ is the support function of the closed convex set $\partial f(x)$, the subdifferential of f at x:

$$\mathrm{cl}(f'(x; -)) = \delta^*(-|\partial f(x)).$$

Sketch of proof. Proposition 15.14 is proven in Rockafellar [Rockafellar (1970)] (Theorem 23.2). We prove the inequality. If we write $z = x + \lambda y$ with $\lambda > 0$, then the subgradient inequality implies

$$f(x + \lambda u) \geq f(x) + \langle z - x, u \rangle = f(x) + \lambda \langle y, u \rangle,$$

so we get

$$\frac{f(x + \lambda y) - f(x)}{\lambda} \geq \langle y, u \rangle.$$

Since the expression on the left tends to $f'(x; y)$ as $\lambda > 0$ tends to zero, we obtain the desired inequality. The second part follows from Corollary 13.2.1 in Rockafellar [Rockafellar (1970)]. □

If f is a proper function on $\mathbb{R}$, then its effective domain being convex is an interval whose relative interior is an open interval (a, b). In Proposition 15.14, we can pick $y = 1$ so $\langle y, u \rangle = u$, and for any $x \in (a, b)$, since the limits $f'_-(x) = -f'(x; -1)$ and $f'_+(x) = f'(x; 1)$ exist, with $f'_-(x) \leq f'_+(x)$, we deduce that $\partial f(x) = [f'_-(x), f'_+(x)]$. The numbers $\alpha \in [f'_-(x), f'_+(x)]$ are the slopes of nonvertical lines in $\mathbb{R}^2$ passing through $(x, f(x))$ that are supporting lines to the epigraph $\mathbf{epi}(f)$ of f.

Example 15.10. If f is the celebrated **ReLU** function (ramp function) from deep learning defined so that

$$\mathrm{ReLU}(x) = \max\{x, 0\} = \begin{cases} 0 & \text{if } x < 0 \\ x & \text{if } x \geq 0, \end{cases}$$

then $\partial\,\mathrm{ReLU}(0) = [0, 1]$. See Figure 15.20. The function ReLU is differentiable for $x \neq 0$, with $\mathrm{ReLU}'(x) = 0$ if $x < 0$ and $\mathrm{ReLU}'(x) = 1$ if $x > 0$.

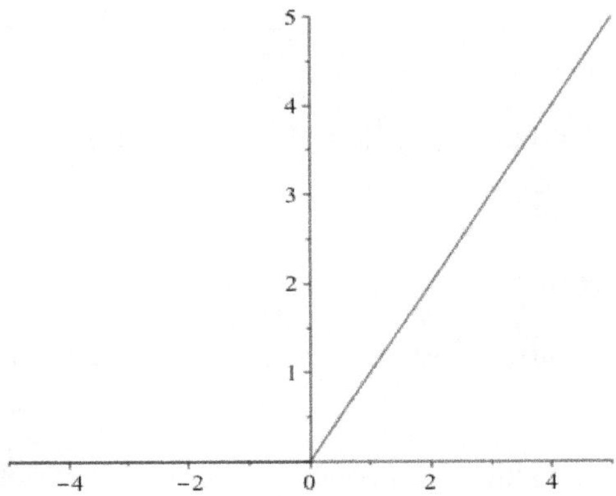

Fig. 15.20 The graph of the ReLU function.

Proposition 15.14 has several interesting consequences.

Proposition 15.15. *Let $f\colon \mathbb{R}^n \to \mathbb{R} \cup \{-\infty, +\infty\}$ be a convex function. For any $x \in \mathbb{R}^n$, if $f(x)$ is finite and if f is subdifferentiable at x, then f is proper. If f is not subdifferentiable at x, then there is some $y \neq 0$ such that*

$$f'(x; y) = -f'(x; -y) = -\infty.$$

Proposition 15.15 is proven in Rockafellar [Rockafellar (1970)] (Theorem 23.3). It confirms that improper convex functions are rather pathological objects, because if a convex function is subdifferentiable for some x such that $f(x)$ is finite, then f must be proper. This is because if $f(x)$ is finite, then the subgradient inequality implies that f majorizes an affine function, which is proper.

The next theorem is one of the most important results about the connection between one-sided directional derivatives and subdifferentials. It sharpens the result of Theorem 15.2.

Theorem 15.3. *Let $f\colon \mathbb{R}^n \to \mathbb{R} \cup \{+\infty\}$ be a proper convex function. For any $x \notin \mathrm{dom}(f)$, we have $\partial f(x) = \emptyset$. For any $x \in \mathbf{relint}(\mathrm{dom}(f))$, we have $\partial f(x) \neq \emptyset$, the map $y \mapsto f'(x; y)$ is convex, closed and proper, and*

$$f'(x; y) = \sup_{u \in \partial f(x)} \langle y, u \rangle = \delta^*(y | \partial f(x)) \quad \textit{for all } y \in \mathbb{R}^n.$$

The subdifferential $\partial f(x)$ is nonempty and bounded (also closed and convex) if and only if $x \in \mathrm{int}(\mathrm{dom}(f))$, in which case $f'(x; y)$ is finite for all $y \in \mathbb{R}^n$.

Theorem 15.3 is proven in Rockafellar [Rockafellar (1970)] (Theorem 23.4). If we write

$$\mathrm{dom}(\partial f) = \{x \in \mathbb{R}^n \mid \partial f(x) \neq \emptyset\},$$

then Theorem 15.3 implies that

$$\mathbf{relint}(\mathrm{dom}(f)) \subseteq \mathrm{dom}(\partial f) \subseteq \mathrm{dom}(f).$$

However, $\mathrm{dom}(\partial f)$ is not necessarily convex as shown by the following counterexample.

Example 15.11. Consider the proper convex function defined on $\mathbb{R}^2$ given by

$$f(x, y) = \max\{g(x), |y|\},$$

where

$$g(x) = \begin{cases} 1 - \sqrt{x} & \text{if } x \geq 0 \\ +\infty & \text{if } x < 0. \end{cases}$$

See Figure 15.21. It is easy to see that $\mathrm{dom}(f) = \{(x, y) \in \mathbb{R}^2 \mid x \geq 0\}$, but $\mathrm{dom}(\partial f) = \{(x, y) \in \mathbb{R}^2 \mid x \geq 0\} - \{(0, y) \mid -1 < y < 1\}$, which is not convex.

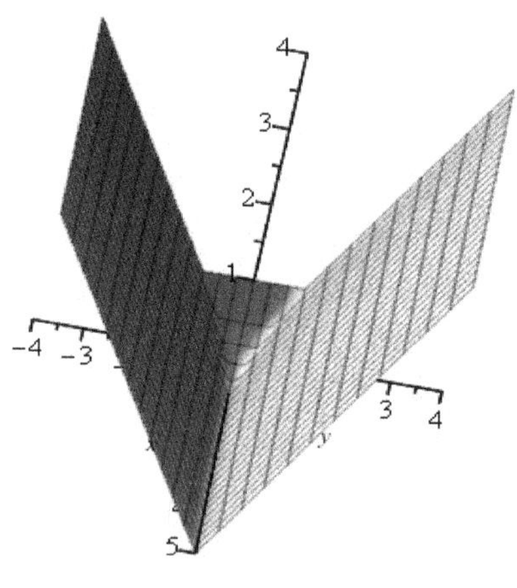

Fig. 15.21 The graph of the function from Example 15.11 with a view along the positive x axis.

The following theorem is important because it tells us when a convex function is differentiable in terms of its subdifferential, as shown in Rockafellar [Rockafellar (1970)] (Theorem 25.1).

Theorem 15.4. *Let f be a convex function on $\mathbb{R}^n$, and let $x \in \mathbb{R}^n$ such that $f(x)$ is finite. If f is differentiable at x then $\partial f(x) = \{\nabla f_x\}$ (where ∇f_x is the gradient of f at x) and we have*

$$f(z) \geq f(x) + \langle z - x, \nabla f_x \rangle \quad \textit{for all } z \in \mathbb{R}^n.$$

Conversely, if $\partial f(x)$ consists of a single vector, then $\partial f(x) = \{\nabla f_x\}$ and f is differentiable at x.

The first direction is easy to prove. Indeed, if f is differentiable at x, then

$$f'(x; y) = \langle y, \nabla f_x \rangle \quad \text{for all } y \in \mathbb{R}^n,$$

so by Proposition 15.14, a vector u is a subgradient at x iff

$$\langle y, \nabla f_x \rangle \geq \langle y, u \rangle \quad \text{for all } y \in \mathbb{R}^n,$$

so $\langle y, \nabla f_x - u \rangle \geq 0$ for all y, which implies that $u = \nabla f_x$.

We obtain the following corollary.

Corollary 15.1. *Let f be a convex function on $\mathbb{R}^n$, and let $x \in \mathbb{R}^n$ such that $f(x)$ is finite. If f is differentiable at x, then f is proper and $x \in \mathrm{int}(\mathrm{dom}(f))$.*

The following theorem shows that proper convex functions are differentiable almost everywhere.

Theorem 15.5. *Let f be a proper convex function on $\mathbb{R}^n$, and let D be the set of vectors where f is differentiable. Then D is a dense subset of $\mathrm{int}(\mathrm{dom}(f))$, and its complement in $\mathrm{int}(\mathrm{dom}(f))$ has measure zero. Furthermore, the gradient map $x \mapsto \nabla f_x$ is continuous on D.*

Theorem 15.5 is proven in Rockafellar [Rockafellar (1970)] (Theorem 25.5).

Remark: If $f\colon (a, b) \to \mathbb{R}$ is a finite convex function on an open interval of $\mathbb{R}$, then the set D where f is differentiable is dense in (a, b), and $(a, b) - D$ is at most countable. The map f' is continuous and nondecreasing on D. See Rockafellar [Rockafellar (1970)] (Theorem 25.3).

We also have the following result showing that in "most cases" the subdifferential $\partial f(x)$ can be constructed from the gradient map; see Rockafellar [Rockafellar (1970)] (Theorem 25.6).

Theorem 15.6. *Let f be a closed proper convex function on $\mathbb{R}^n$. If $\mathrm{int}(\mathrm{dom}(f)) \neq \emptyset$, then for every $x \in \mathrm{dom}(f)$, we have*

$$\partial f(x) = \overline{\mathrm{conv}(S(x))} + N_{\mathrm{dom}(f)}(x)$$

where $N_{\mathrm{dom}(f)}(x)$ is the normal cone to $\mathrm{dom}(f)$ at x, and $S(x)$ is the set of all limits of sequences of the form $\nabla f_{x_1}, \nabla f_{x_2}, \ldots, \nabla f_{x_p}, \ldots$, where $x_1, x_2, \ldots, x_p, \ldots$ is a sequence in $\mathrm{dom}(f)$ converging to x such that each ∇f_{x_p} is defined.

The next two results generalize familiar results about derivatives to subdifferentials.

Proposition 15.16. *Let $f_1, \ldots, f_n$ be proper convex functions on $\mathbb{R}^n$, and let $f = f_1 + \cdots + f_n$. For $x \in \mathbb{R}^n$, we have*

$$\partial f(x) \supseteq \partial f_1(x) + \cdots + \partial f_n(x).$$

*If $\bigcap_{i=1}^n$ **relint**$(\mathrm{dom}(f_i)) \neq \emptyset$, then*

$$\partial f(x) = \partial f_1(x) + \cdots + \partial f_n(x).$$

Proposition 15.16 is proven in Rockafellar [Rockafellar (1970)] (Theorem 23.8).

The next result can be viewed as a generalization of the chain rule.

Proposition 15.17. *Let f be the function given by $f(x) = h(Ax)$ for all $x \in \mathbb{R}^n$, where h is a proper convex function on $\mathbb{R}^m$ and A is an $m \times n$ matrix. Then*

$$\partial f(x) \supseteq A^\top(\partial h(Ax)) \quad \textit{for all } x \in \mathbb{R}^n.$$

If the range of A contains a point of **relint**$(\mathrm{dom}(h))$, *then*

$$\partial f(x) = A^\top(\partial h(Ax)).$$

Proposition 15.17 is proven in Rockafellar [Rockafellar (1970)] (Theorem 23.9).

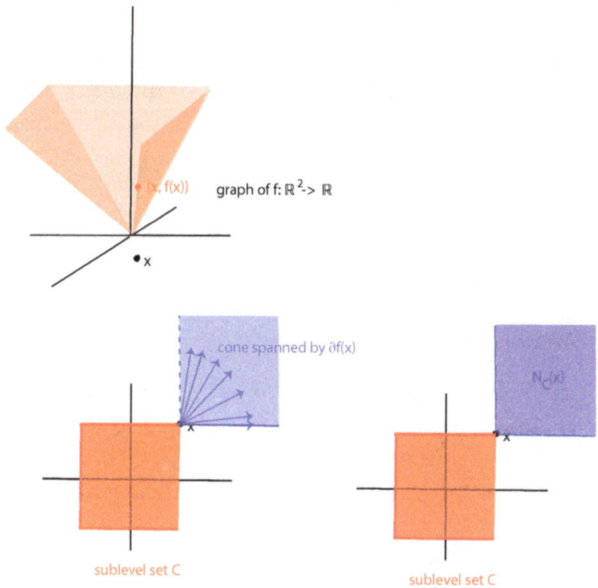

Fig. 15.22 Let f be the proper convex function whose graph in $\mathbb{R}^3$ is the peach polyhedral surface. The sublevel set $C = \{z \in \mathbb{R}^2 \mid f(z) \leq f(x)\}$ is the orange square which is closed on three sides. Then the normal cone $N_C(x)$ is the closure of the convex cone spanned by $\partial f(x)$.

15.4 Additional Properties of Subdifferentials

In general, if $f\colon \mathbb{R}^n \to \mathbb{R}$ is a function (not necessarily convex) and f is differentiable at x, we expect that the gradient ∇f_x of f at x is normal to the level set $\{z \in \mathbb{R}^n \mid f(z) = f(x)\}$ at $f(x)$. An analogous result, as illustrated in Figure 15.22, holds for proper convex functions in terms of subdifferentials.

Proposition 15.18. *Let f be a proper convex function on $\mathbb{R}^n$, and let $x \in \mathbb{R}^n$ be a vector such that f is subdifferentiable at x but f does not achieve its minimum at x. Then the normal cone $N_C(x)$ at x to the sublevel set $C = \{z \in \mathbb{R}^n \mid f(z) \leq f(x)\}$ is the closure of the convex cone spanned by $\partial f(x)$.*

Proposition 15.18 is proven in Rockafellar [Rockafellar (1970)] (Theorem 23.7).

The following result sharpens Proposition 15.8.

Proposition 15.19. *Let f be a closed proper convex function on $\mathbb{R}^n$, and let S be a nonempty closed and bounded subset of* $\mathrm{int}(\mathrm{dom}(f))$. *Then*

$$\partial f(S) = \bigcup_{x \in S} \partial f(x)$$

is nonempty, closed and bounded. If

$$\alpha = \sup_{y \in \partial f(S)} \|y\|_2 < +\infty,$$

then f is Lipschitizan on S, and we have

$$f'(x; z) \leq \alpha \|z\|_2 \qquad \text{for all } x \in S \text{ and all } z \in \mathbb{R}^n$$
$$|f(y) - f(x)| \leq \alpha \|y - z\|_2 \qquad \text{for all } x, y \in S.$$

Proposition 15.17 is proven in Rockafellar [Rockafellar (1970)] (Theorem 24.7).

The subdifferentials of a proper convex function f and its conjugate f^* are closely related. First, we have the following proposition from Rockafellar [Rockafellar (1970)] (Theorem 12.2).

Proposition 15.20. *Let f be convex function on $\mathbb{R}^n$. The conjugate function f^* of f is a closed and convex function, proper iff f is proper. Furthermore,* $(\mathrm{cl}(f))^* = f^*$, *and* $f^{**} = \mathrm{cl}(f)$.

As a corollary of Proposition 15.20, it can be shown that

$$f^*(y) = \sup_{x \in \mathbf{relint}(\mathrm{dom}(f))} (\langle x, y\rangle - f(x)).$$

The following result is proven in Rockafellar [Rockafellar (1970)] (Theorem 23.5).

Proposition 15.21. *For any proper convex function f on $\mathbb{R}^n$ and for any vector $x \in \mathbb{R}^n$, the following conditions on a vector $y \in \mathbb{R}^n$ are equivalent.*

(a) $y \in \partial f(x)$.
(b) The function $\langle z, y\rangle - f(z)$ *achieves its supremum in z at $z = x$.*
(c) $f(x) + f^*(y) \leq \langle x, y\rangle$.
(d) $f(x) + f^*(y) = \langle x, y\rangle$.

If $(\mathrm{cl}(f))(x) = f(x)$, *then there are three more conditions all equivalent to the above conditions.*

(a^*) $x \in \partial f^*(y)$.
(b^*) *The function* $\langle x, z\rangle - f^*(z)$ *achieves its supremum in* z *at* $z = y$.
(a^{**}) $y \in \partial(\mathrm{cl}(f))(x)$.

The following results are corollaries of Proposition 15.21; see Rockafellar [Rockafellar (1970)] (Corollaries 23.5.1, 23.5.2, 23.5.3).

Corollary 15.2. *For any proper convex function* f *on* $\mathbb{R}^n$, *if* f *is closed, then* $y \in \partial f(x)$ *iff* $x \in \partial f^*(y)$, *for all* $x, y \in \mathbb{R}^n$.

Corollary 15.2 states a sort of adjunction property.

Corollary 15.3. *For any proper convex function* f *on* $\mathbb{R}^n$, *if* f *is subdifferentiable at* $x \in \mathbb{R}^n$, *then* $(\mathrm{cl}(f))(x) = f(x)$ *and* $\partial(\mathrm{cl}(f))(x) = \partial f(x)$.

Corollary 15.3 shows that the closure of a proper convex function f agrees with f wherever f is subdifferentiable.

Corollary 15.4. *For any proper convex function* f *on* $\mathbb{R}^n$, *for any nonempty closed convex subset* C *of* $\mathbb{R}^n$, *for any* $y \in \mathbb{R}^n$, *the set* $\partial\delta^*(y|C) = \partial I_C^*(y)$ *consists of the vectors* $x \in \mathbb{R}^n$ *(if any) where the linear form* $z \mapsto \langle z, y\rangle$ *achieves its maximum over* C.

There is a notion of approximate subgradient which turns out to be useful in optimization theory; see Bertsekas [Bertsekas *et al.* (2003); Bertsekas (2015)].

Definition 15.17. Let $f\colon \mathbb{R}^n \to \mathbb{R} \cup \{+\infty\}$ be any proper convex function. For any $\epsilon > 0$, for any $x \in \mathbb{R}^n$, if $f(x)$ is finite, then an *ϵ-subgradient of f at x* is any vector $u \in \mathbb{R}^n$ such that

$$f(z) \geq f(x) - \epsilon + \langle z - x, u\rangle, \quad \text{for all } z \in \mathbb{R}^n.$$

See Figure 15.23. The set of all ϵ-subgradients of f at x is denoted $\partial_\epsilon f(x)$ and is called the *ϵ-subdifferential* of f at x.

The set $\partial_\epsilon f(x)$ can be defined in terms of the conjugate of the function h_x given by

$$h_x(y) = f(x + y) - f(x), \quad \text{for all } y \in \mathbb{R}^n.$$

Proposition 15.22. *Let* $f\colon \mathbb{R}^n \to \mathbb{R} \cup \{+\infty\}$ *be any proper convex function. For any* $\epsilon > 0$, *if* h_x *is given by*

$$h_x(y) = f(x + y) - f(x), \quad \textit{for all } y \in \mathbb{R}^n,$$

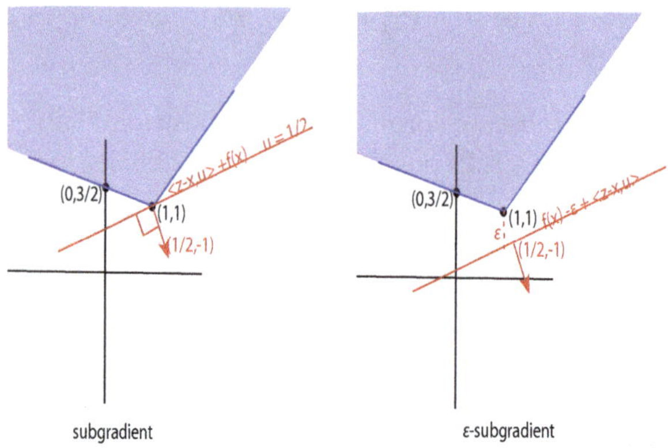

Fig. 15.23 Let $f\colon \mathbb{R} \to \mathbb{R} \cup \{-\infty, +\infty\}$ be the piecewise function defined by $f(x) = x+1$ for $x \geq 1$ and $f(x) = -\frac{1}{2}x + \frac{3}{2}$ for $x < 1$. Its epigraph is the shaded blue region in $\mathbb{R}^2$. The line $\frac{1}{2}(x-1)+1$ (with normal $(\frac{1}{2}, -1)$ is a supporting hyperplane to the graph of $f(x)$ at $(1,1)$ while the line $\frac{1}{2}(x-1)+1-\epsilon$ is the hyperplane associated with the ϵ-subgradient at $x = 1$ and shows that $u = \frac{1}{2} \in \partial_\epsilon f(x)$.

then

$$h_x^*(y) = f^*(y) + f(x) - \langle x, y\rangle \quad \textit{for all } y \in \mathbb{R}^n$$

and

$$\partial_\epsilon f(x) = \{u \in \mathbb{R}^n \mid h_x^*(u) \leq \epsilon\}.$$

Proof. We have

$$\begin{aligned} h_x^*(y) &= \sup_{z\in\mathbb{R}^n} (\langle y, z\rangle - h_x(z)) \\ &= \sup_{z\in\mathbb{R}^n} (\langle y, z\rangle - f(x+z) + f(x)) \\ &= \sup_{x+z\in\mathbb{R}^n} (\langle y, x+z\rangle - f(x+z) - \langle y, x\rangle + f(x)) \\ &= f^*(y) + f(x) - \langle x, y\rangle. \end{aligned}$$

Observe that $u \in \partial_\epsilon f(x)$ iff for every $y \in \mathbb{R}^n$,

$$f(x+y) \geq f(x) - \epsilon + \langle y, u\rangle$$

iff

$$\epsilon \geq \langle y, u\rangle - f(x+y) + f(x) = \langle y, u\rangle - h_x(y).$$

Since by definition

$$h_x^*(u) = \sup_{y \in \mathbb{R}^n} (\langle y, u \rangle - h_x(y)),$$

we conclude that

$$\partial_\epsilon f(x) = \{u \in \mathbb{R}^n \mid h_x^*(u) \leq \epsilon\},$$

as claimed. ☐

Remark: By Fenchel's inequality $h_x^*(y) \geq 0$, and by Proposition 15.21(d), the set of vectors where h_x^* vanishes is $\partial f(x)$.

The equation $\partial_\epsilon f(x) = \{u \in \mathbb{R}^n \mid h_x^*(u) \leq \epsilon\}$ shows that $\partial_\epsilon f(x)$ is a closed convex set. As ϵ gets smaller, the set $\partial_\epsilon f(x)$ decreases, and we have

$$\partial f(x) = \bigcap_{\epsilon > 0} \partial_\epsilon f(x).$$

However $\delta^*(y|\partial_\epsilon f(x)) = I^*_{\partial_\epsilon f(x)}(y)$ does not necessarily decrease to $\delta^*(y|\partial f(x)) = I^*_{\partial f(x)}(y)$ as ϵ decreases to zero. The discrepancy corresponds to the discrepancy between $f'(x; y)$ and $\delta^*(y|\partial f(x)) = I^*_{\partial f(x)}(y)$ and is due to the fact that f is not necessarily closed (see Proposition 15.14) as shown by the following result proven in Rockafellar [Rockafellar (1970)] (Theorem 23.6).

Proposition 15.23. *Let f be a closed and proper convex function, and let $x \in \mathbb{R}^n$ such that $f(x)$ is finite. Then*

$$f'(x; y) = \lim_{\epsilon \downarrow 0} \delta^*(y|\partial_\epsilon f(x)) = \lim_{\epsilon \downarrow 0} I^*_{\partial_\epsilon f(x)}(y) \quad \text{for all } y \in \mathbb{R}^n.$$

The theory of convex functions is rich and we have only given a sample of some of the most significant results that are relevant to optimization theory. There are a few more results regarding the minimum of convex functions that are particularly important due to their applications to optimization theory.

15.5 The Minimum of a Proper Convex Function

Let h be a proper convex function on $\mathbb{R}^n$. The general problem is to study the minimum of h over a nonempty convex set C in $\mathbb{R}^n$, possibly defined by a set of inequality and equality constraints. We already observed that minimizing h over C is equivalent to minimizing the proper convex function f given by

$$f(x) = h(x) + I_C(x) = \begin{cases} h(x) & \text{if } x \in C \\ +\infty & \text{if } x \notin C. \end{cases}$$

Therefore it makes sense to begin by considering the problem of minimizing a proper convex function f over $\mathbb{R}^n$. Of course, minimizing over $\mathbb{R}^n$ is equivalent to minimizing over $\text{dom}(f)$.

Definition 15.18. Let f be a proper convex function on $\mathbb{R}^n$. We denote by $\inf f$ the quantity

$$\inf f = \inf_{x \in \text{dom}(f)} f(x).$$

This is the minimum of the function f over $\mathbb{R}^n$ (it may be equal to $-\infty$).

For every $\alpha \in \mathbb{R}$, we have the sublevel set

$$\text{sublev}_\alpha(f) = \{x \in \mathbb{R}^n \mid f(x) \leq \alpha\}.$$

By Proposition 15.2, we know that the sublevel sets $\text{sublev}_\alpha(f)$ are convex and that

$$\text{dom}(f) = \bigcup_{\alpha \in \mathbb{R}} \text{sublev}_\alpha(f).$$

Observe that $\text{sublev}_\alpha(f) = \emptyset$ if $\alpha < \inf f$. If $\inf f > -\infty$, then for $\alpha = \inf f$, the sublevel set $\text{sublev}_\alpha(f)$ consists of the set of vectors where f achieves it minimum.

Definition 15.19. Let f be a proper convex function on $\mathbb{R}^n$. If $\inf f > -\infty$, then the sublevel set $\text{sublev}_{\inf f}(f)$ is called the *minimum set* of f (this set may be empty). See Figure 15.24.

It is important to determine whether the minimum set is empty or nonempty, or whether it contains a single point. As we noted in Theorem 4.5(2), if f is strictly convex then the minimum set contains at most one point.

In any case, we know from Proposition 15.2 and Proposition 15.3 that the minimum set of f is convex, and closed iff f is closed.

Subdifferentials provide the first criterion for deciding whether a vector $x \in \mathbb{R}^n$ belongs to the minimum set of f. Indeed, the very definition of a

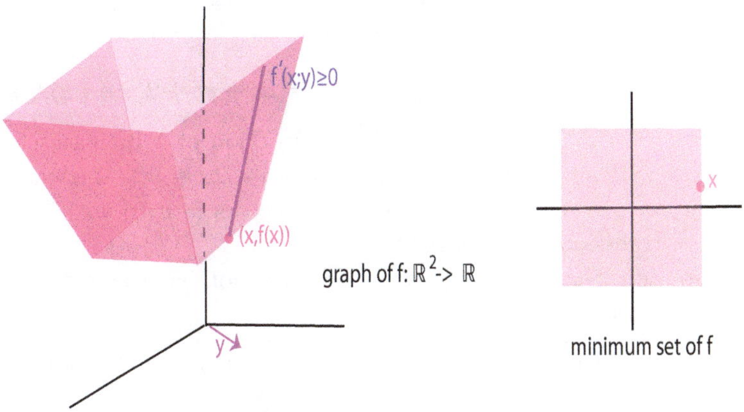

Fig. 15.24 Let f be the proper convex function whose graph is the surface of the upward facing pink trough. The minimum set of f is the light pink square of $\mathbb{R}^2$ which maps to the bottom surface of the trough in $\mathbb{R}^3$. For any x in the minimum set, $f'(x;y) \geq 0$, a fact substantiated by Proposition 15.24.

subgradient says that $x \in \mathbb{R}^n$ belongs to the minimum set of f iff $0 \in \partial f(x)$. Using Proposition 15.14, we obtain the following result.

Proposition 15.24. *Let f be a proper convex function over $\mathbb{R}^n$. A vector $x \in \mathbb{R}^n$ belongs to the minimum set of f iff*

$$0 \in \partial f(x)$$

iff $f(x)$ is finite and

$$f'(x;y) \geq 0 \quad \textit{for all } y \in \mathbb{R}^n.$$

Of course, if f is differentiable at x, then $\partial f(x) = \{\nabla f_x\}$, and we obtain the well-known condition $\nabla f_x = 0$.

There are many ways of expressing the conditions of Proposition 15.24, and the minimum set of f can even be characterized in terms of the conjugate function f^*. The notion of direction of recession plays a key role.

Definition 15.20. Let $f: \mathbb{R}^n \to \mathbb{R} \cup \{+\infty\}$ be any function. A *direction of recession* of f is any non-zero vector $u \in \mathbb{R}^n$ such that for every $x \in \mathrm{dom}(f)$, the function $\lambda \mapsto f(x + \lambda u)$ is nonincreasing (this means that for all $\lambda_1, \lambda_2 \in \mathbb{R}$, if $\lambda_1 < \lambda_2$, then $x + \lambda_1 u \in \mathrm{dom}(f)$, $x + \lambda_2 u \in \mathrm{dom}(f)$, and $f(x + \lambda_2 u) \leq f(x + \lambda_1 u)$).

Example 15.12. Consider the function $f\colon \mathbb{R}^2 \to \mathbb{R}$ given by $f(x, y) = 2x + y^2$. Since

$$f(x + \lambda u, y + \lambda v) = 2(x + \lambda u) + (y + \lambda v)^2 = 2x + y^2 + 2(u + yv)\lambda + v^2\lambda^2,$$

if $v \neq 0$, we see that the above quadratic function of λ increases for $\lambda \geq -(u + yv)/v^2$. If $v = 0$, then the function $\lambda \mapsto 2x + y^2 + 2u\lambda$ decreases to $-\infty$ when λ goes to $+\infty$ if $u < 0$, so all vectors $(-u, 0)$ with $u > 0$ are directions of recession. See Figure 15.25.

The function $f(x, y) = 2x + x^2 + y^2$ does not have any direction of recession, because

$$f(x + \lambda u, y + \lambda v) = 2x + x^2 + y^2 + 2(u + ux + yv)\lambda + (u^2 + v^2)\lambda^2,$$

and since $(u, v) \neq (0, 0)$, we have $u^2 + v^2 > 0$, so as a function of λ, the above quadratic function increases for $\lambda \geq -(u + ux + yv)/(u^2 + v^2)$. See Figure 15.25.

In fact, the above example is typical. For any symmetric positive definite $n \times n$ matrix A and any vector $b \in \mathbb{R}^n$, the quadratic strictly convex function q given by $q(x) = x^\top Ax + b^\top x$ has no directions of recession. For any $u \in \mathbb{R}^n$, with $u \neq 0$, we have

$$\begin{aligned} q(x + \lambda u) &= (x + \lambda u)^\top A(x + \lambda u) + b^\top (x + \lambda u) \\ &= x^\top Ax + b^\top x + (2x^\top Au + b^\top u)\lambda + (u^\top Au)\lambda^2. \end{aligned}$$

Since $u \neq 0$ and A is SPD, we have $u^\top Au > 0$, and the above quadratic function increases for $\lambda \geq -(2x^\top Au + b^\top u)/(2u^\top Au)$.

The above fact yields an important trick of convex optimization. If f is any proper closed and convex function, then for any quadratic strictly convex function q, the function $h = f + q$ is a proper and closed strictly convex function that has a minimum which is attained for a *unique* vector. This trick is at the core of the method of augmented Lagrangians, and in particular ADMM. Surprisingly, a rigorous proof requires the deep theorem below.

One should be careful not to conclude hastily that if a convex function is proper and closed, then $\text{dom}(f)$ and $\text{Im}(f)$ are also closed. Also, a closed and proper convex function may not attain its minimum. For example, the function $f\colon \mathbb{R} \to \mathbb{R} \cup \{+\infty\}$ given by

$$f(x) = \begin{cases} \dfrac{1}{x} & \text{if } x > 0 \\ +\infty & \text{if } x \leq 0 \end{cases}$$

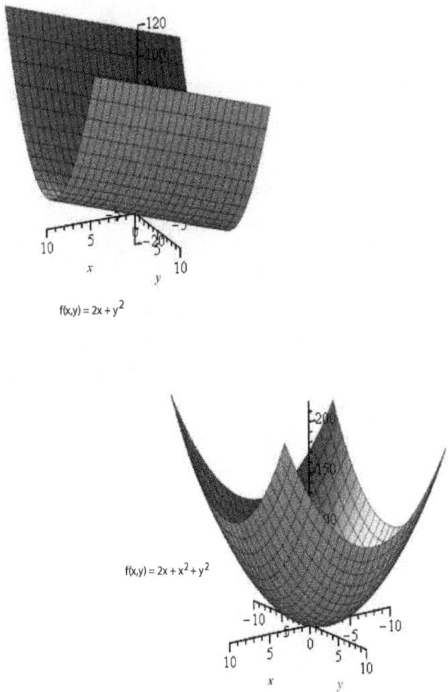

Fig. 15.25 The graphs of the two functions discussed in Example 15.12. The graph of $f(x, y) = 2x + y^2$ slopes "downward" along the negative x-axis, reflecting the fact that $(-u, 0)$ is a direction of recession.

is a proper, closed and convex function, but $\mathrm{dom}(f) = (0, +\infty)$ and $\mathrm{Im}(f) = (0, +\infty)$. Note that $\inf f = 0$ is not attained. The problem is that f has 1 has a direction of recession as evidenced by the graph provided in Figure 15.26.

The following theorem is proven in Rockafellar [Rockafellar (1970)] (Theorem 27.1).

Theorem 15.7. *Let f be a proper and closed convex function over $\mathbb{R}^n$. The following statements hold:*

(1) We have $\inf f = -f^*(0)$. Thus f is bounded below iff $0 \in \mathrm{dom}(f^*)$.

(2) The minimum set of f is equal to $\partial f^*(0)$. Thus the infimum of f is attained (which means that there is some $x \in \mathbb{R}^n$ such that $f(x) = \inf f$) iff f^* is subdifferentiable at 0. This condition holds in particular when $0 \in \mathbf{relint}(\mathrm{dom}(f^*))$. Moreover, $0 \in \mathbf{relint}(\mathrm{dom}(f^*))$ iff every

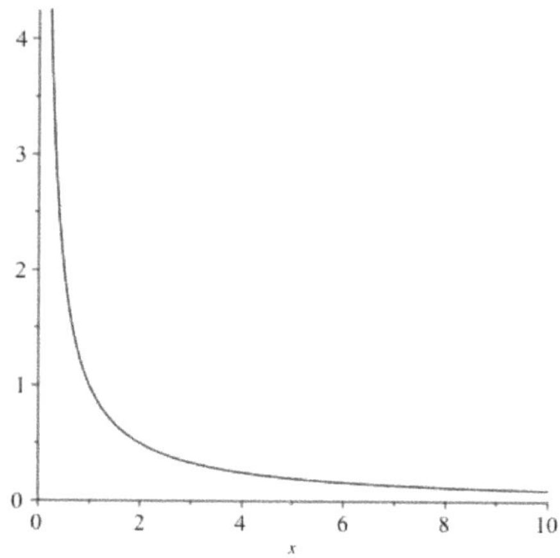

Fig. 15.26 The graph of the partial function $f(x) = \frac{1}{x}$ for $x > 0$. The graph of this function decreases along the x-axis since 1 is a direction of recession.

direction of recession of f is a direction in which f is constant.

(3) *For the infimum of f to be finite but unattained, it is necessary and sufficient that $f^*(0)$ be finite and $(f^*)'(0; y) = -\infty$ for some $y \in \mathbb{R}^n$.*

(4) *The minimum set of f is a nonempty bounded set iff $0 \in \text{int}(\text{dom}(f^*))$. This condition holds iff f has no directions of recession.*

(5) *The minimum set of f consists of a unique vector x iff f^* is differentiable at x and $x = \nabla f_0^*$.*

(6) *For each $\alpha \in \mathbb{R}$, the support function of $\text{sublev}_\alpha(f)$ is the closure of the positively homogeneous convex function generated by $f^* + \alpha$. If f is bounded below, then the support function of the minimum set of f is the closure of the directional derivative map $y \mapsto (f^*)'(0; y)$.*

In view of the importance of Theorem 15.7(4), we state this property as the following corollary.

Corollary 15.5. *Let f be a closed proper convex function on $\mathbb{R}^n$. Then the minimal set of f is a non-empty bounded set iff f has no directions of recession. In particular, if f has no directions of recession, then the minimum* $\inf f$ *of f is finite and attained for some $x \in \mathbb{R}^n$.*

Theorem 15.2 implies the following result which is very important for the design of optimization procedures.

Proposition 15.25. *Let f be a proper and closed convex function over $\mathbb{R}^n$.*

The function h given by $h(x) = f(x) + q(x)$ obtained by adding any strictly convex quadratic function q of the form $q(x) = x^\top Ax + b^\top x$ (where A is symmetric positive definite) is a proper closed strictly convex function such that $\inf h$ is finite, and there is a unique $x^* \in \mathbb{R}^n$ such that h attains its minimum in x^* (that is, $h(x^*) = \inf h$).

Proof. By Theorem 15.2 there is some affine form φ given by $\varphi(x) = c^\top x + \alpha$ (with $\alpha \in \mathbb{R}$) such that $f(x) \geq \varphi(x)$ for all $x \in \mathbb{R}^n$. Then we have

$$h(x) = f(x) + q(x) \geq x^\top Ax + (b^\top + c^\top)x + \alpha \quad \text{for all } x \in \mathbb{R}^n.$$

Since A is symmetric positive definite, by Example 15.12, the quadratic function Q given by $Q(x) = x^\top Ax + (b^\top + c^\top)x + \alpha$ has no directions of recession. Since $h(x) \geq Q(x)$ for all $x \in \mathbb{R}^n$, we claim that h has no directions of recession. Otherwise, there would be some nonzero vector u, such that the function $\lambda \mapsto h(x+\lambda u)$ is nonincreasing for all $x \in \text{dom}(h)$, so $h(x+\lambda u) \leq \beta$ for some β for all λ. But we showed that for λ large enough, the function $\lambda \mapsto Q(x + \lambda u)$ increases like λ^2, so for λ large enough, we will have $Q(x + \lambda u) > \beta$, contradicting the fact that h majorizes Q. By Corollary 15.5, h has a finite minimum x^* which is attained.

If f and g are proper convex functions and if g is strictly convex, then $f+g$ is a proper function. For all $x, y \in \mathbb{R}^n$, for any λ such that $0 < \lambda < 1$, since f is convex and g is strictly convex, we have

$$\begin{aligned} f((1-\lambda)x + \lambda y) &\leq (1-\lambda)f(x) + \lambda f(y) \\ g((1-\lambda)x + \lambda y) &< (1-\lambda)g(x) + \lambda g(y), \end{aligned}$$

so we deduce that

$$f((1-\lambda)x+\lambda y)+g((1-\lambda)x+\lambda y) < ((1-\lambda)(f(x)+g(x))+\lambda(f(x)+g(x))),$$

which shows that $f+g$ is strictly convex. Then, as $f+q$ is strictly convex, it has a unique minimum at x^*. $\square$

We now come back to the problem of minimizing a proper convex function h over a nonempty convex subset C . Here is a nice characterization.

Proposition 15.26. *Let h be a proper convex function on $\mathbb{R}^n$, and let C be a nonempty convex subset of $\mathbb{R}^n$.*

(1) For any $x \in \mathbb{R}^n$, if there is some $y \in \partial h(x)$ such that $-y \in N_C(x)$, that is, $-y$ is normal to C at x, then h attains its minimum on C at x.

(2) If $\mathbf{relint}(\mathrm{dom}(h)) \cap \mathbf{relint}(C) \neq \emptyset$*, then the converse of (1) holds. This means that if* h *attains its minimum on* C *at* x*, then there is some* $y \in \partial h(x)$ *such that* $-y \in N_C(x)$.

Proposition 15.26 is proven in Rockafellar [Rockafellar (1970)] (Theorem 27.4). The proof is actually quite simple.

Proof. (1) By Proposition 15.24, h attains its minimum on C at x iff

$$0 \in \partial(h + I_C)(x).$$

By Proposition 15.16, since

$$\partial(h + I_C)(x) \subseteq \partial h(x) + \partial I_C(x),$$

if $0 \in \partial h(x) + \partial I_C(x)$, then h attains its minimum on C at x. But we saw in Section 15.2 that $\partial I_C(x) = N_C(x)$, the normal cone to C at x. Then the condition $0 \in \partial h(x) + \partial I_C(x)$ says that there is some $y \in \partial h(x)$ such that $y + z = 0$ for some $z \in N_C(x)$, and this is equivalent to $-y \in N_C(x)$.

(2) By definition of I_C, the condition $\mathbf{relint}(\mathrm{dom}(h)) \cap \mathbf{relint}(C) \neq \emptyset$ is the hypothesis of Proposition 15.16 to have

$$\partial(h + I_C)(x) = \partial h(x) + \partial I_C(x),$$

so we deduce that $y \in \partial(h + I_C)(x)$, and by Proposition 15.24, h attains its minimum on C at x. □

Remark: A *polyhedral function* is a convex function whose epigraph is a polyhedron. It is easy to see that Proposition 15.26(2) also holds in the following cases

(1) C is a $\mathcal{H}$-polyhedron and $\mathbf{relint}(\mathrm{dom}(h)) \cap C \neq \emptyset$
(2) h is polyhedral and $\mathrm{dom}(h) \cap \mathbf{relint}(C) \neq \emptyset$.
(3) Both h and C are polyhedral, and $\mathrm{dom}(h) \cap C \neq \emptyset$.

15.6 Generalization of the Lagrangian Framework

Essentially all the results presented in Section 14.3, Section 14.7, Section 14.8, and Section 14.9 about Lagrangians and Lagrangian duality generalize to programs involving a proper and convex objective function J, proper and convex inequality constraints, and affine equality constraints. The extra generality is that it is no longer assumed that these functions are differentiable. This theory is thoroughly discussed in Part VI, Section 28, of

Rockafellar [Rockafellar (1970)], for programs called ordinary convex programs. We do not have the space to even sketch this theory but we will spell out some of the key results.

We will be dealing with programs consisting of an objective function $J\colon \mathbb{R}^n \to \mathbb{R}\cup\{+\infty\}$ which is convex and proper, subject to $m \geq 0$ inequality constraints $\varphi_i(v) \leq 0$, and $p \geq 0$ affine equality constraints $\psi_j(v) = 0$. The constraint functions φ_i are also convex and proper, and we assume that

$$\textbf{relint}(\mathrm{dom}(J)) \subseteq \textbf{relint}(\mathrm{dom}(\varphi_i)), \quad \mathrm{dom}(J) \subseteq \mathrm{dom}(\varphi_i), \quad i = 1, \ldots, m.$$

Such programs are *called ordinary convex programs*. Let

$$U = \mathrm{dom}(J) \cap \{v \in \mathbb{R}^n \mid \varphi_i(v) \leq 0,\ \psi_j(v) = 0,\ 1 \leq i \leq m,\ 1 \leq j \leq p\},$$

be the set of *feasible solutions*. We are seeking elements in $u \in U$ that minimize J over U.

A generalized version of Theorem 14.6 holds under the above hypotheses on J and the constraints φ_i and ψ_j, except that in the KKT conditions, the equation involving gradients must be replaced by the following condition involving subdifferentials:

$$0 \in \partial\left(J + \sum_{i=1}^{m} \lambda_i \varphi_i + \sum_{j=1}^{p} \mu_j \psi_j\right)(u),$$

with $\lambda_i \geq 0$ for $i = 1, \ldots, m$ and $\mu_j \in \mathbb{R}$ for $j = 1, \ldots, p$ (where $u \in U$ and J attains its minimum at u).

The *Lagrangian* $L(v, \lambda, \nu)$ of our problem is defined as follows: Let

$$E_m = \{x \in \mathbb{R}^{m+p} \mid x_i \geq 0,\ 1 \leq i \leq m\}.$$

Then

$$L(v, \lambda, \mu) = \begin{cases} J(v) + \sum_{i=1}^{m} \lambda_i \varphi_i(v) + \sum_{j=1}^{p} \mu_j \psi_j(v) & \text{if } (\lambda, \mu) \in E_m, v \in \mathrm{dom}(J) \\ -\infty & \text{if } (\lambda, \mu) \notin E_m, v \in \mathrm{dom}(J) \\ +\infty & \text{if } v \notin \mathrm{dom}(J). \end{cases}$$

For *fixed values* $(\lambda, \mu) \in \mathbb{R}^m_+ \times \mathbb{R}^p$, we also define the function $h\colon \mathbb{R}^n \to \mathbb{R} \cup \{+\infty\}$ given by

$$h(x) = J(x) + \sum_{i=1}^{m} \lambda_i \varphi_i(x) + \sum_{j=1}^{p} \mu_j \psi_j(x),$$

whose effective domain is $\mathrm{dom}(J)$ (since we are assuming that $\mathrm{dom}(J) \subseteq \mathrm{dom}(\varphi_i)$, $i = 1, \ldots, m$). Thus $h(x) = L(x, \lambda, \mu)$, but h is a *function only of* x, so we denote it differently to avoid confusion (also, technically, $L(x, \lambda, \mu)$ may take the value $-\infty$, but h does not). Since J and the φ_i are proper convex functions and the ψ_j are affine, the function h is a proper convex function.

A proof of a generalized version of Theorem 14.6 can be obtained by putting together Theorem 28.1, Theorem 28.2, and Theorem 28.3, in Rockafellar [Rockafellar (1970)]. For the sake of completeness, we state these theorems. Here is Theorem 28.1.

Theorem 15.8. *(Theorem 28.1, Rockafellar) Let (P) be an ordinary convex program. Let $(\lambda, \mu) \in \mathbb{R}^m_+ \times \mathbb{R}^p$ be Lagrange multipliers such that the infimum of the function $h = J + \sum_{i=1}^m \lambda_i \varphi_i + \sum_{j=1}^p \mu_j \psi_j$ is finite and equal to the optimal value of J over U. Let D be the minimal set of h over $\mathbb{R}^n$, and let $I = \{i \in \{1, \dots, m\} \mid \lambda_i = 0\}$. If D_0 is the subset of D consisting of vectors x such that*

$$\begin{aligned} \varphi_i(x) &\leq 0 && \text{for all } i \in I \\ \varphi_i(x) &= 0 && \text{for all } i \notin I \\ \psi_j(x) &= 0 && \text{for all } j = 1, \dots, p, \end{aligned}$$

then D_0 is the set of minimizers of (P) over U.

And now here is Theorem 28.2.

Theorem 15.9. *(Theorem 28.2, Rockafellar) Let (P) be an ordinary convex program, and let $I \subseteq \{1, \dots, m\}$ be the subset of indices of inequality constraints that are not affine. Assume that the optimal value of (P) is finite, and that (P) has at least one feasible solution $x \in$ **relint**$(\mathrm{dom}(J))$ such that*

$$\varphi_i(x) < 0 \quad \text{for all } i \in I.$$

Then there exist some Lagrange multipliers $(\lambda, \mu) \in \mathbb{R}^m_+ \times \mathbb{R}^p$ (not necessarily unique) such that

(a) The infimum of the function $h = J + \sum_{i=1}^m \lambda_i \varphi_i + \sum_{j=1}^p \mu_j \psi_j$ is finite and equal to the optimal value of J over U.

The hypotheses of Theorem 15.9 are qualification conditions on the constraints, essentially Slater's conditions from Definition 14.6.

Definition 15.21. Let (P) be an ordinary convex program, and let $I \subseteq \{1, \dots, m\}$ be the subset of indices of inequality constraints that are not affine. The constraints are *qualified* is there is a feasible solution $x \in$ **relint**(dom(J)) such that

$$\varphi_i(x) < 0 \quad \text{for all } i \in I.$$

Finally, here is Theorem 28.3 from Rockafellar [Rockafellar (1970)].

Theorem 15.10. *(Theorem 28.3, Rockafellar) Let (P) be an ordinary convex program. If $x \in \mathbb{R}^n$ and $(\lambda, \mu) \in \mathbb{R}^m_+ \times \mathbb{R}^p$, then (λ, μ) and x have the property that*

(a) The infimum of the function $h = J + \sum_{i=1}^m \lambda_i \varphi_i + \sum_{j=1}^p \mu_j \psi_j$ is finite and equal to the optimal value of J over U, and
(b) The vector x is an optimal solution of (P) (so $x \in U$),

iff (x, λ, μ) is a saddle point of the Lagrangian $L(x, \lambda, \mu)$ of (P).
Moreover, this condition holds iff the following KKT conditions hold:

(1) $\lambda \in \mathbb{R}^m_+$, $\varphi_i(x) \leq 0$, and $\lambda_i \varphi_i(x) = 0$ for $i = 1, \ldots, m$.
(2) $\psi_j(x) = 0$ for $j = 1, \ldots p$.
(3) $0 \in \partial J(x) + \sum_{i=1}^m \partial \lambda_i \varphi_i(x) + \sum_{j=1}^p \partial \mu_j \psi_i(x)$.

Observe that by Theorem 15.9, if the optimal value of (P) is finite and if the constraints are qualified, then Condition (a) of Theorem 15.10 holds for (λ, μ). As a consequence we obtain the following corollary of Theorem 15.10 attributed to Kuhn and Tucker, which is one of the main results of the theory. It is a generalized version of Theorem 14.6.

Theorem 15.11. *(Theorem 28.3.1, Rockafellar) Let (P) be an ordinary convex program satisfying the hypothesis of Theorem 15.9, which means that the optimal value of (P) is finite, and that the constraints are qualified. In order that a vector $x \in \mathbb{R}^n$ be an optimal solution to (P), it is necessary and sufficient that there exist Lagrange multipliers $(\lambda, \mu) \in \mathbb{R}^m_+ \times \mathbb{R}^p$ such that (x, λ, μ) is a saddle point of $L(x, \lambda, \mu)$. Equivalently, x is an optimal solution of (P) if and only if there exist Lagrange multipliers $(\lambda, \mu) \in \mathbb{R}^m_+ \times \mathbb{R}^p$, which, together with x, satisfy the KKT conditions from Theorem 15.10.*

Theorem 15.11 has to do with the existence of an optimal solution for (P), but it does not say anything about the optimal value of (P). To establish such a result, we need the notion of dual function.

The *dual function* G is defined by

$$G(\lambda, \mu) = \inf_{v \in \mathbb{R}^n} L(v, \lambda, \mu).$$

It is a concave function (so $-G$ is convex) which may take the values $\pm\infty$. Note that maximizing G, which is equivalent to minimizing $-G$, runs into troubles if $G(\lambda, \mu) = +\infty$ for some λ, μ, but that $G(\lambda, \mu) = -\infty$ does not cause a problem. At first glance, this seems counterintuitive, but remember that G is *concave*, not *convex*. It is $-G$ that is convex, and $-\infty$ and $+\infty$ get flipped.

Then a generalized and stronger version of Theorem 14.7(2) also holds. A proof can be obtained by putting together Corollary 28.3.1, Theorem 28.4, and Corollary 28.4.1, in Rockafellar [Rockafellar (1970)]. For the sake of completeness, we state the following results from Rockafellar [Rockafellar (1970)].

Theorem 15.12. *(Theorem 28.4, Rockafellar) Let (P) be an ordinary convex program with Lagrangian $L(x, \lambda, \mu)$. If the Lagrange multipliers $(\lambda^*, \mu^*) \in \mathbb{R}^m_+ \times \mathbb{R}^p$ and the vector $x^* \in \mathbb{R}^n$ have the property that*

(a) The infimum of the function $h = J + \sum_{i=1}^m \lambda_i^* \varphi_i + \sum_{j=1}^p \mu_j^* \psi_j$ is finite and equal to the optimal value of J over U, and
(b) The vector x^* is an optimal solution of (P) (so $x^* \in U$),

then the saddle value $L(x^*, \lambda^*, \mu^*)$ is the optimal value $J(x^*)$ of (P).

More generally, the Lagrange multipliers $(\lambda^*, \mu^*) \in \mathbb{R}^m_+ \times \mathbb{R}^p$ have Property (a) iff

$$-\infty < \inf_x L(x, \lambda^*, \mu^*) \leq \sup_{\lambda,\mu} \inf_x L(x, \lambda, \mu) = \inf_x \sup_{\lambda,\mu} L(x, \lambda, \mu),$$

in which case, the common value of the extremum value is the optimal value of (P). In particular, if x^* is an optimal solution for (P), then $\sup_{\lambda,\mu} G(\lambda, \mu) = L(x^*, \lambda^*, \mu^*) = J(x^*)$ (zero duality gap).

Observe that Theorem 15.12 gives sufficient Conditions (a) and (b) for the duality gap to be zero. In view of Theorem 15.10, these conditions are equivalent to the fact that (x^*, λ^*, μ^*) is a saddle point of L, or equivalently that the KKT conditions hold.

Again, by Theorem 15.9, if the optimal value of (P) is finite and if the constraints are qualified, then Condition (a) of Theorem 15.12 holds for (λ, μ). Then the following corollary of Theorem 15.12 holds.

Theorem 15.13. *(Theorem 28.4.1, Rockafellar) Let (P) be an ordinary convex program satisfying the hypothesis of Theorem 15.9, which means that the optimal value of (P) is finite, and that the constraints are qualified.*

The Lagrange multipliers $(\lambda, \mu) \in \mathbb{R}^m_+ \times \mathbb{R}^p$ *that have the property that the infimum of the function* $h = J + \sum_{i=1}^m \lambda_i \varphi_i + \sum_{j=1}^p \mu_j \psi_j$ *is finite and equal to the optimal value of* J *over* U *are exactly the vectors where the dual function* G *attains is supremum over* $\mathbb{R}^n$.

Theorem 15.13 is a generalized and stronger version of Theorem 14.7(2). Part (1) of Theorem 14.7 requires J and the φ_i to be differentiable, so it does not generalize.

More results can shown about ordinary convex programs, and another class of programs called *generalized convex programs*. However, we do not need such results for our purposes, in particular to discuss the ADMM method. The interested reader is referred to Rockafellar [Rockafellar (1970)] (Part VI, Sections 28 and 29).

15.7 Summary

The main concepts and results of this chapter are listed below:

- Extended real-valued functions.
- Epigraph ($\mathbf{epi}(f)$).
- Convex and concave (extended real-valued) functions.
- Effective domain ($\text{dom}(f)$).
- Proper and improper convex functions.
- Sublevel sets.
- Lower semi-continuous functions.
- Lower semi-continuous hull; closure of a convex function.
- Relative interior ($\mathbf{relint}(C)$).
- Indicator function.
- Lipschitz condition.
- Affine form, affine hyperplane.
- Half spaces.
- Supporting hyperplane.
- Normal cone at a.
- Subgradient, subgradient inequality, subdifferential.
- Minkowski's supporting hyperplane theorem.
- One-sided directional derivative.
- Support function.
- ReLU function.
- ϵ-subgradient.
- Minimum set of a convex function.

- Direction of recession.
- Ordinary convex programs.
- Set of feasible solutions.
- Lagrangian.
- Saddle point.
- KKT conditions.
- Qualified constraints.
- Duality gap.

15.8 Problems

Problem 15.1. Prove Proposition 15.1.

Problem 15.2. Prove Proposition 15.2.

Problem 15.3. Prove Proposition 15.3.

Problem 15.4. Prove that the convex function defined in Example 15.4 has the property that the limit along any line segment from $(0,0)$ to a point in the open right half-plane is 0.

Problem 15.5. Check that the normal cone to C at a is a convex cone.

Problem 15.6. Prove that $\partial f(x)$ is closed and convex.

Problem 15.7. For Example 15.6, with $f(x) = \|x\|_\infty$, prove that $\partial f(0)$ is the polyhedron

$$\partial f(0) = \text{conv}\{\pm e_1, \ldots, \pm e_n\}.$$

Problem 15.8. For Example 15.7, with

$$f(x) = \begin{cases} -(1-|x|^2)^{1/2} & \text{if } |x| \leq 1 \\ +\infty & \text{otherwise.} \end{cases}$$

prove that f is subdifferentiable (in fact differentiable) at x when $|x| < 1$, but $\partial f(x) = \emptyset$ when $|x| \geq 1$, even though $x \in \text{dom}(f)$ for $|x| = 1$

Problem 15.9. Prove Proposition 15.13.

Problem 15.10. Prove that as a convex function of u, the effective domain of the function $u \mapsto f'(x;u)$ is the convex cone generated by $\text{dom}(f) - x$.

Problem 15.11. Prove Proposition 15.21.

Problem 15.12. Prove Proposition 15.23.

Problem 15.13. Prove that Proposition 15.26(2) also holds in the following cases:

(1) C is a $\mathcal{H}$-polyhedron and $\mathbf{relint}(\mathrm{dom}(h)) \cap C \neq \emptyset$
(2) h is polyhedral and $\mathrm{dom}(h) \cap \mathbf{relint}(C) \neq \emptyset$.
(3) Both h and C are polyhedral, and $\mathrm{dom}(h) \cap C \neq \emptyset$.

Chapter 16

Dual Ascent Methods; ADMM

This chapter is devoted to the presentation of one of the best methods known at the present for solving optimization problems involving equality constraints. In fact, this method can also handle more general constraints, namely, membership in a convex set. It can also be used to solve a range of problems arising in machine learning including *lasso minimization*, *elastic net regression*, *support vector machine (SVM)*, and *ν-SV regression*. In order to obtain a good understanding of this method, called the *alternating direction method of multipliers*, for short *ADMM*, we review two precursors of ADMM, the *dual ascent method* and the *method of multipliers*.

ADMM is not a new method. In fact, it was developed in the 1970's. It has been revived as a very effective method to solve problems in statistical and machine learning dealing with very large data because it is well suited to distributed (convex) optimization. An extensive presentation of ADMM, its variants, and its applications, is given in the excellent paper by Boyd, Parikh, Chu, Peleato and Eckstein [Boyd *et al.* (2010)]. This paper is essentially a book on the topic of ADMM, and our exposition is deeply inspired by it.

In this chapter, we consider the problem of minimizing a convex function J (not necessarily differentiable) under the equality constraints $Ax = b$. In Section 16.1 we discuss the dual ascent method. It is essentially gradient descent applied to the dual function G, but since G is maximized, gradient descent becomes gradient ascent.

In order to make the minimization step of the dual ascent method more robust, one can use the trick of adding the penalty term $(\rho/2)\left\|Au - b\right\|_2^2$ to the Lagrangian. We obtain the *augmented Lagrangian*

$$L_\rho(u, \lambda) = J(u) + \lambda^\top (Au - b) + (\rho/2)\left\|Au - b\right\|_2^2,$$

with $\lambda \in \mathbb{R}^m$, and where $\rho > 0$ is called the *penalty parameter*. We obtain the minimization Problem (P_ρ),

$$\begin{aligned}&\text{minimize} \quad J(u) + (\rho/2)\left\|Au - b\right\|_2^2\\&\text{subject to} \quad Au = b,\end{aligned}$$

which is equivalent to the original problem.

The benefit of adding the penalty term $(\rho/2)\left\|Au - b\right\|_2^2$ is that by Proposition 15.25, Problem (P_ρ) has a unique optimal solution under mild conditions on A. Dual ascent applied to the dual of (P_ρ) is called the *method of multipliers* and is discussed in Section 16.2.

The alternating direction method of multipliers, for short ADMM, combines the decomposability of dual ascent with the superior convergence properties of the method of multipliers. The idea is to split the function J into two independent parts, as $J(x, z) = f(x) + g(z)$, and to consider the Minimization Problem (P_{admm}),

$$\begin{aligned}&\text{minimize} \quad f(x) + g(z)\\&\text{subject to} \quad Ax + Bz = c,\end{aligned}$$

for some $p \times n$ matrix A, some $p \times m$ matrix B, and with $x \in \mathbb{R}^n$, $z \in \mathbb{R}^m$, and $c \in \mathbb{R}^p$. We also assume that f and g are convex. Further conditions will be added later.

As in the method of multipliers, we form the augmented Lagrangian

$$L_\rho(x, z, \lambda) = f(x) + g(z) + \lambda^\top (Ax + Bz - c) + (\rho/2)\left\|Ax + Bz - c\right\|_2^2,$$

with $\lambda \in \mathbb{R}^p$ and for some $\rho > 0$. The major difference with the method of multipliers is that instead of performing a minimization step jointly over x and z, ADMM *first performs an x-minimization step and then a z-minimization step*. Thus x and z are updated in an alternating or sequential fashion, which accounts for the term *alternating direction*. Because the Lagrangian is augmented, some mild conditions on A and B imply that these minimization steps are guaranteed to terminate. ADMM is presented in Section 16.3.

In Section 16.4 we prove the convergence of ADMM under the following assumptions:

(1) The functions $f\colon \mathbb{R} \to \mathbb{R} \cup \{+\infty\}$ and $g\colon \mathbb{R} \to \mathbb{R} \cup \{+\infty\}$ are proper and closed convex functions (see Section 15.1) such that $\textbf{relint}(\text{dom}(f)) \cap \textbf{relint}(\text{dom}(g)) \neq \emptyset$.

(2) The $n \times n$ matrix $A^\top A$ is invertible and the $m \times m$ matrix $B^\top B$ is invertible. Equivalently, the $p \times n$ matrix A has rank n and the $p \times m$ matrix has rank m.

(3) The unaugmented Lagrangian $L_0(x, z, \lambda) = f(x) + g(z) + \lambda^\top(Ax + Bz - c)$ has a saddle point, which means there exists x^*, z^*, λ^* (not necessarily unique) such that

$$L_0(x^*, z^*, \lambda) \leq L_0(x^*, z^*, \lambda^*) \leq L_0(x, z, \lambda^*)$$

for all x, z, λ.

By Theorem 15.10, Assumption (3) is equivalent to the fact that the KKT equations are satisfied by some triple (x^*, z^*, λ^*), namely

$$Ax^* + Bz^* - c = 0 \tag{$*$}$$

and

$$0 \in \partial f(x^*) + \partial g(z^*) + A^\top\lambda^* + B^\top\lambda^*, \tag{$\dagger$}$$

Assumption (3) is also equivalent to Conditions (a) and (b) of Theorem 15.10. In particular, our program has an optimal solution (x^*, z^*). By Theorem 15.12, λ^* is maximizer of the dual function $G(\lambda) = \inf_{x,z} L_0(x, z, \lambda)$ and strong duality holds, that is, $G(\lambda^*) = f(x^*) + g(z^*)$ (the duality gap is zero).

We will show after the proof of Theorem 16.1 that Assumption (2) is actually implied by Assumption (3). This allows us to prove a convergence result stronger than the convergence result proven in Boyd et al. [Boyd *et al.* (2010)] (under the exact same assumptions (1) and (3)). In particular, we prove that *all* of the sequences (x^k), (z^k), and (λ^k) converge to optimal solutions $(\widetilde{x}, \widetilde{z})$, and $\widetilde{\lambda}$. The core of our proof is due to Boyd et al. [Boyd *et al.* (2010)], but there are new steps because we have the stronger hypothesis (2).

In Section 16.5, we discuss stopping criteria.

In Section 16.6 we present some applications of ADMM, in particular, minimization of a proper closed convex function f over a closed convex set C in $\mathbb{R}^n$ and quadratic programming. The second example provides one of the best methods for solving quadratic problems, including the SVM problems discussed in Chapter 18, the elastic net method in Section 19.6, and ν-SV regression in Chapter 20.

Section 16.8 gives applications of ADMM to ℓ^1-norm problems, in particular, lasso regularization, which plays an important role in machine learning.

16.1 Dual Ascent

Our goal is to solve the minimization problem, Problem (P),

$$\begin{aligned} &\text{minimize} && J(u) \\ &\text{subject to} && Au = b, \end{aligned}$$

with affine equality constraints (with A an $m \times n$ matrix and $b \in \mathbb{R}^m$). The Lagrangian $L(u, \lambda)$ of Problem (P) is given by

$$L(u, \lambda) = J(u) + \lambda^\top (Au - b).$$

with $\lambda \in \mathbb{R}^m$. From Proposition 14.13, the dual function $G(\lambda) = \inf_{u \in \mathbb{R}^n} L(u, \lambda)$ is given by

$$G(\lambda) = \begin{cases} -b^\top \lambda - J^*(-A^\top \lambda) & \text{if } -A^\top \lambda \in \mathrm{dom}(J^*), \\ -\infty & \text{otherwise}, \end{cases}$$

for all $\lambda \in \mathbb{R}^m$, where J^* is the conjugate of J. Recall that by Definition 14.11, the *conjugate* f^* of a function $f \colon U \to \mathbb{R}$ defined on a subset U of $\mathbb{R}^n$ is the partial function $f^* \colon \mathbb{R}^n \to \mathbb{R}$ defined by

$$f^*(y) = \sup_{x \in U} (y^\top x - f(x)), \quad y \in \mathbb{R}^n.$$

If the conditions of Theorem 14.7(1) hold, which in our case means that for every $\lambda \in \mathbb{R}^m$, there is a unique $u_\lambda \in \mathbb{R}^n$ such that

$$G(\lambda) = L(u_\lambda, \lambda) = \inf_{u \in \mathbb{R}^n} L(u, \lambda),$$

and that the function $\lambda \mapsto u_\lambda$ is continuous, then G is differentiable. Furthermore, we have

$$\nabla G_\lambda = Au_\lambda - b,$$

and for any solution $\mu = \lambda^*$ of the dual problem

$$\begin{aligned} &\text{maximize} && G(\lambda) \\ &\text{subject to} && \lambda \in \mathbb{R}^m, \end{aligned}$$

the vector $u^* = u_\mu$ is a solution of the primal Problem (P). Furthermore, $J(u^*) = G(\lambda^*)$, that is, the duality gap is zero.

The dual ascent method is essentially gradient descent applied to the dual function G. But since G is maximized, gradient descent becomes gradient ascent. Also, we no longer worry that the minimization problem $\inf_{u \in \mathbb{R}^n} L(u, \lambda)$ has a unique solution, so we denote by u^+ some minimizer of the above problem, namely

$$u^+ = \arg\min_u L(u, \lambda).$$

Given some initial dual variable λ^0, the *dual ascent method* consists of the following two steps:

$$u^{k+1} = \arg\min_u L(u, \lambda^k)$$
$$\lambda^{k+1} = \lambda^k + \alpha^k(Au^{k+1} - b),$$

where $\alpha^k > 0$ is a step size. The first step is used to compute the "new gradient" (indeed, if the minimizer u^{k+1} is unique, then $\nabla G_{\lambda^k} = Au^{k+1} - b$), and the second step is a dual variable update.

Example 16.1. Let us look at a very simple example of the gradient ascent method applied to a problem we first encountered in Section 6.1, namely minimize $J(x, y) = (1/2)(x^2 + y^2)$ subject to $2x - y = 5$. The Lagrangian is

$$L(x, y, \lambda) = \frac{1}{2}(x^2 + y^2) + \lambda(2x - y - 5).$$

See Figure 16.1.

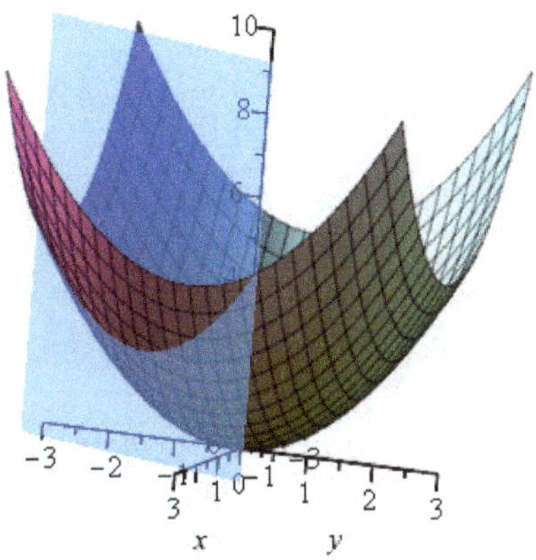

Fig. 16.1 The graph of $J(x, y) = (1/2)(x^2 + y^2)$ is the parabolic surface while the graph of $2x - y = 5$ is the transparent blue plane. The solution to Example 16.1 is apex of the intersection curve, namely the point $(2, -1, \frac{5}{2})$.

The method of Lagrangian duality says first calculate

$$G(\lambda) = \inf_{(x,y)\in\mathbb{R}^2} L(x, y, \lambda).$$

Since

$$J(x, y) = \frac{1}{2}\begin{pmatrix} x & y \end{pmatrix}\begin{pmatrix} 1 & 0 \\ 0 & 1 \end{pmatrix}\begin{pmatrix} x \\ y \end{pmatrix},$$

we observe that $J(x,y)$ is a quadratic function determined by the positive definite matrix $\begin{pmatrix} 1 & 0 \\ 0 & 1 \end{pmatrix}$, and hence to calculate $G(\lambda)$, we must set $\nabla L_{x,y} = 0$. By calculating $\frac{\partial J}{\partial x} = 0$ and $\frac{\partial J}{\partial y} = 0$, we find that $x = -2\lambda$ and $y = \lambda$. Then $G(\lambda) = -5/2\lambda^2 - 5\lambda$, and we must calculate the maximum of $G(\lambda)$ with respect to $\lambda \in \mathbb{R}$. This means calculating $G'(\lambda) = 0$ and obtaining $\lambda = -1$ for the solution of $(x, y, \lambda) = (-2\lambda, \lambda, \lambda) = (2, -1, -1)$.

Instead of solving *directly* for λ in terms of (x, y), the method of dual assent begins with a *numerical* estimate for λ, namely λ^0, and forms the "numerical" Lagrangian

$$L(x, y, \lambda^0) = \frac{1}{2}(x^2 + y^2) + \lambda^0(2x - y - 5).$$

With this numerical value λ^0, we minimize $L(x, y, \lambda^0)$ with respect to (x, y). This calculation will be identical to that used to form $G(\lambda)$ above, and as such, we obtain the iterative step $(x^1, y^1) = (-2\lambda^0, \lambda^0)$. So if we replace λ^0 by λ^k, we have the first step of the dual ascent method, namely

$$u^{k+1} = \begin{pmatrix} x^{k+1} \\ y^{k+1} \end{pmatrix} = \begin{pmatrix} -2 \\ 1 \end{pmatrix} \lambda^k.$$

The second step of the dual ascent method refines the numerical estimate of λ by calculating

$$\lambda^{k+1} = \lambda^k + \alpha^k \left(\begin{pmatrix} 2 & -1 \end{pmatrix} \begin{pmatrix} x^{k+1} \\ y^{k+1} \end{pmatrix} - 5 \right).$$

(Recall that in our original problem the constraint is $2x - y = 5$ or $\begin{pmatrix} 2 & -1 \end{pmatrix} \begin{pmatrix} x \\ y \end{pmatrix} - 5$, so $A = \begin{pmatrix} 2 & -1 \end{pmatrix}$ and $b = 5$.) By simplifying the above equation, we find that

$$\lambda^{k+1} = (1 - \beta)\lambda^k - \beta, \qquad \beta = 5\alpha^k.$$

Back substituting for λ^k in the preceding equation shows that

$$\lambda^{k+1} = (1 - \beta)^{k+1}\lambda^0 + (1 - \beta)^{k+1} - 1.$$

If $0 < \beta \leq 1$, the preceding line implies that λ^{k+1} converges to $\lambda = -1$, which coincides with the answer provided by the original Lagrangian duality method. Observe that if $\beta = 1$ or $\alpha^k = \frac{1}{5}$, the dual ascent method terminates in one step.

With an appropriate choice of α^k, we have $G(\lambda^{k+1}) > G(\lambda^k)$, so the method makes progress. Under certain assumptions, for example, that J is strictly convex and some conditions of the α^k, it can be shown that dual ascent converges to an optimal solution (both for the primal and the dual). However, the main flaw of dual ascent is that the minimization step may diverge. For example, this happens is J is a nonzero affine function of one of its components. The remedy is to add a penalty term to the Lagrangian.

On the positive side, the dual ascent method leads to a decentralized algorithm if the function J is separable. Suppose that u can be split as $u = \sum_{i=1}^N u_i$, with $u_i \in \mathbb{R}^{n_i}$ and $n = \sum_{i=1}^N n_i$, that

$$J(u) = \sum_{i=1}^N J_i(u_i),$$

and that A is split into N blocks A_i (with A_i a $m \times n_i$ matrix) as $A = [A_1 \cdots A_N]$, so that $Au = \sum_{k=1}^N A_i u_i$. Then the Lagrangian can be written as

$$L(u, \lambda) = \sum_{i=1}^N L_i(u_i, \lambda),$$

with

$$L_i(u_i, \lambda) = J_i(u_i) + \lambda^\top \left(A_i u_i - \frac{1}{N} b \right).$$

it follows that the minimization of $L(u, \lambda)$ with respect to the primal variable u can be split into N separate minimization problems that can be solved in parallel. The algorithm then performs the N updates

$$u_i^{k+1} = \arg\min_{u_i} L_i(u_i, \lambda^k)$$

in parallel, and then the step

$$\lambda^{k+1} = \lambda^k + \alpha^k (Au^{k+1} - b).$$

16.2 Augmented Lagrangians and the Method of Multipliers

In order to make the minimization step of the dual ascent method more robust, one can use the trick of adding the penalty term $(\rho/2) \|Au - b\|_2^2$ to the Lagrangian.

Definition 16.1. Given the Optimization Problem (P),

$$\begin{aligned} &\text{minimize} \quad J(u) \\ &\text{subject to} \quad Au = b, \end{aligned}$$

the *augmented Lagrangian* is given by

$$L_\rho(u,\lambda) = J(u) + \lambda^\top (Au - b) + (\rho/2)\,\|Au - b\|_2^2,$$

with $\lambda \in \mathbb{R}^m$, and where $\rho > 0$ is called the *penalty parameter*.

The augmented Lagrangian $L_\rho(u,\lambda)$ can be viewed as the ordinary Lagrangian of the Minimization Problem (P_ρ),

$$\begin{aligned} &\text{minimize} \quad J(u) + (\rho/2)\,\|Au - b\|_2^2 \\ &\text{subject to} \quad Au = b. \end{aligned}$$

The above problem is equivalent to Program (P), since for any feasible solution of (P_ρ), we must have $Au - b = 0$.

The benefit of adding the penalty term $(\rho/2)\,\|Au - b\|_2^2$ is that by Proposition 15.25, Problem (P_ρ) has a unique optimal solution under mild conditions on A.

Dual ascent applied to the dual of (P_ρ) yields the the *method of multipliers*, which consists of the following steps, given some initial λ^0:

$$\begin{aligned} u^{k+1} &= \operatorname*{arg\,min}_u L_\rho(u,\lambda^k) \\ \lambda^{k+1} &= \lambda^k + \rho(Au^{k+1} - b). \end{aligned}$$

Observe that the second step uses the parameter ρ. The reason is that it can be shown that choosing $\alpha^k = \rho$ guarantees that (u^{k+1},λ^{k+1}) satisfies the equation

$$\nabla J_{u^{k+1}} + A^\top \lambda^{k+1} = 0,$$

which means that (u^{k+1},λ^{k+1}) is dual feasible; see Boyd, Parikh, Chu, Peleato and Eckstein [Boyd *et al.* (2010)], Section 2.3.

Example 16.2. Consider the minimization problem

$$\begin{aligned} &\text{minimize} \quad y^2 + 2x \\ &\text{subject to} \quad 2x - y = 0. \end{aligned}$$

See Figure 16.2.

The quadratic function

$$J(x,y) = y^2 + 2x = (x\ y)\begin{pmatrix} 0 & 0 \\ 0 & 1 \end{pmatrix}\begin{pmatrix} x \\ y \end{pmatrix} + (2\ 0)\begin{pmatrix} x \\ y \end{pmatrix}$$

is convex but not strictly convex. Since $y = 2x$, the problem is equivalent to minimizing $y^2 + 2x = 4x^2 + 2x$, whose minimum is achieved for $x = -1/4$ (since setting the derivative of the function $x \mapsto 4x^2 + 2$ yields $8x + 2 = 0$).

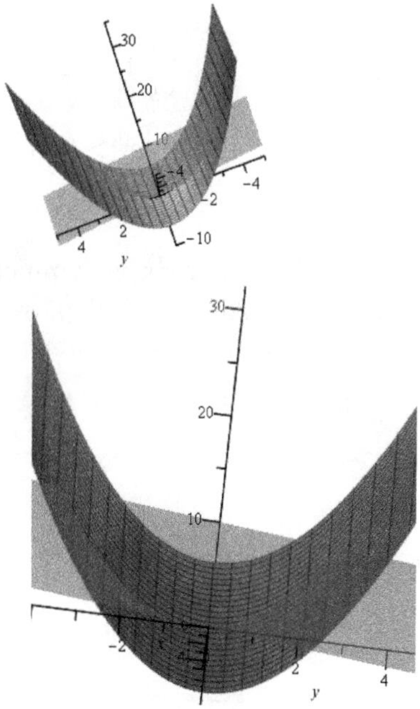

Fig. 16.2 Two views of the graph of $y^2 + 2x$ intersected with the transparent red plane $2x - y = 0$. The solution to Example 16.2 is apex of the intersection curve, namely the point $(-\frac{1}{4}, -\frac{1}{2}, -\frac{15}{16})$.

Thus, the unique minimum of our problem is achieved for $(x = -1/4, y = -1/2)$. The Largrangian of our problem is

$$L(x, y, \lambda) = y^2 + 2x + \lambda(2x - y).$$

If we apply the dual ascent method, minimization of $L(x, y, \lambda)$ with respect to x and y holding λ constant yields the equations

$$2 + 2\lambda = 0$$
$$2y - \lambda = 0,$$

obtained by setting the gradient of L (with respect to x and y) to zero. If $\lambda \neq -1$, the problem has no solution. Indeed, if $\lambda \neq -1$, minimizing $L(x, y, \lambda) = y^2 + 2x + \lambda(2x - y)$ with respect to x and y yields $-\infty$.

The augmented Lagrangian is

$$\begin{aligned} L_\rho(x, y, \lambda) &= y^2 + 2x + \lambda(2x - y) + (\rho/2)(2x - y)^2 \\ &= 2\rho x^2 - 2\rho xy + 2(1 + \lambda)x - \lambda y + \left(1 + \frac{\rho}{2}\right) y^2, \end{aligned}$$

which in matrix form is

$$L_\rho(x, y, \lambda) = (x \; y) \begin{pmatrix} 2\rho^2 & -\rho \\ -\rho & 1 + \dfrac{\rho}{2} \end{pmatrix} \begin{pmatrix} x \\ y \end{pmatrix} + (2(1+\lambda) \; -\lambda) \begin{pmatrix} x \\ y \end{pmatrix}.$$

The trace of the above matrix is $1 + \frac{\rho}{2} + 2\rho^2 > 0$, and the determinant is

$$2\rho^2 \left(1 + \frac{\rho}{2}\right) - \rho^2 = \rho^2(1 + \rho) > 0,$$

since $\rho > 0$. Therefore, the above matrix is symmetric positive definite. Minimizing $L_\rho(x, y, \lambda)$ with respect to x and y, we set the gradient of $L_\rho(x, y, \lambda)$ (with respect to x and y) to zero, and we obtain the equations:

$$\begin{aligned} 2\rho x - \rho y + (1 + \lambda) &= 0 \\ -2\rho x + (2 + \rho)y - \lambda &= 0. \end{aligned}$$

The solution is

$$x = -\frac{1}{4} - \frac{1 + \lambda}{2\rho}, \quad y = -\frac{1}{2}.$$

Thus the steps for the method of multipliers are

$$\begin{aligned} x^{k+1} &= -\frac{1}{4} - \frac{1 + \lambda^k}{2\rho} \\ y^{k+1} &= -\frac{1}{2} \\ \lambda^{k+1} &= \lambda^k + \rho \, (2 \; -1) \begin{pmatrix} -\frac{1}{4} - \frac{1+\lambda^k}{2\rho} \\ -\frac{1}{2} \end{pmatrix}, \end{aligned}$$

and the second step simplifies to

$$\lambda^{k+1} = -1.$$

Consequently, we see that the method converges after two steps for any initial value of λ^0, and we get

$$x = -\frac{1}{4} \quad y = -\frac{1}{2}, \quad \lambda = -1.$$

The method of multipliers also converges for functions J that are not even convex, as illustrated by the next example.

Example 16.3. Consider the minimization problem

$$\begin{aligned} &\text{minimize} \quad 2\beta xy \\ &\text{subject to} \quad 2x - y = 0, \end{aligned}$$

with $\beta > 0$. See Figure 16.3.

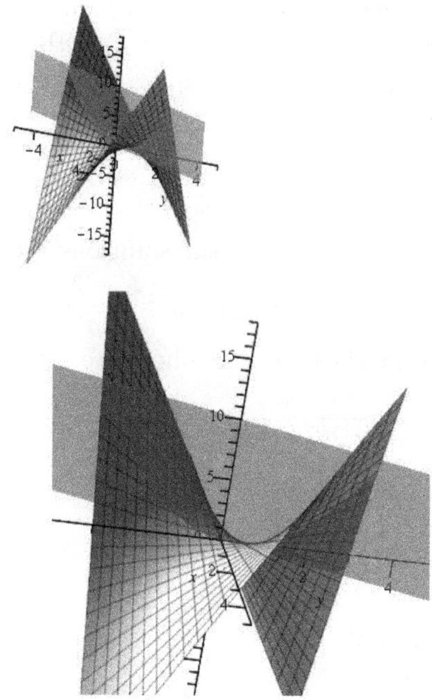

Fig. 16.3 Two views of the graph of the saddle of $2xy$ ($\beta = 1$) intersected with the transparent magenta plane $2x - y = 0$. The solution to Example 16.3 is apex of the intersection curve, namely the point $(0, 0, 0)$.

The quadratic function

$$J(x, y) = 2\beta xy = (x\ y) \begin{pmatrix} 0 & \beta \\ \beta & 0 \end{pmatrix} \begin{pmatrix} x \\ y \end{pmatrix}$$

is not convex because the above matrix is not even positive semidefinite (the eigenvalues of the matrix are $-\beta$ and $+\beta$). The augmented Lagrangian is

$$\begin{aligned} L_\rho(x, y, \lambda) &= 2\beta xy + \lambda(2x - y) + (\rho/2)(2x - y)^2 \\ &= 2\rho x^2 + 2(\beta - \rho)xy + 2\lambda x - \lambda y + \frac{\rho}{2}y^2, \end{aligned}$$

which in matrix form is

$$L_\rho(x, y, \lambda) = (x\ y) \begin{pmatrix} 2\rho & \beta - \rho \\ \beta - \rho & \dfrac{\rho}{2} \end{pmatrix} \begin{pmatrix} x \\ y \end{pmatrix} + (2\lambda\ {-\lambda}) \begin{pmatrix} x \\ y \end{pmatrix}.$$

The trace of the above matrix is $2\rho + \frac{\rho}{2} = \frac{5}{2}\rho > 0$, and the determinant is

$$\rho^2 - (\beta - \rho)^2 = \beta(2\rho - \beta).$$

This determinant is positive if $\rho > \beta/2$, in which case the matrix is symmetric positive definite. Minimizing $L_\rho(x, y, \lambda)$ with respect to x and y, we set the gradient of $L_\rho(x, y, \lambda)$ (with respect to x and y) to zero, and we obtain the equations:

$$\begin{aligned}2\rho x + (\beta - \rho)y + \lambda &= 0\\ 2(\beta - \rho)x + \rho y - \lambda &= 0.\end{aligned}$$

Since we are assuming that $\rho > \beta/2$, the solutions are

$$x = -\frac{\lambda}{2(2\rho - \beta)}, \quad y = \frac{\lambda}{(2\rho - \beta)}.$$

Thus the steps for the method of multipliers are

$$\begin{aligned}x^{k+1} &= -\frac{\lambda^k}{2(2\rho - \beta)}\\ y^{k+1} &= \frac{\lambda^k}{(2\rho - \beta)}\\ \lambda^{k+1} &= \lambda^k + \frac{\rho}{2(2\rho - \beta)}\begin{pmatrix}2 & -1\end{pmatrix}\begin{pmatrix}-\lambda^k\\ 2\lambda^k\end{pmatrix},\end{aligned}$$

and the second step simplifies to

$$\lambda^{k+1} = \lambda^k + \frac{\rho}{2(2\rho - \beta)}(-4\lambda^k),$$

that is,

$$\lambda^{k+1} = -\frac{\beta}{2\rho - \beta}\lambda^k.$$

If we pick $\rho > \beta > 0$, which implies that $\rho > \beta/2$, then

$$\frac{\beta}{2\rho - \beta} < 1,$$

and the method converges for any initial value λ^0 to the solution

$$x = 0, \quad y = 0, \quad \lambda = 0.$$

Indeed, since the constraint $2x - y = 0$ holds, $2\beta xy = 4\beta x^2$, and the minimum of the function $x \mapsto 4\beta x^2$ is achieved for $x = 0$ (since $\beta > 0$).

As an exercise, the reader should verify that dual ascent (with $\alpha^k = \rho$) yields the equations

$$\begin{aligned}x^{k+1} &= \frac{\lambda^k}{2\beta}\\ y^{k+1} &= -\frac{\lambda^k}{\beta}\\ \lambda^{k+1} &= \left(1 + \frac{2\rho}{\beta}\right)\lambda^k,\end{aligned}$$

and so the method diverges, except for $\lambda^0 = 0$, which is the optimal solution.

The method of multipliers converges under conditions that are far more general than the dual ascent. However, the addition of the penalty term has the negative effect that even if J is separable, then the Lagrangian L_ρ is not separable. Thus the basic method of multipliers cannot be used for decomposition and is not parallelizable. The next method deals with the problem of separability.

16.3 ADMM: Alternating Direction Method of Multipliers

The alternating direction method of multipliers, for short ADMM, combines the decomposability of dual ascent with the superior convergence properties of the method of multipliers. It can be viewed as an approximation of the method of multipliers, but it is generally superior.

The idea is to split the function J into two independent parts, as $J(x, z) = f(x) + g(z)$, and to consider the Minimization Problem (P_{admm}),

$$\begin{aligned} &\text{minimize} \quad f(x) + g(z) \\ &\text{subject to} \quad Ax + Bz = c, \end{aligned}$$

for some $p \times n$ matrix A, some $p \times m$ matrix B, and with $x \in \mathbb{R}^n$, $z \in \mathbb{R}^m$, and $c \in \mathbb{R}^p$. We also assume that f and g are convex. Further conditions will be added later.

As in the method of multipliers, we form the *augmented Lagrangian*

$$L_\rho(x, z, \lambda) = f(x) + g(z) + \lambda^\top (Ax + Bz - c) + (\rho/2) \left\| Ax + Bz - c \right\|_2^2,$$

with $\lambda \in \mathbb{R}^p$ and for some $\rho > 0$.

Given some initial values (z^0, λ^0), the *ADMM method* consists of the following iterative steps:

$$\begin{aligned} x^{k+1} &= \arg\min_x L_\rho(x, z^k, \lambda^k) \\ z^{k+1} &= \arg\min_z L_\rho(x^{k+1}, z, \lambda^k) \\ \lambda^{k+1} &= \lambda^k + \rho(Ax^{k+1} + Bz^{k+1} - c). \end{aligned}$$

Instead of performing a minimization step jointly over x and z, as the method of multipliers would in the step

$$(x^{k+1}, z^{k+1}) = \arg\min_{x,z} L_\rho(x, z, \lambda^k),$$

ADMM first performs an x-minimization step, and then a z-minimization step. Thus x and z are updated in an alternating or sequential fashion, which accounts for the term *alternating direction*.

The algorithm state in ADMM is (z^k, λ^k), in the sense that (z^{k+1}, λ^{k+1}) is a function of (z^k, λ^k). The variable x^{k+1} is an auxiliary variable which is used to compute z^{k+1} from (z^k, λ^k). The roles of x and z are not quite symmetric, since the update of x is done before the update of λ. By switching x and z, f and g and A and B, we obtain a variant of ADMM in which the order of the x-update step and the z-update step are reversed.

Example 16.4. Let us reconsider the problem of Example 16.2 to solve it using ADMM. We formulate the problem as

$$\begin{aligned} &\text{minimize} \quad 2x + z^2 \\ &\text{subject to} \quad 2x - z = 0, \end{aligned}$$

with $f(x) = 2x$ and $g(z) = z^2$. The augmented Lagrangian is given by

$$L_\rho(x, z, \lambda) = 2x + z^2 + 2\lambda x - \lambda z + 2\rho x^2 - 2\rho xz + \frac{\rho}{2} z^2.$$

The ADMM steps are as follows. The x-update is

$$x^{k+1} = \arg\min_x \left(2\rho x^2 - 2\rho x z^k + 2\lambda^k x + 2x\right),$$

and since this is a quadratic function in x, its minimum is achieved when the derivative of the above function (with respect to x) is zero, namely

$$x^{k+1} = \frac{1}{2} z^k - \frac{1}{2\rho}\lambda^k - \frac{1}{2\rho}. \tag{1}$$

The z-update is

$$z^{k+1} = \arg\min_z \left(z^2 + \frac{\rho}{2} z^2 - 2\rho x^{k+1} z - \lambda^k z\right),$$

and as for the x-step, the minimum is achieved when the derivative of the above function (with respect to z) is zero, namely

$$z^{k+1} = \frac{2\rho x^{k+1}}{\rho + 2} + \frac{\lambda^k}{\rho + 2}. \tag{2}$$

The λ-update is

$$\lambda^{k+1} = \lambda^k + \rho(2x^{k+1} - z^{k+1}). \tag{3}$$

Substituting the right hand side of (1) for x^{k+1} in (2) yields

$$z^{k+1} = \frac{\rho z^k}{\rho + 2} - \frac{1}{\rho + 2}. \tag{4}$$

Using (2), we obtain

$$2x^{k+1} - z^{k+1} = \frac{4x^{k+1}}{\rho + 2} - \frac{\lambda^k}{\rho + 2}, \tag{5}$$

and then using (3) we get

$$\lambda^{k+1} = \frac{2\lambda^k}{\rho+2} + \frac{4\rho x^{k+1}}{\rho+2}. \tag{6}$$

Substituting the right hand side of (1) for x^{k+1} in (6), we obtain

$$\lambda^{k+1} = \frac{2\rho z^k}{\rho+2} - \frac{2}{\rho+2}. \tag{7}$$

Equation (7) shows that z^k determines λ^{k+1}, and Equation (1) for $k+2$, along with Equation (4), shows that z^k also determines x^{k+2}. In particular, we find that

$$\begin{aligned} x^{k+2} &= \frac{1}{2} z^{k+1} - \frac{1}{2\rho}\lambda^{k+1} - \frac{1}{2\rho} \\ &= \frac{(\rho-2)z^k}{2(\rho+2)} - \frac{1}{\rho+2}. \end{aligned}$$

Thus is suffices to find the limit of the sequence (z^k). Since we already know from Example 16.2 that this limit is $-1/2$, using (4), we write

$$z^{k+1} = -\frac{1}{2} + \frac{\rho z^k}{\rho+2} - \frac{1}{\rho+2} + \frac{1}{2} = -\frac{1}{2} + \frac{\rho}{\rho+2}\left(\frac{1}{2} + z^k\right).$$

By induction, we deduce that

$$z^{k+1} = -\frac{1}{2} + \left(\frac{\rho}{\rho+2}\right)^{k+1} \left(\frac{1}{2} + z^0\right),$$

and since $\rho > 0$, we have $\rho/(\rho+2) < 1$, so the limit of the sequence (z^{k+1}) is indeed $-1/2$, and consequently the limit of (λ^{k+1}) is -1 and the limit of x^{k+2} is $-1/4$.

For ADMM to be practical, the x-minimization step and the z-minimization step have to be doable efficiently.

It is often convenient to write the ADMM updates in terms of the *scaled dual variable* $\mu = (1/\rho)\lambda$. If we define the *residual* as

$$r = Ax + bz - c,$$

then we have

$$\begin{aligned} \lambda^\top r + (\rho/2)\left\|r\right\|_2^2 &= (\rho/2)\left\|r + (1/\rho)\lambda\right\|_2^2 - (1/(2\rho))\left\|\lambda\right\|_2^2 \\ &= (\rho/2)\left\|r + \mu\right\|_2^2 - (\rho/2)\left\|\mu\right\|_2^2. \end{aligned}$$

The *scaled form of ADMM* consists of the following steps:

$$x^{k+1} = \arg\min_x \left(f(x) + (\rho/2) \left\|Ax + Bz^k - c + \mu^k\right\|_2^2 \right)$$

$$z^{k+1} = \arg\min_z \left(g(z) + (\rho/2) \left\|Ax^{k+1} + Bz - c + \mu^k\right\|_2^2 \right)$$

$$\mu^{k+1} = \mu^k + Ax^{k+1} + Bz^{k+1} - c.$$

If we define the *residual* r^k *at step* k as

$$r^k = Ax^k + Bz^k - c = \mu^k - \mu^{k-1} = (1/\rho)(\lambda^k - \lambda^{k-1}),$$

then we see that

$$r = u^0 + \sum_{j=1}^{k} r^j.$$

The formulae in the scaled form are often shorter than the formulae in the unscaled form.

We now discuss the convergence of ADMM.

16.4 Convergence of ADMM ⊛

Let us repeat the steps of ADMM: Given some initial (z^0, λ^0), do:

$$x^{k+1} = \arg\min_x L_\rho(x, z^k, \lambda^k) \qquad (x\text{-update})$$

$$z^{k+1} = \arg\min_z L_\rho(x^{k+1}, z, \lambda^k) \qquad (z\text{-update})$$

$$\lambda^{k+1} = \lambda^k + \rho(Ax^{k+1} + Bz^{k+1} - c). \qquad (\lambda\text{-update})$$

The convergence of ADMM can be proven under the following three assumptions:

(1) The functions $f \colon \mathbb{R} \to \mathbb{R} \cup \{+\infty\}$ and $g \colon \mathbb{R} \to \mathbb{R} \cup \{+\infty\}$ are proper and closed convex functions (see Section 15.1) such that $\textbf{relint}(\text{dom}(f)) \cap \textbf{relint}(\text{dom}(g)) \neq \emptyset$.

(2) The $n \times n$ matrix $A^\top A$ is invertible and the $m \times m$ matrix $B^\top B$ is invertible. Equivalently, the $p \times n$ matrix A has rank n and the $p \times m$ matrix has rank m.

(3) The unaugmented Lagrangian $L_0(x, z, \lambda) = f(x) + g(z) + \lambda^\top(Ax + Bz - c)$ has a saddle point, which means there exists x^*, z^*, λ^* (not necessarily unique) such that

$$L_0(x^*, z^*, \lambda) \leq L_0(x^*, z^*, \lambda^*) \leq L_0(x, z, \lambda^*)$$

for all x, z, λ.

Recall that the augmented Lagrangian is given by

$$L_\rho(x,z,\lambda) = f(x) + g(z) + \lambda^\top(Ax + Bz - c) + (\rho/2)\left\|Ax + Bz - c\right\|_2^2.$$

For z (and λ) fixed, we have

$$\begin{aligned}L_\rho(x,z,\lambda) &= f(x) + g(z) + \lambda^\top(Ax + Bz - c)\\ &\quad + (\rho/2)(Ax + Bz - c)^\top(Ax + Bz - c)\\ &= f(x) + (\rho/2)x^\top A^\top Ax + (\lambda^\top + \rho(Bz - c)^\top)Ax\\ &\quad + g(z) + \lambda^\top(Bz - c) + (\rho/2)(Bz - c)^\top(Bz - c).\end{aligned}$$

Assume that (1) and (2) hold. Since $A^\top A$ is invertible, then it is symmetric positive definite, and by Proposition 15.25 the x-minimization step has a unique solution (the minimization problem succeeds with a unique minimizer).

Similarly, for x (and λ) fixed, we have

$$\begin{aligned}L_\rho(x,z,\lambda) &= f(x) + g(z) + \lambda^\top(Ax + Bz - c)\\ &\quad + (\rho/2)(Ax + Bz - c)^\top(Ax + Bz - c)\\ &= g(z) + (\rho/2)z^\top B^\top Bz + (\lambda^\top + \rho(Ax - c)^\top)Bz\\ &\quad + f(x) + \lambda^\top(Ax - c) + (\rho/2)(Ax - c)^\top(Ax - c).\end{aligned}$$

Since $B^\top B$ is invertible, then it is symmetric positive definite, and by Proposition 15.25 the z-minimization step has a unique solution (the minimization problem succeeds with a unique minimizer).

By Theorem 15.10, Assumption (3) is equivalent to the fact that the KKT equations are satisfied by some triple (x^*, z^*, λ^*), namely

$$Ax^* + Bz^* - c = 0 \tag{$*$}$$

and

$$0 \in \partial f(x^*) + \partial g(z^*) + A^\top\lambda^* + B^\top\lambda^*, \tag{$\dagger$}$$

Assumption (3) is also equivalent to Conditions (a) and (b) of Theorem 15.10. In particular, our program has an optimal solution (x^*, z^*). By Theorem 15.12, λ^* is maximizer of the dual function $G(\lambda) = \inf_{x,z} L_0(x,z,\lambda)$ and strong duality holds, that is, $G(\lambda^*) = f(x^*) + g(z^*)$ (the duality gap is zero).

We will see after the proof of Theorem 16.1 that Assumption (2) is actually implied by Assumption (3). This allows us to prove a convergence result stronger than the convergence result proven in Boyd et al. [Boyd *et al.* (2010)] under the exact same assumptions (1) and (3).

Let p^* be the minimum value of $f + g$ over the convex set $\{(x,z) \in \mathbb{R}^{m+p} \mid Ax + Bz - c = 0\}$, and let (p^k) be the sequence given by $p^k = f(x^k) + g(z^k)$, and recall that $r^k = Ax^k + Bz^k - c$.

Our main goal is to prove the following result.

Theorem 16.1. *Suppose the following assumptions hold:*

(1) The functions $f\colon \mathbb{R} \to \mathbb{R} \cup \{+\infty\}$ and $g\colon \mathbb{R} \to \mathbb{R} \cup \{+\infty\}$ are proper and closed convex functions (see Section 15.1) such that **relint**$(\mathrm{dom}(f)) \cap$ **relint**$(\mathrm{dom}(g)) \neq \emptyset$.

(2) The $n \times n$ matrix $A^\top A$ is invertible and the $m \times m$ matrix $B^\top B$ is invertible. Equivalently, the $p \times n$ matrix A has rank n and the $p \times m$ matrix has rank m. (This assumption is actually redundant, because it is implied by Assumption (3).)

(3) The unaugmented Lagrangian $L_0(x,z,\lambda) = f(x) + g(z) + \lambda^\top(Ax + Bz - c)$ has a saddle point, which means there exists x^*, z^*, λ^* (not necessarily unique) such that

$$L_0(x^*, z^*, \lambda) \leq L_0(x^*, z^*, \lambda^*) \leq L_0(x, z, \lambda^*)$$

for all x, z, λ.

Then for any initial values (z^0, λ^0), the following properties hold:

(1) The sequence (r^k) converges to 0 (residual convergence).

(2) The sequence (p^k) converge to p^* (objective convergence).

(3) The sequences (x^k) and (z^k) converge to an optimal solution $(\widetilde{x}, \widetilde{z})$ of Problem (P_{admm}) and the sequence (λ^k) converges an optimal solution $\widetilde{\lambda}$ of the dual problem (primal and dual variable convergence).

Proof. The core of the proof is due to Boyd et al. [Boyd *et al.* (2010)], but there are new steps because we have the stronger hypothesis (2), which yield the stronger result (3).

The proof consists of several steps. It is not possible to prove directly that the sequences (x^k), (z^k), and (λ^k) converge, so first we prove that the sequence (r^{k+1}) converges to zero, and that the sequences (Ax^{k+1}) and (Bz^{k+1}) also converge.

Step 1. Prove the inequality (A1) below.

Consider the sequence of reals (V^k) given by

$$V^k = (1/\rho) \left\| \lambda^k - \lambda^* \right\|_2^2 + \rho \left\| B(z^k - z^*) \right\|_2^2.$$

It can be shown that the V^k satisfy the following inequality:

$$V^{k+1} \leq V^k - \rho \left\| r^{k+1} \right\|_2^2 - \rho \left\| B(z^{k+1} - z^k) \right\|_2^2. \tag{A1}$$

This is rather arduous. Since a complete proof is given in Boyd et al. [Boyd *et al.* (2010)], we will only provide some of the key steps later.

Inequality (A1) shows that the sequence (V^k) in nonincreasing. If we write these inequalities for $k, k-1, \ldots, 0$, we have

$$\begin{aligned} V^{k+1} &\leq V^k - \rho \left\| r^{k+1} \right\|_2^2 - \rho \left\| B(z^{k+1} - z^k) \right\|_2^2 \\ V^k &\leq V^{k-1} - \rho \left\| r^k \right\|_2^2 - \rho \left\| B(z^k - z^{k-1}) \right\|_2^2 \\ &\vdots \\ V^1 &\leq V^0 - \rho \left\| r^1 \right\|_2^2 - \rho \left\| B(z^1 - z^0) \right\|_2^2, \end{aligned}$$

and by adding up these inequalities, we obtain

$$V^{k+1} \leq V^0 - \rho \sum_{j=0}^{k} \left(\left\| r^{j+1} \right\|_2^2 + \left\| B(z^{j+1} - z^j) \right\|_2^2 \right),$$

which implies that

$$\rho \sum_{j=0}^{k} \left(\left\| r^{j+1} \right\|_2^2 + \left\| B(z^{j+1} - z^j) \right\|_2^2 \right) \leq V_0 - V^{k+1} \leq V^0, \tag{B}$$

since $V^{k+1} \leq V^0$.

Step 2. Prove that the sequence (r^k) converges to 0, and that the sequences (Ax^{k+1}) and (Bz^{k+1}) also converge.

Inequality (B) implies that the series $\sum_{k=1}^{\infty} r^k$ and $\sum_{k=0}^{\infty} B(z^{k+1} - z^k)$ converge absolutely. In particular, the sequence (r^k) converges to 0.

The nth partial sum of the series $\sum_{k=0}^{\infty} B(z^{k+1} - z^k)$ is

$$\sum_{k=0}^{n} B(z^{k+1} - z^k) = B(z^{n+1} - z^0),$$

and since the series $\sum_{k=0}^{\infty} B(z^{k+1} - z^k)$ converges, we deduce that the sequence (Bz^{k+1}) converges. Since $Ax^{k+1} + Bz^{k+1} - c = r^{k+1}$, the convergence of (r^{k+1}) and (Bz^{k+1}) implies that the sequence (Ax^{k+1}) also converges.

Step 3. Prove that the sequences (x^{k+1}) and (z^{k+1}) converge. By Assumption (2), the matrices $A^\top A$ and $B^\top B$ are invertible, so multiplying each vector Ax^{k+1} by $(A^\top A)^{-1}A^\top$, if the sequence (Ax^{k+1}) converges to u, then the sequence (x^{k+1}) converges to $(A^\top A)^{-1}A^\top u$. Similarly, if the sequence (Bz^{k+1}) converges to v, then the sequence (z^{k+1}) converges to $(B^\top B)^{-1}B^\top v$.

Step 4. Prove that the sequence (λ^k) converges.

Recall that

$$\lambda^{k+1} = \lambda^k + \rho r^{k+1}.$$

It follows by induction that

$$\lambda^{k+p} = \lambda^k + \rho(r^{k+1} + \cdots + \rho^{k+p}), \quad p \geq 2.$$

As a consequence, we get

$$\left\|\lambda^{k+p} - \lambda^k\right\| \leq \rho(\left\|r^{k+1}\right\| + \cdots + \left\|r^{k+p}\right\|).$$

Since the series $\sum_{k=1}^{\infty} \left\|r^k\right\|$ converges, the partial sums form a Cauchy sequence, and this immediately implies that for any $\epsilon > 0$ we can find $N > 0$ such that

$$\rho(\left\|r^{k+1}\right\| + \cdots + \left\|r^{k+p}\right\|) < \epsilon, \quad \text{for all } k, p + k \geq N,$$

so the sequence (λ^k) is also a Cauchy sequence, thus it converges.

Step 5. Prove that the sequence (p^k) converges to p^*.

For this, we need two more inequalities. Following Boyd et al. [Boyd *et al.* (2010)], we need to prove that

$$p^{k+1} - p^* \leq -(\lambda^{k+1})^\top r^{k+1} - \rho(B(z^{k+1} - z^k))^\top(-r^{k+1} + B(z^{k+1} - z^*)) \quad \text{(A2)}$$

and

$$p^* - p^{k+1} \leq (\lambda^*)^\top r^{k+1}. \quad \text{(A3)}$$

Since we proved that the sequence (r^k) and $B(z^{k+1} - z^k)$ converge to 0, and that the sequence (λ^{k+1}) converges, from

$$\begin{aligned}(\lambda^{k+1})^\top r^{k+1} + \rho(B(z^{k+1} - z^k))^\top(-r^{k+1} + B(z^{k+1} - z^*)) &\leq p^* - p^{k+1} \\ &\leq (\lambda^*)^\top r^{k+1},\end{aligned}$$

we deduce that in the limit, p^{k+1} converges to p^*.

Step 6. Prove (A3).

Since (x^*, y^*, λ^*) is a saddle point, we have

$$L_0(x^*, z^*, \lambda^*) \leq L_0(x^{k+1}, z^{k+1}, \lambda^*).$$

Since $Ax^* + Bz^* = c$, we have $L_0(x^*, z^*, \lambda^*) = p^*$, and since $p^{k+1} = f(x^{k+1}) + g(z^{k+1})$, we have

$$L_0(x^{k+1}, z^{k+1}, \lambda^*) = p^{k+1} + (\lambda^*)^\top r^{k+1},$$

so we obtain

$$p^* \leq p^{k+1} + (\lambda^*)^\top r^{k+1},$$

which yields (A3).

Step 7. Prove (A2).

By Proposition 15.24, z^{k+1} minimizes $L_\rho(x^{k+1}, z, \lambda^k)$ iff

$$\begin{aligned}0 &\in \partial g(z^{k+1}) + B^\top \lambda^k + \rho B^\top (Ax^{k+1} + Bz^{k+1} - c) \\ &= \partial g(z^{k+1}) + B^\top \lambda^k + \rho B^\top r^{k+1} \\ &= \partial g(z^{k+1}) + B^\top \lambda^{k+1},\end{aligned}$$

since $r^{k+1} = Ax^{k+1} + Bz^{k+1} - c$ and $\lambda^{k+1} = \lambda^k + \rho(Ax^{k+1} + Bz^{k+1} - c)$.

In summary, we have

$$0 \in \partial g(z^{k+1}) + B^\top \lambda^{k+1}, \tag{$\dagger_1$}$$

which shows that z^{k+1} minimizes the function

$$z \mapsto g(z) + (\lambda^{k+1})^\top Bz.$$

Consequently, we have

$$g(z^{k+1}) + (\lambda^{k+1})^\top Bz^{k+1} \leq g(z^*) + (\lambda^{k+1})^\top Bz^*. \tag{B1}$$

Similarly, x^{k+1} minimizes $L_\rho(x, z^k, \lambda^k)$ iff

$$\begin{aligned}0 &\in \partial f(x^{k+1}) + A^\top \lambda^k + \rho A^\top (Ax^{k+1} + Bz^k - c) \\ &= \partial f(x^{k+1}) + A^\top (\lambda^k + \rho r^{k+1} + \rho B(z^k - z^{k+1})) \\ &= \partial f(x^{k+1}) + A^\top \lambda^{k+1} + \rho A^\top B(z^k - z^{k+1})\end{aligned}$$

since $r^{k+1} - Bz^{k+1} = Ax^{k+1} - c$ and $\lambda^{k+1} = \lambda^k + \rho(Ax^{k+1} + Bz^{k+1} - c) = \lambda^k + \rho r^{k+1}$.

Equivalently, the above derivation shows that

$$0 \in \partial f(x^{k+1}) + A^\top (\lambda^{k+1} - \rho B(z^{k+1} - z^k)), \tag{$\dagger_2$}$$

which shows that x^{k+1} minimizes the function

$$x \mapsto f(x) + (\lambda^{k+1} - \rho B(z^{k+1} - z^k))^\top Ax.$$

Consequently, we have

$$\begin{aligned}f(x^{k+1}) &+ (\lambda^{k+1} - \rho B(z^{k+1} - z^k))^\top Ax^{k+1} \\ &\leq f(x^*) + (\lambda^{k+1} - \rho B(z^{k+1} - z^k))^\top Ax^*. \end{aligned} \tag{B2}$$

Adding up Inequalities (B1) and (B2), using the equation $Ax^* + Bz^* = c$, and rearranging, we obtain inequality (A2).

Step 8. Prove that $(x^k), (z^k)$, and (λ^k) converge to optimal solutions.

Recall that (r^k) converges to 0, and that (x^k), (z^k), and (λ^k) converge to limits $\widetilde{x}$, $\widetilde{z}$, and $\widetilde{\lambda}$. Since $r^k = Ax^k + Bz^k - c$, in the limit, we have

$$A\widetilde{x} + B\widetilde{z} - c = 0. \tag{$*_1$}$$

Using ($\dagger_1$), in the limit, we obtain

$$0 \in \partial g(\widetilde{z}) + B^\top \widetilde{\lambda}. \qquad (*_2)$$

Since $(B(z^{k+1} - z^k))$ converges to 0, using ($\dagger_2$), in the limit, we obtain

$$0 \in \partial f(\widetilde{x}) + A^\top \widetilde{\lambda}. \qquad (*_3)$$

From ($*_2$) and ($*_3$), we obtain

$$0 \in \partial f(\widetilde{x}) + \partial g(\widetilde{z}) + A^\top \widetilde{\lambda} + B^\top \widetilde{\lambda}. \qquad (*_4)$$

But ($*_1$) and ($*_4$) are exactly the KKT equations, and by Theorem 15.10, we conclude that $\widetilde{x}, \widetilde{z}, \widetilde{\lambda}$ are optimal solutions.

Step 9. Prove (A1). This is the most tedious step of the proof. We begin by adding up (A2) and (A3), and then perform quite a bit or rewriting and manipulation. The complete derivation can be found in Boyd et al. [Boyd *et al.* (2010)]. □

Remarks:

(1) In view of Theorem 15.11, we could replace Assumption (3) by the slightly stronger assumptions that the optimum value of our program is finite and that the constraints are qualified. Since the constraints are affine, this means that there is some feasible solution in **relint**(dom(f)) ∩ **relint**(dom(g)). These assumptions are more practical than Assumption (3).
(2) Actually, Assumption (3) implies Assumption (2). Indeed, we know from Theorem 15.10 that the existence of a saddle point implies that our program has a finite optimal solution. However, if either $A^\top A$ or $B^\top B$ is not invertible, then Program (P) may not have a finite optimal solution, as shown by the following counterexample.

Example 16.5. Let

$$f(x, y) = x, \quad g(z) = 0, \quad y - z = 0.$$

Then

$$L_\rho(x, y, z, \lambda) = x + \lambda(y - z) + (\rho/2)(y - z)^2,$$

but minimizing over (x, y) with z held constant yields $-\infty$, which implies that the above program has no finite optimal solution. See Figure 16.4.

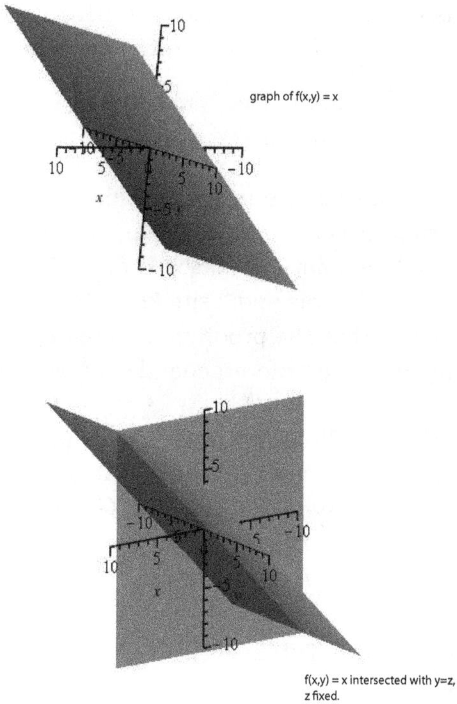

Fig. 16.4 A graphical representation of the Example 16.5. This is an illustration of the x minimization step when z is held fixed. Since the intersection of the two planes is an unbounded line, we "see" that minimizing over x yields $-\infty$.

The problem is that

$$A = \begin{pmatrix} 0 & 1 \end{pmatrix}, \quad B = \begin{pmatrix} -1 \end{pmatrix},$$

but

$$A^\top A = \begin{pmatrix} 0 & 0 \\ 0 & 1 \end{pmatrix}$$

is not invertible.

(3) Proving (A1), (A2), (A3), and the convergence of (r^k) to 0 and of (p^k) to p^* does not require Assumption (2). The proof, using the ingeneous Inequality (A1) (and (B)) is the proof given in Boyd et al. [Boyd *et al.* (2010)]. We were also able to prove that (λ^k), (Ax^k) and (Bz^k) converge without Assumption (2), but to prove that (x^k), (y^k), and (λ^k) converge to optimal solutions, we had to use Assumption (2).

(4) Bertsekas discusses ADMM in [Bertsekas (2015)], Sections 2.2 and 5.4. His formulation of ADMM is slightly different, namely

$$\begin{aligned} &\text{minimize} \quad f(x) + g(z) \\ &\text{subject to} \quad Ax = z. \end{aligned}$$

Bertsekas states a convergence result for this version of ADMM under the hypotheses that either $\text{dom}(f)$ is compact or that $A^\top A$ is invertible, and that a saddle point exists; see Proposition 5.4.1. The proof is given in Bertsekas [Bertsekas and Tsitsiklis (1997)], Section 3.4, Proposition 4.2. It appears that the proof makes use of gradients, so it is not clear that it applies in the more general case where f and g are not differentiable.

(5) Versions of ADMM are discussed in Gabay [Gabay (1983)] (Sections 4 and 5). They are more general than the version discussed here. Some convergence proofs are given, but because Gabay's framework is more general, it is not clear that they apply to our setting. Also, these proofs rely on earlier result by Lions and Mercier, which makes the comparison difficult.

(5) Assumption (2) does not imply that the system $Ax + Bz = c$ has any solution. For example, if

$$A = \begin{pmatrix} 1 \\ 1 \end{pmatrix}, \quad B = \begin{pmatrix} -1 \\ -1 \end{pmatrix}, \quad c = \begin{pmatrix} 1 \\ 0 \end{pmatrix},$$

the system

$$\begin{aligned} x - z &= 1 \\ x - z &= 0 \end{aligned}$$

has no solution. However, since Assumption (3) implies that the program has an optimal solution, it implies that c belongs to the column space of the $p \times (n + m)$ matrix $\begin{pmatrix} A & B \end{pmatrix}$.

Here is an example where ADMM diverges for a problem whose optimum value is $-\infty$.

Example 16.6. Consider the problem given by

$$f(x) = x, \quad g(z) = 0, \quad x - z = 0.$$

Since $f(x) + g(z) = x$, and $x = z$, the variable x is unconstrained and the above function goes to $-\infty$ when x goes to $-\infty$. The augmented Lagrangian is

$$\begin{aligned} L_\rho(x,z,\lambda) &= x + \lambda(x-z) + \frac{\rho}{2}(x-z)^2 \\ &= \frac{\rho}{2}x^2 - \rho xz + \frac{\rho}{2}z^2 + x + \lambda x - \lambda z. \end{aligned}$$

The matrix

$$\begin{pmatrix} \frac{1}{2} & -\frac{1}{2} \\ -\frac{1}{2} & \frac{1}{2} \end{pmatrix}$$

is singular and $L_\rho(x,z,\lambda)$ goes to $-\infty$ in when $(x,z) = t(1,1)$ and t goes to $-\infty$. The ADMM steps are:

$$\begin{aligned} x^{k+1} &= z^k - \frac{1}{\rho}\lambda^k - \frac{1}{\rho} \\ z^{k+1} &= x^{k+1} + \frac{1}{\rho}\lambda^k \\ \lambda^{k+1} &= \lambda^k + \rho(x^{k+1} - z^{k+1}), \end{aligned}$$

and these equations hold for all $k \geq 0$. From the last two equations we deduce that

$$\lambda^{k+1} = \lambda^k + \rho(x^{k+1} - z^{k+1}) = \lambda^k + \rho(-\frac{1}{\rho}\lambda^k) = 0, \quad \text{for all } k \geq 0,$$

so

$$z^{k+2} = x^{k+2} + \frac{1}{\rho}\lambda^{k+1} = x^{k+2}, \quad \text{for all } k \geq 0.$$

Consequently we find that

$$x^{k+3} = z^{k+2} + \frac{1}{\rho}\lambda^{k+2} - \frac{1}{\rho} = x^{k+2} - \frac{1}{\rho}.$$

By induction, we obtain

$$x^{k+3} = x^2 - \frac{k+1}{\rho}, \quad \text{for all } k \geq 0,$$

which shows that x^{k+3} goes to $-\infty$ when k goes to infinity, and since $x^{k+2} = z^{k+2}$, similarly z^{k+3} goes to $-\infty$ when k goes to infinity.

16.5 Stopping Criteria

Going back to Inequality (A2),

$$p^{k+1} - p^* \leq -(\lambda^{k+1})^\top r^{k+1} - \rho(B(z^{k+1} - z^k))^\top(-r^{k+1} + B(z^{k+1} - z^*)), \quad \text{(A2)}$$

using the fact that $Ax^* + Bz^* - c = 0$ and $r^{k+1} = Ax^{k+1} + Bz^{k+1} - c$, we have

$$\begin{aligned} -r^{k+1} + B(z^{k+1} - z^*) &= -Ax^{k+1} - Bz^{k+1} + c + B(z^{k+1} - z^*) \\ &= -Ax^{k+1} + c - Bz^* \\ &= -Ax^{k+1} + Ax^* = -A(x^{k+1} - x^*), \end{aligned}$$

so (A2) can be rewritten as

$$p^{k+1} - p^* \leq -(\lambda^{k+1})^\top r^{k+1} + \rho(B(z^{k+1} - z^k))^\top A(x^{k+1} - x^*),$$

or equivalently as

$$p^{k+1} - p^* \leq -(\lambda^{k+1})^\top r^{k+1} + (x^{k+1} - x^*)^\top \rho A^\top B(z^{k+1} - z^k). \qquad (s_1)$$

We define the *dual residual* as

$$s^{k+1} = \rho A^\top B(z^{k+1} - z^k),$$

the quantity $r^{k+1} = Ax^{k+1} + Bz^{k+1} - c$ being the *primal residual*. Then (s_1) can be written as

$$p^{k+1} - p^* \leq -(\lambda^{k+1})^\top r^{k+1} + (x^{k+1} - x^*)^\top s^{k+1}. \qquad (s)$$

Inequality (s) shows that when the residuals r^k and s^k are small, then p^k is close to p^* from below. Since x^* is unknown, we can't use this inequality, but if we have a guess that $\left\|x^k - x^*\right\| \leq d$, then using Cauchy–Schwarz we obtain

$$p^{k+1} - p^* \leq \left\|\lambda^{k+1}\right\| \left\|r^{k+1}\right\| + d \left\|s^{k+1}\right\|.$$

The above suggests that a reasonable termination criterion is that $\left\|r^k\right\|$ and $\left\|s^k\right\|$ should be small, namely that

$$\left\|r^k\right\| \leq \epsilon^{\text{pri}} \quad \text{and} \quad \left\|s^k\right\| \leq \epsilon^{\text{dual}},$$

for some chosen feasibility tolerances ϵ^{pri} and ϵ^{dual}. Further discussion for choosing these parameters can be found in Boyd et al. [Boyd *et al.* (2010)] (Section 3.3.1).

Various extensions and variations of ADMM are discussed in Boyd et al. [Boyd *et al.* (2010)] (Section 3.4). In order to accelerate convergence of the method, one may choose a different ρ at each step (say ρ^k), although proving the convergence of such a method may be difficult. If we assume that ρ^k becomes constant after a number of iterations, then the proof that we gave still applies. A simple scheme is this:

$$\rho^{k+1} = \begin{cases} \tau^{\text{incr}} \rho^k & \text{if } \left\|r^k\right\| > \mu \left\|s^k\right\| \\ \rho^k / \tau^{\text{decr}} & \text{if } \left\|s^k\right\| > \mu \left\|r^k\right\| \\ \rho^k & \text{otherwise,} \end{cases}$$

where $\tau^{\text{incr}} > 1, \tau^{\text{decr}} > 1$, and $\mu > 1$ are some chosen parameters. Again, we refer the interested reader to Boyd et al. [Boyd *et al.* (2010)] (Section 3.4).

16.6 Some Applications of ADMM

Structure in f, g, A, and B can often be exploited to yield more efficient methods for performing the x-update and the z-update. We focus on the x-update, but the discussion applies just as well to the z-update. Since z and λ are held constant during minimization over x, it is more convenient to use the scaled form of ADMM. Recall that

$$x^{k+1} = \arg\min_x \left(f(x) + (\rho/2) \left\| Ax + Bz^k - c + u^k \right\|_2^2 \right)$$

(here we use u instead of μ), so we can express the x-update step as

$$x^+ = \arg\min_x \left(f(x) + (\rho/2) \left\| Ax - v \right\|_2^2 \right),$$

with $v = -Bz^k + c - u^k$.

Example 16.7. A first simplification arises when $A = I$, in which case the x-update is

$$x^+ = \arg\min_x \left(f(x) + (\rho/2) \left\| x - v \right\|_2^2 \right) = \mathbf{prox}_{f,\rho}(v).$$

The map $v \mapsto \mathbf{prox}_{f,\rho}(v)$ is known as the *proximity operator of f with penalty ρ*. The above minimization is generally referred to as *proximal minimization*.

Example 16.8. When the function f is simple enough, the proximity operator can be computed analytically. This is the case in particular when $f = I_C$, the indicator function of a nonempty closed convex set C. In this case, it is easy to see that

$$x^+ = \arg\min_x \left(I_C(x) + (\rho/2) \left\| x - v \right\|_2^2 \right) = \Pi_C(v),$$

the orthogonal projection of v onto C. In the special case where $C = \mathbb{R}^n_+$ (the first orthant), then

$$x^+ = (v)_+,$$

the vector obtained by setting the negative components of v to zero.

Example 16.9. A second case where simplifications arise is the case where f is a convex quadratic functional of the form

$$f(x) = \frac{1}{2} x^\top P x + q^\top x + r,$$

where P is an $n \times n$ symmetric positive semidefinite matrix, $q \in \mathbb{R}^n$ and $r \in \mathbb{R}$. In this case the gradient of the map

$$x \mapsto f(x) + (\rho/2)\, \|Ax - v\|_2^2 = \frac{1}{2}x^\top Px + q^\top x + r + \frac{\rho}{2}x^\top (A^\top A)x \\ - \rho x^\top A^\top v + \frac{\rho}{2}v^\top v$$

is given by

$$(P + \rho A^\top A)x + q - \rho A^\top v,$$

and since A has rank n, the matrix $A^\top A$ is symmetric positive definite, so we get

$$x^+ = (P + \rho A^\top A))^{-1}(\rho A^\top v - q).$$

Methods from numerical linear algebra can be used so compute x^+ fairly efficiently; see Boyd et al. [Boyd *et al.* (2010)] (Section 4).

Example 16.10. A third case where simplifications arise is the variation of the previous case where f is a convex quadratic functional of the form

$$f(x) = \frac{1}{2}x^\top Px + q^\top x + r,$$

except that f is constrained by equality constraints $Cx = b$, as in Section 14.4, which means that $\mathrm{dom}(f) = \{x \in \mathbb{R}^n \mid Cx = b\}$, and $A = I$. The x-minimization step consists in minimizing the function

$$J(x) = \frac{1}{2}x^\top Px + q^\top x + r + \frac{\rho}{2}x^\top x - \rho x^\top v + \frac{\rho}{2}v^\top v$$

subject to the constraint

$$Cx = b,$$

so by the results of Section 14.4, x^+ is a component of the solution of the KKT-system

$$\begin{pmatrix} P + \rho I & C^\top \\ C & 0 \end{pmatrix} \begin{pmatrix} x^+ \\ y \end{pmatrix} = \begin{pmatrix} -q + \rho v \\ b \end{pmatrix}.$$

The matrix $P + \rho I$ is symmetric positive definite, so the KKT-matrix is invertible.

We can now describe how ADMM is used to solve two common problems of convex optimization.

(1) *Minimization of a proper closed convex function f over a closed convex set C in $\mathbb{R}^n$.* This is the following problem

$$\begin{aligned} &\text{minimize} && f(x) \\ &\text{subject to} && x \in C, \end{aligned}$$

which can be rewritten in AADM form as

$$\begin{aligned} &\text{minimize} && f(x) + I_C(z) \\ &\text{subject to} && x - z = 0. \end{aligned}$$

Using the scaled dual variable $u = \lambda/\rho$, the augmented Lagrangian is

$$L_\rho(x, z, u) = f(x) + I_C(z) + \frac{\rho}{2} \|x - z + u\|_2^2 - \frac{\rho}{2} \|u\|^2 .$$

In view of Example 16.8, the scaled form of ADMM for this problem is

$$\begin{aligned} x^{k+1} &= \arg\min_x \left(f(x) + (\rho/2) \left\|x - z^k + u^k\right\|_2^2\right) \\ z^{k+1} &= \Pi_C(x^{k+1} + u^k) \\ u^{k+1} &= u^k + x^{k+1} - z^{k+1}. \end{aligned}$$

The x-update involves evaluating a proximal operator. Note that the function f need not be differentiable. Of course, these minimizations depend on having efficient computational procedures for the proximal operator and the projection operator.

(2) *Quadratic Programming.* Here the problem is

$$\begin{aligned} &\text{minimize} && \frac{1}{2} x^\top P x + q^\top x + r \\ &\text{subject to} && Ax = b,\ x \geq 0, \end{aligned}$$

where P is an $n \times n$ symmetric positive semidefinite matrix, $q \in \mathbb{R}^n$, $r \in \mathbb{R}$, and A is an $m \times n$ matrix of rank m.

The above program is converted in ADMM form as follows:

$$\begin{aligned} &\text{minimize} && f(x) + g(z) \\ &\text{subject to} && x - z = 0, \end{aligned}$$

with

$$f(x) = \frac{1}{2} x^\top P x + q^\top x + r, \quad \text{dom}(f) = \{x \in \mathbb{R}^n \mid Ax = b\},$$

and

$$g = I_{\mathbb{R}^n_+},$$

the indicator function of the positive orthant $\mathbb{R}^n_+$. In view of Example 16.8 and Example 16.10, the scaled form of ADMM consists of the following steps:

$$
\begin{aligned}
x^{k+1} &= \arg\min_x \left(f(x) + (\rho/2) \left\| x - z^k + u^k \right\|_2^2 \right) \\
z^{k+1} &= (x^{k+1} + u^k)_+ \\
u^{k+1} &= u^k + x^{k+1} - z^{k+1}.
\end{aligned}
$$

The x-update involves solving the KKT equations

$$
\begin{pmatrix} P + \rho I & A^\top \\ A & 0 \end{pmatrix} \begin{pmatrix} x^{k+1} \\ y \end{pmatrix} = \begin{pmatrix} -q + \rho(z^k - u^k) \\ b \end{pmatrix}.
$$

This is an important example because it provides one of the best methods for solving quadratic problems, in particular, the SVM problems discussed in Chapter 18.

We programmed the above method in `Matlab` as the function `qsolve1`, see Appendix B, Section B.1. Here are two examples.

Example 16.11. Consider the quadratic program for which

$$
P_1 = \begin{pmatrix} 4 & 1 & 0 \\ 1 & 4 & 1 \\ 0 & 1 & 4 \end{pmatrix} \qquad q_1 = -\begin{pmatrix} 4 \\ 4 \\ 4 \end{pmatrix}
$$

$$
A_1 = \begin{pmatrix} 1 & 1 & -1 \\ 1 & -1 & -1 \end{pmatrix} \qquad b_1 = \begin{pmatrix} 0 \\ 0 \end{pmatrix}.
$$

We see immediately that the constraints

$$
\begin{aligned}
x + y - z &= 0 \\
x - y - z &= 0
\end{aligned}
$$

imply that $z = x$ and $y = 0$. Then it is easy using calculus to find that the unique minimum is given by $(x, y, z) = (1, 0, 1)$. Running `qsolve1` on P_1, q_1, A_1, b_1 with $\rho = 10$, $tolr = tols = 10^{-12}$ and $iternum = 10000$, we find that after 83 iterations the primal and the dual residuals are less than 10^{-12}, and we get

$$
(x, y, z) = (1.000000000000149, 0.000000000000000, 1.000000000000148).
$$

Example 16.12. Consider the quadratic program for which

$$P_2 = \begin{pmatrix} 4 & 1 & 0 & 0 \\ 1 & 4 & 1 & 0 \\ 0 & 1 & 4 & 1 \\ 0 & 0 & 1 & 4 \end{pmatrix} \qquad q_2 = -\begin{pmatrix} 4 \\ 4 \\ 4 \\ 4 \end{pmatrix}$$

$$A_2 = \begin{pmatrix} 1 & 1 & -1 & 0 \\ 1 & -1 & -1 & 0 \end{pmatrix} \qquad b_2 = \begin{pmatrix} 0 \\ 0 \end{pmatrix}.$$

Again, we see immediately that the constraints imply that $z = x$ and $y = 0$. Then it is easy using calculus to find that the unique minimum is given by $(x, y, z) = (28/31, 0, 28/31, 24/31)$. Running `qsolve1` on P_2, q_2, A_2, b_2 with $\rho = 10$, $tolr = tols = 10^{-12}$ and $iternum = 10000$, we find that after 95 iterations the primal and the dual residuals are less than 10^{-12}, and we get

$$(x, y, z, t) = (0.903225806451495, 0.000000000000000, \\ 0.903225806451495, 0.774193548387264),$$

which agrees with the answer found earlier up to 11 decimals.

As an illustration of the wide range of applications of ADMM we show in the next section how to solve the hard margin SVM (SVM_{h2}) discussed in Section 14.6.

16.7 Solving Hard Margin (SVM_{h2}) Using ADMM

Recall that we would like to solve the following optimization problem (see Section 14.6):

Hard margin SVM (SVM_{h2}):

$$\begin{aligned} &\text{minimize} \quad \frac{1}{2}\|w\|^2 \\ &\text{subject to} \\ &\qquad w^\top u_i - b \geq 1 \qquad i = 1, \ldots, p \\ &\qquad -w^\top v_j + b \geq 1 \qquad j = 1, \ldots, q. \end{aligned}$$

The margin is $\delta = 1/\|w\|$. The separating hyperplane $H_{w,b}$ is the hyperplane of equation $w^\top x - b = 0$, and the margin hyperplanes are the hyperplanes $H_{w,b+1}$ (the blue hyperplane) of equation $w^\top x - b - 1 = 0$ and $H_{w,b-1}$ (the red hyperplane) of equation $w^\top x - b + 1 = 0$. The dual program derived in Section 14.10 is the following program:

Dual of the Hard margin SVM (SVM$_{h2}$):

$$\begin{aligned} &\text{minimize} \quad \frac{1}{2}\begin{pmatrix}\lambda^\top & \mu^\top\end{pmatrix} X^\top X \begin{pmatrix}\lambda \\ \mu\end{pmatrix} - \begin{pmatrix}\lambda^\top & \mu^\top\end{pmatrix} \mathbf{1}_{p+q} \\ &\text{subject to} \\ &\qquad \sum_{i=1}^p \lambda_i - \sum_{j=1}^q \mu_j = 0 \\ &\qquad \lambda \geq 0,\ \mu \geq 0, \end{aligned}$$

where X is the $n \times (p+q)$ matrix given by

$$X = \begin{pmatrix} -u_1 & \cdots & -u_p & v_1 & \cdots & v_q \end{pmatrix}.$$

Then w is determined as follows:

$$w = -X \begin{pmatrix}\lambda \\ \mu\end{pmatrix} = \sum_{i=1}^p \lambda_i u_i - \sum_{j=1}^q \mu_j v_j.$$

To solve the dual using ADMM we need to determine the matrices P, q A and c as in Section 16.6(2). We renamed b as c to avoid a clash since b is already used. We have

$$P = X^\top X, \quad q = -\mathbf{1}_{p+q},$$

and since the only constraint is

$$\sum_{i=1}^p \lambda_i - \sum_{j=1}^q \mu_j = 0,$$

the matrix A is the $1 \times (p+q)$ row vector

$$A = \begin{pmatrix} \mathbf{1}_p^\top & -\mathbf{1}_q^\top \end{pmatrix},$$

and the right-hand side c is the scalar

$$c = 0.$$

Obviously the matrix A has rank 1. We obtain b using any i_0 such that $\lambda_{i_0} > 0$ and any j_0 such that $\mu_{j_0} > 0$. Since the corresponding constraints are active, we have

$$w^\top u_{i_0} - b = 1, \quad -w^\top v_{j_0} + b = 1,$$

so we obtain

$$b = w^\top (u_{i_0} + v_{j_0})/2.$$

For improved numerical stability, we can average over the sets of indices defined as $I_{\lambda>0} = \{i \in \{1,\ldots,p\} \mid \lambda_i > 0\}$ and $I_{\mu>0} = \{j \in \{1,\ldots,q\} \mid \mu_j > 0\}$. We obtain

$$b = w^\top \left(\left(\sum_{i \in I_{\lambda>0}} u_i \right) / |I_{\lambda>0}| + \left(\sum_{j \in I_{\mu>0}} v_j \right) / |I_{\mu>0}| \right) /2.$$

The Matlab programs implementing the above method are given in Appendix B, Section B.1. This should convince the reader that there is very little gap between theory and practice, although it is quite consuming to tune the tolerance parameters needed to deal with floating-point arithmetic.

Figure 16.5 shows the result of running the Matlab program implementing the above method using ADMM on two sets of points of 50 points each generated at random using the following Matlab code.

```
u14 = 10.1*randn(2,50)+18;
v14 = -10.1*randn(2,50)-18;
```

The function SVMhard2 is called with $\rho = 10$ as follows

```
[lamb,mu,w] = SVMhard2(10,u14,v14)
```

and produces the output shown in Figure 16.5. Observe that there is one blue support vector and two red support vectors.

16.8 Applications of ADMM to ℓ^1-Norm Problems

Another important application of ADMM is to ℓ^1-norm minimization problems, especially lasso minimization, discussed below and in Section 19.4. This involves the special case of ADMM where $f(x) = \tau \|x\|_1$ and $A = I$. In particular, in the one-dimensional case, we need to solve the minimization problem: find

$$x^* = \arg\min_x \left(\tau|x| + (\rho/2)(x-v)^2 \right),$$

with $x, v \in \mathbb{R}$, and $\rho, \tau > 0$. Let $c = \tau/\rho$ and write

$$f(x) = \frac{\tau}{2c}\left(2c|x| + (x-v)^2 \right).$$

Minimizing f over x is equivalent to minimizing

$$g(x) = 2c|x| + (x-v)^2 = 2c|x| + x^2 - 2xv + v^2,$$

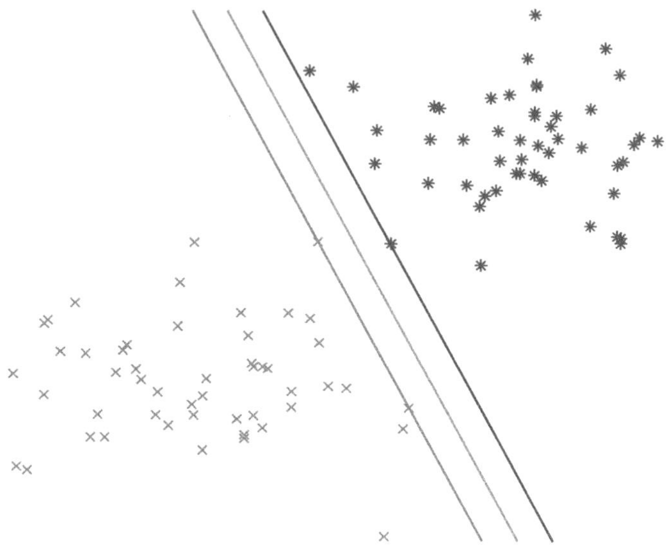

Fig. 16.5 An example of hard margin SVM.

which is equivalent to minimizing

$$h(x) = x^2 + 2(c|x| - xv)$$

over x. If $x \geq 0$, then

$$h(x) = x^2 + 2(cx - xv) = x^2 + 2(c - v)x = (x - (v - c))^2 - (v - c)^2.$$

If $v - c > 0$, that is, $v > c$, since $x \geq 0$, the function $x \mapsto (x - (v - c))^2$ has a minimum for $x = v - c > 0$, else if $v - c \leq 0$, then the function $x \mapsto (x - (v - c))^2$ has a minimum for $x = 0$.

If $x \leq 0$, then

$$h(x) = x^2 + 2(-cx - xv) = x^2 - 2(c + v)x = (x - (v + c))^2 - (v + c)^2.$$

if $v+c < 0$, that is, $v < -c$, since $x \leq 0$, the function $x \mapsto (x-(v+c))^2$ has a minimum for $x = v+c$, else if $v+c \geq 0$, then the function $x \mapsto (x-(v+c))^2$ has a minimum for $x = 0$.

In summary, $\inf_x h(x)$ is the function of v given by

$$S_c(v) = \begin{cases} v - c & \text{if } v > c \\ 0 & \text{if } |v| \leq c \\ v + c & \text{if } v < -c. \end{cases}$$

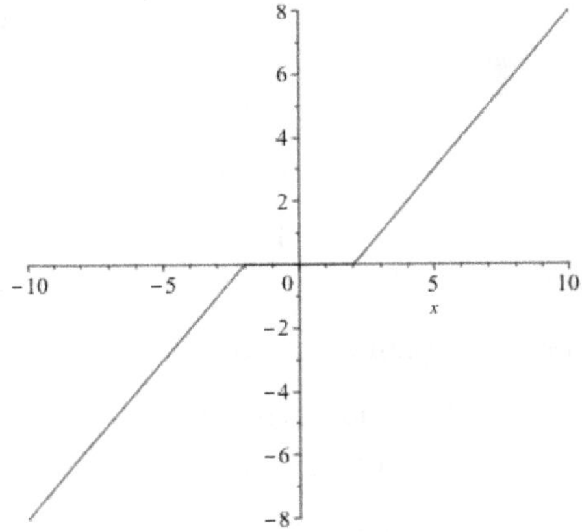

Fig. 16.6 The graph of S_c (when $c = 2$).

The function S_c is known as a *soft thresholding operator*. The graph of S_c shown in Figure 16.6.

One can check that

$$S_c(v) = (v - c)_+ - (-v - c)_+,$$

and also

$$S_c(v) = (1 - c/|v|)_+ v, \quad v \neq 0,$$

which shows that S_c is a *shrinkage operator* (it moves a point toward zero).

The operator S_c is extended to vectors in $\mathbb{R}^n$ component wise, that is, if $x = (x_1, \ldots, x_n)$, then

$$S_c(x) = (S_c(x_1), \ldots, S_c(x_n)).$$

We now consider several ℓ^1-norm problems.

(1) *Least absolute deviation.*
This is the problem of minimizing $\|Ax - b\|_1$, rather than $\|Ax - b\|_2$. Least absolute deviation is more robust than least squares fit because it deals better with outliers. The problem can be formulated in ADMM form as follows:

$$\begin{aligned} &\text{minimize} \quad \|z\|_1 \\ &\text{subject to} \quad Ax - z = b, \end{aligned}$$

with $f = 0$ and $g = \| \|_1$. As usual, we assume that A is an $m \times n$ matrix of rank n, so that $A^\top A$ is invertible. ADMM (in scaled form) can be expressed as

$$\begin{aligned} x^{k+1} &= (A^\top A)^{-1} A^\top (b + z^k - u^k) \\ z^{k+1} &= S_{1/\rho}(Ax^{k+1} - b + u^k) \\ u^{k+1} &= u^k + Ax^{k+1} - z^{k+1} - b. \end{aligned}$$

(2) *Basis pursuit.*
This is the following minimization problem:

$$\begin{aligned} &\text{minimize} \quad \|x\|_1 \\ &\text{subject to} \quad Ax = b, \end{aligned}$$

where A is an $m \times n$ matrix of rank $m < n$, and $b \in \mathbb{R}^m$, $x \in \mathbb{R}^n$. The problem is to find a sparse solution to an underdetermined linear system, which means a solution x with many zero coordinates. This problem plays a central role in compressed sensing and statistical signal processing.
Basis pursuit can be expressed in ADMM form as the problem

$$\begin{aligned} &\text{minimize} \quad I_C(x) + \|z\|_1 \\ &\text{subject to} \quad x - z = 0, \end{aligned}$$

with $C = \{x \in \mathbb{R}^n \mid Ax = b\}$. It is easy to see that the ADMM procedure (in scaled form) is

$$\begin{aligned} x^{k+1} &= \Pi_C(z^k - u^k) \\ z^{k+1} &= S_{1/\rho}(x^{k+1} + u^k) \\ u^{k+1} &= u^k + x^{k+1} - z^{k+1}, \end{aligned}$$

where Π_C is the orthogonal projection onto the subspace C. In fact, it is not hard to show that

$$x^{k+1} = (I - A^\top (AA^\top)^{-1} A)(z^k - u^k) + A^\top (AA^\top)^{-1} b.$$

In some sense, an ℓ^1-minimization problem is reduced to a sequence of ℓ^2-norm problems. There are ways of improving the efficiency of the method; see Boyd et al. [Boyd *et al.* (2010)] (Section 6.2)

(3) *General ℓ^1-regularized loss minimization.*
This is the following minimization problem:

$$\text{minimize} \quad l(x) + \tau \|x\|_1,$$

where l is any proper closed and convex loss function, and $\tau > 0$. We convert the problem to the ADMM problem:

$$\begin{aligned} &\text{minimize} \quad l(x) + \tau \|z\|_1 \\ &\text{subject to} \quad x - z = 0. \end{aligned}$$

The ADMM procedure (in scaled form) is

$$\begin{aligned} x^{k+1} &= \arg\min_x \left(l(x) + (\rho/2) \left\| x - z^k + u^k \right\|_2^2 \right) \\ z^{k+1} &= S_{\tau/\rho}(x^{k+1} + u^k) \\ u^{k+1} &= u^k + x^{k+1} - z^{k+1}. \end{aligned}$$

The x-update is a proximal operator evaluation. In general, one needs to apply a numerical procedure to compute x^{k+1}, for example, a version of Newton's method. The special case where $l(x) = (1/2) \|Ax - b\|_2^2$ is particularly important.

(4) *Lasso regularization.*

This is the following minimization problem:

$$\text{minimize} \quad (1/2) \|Ax - b\|_2^2 + \tau \|x\|_1 .$$

This is a linear regression with the regularizing term $\tau \|x\|_1$ instead of $\tau \|x\|_2$, to encourage a sparse solution. This method was first proposed by Tibshirani arround 1996, under the name *lasso*, which stands for "least absolute selection and shrinkage operator." This method is also known as *ℓ^1-regularized regression*, but this is not as cute as "lasso," which is used predominantly. This method is discussed extensively in Hastie, Tibshirani, and Wainwright [Hastie *et al.* (2015)].

The lasso minimization is converted to the following problem in ADMM form:

$$\begin{aligned} &\text{minimize} \quad \|Ax - b\|_2^2 + \tau \|z\|_1 \\ &\text{subject to} \quad x - z = 0. \end{aligned}$$

Then the ADMM procedure (in scaled form) is

$$\begin{aligned} x^{k+1} &= (A^\top A + \rho I)^{-1}(A^\top b + \rho(z^k - u^k)) \\ z^{k+1} &= S_{\tau/\rho}(x^{k+1} + u^k) \\ u^{k+1} &= u^k + x^{k+1} - z^{k+1}. \end{aligned}$$

Since $\rho > 0$, the matrix $A^\top A + \rho I$ is symmetric positive definite. Note that the x-update looks like a *ridge regression step* (see Section 19.1). There are various generalizations of lasso.

(5) *Generalized Lasso regularization.*
This is the following minimization problem:

$$\text{minimize} \quad (1/2)\,\|Ax - b\|_2^2 + \tau\,\|Fx\|_1\,,$$

where A is an $m \times n$ matrix, F is a $p \times n$ matrix, and either A has rank n or F has rank n. This problem is converted to the ADMM problem

$$\begin{aligned} &\text{minimize} \quad \|Ax - b\|_2^2 + \tau\,\|z\|_1 \\ &\text{subject to} \quad Fx - z = 0, \end{aligned}$$

and the corresponding ADMM procedure (in scaled form) is

$$\begin{aligned} x^{k+1} &= (A^\top A + \rho F^\top F)^{-1}(A^\top b + \rho F^\top (z^k - u^k)) \\ z^{k+1} &= S_{\tau/\rho}(Fx^{k+1} + u^k) \\ u^{k+1} &= u^k + Fx^{k+1} - z^{k+1}. \end{aligned}$$

(6) *Group Lasso.*
This a generalization of (3). Here we assume that x is split as $x = (x_1, \ldots, x_N)$, with $x_i \in \mathbb{R}^{n_i}$ and $n_1 + \cdots + x_N = n$, and the regularizing term $\|x\|_1$ is replaced by $\sum_{i=1}^N \|x_i\|_2$. When $n_i = 1$, this reduces to (3). The z-update of the ADMM procedure needs to modified. We define the soft thresholding operator $\mathcal{S}_c\colon \mathbb{R}^m \to \mathbb{R}^m$ given by

$$\mathcal{S}_c(v) = \left(1 - \frac{c}{\|v\|_2}\right)_+ v,$$

with $\mathcal{S}_c(0) = 0$. Then the z-update consists of the N updates

$$z_i^{k+1} = \mathcal{S}_{\tau/\rho}(x_i^{k+1} + u^k), \quad i = 1, \ldots, N.$$

The method can be extended to deal with overlapping groups; see Boyd et al. [Boyd *et al.* (2010)] (Section 6.4).

There are many more applications of ADMM discussed in Boyd et al. [Boyd *et al.* (2010)], including consensus and sharing. See also Strang [Strang (2019)] for a brief overview.

16.9 Summary

The main concepts and results of this chapter are listed below:

- Dual ascent.
- Augmented Lagrangian.
- Penalty parameter.

- Method of multipliers.
- ADMM (alternating direction method of multipliers).
- x-update, z-update, λ-update.
- Scaled form of ADMM.
- Residual, dual residual.
- Stopping criteria.
- Proximity operator, proximal minimization.
- Quadratic programming.
- KKT equations.
- Soft thresholding operator.
- Shrinkage operator.
- Least absolute deviation.
- Basis pursuit.
- General ℓ^1-regularized loss minimization.
- Lasso regularization.
- Generalized lasso regularization.
- Group lasso.

16.10 Problems

Problem 16.1. In the method of multipliers described in Section 16.2, prove that choosing $\alpha^k = \rho$ guarantees that (u^{k+1}, λ^{k+1}) satisfies the equation

$$\nabla J_{u^{k+1}} + A^\top \lambda^{k+1} = 0.$$

Problem 16.2. Prove that the Inequality (A1) follows from the Inequalities (A2) and (A3) (see the proof of Theorem 16.1). For help consult Appendix A of Boyd et al. [Boyd *et al.* (2010)].

Problem 16.3. Consider Example 16.8. Prove that if $f = I_C$, the indicator function of a nonempty closed convex set C, then

$$x^+ = \arg\min_x \left(I_C(x) + (\rho/2) \left\|x - v\right\|_2^2 \right) = \Pi_C(v),$$

the orthogonal projection of v onto C. In the special case where $C = \mathbb{R}^n_+$ (the first orthant), then

$$x^+ = (v)_+,$$

the vector obtained by setting the negative components of v to zero.

Problem 16.4. Prove that the soft thresholding operator S_c from Section 16.8 satisfies the equations

$$S_c(v) = (v - c)_+ - (-v - c)_+,$$

and

$$S_c(v) = (1 - c/|v|)_+ v, \quad v \neq 0.$$

Problem 16.5. Rederive the formula

$$S_c(v) = \begin{cases} v - c & \text{if } v > c \\ 0 & \text{if } |v| \leq c \\ v + c & \text{if } v < -c \end{cases}$$

using subgradients.

Problem 16.6. In basis pursuit (see Section 16.8 (2)) prove that

$$x^{k+1} = (I - A^\top (AA^\top)^{-1} A)(z^k - u^k) + A^\top (AA^\top)^{-1} b.$$

Problem 16.7. Implement (in `Matlab`) ADMM applied to lasso regularization as described in Section 16.6 (4). The stopping criterion should be based on feasibility tolerances ϵ^{pri} and ϵ^{dual}, say 10^{-4}, and on a maximum number of iteration steps, say 10000. There is a build in `Matlab` function **`wthresh`** implementing soft thresholding. You may use the `Matlab` command **`randn`** to create a random data set X and a random response vector y (see the help menu in `Matlab` under **`lasso`**). Try various values of ρ and τ. You will observe that the choice of ρ greatly affects the rate of convergence of the procedure.

PART 4

Applications to Machine Learning

Chapter 17

Positive Definite Kernels

This chapter is an introduction to positive definite kernels and the use of kernel functions in machine learning. Let X be a nonempty set. If the set X represents a set of highly nonlinear data, it may be advantageous to map X into a space F of much higher dimension called the *feature space*, using a function $\varphi\colon X \to F$ called a *feature map*. This idea is that φ "unwinds" the description of the objects in F in an attempt to make it linear. The space F is usually a vector space equipped with an inner product $\langle -, - \rangle$. If F is infinite dimensional, then we assume that it is a Hilbert space.

Many algorithms that analyze or classify data make use of the inner products $\langle \varphi(x), \varphi(y) \rangle$, where $x, y \in X$. These algorithms make use of the function $\kappa\colon X \times X \to \mathbb{C}$ given by

$$\kappa(x, y) = \langle \varphi(x), \varphi(y) \rangle, \qquad x, y \in X,$$

called a *kernel function*.

The kernel trick is to pretend that we have a feature embedding $\varphi\colon X \to F$ (actually unknown), but to only use inner products $\langle \varphi(x), \varphi(y) \rangle$ that can be evaluated using the original data through the known kernel function κ. It turns out that the functions of the form κ as above can be defined in terms of a condition which is reminiscent of positive semidefinite matrices (see Definition 17.2). Furthermore, every function satisfying Definition 17.2 arises from a suitable feature map into a Hilbert space; see Theorem 17.1.

We illustrate the kernel methods on kernel PCA (see Section 17.4).

17.1 Feature Maps and Kernel Functions

Definition 17.1. Let X be a nonempty set, let H be a (complex) Hilbert space, and let $\varphi\colon X \to H$ be a function called a *feature map*. The function

$\kappa\colon X \times X \to \mathbb{C}$ given by

$$\kappa(x, y) = \langle \varphi(x), \varphi(y) \rangle, \qquad x, y \in X,$$

is called a *kernel function*.

Remark: A *feature map* is often called a *feature embedding*, but this terminology is a bit misleading because it suggests that such a map is injective, which is not necessarily the case. Unfortunately this terminology is used by most people.

Example 17.1. Suppose we have two feature maps $\varphi_1\colon X \to \mathbb{R}^{n_1}$ and $\varphi_2\colon X \to \mathbb{R}^{n_2}$, and let $\kappa_1(x, y) = \langle \varphi_1(x), \varphi_1(y) \rangle$ and $\kappa_2(x, y) = \langle \varphi_2(x), \varphi_2(y) \rangle$ be the corresponding kernel functions (where $\langle -, - \rangle$ is the standard inner product on $\mathbb{R}^n$). Define the feature map $\varphi\colon X \to \mathbb{R}^{n_1+n_2}$ by

$$\varphi(x) = (\varphi_1(x), \varphi_2(x)),$$

an $(n_1 + n_2)$-tuple. We have

$$\begin{aligned}
\langle \varphi(x), \varphi(y) \rangle &= \langle (\varphi_1(x), \varphi_2(x)), (\varphi_1(y), \varphi_2(y)) \rangle \\
&= \langle \varphi_1(x), \varphi_1(y) \rangle + \langle \varphi_2(x), \varphi_2(y) \rangle \\
&= \kappa_1(x, y) + \kappa_2(x, y),
\end{aligned}$$

which shows that the map κ given by

$$\kappa(x, y) = \kappa_1(x, y) + \kappa_2(x, y)$$

is the kernel function corresponding to the feature map $\varphi\colon X \to \mathbb{R}^{n_1+n_2}$.

Example 17.2. Let X be a subset of $\mathbb{R}^2$, and let $\varphi_1\colon X \to \mathbb{R}^3$ be the map given by

$$\varphi_1(x_1, x_2) = (x_1^2, x_2^2, \sqrt{2}x_1x_2).$$

Figure 17.1 illustrates $\varphi_1\colon X \to \mathbb{R}^3$ when $X = \{((x_1, x_2) \mid -10 \leq x_1 \leq 10, -10 \leq x_2 \leq 10\}$.

Observe that linear relations in the feature space $H = \mathbb{R}^3$ correspond to quadratic relations in the input space (of data). We have

$$\begin{aligned}
\langle \varphi_1(x), \varphi_1(y) \rangle &= \langle (x_1^2, x_2^2, \sqrt{2}x_1x_2), (y_1^2, y_2^2, \sqrt{2}y_1y_2) \rangle \\
&= x_1^2y_1^2 + x_2^2y_2^2 + 2x_1x_2y_1y_2 \\
&= (x_1y_1 + x_2y_2)^2 = \langle x, y \rangle^2,
\end{aligned}$$

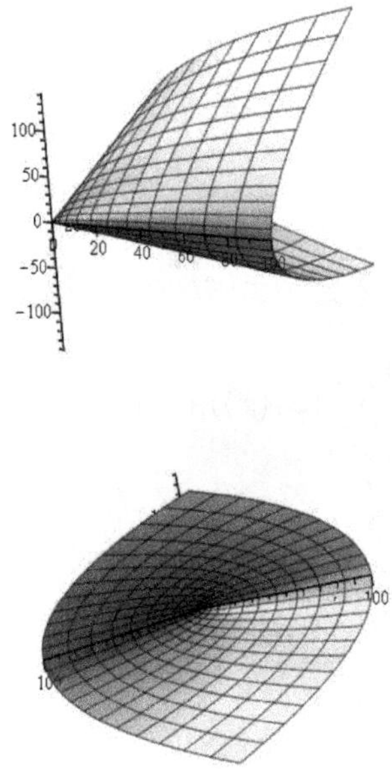

Fig. 17.1 The parametric surface $\varphi_1(x_1, x_2) = (x_1^2, x_2^2, \sqrt{2}x_1x_2)$ where $-10 \leq x_1 \leq 10$ and $-10 \leq x_2 \leq 10$.

where $\langle x, y\rangle$ is the usual inner product on $\mathbb{R}^2$. Hence the function

$$\kappa(x, y) = \langle x, y\rangle^2$$

is a kernel function associated with the feature space $\mathbb{R}^3$.

If we now consider the map $\varphi_2 \colon X \to \mathbb{R}^4$ given by

$$\varphi_2(x_1, x_2) = (x_1^2, x_2^2, x_1x_2, x_1x_2),$$

we check immediately that

$$\langle \varphi_2(x), \varphi_2(y)\rangle = \kappa(x, y) = \langle x, y\rangle^2,$$

which shows that the same kernel can arise from different maps into different feature spaces.

Example 17.3. Example 17.2 can be generalized as follows. Suppose we have a feature map $\varphi_1 \colon X \to \mathbb{R}^n$ and let $\kappa_1(x, y) = \langle \varphi_1(x), \varphi_1(y)\rangle$ be the corresponding kernel function (where $\langle -, -\rangle$ is the standard inner product on $\mathbb{R}^n$). Define the feature map $\varphi \colon X \to \mathbb{R}^n \times \mathbb{R}^n$ by its n^2 components

$$\varphi(x)_{(i,j)} = (\varphi_1(x))_i(\varphi_1(x))_j, \qquad 1 \leq i, j \leq n,$$

with the inner product on $\mathbb{R}^n \times \mathbb{R}^n$ given by

$$\langle u, v\rangle = \sum_{i,j=1}^{n} u_{(i,j)} v_{(i,j)}.$$

Then we have

$$\begin{aligned}\langle \varphi(x), \varphi(y)\rangle &= \sum_{i,j=1}^{n} \varphi_{(i,j)}(x)\varphi_{(i,j)}(y) \\ &= \sum_{i,j=1}^{n} (\varphi_1(x))_i(\varphi_1(x))_j(\varphi_1(y))_i(\varphi_1(y))_j \\ &= \sum_{i=1}^{n}(\varphi_1(x))_i(\varphi_1(y))_i \sum_{j=1}^{n}(\varphi_1(x))_j(\varphi_1(y))_j \\ &= (\kappa_1(x,y))^2.\end{aligned}$$

Thus the map κ given by $\kappa(x, y) = (\kappa_1(x, y))^2$ is a kernel map associated with the feature map $\varphi\colon X \to \mathbb{R}^n \times \mathbb{R}^n$. The feature map φ is a direct generalization of the feature map φ_2 of Example 17.2.

The above argument is immediately adapted to show that if $\varphi_1\colon X \to \mathbb{R}^{n_1}$ and $\varphi_2\colon X \to \mathbb{R}^{n_2}$ are two feature maps and if $\kappa_1(x, y) = \langle \varphi_1(x), \varphi_1(y)\rangle$ and $\kappa_2(x, y) = \langle \varphi_2(x), \varphi_2(y)\rangle$ are the corresponding kernel functions, then the map defined by

$$\kappa(x, y) = \kappa_1(x, y)\kappa_2(x, y)$$

is a kernel function for the feature space $\mathbb{R}^{n_1} \times \mathbb{R}^{n_2}$ and the feature map

$$\varphi(x)_{(i,j)} = (\varphi_1(x))_i(\varphi_2(x))_j, \qquad 1 \leq i \leq n_1,\ 1 \leq j \leq n_2.$$

Example 17.4. Note that the feature map $\varphi\colon X \to \mathbb{R}^n \times \mathbb{R}^n$ is not very economical because if $i \neq j$ then the components $\varphi_{(i,j)}(x)$ and $\varphi_{(j,i)}(x)$ are both equal to $(\varphi_1(x))_i(\varphi_1(x))_j$. Therefore we can define the more economical embedding $\varphi'\colon X \to \mathbb{R}^{\binom{n+1}{2}}$ given by

$$\varphi'(x)_{(i,j)} = \begin{cases} (\varphi_1(x))_i^2 & i = j, \\ \sqrt{2}(\varphi_1(x))_i(\varphi_1(x))_j & i < j, \end{cases}$$

where the pairs (i, j) with $1 \leq i \leq j \leq n$ are ordered lexicographically. The feature map φ is a direct generalization of the feature map φ_1 of Example 17.2.

Observe that φ' can also be defined in the following way which makes it easier to come up with the generalization to any power:

$$\varphi'_{(i_1,\ldots,i_n)}(x) = \binom{2}{i_1 \cdots i_n}^{1/2} (\varphi_1(x))_1^{i_1} (\varphi_1(x))_1^{i_2} \cdots (\varphi_1(x))_1^{i_n},$$
$$i_1 + i_2 + \cdots + i_n = 2,\ i_j \in \mathbb{N},$$

where the n-tuples $(i_1, \ldots, i_n)$ are ordered lexicographically. Recall that for any $m \geq 1$ and any $(i_1, \ldots, i_n) \in \mathbb{N}^m$ such that $i_1 + i_2 + \cdots + i_n = m$, we have

$$\binom{m}{i_1 \cdots i_n} = \frac{m!}{i_1! \cdots i_n!}.$$

More generally, for any $m \geq 2$, using the multinomial theorem, we can define a feature embedding $\varphi\colon X \to \mathbb{R}^{\binom{n+m-1}{m}}$ defining the kernel function κ given by $\kappa(x, y) = (\kappa_1(x, y))^m$, with φ given by

$$\varphi_{(i_1,\ldots,i_n)}(x) = \binom{m}{i_1 \cdots i_n}^{1/2} (\varphi_1(x))_1^{i_1} (\varphi_1(x))_1^{i_2} \cdots (\varphi_1(x))_1^{i_n},$$
$$i_1 + i_2 + \cdots + i_n = m,\ i_j \in \mathbb{N},$$

where the n-tuples $(i_1, \ldots, i_n)$ are ordered lexicographically.

Example 17.5. For any positive real constant $R > 0$, the constant function $\kappa(x, y) = R$ is a kernel function corresponding to the feature map $\varphi\colon X \to \mathbb{R}$ given by $\varphi(x, y) = \sqrt{R}$.

By definition, the function $\kappa'_1\colon \mathbb{R}^n \to \mathbb{R}$ given by $\kappa'_1(x, y) = \langle x, y \rangle$ is a kernel function (the feature map is the identity map from $\mathbb{R}^n$ to itself). We just saw that for any positive real constant $R > 0$, the constant $\kappa'_2(x, y) = R$ is a kernel function. By Example 17.1, the function $\kappa'_3(x, y) = \kappa'_1(x, y) + \kappa'_2(x, y)$ is a kernel function, and for any integer $d \geq 1$, by Example 17.4, the function κ_d given by

$$\kappa_d(x, y) = (\kappa'_3(x, y))^d = (\langle x, y \rangle + R)^d,$$

is a kernel function on $\mathbb{R}^n$. By the binomial formula,

$$\kappa_d(x, y) = \sum_{m=0}^{d} R^{d-m} \langle x, y \rangle^m.$$

By Example 17.1, the feature map of this kernel function is the concatenation of the features of the $d + 1$ kernel maps $R^{d-m}\langle x, y \rangle^m$. By Example 17.3, the components of the feature map of the kernel map $R^{d-m}\langle x, y \rangle^m$ are reweightings of the functions

$$\varphi_{(i_1,\ldots,i_n)}(x) = x_1^{i_1} x_2^{i_2} \cdots x_n^{i_n}, \quad i_1 + i_2 + \cdots + i_n = m,$$

with $(i_1, \dots, i_n) \in \mathbb{N}^n$. Thus the components of the feature map of the kernel function κ_d are reweightings of the functions

$$\varphi_{(i_1,\dots,i_n)}(x) = x_1^{i_1} x_2^{i_2} \cdots x_n^{i_n}, \quad i_1 + i_2 + \cdots + i_n \leq d,$$

with $(i_1, \dots, i_n) \in \mathbb{N}^n$. It is easy to see that the dimension of this feature space is $\binom{m+d}{d}$.

There are a number of variations of the polynomial kernel κ_d; all-subsets embedding kernels, ANOVA kernels; see Shawe-Taylor and Christianini [Shawe-Taylor and Cristianini (2004)], Chapter III.

In the next example the set X is not a vector space.

Example 17.6. Let D be a finite set and let $X = 2^D$ be its power set. If $|D| = n$, let $H = \mathbb{R}^X \cong \mathbb{R}^{2^n}$. We are assuming that the subsets of D are enumerated in some fashion so that each coordinate of $\mathbb{R}^{2^n}$ corresponds to one of these subsets. For example, if $D = \{1, 2, 3, 4\}$, let

$$\begin{array}{llll} U_1 = \emptyset & U_2 = \{1\} & U_3 = \{2\} & U_4 = \{3\} \\ U_5 = \{4\} & U_6 = \{1,2\} & U_7 = \{1,3\} & U_8 = \{1,4\} \\ U_9 = \{2,3\} & U_{10} = \{2,4\} & U_{11} = \{3,4\} & U_{12} = \{1,2,3\} \\ U_{13} = \{1,2,4\} & U_{14} = \{1,3,4\} & U_{15} = \{2,3,4\} & U_{16} = \{1,2,3,4\}. \end{array}$$

Let $\varphi \colon X \to H$ be the feature map defined as follows: for any subsets $A, U \in X$,

$$\varphi(A)_U = \begin{cases} 1 & \text{if } U \subseteq A \\ 0 & \text{otherwise.} \end{cases}$$

For example, if $A_1 = \{1, 2, 3\}$, we obtain the vector

$$\varphi(\{1, 2, 3\}) = (1, 1, 1, 1, 0, 1, 1, 0, 1, 0, 0, 1, 0, 0, 0, 0),$$

and if $A_2 = \{2, 3, 4\}$, we obtain the vector

$$\varphi(\{2, 3, 4\}) = (1, 0, 1, 1, 1, 0, 0, 0, 1, 1, 1, 0, 0, 0, 1, 0).$$

For any two subsets A_1 and A_2 of D, it is easy to check that

$$\langle \varphi(A_1), \varphi(A_2) \rangle = 2^{|A_1 \cap A_2|},$$

the number of common subsets of A_1 and A_2. For example, $A_1 \cap A_2 = \{2, 3\}$, and

$$\langle \varphi(A_1), \varphi(A_2) \rangle = 4.$$

Therefore, the function $\kappa \colon X \times X \to \mathbb{R}$ given by

$$\kappa(A_1, A_2) = 2^{|A_1 \cap A_2|}, \qquad A_1, A_2 \subseteq D$$

is a kernel function.

Kernels on collections of sets can be defined in terms of measures.

Example 17.7. Let $(D, \mathcal{A})$ be a measurable space, where D is a nonempty set and $\mathcal{A}$ is a σ-algebra on D (the measurable sets). Let X be a subset of $\mathcal{A}$. If μ is a positive measure on $(D, \mathcal{A})$ and if μ is finite, which means that $\mu(D)$ is finite, then we can define the map $\kappa_1 \colon X \times X \to \mathbb{R}$ given by

$$\kappa_1(A_1, A_2) = \mu(A_1 \cap A_2), \qquad A_1, A_2 \in X.$$

We can show that κ is a kernel function as follows. Let $H = \mathrm{L}^2_\mu(D, \mathcal{A}, \mathbb{R})$ be the Hilbert space of μ-square-integrable functions with the inner product

$$\langle f, g \rangle = \int_D f(s) g(s)\, d\mu(s),$$

and let $\varphi \colon X \to H$ be the feature embedding given by

$$\varphi(A) = \chi_A, \qquad A \in X,$$

the characteristic function of A. Then we have

$$\begin{aligned}
\kappa_1(A_1, A_2) = \mu(A_1 \cap A_2) &= \int_D \chi_{A_1 \cap A_2}(s)\, d\mu(s) \\
&= \int_D \chi_{A_1}(s) \chi_{A_2}(s)\, d\mu(s) = \langle \chi_{A_1}, \chi_{A_2} \rangle \\
&= \langle \varphi(A_1), \varphi(A_2) \rangle.
\end{aligned}$$

The above kernel is called the *intersection kernel*. If we assume that μ is normalized so that $\mu(D) = 1$, then we also have the *union complement kernel*:

$$\kappa_2(A_1, A_2) = \mu(\overline{A_1} \cap \overline{A_2}) = 1 - \mu(A_1 \cup A_2).$$

The sum κ_3 of the kernels κ_1 and κ_2 is the *agreement kernel*:

$$\kappa_s(A_1, A_2) = 1 - \mu(A_1 - A_2) - \mu(A_2 - A_1).$$

Many other kinds of kernels can be designed, in particular, graph kernels. For comprehensive presentations of kernels, see Schölkopf and Smola [Schölkopf and Smola (2002)] and Shawe-Taylor and Christianini [Shawe-Taylor and Cristianini (2004)].

Kernel functions have the following important property.

Proposition 17.1. *Let X be any nonempty set, let H be any (complex) Hilbert space, let $\varphi\colon X \to H$ be any function, and let $\kappa\colon X \times X \to \mathbb{C}$ be the kernel given by*

$$\kappa(x, y) = \langle \varphi(x), \varphi(y) \rangle, \qquad x, y \in X.$$

For any finite subset $S = \{x_1, \ldots, x_p\}$ of X, if K_S is the $p \times p$ matrix

$$K_S = (\kappa(x_j, x_i))_{1 \leq i,j \leq p} = (\langle \varphi(x_j), \varphi(x_i) \rangle)_{1 \leq i,j \leq p},$$

then we have

$$u^* K_S\, u \geq 0, \qquad \textit{for all } u \in \mathbb{C}^p.$$

Proof. We have

$$\begin{aligned} u^* K_S\, u = u^\top K_S^\top\, \overline{u} &= \sum_{i,j=1}^p \kappa(x_i, x_j) u_i \overline{u_j} \\ &= \sum_{i,j=1}^p \langle \varphi(x), \varphi(y) \rangle u_i \overline{u_j} \\ &= \left\langle \sum_{i=1}^p u_i \varphi(x_i), \sum_{j=1}^p u_j \varphi(x_j) \right\rangle = \left\| \sum_{i=1}^p u_i \varphi(x_i) \right\|^2 \geq 0, \end{aligned}$$

as claimed. □

17.2 Basic Properties of Positive Definite Kernels

Proposition 17.1 suggests a second approach to kernel functions which does not assume that a feature space and a feature map are provided. We will see in Section 17.3 that the two approaches are equivalent. The second approach is useful in practice because it is often difficult to define a feature space and a feature map in a simple manner.

Definition 17.2. Let X be a nonempty set. A function $\kappa\colon X \times X \to \mathbb{C}$ is a *positive definite kernel* if for every finite subset $S = \{x_1, \ldots, x_p\}$ of X, if K_S is the $p \times p$ matrix

$$K_S = (\kappa(x_j, x_i))_{1 \leq i,j \leq p}$$

called a *Gram matrix*, then we have

$$u^* K_S\, u = \sum_{i,j=1}^p \kappa(x_i, x_j) u_i \overline{u_j} \geq 0, \qquad \text{for all } u \in \mathbb{C}^p.$$

Observe that Definition 17.2 does not require that $u^* K_S u > 0$ if $u \neq 0$, so the terminology *positive definite* is a bit abusive, and it would be more appropriate to use the terminology *positive semidefinite*. However, it seems customary to use the term *positive definite kernel*, or even *positive kernel*.

Proposition 17.2. *Let $\kappa\colon X \times X \to \mathbb{C}$ be a positive definite kernel. Then $\kappa(x,x) \geq 0$ for all $x \in X$, and for any finite subset $S = \{x_1, \dots, x_p\}$ of X, the $p \times p$ matrix K_S given by*

$$K_S = (\kappa(x_j, x_i))_{1 \leq i,j \leq p}$$

is Hermitian, that is, $K_S^* = K_S$.

Proof. The first property is obvious by choosing $S = \{x\}$. To prove that K_S is Hermitian, observe that we have

$$(u+v)^* K_S (u+v) = u^* K_S u + u^* K_S v + v^* K_S u + v^* K_S v,$$

and since $(u+v)^* K_S (u+v), u^* K_S u, v^* K_S v \geq 0$, we deduce that

$$2A = u^* K_S v + v^* K_S u \tag{1}$$

must be real. By replacing u by iu, we see that

$$2B = -iu^* K_S v + iv^* K_S u \tag{2}$$

must also be real. By multiplying Equation (2) by i and adding it to Equation (1) we get

$$u^* K_S v = A + iB. \tag{3}$$

By subtracting Equation (3) from Equation (1) we get

$$v^* K_S u = A - iB.$$

Then

$$u^* K_S^* v = \overline{v^* K_S u} = \overline{A - iB} = A + iB = u^* K_S v,$$

for all $u, v \in \mathbb{C}^*$, which implies $K_S^* = K_S$. ☐

If the map $\kappa\colon X \times X \to \mathbb{R}$ is real-valued, then we have the following criterion for κ to be a positive definite kernel that only involves real vectors.

Proposition 17.3. *If $\kappa\colon X \times X \to \mathbb{R}$, then κ is a positive definite kernel iff for any finite subset $S = \{x_1, \dots, x_p\}$ of X, the $p \times p$ real matrix K_S given by*

$$K_S = (\kappa(x_k, x_j))_{1 \leq j,k \leq p}$$

is symmetric, that is, $K_S^\top = K_S$, and

$$u^\top K_S u = \sum_{j,k=1}^{p} \kappa(x_j, x_k) u_j u_k \geq 0, \qquad \text{for all } u \in \mathbb{R}^p.$$

Proof. If κ is a real-valued positive definite kernel, then the proposition is a trivial consequence of Proposition 17.2.

For the converse assume that κ is symmetric and that it satisfies the second condition of the proposition. We need to show that κ is a positive definite kernel with respect to complex vectors. If we write $u_k = a_k + ib_k$, then

$$\begin{aligned} u^* K_S\, u &= \sum_{j,k=1}^{p} \kappa(x_j, x_k)(a_j + ib_j)(a_k - ib_k) \\ &= \sum_{j,k=1}^{p} (a_j a_k + b_j b_k)\kappa(x_j, x_k) + i \sum_{j,k=1}^{p} (b_j a_k - a_j b_k)\kappa(x_j, x_k) \\ &= \sum_{j,k=1}^{p} (a_j a_k + b_j b_k)\kappa(x_j, x_k) \\ &\quad + i \sum_{1 \le j < k \le p} b_j a_k(\kappa(x_j, x_k) - \kappa(x_k, x_j)). \end{aligned}$$

Thus $u^* K_S u$ is real iff K_S is symmetric. ☐

Consequently we make the following definition.

Definition 17.3. Let X be a nonempty set. A function $\kappa\colon X \times X \to \mathbb{R}$ is a *(real) positive definite kernel* if $\kappa(x, y) = \kappa(y, x)$ for all $x, y \in X$, and for every finite subset $S = \{x_1, \dots, x_p\}$ of X, if K_S is the $p \times p$ real symmetric matrix

$$K_S = (\kappa(x_i, x_j))_{1 \le i,j \le p},$$

then we have

$$u^\top K_S\, u = \sum_{i,j=1}^{p} \kappa(x_i, x_j) u_i u_j \ge 0, \qquad \text{for all } u \in \mathbb{R}^p.$$

Among other things, the next proposition shows that a positive definite kernel satisfies the Cauchy–Schwarz inequality.

Proposition 17.4. *A Hermitian* 2×2 *matrix*

$$A = \begin{pmatrix} a & \overline{b} \\ b & d \end{pmatrix}$$

is positive semidefinite if and only if $a \ge 0$, $d \ge 0$, *and* $ad - |b|^2 \ge 0$.

Let $\kappa\colon X \times X \to \mathbb{C}$ *be a positive definite kernel. For all* $x, y \in X$, *we have*

$$|\kappa(x, y)|^2 \le \kappa(x, x)\kappa(y, y).$$

Proof. For all $x, y \in \mathbb{C}$, we have

$$\begin{aligned}(\overline{x}\ \overline{y}) \begin{pmatrix} a & \overline{b} \\ b & d \end{pmatrix} \begin{pmatrix} x \\ y \end{pmatrix} &= (\overline{x}\ \overline{y}) \begin{pmatrix} ax + \overline{b}y \\ bx + dy \end{pmatrix} \\ &= a|x|^2 + bx\overline{y} + \overline{bx\overline{y}} + d|y|^2.\end{aligned}$$

If A is positive semidefinite, then we already know that $a \geq 0$ and $d \geq 0$. If $a = 0$, then we must have $b = 0$, since otherwise we can make $bx\overline{y} + \overline{bx\overline{y}}$, which is twice the real part of $bx\overline{y}$, as negative as we want. In this case, $ad - |b|^2 = 0$.

If $a > 0$, then

$$a|x|^2 + bx\overline{y} + \overline{bx\overline{y}} + d|y|^2 = a\left|x + \frac{\overline{b}}{a}y\right|^2 + \frac{|y|^2}{a}(ad - |b|^2).$$

If $ad - |b|^2 < 0$, we can pick $y \neq 0$ and $x = -(\overline{b}y)/a$, so that the above expression is negative. Therefore, $ad - |b|^2 \geq 0$. The converse is trivial.

If $x = y$, the inequality $|\kappa(x,y)|^2 \leq \kappa(x,x)\kappa(y,y)$ is trivial. If $x \neq y$, the inequality follows by applying the criterion for being positive semidefinite to the matrix

$$\begin{pmatrix} \kappa(x,x) & \overline{\kappa(x,y)} \\ \kappa(x,y) & \kappa(y,y) \end{pmatrix},$$

as claimed. ☐

The following property due to I. Schur (1911) shows that the pointwise product of two positive definite kernels is also a positive definite kernel.

Proposition 17.5. *(I. Schur) If $\kappa_1 \colon X \times X \to \mathbb{C}$ and $\kappa_2 \colon X \times X \to \mathbb{C}$ are two positive definite kernels, then the function $\kappa \colon X \times X \to \mathbb{C}$ given by $\kappa(x,y) = \kappa_1(x,y)\kappa_2(x,y)$ for all $x, y \in X$ is also a positive definite kernel.*

Proof. It suffices to prove that if $A = (a_{jk})$ and $B = (b_{jk})$ are two Hermitian positive semidefinite $p \times p$ matrices, then so is their pointwise product $C = A \circ B = (a_{jk}b_{jk})$ (also known as Hadamard or Schur product). Recall that a Hermitian positive semidefinite matrix A can be diagonalized as $A = U\Lambda U^*$, where Λ is a diagonal matrix with nonnegative entries and U is a unitary matrix. Let $\Lambda^{1/2}$ be the diagonal matrix consisting of the positive square roots of the diagonal entries in Λ. Then we have

$$A = U\Lambda U^* = U\Lambda^{1/2}\Lambda^{1/2}U^* = U\Lambda^{1/2}(U\Lambda^{1/2})^*.$$

Thus if we set $R = U\Lambda^{1/2}$, we have

$$A = RR^*,$$

which means that

$$a_{jk} = \sum_{h=1}^{p} r_{jh}\overline{r_{kh}}.$$

Then for any $u \in \mathbb{C}^p$, we have

$$\begin{aligned} u^*(A \circ B)u &= \sum_{j,k=1}^{p} a_{jk} b_{jk} u_j \overline{u_k} \\ &= \sum_{j,k=1}^{p} \sum_{h=1}^{p} r_{jh}\overline{r_{kh}} b_{jk} u_j \overline{u_k} \\ &= \sum_{h=1}^{p} \sum_{j,k=1}^{p} b_{jk} u_j r_{jh} \overline{u_k r_{kh}}. \end{aligned}$$

Since B is positive semidefinite, for each fixed h, we have

$$\sum_{j,k=1}^{p} b_{jk} u_j r_{jh} \overline{u_k r_{kh}} = \sum_{j,k=1}^{p} b_{jk} z_j \overline{z_k} \geq 0,$$

as we see by letting $z = (u_1 r_{1h}, \ldots, u_p r_{ph})$, □

In contrast, the ordinary product AB of two symmetric positive semidefinite matrices A and B may not be symmetric positive semidefinite; see Section 7.9 in Volume I for an example.

Here are other ways of obtaining new positive definite kernels from old ones.

Proposition 17.6. *Let $\kappa_1 \colon X \times X \to \mathbb{C}$ and $\kappa_2 \colon X \times X \to \mathbb{C}$ be two positive definite kernels, $f \colon X \to \mathbb{C}$ be a function, $\psi \colon X \to \mathbb{R}^N$ be a function, $\kappa_3 \colon \mathbb{R}^N \times \mathbb{R}^N \to \mathbb{C}$ be a positive definite kernel, and $a \in \mathbb{R}$ be any positive real number. Then the following functions are positive definite kernels:*

(1) $\kappa(x, y) = \kappa_1(x, y) + \kappa_2(x, y)$.
(2) $\kappa(x, y) = a\kappa_1(x, y)$.
(3) $\kappa(x, y) = f(x)\overline{f(y)}$.
(4) $\kappa(x, y) = \kappa_3(\psi(x), \psi(y))$.
(5) If B is a symmetric positive semidefinite $n \times n$ matrix, then the map $\kappa \colon \mathbb{R}^n \times \mathbb{R}^n \to \mathbb{R}$ given by

$$\kappa(x, y) = x^\top B y$$

is a positive definite kernel.

Proof. (1) For every finite subset $S = \{x_1, \ldots, x_p\}$ of X, if K_1 is the $p \times p$ matrix

$$K_1 = (\kappa_1(x_k, x_j))_{1 \leq j,k \leq p}$$

and if K_2 is the $p \times p$ matrix

$$K_2 = (\kappa_2(x_k, x_j))_{1 \leq j,k \leq p},$$

then for any $u \in \mathbb{C}^p$, we have

$$u^*(K_1 + K_2)u = u^* K_1 u + u^* K_2 u \geq 0,$$

since $u^* K_1 u \geq 0$ and $u^* K_2 u \geq 0$ because κ_2 and κ_2 are positive definite kernels, which means that K_1 and K_2 are positive semidefinite.

(2) We have

$$u^*(aK_1)u = au^* K_1 u \geq 0,$$

since $a > 0$ and $u^* K_1 u \geq 0$.

(3) For every finite subset $S = \{x_1, \ldots, x_p\}$ of X, if K is the $p \times p$ matrix

$$K = (\kappa(x_k, x_j))_{1 \leq j,k \leq p} = (\overline{f(x_k)} f(x_j))_{1 \leq j,k \leq p}$$

then we have

$$\begin{aligned} u^* K u = u^\top K^\top \overline{u} = \sum_{j,k=1}^{p} \kappa(x_j, x_k) u_j \overline{u_k} &= \sum_{j,k=1}^{p} u_j f(x_j) \overline{u_k f(x_k)} \\ &= \left| \sum_{j=1}^{p} u_j f(x_j) \right|^2 \geq 0. \end{aligned}$$

(4) For every finite subset $S = \{x_1, \ldots, x_p\}$ of X, the $p \times p$ matrix K given by

$$K = (\kappa(x_k, x_j))_{1 \leq j,k \leq p} = (\kappa_3(\psi(x_k), \psi(x_j)))_{1 \leq j,k \leq p}$$

is symmetric positive semidefinite since κ_3 is a positive definite kernel.

(5) As in the proof of Proposition 17.5 (adapted to the real case) there is a matrix R such that

$$B = RR^\top,$$

so

$$\kappa(x, y) = x^\top B y = x^\top R R^\top y = (R^\top x)^\top R^\top y = \langle R^\top x, R^\top y \rangle,$$

so κ is the kernel function given by the feature map $\varphi(x) = R^\top x$ from $\mathbb{R}^n$ to itself, and by Proposition 17.1, it is a symmetric positive definite kernel. $\square$

Proposition 17.7. *Let $\kappa_1 \colon X \times X \to \mathbb{C}$ be a positive definite kernel, and let $p(z)$ be a polynomial with nonnegative coefficients. Then the following functions κ defined below are also positive definite kernels.*

(1) $\kappa(x, y) = p(\kappa_1(x, y))$.
(2) $\kappa(x, y) = e^{\kappa_1(x,y)}$.
(3) If X is real Hilbert space with inner product $\langle -, - \rangle_X$ and corresponding norm $\| \, \|_X$,

$$\kappa(x, y) = e^{-\frac{\|x-y\|_X^2}{2\sigma^2}}$$

for any $\sigma > 0$.

Proof. (1) If $p(z) = a_m z^m + \cdots + a_1 z + a_0$, then

$$p(\kappa_1(x, y)) = a_m \kappa_1(x, y)^m + \cdots + a_1 \kappa_1(x, y) + a_0.$$

Since $a_k \geq 0$ for $k = 0, \ldots, m$, by Proposition 17.5 and Proposition 17.6(2), each function $a_k \kappa_i(x, y)^k$ with $1 \leq k \leq m$ is a positive definite kernel, by Proposition 17.6(3) with $f(x) = \sqrt{a_0}$, the constant function a_0 is a positive definite kernel, and by Proposition 17.6(1), $p(\kappa_1(x, y))$ is a positive definite kernel.

(2) We have

$$e^{\kappa_1(x,y)} = \sum_{k=0}^{\infty} \frac{\kappa_1(x, y)^k}{k!}.$$

By (1), the partial sums

$$\sum_{k=0}^{m} \frac{\kappa_1(x, y)^k}{k!}$$

are positive definite kernels, and since $e^{\kappa_1(x,y)}$ is the (uniform) pointwise limit of positive definite kernels, it is also a positive definite kernel.

(3) By Proposition 17.6(2), since the map $(x, y) \mapsto \langle x, y \rangle_X$ is obviously a positive definite kernel (the feature map is the identity) and since $\sigma \neq 0$, the function $(x, y) \mapsto \langle x, y \rangle_X / \sigma^2$ is a positive definite kernel (by Proposition 17.6(2)), so by (2),

$$\kappa_1(x, y) = e^{\frac{\langle x, y \rangle_X}{\sigma^2}}$$

is a positive definite kernel. Let $f \colon X \to \mathbb{R}$ be the function given by

$$f(x) = e^{-\frac{\|x\|^2}{2\sigma^2}}.$$

Then by Proposition 17.6(3),

$$\kappa_2(x,y) = f(x)f(y) = e^{-\frac{\|x\|^2}{2\sigma^2}} e^{-\frac{\|y\|^2}{2\sigma^2}} = e^{-\frac{\|x\|_X^2 + \|y\|_X^2}{2\sigma^2}}$$

is a positive definite kernel. By Proposition 17.5, the function $\kappa_1\kappa_2$ is a positive definite kernel, that is

$$\kappa_1(x,y)\kappa_2(x,y) = e^{\frac{\langle x,y\rangle_X}{\sigma^2}} e^{-\frac{\|x\|_X^2 + \|y\|_X^2}{2\sigma^2}} = e^{\frac{\langle x,y\rangle_X}{\sigma^2} - \frac{\|x\|_X^2 + \|y\|_X^2}{2\sigma^2}} = e^{-\frac{\|x-y\|_X^2}{2\sigma^2}}$$

is a positive definite kernel. □

Definition 17.4. The positive definite kernel

$$\kappa(x,y) = e^{-\frac{\|x-y\|_X^2}{2\sigma^2}}$$

is called a *Gaussian kernel.*

This kernel requires a feature map in an infinite-dimensional space because it is an infinite sum of distinct kernels.

Remark: If κ_1 is a positive definite kernel, the proof of Proposition 17.7(3) is immediately adapted to show that

$$\kappa(x,y) = e^{-\frac{\kappa_1(x,x)+\kappa_1(y,y)-2\kappa_1(x,y)}{2\sigma^2}}$$

is a positive definite kernel.

Next we prove that every positive definite kernel arises from a feature map in a Hilbert space which is a function space.

17.3 Hilbert Space Representation of a Positive Definite Kernel ⊛

The following result shows how to construct a so-called *reproducing kernel Hilbert space*, for short RKHS, from a positive definite kernel.

Theorem 17.1. *Let $\kappa\colon X \times X \to \mathbb{C}$ be a positive definite kernel on a nonempty set X. For every $x \in X$, let $\kappa_x\colon X \to \mathbb{C}$ be the function given by*

$$\kappa_x(y) = \kappa(x,y), \qquad y \in X.$$

Let H_0 be the subspace of the vector space $\mathbb{C}^X$ of functions from X to $\mathbb{C}$ spanned by the family of functions $(\kappa_x)_{\in X}$, and let $\varphi\colon X \to H_0$ be the map given by $\varphi(x) = \kappa_x$. There is a Hermitian inner product $\langle -, -\rangle$ on H_0 such that

$$\kappa(x,y) = \langle \varphi(x), \varphi(y)\rangle, \qquad \text{for all } x, y \in X.$$

The completion H of H_0 is a Hilbert space, and the map $\eta\colon H \to \mathbb{C}^X$ given by

$$\eta(f)(x) = \langle f, \kappa_x \rangle, \qquad x \in X,$$

is linear and injective, so H can be identified with a subspace of $\mathbb{C}^X$. We also have

$$\kappa(x, y) = \langle \varphi(x), \varphi(y) \rangle, \qquad \text{for all } x, y \in X.$$

For all $f \in H_0$ and all $x \in X$,

$$\langle f, \kappa_x \rangle = f(x), \tag{$*$}$$

a property known as the **reproducing property**.

Proof. *Step 1.* Define a candidate inner product.

For any two linear combinations $f = \sum_{j=1}^{p} \alpha_j \kappa_{x_j}$ and $g = \sum_{k=1}^{q} \beta_k \kappa_{y_k}$ in H_0, with $x_j, y_k \in X$ and $\alpha_j, \beta_k \in \mathbb{C}$, define $\langle f, g \rangle$ by

$$\langle f, g \rangle = \sum_{j=1}^{p} \sum_{k=1}^{q} \alpha_j \overline{\beta_k} \kappa(x_j, y_k). \tag{$\dagger$}$$

At first glance, the above expression appears to depend on the expression of f and g as linear combinations, but since $\kappa(x_j, y_k) = \overline{\kappa(y_k, x_j)}$, observe that

$$\sum_{k=1}^{q} \overline{\beta_k} f(y_k) = \sum_{j=1}^{p} \sum_{k=1}^{q} \alpha_j \overline{\beta_k} \kappa(x_j, y_k) = \sum_{j=1}^{p} \alpha_j \overline{g(x_j)}, \tag{$*$}$$

and since the first and the third term are equal for all linear combinations representing f and g, we conclude that ($\dagger$) depends only on f and g and not on their representation as a linear combination.

Step 2. Prove that the inner product defined by ($\dagger$) is Hermitian semidefinite.

Obviously ($\dagger$) defines a Hermitian sequilinear form. For every $f \in H_0$, we have

$$\langle f, f \rangle = \sum_{j,k=1}^{p} \alpha_j \overline{\alpha_k} \kappa(x_j, x_k) \geq 0,$$

since κ is a positive definite kernel.

Step 3. Prove that the inner product defined by ($\dagger$) is positive definite.

For any finite subset $\{f_1, \ldots, f_n\}$ of H_0 and any $z \in \mathbb{C}^n$, we have

$$\sum_{j,k=1}^{n} \langle f_j, f_k \rangle z_j \overline{z_k} = \left\langle \sum_{j=1}^{n} z_j f_j, \sum_{j=1}^{n} z_j f_j \right\rangle \geq 0,$$

which shows that the map $(f, g) \mapsto \langle f, g \rangle$ from $H_0 \times H_0$ to $\mathbb{C}$ is a positive definite kernel.

Observe that for all $f \in H_0$ and all $x \in X$, (†) implies that

$$\langle f, \kappa_x \rangle = \sum_{j=1}^{k} \alpha_j \kappa(x_j, x) = f(x),$$

a property known as the *reproducing property*. The above implies that

$$\langle \kappa_x, \kappa_y \rangle = \kappa(x, y). \tag{$**$}$$

By Proposition 17.4 applied to the positive definite kernel $(f, g) \mapsto \langle f, g \rangle$, we have

$$|\langle f, \kappa_x \rangle|^2 \leq \langle f, f \rangle \langle \kappa_x, \kappa_x \rangle,$$

that is,

$$|f(x)|^2 \leq \langle f, f \rangle \kappa(x, x),$$

so $\langle f, f \rangle = 0$ implies that $f(x) = 0$ for all $x \in X$, which means that $\langle -, - \rangle$ as defined by (†) is positive definite. Therefore, $\langle -, - \rangle$ is a Hermitian inner product on H_0, and by ($**$) and since $\varphi(x) = \kappa_x$, we have

$$\kappa(x, y) = \langle \varphi(x), \varphi(y) \rangle, \qquad \text{for all } x, y \in X.$$

Step 4. Define the embedding η.

Let H be the Hilbert space which is the completion of H_0, so that H_0 is dense in H. The map $\eta\colon H \to \mathbb{C}^X$ given by

$$\eta(f)(x) = \langle f, \kappa_x \rangle$$

is obviously linear, and it is injective because the family $(\kappa_x)_{x \in X}$ spans H_0 which is dense in H, thus it is also dense in H, so if $\langle f, \kappa_x \rangle = 0$ for all $x \in X$, then $f = 0$. □

Corollary 17.1. *If we identify a function $f \in H$ with the function $\eta(f)$, then we have the reproducing property*

$$\langle f, \kappa_x \rangle = f(x), \qquad \textit{for all } f \in H \textit{ and all } x \in X.$$

If X is finite, then $\mathbb{C}^X$ is finite-dimensional. If X is a separable topological space and if κ is continuous, then it can be shown that H is a separable Hilbert space.

Also, if $\kappa\colon X \times X \to \mathbb{R}$ is a real symmetric positive definite kernel, then we see immediately that Theorem 17.1 holds with H_0 a real Euclidean space and H a real Hilbert space.

⊛ **Remark:** If $X = G$, where G is a locally compact group, then a function $p\colon G \to \mathbb{C}$ (not necessarily continuous) is *positive semidefinite* if for all $s_1, \ldots, s_n \in G$ and all $\xi_1, \ldots, \xi_n \in \mathbb{C}$, we have

$$\sum_{j,k=1}^{n} p(s_j^{-1} s_k)\xi_k \overline{\xi_j} \geq 0.$$

So if we define $\kappa\colon G \times G \to \mathbb{C}$ by

$$\kappa(s,t) = p(t^{-1}s),$$

then κ is a positive definite kernel on G. If p is continuous, then it is known that p arises from a unitary representation $U\colon G \to \mathbf{U}(H)$ of the group G in a Hilbert space H with inner product $\langle -, - \rangle$ (a homomorphism with a certain continuity property), in the sense that there is some vector $x_0 \in H$ such that

$$p(s) = \langle U(s)(x_0), x_0 \rangle, \qquad \text{for all } s \in G.$$

Since the $U(s)$ are unitary operators on H,

$$\begin{aligned} p(t^{-1}s) &= \langle U(t^{-1}s)(x_0), x_0 \rangle = \langle U(t^{-1})(U(s)(x_0)), x_0 \rangle \\ &= \langle U(t)^*(U(s)(x_0)), x_0 \rangle = \langle U(s)(x_0), U(t)(x_0) \rangle, \end{aligned}$$

which shows that

$$\kappa(s,t) = \langle U(s)(x_0), U(t)(x_0) \rangle,$$

so the map $\varphi\colon G \to H$ given by

$$\varphi(s) = U(s)(x_0)$$

is a feature map into the feature space H. This theorem is due to Gelfand and Raikov (1943).

The proof of Theorem 17.1 is essentially identical to part of Godement's proof of the above result about the correspondence between functions of positive type and unitary representations; see Helgason [Helgason (2000)], Chapter IV, Theorem 1.5. Theorem 17.1 is a little more general since it does not assume that X is a group, but when G is a group, the feature map arises from a unitary representation.

17.4 Kernel PCA

As an application of kernel functions, we discuss a generalization of the method of principal component analysis (PCA). Suppose we have a set of data $S = \{x_1, \ldots, x_n\}$ in some input space $\mathcal{X}$, and pretend that we have an embedding $\varphi \colon \mathcal{X} \to F$ of $\mathcal{X}$ in a (real) feature space $(F, \langle -, - \rangle)$, but that we only have access to the kernel function $\kappa(x, y) = \langle \varphi(x), \varphi(y) \rangle$. We would like to do PCA analysis on the set $\varphi(S) = \{\varphi(x_1), \ldots, \varphi(x_n)\}$.

There are two obstacles:

(1) We need to center the data and compute the inner products of pairs of centered data. More precisely, if the centroid of $\varphi(S)$ is
$$\mu = \frac{1}{n}(\varphi(x_1) + \cdots + \varphi(x_n)),$$
then we need to compute the inner products $\langle \varphi(x) - \mu, \varphi(y) - \mu \rangle$.

(2) Let us assume that $F = \mathbb{R}^d$ with the standard Euclidean inner product and that the data points $\varphi(x_i)$ are expressed as *row vectors* X_i of an $n \times d$ matrix X (as it is customary). Then the inner products $\kappa(x_i, x_j) = \langle \varphi(x_i), \varphi(x_j) \rangle$ are given by the *kernel matrix* $\mathbf{K} = XX^\top$. Be aware that with this representation, in the expression $\langle \varphi(x_i), \varphi(x_j) \rangle$, $\varphi(x_i)$ is a d-dimensional column vector, while $\varphi(x_i) = X_i^\top$. However, the jth component $(Y_k)_j$ of the principal component Y_k (viewed as a n-dimensional column vector) is given by the projection of $\widehat{X}_j = X_j - \mu$ onto the direction u_k (viewing μ as a d-dimensional row vector), which is a unit eigenvector of the matrix $(X - \mu)^\top (X - \mu)$ (where $\widehat{X} = X - \mu$ is the matrix whose jth row is $\widehat{X}_j = X_j - \mu$), is given by the inner product
$$\langle X_j - \mu, u_k \rangle = (Y_k)_j;$$
see Definition 21.2 (Vol. I) and Theorem 21.11 (Vol. I). The problem is that we know what the matrix $(X - \mu)(X - \mu)^\top$ is from (1), because it can be expressed in terms of $\mathbf{K}$, but we don't know what $(X - \mu)^\top (X - \mu)$ is because we don't have access to $\widehat{X} = X - \mu$.

Both difficulties are easily overcome. For (1) we have
$$\begin{aligned}\langle \varphi(x) - \mu, \varphi(y) - \mu \rangle &= \left\langle \varphi(x) - \frac{1}{n}\sum_{k=1}^{n} \varphi(x_k), \varphi(y) - \frac{1}{n}\sum_{k=1}^{n} \varphi(x_k) \right\rangle \\ &= \kappa(x, y) - \frac{1}{n}\sum_{i=1}^{n} \kappa(x, x_i) - \frac{1}{n}\sum_{j=1}^{n} \kappa(x_j, y) + \frac{1}{n^2}\sum_{i,j=1}^{n} \kappa(x_i, x_j).\end{aligned}$$

For (2), if $\mathbf{K}$ is the kernel matrix $\mathbf{K} = (\kappa(x_i, x_j))$, then the kernel matrix $\widehat{\mathbf{K}}$ corresponding to the kernel function $\widehat{\kappa}$ given by

$$\widehat{\kappa}(x, y) = \langle \varphi(x) - \mu, \varphi(y) - \mu \rangle$$

can be expressed in terms of $\mathbf{K}$. Let $\mathbf{1}$ be the column vector (of dimension n) whose entries are all 1. Then $\mathbf{1}\mathbf{1}^\top$ is the $n \times n$ matrix whose entries are all 1. If A is an $n \times n$ matrix, then $\mathbf{1}^\top A$ is the row vector consisting of the sums of the columns of A, $A\mathbf{1}$ is the column vector consisting of the sums of the rows of A, and $\mathbf{1}^\top A\mathbf{1}$ is the sum of all the entries in A. Then it is easy to see that the kernel matrix corresponding to the kernel function $\widehat{\kappa}$ is given by

$$\widehat{\mathbf{K}} = \mathbf{K} - \frac{1}{n}\mathbf{1}\mathbf{1}^\top\mathbf{K} - \frac{1}{n}\mathbf{K}\mathbf{1}\mathbf{1}^\top + \frac{1}{n^2}(\mathbf{1}^\top\mathbf{K}\mathbf{1})\mathbf{1}\mathbf{1}^\top.$$

Suppose $\widehat{X} = X - \mu$ has rank r. To overcome the second problem, note that if

$$\widehat{X} = VDU^\top$$

is an SVD for $\widehat{X}$, then

$$\widehat{X}^\top = UD^\top V^\top$$

is an SVD for $\widehat{X}^\top$, and the $r \times r$ submatrix of $D^\top$ consisting of the first r rows and r columns of $D^\top$ (and D), is the diagonal Σ_r matrix consisting of the singular values $\sigma_1 \geq \cdots \geq \sigma_r$ of $\widehat{X}$, so we can express the matrix U_r consisting of the first r columns u_k of U in terms of the matrix V_r consisting of the first r columns v_k of V $(1 \leq k \leq r)$ as

$$U_r = \widehat{X}^\top V_r \Sigma_r^{-1}.$$

Furthermore, $\sigma_1^2 \geq \cdots \geq \sigma_r^2$ are the nonzero eigenvalues of $\widehat{\mathbf{K}} = \widehat{X}\widehat{X}^\top$, and the columns of V_r are corresponding unit eigenvectors of $\widehat{\mathbf{K}}$. From

$$U_r = \widehat{X}^\top V_r \Sigma_r^{-1}$$

the kth column u_k of U_r (which is a unit eigenvector of $\widehat{X}^\top\widehat{X}$ associated with the eigenvalue σ_k^2) is given by

$$u_k = \sum_{i=1}^n \sigma_k^{-1}(v_k)_i \widehat{X}_i^\top = \sum_{i=1}^n \sigma_k^{-1}(v_k)_i \widehat{\varphi(x_i)}, \quad 1 \leq k \leq r,$$

so the projection of $\widehat{\varphi(x)}$ onto u_k is given by

$$\begin{aligned}\langle \widehat{\varphi(x)}, u_k \rangle &= \left\langle \widehat{\varphi(x)}, \sum_{i=1}^n \sigma_k^{-1}(v_k)_i \widehat{\varphi(x_i)} \right\rangle \\ &= \sum_{i=1}^n \sigma_k^{-1}(v_k)_i \left\langle \widehat{\varphi(x)}, \widehat{\varphi(x_i)} \right\rangle = \sum_{i=1}^n \sigma_k^{-1}(v_k)_i \widehat{\kappa}(x, x_i).\end{aligned}$$

Therefore, the jth component of the principal component Y_k in the principal direction u_k is given by

$$(Y_k)_j = \langle X_j - \mu, u_k \rangle = \sum_{i=1}^{n} \sigma_k^{-1} (v_k)_i \widehat{\kappa}(x_j, x_i) = \sum_{i=1}^{n} \sigma_k^{-1} (v_k)_i \widehat{\mathbf{K}}_{ij}.$$

The generalization of kernel PCA to a general embedding $\varphi\colon \mathcal{X} \to F$ of $\mathcal{X}$ in a (real) feature space $(F, \langle -, - \rangle)$ (where F is not restricted to be equal to $\mathbb{R}^d$) with the kernel matrix $\mathbf{K}$ given by

$$\mathbf{K}_{ij} = \langle \varphi(x_i), \varphi(x_j) \rangle,$$

goes as follows.

- Let r be the rank of $\widehat{\mathbf{K}}$, where
$$\widehat{\mathbf{K}} = \mathbf{K} - \frac{1}{n}\mathbf{1}\mathbf{1}^\top \mathbf{K} - \frac{1}{n}\mathbf{K}\mathbf{1}\mathbf{1}^\top + \frac{1}{n^2}(\mathbf{1}^\top \mathbf{K}\mathbf{1})\mathbf{1}\mathbf{1}^\top,$$
let $\sigma_1^2 \geq \cdots \geq \sigma_r^2$ be the nonzero eigenvalues of $\widehat{\mathbf{K}}$, and let $v_1, \ldots, v_r$ be corresponding unit eigenvectors. The notation
$$\alpha_k = \sigma_k^{-1} v_k$$
is often used, where the α_k are called the *dual variables*.
- The column vector Y_k $(1 \leq k \leq r)$ defined by
$$Y_k = \left(\sum_{i=1}^{n} (\alpha_k)_i \widehat{\mathbf{K}}_{ij} \right)_{j=1}^{n}$$
is called the *kth kernel principal component* (for short *kth kernel PCA*) of the data set $S = \{x_1, \ldots, x_n\}$ in the direction $u_k = \sum_{i=1}^{n} \sigma_k^{-1} (v_k)_i \widehat{X}_i^\top$ (even though the matrix $\widehat{X}$ is *not known*).

17.5 Summary

The main concepts and results of this chapter are listed below:

- Feature map, feature embedding, feature space.
- Kernel function.
- Positive definite kernel, real positive definite kernel.
- Gram matrix.
- Hadamard product, Schur product.
- Gaussian kernel.
- Reproducing kernel Hilbert space (RKHS).
- Reproducing property.
- Intersection kernel, union complement kernel, agreement kernel.
- Kernel PCA.
- k-th kernel PCA.

17.6 Problems

Problem 17.1. Referring back to Example 17.3, prove that if $\varphi_1 \colon X \to \mathbb{R}^{n_1}$ and $\varphi_2 \colon X \to \mathbb{R}^{n_2}$ are two feature maps and if $\kappa_1(x, y) = \langle \varphi_1(x), \varphi_1(y) \rangle$ and $\kappa_2(x, y) = \langle \varphi_2(x), \varphi_2(y) \rangle$ are the corresponding kernel functions, then the map defined by

$$\kappa(x, y) = \kappa_1(x, y)\kappa_2(x, y)$$

is a kernel function, for the feature space $\mathbb{R}^{n_1} \times \mathbb{R}^{n_2}$ and the feature map

$$\varphi(x)_{(i,j)} = (\varphi_1(x))_i(\varphi_2(x))_j, \qquad 1 \leq i \leq n_1,\ 1 \leq j \leq n_2.$$

Problem 17.2. Referring back to Example 17.3, prove that the feature embedding $\varphi \colon X \to \mathbb{R}^{\binom{n+m-1}{m}}$ given by

$$\varphi_{(i_1,\ldots,i_n)}(x) = \binom{m}{i_1 \cdots i_n}^{1/2} (\varphi_1(x))_1^{i_1} (\varphi_1(x))_1^{i_2} \cdots (\varphi_1(x))_1^{i_n},$$
$$i_1 + i_2 + \cdots + i_n = m,\ i_j \in \mathbb{N},$$

where the n-tuples $(i_1, \ldots, i_n)$ are ordered lexicographically, defines the kernel function κ given by $\kappa(x, y) = (\kappa_1(x, y))^m$.

Problem 17.3. In Example 17.6, prove that for any two subsets A_1 and A_2 of D,

$$\langle \varphi(A_1), \varphi(A_2) \rangle = 2^{|A_1 \cap A_2|},$$

the number of common subsets of A_1 and A_2.

Problem 17.4. Prove that the pointwise limit of positive definite kernels is also a positive definite kernel.

Problem 17.5. Prove that if κ_1 is a positive definite kernel, then

$$\kappa(x, y) = e^{-\frac{\kappa_1(x,x) + \kappa_1(y,y) - 2\kappa_1(x,y)}{2\sigma^2}}$$

is a positive definite kernel.

Chapter 18

Soft Margin Support Vector Machines

In Sections 14.5 and 14.6 we considered the problem of separating two nonempty disjoint finite sets of p *blue* points $\{u_i\}_{i=1}^p$ and q *red* points $\{v_j\}_{j=1}^q$ in $\mathbb{R}^n$. The goal is to find a hyperplane H of equation $w^\top x - b = 0$ (where $w \in \mathbb{R}^n$ is a nonzero vector and $b \in \mathbb{R}$), such that all the blue points u_i are in one of the two open half-spaces determined by H, and all the red points v_j are in the other open half-space determined by H. SVM picks a hyperplane which maximizes the minimum distance from these points to the hyperplane. See Figure 18.1.

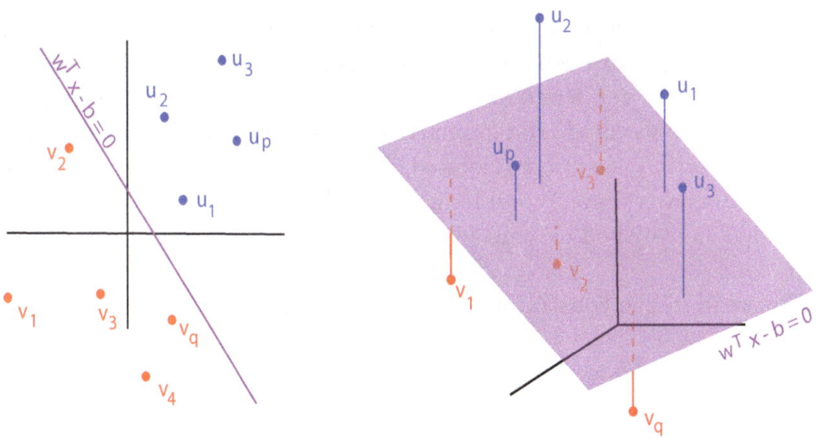

Fig. 18.1 Two examples of the SVM separation problem. The left figure is SVM in $\mathbb{R}^2$, while the right figure is SVM in $\mathbb{R}^3$.

In this chapter we return to the problem of separating two disjoint sets of points, $\{u_i\}_{i=1}^p$ and $\{v_j\}_{j=1}^q$, but this time we do not assume that these two sets are separable. To cope with nonseparability, we allow points to invade the safety zone around the separating hyperplane, and even points on the wrong side of the hyperplane. Such a method is called *soft margin support vector machine*. We discuss variations of this method, including ν-SV classification. In each case we present a careful derivation of the dual.

If the sets of points $\{u_1, \ldots, u_p\}$ and $\{v_1, \ldots, v_q\}$ are not linearly separable (with $u_i, v_j \in \mathbb{R}^n$), we can use a trick from linear programming which is to introduce nonnegative "slack variables" $\epsilon = (\epsilon_1, \ldots, \epsilon_p) \in \mathbb{R}^p$ and $\xi = (\xi_1, \ldots, \xi_q) \in \mathbb{R}^q$ to relax the "hard" constraints

$$\begin{aligned} w^\top u_i - b &\geq \delta & i &= 1, \ldots, p \\ -w^\top v_j + b &\geq \delta & j &= 1, \ldots, q \end{aligned}$$

of Problem (SVM$_{h1}$) from Section 14.5 to the "soft" constraints

$$\begin{aligned} w^\top u_i - b &\geq \delta - \epsilon_i, & \epsilon_i &\geq 0 & i &= 1, \ldots, p \\ -w^\top v_j + b &\geq \delta - \xi_j, & \xi_j &\geq 0 & j &= 1, \ldots, q. \end{aligned}$$

Recall that $w \in \mathbb{R}^n$ and $b, \delta \in \mathbb{R}$.

If $\epsilon_i > 0$, the point u_i may be misclassified, in the sense that it can belong to the margin (the slab), or even to the wrong half-space classifying the negative (red) points. See Figures 18.5 (2) and (3). Similarly, if $\xi_j > 0$, the point v_j may be misclassified, in the sense that it can belong to the margin (the slab), or even to the wrong half-space classifying the positive (blue) points. We can think of ϵ_i as a measure of how much the constraint $w^\top u_i - b \geq \delta$ is violated, and similarly of ξ_j as a measure of how much the constraint $-w^\top v_j + b \geq \delta$ is violated. If $\epsilon = 0$ and $\xi = 0$, then we recover the original constraints. By making ϵ and ξ large enough, these constraints can always be satisfied. We add the constraint $w^\top w \leq 1$ and we minimize $-\delta$.

If instead of the constraints of Problem (SVM$_{h1}$) we use the hard constraints

$$\begin{aligned} w^\top u_i - b &\geq 1 & i &= 1, \ldots, p \\ -w^\top v_j + b &\geq 1 & j &= 1, \ldots, q \end{aligned}$$

of Problem (SVM$_{h2}$) (see Example 14.6), then we relax to the soft constraints

$$\begin{aligned} w^\top u_i - b &\geq 1 - \epsilon_i, & \epsilon_i &\geq 0 & i &= 1, \ldots, p \\ -w^\top v_j + b &\geq 1 - \xi_j, & \xi_j &\geq 0 & j &= 1, \ldots, q. \end{aligned}$$

In this case there is no constraint on w, but we minimize $(1/2)w^\top w$.

Ideally we would like to find a separating hyperplane that *minimizes the number of misclassified points*, which means that the variables ϵ_i and ξ_j should be as small as possible, but there is a trade-off in maximizing the margin (the thickness of the slab), and minimizing the number of misclassified points. This is reflected in the choice of the objective function, and there are several options, depending on whether we minimize a linear function of the variables ϵ_i and ξ_j, or a quadratic functions of these variables, or whether we include the term $(1/2)b^2$ in the objective function. These methods are known as *support vector classification* algorithms (for short *SVC* algorithms).

SVC algorithms seek an "optimal" separating hyperplane H of equation $w^\top x - b = 0$. If some new data $x \in \mathbb{R}^n$ comes in, we can classify it by determining in which of the two half spaces determined by the hyperplane H they belong by computing the sign of the quantity $w^\top x - b$. The function $\mathrm{sgn}\colon \mathbb{R} \to \{-1, 1\}$ is given by

$$\mathrm{sgn}(x) = \begin{cases} +1 & \text{if } x \geq 0 \\ -1 & \text{if } x < 0. \end{cases}$$

Then we define the *(binary) classification function* associated with the hyperplane H of equation $w^\top x - b = 0$ as

$$f(x) = \mathrm{sgn}(w^\top x - b).$$

Remarkably, all the known optimization problems for finding this hyperplane share the property that the weight vector w and the constant b are given by expressions that *only involves inner products of the input data points u_i and v_j*, and so does the classification function

$$f(x) = \mathrm{sgn}(w^\top x - b).$$

This is a key fact that allows a far reaching generalization of the support vector machine using the method of *kernels*.

The method of kernels consists in assuming that the input space $\mathbb{R}^n$ is embedded in a larger (possibly infinite dimensional) Euclidean space F (with an inner product $\langle -, - \rangle$) usually called a *feature space*, using a function

$$\varphi\colon \mathbb{R}^n \to F$$

called a *feature map*. The function $\kappa\colon \mathbb{R}^n \times \mathbb{R}^n \to \mathbb{R}$ given by

$$\kappa(x, y) = \langle \varphi(x), \varphi(y) \rangle$$

is the kernel function associated with the embedding φ; see Chapter 17. The idea is that the feature map φ "unwinds" the input data, making it somehow more linear in the higher dimensional space F. Now even if we don't know what the feature space F is and what the embedding map φ is, we can pretend to solve our separation problem in F for the embedded data points $\varphi(u_i)$ and $\varphi(v_j)$. Thus we seek a hyperplane H of equation

$$\langle w, \zeta \rangle - b = 0, \quad \zeta \in F,$$

in the feature space F, to attempt to separate the points $\varphi(u_i)$ and the points $\varphi(v_j)$. As we said, it turns out that w and b are given by expression involving only the inner products $\kappa(u_i, u_j) = \langle \varphi(u_i), \varphi(u_j) \rangle, \kappa(u_i, v_j) = \langle \varphi(u_i), \varphi(v_j) \rangle$, and $\kappa(v_i, v_j) = \langle \varphi(v_i), \varphi(v_j) \rangle$, which form the symmetric $(p+q) \times (p+q)$ matrix $\mathbf{K}$ (a kernel matrix) given by

$$\mathbf{K}_{ij} = \begin{cases} \kappa(u_i, u_j) & 1 \leq i \leq p,\ 1 \leq j \leq q \\ -\kappa(u_i, v_{j-p}) & 1 \leq i \leq p,\ p+1 \leq j \leq p+q \\ -\kappa(v_{i-p}, u_j) & p+1 \leq i \leq p+q,\ 1 \leq j \leq p \\ \kappa(v_{i-p}, v_{j-q}) & p+1 \leq i \leq p+q,\ p+1 \leq j \leq p+q. \end{cases}$$

For example, if $p = 2$ and $q = 3$, we have the matrix

$$\mathbf{K} = \begin{pmatrix} \kappa(u_1, u_1) & \kappa(u_1, u_2) & -\kappa(u_1, v_1) & -\kappa(u_1, v_2) & -\kappa(u_1, v_3) \\ \kappa(u_2, u_1) & \kappa(u_2, u_2) & -\kappa(u_2, v_1) & -\kappa(u_2, v_2) & -\kappa(u_2, v_3) \\ -\kappa(v_1, u_1) & -\kappa(v_1, u_2) & \kappa(v_1, v_1) & \kappa(v_1, v_2) & \kappa(v_1, v_3) \\ -\kappa(v_2, u_1) & -\kappa(v_2, u_2) & \kappa(v_2, v_1) & \kappa(v_2, v_2) & \kappa(v_2, v_3) \\ -\kappa(v_3, u_1) & -\kappa(v_3, u_2) & \kappa(v_3, v_1) & \kappa(v_3, v_2) & \kappa(v_3, v_3) \end{pmatrix}.$$

Then the classification function

$$f(x) = \operatorname{sgn}(\langle w, \varphi(x) \rangle - b)$$

for points in the original data space $\mathbb{R}^n$ is also expressed solely in terms of the matrix $\mathbf{K}$ and the inner products $\kappa(u_i, x) = \langle \varphi(u_i), \varphi(x) \rangle$ and $\kappa(v_j, x) = \langle \varphi(v_j), \varphi(x) \rangle$. As a consequence, in the original data space $\mathbb{R}^n$, the hypersurface

$$\mathcal{S} = \{x \in \mathbb{R}^n \mid \langle w, \varphi(x) \rangle - b = 0\}$$

separates the data points u_i and v_j, but it is not an affine subspace of $\mathbb{R}^n$. The classification function f tells us on which "side" of $\mathcal{S}$ is a new data point $x \in \mathbb{R}^n$. Thus, we managed to separate the data points u_i and v_j that are not separable by an affine hyperplane, by a *nonaffine hypersurface* $\mathcal{S}$, by assuming that an embedding $\varphi\colon \mathbb{R}^n \to F$ exists, even though we don't know what it is, but having access to F through the kernel function $\kappa\colon \mathbb{R}^n \times \mathbb{R}^n \to \mathbb{R}$ given by the inner products $\kappa(x, y) = \langle \varphi(x), \varphi(y)\rangle$.

In practice the art of using the kernel method is to choose the right kernel (as the knight says in Indiana Jones, to "choose wisely.").

The method of kernels is very flexible. It also applies to the soft margin versions of SVM, but also to regression problems, to principal component analysis (PCA), and to other problems arising in machine learning.

We discussed the method of kernels in Chapter 17. Other comprehensive presentations of the method of kernels are found in Schölkopf and Smola [Schölkopf and Smola (2002)] and Shawe-Taylor and Christianini [Shawe-Taylor and Cristianini (2004)]. See also Bishop [Bishop (2006)].

We first consider the soft margin SVM arising from Problem (SVM$_{h1}$).

18.1 Soft Margin Support Vector Machines; (SVM$_{s1}$)

In this section we derive the dual function G associated with the following version of the soft margin SVM coming from Problem (SVM$_{h1}$), where the maximization of the margin δ has been replaced by the minimization of $-\delta$, and where we added a "regularizing term" $K\Big(\sum_{i=1}^p \epsilon_i + \sum_{j=1}^q \xi_j\Big)$ whose purpose is to make $\epsilon \in \mathbb{R}^p$ and $\xi \in \mathbb{R}^q$ *sparse* (that is, try to make ϵ_i and ξ_j have as many zeros as possible), where $K > 0$ is a fixed constant that can be adjusted to determine the influence of this regularizing term. If the primal problem (SVM$_{s1}$) has an optimal solution $(w, \delta, b, \epsilon, \xi)$, we attempt to use the dual function G to obtain it, but we will see that with this particular formulation of the problem, the constraint $w^\top w \leq 1$ causes troubles even though it is convex.

Soft margin SVM (SVM$_{s1}$):

$$\begin{aligned}
&\text{minimize} \quad -\delta + K\Big(\sum_{i=1}^p \epsilon_i + \sum_{j=1}^q \xi_j\Big)\\
&\text{subject to}\\
&\qquad w^\top u_i - b \geq \delta - \epsilon_i, \quad \epsilon_i \geq 0 \qquad i = 1, \ldots, p\\
&\qquad -w^\top v_j + b \geq \delta - \xi_j, \quad \xi_j \geq 0 \qquad j = 1, \ldots, q\\
&\qquad w^\top w \leq 1.
\end{aligned}$$

It is customary to write $\ell = p+q$. Figure 18.2 illustrates the correct margin half space associated with $w^\top x - b - \delta = 0$ while Figure 18.3 illustrates the

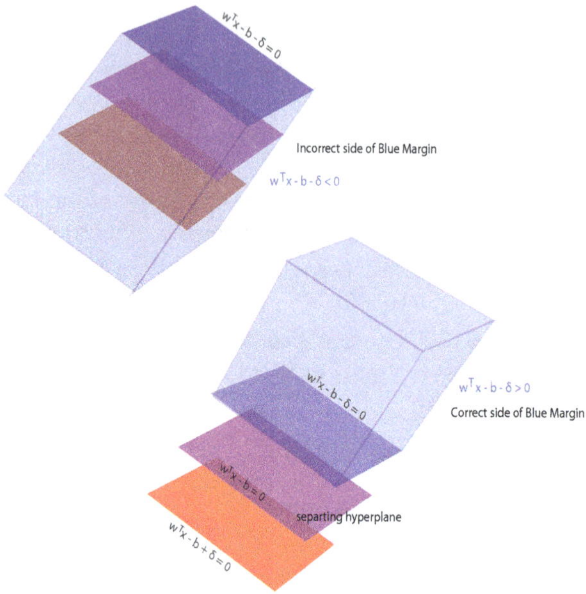

Fig. 18.2 The blue margin half space associated with $w^\top x - b - \delta = 0$.

correct margin half space associated with $w^\top x - b + \delta = 0$. Ideally, all the points should be contained in one of the two correct shifted margin regions described by affine constraints $w^\top u_i - b \geq \delta - \epsilon_i$, or $-w^\top v_j + b \geq \delta - \xi_j$.

For this problem, the primal problem may have an optimal solution $(w, \delta, b, \epsilon, \xi)$ with $\|w\| = 1$ and $\delta > 0$, but if the sets of points are not linearly separable then an optimal solution of the dual may not yield w.

The objective function of our problem is affine and the only nonaffine constraint $w^\top w \leq 1$ is convex. This constraint is qualified because for any $w \neq 0$ such that $w^\top w < 1$ and for any $\delta > 0$ and any b we can pick ϵ and ξ large enough so that the constraints are satisfied. Consequently, by Theorem 14.5(2) *if* the primal problem (SVM$_{s1}$) has an optimal solution, *then* the dual problem has a solution too, and the duality gap is zero.

Unfortunately this does not imply that an optimal solution of the dual yields an optimal solution of the primal because the hypotheses of Theorem 14.5(1) fail to hold. In general, there may not be a unique vector $(w, \epsilon, \xi, b, \delta)$ such that

$$\inf_{w,\epsilon,\xi,b,\delta} L(w, \epsilon, \xi, b, \delta, \lambda, \mu, \alpha, \beta, \gamma) = G(\lambda, \mu, \alpha, \beta, \gamma).$$

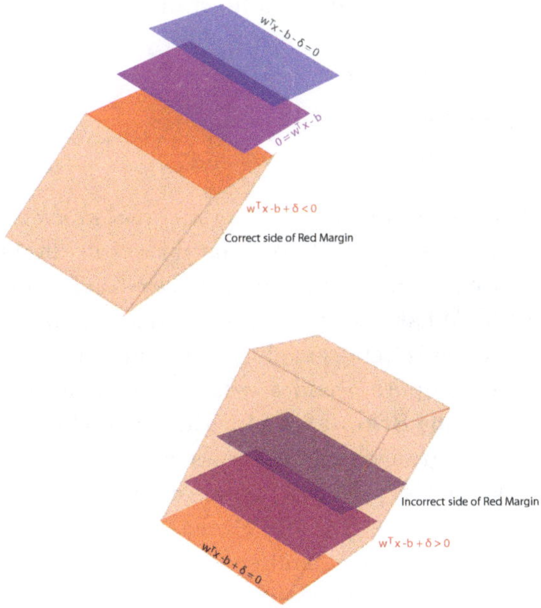

Fig. 18.3 The red margin half space associated with $w^\top x - b + \delta = 0$.

If the sets $\{u_i\}$ and $\{v_j\}$ are *not* linearly separable, then the dual problem may have a solution for which $\gamma = 0$,

$$\sum_{i=1}^{p} \lambda_i = \sum_{j=1}^{q} \mu_j = \frac{1}{2},$$

and

$$\sum_{i=1}^{p} \lambda_i u_i = \sum_{j=1}^{q} \mu_j v_j,$$

so that the dual function $G(\lambda, \mu, \alpha, \beta, \gamma)$, which is a *partial function*, is defined and has the value $G(\lambda, \mu, \alpha, \beta, 0) = 0$. Such a pair (λ, μ) corresponds to the coefficients of two convex combinations

$$\sum_{i=1}^{p} 2\lambda_i u_i = \sum_{j=1}^{q} 2\mu_j v_j$$

which correspond to the *same point* in the (nonempty) intersection of the convex hulls $\mathrm{conv}(u_1, \ldots, u_p)$ and $\mathrm{conv}(v_1, \ldots, v_q)$. It turns out that the only connection between w and the dual function is the equation

$$2\gamma w = \sum_{i=1}^{p} \lambda_i u_i - \sum_{j=1}^{q} \mu_j v_j,$$

and when $\gamma = 0$ this is equation is $0 = 0$, *so the dual problem is useless to determine* w. This point seems to have been missed in the literature (for example, in Shawe-Taylor and Christianini [Shawe-Taylor and Cristianini (2004)], Section 7.2). What the dual problem does show is that $\delta \geq 0$. However, if $\gamma \neq 0$, then w is determined by any solution (λ, μ) of the dual.

It still remains to compute δ and b, which can be done under a mild hypothesis that we call the **Standard Margin Hypothesis.**

Let $\lambda \in \mathbb{R}^p_+$ be the Lagrange multipliers associated with the inequalities $w^\top u_i - b \geq \delta - \epsilon_i$, let $\mu \in \mathbb{R}^q_+$ be the Lagrange multipliers are associated with the inequalities $-w^\top v_j + b \geq \delta - \xi_j$, let $\alpha \in \mathbb{R}^p_+$ be the Lagrange multipliers associated with the inequalities $\epsilon_i \geq 0$, $\beta \in \mathbb{R}^q_+$ be the Lagrange multipliers associated with the inequalities $\xi_j \geq 0$, and let $\gamma \in \mathbb{R}^+$ be the Lagrange multiplier associated with the inequality $w^\top w \leq 1$.

The linear constraints are given by the $2(p+q) \times (n+p+q+2)$ matrix given in block form by

$$C = \begin{pmatrix} X^\top & -I_{p+q} & \begin{matrix} \mathbf{1}_p \\ -\mathbf{1}_q \end{matrix} & \mathbf{1}_{p+q} \\ 0_{p+q,n} & -I_{p+q} & 0_{p+q} & 0_{p+q} \end{pmatrix},$$

where X is the $n \times (p+q)$ matrix

$$X = \begin{pmatrix} -u_1 & \cdots & -u_p & v_1 & \cdots & v_q \end{pmatrix},$$

and the linear constraints are expressed by

$$\begin{pmatrix} X^\top & -I_{p+q} & \begin{matrix} \mathbf{1}_p \\ -\mathbf{1}_q \end{matrix} & \mathbf{1}_{p+q} \\ 0_{p+q,n} & -I_{p+q} & 0_{p+q} & 0_{p+q} \end{pmatrix} \begin{pmatrix} w \\ \epsilon \\ \xi \\ b \\ \delta \end{pmatrix} \leq \begin{pmatrix} 0_{p+q} \\ 0_{p+q} \end{pmatrix}.$$

More explicitly, C is the following matrix:

$$C = \begin{pmatrix} -u_1^\top & -1 & \cdots & 0 & 0 & \cdots & 0 & 1 & 1 \\ \vdots & \vdots & \ddots & \vdots & \vdots & \ddots & \vdots & \vdots & \vdots \\ -u_p^\top & 0 & \cdots & -1 & 0 & \cdots & 0 & 1 & 1 \\ v_1^\top & 0 & \cdots & 0 & -1 & \cdots & 0 & -1 & 1 \\ \vdots & \vdots & \ddots & \vdots & \vdots & \ddots & \vdots & \vdots & \vdots \\ v_q^\top & 0 & \cdots & 0 & 0 & \cdots & -1 & -1 & 1 \\ 0 & -1 & \cdots & 0 & 0 & \cdots & 0 & 0 & 0 \\ \vdots & \vdots & \ddots & \vdots & \vdots & \ddots & \vdots & \vdots & \vdots \\ 0 & 0 & \cdots & -1 & 0 & \cdots & 0 & 0 & 0 \\ 0 & 0 & \cdots & 0 & -1 & \cdots & 0 & 0 & 0 \\ \vdots & \vdots & \ddots & \vdots & \vdots & \ddots & \vdots & \vdots & \vdots \\ 0 & 0 & \cdots & 0 & 0 & \cdots & -1 & 0 & 0 \end{pmatrix}.$$

The objective function is given by

$$J(w, \epsilon, \xi, b, \delta) = -\delta + K \begin{pmatrix} \epsilon^\top & \xi^\top \end{pmatrix} \mathbf{1}_{p+q}.$$

The Lagrangian $L(w, \epsilon, \xi, b, \delta, \lambda, \mu, \alpha, \beta, \gamma)$ with $\lambda, \alpha \in \mathbb{R}^p_+$, $\mu, \beta \in \mathbb{R}^q_+$, and $\gamma \in \mathbb{R}^+$ is given by

$$L(w, \epsilon, \xi, b, \delta, \lambda, \mu, \alpha, \beta, \gamma) = -\delta + K \begin{pmatrix} \epsilon^\top & \xi^\top \end{pmatrix} \mathbf{1}_{p+q} + \begin{pmatrix} w^\top & \begin{pmatrix} \epsilon^\top & \xi^\top \end{pmatrix} & b & \delta \end{pmatrix} C^\top \begin{pmatrix} \lambda \\ \mu \\ \alpha \\ \beta \end{pmatrix} + \gamma(w^\top w - 1).$$

Since

$$\begin{pmatrix} w^\top & \begin{pmatrix} \epsilon^\top & \xi^\top \end{pmatrix} & b & \delta \end{pmatrix} C^\top \begin{pmatrix} \lambda \\ \mu \\ \alpha \\ \beta \end{pmatrix} = w^\top X \begin{pmatrix} \lambda \\ \mu \end{pmatrix} - \epsilon^\top(\lambda + \alpha) - \xi^\top(\mu + \beta) + b(\mathbf{1}_p^\top \lambda - \mathbf{1}_q^\top \mu) + \delta(\mathbf{1}_p^\top \lambda + \mathbf{1}_q^\top \mu),$$

the Lagrangian can be written as

$$\begin{aligned} L(w, \epsilon, \xi, b, \delta, \lambda, \mu, \alpha, \beta, \gamma) &= -\delta + K(\epsilon^\top \mathbf{1}_p + \xi^\top \mathbf{1}_q) + w^\top X \begin{pmatrix} \lambda \\ \mu \end{pmatrix} \\ &\quad + \gamma(w^\top w - 1) - \epsilon^\top(\lambda + \alpha) - \xi^\top(\mu + \beta) + b(\mathbf{1}_p^\top \lambda - \mathbf{1}_q^\top \mu) + \delta(\mathbf{1}_p^\top \lambda + \mathbf{1}_q^\top \mu) \\ &= (\mathbf{1}_p^\top \lambda + \mathbf{1}_q^\top \mu - 1)\delta + w^\top X \begin{pmatrix} \lambda \\ \mu \end{pmatrix} + \gamma(w^\top w - 1) \\ &\quad + \epsilon^\top(K\mathbf{1}_p - (\lambda + \alpha)) + \xi^\top(K\mathbf{1}_q - (\mu + \beta)) + b(\mathbf{1}_p^\top \lambda - \mathbf{1}_q^\top \mu). \end{aligned}$$

To find the dual function $G(\lambda, \mu, \alpha, \beta, \gamma)$ we minimize $L(w, \epsilon, \xi, b, \delta, \lambda, \mu, \alpha, \beta, \gamma)$ with respect to w, ϵ, ξ, b, and δ. Since the Lagrangian is convex and $(w, \epsilon, \xi, b, \delta) \in \mathbb{R}^n \times \mathbb{R}^p \times \mathbb{R}^q \times \mathbb{R} \times \mathbb{R}$, a convex open set, by Theorem 4.5, the Lagrangian has a minimum in $(w, \epsilon, \xi, b, \delta)$ iff $\nabla L_{w,\epsilon,\xi,b,\delta} = 0$, so we compute the gradient with respect to $w, \epsilon, \xi, b, \delta$, and we get

$$\nabla L_{w,\epsilon,\xi,b,\delta} = \begin{pmatrix} X\begin{pmatrix}\lambda\\ \mu\end{pmatrix} + 2\gamma w \\ K\mathbf{1}_p - (\lambda + \alpha) \\ K\mathbf{1}_q - (\mu + \beta) \\ \mathbf{1}_p^\top \lambda - \mathbf{1}_q^\top \mu \\ \mathbf{1}_p^\top \lambda + \mathbf{1}_q^\top \mu - 1 \end{pmatrix}.$$

By setting $\nabla L_{w,\epsilon,\xi,b,\delta} = 0$ we get the equations

$$\begin{aligned} 2\gamma w &= -X\begin{pmatrix}\lambda\\ \mu\end{pmatrix} \\ \lambda + \alpha &= K\mathbf{1}_p \\ \mu + \beta &= K\mathbf{1}_q \\ \mathbf{1}_p^\top \lambda &= \mathbf{1}_q^\top \mu \\ \mathbf{1}_p^\top \lambda + \mathbf{1}_q^\top \mu &= 1. \end{aligned} \qquad (*_w)$$

The second and third equations are equivalent to the inequalities

$$0 \leq \lambda_i, \mu_j \leq K, \quad i = 1, \ldots, p,\ j = 1, \ldots, q,$$

often called *box constraints*, and the fourth and fifth equations yield

$$\mathbf{1}_p^\top \lambda = \mathbf{1}_q^\top \mu = \frac{1}{2}.$$

First let us consider the singular case $\gamma = 0$. In this case, $(*_w)$ implies that

$$X\begin{pmatrix}\lambda\\ \mu\end{pmatrix} = 0,$$

and the term $\gamma(w^\top w - 1)$ is missing from the Lagrangian, which in view of the other four equations above reduces to

$$L(w, \epsilon, \xi, b, \delta, \lambda, \mu, \alpha, \beta, 0) = w^\top X\begin{pmatrix}\lambda\\ \mu\end{pmatrix} = 0.$$

In summary, we proved that if $\gamma = 0$, then

$$G(\lambda, \mu, \alpha, \beta, 0) = \begin{cases} 0 & \text{if } \begin{cases} \sum_{i=1}^p \lambda_i = \sum_{j=1}^q \mu_j = \frac{1}{2} \\ 0 \leq \lambda_i \leq K,\ i = 1, \ldots, p \\ 0 \leq \mu_j \leq K,\ j = 1, \ldots, q \end{cases} \\ -\infty & \text{otherwise} \\ & \text{and } \sum_{i=1}^p \lambda_i u_i - \sum_{j=1}^q \mu_j v_j = 0. \end{cases}$$

Geometrically, (λ, μ) corresponds to the coefficients of two convex combinations

$$\sum_{i=1}^{p} 2\lambda_i u_i = \sum_{j=1}^{q} 2\mu_j v_j$$

which correspond to the *same point* in the intersection of the convex hulls $\mathrm{conv}(u_1, \ldots, u_p)$ and $\mathrm{conv}(v_1, \ldots, v_q)$ iff the sets $\{u_i\}$ and $\{v_j\}$ are *not linearly separable*. If the sets $\{u_i\}$ and $\{v_j\}$ are *linearly separable*, then the convex hulls $\mathrm{conv}(u_1, \ldots, u_p)$ and $\mathrm{conv}(v_1, \ldots, v_q)$ are disjoint, which implies that $\gamma > 0$.

Let us now assume that $\gamma > 0$. Plugging back w from equation $(*_w)$ into the Lagrangian, after simplifications we get

$$\begin{aligned} G(\lambda, \mu, \alpha, \beta, \gamma) &= -\frac{1}{2\gamma} \begin{pmatrix} \lambda^\top & \mu^\top \end{pmatrix} X^\top X \begin{pmatrix} \lambda \\ \mu \end{pmatrix} + \frac{\gamma}{4\gamma^2} \begin{pmatrix} \lambda^\top & \mu^\top \end{pmatrix} X^\top X \begin{pmatrix} \lambda \\ \mu \end{pmatrix} - \gamma \\ &= -\frac{1}{4\gamma} \begin{pmatrix} \lambda^\top & \mu^\top \end{pmatrix} X^\top X \begin{pmatrix} \lambda \\ \mu \end{pmatrix} - \gamma, \end{aligned}$$

so if $\gamma > 0$ the dual function is independent of α, β and is given by

$$\begin{aligned} &G(\lambda, \mu, \alpha, \beta, \gamma) \\ &= \begin{cases} -\frac{1}{4\gamma} \begin{pmatrix} \lambda^\top & \mu^\top \end{pmatrix} X^\top X \begin{pmatrix} \lambda \\ \mu \end{pmatrix} - \gamma & \text{if } \begin{cases} \sum_{i=1}^p \lambda_i = \sum_{j=1}^q \mu_j = \frac{1}{2} \\ 0 \leq \lambda_i \leq K, \; i = 1, \ldots, p \\ 0 \leq \mu_j \leq K, \; j = 1, \ldots, q \end{cases} \\ -\infty & \text{otherwise.} \end{cases} \end{aligned}$$

Since $X^\top X$ is symmetric positive semidefinite and $\gamma \geq 0$, obviously

$$G(\lambda, \mu, \alpha, \beta, \gamma) \leq 0$$

for all $\gamma > 0$.

The dual program is given by

$$\begin{aligned}
&\text{maximize} \quad -\frac{1}{4\gamma}\begin{pmatrix}\lambda^\top & \mu^\top\end{pmatrix} X^\top X \begin{pmatrix}\lambda \\ \mu\end{pmatrix} - \gamma \quad \text{if } \gamma > 0 \\
&\qquad\qquad\quad 0 \quad \text{if} \quad \gamma = 0 \\
&\text{subject to} \\
&\qquad \sum_{i=1}^{p} \lambda_i - \sum_{j=1}^{q} \mu_j = 0 \\
&\qquad \sum_{i=1}^{p} \lambda_i + \sum_{j=1}^{q} \mu_j = 1 \\
&\qquad 0 \leq \lambda_i \leq K, \quad i = 1, \ldots, p \\
&\qquad 0 \leq \mu_j \leq K, \quad j = 1, \ldots, q.
\end{aligned}$$

Also, if $\gamma = 0$, then $X \begin{pmatrix}\lambda \\ \mu\end{pmatrix} = 0$.

Maximizing with respect to $\gamma > 0$ by setting $\frac{\partial}{\partial\gamma} G(\lambda, \mu, \alpha, \beta, \gamma) = 0$ yields

$$\gamma^2 = \frac{1}{4}\begin{pmatrix}\lambda^\top & \mu^\top\end{pmatrix} X^\top X \begin{pmatrix}\lambda \\ \mu\end{pmatrix},$$

so we obtain

$$G(\lambda, \mu) = -\left(\begin{pmatrix}\lambda^\top & \mu^\top\end{pmatrix} X^\top X \begin{pmatrix}\lambda \\ \mu\end{pmatrix}\right)^{1/2}.$$

Finally, since $G(\lambda, \mu) = 0$ and $X \begin{pmatrix}\lambda \\ \mu\end{pmatrix} = 0$ if $\gamma = 0$, the dual program is equivalent to the following minimization program:

Dual of Soft margin SVM (SVM_{s1}):

$$\begin{aligned}
&\text{minimize} \quad \begin{pmatrix}\lambda^\top & \mu^\top\end{pmatrix} X^\top X \begin{pmatrix}\lambda \\ \mu\end{pmatrix} \\
&\text{subject to} \\
&\qquad \sum_{i=1}^{p} \lambda_i - \sum_{j=1}^{q} \mu_j = 0 \\
&\qquad \sum_{i=1}^{p} \lambda_i + \sum_{j=1}^{q} \mu_j = 1 \\
&\qquad 0 \leq \lambda_i \leq K, \quad i = 1, \ldots, p \\
&\qquad 0 \leq \mu_j \leq K, \quad j = 1, \ldots, q.
\end{aligned}$$

Observe that the constraints imply that K must be chosen so that

$$K \geq \max\left\{\frac{1}{2p}, \frac{1}{2q}\right\}.$$

If $(w, \delta, b, \epsilon, \xi)$ is an optimal solution of Problem (SVM$_{s1}$), then the complementary slackness conditions yield a classification of the points u_i and v_j in terms of the values of λ and μ. Indeed, we have $\epsilon_i\alpha_i = 0$ for $i = 1, \ldots, p$ and $\xi_j\beta_j = 0$ for $j = 1, \ldots, q$. Also, if $\lambda_i > 0$, then corresponding constraint is active, and similarly if $\mu_j > 0$. Since $\lambda_i + \alpha_i = K$, it follows that $\epsilon_i\alpha_i = 0$ iff $\epsilon_i(K - \lambda_i) = 0$, and since $\mu_j + \beta_j = K$, we have $\xi_j\beta_j = 0$ iff $\xi_j(K - \mu_j) = 0$. Thus if $\epsilon_i > 0$, then $\lambda_i = K$, and if $\xi_j > 0$, then $\mu_j = K$. Consequently, if $\lambda_i < K$, then $\epsilon_i = 0$ and u_i is correctly classified, and similarly if $\mu_j < K$, then $\xi_j = 0$ and v_j is correctly classified. We have the following classification:

(1) If $0 < \lambda_i < K$, then $\epsilon_i = 0$ and the i-th inequality is active, so

$$w^\top u_i - b - \delta = 0.$$

This means that u_i is on the blue margin (the hyperplane $H_{w,b+\delta}$ of equation $w^\top x = b + \delta$) and is classified correctly.
Similarly, if $0 < \mu_j < K$, then $\xi_j = 0$ and

$$w^\top v_j - b + \delta = 0,$$

so v_j is on the red margin (the hyperplane $H_{w,b-\delta}$ of equation $w^\top x = b - \delta$) and is classified correctly. See Figure 18.4.

(2) If $\lambda_i = K$, then the i-th inequality is active, so

$$w^\top u_i - b - \delta = -\epsilon_i.$$

If $\epsilon_i = 0$, then the point u_i is on the blue margin. If $\epsilon_i > 0$, then u_i is within the open half space bounded by the blue margin hyperplane $H_{w,b+\delta}$ and containing the separating hyperplane $H_{w,b}$; if $\epsilon_i \leq \delta$, then u_i is classified correctly, and if $\epsilon_i > \delta$, then u_i is misclassified (u_i lies on the wrong side of the separating hyperplane, the red side). See Figure 18.5.
Similarly, if $\mu_j = K$, then

$$w^\top v_j - b + \delta = \xi_j.$$

If $\xi_j = 0$, then the point v_j is on the red margin. If $\xi_j > 0$, then v_j is within the open half space bounded by the red margin hyperplane $H_{w,b-\delta}$ and containing the separating hyperplane $H_{w,b}$; if $\xi_j \leq \delta$, then v_j is classified correctly, and if $\xi_j > \delta$, then v_j is misclassified (v_j lies on the wrong side of the separating hyperplane, the blue side). See Figure 18.5.

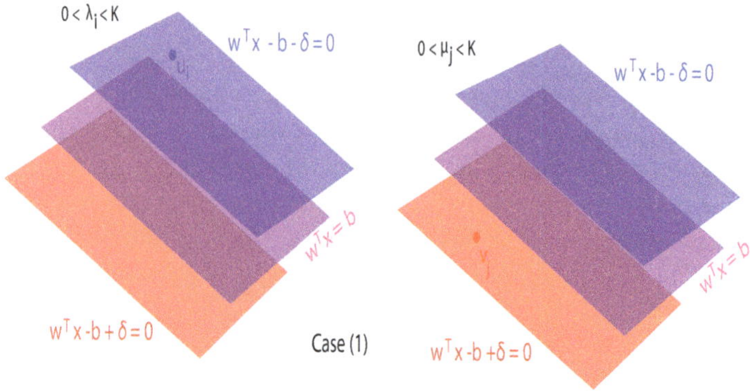

Fig. 18.4 When $0 < \lambda_i < K$, u_i is contained within the blue margin hyperplane. When $0 < \mu_j < K$, v_j is contained within the red margin hyperplane.

(3) If $\lambda_i = 0$, then $\epsilon_i = 0$ and the i-th inequality may or may not be active, so

$$w^\top u_i - b - \delta \geq 0.$$

Thus u_i is in the closed half space on the blue side bounded by the blue margin hyperplane $H_{w,b+\delta}$ (of course, classified correctly).
Similarly, if $\mu_j = 0$, then

$$w^\top v_j - b + \delta \leq 0$$

and v_j is in the closed half space on the red side bounded by the red margin hyperplane $H_{w,b-\delta}$ (of course, classified correctly). See Figure 18.6.

Definition 18.1. The vectors u_i on the blue margin $H_{w,b+\delta}$ and the vectors v_j on the red margin $H_{w,b-\delta}$ are called *support vectors*. Support vectors correspond to vectors u_i for which $w^\top u_i - b - \delta = 0$ (which implies $\epsilon_i = 0$), and vectors v_j for which $w^\top v_j - b + \delta = 0$ (which implies $\xi_j = 0$). Support vectors u_i such that $0 < \lambda_i < K$ and support vectors v_j such that $0 < \mu_j < K$ are *support vectors of type 1*. Support vectors of type 1 play a special role so we denote the sets of indices associated with them by

$$\begin{aligned} I_\lambda &= \{i \in \{1, \ldots, p\} \mid 0 < \lambda_i < K\} \\ I_\mu &= \{j \in \{1, \ldots, q\} \mid 0 < \mu_j < K\}. \end{aligned}$$

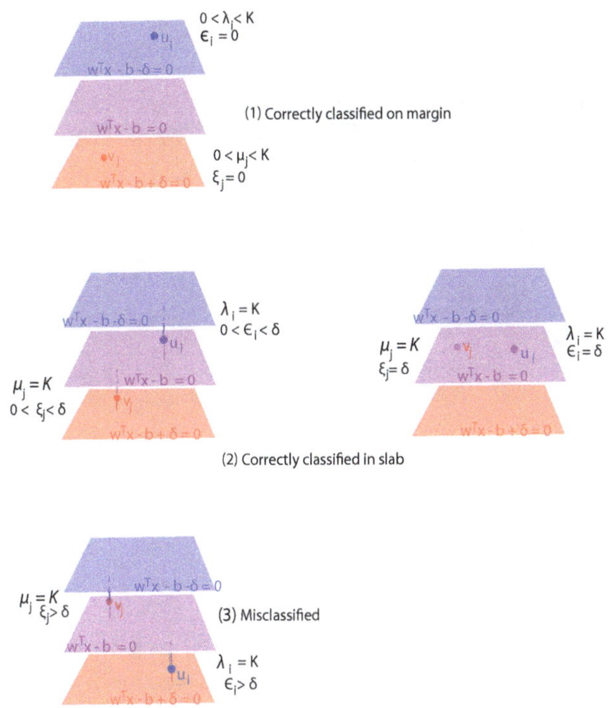

Fig. 18.5 Figure (1) illustrates the case of u_i contained in the margin and occurs when $\epsilon_i = 0$. Figure (1) also illustrates the case of v_j contained in the margin when $\xi_j = 0$. The left illustration of Figure (2) is when u_i is inside the margin yet still on the correct side of the separating hyperplane $w^\top x - b = 0$. Similarly, v_j is inside the margin on the correct side of the separating hyperplane. The right illustration depicts u_i and v_j on the separating hyperplane. Figure (3) illustrations a misclassification of u_i and v_j.

We denote their cardinalities by $numsvl_1 = |I_\lambda|$ and $numsvm_1 = |I_\mu|$. Support vectors u_i such that $\lambda_i = K$ and support vectors v_j such that $\mu_j = K$ are *support vectors of type 2*. Those support vectors u_i such that $\lambda_i = 0$ and those support vectors v_j such that $\mu_j = 0$ are called *exceptional support vectors*.

The vectors u_i for which $\lambda_i = K$ and the vectors v_j for which $\mu_j = K$ are said to *fail the margin*. The sets of indices associated with the vectors failing the margin are denoted by

$$K_\lambda = \{i \in \{1, \ldots, p\} \mid \lambda_i = K\}$$
$$K_\mu = \{j \in \{1, \ldots, q\} \mid \mu_j = K\}.$$

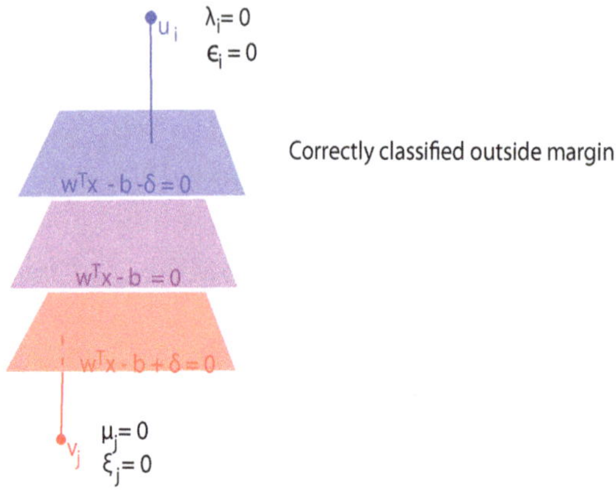

Fig. 18.6 When $\lambda_i = 0$, u_i is correctly classified outside the blue margin. When $\mu_j = 0$, v_j is correctly classified outside the red margin.

We denote their cardinalities by $p_f = |K_\lambda|$ and $q_f = |K_\mu|$.

Vectors u_i such that $\lambda_i > 0$ and vectors v_j such that $\mu_j > 0$ are said to *have margin at most* δ. The sets of indices associated with these vectors are denoted by

$$\begin{aligned} I_{\lambda>0} &= \{i \in \{1, \ldots, p\} \mid \lambda_i > 0\} \\ I_{\mu>0} &= \{j \in \{1, \ldots, q\} \mid \mu_j > 0\}. \end{aligned}$$

We denote their cardinalities by $p_m = |I_{\lambda>0}|$ and $q_m = |I_{\mu>0}|$.

Obviously, $I_{\lambda>0} = I_\lambda \cup K_\lambda$ and $I_{\mu>0} = I_\mu \cup K_\mu$, so $p_f \leq p_m$ and $q_f \leq q_m$. *Intuitively a blue point that fails the margin is on the wrong side of the blue margin and a red point that fails the margin is on the wrong side of the red margin.* The points in $I_{\lambda>0}$ not in K_λ are on the blue margin and the points in $I_{\mu>0}$ not in K_μ are on the red margin. There are $p - p_m$ points u_i classified correctly on the blue side and outside the δ-slab and there are $q - q_m$ points v_j classified correctly on the red side and outside the δ-slab.

It is easy to show that we have the following bounds on K:

$$\max\left\{\frac{1}{2p_m}, \frac{1}{2q_m}\right\} \leq K \leq \min\left\{\frac{1}{2p_f}, \frac{1}{2q_f}\right\}.$$

These inequalities restrict the choice of K quite heavily.

It will also be useful to understand how points are classified in terms of ϵ_i (or ξ_j).

(1) If $\epsilon_i > 0$, then by complementary slackness $\lambda_i = K$, so the ith equation is active and by (2) above,

$$w^\top u_i - b - \delta = -\epsilon_i.$$

Since $\epsilon_i > 0$, the point u_i is within the open half space bounded by the blue margin hyperplane $H_{w,b+\delta}$ and containing the separating hyperplane $H_{w,b}$; if $\epsilon_i \leq \delta$, then u_i is classified correctly, and if $\epsilon_i > \delta$, then u_i is misclassified.
Similarly, if $\xi_j > 0$, then v_j is within the open half space bounded by the red margin hyperplane $H_{w,b-\delta}$ and containing the separating hyperplane $H_{w,b}$; if $\xi_j \leq \delta$, then v_j is classified correctly, and if $\xi_j > \delta$, then v_j is misclassified.

(2) If $\epsilon_i = 0$, then the point u_i is correctly classified. If $\lambda_i = 0$, then by (3) above, u_i is in the closed half space on the blue side bounded by the blue margin hyperplane $H_{w,b+\delta}$. If $\lambda_i > 0$, then by (1) and (2) above, the point u_i is on the blue margin.
Similarly, if $\xi_j = 0$, then the point v_j is correctly classified. If $\mu_j = 0$, then v_j is in the closed half space on the red side bounded by the red margin hyperplane $H_{w,b-\delta}$, and if $\mu_j > 0$, then the point v_j is on the red margin.

It shown in Section 18.2 how the dual program is solved using ADMM from Section 16.6. If the primal problem is solvable, this yields solutions for λ and μ.

If the optimal value is 0, then $\gamma = 0$ and $X \begin{pmatrix} \lambda \\ \mu \end{pmatrix} = 0$, so in this case it is not possible to determine w. However, if the optimal value is > 0, then once a solution for λ and μ is obtained, by $(*_w)$, we have

$$\gamma = \frac{1}{2}\left(\begin{pmatrix} \lambda^\top & \mu^\top \end{pmatrix} X^\top X \begin{pmatrix} \lambda \\ \mu \end{pmatrix} \right)^{1/2}$$

$$w = \frac{1}{2\gamma}\left(\sum_{i=1}^{p} \lambda_i u_i - \sum_{j=1}^{q} \mu_j v_j \right),$$

so we get

$$w = \frac{\sum_{i=1}^{p} \lambda_i u_i - \sum_{j=1}^{q} \mu_j v_j}{\left(\begin{pmatrix} \lambda^\top & \mu^\top \end{pmatrix} X^\top X \begin{pmatrix} \lambda \\ \mu \end{pmatrix} \right)^{1/2}},$$

which is the result of making $\sum_{i=1}^{p} \lambda_i u_i - \sum_{j=1}^{q} \mu_j v_j$ a unit vector, since

$$X = \begin{pmatrix} -u_1 & \cdots & -u_p & v_1 & \cdots & v_q \end{pmatrix}.$$

It remains to find b and δ, which are not given by the dual program and for this we use the complementary slackness conditions.

The equations

$$\sum_{i=1}^{p} \lambda_i = \sum_{j=1}^{q} \mu_j = \frac{1}{2}$$

imply that there is some i_0 such that $\lambda_{i_0} > 0$ and some j_0 such that $\mu_{j_0} > 0$, but a priori, nothing prevents the situation where $\lambda_i = K$ for all nonzero λ_i or $\mu_j = K$ for all nonzero μ_j. If this happens, we can rerun the optimization method with a larger value of K. If the following mild hypothesis holds, then b and δ can be found.

Standard Margin Hypothesis for (SVM$_{s1}$). There is some index i_0 such that $0 < \lambda_{i_0} < K$ and there is some index j_0 such that $0 < \mu_{j_0} < K$. This means that some u_{i_0} is a support vector of type 1 on the blue margin, and some v_{j_0} is a support of type 1 on the red margin.

If the **Standard Margin Hypothesis** for (SVM$_{s1}$) holds, then $\epsilon_{i_0} = 0$ and $\mu_{j_0} = 0$, and then we have the active equations

$$w^\top u_{i_0} - b = \delta \quad \text{and} \quad -w^\top v_{j_0} + b = \delta,$$

and we obtain the values of b and δ as

$$\begin{aligned} b &= \frac{1}{2}(w^\top u_{i_0} + w^\top v_{j_0}) \\ \delta &= \frac{1}{2}(w^\top u_{i_0} - w^\top v_{j_0}). \end{aligned}$$

Due to numerical instability, when writing a computer program it is preferable to compute the lists of indices I_λ and I_μ given by

$$\begin{aligned} I_\lambda &= \{i \in \{1, \ldots, p\} \mid 0 < \lambda_i < K\} \\ I_\mu &= \{j \in \{1, \ldots, q\} \mid 0 < \mu_j < K\}. \end{aligned}$$

Then it is easy to see that we can compute b and δ using the following averaging formulae:

$$b = w^\top \left(\left(\sum_{i \in I_\lambda} u_i \right) / |I_\lambda| + \left(\sum_{j \in I_\mu} v_j \right) / |I_\mu| \right) /2$$

$$\delta = w^\top \left(\left(\sum_{i \in I_\lambda} u_i \right) / |I_\lambda| - \left(\sum_{j \in I_\mu} v_j \right) / |I_\mu| \right) /2.$$

As we said earlier, the hypotheses of Theorem 14.5(2) hold, so *if* the primal problem (SVM$_{s1}$) has an optimal solution with $w \neq 0$, *then* the dual problem has a solution too, and the duality gap is zero. Therefore, for optimal solutions we have

$$L(w, \epsilon, \xi, b, \delta, \lambda, \mu, \alpha, \beta, \gamma) = G(\lambda, \mu, \alpha, \beta, \gamma),$$

which means that

$$-\delta + K\left(\sum_{i=1}^p \epsilon_i + \sum_{j=1}^q \xi_j \right) = -\left(\begin{pmatrix} \lambda^\top & \mu^\top \end{pmatrix} X^\top X \begin{pmatrix} \lambda \\ \mu \end{pmatrix} \right)^{1/2},$$

so we get

$$\delta = K\left(\sum_{i=1}^p \epsilon_i + \sum_{j=1}^q \xi_j \right) + \left(\begin{pmatrix} \lambda^\top & \mu^\top \end{pmatrix} X^\top X \begin{pmatrix} \lambda \\ \mu \end{pmatrix} \right)^{1/2}.$$

Therefore, we confirm that $\delta \geq 0$.

It is important to note that the objective function of the dual program

$$-G(\lambda, \mu) = \left(\begin{pmatrix} \lambda^\top & \mu^\top \end{pmatrix} X^\top X \begin{pmatrix} \lambda \\ \mu \end{pmatrix} \right)^{1/2}$$

only involves the inner products of the u_i and the v_j through the matrix $X^\top X$, and similarly, the equation of the optimal hyperplane can be written as

$$\sum_{i=1}^p \lambda_i u_i^\top x - \sum_{j=1}^q \mu_j v_j^\top x - \left(\begin{pmatrix} \lambda^\top & \mu^\top \end{pmatrix} X^\top X \begin{pmatrix} \lambda \\ \mu \end{pmatrix} \right)^{1/2} b = 0,$$

an expression that only involves inner products of x with the u_i and the v_j and inner products of the u_i and the v_j.

As explained at the beginning of this chapter, this is a key fact that allows a generalization of the support vector machine using the method of *kernels*. We can define the following "kernelized" version of Problem (SVM$_{s1}$):

Soft margin kernel SVM (SVM_{s1}):

$$\begin{aligned}
&\text{minimize} \quad -\delta + K\Big(\sum_{i=1}^{p} \epsilon_i + \sum_{j=1}^{q} \xi_j\Big) \\
&\text{subject to} \\
&\qquad \langle w, \varphi(u_i)\rangle - b \geq \delta - \epsilon_i, \quad \epsilon_i \geq 0 \qquad i = 1, \ldots, p \\
&\qquad -\langle w, \varphi(v_j)\rangle + b \geq \delta - \xi_j, \quad \xi_j \geq 0 \qquad j = 1, \ldots, q \\
&\qquad \langle w, w\rangle \leq 1.
\end{aligned}$$

Tracing through the computation that led us to the dual program with u_i replaced by $\varphi(u_i)$ and v_j replaced by $\varphi(v_j)$, we find the following version of the dual program:

Dual of Soft margin kernel SVM (SVM_{s1}):

$$\begin{aligned}
&\text{minimize} \quad (\lambda^\top \ \mu^\top)\, \mathbf{K} \begin{pmatrix} \lambda \\ \mu \end{pmatrix} \\
&\text{subject to} \\
&\qquad \sum_{i=1}^{p} \lambda_i - \sum_{j=1}^{q} \mu_j = 0 \\
&\qquad \sum_{i=1}^{p} \lambda_i + \sum_{j=1}^{q} \mu_j = 1 \\
&\qquad 0 \leq \lambda_i \leq K, \quad i = 1, \ldots, p \\
&\qquad 0 \leq \mu_j \leq K, \quad j = 1, \ldots, q,
\end{aligned}$$

where $\mathbf{K}$ is the $\ell \times \ell$ kernel symmetric matrix (with $\ell = p + q$) given by

$$\mathbf{K}_{ij} = \begin{cases} \kappa(u_i, u_j) & 1 \leq i \leq p,\ 1 \leq j \leq q \\ -\kappa(u_i, v_{j-p}) & 1 \leq i \leq p,\ p+1 \leq j \leq p+q \\ -\kappa(v_{i-p}, u_j) & p+1 \leq i \leq p+q,\ 1 \leq j \leq p \\ \kappa(v_{i-p}, v_{j-q}) & p+1 \leq i \leq p+q,\ p+1 \leq j \leq p+q. \end{cases}$$

We also find that

$$w = \frac{\displaystyle\sum_{i=1}^{p} \lambda_i \varphi(u_i) - \sum_{j=1}^{q} \mu_j \varphi(v_j)}{\left((\lambda^\top \ \mu^\top)\, K \begin{pmatrix} \lambda \\ \mu \end{pmatrix}\right)^{1/2}}.$$

Under the Standard Margin Hypothesis, there is some index i_0 such that $0 < \lambda_{i_0} < K$ and there is some index j_0 such that $0 < \mu_{j_0} < K$, and we obtain the value of b and δ as

$$b = \frac{1}{2}(\langle w, \varphi(u_{i_0}\rangle + \langle w, \varphi(v_{j_0})\rangle)$$
$$\delta = \frac{1}{2}(\langle w, \varphi(u_{i_0})\rangle - \langle w, \varphi(v_{j_0})\rangle).$$

Using the above value for w, we obtain

$$b = \frac{\sum_{i=1}^{p} \lambda_i(\kappa(u_i, u_{i_0}) + \kappa(u_i, v_{j_0})) - \sum_{j=1}^{q} \mu_j(\kappa(v_j, u_{i_0}) + \kappa(v_j, v_{j_0}))}{2\left(\begin{pmatrix}\lambda^\top & \mu^\top\end{pmatrix} K \begin{pmatrix}\lambda \\ \mu\end{pmatrix}\right)^{1/2}}.$$

It follows that the classification function

$$f(x) = \operatorname{sgn}(\langle w, \varphi(x)\rangle - b)$$

is given by

$$f(x) = \operatorname{sgn}\Bigg(\sum_{i=1}^{p} \lambda_i(2\kappa(u_i, x) - \kappa(u_i, u_{i_0}) - \kappa(u_i, v_{j_0})) \\ - \sum_{j=1}^{q} \mu_j(2\kappa(v_j, x) - \kappa(v_j, u_{i_0}) - \kappa(v_j, v_{j_0}))\Bigg),$$

which is solely expressed in terms of the kernel κ.

Kernel methods for SVM are discussed in Schölkopf and Smola [Schölkopf and Smola (2002)] and Shawe-Taylor and Christianini [Shawe-Taylor and Cristianini (2004)].

18.2 Solving SVM (SVM$_{s1}$) Using ADMM

In order to solve (SVM$_{s1}$) using ADMM we need to write the matrix corresponding to the constraints in equational form,

$$\sum_{i=1}^{p} \lambda_i - \sum_{j=1}^{q} \mu_j = 0$$
$$\sum_{i=1}^{p} \lambda_i + \sum_{j=1}^{q} \mu_j = 1$$
$$\lambda_i + \alpha_i = K, \quad i = 1, \ldots, p$$
$$\mu_j + \beta_j = K, \quad j = 1, \ldots, q.$$

This is the $(p+q+2) \times 2(p+q)$ matrix A given by

$$A = \begin{pmatrix} \mathbf{1}_p^\top & -\mathbf{1}_q^\top & 0_p^\top & 0_q^\top \\ \mathbf{1}_p^\top & \mathbf{1}_q^\top & 0_p^\top & 0_q^\top \\ I_p & 0_{p,q} & I_q & 0_{p,q} \\ 0_{q,p} & I_q & 0_{q,p} & I_q \end{pmatrix}.$$

We leave it as an exercise to prove that A has rank $p+q+2$. The right-hand side is

$$c = \begin{pmatrix} 0 \\ 1 \\ K\mathbf{1}_{p+q} \end{pmatrix}.$$

The symmetric positive semidefinite $(p+q) \times (p+q)$ matrix P defining the quadratic functional is

$$P = 2X^\top X, \quad \text{with} \quad X = \begin{pmatrix} -u_1 & \cdots & -u_p & v_1 & \cdots & v_q \end{pmatrix},$$

and

$$q = 0_{p+q}.$$

Since there are $2(p+q)$ Lagrange multipliers $(\lambda, \mu, \alpha, \beta)$, the $(p+q) \times (p+q)$ matrix $X^\top X$ must be augmented with zero's to make it a $2(p+q) \times 2(p+q)$ matrix P_a given by

$$P_a = \begin{pmatrix} X^\top X & 0_{p+q,p+q} \\ 0_{p+q,p+q} & 0_{p+q,p+q} \end{pmatrix},$$

and similarly q is augmented with zeros as the vector $q_a = 0_{2(p+q)}$.

Since the constraint $w^\top w \leq 1$ causes troubles, we trade it for a different objective function in which $-\delta$ is replaced by $(1/2)\,\|w\|_2^2$. This way we are left with purely affine constraints. In the next section we discuss a generalization of Problem (SVM$_{h2}$) obtained by adding a linear regularizing term.

18.3 Soft Margin Support Vector Machines; (SVM$_{s2}$)

In this section we consider the generalization of Problem (SVM$_{h2}$) where we minimize $(1/2)w^\top w$ by adding the "regularizing term" $K\Big(\sum_{i=1}^p \epsilon_i + \sum_{j=1}^q \xi_j,\Big)$ for some $K > 0$. Recall that the margin δ is given by $\delta = 1/\,\|w\|$.

Soft margin SVM (SVM_{s2}):

$$\begin{aligned}
&\text{minimize} \quad \frac{1}{2}w^\top w + K\left(\epsilon^\top\ \xi^\top\right)\mathbf{1}_{p+q} \\
&\text{subject to} \\
&\qquad w^\top u_i - b \geq 1 - \epsilon_i, \quad \epsilon_i \geq 0 \qquad i = 1,\ldots,p \\
&\qquad - w^\top v_j + b \geq 1 - \xi_j, \quad \xi_j \geq 0 \qquad j = 1,\ldots,q.
\end{aligned}$$

This is the classical problem discussed in all books on machine learning or pattern analysis, for instance Vapnik [Vapnik (1998)], Bishop [Bishop (2006)], and Shawe-Taylor and Christianini [Shawe-Taylor and Cristianini (2004)]. The trivial solution where all variables are 0 is ruled out because of the presence of the 1 in the inequalities, but it is not clear that if (w, b, ϵ, ξ) is an optimal solution, then $w \neq 0$.

We prove that if the primal problem has an optimal solution (w, ϵ, ξ, b) with $w \neq 0$, then w is determined by any optimal solution (λ, μ) of the dual. We also prove that there is some i for which $\lambda_i > 0$ and some j for which $\mu_j > 0$. Under a mild hypothesis that we call the **Standard Margin Hypothesis**, b can be found.

Note that this framework is still somewhat sensitive to outliers because the penalty for misclassification is linear in ϵ and ξ.

First we write the constraints in matrix form. The $2(p+q)\times(n+p+q+1)$ matrix C is written in block form as

$$C = \begin{pmatrix} X^\top & -I_{p+q} & \begin{matrix}\mathbf{1}_p \\ -\mathbf{1}_q\end{matrix} \\ 0_{p+q,n} & -I_{p+q} & 0_{p+q} \end{pmatrix},$$

where X is the $n \times (p+q)$ matrix

$$X = \begin{pmatrix} -u_1 & \cdots & -u_p & v_1 & \cdots & v_q \end{pmatrix},$$

and the constraints are expressed by

$$\begin{pmatrix} X^\top & -I_{p+q} & \begin{matrix}\mathbf{1}_p \\ -\mathbf{1}_q\end{matrix} \\ 0_{p+q,n} & -I_{p+q} & 0_{p+q} \end{pmatrix} \begin{pmatrix} w \\ \epsilon \\ \xi \\ b \end{pmatrix} \leq \begin{pmatrix} -\mathbf{1}_{p+q} \\ 0_{p+q} \end{pmatrix}.$$

The objective function $J(w, \epsilon, \xi, b)$ is given by

$$J(w, \epsilon, \xi, b) = \frac{1}{2}w^\top w + K\left(\epsilon^\top\ \xi^\top\right)\mathbf{1}_{p+q}.$$

The Lagrangian $L(w, \epsilon, \xi, b, \lambda, \mu, \alpha, \beta)$ with $\lambda, \alpha \in \mathbb{R}^p_+$ and with $\mu, \beta \in \mathbb{R}^q_+$ is given by

$$L(w, \epsilon, \xi, b, \lambda, \mu, \alpha, \beta) = \frac{1}{2} w^\top w + K \begin{pmatrix} \epsilon^\top & \xi^\top \end{pmatrix} \mathbf{1}_{p+q} \\ + \begin{pmatrix} w^\top & \begin{pmatrix} \epsilon^\top & \xi^\top \end{pmatrix} & b \end{pmatrix} C^\top \begin{pmatrix} \lambda \\ \mu \\ \alpha \\ \beta \end{pmatrix} + \begin{pmatrix} \mathbf{1}_{p+q}^\top & 0_{p+q}^\top \end{pmatrix} \begin{pmatrix} \lambda \\ \mu \\ \alpha \\ \beta \end{pmatrix}.$$

Since

$$\begin{pmatrix} w^\top & \begin{pmatrix} \epsilon^\top & \xi^\top \end{pmatrix} & b \end{pmatrix} C^\top \begin{pmatrix} \lambda \\ \mu \\ \alpha \\ \beta \end{pmatrix} = \begin{pmatrix} w^\top & \begin{pmatrix} \epsilon^\top & \xi^\top \end{pmatrix} & b \end{pmatrix} \begin{pmatrix} X & 0_{n,p+q} \\ -I_{p+q} & -I_{p+q} \\ \mathbf{1}_p^\top \; -\mathbf{1}_q^\top & 0_{p+q}^\top \end{pmatrix} \begin{pmatrix} \lambda \\ \mu \\ \alpha \\ \beta \end{pmatrix},$$

we get

$$\begin{aligned} \begin{pmatrix} w^\top & \begin{pmatrix} \epsilon^\top & \xi^\top \end{pmatrix} & b \end{pmatrix} C^\top \begin{pmatrix} \lambda \\ \mu \\ \alpha \\ \beta \end{pmatrix} &= \begin{pmatrix} w^\top & \begin{pmatrix} \epsilon^\top & \xi^\top \end{pmatrix} & b \end{pmatrix} \begin{pmatrix} X \begin{pmatrix} \lambda \\ \mu \end{pmatrix} \\ -\begin{pmatrix} \lambda + \alpha \\ \mu + \beta \end{pmatrix} \\ \mathbf{1}_p^\top \lambda - \mathbf{1}_q^\top \mu \end{pmatrix} \\ &= w^\top X \begin{pmatrix} \lambda \\ \mu \end{pmatrix} - \epsilon^\top (\lambda + \alpha) - \xi^\top (\mu + \beta) + b(\mathbf{1}_p^\top \lambda - \mathbf{1}_q^\top \mu), \end{aligned}$$

and since

$$\begin{pmatrix} \mathbf{1}_{p+q}^\top & 0_{p+q}^\top \end{pmatrix} \begin{pmatrix} \lambda \\ \mu \\ \alpha \\ \beta \end{pmatrix} = \mathbf{1}_{p+q}^\top \begin{pmatrix} \lambda \\ \mu \end{pmatrix} = \begin{pmatrix} \lambda^\top & \mu^\top \end{pmatrix} \mathbf{1}_{p+q},$$

the Lagrangian can be rewritten as

$$L(w, \epsilon, \xi, b, \lambda, \mu, \alpha, \beta) = \frac{1}{2} w^\top w + w^\top X \begin{pmatrix} \lambda \\ \mu \end{pmatrix} + \epsilon^\top (K\mathbf{1}_p - (\lambda + \alpha)) \\ + \xi^\top (K\mathbf{1}_q - (\mu + \beta)) + b(\mathbf{1}_p^\top \lambda - \mathbf{1}_q^\top \mu) + \begin{pmatrix} \lambda^\top & \mu^\top \end{pmatrix} \mathbf{1}_{p+q}.$$

To find the dual function $G(\lambda, \mu, \alpha, \beta)$ we minimize $L(w, \epsilon, \xi, b, \lambda, \mu, \alpha, \beta)$ with respect to w, ϵ, ξ and b. Since the Lagrangian is convex and $(w, \epsilon, \xi, b) \in \mathbb{R}^n \times \mathbb{R}^p \times \mathbb{R}^q \times \mathbb{R}$, a convex open set, by Theorem 4.5, the Lagrangian has a minimum in (w, ϵ, ξ, b) iff $\nabla L_{w,\epsilon,\xi,b} = 0$, so we compute its gradient with respect to w, ϵ, ξ and b, and we get

$$\nabla L_{w,\epsilon,\xi,b} = \begin{pmatrix} w + X\begin{pmatrix}\lambda \\ \mu\end{pmatrix} \\ K\mathbf{1}_p - (\lambda + \alpha) \\ K\mathbf{1}_q - (\mu + \beta) \\ \mathbf{1}_p^\top \lambda - \mathbf{1}_q^\top \mu \end{pmatrix}.$$

By setting $\nabla L_{w,\epsilon,\xi,b} = 0$ we get the equations

$$w = -X\begin{pmatrix}\lambda \\ \mu\end{pmatrix} \qquad (*_w)$$

$$\begin{aligned}\lambda + \alpha &= K\mathbf{1}_p \\ \mu + \beta &= K\mathbf{1}_q \\ \mathbf{1}_p^\top \lambda &= \mathbf{1}_q^\top \mu.\end{aligned}$$

The first and the fourth equation are identical to the equations $(*_1)$ and $(*_2)$ that we obtained in Example 14.10. Since $\lambda, \mu, \alpha, \beta \geq 0$, the second and the third equation are equivalent to the box constraints

$$0 \leq \lambda_i, \mu_j \leq K, \quad i = 1, \ldots, p,\ j = 1, \ldots, q.$$

Using the equations that we just derived, after simplifications we get

$$G(\lambda, \mu, \alpha, \beta) = -\frac{1}{2}\left(\lambda^\top\ \mu^\top\right) X^\top X \begin{pmatrix}\lambda \\ \mu\end{pmatrix} + \left(\lambda^\top\ \mu^\top\right) \mathbf{1}_{p+q},$$

which is independent of α and β and is identical to the dual function obtained in $(*_4)$ of Example 14.10. To be perfectly rigorous,

$$G(\lambda, \mu) = -\frac{1}{2}\left(\lambda^\top\ \mu^\top\right) X^\top X \begin{pmatrix}\lambda \\ \mu\end{pmatrix} + \left(\lambda^\top\ \mu^\top\right) \mathbf{1}_{p+q}$$

$$\text{if} \quad \begin{cases} \sum_{i=1}^p \lambda_i = \sum_{j=1}^q \mu_j \\ 0 \leq \lambda_i \leq K,\ i = 1, \ldots, p \\ 0 \leq \mu_j \leq K,\ j = 1, \ldots, q \end{cases}$$

$$-\infty \quad \text{otherwise.}$$

As in Example 14.10, the the dual program can be formulated as

$$\text{maximize} \quad -\frac{1}{2}\left(\lambda^\top\ \mu^\top\right) X^\top X \begin{pmatrix}\lambda \\ \mu\end{pmatrix} + \left(\lambda^\top\ \mu^\top\right) \mathbf{1}_{p+q}$$

subject to

$$\begin{aligned} &\sum_{i=1}^p \lambda_i - \sum_{j=1}^q \mu_j = 0 \\ &0 \leq \lambda_i \leq K, \quad i = 1, \ldots, p \\ &0 \leq \mu_j \leq K, \quad j = 1, \ldots, q, \end{aligned}$$

or equivalently

Dual of Soft margin SVM (SVM_{s2}):

$$\begin{aligned} &\text{minimize} \quad \frac{1}{2}\begin{pmatrix}\lambda^\top & \mu^\top\end{pmatrix} X^\top X \begin{pmatrix}\lambda \\ \mu\end{pmatrix} - \begin{pmatrix}\lambda^\top & \mu^\top\end{pmatrix} \mathbf{1}_{p+q} \\ &\text{subject to} \\ &\qquad \sum_{i=1}^p \lambda_i - \sum_{j=1}^q \mu_j = 0 \\ &\qquad 0 \leq \lambda_i \leq K, \quad i = 1, \ldots, p \\ &\qquad 0 \leq \mu_j \leq K, \quad j = 1, \ldots, q. \end{aligned}$$

If (w, ϵ, ξ, b) is an optimal solution of Problem (SVM_{s2}), then the complementary slackness conditions yield a classification of the points u_i and v_j in terms of the values of λ and μ. Indeed, we have $\epsilon_i \alpha_i = 0$ for $i = 1, \ldots, p$ and $\xi_j \beta_j = 0$ for $j = 1, \ldots, q$. Also, if $\lambda_i > 0$, then corresponding constraint is active, and similarly if $\mu_j > 0$. Since $\lambda_i + \alpha_i = K$, it follows that $\epsilon_i \alpha_i = 0$ iff $\epsilon_i(K - \lambda_i) = 0$, and since $\mu_j + \beta_j = K$, we have $\xi_j \beta_j = 0$ iff $\xi_j(K - \mu_j) = 0$. Thus if $\epsilon_i > 0$, then $\lambda_i = K$, and if $\xi_j > 0$, then $\mu_j = K$. Consequently, if $\lambda_i < K$, then $\epsilon_i = 0$ and u_i is correctly classified, and similarly if $\mu_j < K$, then $\xi_j = 0$ and v_j is correctly classified.

We have a classification of the points u_i and v_j in terms of λ and μ obtained from the classification given in Section 18.1 by replacing δ with 1. Since it is so similar, it is omitted. Let us simply recall that the vectors u_i on the blue margin and the vectors v_j on the red margin are called *support vectors*; these are the vectors u_i for which $w^\top u_i - b - 1 = 0$ (which implies $\epsilon_i = 0$), and the vectors v_j for which $w^\top v_j - b + 1 = 0$ (which implies $\xi_j = 0$). Those support vectors u_i such that $\lambda_i = 0$ and those support vectors such that $\mu_j = 0$ are called *exceptional support vectors*.

We also have the following classification of the points u_i and v_j terms of ϵ_i (or ξ_j) obtained by replacing δ with 1.

(1) If $\epsilon_i > 0$, then by complementary slackness $\lambda_i = K$, so the ith equation is active and by (2) above,
$$w^\top u_i - b - 1 = -\epsilon_i.$$
Since $\epsilon_i > 0$, the point u_i is within the open half space bounded by the blue margin hyperplane $H_{w,b+1}$ and containing the separating hyperplane $H_{w,b}$; if $\epsilon_i \leq 1$, then u_i is classified correctly, and if $\epsilon_i > 1$, then u_i is misclassified.

Similarly, if $\xi_j > 0$, then v_j is within the open half space bounded by the red margin hyperplane $H_{w,b-1}$ and containing the separating hyperplane $H_{w,b}$; if $\xi_j \leq 1$, then v_j is classified correctly, and if $\xi_j > 1$, then v_j is misclassified.

(2) If $\epsilon_i = 0$, then the point u_i is correctly classified. If $\lambda_i = 0$, then by (3) above, u_i is in the closed half space on the blue side bounded by the blue margin hyperplane $H_{w,b+\eta}$. If $\lambda_i > 0$, then by (1) and (2) above, the point u_i is on the blue margin.
Similarly, if $\xi_j = 0$, then the point v_j is correctly classified. If $\mu_j = 0$, then v_j is in the closed half space on the red side bounded by the red margin hyperplane $H_{w,b-\eta}$, and if $\mu_j > 0$, then the point v_j is on the red margin. See Figure 18.5 (3).

Vectors u_i for which $\lambda_i = K$ and vectors v_j such that $\xi_j = K$ are said to *fail the margin*.

It is shown in Section 18.4 how the dual program is solved using ADMM from Section 16.6. If the primal problem is solvable, this yields solutions for λ and μ.

Remark. The hard margin Problem (SVM$_{h2}$) corresponds to the special case of Problem (SVM$_{s2}$) in which $\epsilon = 0$, $\xi = 0$, and $K = +\infty$. Indeed, in Problem (SVM$_{h2}$) the terms involving ϵ and ξ are missing from the Lagrangian and the effect is that the box constraints are missing; we simply have $\lambda_i \geq 0$ and $\mu_j \geq 0$.

We can use the dual program to solve the primal. Once $\lambda \geq 0, \mu \geq 0$ have been found, w is given by

$$w = -X \begin{pmatrix} \lambda \\ \mu \end{pmatrix} = \sum_{i=1}^{p} \lambda_i u_i - \sum_{j=1}^{q} \mu_j v_j.$$

To find b we use the complementary slackness conditions.

If the primal has a solution $w \neq 0$, then the equation

$$w = \sum_{i=1}^{p} \lambda_i u_i - \sum_{j=1}^{q} \mu_j v_j$$

implies that either there is some index i_0 such that $\lambda_{i_0} > 0$ or there is some index j_0 such that $\mu_{j_0} > 0$. The constraint

$$\sum_{i=1}^{p} \lambda_i - \sum_{j=1}^{q} \mu_j = 0$$

implies that there is some index i_0 such that $\lambda_{i_0} > 0$ and there is some index j_0 such that $\mu_{j_0} > 0$. However, a priori, nothing prevents the situation where $\lambda_i = K$ for all nonzero λ_i or $\mu_j = K$ for all nonzero μ_j. If this happens, we can rerun the optimization method with a larger value of K. Observe that the equation

$$\sum_{i=1}^{p} \lambda_i - \sum_{j=1}^{q} \mu_j = 0$$

implies that if there is some index i_0 such that $0 < \lambda_{i_0} < K$, then there is some index j_0 such that $0 < \mu_{j_0} < K$, and vice-versa. If the following mild hypothesis holds, then b can be found.

Standard Margin Hypothesis for (SVM_{s2}). There is some index i_0 such that $0 < \lambda_{i_0} < K$ and there is some index j_0 such that $0 < \mu_{j_0} < K$. This means that some u_{i_0} is a support vector of type 1 on the blue margin, and some v_{j_0} is a support vector of type 1 on the red margin.

If the **Standard Margin Hypothesis** for (SVM_{s2}) holds, then $\epsilon_{i_0} = 0$ and $\mu_{j_0} = 0$, and then we have the active equations

$$w^\top u_{i_0} - b = 1 \quad \text{and} \quad -w^\top v_{j_0} + b = 1,$$

and we obtain

$$b = \frac{1}{2}(w^\top u_{i_0} + w^\top v_{j_0}).$$

Due to numerical instability, when writing a computer program it is preferable to compute the lists of indices I_λ and I_μ given by

$$\begin{aligned} I_\lambda &= \{i \in \{1, \ldots, p\} \mid 0 < \lambda_i < K\} \\ I_\mu &= \{j \in \{1, \ldots, q\} \mid 0 < \mu_j < K\}. \end{aligned}$$

Then it is easy to see that we can compute b using the following averaging formula

$$b = w^\top \left(\Big(\sum_{i \in I_\lambda} u_i \Big) / |I_\lambda| + \Big(\sum_{j \in I_\mu} v_j \Big) / |I_\mu| \right) /2.$$

Recall that $\delta = 1/\left\| w \right\|$.

Remark: There is a cheap version of Problem (SVM_{s2}) which consists in dropping the term $(1/2)w^\top w$ from the objective function:

Soft margin classifier (SVM_{s2l}):

$$\begin{aligned}
&\text{minimize} \quad \sum_{i=1}^{p} \epsilon_i + \sum_{j=1}^{q} \xi_j \\
&\text{subject to} \\
&\qquad w^\top u_i - b \geq 1 - \epsilon_i, \quad \epsilon_i \geq 0 \qquad i = 1, \ldots, p \\
&\qquad - w^\top v_j + b \geq 1 - \xi_j, \quad \xi_j \geq 0 \qquad j = 1, \ldots, q.
\end{aligned}$$

The above program is a linear program that minimizes the number of misclassified points but does not care about enforcing a minimum margin. An example of its use is given in Boyd and Vandenberghe; see [Boyd and Vandenberghe (2004)], Section 8.6.1.

The "kernelized" version of Problem (SVM_{s2}) is the following:

Soft margin kernel SVM (SVM_{s2}):

$$\begin{aligned}
&\text{minimize} \quad \frac{1}{2}\langle w, w\rangle + K \left(\epsilon^\top \; \xi^\top\right) \mathbf{1}_{p+q} \\
&\text{subject to} \\
&\qquad \langle w, \varphi(u_i)\rangle - b \geq 1 - \epsilon_i, \quad \epsilon_i \geq 0 \qquad i = 1, \ldots, p \\
&\qquad - \langle w, \varphi(v_j)\rangle + b \geq 1 - \xi_j, \quad \xi_j \geq 0 \qquad j = 1, \ldots, q.
\end{aligned}$$

Redoing the computation of the dual function, we find that the dual program is given by

Dual of Soft margin kernel SVM (SVM_{s2}):

$$\begin{aligned}
&\text{minimize} \quad \frac{1}{2}\left(\lambda^\top \; \mu^\top\right) \mathbf{K} \begin{pmatrix} \lambda \\ \mu \end{pmatrix} - \left(\lambda^\top \; \mu^\top\right) \mathbf{1}_{p+q} \\
&\text{subject to} \\
&\qquad \sum_{i=1}^{p} \lambda_i - \sum_{j=1}^{q} \mu_j = 0 \\
&\qquad 0 \leq \lambda_i \leq K, \quad i = 1, \ldots, p \\
&\qquad 0 \leq \mu_j \leq K, \quad j = 1, \ldots, q,
\end{aligned}$$

where $\mathbf{K}$ is the $\ell \times \ell$ kernel symmetric matrix (with $\ell = p + q$) given at the end of Section 18.1. We also find that

$$w = \sum_{i=1}^{p} \lambda_i \varphi(u_i) - \sum_{j=1}^{q} \mu_j \varphi(v_j),$$

so

$$b = \frac{1}{2}\left(\sum_{i=1}^{p} \lambda_i(\kappa(u_i, u_{i_0}) + \kappa(u_i, v_{j_0})) - \sum_{j=1}^{q} \mu_j(\kappa(v_j, u_{i_0}) + \kappa(v_j, v_{j_0}))\right),$$

and the classification function

$$f(x) = \mathrm{sgn}(\langle w, \varphi(x)\rangle - b)$$

is given by

$$\begin{aligned} f(x) = \mathrm{sgn}\Bigg(&\sum_{i=1}^{p} \lambda_i(2\kappa(u_i, x) - \kappa(u_i, u_{i_0}) - \kappa(u_i, v_{j_0})) \\ &- \sum_{j=1}^{q} \mu_j(2\kappa(v_j, x) - \kappa(v_j, u_{i_0}) - \kappa(v_j, v_{j_0}))\Bigg). \end{aligned}$$

18.4 Solving SVM ($\mathbf{SVM_{s2}}$) Using ADMM

In order to solve (SVM_{s2}) using ADMM we need to write the matrix corresponding to the constraints in equational form,

$$\begin{aligned} \sum_{i=1}^{p} \lambda_i - \sum_{j=1}^{q} \mu_j &= 0 \\ \lambda_i + \alpha_i &= K, \quad i = 1, \ldots, p \\ \mu_j + \beta_j &= K, \quad j = 1, \ldots, q. \end{aligned}$$

This is the $(p + q + 1) \times 2(p + q)$ matrix A given by

$$A = \begin{pmatrix} \mathbf{1}_p^\top & -\mathbf{1}_q^\top & 0_p^\top & 0_q^\top \\ I_p & 0_{p,q} & I_q & 0_{p,q} \\ 0_{q,p} & I_q & 0_{q,p} & I_q \end{pmatrix}.$$

We leave it as an exercise to prove that A has rank $p+q+1$. The right-hand side is

$$c = \begin{pmatrix} 0 \\ K\mathbf{1}_{p+q} \end{pmatrix}.$$

The symmetric positive semidefinite $(p+q) \times (p+q)$ matrix P defining the quadratic functional is

$$P = X^\top X, \quad \text{with} \quad X = \begin{pmatrix} -u_1 & \cdots & -u_p & v_1 & \cdots & v_q \end{pmatrix},$$

and

$$q = -\mathbf{1}_{p+q}.$$

Since there are $2(p+q)$ Lagrange multipliers $(\lambda, \mu, \alpha, \beta)$, the $(p+q)\times(p+q)$ matrix $X^\top X$ must be augmented with zero's to make it a $2(p+q)\times 2(p+q)$ matrix P_a given by

$$P_a = \begin{pmatrix} X^\top X & 0_{p+q,p+q} \\ 0_{p+q,p+q} & 0_{p+q,p+q} \end{pmatrix},$$

and similarly q is augmented with zeros as the vector

$$q_a = \begin{pmatrix} -\mathbf{1}_{p+q} \\ 0_{p+q}. \end{pmatrix}$$

18.5 Soft Margin Support Vector Machines; ($\mathrm{SVM}_{s2'}$)

In this section we consider a generalization of Problem (SVM_{s2}) for a version of the soft margin SVM coming from Problem (SVM_{h2}) by adding an extra degree of freedom, namely instead of the margin $\delta = 1/\|w\|$, we use the margin $\delta = \eta/\|w\|$ where η is some positive constant that we wish to maximize. To do so, we add a term $-K_m\eta$ to the objective function $(1/2)w^\top w$ as well as the "regularizing term" $K_s\left(\sum_{i=1}^p \epsilon_i + \sum_{j=1}^q \xi_j\right)$ whose purpose is to make ϵ and ξ sparse, where $K_m > 0$ (m refers to margin) and $K_s > 0$ (s refers to sparse) are fixed constants that can be adjusted to determine the influence of η and the regularizing term.

Soft margin SVM ($\mathrm{SVM}_{s2'}$):

$$\text{minimize} \quad \frac{1}{2}w^\top w - K_m\eta + K_s\left(\epsilon^\top \; \xi^\top\right)\mathbf{1}_{p+q}$$

subject to

$$\begin{aligned} w^\top u_i - b &\geq \eta - \epsilon_i, \quad & \epsilon_i \geq 0 \qquad & i = 1, \ldots, p \\ -w^\top v_j + b &\geq \eta - \xi_j, \quad & \xi_j \geq 0 \qquad & j = 1, \ldots, q \\ \eta &\geq 0. & & \end{aligned}$$

This version of the SVM problem was first discussed in Schölkopf, Smola, Williamson, and Bartlett [Schölkopf *et al.* (2000)] under the name of ν-*SVC* (or ν-*SVM*), and also used in Schölkopf, Platt, Shawe-Taylor, and Smola [Schölkopf *et al.* (2001)]. The ν-SVC method is also presented in Schölkopf and Smola [Schölkopf and Smola (2002)] (which contains much more). The difference between the ν-SVC method and the method presented in Section 18.3, sometimes called the C-SVM method, was thoroughly investigated by Chan and Lin [Chang and Chih-Jen (2001)].

For this problem it is no longer clear that if $(w, \eta, b, \epsilon, \xi)$ is an optimal solution, then $w \neq 0$ and $\eta > 0$. In fact, if the sets of points are not linearly separable and if K_s is chosen too big, Problem $(\mathrm{SVM}_{s2'})$ may fail to have an optimal solution.

We show that in order for the problem to have a solution we must pick K_m and K_s so that

$$K_m \leq \min\{2pK_s, 2qK_s\}.$$

If we define ν by

$$\nu = \frac{K_m}{(p+q)K_s},$$

then $K_m \leq \min\{2pK_s, 2qK_s\}$ is equivalent to

$$\nu \leq \min\left\{\frac{2p}{p+q}, \frac{2q}{p+q}\right\} \leq 1.$$

The reason for introducing ν is that $\nu(p+q)/2$ can be interpreted as the maximum number of points failing to achieve the margin $\delta = \eta/\left\|w\right\|$. We will show later that if the points u_i and v_j are not separable, then we must pick ν so that $\nu \geq 2/(p+q)$ for the method to have a solution for which $w \neq 0$ and $\eta > 0$.

The objective function of our problem is convex and the constraints are affine. Consequently, by Theorem 14.5(2) *if* the Primal Problem $(\mathrm{SVM}_{s2'})$ has an optimal solution, *then* the dual problem has a solution too, and the duality gap is zero. This does not immediately imply that an optimal solution of the dual yields an optimal solution of the primal because the hypotheses of Theorem 14.5(1) fail to hold.

We show that if the primal problem has an optimal solution $(w, \eta, \epsilon, \xi, b)$ with $w \neq 0$, then any optimal solution of the dual problem determines λ and μ, which in turn determine w *via* the equation

$$w = -X\begin{pmatrix}\lambda \\ \mu\end{pmatrix} = \sum_{i=1}^{p} \lambda_i u_i - \sum_{j=1}^{q} \mu_j v_j, \qquad (*_w)$$

and $\eta \geq 0$.

It remains to determine b, η, ϵ and ξ. The solution of the dual does not determine b, η, ϵ, ξ directly, and we are not aware of necessary and sufficient conditions that ensure that they can be determined. The best we can do is to use the KKT conditions.

The simplest sufficient condition is what we call the

Standard Margin Hypothesis for (SVM$_{s2'}$): There is some i_0 such that $0 < \lambda_{i_0} < K_s$, and there is some μ_{j_0} such that $0 < \mu_{j_0} < K_s$. This means that there is some support vector u_{i_0} of type 1 *and* there is some support vector v_{j_0} of type 1.

In this case, then by complementary slackness, it can be shown that $\epsilon_{i_0} = 0$, $\xi_{i_0} = 0$, and the corresponding inequalities are active, that is we have the equations

$$w^\top u_{i_0} - b = \eta, \quad -w^\top v_{j_0} + b = \eta,$$

so we can solve for b and η. Then since by complementary slackness, if $\epsilon_i > 0$, then $\lambda_i = K_s$ and if $\xi_j > 0$, then $\mu_j = K_s$, all inequalities corresponding to such $\epsilon_i > 0$ and $\mu_j > 0$ are active, and we can solve for ϵ_i and ξ_j.

The linear constraints are given by the $(2(p+q)+1) \times (n+p+q+2)$ matrix given in block form by

$$C = \begin{pmatrix} X^\top & -I_{p+q} & \begin{matrix} \mathbf{1}_p \\ -\mathbf{1}_q \end{matrix} & \mathbf{1}_{p+q} \\ 0_{p+q,n} & -I_{p+q} & 0_{p+q} & 0_{p+q} \\ 0_n^\top & 0_{p+q}^\top & 0 & -1 \end{pmatrix},$$

where X is the $n \times (p+q)$ matrix

$$X = \begin{pmatrix} -u_1 & \cdots & -u_p & v_1 & \cdots & v_q \end{pmatrix},$$

and the linear constraints are expressed by

$$\begin{pmatrix} X^\top & -I_{p+q} & \begin{matrix} \mathbf{1}_p \\ -\mathbf{1}_q \end{matrix} & \mathbf{1}_{p+q} \\ 0_{p+q,n} & -I_{p+q} & 0_{p+q} & 0_{p+q} \\ 0_n^\top & 0_{p+q}^\top & 0 & -1 \end{pmatrix} \begin{pmatrix} w \\ \epsilon \\ \xi \\ b \\ \eta \end{pmatrix} \leq \begin{pmatrix} 0_{p+q} \\ 0_{p+q} \\ 0 \end{pmatrix}.$$

The objective function is given by

$$J(w, \epsilon, \xi, b, \eta) = \frac{1}{2} w^\top w - K_m \eta + K_s \begin{pmatrix} \epsilon^\top & \xi^\top \end{pmatrix} \mathbf{1}_{p+q}.$$

The Lagrangian $L(w, \epsilon, \xi, b, \eta, \lambda, \mu, \alpha, \beta, \gamma)$ with $\lambda, \alpha \in \mathbb{R}^p_+$, $\mu, \beta \in \mathbb{R}^q_+$, and $\gamma \in \mathbb{R}_+$ is given by

$$\begin{aligned} L(w, \epsilon, \xi, b, \eta, \lambda, \mu, \alpha, \beta, \gamma) &= \frac{1}{2} w^\top w - K_m \eta + K_s \begin{pmatrix} \epsilon^\top & \xi^\top \end{pmatrix} \mathbf{1}_{p+q} \\ &\quad + \begin{pmatrix} w^\top & \begin{pmatrix} \epsilon^\top & \xi^\top \end{pmatrix} & b & \eta \end{pmatrix} C^\top \begin{pmatrix} \lambda \\ \mu \\ \alpha \\ \beta \\ \gamma \end{pmatrix}. \end{aligned}$$

Since

$$\begin{pmatrix} w^\top & (\epsilon^\top \ \xi^\top) & b & \eta \end{pmatrix} C^\top \begin{pmatrix} \lambda \\ \mu \\ \alpha \\ \beta \\ \gamma \end{pmatrix} = w^\top X \begin{pmatrix} \lambda \\ \mu \end{pmatrix} - \epsilon^\top(\lambda + \alpha) - \xi^\top(\mu + \beta)$$

$$+ b(\mathbf{1}_p^\top \lambda - \mathbf{1}_q^\top \mu) + \eta(\mathbf{1}_p^\top \lambda + \mathbf{1}_q^\top \mu) - \gamma\eta,$$

the Lagrangian can be written as

$$\begin{aligned} L(w, \epsilon, \xi, b, \eta, \lambda, \mu, \alpha, \beta, \gamma) &= \frac{1}{2} w^\top w + K_s(\epsilon^\top \mathbf{1}_p + \xi^\top \mathbf{1}_q) + w^\top X \begin{pmatrix} \lambda \\ \mu \end{pmatrix} \\ - K_m\eta - \epsilon^\top(\lambda + \alpha) &- \xi^\top(\mu + \beta) + b(\mathbf{1}_p^\top \lambda - \mathbf{1}_q^\top \mu) + \eta(\mathbf{1}_p^\top \lambda + \mathbf{1}_q^\top \mu) - \gamma\eta \\ &= \frac{1}{2} w^\top w + w^\top X \begin{pmatrix} \lambda \\ \mu \end{pmatrix} + (\mathbf{1}_p^\top \lambda + \mathbf{1}_q^\top \mu - K_m - \gamma)\eta \\ &+ \epsilon^\top(K_s \mathbf{1}_p - (\lambda + \alpha)) + \xi^\top(K_s \mathbf{1}_q - (\mu + \beta)) + b(\mathbf{1}_p^\top \lambda - \mathbf{1}_q^\top \mu). \end{aligned}$$

To find the dual function $G(\lambda, \mu, \alpha, \beta, \gamma)$ we minimize $L(w, \epsilon, \xi, b, \eta, \lambda, \mu, \alpha, \beta, \gamma)$ with respect to w, ϵ, ξ, b, and η. Since the Lagrangian is convex and $(w, \epsilon, \xi, b, \eta) \in \mathbb{R}^n \times \mathbb{R}^p \times \mathbb{R}^q \times \mathbb{R} \times \mathbb{R}$, a convex open set, by Theorem 4.5, the Lagrangian has a minimum in $(w, \epsilon, \xi, b, \eta)$ iff $\nabla L_{w,\epsilon,\xi,b,\eta} = 0$, so we compute its gradient with respect to $w, \epsilon, \xi, b, \eta$, and we get

$$\nabla L_{w,\epsilon,\xi,b,\eta} = \begin{pmatrix} X \begin{pmatrix} \lambda \\ \mu \end{pmatrix} + w \\ K_s \mathbf{1}_p - (\lambda + \alpha) \\ K_s \mathbf{1}_q - (\mu + \beta) \\ \mathbf{1}_p^\top \lambda - \mathbf{1}_q^\top \mu \\ \mathbf{1}_p^\top \lambda + \mathbf{1}_q^\top \mu - K_m - \gamma \end{pmatrix}.$$

By setting $\nabla L_{w,\epsilon,\xi,b,\eta} = 0$ we get the equations

$$w = -X \begin{pmatrix} \lambda \\ \mu \end{pmatrix} \qquad (*_w)$$

$$\begin{aligned} \lambda + \alpha &= K_s \mathbf{1}_p \\ \mu + \beta &= K_s \mathbf{1}_q \\ \mathbf{1}_p^\top \lambda &= \mathbf{1}_q^\top \mu, \end{aligned}$$

and

$$\mathbf{1}_p^\top \lambda + \mathbf{1}_q^\top \mu = K_m + \gamma. \qquad (*_\gamma)$$

The second and third equations are equivalent to the box constraints

$$0 \leq \lambda_i, \mu_j \leq K_s, \quad i = 1, \ldots, p,\ j = 1, \ldots, q,$$

and since $\gamma \geq 0$ equation $(*_\gamma)$ is equivalent to

$$\mathbf{1}_p^\top \lambda + \mathbf{1}_q^\top \mu \geq K_m.$$

Plugging back w from $(*_w)$ into the Lagrangian, after simplifications we get

$$\begin{aligned} G(\lambda, \mu, \alpha, \beta) &= \frac{1}{2} \begin{pmatrix} \lambda^\top & \mu^\top \end{pmatrix} X^\top X \begin{pmatrix} \lambda \\ \mu \end{pmatrix} - \begin{pmatrix} \lambda^\top & \mu^\top \end{pmatrix} X^\top X \begin{pmatrix} \lambda \\ \mu \end{pmatrix} \\ &= -\frac{1}{2} \begin{pmatrix} \lambda^\top & \mu^\top \end{pmatrix} X^\top X \begin{pmatrix} \lambda \\ \mu \end{pmatrix}, \end{aligned}$$

so the dual function is independent of α, β and is given by

$$G(\lambda, \mu) = -\frac{1}{2} \begin{pmatrix} \lambda^\top & \mu^\top \end{pmatrix} X^\top X \begin{pmatrix} \lambda \\ \mu \end{pmatrix}.$$

The dual program is given by

$$\begin{aligned} &\text{maximize} \quad -\frac{1}{2} \begin{pmatrix} \lambda^\top & \mu^\top \end{pmatrix} X^\top X \begin{pmatrix} \lambda \\ \mu \end{pmatrix} \\ &\text{subject to} \\ &\qquad \sum_{i=1}^p \lambda_i - \sum_{j=1}^q \mu_j = 0 \\ &\qquad \sum_{i=1}^p \lambda_i + \sum_{j=1}^q \mu_j \geq K_m \\ &\qquad 0 \leq \lambda_i \leq K_s, \quad i = 1, \ldots, p \\ &\qquad 0 \leq \mu_j \leq K_s, \quad j = 1, \ldots, q. \end{aligned}$$

Finally, the dual program is equivalent to the following minimization program:

Dual of Soft margin SVM ($\mathrm{SVM}_{s2'}$):

$$\begin{aligned}
\text{minimize} \quad & \frac{1}{2}\begin{pmatrix}\lambda^\top & \mu^\top\end{pmatrix} X^\top X \begin{pmatrix}\lambda \\ \mu\end{pmatrix} \\
\text{subject to} \quad & \\
& \sum_{i=1}^p \lambda_i - \sum_{j=1}^q \mu_j = 0 \\
& \sum_{i=1}^p \lambda_i + \sum_{j=1}^q \mu_j \geq K_m \\
& 0 \leq \lambda_i \leq K_s, \quad i = 1, \ldots, p \\
& 0 \leq \mu_j \leq K_s, \quad j = 1, \ldots, q.
\end{aligned}$$

If $(w, \eta, \epsilon, \xi, b)$ is an optimal solution of Problem ($\mathrm{SVM}_{s2'}$) with $w \neq 0$ and $\eta \neq 0$, then the complementary slackness conditions yield a classification of the points u_i and v_j in terms of the values of λ and μ. Indeed, we have $\epsilon_i \alpha_i = 0$ for $i = 1, \ldots, p$ and $\xi_j \beta_j = 0$ for $j = 1, \ldots, q$. Also, if $\lambda_i > 0$, then the corresponding constraint is active, and similarly if $\mu_j > 0$. Since $\lambda_i + \alpha_i = K_s$, it follows that $\epsilon_i \alpha_i = 0$ iff $\epsilon_i(K_s - \lambda_i) = 0$, and since $\mu_j + \beta_j = K_s$, we have $\xi_j \beta_j = 0$ iff $\xi_j(K_s - \mu_j) = 0$. Thus if $\epsilon_i > 0$, then $\lambda_i = K_s$, and if $\xi_j > 0$, then $\mu_j = K_s$. Consequently, if $\lambda_i < K_s$, then $\epsilon_i = 0$ and u_i is correctly classified, and similarly if $\mu_j < K_s$, then $\xi_j = 0$ and v_j is correctly classified.

We have the following classification which is basically the classification given in Section 18.1 obtained by replacing δ with η (recall that $\eta > 0$ and $\delta = \eta / \|w\|$).

(1) If $0 < \lambda_i < K_s$, then $\epsilon_i = 0$ and the i-th inequality is active, so

$$w^\top u_i - b - \eta = 0.$$

This means that u_i is on the blue margin (the hyperplane $H_{w,b+\eta}$ of equation $w^\top x = b + \eta$) and is classified correctly.
Similarly, if $0 < \mu_j < K_s$, then $\xi_j = 0$ and

$$w^\top v_j - b + \eta = 0,$$

so v_j is on the red margin (the hyperplane $H_{w,b-\eta}$ of equation $w^\top x = b - \eta$) and is classified correctly.

(2) If $\lambda_i = K_s$, then the i-th inequality is active, so

$$w^\top u_i - b - \eta = -\epsilon_i.$$

If $\epsilon_i = 0$, then the point u_i is on the blue margin. If $\epsilon_i > 0$, then u_i is within the open half space bounded by the blue margin hyperplane $H_{w,b+\eta}$ and containing the separating hyperplane $H_{w,b}$; if $\epsilon_i \leq \eta$, then u_i is classified correctly, and if $\epsilon_i > \eta$, then u_i is misclassified (u_i lies on the wrong side of the separating hyperplane, the red side).
Similarly, if $\mu_j = K_s$, then

$$w^\top v_j - b + \eta = \xi_j.$$

If $\xi_j = 0$, then the point v_j is on the red margin. If $\xi_j > 0$, then v_j is within the open half space bounded by the red margin hyperplane $H_{w,b-\eta}$ and containing the separating hyperplane $H_{w,b}$; if $\xi_j \leq \eta$, then v_j is classified correctly, and if $\xi_j > \eta$, then v_j is misclassified (v_j lies on the wrong side of the separating hyperplane, the blue side).

(3) If $\lambda_i = 0$, then $\epsilon_i = 0$ and the i-th inequality may or may not be active, so

$$w^\top u_i - b - \eta \geq 0.$$

Thus u_i is in the closed half space on the blue side bounded by the blue margin hyperplane $H_{w,b+\eta}$ (of course, classified correctly).
Similarly, if $\mu_j = 0$, then

$$w^\top v_j - b + \eta \leq 0$$

and v_j is in the closed half space on the red side bounded by the red margin hyperplane $H_{w,b-\eta}$ (of course, classified correctly).

Definition 18.2. The vectors u_i on the blue margin $H_{w,b+\eta}$ and the vectors v_j on the red margin $H_{w,b-\eta}$ are called *support vectors*. Support vectors correspond to vectors u_i for which $w^\top u_i - b - \eta = 0$ (which implies $\epsilon_i = 0$), and vectors v_j for which $w^\top v_j - b + \eta = 0$ (which implies $\xi_j = 0$). Support vectors u_i such that $0 < \lambda_i < K_s$ and support vectors v_j such that $0 < \mu_j < K_s$ are *support vectors of type 1*. Support vectors of type 1 play a special role so we denote the sets of indices associated with them by

$$\begin{aligned} I_\lambda &= \{i \in \{1, \dots, p\} \mid 0 < \lambda_i < K_s\} \\ I_\mu &= \{j \in \{1, \dots, q\} \mid 0 < \mu_j < K_s\}. \end{aligned}$$

We denote their cardinalities by $numsvl_1 = |I_\lambda|$ and $numsvm_1 = |I_\mu|$. Support vectors u_i such that $\lambda_i = K_s$ and support vectors v_j such that $\mu_j = K_s$ are *support vectors of type 2*. Those support vectors u_i such that $\lambda_i = 0$ and those support vectors v_j such that $\mu_j = 0$ are called *exceptional support vectors*.

The vectors u_i for which $\lambda_i = K_s$ and the vectors v_j for which $\mu_j = K_s$ are said to *fail the margin*. The sets of indices associated with the vectors failing the margin are denoted by

$$K_\lambda = \{i \in \{1, \dots, p\} \mid \lambda_i = K_s\}$$
$$K_\mu = \{j \in \{1, \dots, q\} \mid \mu_j = K_s\}.$$

We denote their cardinalities by $p_f = |K_\lambda|$ and $q_f = |K_\mu|$.

It will also be useful to understand how points are classified in terms of ϵ_i (or ξ_j).

(1) If $\epsilon_i > 0$, then by complementary slackness $\lambda_i = K_s$, so the ith equation is active and by (2) above,

$$w^\top u_i - b - \eta = -\epsilon_i.$$

Since $\epsilon_i > 0$, the point u_i is strictly within the open half space bounded by the blue margin hyperplane $H_{w,b+\eta}$ and containing the separating hyperplane $H_{w,b}$ (excluding the blue hyperplane $H_{w,b+\eta}$); if $\epsilon_i \leq \eta$, then u_i is classified correctly, and if $\epsilon_i > \eta$, then u_i is misclassified.
Similarly, if $\xi_j > 0$, then v_j is strictly within the open half space bounded by the red margin hyperplane $H_{w,b-\eta}$ and containing the separating hyperplane $H_{w,b}$ (excluding the red hyperplane $H_{w,b-\eta}$); if $\xi_j \leq \eta$, then v_j is classified correctly, and if $\xi_j > \eta$, then v_j is misclassified.

(2) If $\epsilon_i = 0$, then the point u_i is correctly classified. If $\lambda_i = 0$, then by (3) above, u_i is in the closed half space on the blue side bounded by the blue margin hyperplane $H_{w,b+\eta}$. If $\lambda_i > 0$, then by (1) and (2) above, the point u_i is on the blue margin.
Similarly, if $\xi_j = 0$, then the point v_j is correctly classified. If $\mu_j = 0$, then v_j is in the closed half space on the red side bounded by the red margin hyperplane $H_{w,b-\eta}$, and if $\mu_j > 0$, then the point v_j is on the red margin.

Also observe that if $\lambda_i > 0$, then u_i is in the closed half space bounded by the blue hyperplane $H_{w,b+\eta}$ and containing the separating hyperplane $H_{w,b}$ (including the blue hyperplane $H_{w,b+\eta}$).

Similarly, if $\mu_j > 0$, then v_j is in the closed half space bounded by the red hyperplane $H_{w,b+\eta}$ and containing the separating hyperplane $H_{w,b}$ (including the red hyperplane $H_{w,b+\eta}$).

Definition 18.3. Vectors u_i such that $\lambda_i > 0$ and vectors v_j such that $\mu_j > 0$ are said to *have margin at most* δ. The sets of indices associated with these vectors are denoted by

$$I_{\lambda>0} = \{i \in \{1, \ldots, p\} \mid \lambda_i > 0\}$$
$$I_{\mu>0} = \{j \in \{1, \ldots, q\} \mid \mu_j > 0\}.$$

We denote their cardinalities by $p_m = |I_{\lambda>0}|$ and $q_m = |I_{\mu>0}|$.

Vectors u_i such that $\epsilon_i > 0$ and vectors v_j such that $\xi_j > 0$ are said to *strictly fail the margin*. The corresponding sets of indices are denoted by

$$E_\lambda = \{i \in \{1, \ldots, p\} \mid \epsilon_i > 0\}$$
$$E_\mu = \{j \in \{1, \ldots, q\} \mid \xi_j > 0\}.$$

We write $p_{sf} = |E_\lambda|$ and $q_{sf} = |E_\mu|$.

We have the inclusions $E_\lambda \subseteq K_\lambda$ and $E_\mu \subseteq K_\mu$. The difference between the first sets and the second sets is that the second sets may contain support vectors such that $\lambda_i = K_s$ and $\epsilon_i = 0$, or $\mu_j = K_s$ and $\xi_j = 0$. We also have the equations $I_\lambda \cup K_\lambda = I_{\lambda>0}$ and $I_\mu \cup K_\mu = I_{\mu>0}$, and the inequalities $p_{sf} \leq p_f \leq p_m$ and $q_{sf} \leq q_f \leq q_m$.

It is shown in Section 18.8 how the dual program is solved using ADMM from Section 16.6. If the primal problem is solvable, this yields solutions for λ and μ. Once a solution for λ and μ is obtained, we have

$$w = -X \begin{pmatrix} \lambda \\ \mu \end{pmatrix} = \sum_{i=1}^{p} \lambda_i u_i - \sum_{j=1}^{q} \mu_j v_j.$$

As we said earlier, the hypotheses of Theorem 14.5(2) hold, so *if* the primal problem (SVM$_{s2'}$) has an optimal solution with $w \neq 0$, *then* the dual problem has a solution too, and the duality gap is zero. Therefore, for optimal solutions we have

$$L(w, \epsilon, \xi, b, \eta, \lambda, \mu, \alpha, \beta, \gamma) = G(\lambda, \mu, \alpha, \beta, \gamma),$$

which means that

$$\frac{1}{2} w^\top w - K_m \eta + K_s \left(\sum_{i=1}^{p} \epsilon_i + \sum_{j=1}^{q} \xi_j \right) = -\frac{1}{2} \begin{pmatrix} \lambda^\top & \mu^\top \end{pmatrix} X^\top X \begin{pmatrix} \lambda \\ \mu \end{pmatrix},$$

and since

$$w = -X \begin{pmatrix} \lambda \\ \mu \end{pmatrix},$$

we get

$$\frac{1}{2} \begin{pmatrix} \lambda^\top & \mu^\top \end{pmatrix} X^\top X \begin{pmatrix} \lambda \\ \mu \end{pmatrix} - K_m \eta + K_s \left(\sum_{i=1}^{p} \epsilon_i + \sum_{j=1}^{q} \xi_j \right) = -\frac{1}{2} \begin{pmatrix} \lambda^\top & \mu^\top \end{pmatrix} X^\top X \begin{pmatrix} \lambda \\ \mu \end{pmatrix},$$

which yields

$$\eta = \frac{K_s}{K_m}\left(\sum_{i=1}^{p} \epsilon_i + \sum_{j=1}^{q} \xi_j\right) + \frac{1}{K_m}\left(\lambda^\top\ \mu^\top\right) X^\top X \begin{pmatrix}\lambda\\ \mu\end{pmatrix}. \qquad (*)$$

Therefore, we confirm that $\eta \geq 0$.

Remarks: Since we proved that if the Primal Problem (SVM$_{s2'}$) has an optimal solution with $w \neq 0$, then $\eta \geq 0$, one might wonder why the constraint $\eta \geq 0$ was included. If we delete this constraint, it is easy to see that the only difference is that instead of the equation

$$\mathbf{1}_p^\top \lambda + \mathbf{1}_q^\top \mu = K_m + \gamma \qquad (*_1)$$

we obtain the equation

$$\mathbf{1}_p^\top \lambda + \mathbf{1}_q^\top \mu = K_m. \qquad (*_2)$$

If $\eta > 0$, then by complementary slackness $\gamma = 0$, in which case $(*_1)$ and $(*_2)$ are equivalent. But if $\eta = 0$, then γ could be strictly positive.

The option to omit the constraint $\eta \geq 0$ in the primal is slightly advantageous because then the dual involves $2(p+q)$ instead of $2(p+q)+1$ Lagrange multipliers, so the constraint matrix is a $(p+q+2) \times 2(p+q)$ matrix instead of a $(p+q+2) \times (2(p+q)+1)$ matrix and the matrix defining the quadratic functional is a $2(p+q) \times 2(p+q)$ matrix instead of a $(2(p+q)+1) \times (2(p+q)+1)$ matrix; see Section 18.8.

Under the **Standard Margin Hypothesis** for (SVM$_{s2'}$), there is some i_0 such that $0 < \lambda_{i_0} < K_s$ and some j_0 such that $0 < \mu_{j_0} < K_s$, and by the complementary slackness conditions $\epsilon_{i_0} = 0$ and $\xi_{j_0} = 0$, so we have the two active constraints

$$w^\top u_{i_0} - b = \eta, \quad -w^\top v_{j_0} + b = \eta,$$

and we can solve for b and η and we get

$$b = \frac{w^\top u_{i_0} + w^\top v_{j_0}}{2}$$
$$\eta = \frac{w^\top u_{i_0} - w^\top v_{j_0}}{2}$$
$$\delta = \frac{\eta}{\|w\|}.$$

Due to numerical instability, when writing a computer program it is preferable to compute the lists of indices I_λ and I_μ given by

$$I_\lambda = \{i \in \{1, \ldots, p\} \mid 0 < \lambda_i < K_s\}$$
$$I_\mu = \{j \in \{1, \ldots, q\} \mid 0 < \mu_j < K_s\}.$$

Then it is easy to see that we can compute b and η using the following averaging formulae:

$$b = w^\top \left(\Big(\sum_{i \in I_\lambda} u_i \Big) / |I_\lambda| + \Big(\sum_{j \in I_\mu} v_j \Big) / |I_\mu| \right) /2$$

$$\eta = w^\top \left(\Big(\sum_{i \in I_\lambda} u_i \Big) / |I_\lambda| - \Big(\sum_{j \in I_\mu} v_j \Big) / |I_\mu| \right) /2.$$

The "kernelized" version of Problem $(\mathrm{SVM}_{s2'})$ is the following:

Soft margin kernel SVM $(\mathrm{SVM}_{s2'})$:

$$\begin{aligned}
&\text{minimize} \quad \frac{1}{2}\langle w, w\rangle - K_m \eta + K_s \left(\epsilon^\top \; \xi^\top\right) \mathbf{1}_{p+q}\\
&\text{subject to}\\
&\qquad \langle w, \varphi(u_i)\rangle - b \geq \eta - \epsilon_i, \quad \epsilon_i \geq 0 \qquad i = 1, \ldots, p\\
&\qquad - \langle w, \varphi(v_j)\rangle + b \geq \eta - \xi_j, \quad \xi_j \geq 0 \qquad j = 1, \ldots, q\\
&\qquad \eta \geq 0.
\end{aligned}$$

Tracing through the derivation of the dual program we obtain

Dual of the Soft margin kernel SVM $(\mathrm{SVM}_{s2'})$:

$$\begin{aligned}
&\text{minimize} \quad \frac{1}{2} \left(\lambda^\top \; \mu^\top\right) \mathbf{K} \begin{pmatrix} \lambda \\ \mu \end{pmatrix}\\
&\text{subject to}\\
&\qquad \sum_{i=1}^p \lambda_i - \sum_{j=1}^q \mu_j = 0\\
&\qquad \sum_{i=1}^p \lambda_i + \sum_{j=1}^q \mu_j \geq K_m\\
&\qquad 0 \leq \lambda_i \leq K_s, \quad i = 1, \ldots, p\\
&\qquad 0 \leq \mu_j \leq K_s, \quad j = 1, \ldots, q,
\end{aligned}$$

where $\mathbf{K}$ is the kernel matrix of Section 18.1.

As in Section 18.3, we obtain

$$w = \sum_{i=1}^p \lambda_i \varphi(u_i) - \sum_{j=1}^q \mu_j \varphi(v_j),$$

so

$$b = \frac{1}{2}\bigg(\sum_{i=1}^{p} \lambda_i(\kappa(u_i, u_{i_0}) + \kappa(u_i, v_{j_0})) - \sum_{j=1}^{q} \mu_j(\kappa(v_j, u_{i_0}) + \kappa(v_j, v_{j_0}))\bigg),$$

and the classification function

$$f(x) = \mathrm{sgn}(\langle w, \varphi(x)\rangle - b)$$

is given by

$$\begin{aligned} f(x) = \mathrm{sgn}\bigg(&\sum_{i=1}^{p} \lambda_i(2\kappa(u_i, x) - \kappa(u_i, u_{i_0}) - \kappa(u_i, v_{j_0})) \\ &- \sum_{j=1}^{q} \mu_j(2\kappa(v_j, x) - \kappa(v_j, u_{i_0}) - \kappa(v_j, v_{j_0}))\bigg). \end{aligned}$$

18.6 Classification of the Data Points in Terms of ν ($\mathrm{SVM}_{s2'}$)

For a finer classification of the points it turns out to be convenient to consider the ratio

$$\nu = \frac{K_m}{(p+q)K_s}.$$

First note that in order for the constraints to be satisfied, some relationship between K_s and K_m must hold. In addition to the constraints

$$0 \le \lambda_i \le K_s, \quad 0 \le \mu_j \le K_s,$$

we also have the constraints

$$\sum_{i=1}^{p} \lambda_i = \sum_{j=1}^{q} \mu_j$$

$$\sum_{i=1}^{p} \lambda_i + \sum_{j=1}^{q} \mu_j \ge K_m$$

which imply that

$$\sum_{i=1}^{p} \lambda_i \ge \frac{K_m}{2} \quad \text{and} \quad \sum_{j=1}^{q} \mu_j \ge \frac{K_m}{2}. \tag{$\dagger$}$$

Since λ, μ are all nonnegative, if $\lambda_i = K_s$ for all i and if $\mu_j = K_s$ for all j, then

$$\frac{K_m}{2} \le \sum_{i=1}^{p} \lambda_i \le pK_s \quad \text{and} \quad \frac{K_m}{2} \le \sum_{j=1}^{q} \mu_j \le qK_s,$$

so these constraints are not satisfied unless $K_m \leq \min\{2pK_s, 2qK_s\}$, so we assume that $K_m \leq \min\{2pK_s, 2qK_s\}$. The equations in (†) also imply that there is some i_0 such that $\lambda_{i_0} > 0$ and some j_0 such that $\mu_{j_0} > 0$, and so $p_m \geq 1$ and $q_m \geq 1$.

For a finer classification of the points we find it convenient to define $\nu > 0$ such that

$$\nu = \frac{K_m}{(p+q)K_s},$$

so that the objective function $J(w, \epsilon, \xi, b, \eta)$ is given by

$$J(w, \epsilon, \xi, b, \eta) = \frac{1}{2} w^\top w + (p+q)K_s \left(-\nu\eta + \frac{1}{p+q} \begin{pmatrix} \epsilon^\top & \xi^\top \end{pmatrix} \mathbf{1}_{p+q} \right).$$

Observe that the condition $K_m \leq \min\{2pK_s, 2qK_s\}$ is equivalent to

$$\nu \leq \min\left\{ \frac{2p}{p+q}, \frac{2q}{p+q} \right\} \leq 1.$$

Since we obtain an equivalent problem by rescaling by a common positive factor, theoretically it is convenient to normalize K_s as

$$K_s = \frac{1}{p+q},$$

in which case $K_m = \nu$. This method is called the *ν-support vector machine*. Actually, to program the method, it may be more convenient assume that K_s is arbitrary. This helps in avoiding λ_i and μ_j to become to small when $p + q$ is relatively large.

The equations (†) and the box inequalities

$$0 \leq \lambda_i \leq K_s, \quad 0 \leq \mu_j \leq K_s$$

also imply the following facts:

Proposition 18.1. *If Problem* (SVM$_{s2'}$) *has an optimal solution with $w \neq 0$ and $\eta > 0$, then the following facts hold:*

(1) Let p_f be the number of points u_i such that $\lambda_i = K_s$, and let q_f the number of points v_j such that $\mu_j = K_s$. Then $p_f, q_f \leq \nu(p+q)/2$.
(2) Let p_m be the number of points u_i such that $\lambda_i > 0$, and let q_m the number of points v_j such that $\mu_j > 0$. Then $p_m, q_m \geq \nu(p+q)/2$. We have $p_m \geq 1$ and $q_m \geq 1$.
(3) If $p_f \geq 1$ or $q_f \geq 1$, then $\nu \geq 2/(p+q)$.

Proof. (1) Recall that for an optimal solution with $w \neq 0$ and $\eta > 0$, we have $\gamma = 0$, so by $(*_\gamma)$ we have the equations

$$\sum_{i=1}^{p} \lambda_i = \frac{K_m}{2} \quad \text{and} \quad \sum_{j=1}^{q} \mu_j = \frac{K_m}{2}.$$

The point u_i fails to achieve the margin iff $\lambda_i = K_s = K_m/(\nu(p+q))$, so if there are p_f such points then

$$\frac{K_m}{2} = \sum_{i=1}^{p} \lambda_i \geq \frac{K_m p_f}{\nu(p+q)},$$

so

$$p_f \leq \frac{\nu(p+q)}{2}.$$

A similar reasoning applies if v_j fails to achieve the margin δ with $\sum_{i=1}^{p} \lambda_i$ replaced by $\sum_{j=1}^{q} \mu_j$.

(2) A point u_i has margin at most δ iff $\lambda_i > 0$. If

$$I_{\lambda>0} = \{i \in \{1, \ldots, p\} \mid \lambda_i > 0\} \quad \text{and} \quad p_m = |I_{\lambda>0}|,$$

then

$$\frac{K_m}{2} = \sum_{i=1}^{p} \lambda_i = \sum_{i \in I_{\lambda>0}} \lambda_i,$$

and since $\lambda_i \leq K_s = K_m/(\nu(p+q))$, we have

$$\frac{K_m}{2} = \sum_{i \in I_{\lambda>0}} \lambda_i \leq \frac{K_m p_m}{\nu(p+q)},$$

which yields

$$p_m \geq \frac{\nu(p+q)}{2}.$$

A similar reasoning applies if a point v_j has margin at most δ. We already observed that (†) implies that $p_m \geq 1$ and $q_m \geq 1$.

(3) This follows immediately from (1). ☐

Observe that $p_f = q_f = 0$ means that there are no points in the open slab containing the separating hyperplane, namely, the points u_i and the points v_j are separable. So if the points u_i and the points v_j are not separable, then we must pick ν such that $2/(p+q) \leq \nu \leq \min\{2p/(p+q), 2q/(p+q)\}$ for the method to succeed. Otherwise, the method is trying to produce a solution where $w = 0$ and $\eta = 0$, and it does not converge (γ

is nonzero). Actually, Proposition 18.1 yields more accurate bounds on ν for the method to converge, namely

$$\max\left\{\frac{2p_f}{p+q}, \frac{2q_f}{p+q}\right\} \leq \nu \leq \min\left\{\frac{2p_m}{p+q}, \frac{2q_m}{p+q}\right\}.$$

By a previous remark, $p_f \leq p_m$ and $q_f \leq q_m$, the first inequality being strict if there is some i such that $0 < \lambda_i < K$, and the second inequality being strict if there is some j such that $0 < \mu_j < K$. This will be the case under the **Standard Margin Hypothesis**.

Observe that a small value of ν keeps p_f and q_f small, which is achieved if the δ-slab is narrow (to avoid having points on the wrong sides of the margin hyperplanes). A large value of ν allows p_m and q_m to be fairly large, which is achieved if the δ-slab is wide. Thus the smaller ν is, the narrower the δ-slab is, and the larger ν is, the wider the δ-slab is. This is the opposite of the behavior that we witnessed in ν-regression (see Section 20.1).

18.7 Existence of Support Vectors for $(\mathrm{SVM}_{s2'})$

We now consider the issue of the existence of support vectors. We will show that in the "generic case" there is always some blue support vector and some red support vector. The term generic has to do with the choice of ν and will be explained below.

Given any real numbers u, v, x, y, if $\max\{u, v\} < \min\{x, y\}$, then $u < x$ and $v < y$. This is because $u, v \leq \max\{u, v\} < \min\{x, y\} \leq x, y$. Consequently, since by Proposition 18.1, $\max\{2p_f/(p+q), 2q_f/(p+q)\} \leq \nu$, if $\nu < \min\{2p/(p+q), 2q/(p+q)\}$, then $p_f < p$ and $q_f < q$, and since $p_{sf} \leq p_f$ and $q_{sf} \leq q_f$, we also have $p_{sf} < p$ and $q_{sf} < q$. This implies that there are constraints corresponding to some $i \notin E_\lambda$ (in which case $\epsilon_i = 0$) and to some $j \notin E_\mu$ (in which case $\xi_j = 0$), of the form

$$\begin{aligned} w^\top u_i - b &\geq \eta & i &\notin E_\lambda \\ -w^\top v_j + b &\geq \eta & j &\notin E_\mu. \end{aligned}$$

If $w^\top u_i - b = \eta$ for some $i \notin E_\lambda$ and $-w^\top v_j + b = \eta$ for some $j \notin E_\mu$, then we have a blue support vector and a red support vector. Otherwise, we show how to modify b and θ to obtain an optimal solution with a blue support vector and a red support vector.

Proposition 18.2. *For every optimal solution $(w, b, \eta, \epsilon, \xi)$ of Problem $(\mathrm{SVM}_{s2'})$ with $w \neq 0$ and $\eta > 0$, if*

$$\nu < \min\{2p/(p+q), 2q/(p+q)\}$$

and if either no u_i is a support vector or no v_j is a support vector, then there is another optimal solution (for the same w) with some i_0 such that $\epsilon_{i_0} = 0$ and $w^\top u_{i_0} - b = \eta$, and there is some j_0 such that $\xi_{j_0} = 0$ and $-w^\top v_{j_0} + b = \eta$; in other words, some u_{i_0} and some v_{j_0} is a support vector; in particular, $p_{sf} < p$ and $q_{sf} < q$.

Proof. We just explained that $p_{sf} < p$ and $q_{sf} < q$, so the following constraints hold:

$$\begin{aligned} w^\top u_i - b &= \eta - \epsilon_i & \epsilon_i &> 0 & i &\in E_\lambda \\ -w^\top v_j + b &= \eta - \xi_j & \xi_j &> 0 & j &\in E_\mu \\ w^\top u_i - b &\geq \eta & & & i &\notin E_\lambda \\ -w^\top v_j + b &\geq \eta & & & j &\notin E_\mu, \end{aligned}$$

where there is some $i \notin E_\lambda$ and some $j \notin E_\mu$.

If our optimal solution does not have a blue support vector and a red support vector, then either $w^\top u_i - b > \eta$ for all $i \notin E_\lambda$ or $-w^\top v_j + b > \eta$ for all $j \notin E_\mu$.

Case 1. We have

$$\begin{aligned} w^\top u_i - b &> \eta & i &\notin E_\lambda \\ -w^\top v_j + b &\geq \eta & j &\notin E_\mu. \end{aligned}$$

There are two subcases.

Case 1a. Assume that there is some $j \notin E_\mu$ such that $-w^\top v_j + b = \eta$. Our strategy is to increase η and b by a small amount θ in such a way that some inequality becomes an equation for some $i \notin E_\lambda$. Geometrically, this amounts to raising the separating hyperplane $H_{w,b}$ and increasing the width of the slab, keeping the red margin hyperplane unchanged. See Figure 18.7.

Let us pick θ such that

$$\theta = (1/2)\min\{w^\top u_i - b - \eta \mid i \notin E_\lambda\}.$$

Our hypotheses imply that $\theta > 0$. We can write

$$\begin{aligned} w^\top u_i - (b + \theta) &= \eta + \theta - (\epsilon_i + 2\theta) & \epsilon_i &> 0 & i &\in E_\lambda \\ -w^\top v_j + b + \theta &= \eta + \theta - \xi_j & \xi_j &> 0 & j &\in E_\mu \\ w^\top u_i - (b + \theta) &\geq \eta + \theta & & & i &\notin E_\lambda \\ -w^\top v_j + b + \theta &\geq \eta + \theta & & & j &\notin E_\mu. \end{aligned}$$

By hypothesis

$$-w^\top v_j + b + \theta = \eta + \theta \quad \text{for some } j \notin E_\mu,$$

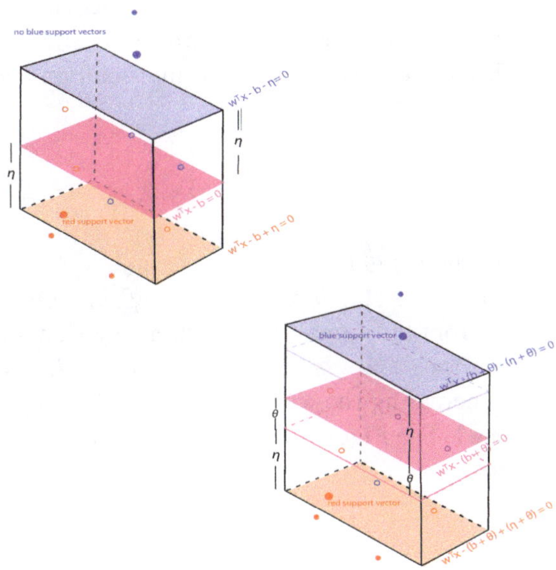

Fig. 18.7 In this illustration points with errors are denoted by open circles. In the original, upper left configuration, there is no blue support vector. By raising the pink separating hyperplane and increasing the margin, we end up with a blue support vector.

and by the choice of θ,

$$w^\top u_i - (b+\theta) = \eta + \theta \quad \text{for some } i \notin E_\lambda.$$

The new value of the objective function is

$$\begin{aligned}\omega(\theta) &= \frac{1}{2} w^\top w - \nu(\eta+\theta) + \frac{1}{p+q}\Big(\sum_{i \in E_\lambda} (\epsilon_i + 2\theta) + \sum_{j \in E_\mu} \xi_j\Big) \\ &= \frac{1}{2} w^\top w - \nu\eta + \frac{1}{p+q}\Big(\sum_{i \in E_\lambda} \epsilon_i + \sum_{j \in E_\mu} \xi_j\Big) - \Big(\nu - \frac{2p_{sf}}{p+q}\Big)\theta.\end{aligned}$$

By Proposition 18.1 we have

$$\max\left\{\frac{2p_f}{p+q}, \frac{2q_f}{p+q}\right\} \leq \nu$$

and $p_{sf} \leq p_f$ and $q_{sf} \leq q_f$, which implies that

$$\nu - \frac{2p_{sf}}{p+q} \geq 0, \tag{$*_1$}$$

and so $\omega(\theta) \leq \omega(0)$. If inequality $(*_1)$ is strict, then this contradicts the optimality of the original solution. Therefore, $\nu = 2p_{sf}/(p+q)$, $\omega(\theta) = \omega(0)$, and $(w, b+\theta, \eta+\theta, \epsilon+2\theta, \xi)$ is an optimal solution such that

$$\begin{aligned} w^\top u_i - (b+\theta) &= \eta + \theta \\ -w^\top v_j + b + \theta &= \eta + \theta \end{aligned}$$

for some $i \notin E_\lambda$ and some $j \notin E_\mu$.

Case 1b. We have $-w^\top v_j + b > \eta$ for all $j \notin E_\mu$. Our strategy is to increase η and the errors by a small θ in such a way that some inequality becomes an equation for some $i \notin E_\lambda$ or for some $j \notin E_\mu$. Geometrically, this corresponds to increasing the width of the slab, keeping the separating hyperplane unchanged. See Figures 18.8 and 18.9. Then we are reduced to *Case 1a* or *Case 2a*.

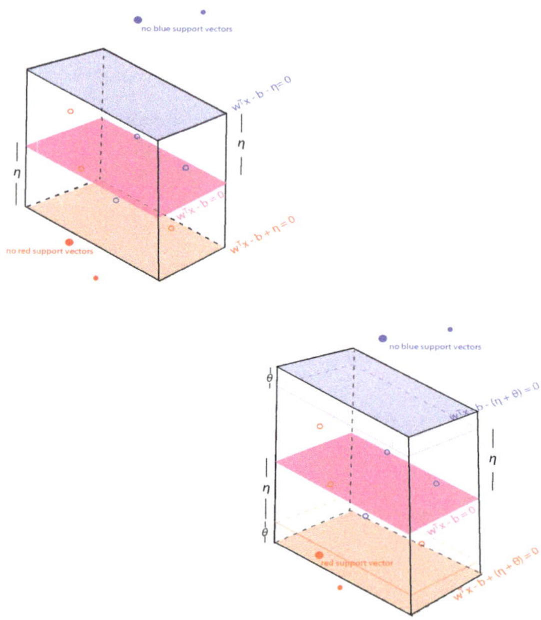

Fig. 18.8 In this illustration points with errors are denoted by open circles. In the original, upper left configuration, there is no blue support vector and no red support vector. By increasing the margin, we end up with a red support vector and reduce to Case 1a.

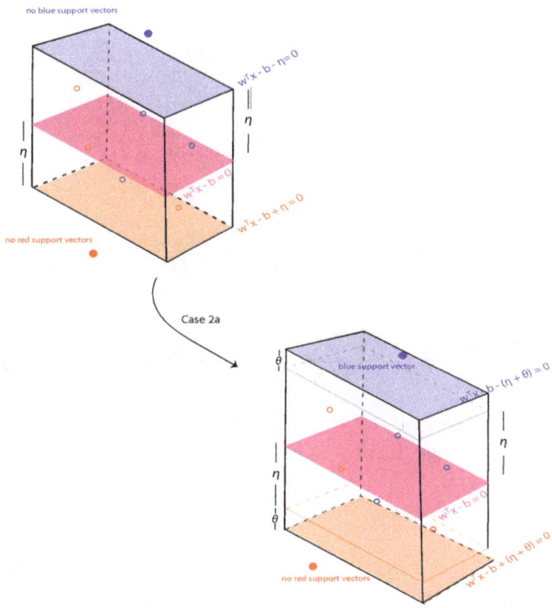

Fig. 18.9 In this illustration points with errors are denoted by open circles. In the original, upper left configuration, there is no blue support vector and no red support vector. By increasing the margin, we end up with a blue support vector and reduce to Case 2a.

We have

$$\begin{aligned}
w^\top u_i - b &= \eta - \epsilon_i & \epsilon_i &> 0 & i &\in E_\lambda \\
-w^\top v_j + b &= \eta - \xi_j & \xi_j &> 0 & j &\in E_\mu \\
w^\top u_i - b &> \eta & & & i &\notin E_\lambda \\
-w^\top v_j + b &> \eta & & & j &\notin E_\mu.
\end{aligned}$$

Let us pick θ such that

$$\theta = \min\{w^\top u_i - b - \eta, \; -w^\top v_j + b - \eta \mid i \notin E_\lambda, j \notin E_\mu\}.$$

Our hypotheses imply that $\theta > 0$. We can write

$$\begin{aligned}
w^\top u_i - b &= \eta + \theta - (\epsilon_i + \theta) & \epsilon_i &> 0 & i &\in E_\lambda \\
-w^\top v_j + b &= \eta + \theta - (\xi_j + \theta) & \xi_j &> 0 & j &\in E_\mu \\
w^\top u_i - b &\geq \eta + \theta & & & i &\notin E_\lambda \\
-w^\top v_j + b &\geq \eta + \theta & & & j &\notin E_\mu,
\end{aligned}$$

and by the choice of θ, either

$$w^\top u_i - b = \eta + \theta \quad \text{for some } i \notin E_\lambda$$

or

$$-w^\top v_j + b = \eta + \theta \quad \text{for some } j \notin E_\mu.$$

The new value of the objective function is

$$\begin{aligned}
\omega(\theta) &= \frac{1}{2}w^\top w - \nu(\eta + \theta) + \frac{1}{p+q}\bigg(\sum_{i\in E_\lambda}(\epsilon_i + \theta) + \sum_{j \in E_\mu}(\xi_j + \theta)\bigg) \\
&= \frac{1}{2}w^\top w - \nu\eta + \frac{1}{p+q}\bigg(\sum_{i\in E_\lambda}\epsilon_i + \sum_{j\in E_\mu}\xi_j\bigg) - \left(\nu - \frac{p_{sf} + q_{sf}}{p+q}\right)\theta.
\end{aligned}$$

Since $\max\{2p_f/(p+q), 2q_f/(p+q)\} \leq \nu$ implies that $(p_f + q_f)/(p+q) \leq \nu$ and $p_{sf} \leq p_f$, $q_{sf} \leq q_f$, we have

$$\nu - \frac{p_{sf} + q_{sf}}{p+q} \geq 0, \qquad (*_2)$$

and so $\omega(\theta) \leq \omega(0)$. If inequality $(*_2)$ is strict, then this contradicts the optimality of the original solution. Therefore, $\nu = (p_{sf} + q_{sf})/(p+q)$, $\omega(\theta) = \omega(0)$ and $(w, b, \eta + \theta, \epsilon + \theta, \xi + \theta)$ is an optimal solution such that either

$$w^\top u_i - b = \eta + \theta \quad \text{for some } i \notin E_\lambda$$

or

$$-w^\top v_j + b = \eta + \theta \quad \text{for some } j \notin E_\mu.$$

We are now reduced to *Case 1a* or *Case 2a*.

Case 2. We have

$$\begin{aligned}
w^\top u_i - b &\geq \eta & i &\notin E_\lambda \\
-w^\top v_j + b &> \eta & j &\notin E_\mu.
\end{aligned}$$

There are two subcases.

Case 2a. Assume that there is some $i \notin E_\lambda$ such that $w^\top u_i - b = \eta$. Our strategy is to increase η and decrease b by a small amount θ in such a way that some inequality becomes an equation for some $j \notin E_\mu$. Geometrically, this amounts to lowering the separating hyperplane $H_{w,b}$ and increasing the width of the slab, keeping the blue margin hyperplane unchanged. See Figure 18.10.

Let us pick θ such that

$$\theta = (1/2)\min\{-w^\top v_j + b - \eta \mid j \notin E_\mu\}.$$

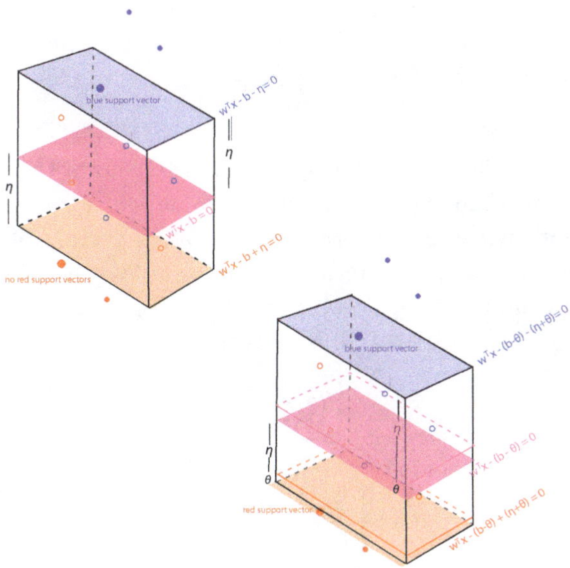

Fig. 18.10 In this illustration points with errors are denoted by open circles. In the original, upper left configuration, there is no red support vector. By lowering the pink separating hyperplane and increasing the margin, we end up with a red support vector.

Our hypotheses imply that $\theta > 0$. We can write

$$\begin{aligned}
w^\top u_i - (b - \theta) &= \eta + \theta - \epsilon_i & \epsilon_i &> 0 & i &\in E_\lambda\\
-w^\top v_j + b - \theta &= \eta + \theta - (\xi_j + 2\theta) & \xi_j &> 0 & j &\in E_\mu\\
w^\top u_i - (b - \theta) &\geq \eta + \theta & & & i &\notin E_\lambda\\
-w^\top v_j + b - \theta &\geq \eta + \theta & & & j &\notin E_\mu.
\end{aligned}$$

By hypothesis

$$w^\top u_i - (b - \theta) = \eta + \theta \quad \text{for some } i \notin E_\lambda,$$

and by the choice of θ,

$$-w^\top v_j + b - \theta = \eta + \theta \quad \text{for some } j \notin E_\mu.$$

The new value of the objective function is

$$\begin{aligned}
\omega(\theta) &= \frac{1}{2} w^\top w - \nu(\eta + \theta) + \frac{1}{p+q}\Bigg(\sum_{i \in E_\lambda} \epsilon_i + \sum_{j \in E_\mu} (\xi_j + 2\theta)\Bigg)\\
&= \frac{1}{2} w^\top w - \nu\eta + \frac{1}{p+q}\Bigg(\sum_{i \in E_\lambda} \epsilon_i + \sum_{j \in E_\mu} \xi_j\Bigg) - \left(\nu - \frac{2q_{sf}}{p+q}\right)\theta.
\end{aligned}$$

The rest of the proof is similar to Case 1a with p_{sf} replaced by q_{sf}.

Case 2b. We have $w^\top u_i - b > \eta$ for all $i \notin E_\lambda$. Since by hypothesis $-w^\top v_j + b > \eta$ for all $j \notin E_\mu$, Case 2b is identical to Case 1b, and we are done. □

A subtle point here is that Proposition 18.2 shows that if there is an optimal solution, then there is one with a blue and a red support vector, but it does not guarantee that these are support vectors of type 1. Since the dual program does not determine b and η unless these support vectors are of type 1, from a practical point of view this proposition is not helpful.

The proof of Proposition 18.2 reveals that there are three critical values for ν:

$$\frac{2p_{sf}}{p+q}, \frac{2q_{sf}}{p+q}, \frac{p_{sf}+q_{sf}}{p+q}.$$

These values can be avoided by requiring the strict inequality

$$\max\left\{\frac{2p_{sf}}{p+q}, \frac{2q_{sf}}{p+q}\right\} < \nu.$$

Then the following corollary holds.

Theorem 18.1. *For every optimal solution* $(w, b, \eta, \epsilon, \xi)$ *of Problem* ($\mathrm{SVM}_{s2'}$) *with* $w \neq 0$ *and* $\eta > 0$, *if*

$$\max\{2p_f/(p+q), 2q_f/(p+q)\} < \nu < \min\{2p/(p+q), 2q/(p+q)\},$$

then some u_{i_0} *and some* v_{j_0} *is a support vector.*

Proof. We proceed by contradiction. Suppose that for every optimal solution with $w \neq 0$ and $\eta > 0$ no u_i is a blue support vector or no v_j is a red support vector. Since $\nu < \min\{2p/(p+q), 2q/(p+q)\}$, Proposition 18.2 holds, so there is another optimal solution. But since the critical values of ν are avoided, the proof of Proposition 18.2 shows that the value of the objective function for this new optimal solution is strictly smaller than the original optimal value, a contradiction. □

We also have the following proposition that gives a sufficient condition implying that η and b can be found in terms of an optimal solution (λ, μ) of the dual.

Proposition 18.3. *If* $(w, b, \eta, \epsilon, \xi)$ *is an optimal solution of Problem* ($\mathrm{SVM}_{s2'}$) *with* $w \neq 0$ *and* $\eta > 0$, *if*

$$\max\{2p_f/(p+q), 2q_f/(p+q)\} < \nu < \min\{2p/(p+q), 2q/(p+q)\},$$

then η *and* b *can always be determined from an optimal solution* (λ, μ) *of the dual in terms of a single support vector.*

Proof. By Theorem 18.1 some u_{i_0} and some v_{j_0} is a support vector. As we already explained, Problem $(\mathrm{SVM}_{s2'})$ satisfies the conditions for having a zero duality gap. Therefore, for optimal solutions we have

$$L(w, \epsilon, \xi, b, \eta, \lambda, \mu, \alpha, \beta) = G(\lambda, \mu, \alpha, \beta),$$

which means that

$$\frac{1}{2} w^\top w - \nu\eta + \frac{1}{p+q}\left(\sum_{i=1}^{p} \epsilon_i + \sum_{j=1}^{q} \xi_j\right) = -\frac{1}{2}\begin{pmatrix}\lambda^\top & \mu^\top\end{pmatrix} X^\top X \begin{pmatrix}\lambda \\ \mu\end{pmatrix},$$

and since

$$w = -X\begin{pmatrix}\lambda \\ \mu\end{pmatrix},$$

we get

$$\frac{1}{p+q}\left(\sum_{i=1}^{p} \epsilon_i + \sum_{j=1}^{q} \xi_j\right) = \nu\eta - \begin{pmatrix}\lambda^\top & \mu^\top\end{pmatrix} X^\top X \begin{pmatrix}\lambda \\ \mu\end{pmatrix}. \qquad (*)$$

Let $K_\lambda = \{i \in \{1, \ldots, p\} \mid \lambda_i = K_s\}$ and $K_\mu = \{j \in \{1, \ldots, q\} \mid \mu_j = K_s\}$. By definition, $p_f = |K_\lambda|$ and $q_f = |K_\mu|$ (here we assuming that $K_s = 1/(p+q)$). By complementary slackness the following equations are active:

$$\begin{aligned} w^\top u_i - b &= \eta - \epsilon_i & i &\in K_\lambda \\ -w^\top v_j + b &= \eta - \xi_j & j &\in K_\mu. \end{aligned}$$

But $(*)$ can be written as

$$\frac{1}{p+q}\left(\sum_{i \in K_\lambda} \epsilon_i + \sum_{j \in K_\mu} \xi_j\right) = \nu\eta - \begin{pmatrix}\lambda^\top & \mu^\top\end{pmatrix} X^\top X \begin{pmatrix}\lambda \\ \mu\end{pmatrix}, \qquad (**)$$

and since

$$\begin{aligned} \epsilon_i &= \eta - w^\top u_i + b & i &\in K_\lambda \\ \xi_j &= \eta + w^\top v_j - b & j &\in K_\mu, \end{aligned}$$

by substituting in the Equation $(**)$ we get

$$\left(\nu - \frac{p_f + q_f}{p+q}\right)\eta = \frac{p_f - q_f}{p+q} b + \frac{1}{p+q} w^\top\left(\sum_{i \in K_\mu} v_j - \sum_{i \in K_\lambda} u_i\right) + \begin{pmatrix}\lambda^\top & \mu^\top\end{pmatrix} X^\top X \begin{pmatrix}\lambda \\ \mu\end{pmatrix}.$$

We also know that $w^\top u_{i_0} - b = \eta$ and $-w^\top v_{j_0} + b = \eta$ for some i_0 and some j_0. In the first case $b = -\eta + w^\top u_{i_0}$, and by substituting b in the above equation we get the equation

$$\left(\nu - \frac{p_f + q_f}{p+q}\right)\eta = -\frac{p_f - q_f}{p+q}\eta + \frac{p_f - q_f}{p+q} w^\top u_{i_0} + \frac{1}{p+q} w^\top \left(\sum_{i\in K_\mu} v_j - \sum_{i\in K_\lambda} u_i\right) + (\lambda^\top \; \mu^\top)\, X^\top X \begin{pmatrix}\lambda \\ \mu\end{pmatrix},$$

that is,

$$\left(\nu - \frac{2q_f}{p+q}\right)\eta = \frac{p_f - q_f}{p+q} w^\top u_{i_0} + \frac{1}{p+q} w^\top \left(\sum_{i\in K_\mu} v_j - \sum_{i\in K_\lambda} u_i\right) + (\lambda^\top \; \mu^\top)\, X^\top X \begin{pmatrix}\lambda \\ \mu\end{pmatrix}.$$

In the second case $b = \eta + w^\top v_{j_0}$, and we get the equation

$$\left(\nu - \frac{p_f + q_f}{p+q}\right)\eta = \frac{p_f - q_f}{p+q}\eta + \frac{p_f - q_f}{p+q} w^\top v_{j_0} + \frac{1}{p+q} w^\top \left(\sum_{i\in K_\mu} v_j - \sum_{i\in K_\lambda} u_i\right) + (\lambda^\top \; \mu^\top)\, X^\top X \begin{pmatrix}\lambda \\ \mu\end{pmatrix},$$

that is,

$$\left(\nu - \frac{2p_f}{p+q}\right)\eta = \frac{p_f - q_f}{p+q} w^\top v_{j_0} + \frac{1}{p+q} w^\top \left(\sum_{i\in K_\mu} v_j - \sum_{i\in K_\lambda} u_i\right) + (\lambda^\top \; \mu^\top)\, X^\top X \begin{pmatrix}\lambda \\ \mu\end{pmatrix}.$$

We need to choose ν such that $2p_f/(p+q) - \nu \neq 0$ and $2q_f/(p+q) - \nu \neq 0$. Since by Proposition 18.1, we have $\max\{2p_f/(p+q), 2q_f/(p+q)\} \leq \nu$, it suffices to pick ν such that $\max\{2p_f/(p+q), 2q_f/(p+q)\} < \nu$. If this condition is satisfied we can solve for η, and then we find b from either $b = -\eta + w^\top u_{i_0}$ or $b = \eta + w^\top v_{j_0}$. ☐

Remark: Of course the hypotheses of the proposition imply that $w^\top u_{i_0} - b = \eta$ and $-w^\top v_{j_0} + b = \eta$ for some i_0 and some j_0. Thus we can also compute b and η using the formulae

$$b = \frac{w^\top (u_{i_0} + v_{j_0})}{2}$$
$$\eta = \frac{w^\top (u_{i_0} - v_{j_0})}{2}.$$

The interest of Proposition 18.3 lies in the fact that it allows us to compute b and η knowing only a *single* support vector.

In practice we can only find support vectors of type 1 so Proposition 18.3 is useful if we can only find some blue support vector of type 1 *or* some red support vector of type 1.

As earlier, if we define I_λ and I_μ as

$$\begin{aligned} I_\lambda &= \{i \in \{1, \ldots, p\} \mid 0 < \lambda_i < K_s\} \\ I_\mu &= \{j \in \{1, \ldots, q\} \mid 0 < \mu_j < K_s\}, \end{aligned}$$

then we have the following cases to compute η and b.

(1) If $I_\lambda \neq \emptyset$ and $I_\mu \neq \emptyset$, then

$$\begin{aligned} b &= w^\top \left(\Big(\sum_{i \in I_\lambda} u_i \Big) / |I_\lambda| + \Big(\sum_{j \in I_\mu} v_j \Big) / |I_\mu| \right) /2 \\ \eta &= w^\top \left(\Big(\sum_{i \in I_\lambda} u_i \Big) / |I_\lambda| - \Big(\sum_{j \in I_\mu} v_j \Big) / |I_\mu| \right) /2. \end{aligned}$$

(2) If $I_\lambda \neq \emptyset$ and $I_\mu = \emptyset$, then

$$\begin{aligned} b &= -\eta + w^\top \Big(\sum_{i \in I_\lambda} u_i \Big) / |I_\lambda| \\ ((p+q)\nu - 2q_f)\eta &= (p_f - q_f) w^\top \Big(\sum_{i \in I_\lambda} u_i \Big) / |I_\lambda| \\ &\quad + w^\top \Big(\sum_{i \in K_\mu} v_j - \sum_{i \in K_\lambda} u_i \Big) \\ &\quad + (p+q) \begin{pmatrix} \lambda^\top & \mu^\top \end{pmatrix} X^\top X \begin{pmatrix} \lambda \\ \mu \end{pmatrix}. \end{aligned}$$

(3) If $I_\lambda = \emptyset$ and $I_\mu \neq \emptyset$, then

$$\begin{aligned} b &= \eta + w^\top \Big(\sum_{j \in I_\mu} v_j \Big) / |I_\mu| \\ ((p+q)\nu - 2p_f)\eta &= (p_f - q_f) w^\top \Big(\sum_{j \in I_\mu} v_j \Big) / |I_\mu| \\ &\quad + w^\top \Big(\sum_{i \in K_\mu} v_j - \sum_{i \in K_\lambda} u_i \Big) \\ &\quad + (p+q) \begin{pmatrix} \lambda^\top & \mu^\top \end{pmatrix} X^\top X \begin{pmatrix} \lambda \\ \mu \end{pmatrix}. \end{aligned}$$

The above formulae correspond to $K_s = 1/(p+q)$. In general we have to replace the rightmost $(p+q)$ by $1/K_s$.

We have examples where there is a single support vector of type 1 and $\nu = 2q_f/(p+q)$, so the above method fails. Curiously, perturbing ν slightly yields a solution with some blue support vector of type 1 and some red support vector of type 1, and so we have not yet found an example where the above method succeeds with a single support vector of type 1. This suggests to conduct some perturbation analysis but it appears to be nontrivial.

Among its advantages, the support vector machinery is conducive to finding interesting statistical bounds in terms of the *VC dimension*, a notion invented by Vapnik and Chernovenkis. We will not go into this here and instead refer the reader to Vapnik [Vapnik (1998)] (especially, Chapter 4 and Chapters 9–13).

18.8 Solving SVM ($\mathrm{SVM}_{s2'}$) Using ADMM

In order to solve ($\mathrm{SVM}_{s2'}$) using ADMM we need to write the matrix corresponding to the constraints in equational form,

$$\begin{aligned}
\sum_{i=1}^{p} \lambda_i - \sum_{j=1}^{q} \mu_j &= 0 \\
\sum_{i=1}^{p} \lambda_i + \sum_{j=1}^{q} \mu_j - \gamma &= K_m \\
\lambda_i + \alpha_i &= K_s, \quad i = 1, \ldots, p \\
\mu_j + \beta_j &= K_s, \quad j = 1, \ldots, q,
\end{aligned}$$

with $K_m = (p+q)K_s\nu$. This is the $(p+q+2) \times (2(p+q)+1)$ matrix A given by

$$A = \begin{pmatrix} \mathbf{1}_p^\top & -\mathbf{1}_q^\top & 0_p^\top & 0_q^\top & 0 \\ \mathbf{1}_p^\top & \mathbf{1}_q^\top & 0_p^\top & 0_q^\top & -1 \\ I_p & 0_{p,q} & I_q & 0_{p,q} & 0_p \\ 0_{q,p} & I_q & 0_{q,p} & I_q & 0_q \end{pmatrix}.$$

Observe the remarkable analogy with the matrix arising in ν-regression in Section 20.3, except that $p = q = m$ and that -1 is replaced by $+1$. We leave it as an exercise to prove that A has rank $p+q+2$. The right-hand side is

$$c = \begin{pmatrix} 0 \\ K_m \\ K_s \mathbf{1}_{p+q} \end{pmatrix}.$$

The symmetric positive semidefinite $(p+q) \times (p+q)$ matrix P defining the quadratic functional is

$$P = X^\top X, \quad \text{with} \quad X = \begin{pmatrix} -u_1 & \cdots & -u_p & v_1 & \cdots & v_q \end{pmatrix},$$

and

$$q = 0_{p+q}.$$

Since there are $2(p+q)+1$ Lagrange multipliers $(\lambda, \mu, \alpha, \beta, \gamma)$, the $(p+q) \times (p+q)$ matrix $X^\top X$ must be augmented with zero's to make it a $(2(p+q)+1) \times (2(p+q)+1)$ matrix P_a given by

$$P_a = \begin{pmatrix} X^\top X & 0_{p+q,p+q+1} \\ 0_{p+q+1,p+q} & 0_{p+q+1,p+q+1} \end{pmatrix},$$

and similarly q is augmented with zeros as the vector $q_a = 0_{2(p+q)+1}$.

As we mentioned in Section 18.5, since $\eta \geq 0$ for an optimal solution, we can drop the constraint $\eta \geq 0$ from the primal problem. In this case there are $2(p+q)$ Lagrange multipliers $(\lambda, \mu, \alpha, \beta)$. It is easy to see that the objective function of the dual is unchanged and the set of constraints is

$$\begin{aligned}
\sum_{i=1}^p \lambda_i - \sum_{j=1}^q \mu_j &= 0 \\
\sum_{i=1}^p \lambda_i + \sum_{j=1}^q \mu_j &= K_m \\
\lambda_i + \alpha_i &= K_s, \quad i = 1, \ldots, p \\
\mu_j + \beta_j &= K_s, \quad j = 1, \ldots, q,
\end{aligned}$$

with $K_m = (p+q)K_s\nu$. The constraint matrix corresponding to this system of equations is the $(p+q+2) \times 2(p+q)$ matrix A_2 given by

$$A_2 = \begin{pmatrix} \mathbf{1}_p^\top & -\mathbf{1}_q^\top & 0_p^\top & 0_q^\top \\ \mathbf{1}_p^\top & \mathbf{1}_q^\top & 0_p^\top & 0_q^\top \\ I_p & 0_{p,q} & I_q & 0_{p,q} \\ 0_{q,p} & I_q & 0_{q,p} & I_q \end{pmatrix}.$$

We leave it as an exercise to prove that A_2 has rank $p+q+2$. The right-hand side is

$$c_2 = \begin{pmatrix} 0 \\ K_m \\ K_s \mathbf{1}_{p+q} \end{pmatrix}.$$

The symmetric positive semidefinite $(p+q) \times (p+q)$ matrix P defining the quadratic functional is

$$P = X^\top X, \quad \text{with} \quad X = \begin{pmatrix} -u_1 & \cdots & -u_p & v_1 & \cdots & v_q \end{pmatrix},$$

and

$$q = 0_{p+q}.$$

Since there are $2(p+q)$ Lagrange multipliers the $(p+q) \times (p+q)$ matrix $X^\top X$ must be augmented with zero's to make it a $2(p+q) \times 2(p+q)$ matrix P_{2a} given by

$$P_{2a} = \begin{pmatrix} X^\top X & 0_{p+q,p+q} \\ 0_{p+q,p+q} & 0_{p+q,p+q} \end{pmatrix},$$

and similarly q is augmented with zeros as the vector $q_{2a} = 0_{2(p+q)}$.

The `Matlab` programs implementing the above method are given in Appendix B, Section B.2. We ran our program on two sets of 30 points each generated at random using the following code which calls the function `runSVMs2pbv3`:

```
rho = 10;
u16 = 10.1*randn(2,30)+7 ;
v16 = -10.1*randn(2,30)-7;
[~,~,~,~,~,~,w3] = runSVMs2pbv3(0.37,rho,u16,v16,1/60)
```

We picked $K = 1/60$ and various values of ν starting with $\nu = 0.37$, which appears to be the smallest value for which the method converges; see Figure 18.11.

In this example, $p_f = 10, q_f = 11, p_m = 12, q_m = 12$. The quadratic solver converged after 8121 steps to reach primal and dual residuals smaller than 10^{-10}.

Reducing ν below $\nu = 0.37$ has the effect that p_f, q_f, p_m, q_m decrease but the following situation arises. Shrinking η a little bit has the effect that $p_f = 9, q_f = 10, p_m = 10, q_m = 11$. Then $\max\{p_f, q_f\} = \min\{p_m, q_m\} = 10$, so the only possible value for ν is $\nu = 20/60 = 1/3 = 0.3333333\cdots$. When we run our program with $\nu = 1/3$, it returns a value of η less than

10^{-13} and a value of w whose components are also less than 10^{-13}. This is probably due to numerical precision. Values of ν less than $1/3$ cause the same problem. It appears that the geometry of the problem constrains the values of p_f, q_f, p_m, q_m in such a way that it has no solution other than $w = 0$ and $\eta = 0$.

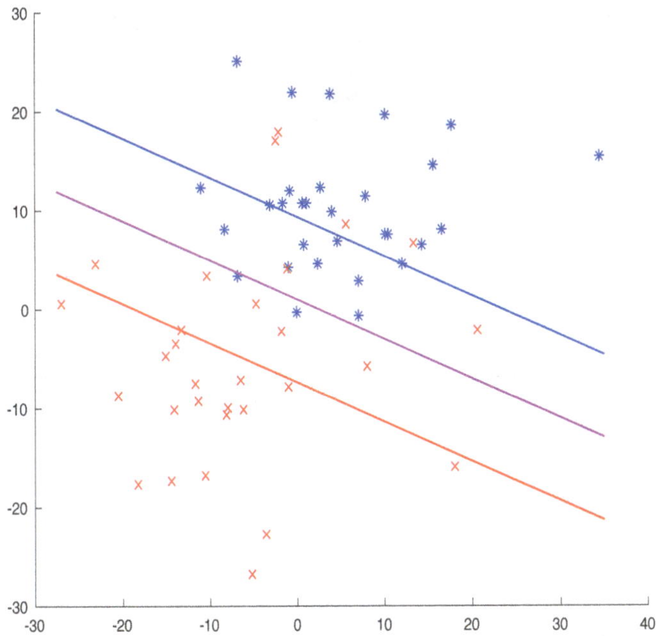

Fig. 18.11 Running (SVM$_{s2'}$) on two sets of 30 points; $\nu = 0.37$.

Figure 18.12 shows the result of running the program with $\nu = 0.51$. We have $p_f = 15, q_f = 16, p_m = 16, q_m = 16$. Interestingly, for $\nu = 0.5$, we run into the singular situation where there is only one support vector and $\nu = 2p_f/(p+q)$.

Next Figure 18.13 shows the result of running the program with $\nu = 0.71$. We have $p_f = 21, q_f = 21, p_m = 22, q_m = 23$. Interestingly, for $\nu = 0.7$, we run into the singular situation where there are no support vectors.

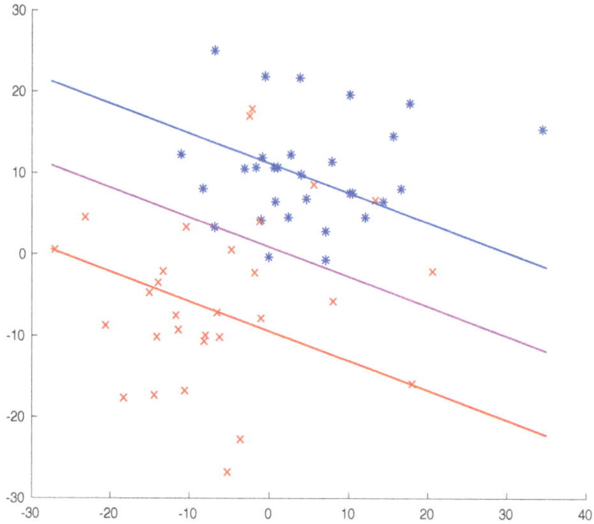

Fig. 18.12 Running ($SVM_{s2'}$) on two sets of 30 points; $\nu = 0.51$.

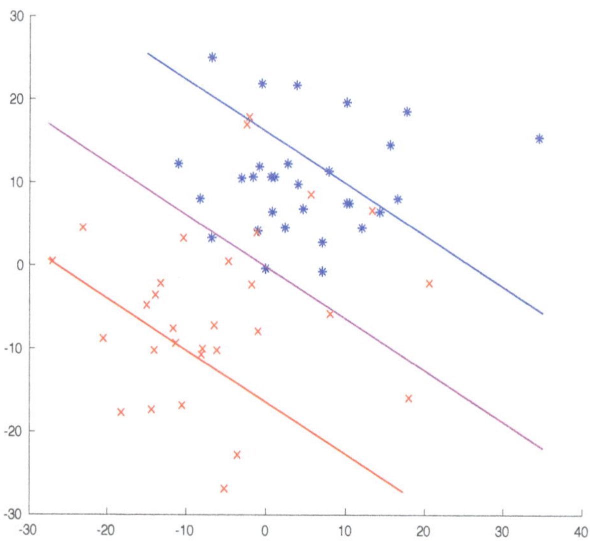

Fig. 18.13 Running ($SVM_{s2'}$) on two sets of 30 points; $\nu = 0.71$.

For our next to the last run, Figure 18.14 shows the result of running the program with $\nu = 0.95$. We have $p_f = 28, q_f = 28, p_m = 29, q_m = 29$.

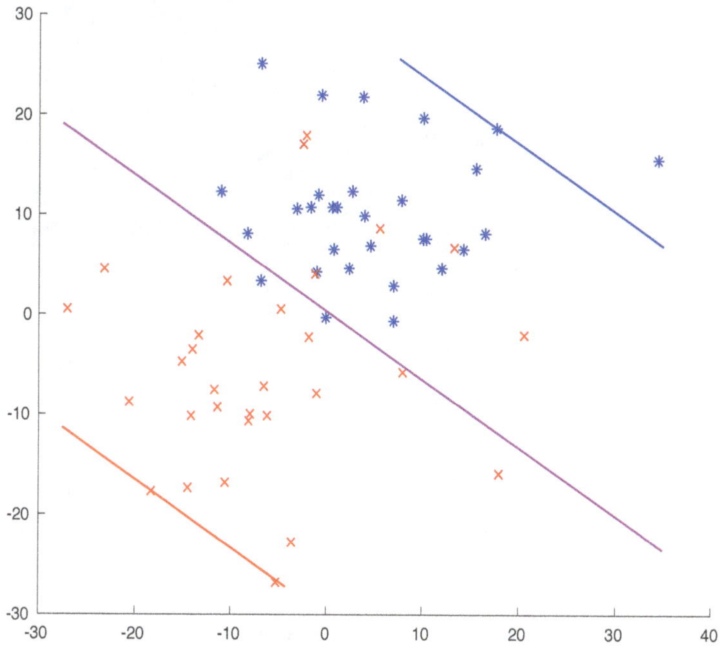

Fig. 18.14 Running (SVM$_{s2'}$) on two sets of 30 points; $\nu = 0.95$.

Figure 18.15 shows the result of running the program with $\nu = 0.97$. We have $p_f = 29, q_f = 29, p_m = 30, q_m = 30$, which shows that the largest margin has been achieved. However, after 80000 iterations the dual residual is less than 10^{-12} but the primal residual is approximately 10^{-4} (our tolerance for convergence is 10^{-10}, which is quite high). Nevertheless the result is visually very good.

18.9 Soft Margin Support Vector Machines; (SVM$_{s3}$)

In this section we consider a variation of Problem (SVM$_{s2'}$) by adding the term $(1/2)b^2$ to the objective function. The result is that in minimizing the Lagrangian to find the dual function G, not just w but also b is determined and η is determined under a mild condition on ν. We also suppress the

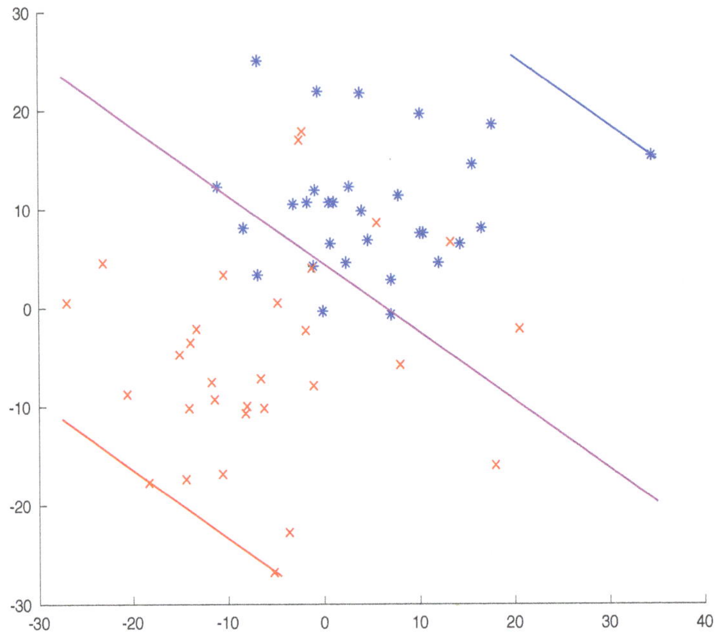

Fig. 18.15 Running ($\text{SVM}_{s2'}$) on two sets of 30 points; $\nu = 0.97$.

constraint $\eta \geq 0$ which turns out to be redundant.

Soft margin SVM (SVM_{s3}):

$$\begin{aligned}
&\text{minimize} \quad \frac{1}{2}w^\top w + \frac{1}{2}b^2 + (p+q)K_s\left(-\nu\eta + \frac{1}{p+q}\left(\epsilon^\top \; \xi^\top\right)\mathbf{1}_{p+q}\right)\\
&\text{subject to}\\
&\qquad w^\top u_i - b \geq \eta - \epsilon_i, \quad \epsilon_i \geq 0 \qquad i = 1, \ldots, p\\
&\qquad -w^\top v_j + b \geq \eta - \xi_j, \quad \xi_j \geq 0 \qquad j = 1, \ldots, q.
\end{aligned}$$

To simplify the presentation we assume that $K_s = 1/(p+q)$. When writing a computer program it is more convenient to assume that K_s is arbitrary. In this case, ν needs to be replaced by $(p+q)K_s\nu$ in all the formulae.

The Lagrangian $L(w, \epsilon, \xi, b, \eta, \lambda, \mu, \alpha, \beta)$ with $\lambda, \alpha \in \mathbb{R}^p_+$, $\mu, \beta \in \mathbb{R}^q_+$ is given by

$$\begin{aligned}
L(w, \epsilon, \xi, b, \eta, \lambda, \mu, \alpha, \beta) &= \frac{1}{2} w^\top w + w^\top X \begin{pmatrix} \lambda \\ \mu \end{pmatrix} + \frac{b^2}{2} + K_s(\epsilon^\top \mathbf{1}_p + \xi^\top \mathbf{1}_q) \\
&\quad - \epsilon^\top(\lambda + \alpha) - \nu\eta - \xi^\top(\mu + \beta) + b(\mathbf{1}_p^\top \lambda - \mathbf{1}_q^\top \mu) + \eta(\mathbf{1}_p^\top \lambda + \mathbf{1}_q^\top \mu) \\
&= \frac{1}{2} w^\top w + w^\top X \begin{pmatrix} \lambda \\ \mu \end{pmatrix} + \frac{b^2}{2} + b(\mathbf{1}_p^\top \lambda - \mathbf{1}_q^\top \mu) + \eta(\mathbf{1}_p^\top \lambda + \mathbf{1}_q^\top \mu - \nu) \\
&\qquad + \epsilon^\top(K_s \mathbf{1}_p - (\lambda + \alpha)) + \xi^\top(K_s \mathbf{1}_q - (\mu + \beta)).
\end{aligned}$$

To find the dual function $G(\lambda, \mu, \alpha, \beta)$, we minimize $L(w, \epsilon, \xi, b, \eta, \lambda, \mu, \alpha, \beta)$ with respect to w, ϵ, ξ, b, and η. Since the Lagrangian is convex and $(w, \epsilon, \xi, b, \eta) \in \mathbb{R}^n \times \mathbb{R}^p \times \mathbb{R}^q \times \mathbb{R} \times \mathbb{R}$, a convex open set, by Theorem 4.5, the Lagrangian has a minimum in $(w, \epsilon, \xi, b, \eta)$ iff $\nabla L_{w,\epsilon,\xi,b,\eta} = 0$, so we compute its gradient with respect to $w, \epsilon, \xi, b, \eta$, and we get

$$\nabla L_{w,\epsilon,\xi,b,\eta} = \begin{pmatrix} X \begin{pmatrix} \lambda \\ \mu \end{pmatrix} + w \\ K_s \mathbf{1}_p - (\lambda + \alpha) \\ K_s \mathbf{1}_q - (\mu + \beta) \\ b + \mathbf{1}_p^\top \lambda - \mathbf{1}_q^\top \mu \\ \mathbf{1}_p^\top \lambda + \mathbf{1}_q^\top \mu - \nu \end{pmatrix}.$$

By setting $\nabla L_{w,\epsilon,\xi,b,\eta} = 0$ we get the equations

$$w = -X \begin{pmatrix} \lambda \\ \mu \end{pmatrix} \tag{$*_w$}$$

$$\begin{aligned}
\lambda + \alpha &= K_s \mathbf{1}_p \\
\mu + \beta &= K_s \mathbf{1}_q \\
\mathbf{1}_p^\top \lambda + \mathbf{1}_q^\top \mu &= \nu,
\end{aligned}$$

and

$$b = -(\mathbf{1}_p^\top \lambda - \mathbf{1}_q^\top \mu). \tag{$*_b$}$$

The second and third equations are equivalent to the box constraints

$$0 \leq \lambda_i, \mu_j \leq K_s, \quad i = 1, \ldots, p,\ j = 1, \ldots, q.$$

Since we assumed that the primal problem has an optimal solution with $w \neq 0$, we have

$$X \begin{pmatrix} \lambda \\ \mu \end{pmatrix} \neq 0.$$

Plugging back w from $(*_w)$ and b from $(*_b)$ into the Lagrangian, we get

$$\begin{aligned}
G(\lambda, \mu, \alpha, \beta) &= \frac{1}{2}\begin{pmatrix}\lambda^\top & \mu^\top\end{pmatrix} X^\top X \begin{pmatrix}\lambda \\ \mu\end{pmatrix} - \begin{pmatrix}\lambda^\top & \mu^\top\end{pmatrix} X^\top X \begin{pmatrix}\lambda \\ \mu\end{pmatrix} + \frac{1}{2}b^2 - b^2 \\
&= -\frac{1}{2}\begin{pmatrix}\lambda^\top & \mu^\top\end{pmatrix} X^\top X \begin{pmatrix}\lambda \\ \mu\end{pmatrix} - \frac{1}{2}b^2 \\
&= -\frac{1}{2}\begin{pmatrix}\lambda^\top & \mu^\top\end{pmatrix} \left(X^\top X + \begin{pmatrix}\mathbf{1}_p\mathbf{1}_p^\top & -\mathbf{1}_p\mathbf{1}_q^\top \\ -\mathbf{1}_q\mathbf{1}_p^\top & \mathbf{1}_q\mathbf{1}_q^\top\end{pmatrix}\right) \begin{pmatrix}\lambda \\ \mu\end{pmatrix},
\end{aligned}$$

so the dual function is independent of α, β and is given by

$$G(\lambda, \mu) = -\frac{1}{2}\begin{pmatrix}\lambda^\top & \mu^\top\end{pmatrix} \left(X^\top X + \begin{pmatrix}\mathbf{1}_p\mathbf{1}_p^\top & -\mathbf{1}_p\mathbf{1}_q^\top \\ -\mathbf{1}_q\mathbf{1}_p^\top & \mathbf{1}_q\mathbf{1}_q^\top\end{pmatrix}\right) \begin{pmatrix}\lambda \\ \mu\end{pmatrix}.$$

The dual program is given by

$$\begin{aligned}
&\text{maximize} \quad -\frac{1}{2}\begin{pmatrix}\lambda^\top & \mu^\top\end{pmatrix} \left(X^\top X + \begin{pmatrix}\mathbf{1}_p\mathbf{1}_p^\top & -\mathbf{1}_p\mathbf{1}_q^\top \\ -\mathbf{1}_q\mathbf{1}_p^\top & \mathbf{1}_q\mathbf{1}_q^\top\end{pmatrix}\right) \begin{pmatrix}\lambda \\ \mu\end{pmatrix} \\
&\text{subject to} \\
&\qquad \sum_{i=1}^p \lambda_i + \sum_{j=1}^q \mu_j = \nu \\
&\qquad 0 \leq \lambda_i \leq K_s, \quad i = 1, \ldots, p \\
&\qquad 0 \leq \mu_j \leq K_s, \quad j = 1, \ldots, q.
\end{aligned}$$

Finally, the dual program is equivalent to the following minimization program:

Dual of the Soft margin SVM (SVM_{s3}):

$$\begin{aligned}
&\text{minimize} \quad \frac{1}{2}\begin{pmatrix}\lambda^\top & \mu^\top\end{pmatrix} \left(X^\top X + \begin{pmatrix}\mathbf{1}_p\mathbf{1}_p^\top & -\mathbf{1}_p\mathbf{1}_q^\top \\ -\mathbf{1}_q\mathbf{1}_p^\top & \mathbf{1}_q\mathbf{1}_q^\top\end{pmatrix}\right) \begin{pmatrix}\lambda \\ \mu\end{pmatrix} \\
&\text{subject to} \\
&\qquad \sum_{i=1}^p \lambda_i + \sum_{j=1}^q \mu_j = \nu \\
&\qquad 0 \leq \lambda_i \leq K_s, \quad i = 1, \ldots, p \\
&\qquad 0 \leq \mu_j \leq K_s, \quad j = 1, \ldots, q.
\end{aligned}$$

The classification of the points u_i and v_j in terms of the values of λ and μ and Definition 18.2 and Definition 18.3 are unchanged.

It is shown in Section 18.12 how the dual program is solved using ADMM from Section 16.6. If the primal problem is solvable, this yields solutions for λ and μ. Once a solution for λ and μ is obtained, we have

$$\begin{aligned} w &= -X\begin{pmatrix}\lambda\\ \mu\end{pmatrix} = \sum_{i=1}^{p}\lambda_i u_i - \sum_{j=1}^{q}\mu_j v_j \\ b &= -(\mathbf{1}_p^\top\lambda - \mathbf{1}_q^\top\mu) = -\sum_{i=1}^{p}\lambda_i + \sum_{j=1}^{q}\mu_j. \end{aligned}$$

We can compute η using duality. As we said earlier, the hypotheses of Theorem 14.5(2) hold, so *if* the primal problem (SVM$_{s3}$) has an optimal solution with $w \neq 0$, *then* the dual problem has a solution too, and the duality gap is zero. Therefore, for optimal solutions we have

$$L(w,\epsilon,\xi,b,\eta,\lambda,\mu,\alpha,\beta) = G(\lambda,\mu,\alpha,\beta),$$

which means that

$$\begin{aligned} &\frac{1}{2}w^\top w + \frac{b^2}{2} - (p+q)K_s\nu\eta + K_s\left(\sum_{i=1}^{p}\epsilon_i + \sum_{j=1}^{q}\xi_j\right) \\ &\qquad = -\frac{1}{2}\begin{pmatrix}\lambda^\top & \mu^\top\end{pmatrix}\left(X^\top X + \begin{pmatrix}\mathbf{1}_p\mathbf{1}_p^\top & -\mathbf{1}_p\mathbf{1}_q^\top\\ -\mathbf{1}_q\mathbf{1}_p^\top & \mathbf{1}_q\mathbf{1}_q^\top\end{pmatrix}\right)\begin{pmatrix}\lambda\\ \mu\end{pmatrix}. \end{aligned}$$

We can use the above equation to determine η.

Since

$$\frac{1}{2}w^\top w + \frac{b^2}{2} = \frac{1}{2}\begin{pmatrix}\lambda^\top & \mu^\top\end{pmatrix}\left(X^\top X + \begin{pmatrix}\mathbf{1}_p\mathbf{1}_p^\top & -\mathbf{1}_p\mathbf{1}_q^\top\\ -\mathbf{1}_q\mathbf{1}_p^\top & \mathbf{1}_q\mathbf{1}_q^\top\end{pmatrix}\right)\begin{pmatrix}\lambda\\ \mu\end{pmatrix},$$

we get

$$\begin{aligned} (p+q)K_s\nu\eta &= K_s\left(\sum_{i=1}^{p}\epsilon_i + \sum_{j=1}^{q}\xi_j\right) \\ &\quad + \begin{pmatrix}\lambda^\top & \mu^\top\end{pmatrix}\left(X^\top X + \begin{pmatrix}\mathbf{1}_p\mathbf{1}_p^\top & -\mathbf{1}_p\mathbf{1}_q^\top\\ -\mathbf{1}_q\mathbf{1}_p^\top & \mathbf{1}_q\mathbf{1}_q^\top\end{pmatrix}\right)\begin{pmatrix}\lambda\\ \mu\end{pmatrix}. \qquad (*) \end{aligned}$$

Since

$$X^\top X + \begin{pmatrix}\mathbf{1}_p\mathbf{1}_p^\top & -\mathbf{1}_p\mathbf{1}_q^\top\\ -\mathbf{1}_q\mathbf{1}_p^\top & \mathbf{1}_q\mathbf{1}_q^\top\end{pmatrix}$$

is positive semidefinite, we confirm that $\eta \geq 0$.

Since nonzero ϵ_i and ξ_j may only occur for vectors u_i and v_j that fail the margin, namely $\lambda_i = K_s$, $\mu_j = K_s$, the corresponding constraints are active and we can solve for ϵ_i and ξ_j in terms of b and η. Let K_λ and K_μ be the sets of indices corresponding to points failing the margin,

$$\begin{aligned} K_\lambda &= \{i \in \{1, \ldots, p\} \mid \lambda_i = K_s\} \\ K_\mu &= \{j \in \{1, \ldots, q\} \mid \mu_j = K_s\}. \end{aligned}$$

By definition $p_f = |K_\lambda|, q_f = |K_\mu|$. Then for every $i \in K_\lambda$ we have

$$\epsilon_i = \eta + b - w^\top u_i$$

and for every $j \in K_\mu$ we have

$$\xi_j = \eta - b + w^\top v_j.$$

Using the above formulae we obtain

$$\begin{aligned} \sum_{i=1}^p \epsilon_i + \sum_{j=1}^q \xi_j &= \sum_{i \in K_\lambda} \epsilon_i + \sum_{j \in K_\mu} \xi_j \\ &= \sum_{i \in K_\lambda} (\eta + b - w^\top u_i) + \sum_{j \in K_\mu} (\eta - b + w^\top v_j) \\ &= (p_f + q_f)\eta + (p_f - q_f)b + w^\top \Bigg(\sum_{j \in K_\mu} v_j - \sum_{i \in K_\lambda} u_i \Bigg) \end{aligned}$$

Substituting this expression in $(*)$ we obtain

$$\begin{aligned} (p+q)K_s\nu\eta &= K_s \Bigg(\sum_{i=1}^p \epsilon_i + \sum_{j=1}^q \xi_j \Bigg) \\ &\quad + \begin{pmatrix} \lambda^\top & \mu^\top \end{pmatrix} \left(X^\top X + \begin{pmatrix} \mathbf{1}_p \mathbf{1}_p^\top & -\mathbf{1}_p \mathbf{1}_q^\top \\ -\mathbf{1}_q \mathbf{1}_p^\top & \mathbf{1}_q \mathbf{1}_q^\top \end{pmatrix} \right) \begin{pmatrix} \lambda \\ \mu \end{pmatrix} \\ &= K_s \Bigg((p_f + q_f)\eta + (p_f - q_f)b + w^\top \Bigg(\sum_{j \in K_\mu} v_j - \sum_{i \in K_\lambda} u_i \Bigg) \Bigg) \\ &\quad + \begin{pmatrix} \lambda^\top & \mu^\top \end{pmatrix} \left(X^\top X + \begin{pmatrix} \mathbf{1}_p \mathbf{1}_p^\top & -\mathbf{1}_p \mathbf{1}_q^\top \\ -\mathbf{1}_q \mathbf{1}_p^\top & \mathbf{1}_q \mathbf{1}_q^\top \end{pmatrix} \right) \begin{pmatrix} \lambda \\ \mu \end{pmatrix}, \end{aligned}$$

which yields

$$\begin{aligned} ((p+q)\nu - p_f - q_f)\eta &= (p_f - q_f)b + w^\top \Bigg(\sum_{j \in K_\mu} v_j - \sum_{i \in K_\lambda} u_i \Bigg) \\ &\quad + \frac{1}{K_s} \begin{pmatrix} \lambda^\top & \mu^\top \end{pmatrix} \left(X^\top X + \begin{pmatrix} \mathbf{1}_p \mathbf{1}_p^\top & -\mathbf{1}_p \mathbf{1}_q^\top \\ -\mathbf{1}_q \mathbf{1}_p^\top & \mathbf{1}_q \mathbf{1}_q^\top \end{pmatrix} \right) \begin{pmatrix} \lambda \\ \mu \end{pmatrix}. \end{aligned}$$

We show in Proposition 18.4 that $p_f + q_f \leq (p+q)\nu$, so if $\nu > (p_f + q_f)/(p+q)$, we can solve for η in terms of b, w, and λ, μ. But b and w are expressed in terms of λ, μ as

$$w = -X \begin{pmatrix} \lambda \\ \mu \end{pmatrix}$$

$$b = -\sum_{i=1}^{p} \lambda_i + \sum_{j=1}^{q} \mu_j = -\mathbf{1}_p^\top \lambda + \mathbf{1}_q^\top \mu$$

so η is also expressed in terms of λ, μ.

The condition $\nu > (p_f+q_f)/(p+q)$ cannot be satisfied if $p_f+q_f = p+q$, but in this case all points fail the margin, which indicates that δ is too big, so we reduce ν and try again.

Remark: The equation

$$\sum_{i=1}^{p} \lambda_i + \sum_{j=1}^{q} \mu_j = \nu$$

implies that either there is some i_0 such that $\lambda_{i_0} > 0$ *or* there is some j_0 such that $\mu_{j_0} > 0$, which implies that $p_m + q_m \geq 1$.

Another way to compute η is to assume the Standard Margin Hypothesis for (SVM$_{s3}$). Under the **Standard Margin Hypothesis** for (SVM$_{s3}$), either there is some i_0 such that $0 < \lambda_{i_0} < K_s$ *or* there is some j_0 such that $0 < \mu_{j_0} < K_s$, in other words, there is some support vector of type 1. By the complementary slackness conditions $\epsilon_{i_0} = 0$ or $\xi_{j_0} = 0$, so we have

$$w^\top u_{i_0} - b = \eta, \quad \text{or} \quad -w^\top v_{j_0} + b = \eta,$$

and we can solve for η.

Due to numerical instability, when writing a computer program it is preferable to compute the lists of indices I_λ and I_μ given by

$$I_\lambda = \{i \in \{1, \dots, p\} \mid 0 < \lambda_i < K_s\}$$
$$I_\mu = \{j \in \{1, \dots, q\} \mid 0 < \mu_j < K_s\}.$$

Then it is easy to see that we can compute η using the following averaging formulae: If $I_\lambda \neq \emptyset$, then

$$\eta = w^\top \Bigg(\sum_{i \in I_\lambda} u_i\Bigg) / |I_\lambda| - b,$$

and if $I_\mu \neq \emptyset$, then

$$\eta = b - w^\top \Bigg(\sum_{j \in I_\mu} v_j\Bigg) / |I_\mu|.$$

Theoretically the condition $\nu > (p_f + q_f)/(p + q)$ is less restrictive that the **Standard Margin Hypothesis** but in practice we have never observed an example for which $\nu > (p_f + q_f)/(p + q)$ and yet the **Standard Margin Hypothesis** fails.

The "kernelized" version of Problem (SVM_{s3}) is the following:

Soft margin kernel SVM (SVM_{s3}):

$$\begin{aligned}
&\text{minimize} \quad \frac{1}{2}\langle w, w\rangle + \frac{1}{2}b^2 - \nu\eta + K_s \left(\epsilon^\top \; \xi^\top\right) \mathbf{1}_{p+q} \\
&\text{subject to} \\
&\qquad \langle w, \varphi(u_i)\rangle - b \geq \eta - \epsilon_i, \quad \epsilon_i \geq 0 \qquad i = 1, \ldots, p \\
&\qquad -\langle w, \varphi(v_j)\rangle + b \geq \eta - \xi_j, \quad \xi_j \geq 0 \qquad j = 1, \ldots, q,
\end{aligned}$$

with $K_s = 1/(p + q)$.

Tracing through the derivation of the dual program, we obtain

Dual of the Soft margin kernel SVM (SVM_{s3}):

$$\begin{aligned}
&\text{minimize} \quad \frac{1}{2}\left(\lambda^\top \; \mu^\top\right)\left(\mathbf{K} + \begin{pmatrix} \mathbf{1}_p\mathbf{1}_p^\top & -\mathbf{1}_p\mathbf{1}_q^\top \\ -\mathbf{1}_q\mathbf{1}_p^\top & \mathbf{1}_q\mathbf{1}_q^\top \end{pmatrix}\right)\begin{pmatrix} \lambda \\ \mu \end{pmatrix} \\
&\text{subject to} \\
&\qquad \sum_{i=1}^{p} \lambda_i + \sum_{j=1}^{q} \mu_j = \nu \\
&\qquad 0 \leq \lambda_i \leq K_s, \quad i = 1, \ldots, p \\
&\qquad 0 \leq \mu_j \leq K_s, \quad j = 1, \ldots, q,
\end{aligned}$$

where $\mathbf{K}$ is the kernel matrix of Section 18.1.

We obtain

$$\begin{aligned}
w &= \sum_{i=1}^{p} \lambda_i \varphi(u_i) - \sum_{j=1}^{q} \mu_j \varphi(v_j) \\
b &= -\sum_{i=1}^{p} \lambda_i + \sum_{j=1}^{q} \mu_j.
\end{aligned}$$

The classification function

$$f(x) = \mathrm{sgn}(\langle w, \varphi(x)\rangle - b)$$

is given by

$$f(x) = \mathrm{sgn}\left(\sum_{i=1}^{p} \lambda_i(\kappa(u_i, x) + 1) - \sum_{j=1}^{q} \mu_j(\kappa(v_j, x) + 1)\right).$$

18.10 Classification of the Data Points in Terms of ν (SVM$_{s3}$)

The equations (†) and the box inequalities

$$0 \leq \lambda_i \leq K_s, \quad 0 \leq \mu_j \leq K_s$$

also imply the following facts (recall that $\delta = \eta / \|w\|$):

Proposition 18.4. *If Problem* (SVM$_{s3}$) *has an optimal solution with $w \neq 0$ and $\eta > 0$ then the following facts hold:*

(1) Let p_f be the number of points u_i such that $\lambda_i = K_s$, and let q_f the number of points v_j such that $\mu_j = K_s$. Then $p_f + q_f \leq (p+q)\nu$.
(2) Let p_m be the number of points u_i such that $\lambda_i > 0$, and let q_m the number of points v_j such that $\mu_j > 0$. Then $p_m + q_m \geq (p+q)\nu$. We have $p_m + q_m \geq 1$.
(3) If $p_f \geq 1$ or $q_f \geq 1$, then $\nu \geq 1/(p+q)$.

Proof. (1) Recall that for an optimal solution with $w \neq 0$ and $\eta > 0$ we have the equation

$$\sum_{i=1}^{p} \lambda_i + \sum_{j=1}^{q} \mu_j = \nu.$$

Since there are p_f points u_i such that $\lambda_i = K_s = 1/(p+q)$ and q_f points v_j such that $\mu_j = K_s = 1/(p+q)$, we have

$$\nu = \sum_{i=1}^{p} \lambda_i + \sum_{j=1}^{q} \mu_j \geq \frac{p_f + q_f}{p+q},$$

so

$$p_f + q_f \leq \nu(p+q).$$

(2) If

$$I_{\lambda>0} = \{i \in \{1, \dots, p\} \mid \lambda_i > 0\} \quad \text{and} \quad p_m = |I_{\lambda>0}|$$

and

$$I_{\mu>0} = \{j \in \{1, \dots, q\} \mid \mu_j > 0\} \quad \text{and} \quad q_m = |I_{\mu>0}|,$$

then

$$\nu = \sum_{i=1}^{p} \lambda_i + \sum_{j=1}^{q} \mu_j = \sum_{i \in I_{\lambda>0}} \lambda_i + \sum_{j \in I_{\mu>0}} \mu_j,$$

and since $\lambda_i, \mu_j \leq K_s = 1/(p+q)$, we have

$$\nu = \sum_{i \in I_{\lambda>0}} \lambda_i + \sum_{j \in I_{\mu>0}} \mu_j \leq \frac{p_m + q_m}{p+q},$$

which yields

$$p_m + q_m \geq \nu(p+q).$$

We already noted earlier that $p_m + q_m \geq 1$.

(3) This follows immediately from (1). ☐

Note that if ν is chosen so that $\nu < 1/(p+q)$, then $p_f = q_f = 0$, which means that none of the data points are misclassified; in other words, the u_is and v_js are linearly separable. Thus we see that if the u_is and v_js are not linearly separable we must pick ν such that $1/(p+q) \leq \nu \leq 1$ for the method to succeed. In fact, by Proposition 18.4, we must choose ν so that

$$\frac{p_f + q_f}{p+q} \leq \nu \leq \frac{p_m + q_m}{p+q}.$$

Furthermore, in order to be able to determine b, we must have the strict inequality

$$\frac{p_f + q_f}{p+q} < \nu.$$

18.11 Existence of Support Vectors for (SVM$_{s3}$)

The following proposition is the version of Proposition 18.2 for Problem (SVM$_{s3}$).

Proposition 18.5. *For every optimal solution $(w, b, \eta, \epsilon, \xi)$ of Problem (SVM$_{s3}$) with $w \neq 0$ and $\eta > 0$, if $\nu < 1$ and if no u_i is a support vector and no v_j is a support vector, then there is another optimal solution such that some u_{i_0} or some v_{j_0} is a support vector.*

Proof. We may assume that $K_s = 1/(p+q)$ and we proceed by contradiction. Thus we assume that for all $i \in \{1, \ldots, p\}$, if $\epsilon_i = 0$, then the constraint $w^\top u_i - b \geq \eta$ is not active, namely $w^\top u_i - b > \eta$, *and* for all $j \in \{1, \ldots, q\}$, if $\xi_j = 0$, then the constraint $-w^\top v_j + b \geq \eta$ is not active, namely $-w^\top v_j + b > \eta$.

Let $E_\lambda = \{i \in \{1, \ldots, p\} \mid \epsilon_i > 0\}$ and let $E_\mu = \{j \in \{1, \ldots, q\} \mid \xi_j > 0\}$. By definition, $p_{sf} = |E_\lambda|$, $q_{sf} = |E_\mu|$, $p_{sf} \leq p_f$ and $q_{sf} \leq q_f$, so by Proposition 18.1,

$$\frac{p_{sf} + q_{sf}}{p+q} \leq \frac{p_f + q_f}{p+q} \leq \nu.$$

Therefore, if $\nu < 1$, then $p_{sf} + q_{sf} < p + q$, which implies that either there is some $i \notin E_\lambda$ such that $\epsilon_i = 0$ or there is some $j \notin E_\mu$ such that $\xi_j = 0$.

By complementary slackness all the constraints for which $i \in E_\lambda$ and $j \in E_\mu$ are active, so our hypotheses are

$$\begin{aligned} w^\top u_i - b &= \eta - \epsilon_i & \epsilon_i &> 0 & i &\in E_\lambda \\ -w^\top v_j + b &= \eta - \xi_j & \xi_j &> 0 & j &\in E_\mu \\ w^\top u_i - b &> \eta & & & i &\notin E_\lambda \\ -w^\top v_j + b &> \eta & & & j &\notin E_\mu, \end{aligned}$$

and either there is some $i \notin E_\lambda$ or there is some $j \notin E_\mu$. Our strategy, as illustrated in Figures 18.8 and 18.9, is to increase the width η of the slab keeping the separating hyperplane unchanged. Let us pick θ such that

$$\theta = \min\{w^\top u_i - b - \eta,\ -w^\top v_j + b - \eta \mid i \notin E_\lambda, j \notin E_\mu\}.$$

Our hypotheses imply that $\theta > 0$. We can write

$$\begin{aligned} w^\top u_i - b &= \eta + \theta - (\epsilon_i + \theta) & \epsilon_i + \theta &> 0 & i &\in E_\lambda \\ -w^\top v_j + b &= \eta + \theta - (\xi_j + \theta) & \xi_j + \theta &> 0 & j &\in E_\mu \\ w^\top u_i - b &\geq \eta + \theta & & & i &\notin E_\lambda \\ -w^\top v_j + b &\geq \eta + \theta & & & j &\notin E_\mu, \end{aligned}$$

and by the choice of θ, either

$$w^\top u_i - b = \eta + \theta \quad \text{for some } i \notin E_\lambda$$

or

$$-w^\top v_j + b = \eta + \theta \quad \text{for some } j \notin E_\mu.$$

The original value of the objective function is

$$\omega(0) = \frac{1}{2} w^\top w + \frac{1}{2} b^2 - \nu\eta + \frac{1}{p+q}\left(\sum_{i \in E_\lambda} \epsilon_i + \sum_{j \in E_\mu} \xi_j\right),$$

and the new value is

$$\begin{aligned} \omega(\theta) &= \frac{1}{2} w^\top w + \frac{1}{2} b^2 - \nu(\eta + \theta) + \frac{1}{p+q}\left(\sum_{i \in E_\lambda} (\epsilon_i + \theta) + \sum_{j \in E_\mu} (\xi_j + \theta)\right) \\ &= \frac{1}{2} w^\top w + \frac{1}{2} b^2 - \nu\eta + \frac{1}{p+q}\left(\sum_{i \in E_\lambda} \epsilon_i + \sum_{j \in E_\mu} \xi_j\right) - \left(\nu - \frac{p_{sf} + q_{sf}}{p+q}\right)\theta. \end{aligned}$$

By Proposition 18.1,

$$\frac{p_{sf} + q_{sf}}{p+q} \leq \frac{p_f + q_f}{p+q} \leq \nu,$$

so

$$\nu - \frac{p_{sf} + q_{sf}}{p+q} \geq 0,$$

and so $\omega(\theta) \leq \omega(0)$. If the inequality is strict, then this contradicts the optimality of the original solution. Therefore, $\omega(\theta) = \omega(0)$ and $(w, b, \eta + \theta, \epsilon + \theta, \xi + \theta)$ is an optimal solution such that either

$$w^\top u_i - b = \eta + \theta \quad \text{for some } i \notin E_\lambda$$

or

$$-w^\top v_j + b = \eta + \theta \quad \text{for some } j \notin E_\mu,$$

as desired. □

Proposition 18.5 cannot be strengthened to claim that there is some support vector u_{i_0} *and* some support vector v_{j_0}. We found examples for which the above condition fails for ν large enough.

The proof of Proposition 18.5 reveals that $(p_{sf} + q_{sf})/(p+q)$ is a critical value for ν. if this value is avoided we have the following corollary.

Theorem 18.2. *For every optimal solution* $(w, b, \eta, \epsilon, \xi)$ *of Problem* (SVM$_{s3}$) *with* $w \neq 0$ *and* $\eta > 0$, *if*

$$(p_{sf} + q_{sf})/(p+q) < \nu < 1,$$

then some u_{i_0} *or some* v_{j_0} *is a support vector.*

The proof proceeds by contradiction using Proposition 18.5 (for a very similar proof, see the proof of Theorem 18.1).

18.12 Solving SVM (SVM$_{s3}$) Using ADMM

In order to solve (SVM$_{s3}$) using ADMM we need to write the matrix corresponding to the constraints in equational form,

$$\begin{aligned}\sum_{i=1}^{p} \lambda_i + \sum_{j=1}^{q} \mu_j &= K_m \\ \lambda_i + \alpha_i &= K_s, \quad i = 1, \ldots, p \\ \mu_j + \beta_j &= K_s, \quad j = 1, \ldots, q\end{aligned}$$

with $K_m = (p+q)K_s\nu$. This is the $(p+q+1) \times 2(p+q)$ matrix A given by

$$A = \begin{pmatrix} \mathbf{1}_p^\top & \mathbf{1}_q^\top & 0_p^\top & 0_q^\top \\ I_p & 0_{p,q} & I_q & 0_{p,q} \\ 0_{q,p} & I_q & 0_{q,p} & I_q \end{pmatrix}.$$

We leave it as an exercise to prove that A has rank $p+q+1$. The right-hand side is

$$c = \begin{pmatrix} K_m \\ K_s \mathbf{1}_{p+q} \end{pmatrix}.$$

The symmetric positive semidefinite $(p+q) \times (p+q)$ matrix P defining the quadratic functional is

$$P = X^\top X + \begin{pmatrix} \mathbf{1}_p \mathbf{1}_p^\top & -\mathbf{1}_p \mathbf{1}_q^\top \\ -\mathbf{1}_q \mathbf{1}_p^\top & \mathbf{1}_q \mathbf{1}_q^\top \end{pmatrix}, \quad \text{with} \quad X = \begin{pmatrix} -u_1 & \cdots & -u_p & v_1 & \cdots & v_q \end{pmatrix},$$

and

$$q = 0_{p+q}.$$

Since there are $2(p+q)$ Lagrange multipliers $(\lambda, \mu, \alpha, \beta)$, the $(p+q) \times (p+q)$ matrix P must be augmented with zero's to make it a $2(p+q) \times 2(p+q)$ matrix P_a given by

$$P_a = \begin{pmatrix} P & 0_{p+q,p+q} \\ 0_{p+q,p+q} & 0_{p+q,p+q} \end{pmatrix},$$

and similarly q is augmented with zeros as the vector

$$q_a = 0_{2(p+q)}.$$

The `Matlab` programs implementing the above method are given in Appendix B, Section B.3. We ran our program on the same input data points used in Section 18.8, namely

```
u16 = 10.1*randn(2,30)+7 ;
v16 = -10.1*randn(2,30)-7;
[~,~,~,~,~,~,w3] = runSVMs3b(0.365,rho,u16,v16,1/60)
```

We picked $K = 1/60$ and various values of ν starting with $\nu = 0.365$, which appears to be the smallest value for which the method converges; see Figure 18.16.

We have $p_f = 10, q_f = 10, p_m = 12$ and $q_m = 11$, as opposed to $p_f = 10, q_f = 11, p_m = 12, q_m = 12$, which was obtained by running (SVM$_{s2'}$); see Figure 18.11. A slightly narrower margin is achieved.

Next we ran our program with $\nu = 0.5$, see Figure 18.17. We have $p_f = 13, q_f = 16, p_m = 14$ and $q_m = 17$.

We also ran our program with $\nu = 0.71$, see Figure 18.18. We have $p_f = 21, q_f = 21, p_m = 22$ and $q_m = 22$. The value $\nu = 0.7$ is a singular value for which there are no support vectors and $\nu = (p_f + q_f)/(p+q)$.

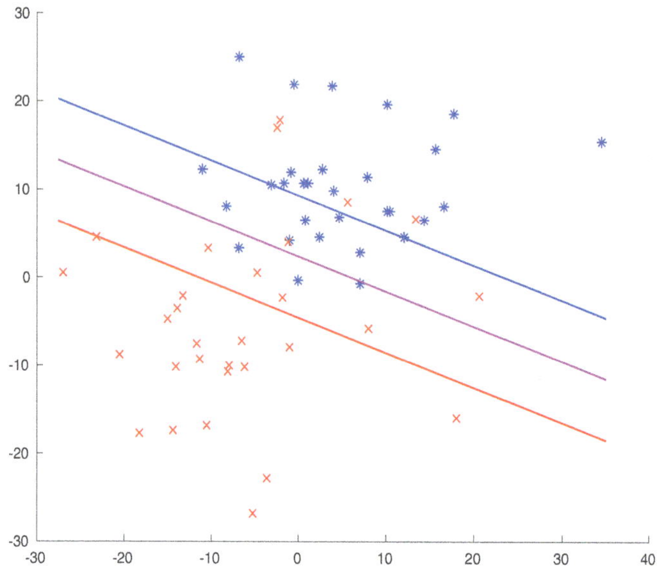

Fig. 18.16 Running (SVM_{s3}) on two sets of 30 points; $\nu = 0.365$.

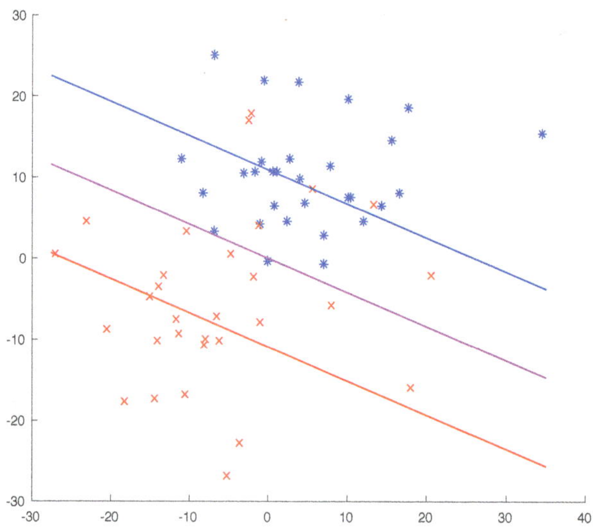

Fig. 18.17 Running (SVM_{s3}) on two sets of 30 points; $\nu = 0.5$.

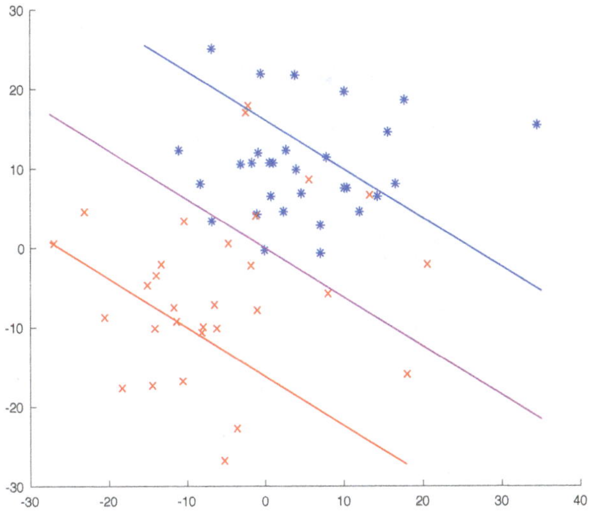

Fig. 18.18 Running (SVM$_{s3}$) on two sets of 30 points; $\nu = 0.71$.

Finally we ran our program with $\nu = 0.98$, see Figure 18.19. We have $p_f = 28, q_f = 30, p_m = 29$ and $q_m = 30$.

Because the term $(1/2)b^2$ is added to the objective function to be minimized, it turns out that (SVM$_{s3}$) yields values of b and η that are smaller than the values returned by (SVM$_{s2'}$). This is the reason why a smaller margin width could be obtained for $\nu = 0.365$. On the other hand, (SVM$_{s3}$) is unable to achieve as big a margin as (SVM$_{s2'}$) for values of $\nu \geq 0.97$, because the separating line produced by (SVM$_{s3}$) is lower than the the separating line produced by (SVM$_{s2'}$).

18.13 Soft Margin SVM; (SVM$_{s4}$)

In this section we consider the version of Problem (SVM$_{s2'}$) in which instead of using the function $K\left(\sum_{i=1}^{p} \epsilon_i + \sum_{j=1}^{q} \xi_j\right)$ as a regularizing function we use the quadratic function $K(\|\epsilon\|_2^2 + \|\xi\|_2^2)$.

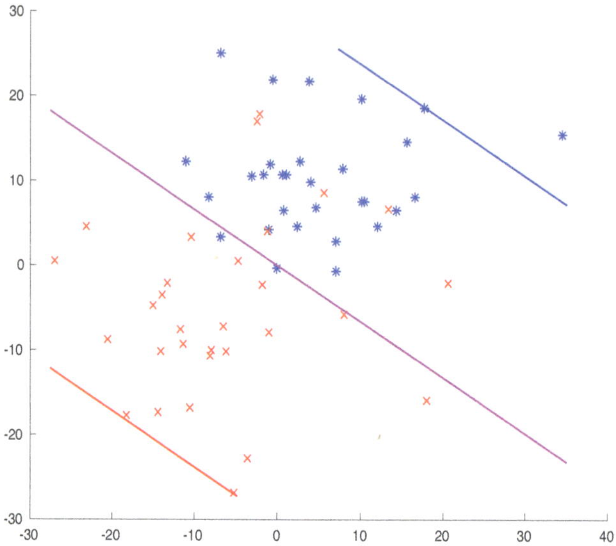

Fig. 18.19 Running (SVM$_{s3}$) on two sets of 30 points; $\nu = 0.98$.

Soft margin SVM (SVM$_{s4}$):

$$\begin{aligned} &\text{minimize} \quad \frac{1}{2} w^\top w + (p+q)K_s \left(-\nu\eta + \frac{1}{p+q}(\epsilon^\top \epsilon + \xi^\top \xi) \right) \\ &\text{subject to} \\ &\qquad w^\top u_i - b \geq \eta - \epsilon_i, \qquad i = 1, \ldots, p \\ &\qquad - w^\top v_j + b \geq \eta - \xi_j, \qquad j = 1, \ldots, q \\ &\qquad \eta \geq 0, \end{aligned}$$

where ν and K_s are two given positive constants. As we saw earlier, theoretically, it is convenient to pick $K_s = 1/(p+q)$. When writing a computer program, it is preferable to assume that K_s is arbitrary. In this case ν needs to be replaced by $(p+q)K_s\nu$ in all the formulae obtained with $K_s = 1/(p+q)$.

The new twist with this formulation of the problem is that if $\epsilon_i < 0$, then the corresponding inequality $w^\top u_i - b \geq \eta - \epsilon_i$ implies the inequality $w^\top u_i - b \geq \eta$ obtained by setting ϵ_i to zero while reducing the value of $\|\epsilon\|^2$, and similarly if $\xi_j < 0$, then the corresponding inequality $-w^\top v_j + b \geq \eta - \xi_j$

implies the inequality $-w^\top v_j + b \geq \eta$ obtained by setting ξ_j to zero while reducing the value of $\|\xi\|^2$. Therefore, if (w, b, ϵ, ξ) is an optimal solution of Problem (SVM$_{s4}$), it is not necessary to restrict the slack variables ϵ_i and ξ_j to the nonnegative, which simplifies matters a bit. In fact, we will see that for an optimal solution, $\epsilon = \lambda/(2K_s)$ and $\xi = \mu/(2K_s)$. The variable η can also be determined by expressing that the duality gap is zero.

One of the advantages of this methods is that ϵ is determined by λ, ξ is determined by μ, and η and b are determined by λ and μ. This method *does not* require support vectors to compute b. We can omit the constraint $\eta \geq 0$, because for an optimal solution it can be shown using duality that $\eta \geq 0$; see Section 18.14.

A drawback of Program (SVM$_{s4}$) is that for fixed K_s, the quantity $\delta = \eta/\|w\|$ and the hyperplanes $H_{w,b}, H_{w,b+\eta}$ and $H_{w,b-\eta}$ are independent of ν. This will be shown in Theorem 18.3. Thus this method is less flexible than (SVM$_{s2'}$) and (SVM$_{s3}$).

The Lagrangian is given by

$$\begin{aligned}
L(w, \epsilon, \xi, b, \eta, \lambda, \mu, \gamma) &= \frac{1}{2}w^\top w - \nu\eta + K_s(\epsilon^\top\epsilon + \xi^\top\xi) + w^\top X \begin{pmatrix}\lambda\\ \mu\end{pmatrix} - \gamma\eta \\
&\quad - \epsilon^\top\lambda - \xi^\top\mu + b(\mathbf{1}_p^\top\lambda - \mathbf{1}_q^\top\mu) + \eta(\mathbf{1}_p^\top\lambda + \mathbf{1}_q^\top\mu) \\
&= \frac{1}{2}w^\top w + w^\top X \begin{pmatrix}\lambda\\ \mu\end{pmatrix} + \eta(\mathbf{1}_p^\top\lambda + \mathbf{1}_q^\top\mu - \nu - \gamma) \\
&\quad + K_s(\epsilon^\top\epsilon + \xi^\top\xi) - \epsilon^\top\lambda - \xi^\top\mu + b(\mathbf{1}_p^\top\lambda - \mathbf{1}_q^\top\mu).
\end{aligned}$$

To find the dual function $G(\lambda, \mu, \gamma)$ we minimize $L(w, \epsilon, \xi, b, \eta, \lambda, \mu, \gamma)$ with respect to w, ϵ, ξ, b, and η. Since the Lagrangian is convex and $(w, \epsilon, \xi, b, \eta) \in \mathbb{R}^n \times \mathbb{R}^p \times \mathbb{R}^q \times \mathbb{R} \times \mathbb{R}$, a convex open set, by Theorem 4.5, the Lagrangian has a minimum in $(w, \epsilon, \xi, b, \eta)$ iff $\nabla L_{w,\epsilon,\xi,b,\eta} = 0$, so we compute $\nabla L_{w,\epsilon,\xi,b,\eta}$. The gradient $\nabla L_{w,\epsilon,\xi,b,\eta}$ is given by

$$\nabla L_{w,\epsilon,\xi,b,\eta} = \begin{pmatrix} w + X\begin{pmatrix}\lambda\\ \mu\end{pmatrix} \\ 2K_s\epsilon - \lambda \\ 2K_s\xi - \mu \\ \mathbf{1}_p^\top\lambda - \mathbf{1}_q^\top\mu \\ \mathbf{1}_p^\top\lambda + \mathbf{1}_q^\top\mu - \nu - \gamma \end{pmatrix}.$$

By setting $\nabla L_{w,\epsilon,\xi,b,\eta} = 0$ we get the equations

$$w = -X\begin{pmatrix}\lambda\\ \mu\end{pmatrix}, \qquad (*_w)$$

$$\begin{aligned}2K_s\epsilon &= \lambda\\ 2K_s\xi &= \mu\\ \mathbf{1}_p^\top\lambda &= \mathbf{1}_q^\top\mu\\ \mathbf{1}_p^\top\lambda + \mathbf{1}_q^\top\mu &= \nu + \gamma.\end{aligned}$$

The last two equations are identical to the last two equations obtained in Problem (SVM$_{s2'}$). We can use the other equations to obtain the following expression for the dual function $G(\lambda, \mu, \gamma)$,

$$\begin{aligned}G(\lambda,\mu,\gamma) &= -\frac{1}{4K_s}(\lambda^\top\lambda + \mu^\top\mu) - \frac{1}{2}\begin{pmatrix}\lambda^\top & \mu^\top\end{pmatrix} X^\top X\begin{pmatrix}\lambda\\ \mu\end{pmatrix}\\ &= -\frac{1}{2}\begin{pmatrix}\lambda^\top & \mu^\top\end{pmatrix}\left(X^\top X + \frac{1}{2K_s}I_{p+q}\right)\begin{pmatrix}\lambda\\ \mu\end{pmatrix}.\end{aligned}$$

Consequently the dual program is equivalent to the minimization program

Dual of the Soft margin SVM (SVM$_{s4}$):

$$\begin{aligned}&\text{minimize} \quad \frac{1}{2}\begin{pmatrix}\lambda^\top & \mu^\top\end{pmatrix}\left(X^\top X + \frac{1}{2K_s}I_{p+q}\right)\begin{pmatrix}\lambda\\ \mu\end{pmatrix}\\ &\text{subject to}\\ &\qquad \sum_{i=1}^p \lambda_i - \sum_{j=1}^q \mu_j = 0\\ &\qquad \sum_{i=1}^p \lambda_i + \sum_{j=1}^q \mu_j \geq \nu\\ &\qquad \lambda_i \geq 0, \quad i = 1,\ldots,p\\ &\qquad \mu_j \geq 0, \quad j = 1,\ldots,q.\end{aligned}$$

The above program is similar to the program that was obtained for Problem (SVM$_{s2'}$) but the matrix $X^\top X$ is replaced by the matrix $X^\top X + (1/2K_s)I_{p+q}$, which is positive definite since $K_s > 0$, and also the inequalities $\lambda_i \leq K_s$ and $\mu_j \leq K_s$ no longer hold.

It is shown in Section 18.14 how the dual program is solved using ADMM from Section 16.6. If the primal problem is solvable, this yields solutions for λ and μ. We obtain w from λ and μ, as in Problem (SVM$_{s2'}$); namely,

$$w = -X\begin{pmatrix}\lambda \\ \mu\end{pmatrix} = \sum_{i=1}^{p} \lambda_i u_i - \sum_{j=1}^{q} \mu_j v_j.$$

Since the variables ϵ_i and ξ_j are not restricted to be nonnegative we no longer have complementary slackness conditions involving them, but we know that

$$\epsilon = \frac{\lambda}{2K_s}, \quad \xi = \frac{\mu}{2K_s}.$$

Also since the constraints

$$\sum_{i=1}^{p} \lambda_i \geq \frac{\nu}{2} \quad \text{and} \quad \sum_{j=1}^{q} \mu_j \geq \frac{\nu}{2}$$

imply that there is some i_0 such that $\lambda_{i_0} > 0$ and some j_0 such that $\mu_{j_0} > 0$, we have $\epsilon_{i_0} > 0$ and $\xi_{j_0} > 0$, which means that at least two points are misclassified, so Problem (SVM$_{s4}$) should only be used when the sets $\{u_i\}$ and $\{v_j\}$ are *not* linearly separable.

Because $\epsilon_i = \lambda_i/(2K_s)$, $\xi_j = \mu_j/(2K_s)$, and there is no upper bound K_s on λ_i and μ_j, the classification of the points is simpler than in the previous cases.

(1) If $\lambda_i = 0$, then $\epsilon_i = 0$ and the inequality $w^\top u_i - b - \eta \geq 0$ holds. If equality holds then u_i is a support vector on the blue margin (the hyperplane $H_{w,b+\eta}$). Otherwise u_i is in the blue open half-space bounded by the margin hyperplane $H_{w,b+\eta}$ (not containing the separating hyperplane $H_{w,b}$). See Figure 18.20.
Similarly, if $\mu_j = 0$, then $\xi_j = 0$ and the inequality $-w^\top v_j + b - \eta \geq$ holds. If equality holds then v_j is a support vector on the red margin (the hyperplane $H_{w,b-\eta}$). Otherwise v_j is in the red open half-space bounded by the margin hyperplane $H_{w,b-\eta}$ (not containing the separating hyperplane $H_{w,b}$). See Figure 18.20.

(2) If $\lambda_i > 0$, then $\epsilon_i = \lambda_i/(2K_s) > 0$. The corresponding constraint is active, so we have

$$w^\top u_i - b = \eta - \epsilon_i.$$

If $\epsilon_i \leq \eta$, then the points u_i is inside the slab bounded by the blue margin hyperplane $H_{w,b+\eta}$ and the separating hyperplane $H_{w,b}$. If

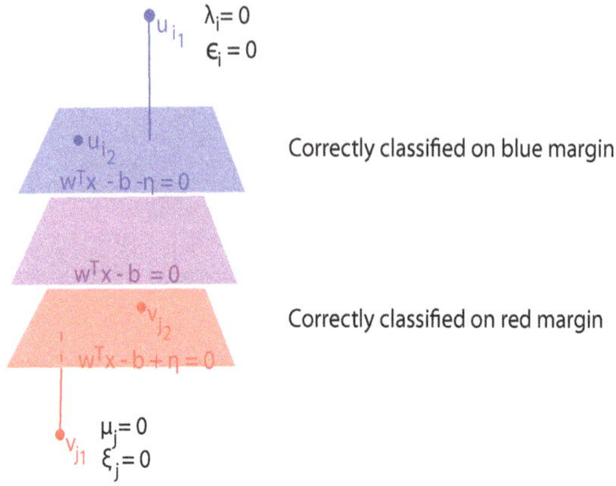

Fig. 18.20 When $\lambda_i = 0$, u_i is correctly classified on or outside the blue margin. When $\mu_j = 0$, v_j is correctly classified on or outside outside the red margin.

$\epsilon_i > \eta$, then the point u_i belongs to the open half-space bounded by the separating hyperplane and containing the red margin hyperplane (the red side); it is misclassified. See Figure 18.21.

Similarly, if $\mu_j > 0$, then $\xi_j = \mu_j/(2K_s) > 0$. The corresponding constraint is active, so we have

$$-w^\top v_j + b = \eta - \xi_j.$$

If $\xi_j \leq \eta$, then the points v_j is inside the slab bounded by the red margin hyperplane $H_{w,b-\eta}$ and the separating hyperplane $H_{w,b}$. If $\xi_j > \eta$, then the point v_j belongs to the open half-space bounded by the separating hyperplane and containing the blue margin hyperplane (the blue side); it is misclassified. See Figure 18.21.

We can use the fact that the duality gap is 0 to find η. We have

$$\frac{1}{2}w^\top w - \nu\eta + K_s(\epsilon^\top \epsilon + \xi^\top \xi) = -\frac{1}{2}\begin{pmatrix}\lambda^\top & \mu^\top\end{pmatrix}\left(X^\top X + \frac{1}{2K_s}I_{p+q}\right)\begin{pmatrix}\lambda \\ \mu\end{pmatrix},$$

and since

$$w = -X\begin{pmatrix}\lambda \\ \mu\end{pmatrix}$$

we get

$$\begin{aligned}\nu\eta &= K_s(\epsilon^\top \epsilon + \xi^\top \xi) + \begin{pmatrix}\lambda^\top & \mu^\top\end{pmatrix}\left(X^\top X + \frac{1}{4K_s}I_{p+q}\right)\begin{pmatrix}\lambda \\ \mu\end{pmatrix} \\ &= \begin{pmatrix}\lambda^\top & \mu^\top\end{pmatrix}\left(X^\top X + \frac{1}{2K_s}I_{p+q}\right)\begin{pmatrix}\lambda \\ \mu\end{pmatrix}.\end{aligned}$$

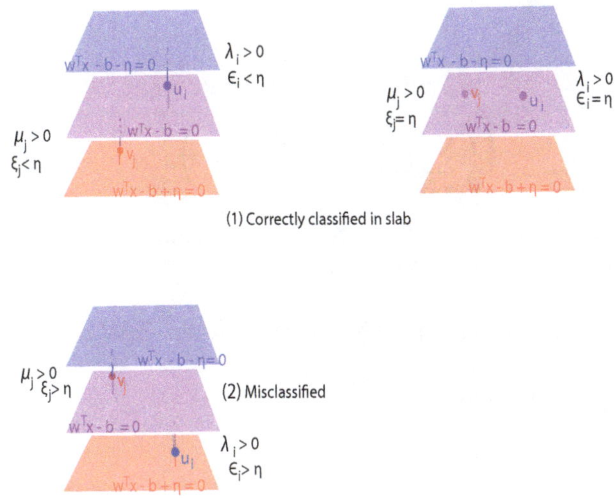

Fig. 18.21 The classification of points for SVM$_{s4}$ when the Lagrange multipliers are positive. The left illustration of Figure (1) is when u_i is inside the margin yet still on the correct side of the separating hyperplane $w^\top x - b = 0$. Similarly, v_j is inside the margin on the correct side of the separating hyperplane. The right illustration depicts u_i and v_j on the separating hyperplane. Figure (2) illustrations a misclassification of u_i and v_j.

The above confirms that at optimality we have $\eta \geq 0$.

Remark: If we do not assume that $K_s = 1/(p+q)$, then the above formula must be replaced by

$$(p+q)K_s\nu\eta = (\lambda^\top \; \mu^\top)\left(X^\top X + \frac{1}{2K_s}I_{p+q}\right)\begin{pmatrix}\lambda\\ \mu\end{pmatrix}.$$

Since η is determined independently of the existence of support vectors, the margin hyperplane $H_{w,b+\eta}$ may not contain any point u_i and the margin hyperplane $H_{w,b-\eta}$ may not contain any point v_j.

We can solve for b using some active constraint corresponding to any i_0 such that $\lambda_{i_0} > 0$ and any j_0 such that $\mu_{j_0} > 0$ (by a previous remark, the constraints imply that such i_0 and j_0 must exist). To improve numerical stability we average over the following sets of indices. Let I_λ and I_μ be the set of indices given by

$$I_\lambda = \{i \in \{1, \ldots, p\} \mid \lambda_i > 0\}$$
$$I_\mu = \{j \in \{1, \ldots, q\} \mid \mu_j > 0\},$$

and let $p_m = |I_\lambda|$ and $q_m = |I_\mu|$. We obtain the formula

$$b = w^\top \Big(\Big(\sum_{i \in I_\lambda} u_i \Big) / p_m + \Big(\sum_{j \in I_\mu} v_j \Big) / q_m \Big) / 2$$
$$+ \Big(\Big(\sum_{i \in I_\lambda} \epsilon_i \Big) / p_m - \Big(\sum_{j \in I_\mu} \xi_j \Big) / q_m \Big) / 2.$$

We now prove that for a fixed K_s, the solution to Problem (SVM$_{s4}$) is unique and independent of the value of ν.

Theorem 18.3. *For K_s and ν fixed, if Problem* (SVM$_{s4}$) *succeeds, then it has a unique solution. If Problem* (SVM$_{s4}$) *succeeds and returns $(\lambda, \mu, \eta, w, b)$ for the value ν and $(\lambda^\kappa, \mu^\kappa, \eta^\kappa, w^\kappa, b^\kappa)$ for the value $\kappa\nu$ with $\kappa > 0$, then*

$$\lambda^\kappa = \kappa\lambda, \quad \mu^\kappa = \kappa\mu, \quad \eta^\kappa = \kappa\eta, \quad w^\kappa = \kappa w, \quad b^\kappa = \kappa b.$$

As a consequence, $\delta = \eta / \|w\| = \eta^\kappa / \|w^\kappa\| = \delta^\kappa$, and the hyperplanes $H_{w,b}, H_{w,b+\eta}$ and $H_{w,b-\eta}$ are independent of ν.

Proof. We already observed that for an optimal solution with $\eta > 0$, we have $\gamma = 0$. This means that (λ, μ) is a solution of the problem

$$\text{minimize} \quad \frac{1}{2} \begin{pmatrix} \lambda^\top & \mu^\top \end{pmatrix} \Big(X^\top X + \frac{1}{2K_s} I_{p+q} \Big) \begin{pmatrix} \lambda \\ \mu \end{pmatrix}$$

subject to

$$\sum_{i=1}^p \lambda_i - \sum_{j=1}^q \mu_j = 0$$
$$\sum_{i=1}^p \lambda_i + \sum_{j=1}^q \mu_j = \nu$$
$$\lambda_i \geq 0, \quad i = 1, \ldots, p$$
$$\mu_j \geq 0, \quad j = 1, \ldots, q.$$

Since $K_s > 0$ and $X^\top X$ is symmetric positive semidefinite, the matrix $P = X^\top X + \frac{1}{2K_s} I_{p+q}$ is symmetric positive definite. Let $\Omega = \mathbb{R}^{p+q}$ and let U be the convex set given by

$$U = \left\{ \begin{pmatrix} \lambda \\ \mu \end{pmatrix} \in \mathbb{R}_+^{p+q} \;\middle|\; \begin{pmatrix} \mathbf{1}_p^\top & -\mathbf{1}_q^\top \\ \mathbf{1}_p^\top & \mathbf{1}_q^\top \end{pmatrix} \begin{pmatrix} \lambda \\ \mu \end{pmatrix} = \begin{pmatrix} 0 \\ (p+q)K_s\nu \end{pmatrix} \right\}.$$

Since the matrix P is symmetric positive definite, the functional

$$F(\lambda,\mu) = -G(\lambda,\mu) = \frac{1}{2}\begin{pmatrix}\lambda^\top & \mu^\top\end{pmatrix} P \begin{pmatrix}\lambda \\ \mu\end{pmatrix}$$

is strictly convex and U is convex, so by Theorem 4.5(2,4), if it has a minimum, then it is unique. Consider the convex set

$$U^\kappa = \left\{ \begin{pmatrix}\lambda \\ \mu\end{pmatrix} \in \mathbb{R}_+^{p+q} \,\middle|\, \begin{pmatrix}\mathbf{1}_p^\top & -\mathbf{1}_q^\top \\ \mathbf{1}_p^\top & \mathbf{1}_q^\top\end{pmatrix}\begin{pmatrix}\lambda \\ \mu\end{pmatrix} = \begin{pmatrix}0 \\ (p+q)K_s\kappa\nu\end{pmatrix} \right\}.$$

Observe that

$$\kappa U = \left\{ \begin{pmatrix}\kappa\lambda \\ \kappa\mu\end{pmatrix} \in \mathbb{R}_+^{p+q} \,\middle|\, \begin{pmatrix}\mathbf{1}_p^\top & -\mathbf{1}_q^\top \\ \mathbf{1}_p^\top & \mathbf{1}_q^\top\end{pmatrix}\begin{pmatrix}\kappa\lambda \\ \kappa\mu\end{pmatrix} = \begin{pmatrix}0 \\ (p+q)K_s\kappa\nu\end{pmatrix} \right\} = U^\kappa.$$

By Theorem 4.5(3), $(\lambda,\mu) \in U$ is a minimum of F over U iff

$$dF_{\lambda,\mu}\begin{pmatrix}\lambda' - \lambda \\ \mu' - \mu\end{pmatrix} \geq 0 \quad \text{for all} \quad \begin{pmatrix}\lambda' \\ \mu'\end{pmatrix} \in U.$$

Since

$$dF_{\lambda,\mu}\begin{pmatrix}\lambda' - \lambda \\ \mu' - \mu\end{pmatrix} = \begin{pmatrix}\lambda^\top & \mu^\top\end{pmatrix} P \begin{pmatrix}\lambda' - \lambda \\ \mu' - \mu\end{pmatrix}$$

the above conditions are equivalent to

$$\begin{aligned}
&\begin{pmatrix}\lambda^\top & \mu^\top\end{pmatrix} P \begin{pmatrix}\lambda' - \lambda \\ \mu' - \mu\end{pmatrix} \geq 0 \\
&\begin{pmatrix}\mathbf{1}_p^\top & -\mathbf{1}_q^\top \\ \mathbf{1}_p^\top & \mathbf{1}_q^\top\end{pmatrix}\begin{pmatrix}\lambda \\ \mu\end{pmatrix} = \begin{pmatrix}0 \\ (p+q)K_s\nu\end{pmatrix} \\
&\qquad\qquad \lambda, \lambda' \in \mathbb{R}_+^p,\ \mu, \mu' \in \mathbb{R}_+^q.
\end{aligned}$$

Since $\kappa > 0$, by multiplying the above inequality by κ^2 and the equations by κ, the following conditions hold:

$$\begin{aligned}
&\begin{pmatrix}\kappa\lambda^\top & \kappa\mu^\top\end{pmatrix} P \begin{pmatrix}\kappa\lambda' - \kappa\lambda \\ \kappa\mu' - \kappa\mu\end{pmatrix} \geq 0 \\
&\begin{pmatrix}\mathbf{1}_p^\top & -\mathbf{1}_q^\top \\ \mathbf{1}_p^\top & \mathbf{1}_q^\top\end{pmatrix}\begin{pmatrix}\kappa\lambda \\ \kappa\mu\end{pmatrix} = \begin{pmatrix}0 \\ (p+q)K_s\kappa\nu\end{pmatrix} \\
&\qquad\qquad \kappa\lambda, \kappa\lambda' \in \mathbb{R}_+^p,\ \kappa\mu, \kappa\mu' \in \mathbb{R}_+^q.
\end{aligned}$$

By Theorem 4.5(3), $(\kappa\lambda, \kappa\mu) \in U^\kappa$ is a minimum of F over U^κ, and because F is strictly convex and U^κ is convex, if F has a minimum over U^κ, then $(\kappa\lambda, \kappa\mu) \in U^\kappa$ is the unique minimum. Therefore, $\lambda^\kappa = \kappa\lambda$, $\mu^\kappa = \kappa\mu$.

Since w is given by the equation

$$w = -X \begin{pmatrix} \lambda \\ \mu \end{pmatrix}$$

and since we just showed that $\lambda^\kappa = \kappa\lambda$, $\mu^\kappa = \kappa\mu$, we deduce that $w^\kappa = \kappa w$.

We showed earlier that η is given by the equation

$$(p+q)K_s\nu\eta = \begin{pmatrix}\lambda^\top & \mu^\top\end{pmatrix} \left(X^\top X + \frac{1}{2K_s} I_{p+q}\right) \begin{pmatrix} \lambda \\ \mu \end{pmatrix}.$$

If we replace ν by $\kappa\nu$, since λ is replaced by $\kappa\lambda$ and μ by $\kappa\nu$, we see that $\eta^\kappa = \kappa\eta$. Finally, b is given by the equation

$$b = \frac{w^\top(u_{i_0} + v_{j_0}) + \epsilon_{i_0} - \xi_{j_0}}{2}$$

for and i_0 such that $\lambda_{i_0} > 0$ and any j_0 such that $\mu_{j_0} > 0$. If λ is replaced by $\kappa\lambda$ and μ by $\kappa\mu$, since $\epsilon = \lambda/(2K_s)$ and $\xi = \mu/(2K_s)$, we see that ϵ is replaced by $\kappa\epsilon$ and ξ by $\kappa\xi$, so $b^\kappa = \kappa b$.

Since $w^\kappa = \kappa w$ and $\eta^\kappa = \kappa\eta$ we obtain $\delta = \eta/\|w\| = \eta^\kappa/\|w^\kappa\| = \delta^\kappa$. Since $w^\kappa = \kappa w$, $\eta^\kappa = \kappa\eta$ and $b^\kappa = \kappa b$, the normalized equations of the hyperplanes $H_{w,b}, H_{w,b+\eta}$ and $H_{w,b-\eta}$ (obtained by dividing by $\|w\|$) are all identical, so the hyperplanes $H_{w,b}, H_{w,b+\eta}$ and $H_{w,b-\eta}$ are independent of ν. □

The width of the slab is controlled by K. The larger K is the smaller is the width of the slab. Theoretically, since this method does not rely on support vectors to compute b, it cannot fail if a solution exists, but in practice the quadratic solver does not converge for values of K that are too large. However, the method handles very small values of K, which can yield slabs of excessive width.

The "kernelized" version of Problem (SVM$_{s4}$) is the following:

Soft margin kernel SVM (SVM$_{s4}$):

$$\begin{aligned} &\text{minimize} \quad \frac{1}{2}\langle w, w\rangle - \nu\eta + K_s(\epsilon^\top\epsilon + \xi^\top\xi) \\ &\text{subject to} \\ &\qquad \langle w, \varphi(u_i)\rangle - b \geq \eta - \epsilon_i, \qquad i = 1, \ldots, p \\ &\qquad -\langle w, \varphi(v_j)\rangle + b \geq \eta - \xi_j, \qquad j = 1, \ldots, q \\ &\qquad \eta \geq 0, \end{aligned}$$

with $K_s = 1/(p+q)$.

By going over the derivation of the dual program, we obtain

Dual of the Soft margin kernel SVM (SVM$_{s4}$):

$$
\begin{aligned}
&\text{minimize} \quad \frac{1}{2}\begin{pmatrix}\lambda^\top & \mu^\top\end{pmatrix}\left(\mathbf{K} + \frac{1}{2K_s} I_{p+q}\right)\begin{pmatrix}\lambda \\ \mu\end{pmatrix} \\
&\text{subject to} \\
&\qquad \sum_{i=1}^p \lambda_i - \sum_{j=1}^q \mu_j = 0 \\
&\qquad \sum_{i=1}^p \lambda_i + \sum_{j=1}^q \mu_j \geq \nu \\
&\qquad \lambda_i \geq 0, \quad i = 1, \ldots, p \\
&\qquad \mu_j \geq 0, \quad j = 1, \ldots, q,
\end{aligned}
$$

where $\mathbf{K}$ is the kernel matrix of Section 18.1. Then w, b, and $f(x)$ are obtained exactly as in Section 18.5.

18.14 Solving SVM (SVM$_{s4}$) Using ADMM

In order to solve (SVM$_{s4}$) using ADMM we need to write the matrix corresponding to the constraints in equational form,

$$
\begin{aligned}
\sum_{i=1}^p \lambda_i - \sum_{j=1}^q \mu_j &= 0 \\
\sum_{i=1}^p \lambda_i + \sum_{j=1}^q \mu_j - \gamma &= K_m,
\end{aligned}
$$

with $K_m = (p+q)K_s\nu$. This is the $2 \times (p+q+1)$ matrix A given by

$$
A = \begin{pmatrix} \mathbf{1}_p^\top & -\mathbf{1}_q^\top & 0 \\ \mathbf{1}_p^\top & \mathbf{1}_q^\top & -1 \end{pmatrix}.
$$

We leave it as an exercise to prove that A has rank 2. The right-hand side is

$$
c = \begin{pmatrix} 0 \\ K_m \end{pmatrix}.
$$

The symmetric positive semidefinite $(p+q) \times (p+q)$ matrix P defining the quadratic functional is

$$
P = X^\top X + \frac{1}{2K_s} I_{p+q}, \quad \text{with} \quad X = \begin{pmatrix} -u_1 & \cdots & -u_p & v_1 & \cdots & v_q \end{pmatrix},
$$

and

$$q = 0_{p+q}.$$

Since there are $p+q+1$ Lagrange multipliers (λ, μ, γ), the $(p+q) \times (p+q)$ matrix P must be augmented with zero's to make it a $(p+q+1)\times(p+q+1)$ matrix P_a given by

$$P_a = \begin{pmatrix} X^\top X & 0_{p+q} \\ 0_{p+q}^\top & 0 \end{pmatrix},$$

and similarly q is augmented with zeros as the vector $q_a = 0_{p+q+1}$.

As in Section 18.8, since $\eta \geq 0$ for an optimal solution, we can drop the constraint $\eta \geq 0$ from the primal problem. In this case, there are $p+q$ Lagrange multipliers (λ, μ). It is easy to see that the objective function of the dual is unchanged and the set of constraints is

$$\sum_{i=1}^{p} \lambda_i - \sum_{j=1}^{q} \mu_j = 0$$
$$\sum_{i=1}^{p} \lambda_i + \sum_{j=1}^{q} \mu_j = K_m,$$

with $K_m = (p+q)K_s\nu$. The matrix corresponding to the above equations is the $2 \times (p+q)$ matrix A_2 given by

$$A_2 = \begin{pmatrix} \mathbf{1}_p^\top & -\mathbf{1}_q^\top \\ \mathbf{1}_p^\top & \mathbf{1}_q^\top \end{pmatrix}.$$

We leave it as an exercise to prove that A_2 has rank 2. The right-hand side is

$$c_2 = \begin{pmatrix} 0 \\ K_m \end{pmatrix}.$$

The symmetric positive semidefinite $(p+q) \times (p+q)$ matrix P defining the quadratic functional is

$$P = X^\top X + \frac{1}{2K_s} I_{p+q}, \quad \text{with} \quad X = \begin{pmatrix} -u_1 & \cdots & -u_p & v_1 & \cdots & v_q \end{pmatrix},$$

and

$$q = 0_{p+q}.$$

Since there are $p+q$ Lagrange multipliers (λ, μ), the $(p+q)\times(p+q)$ matrix P need not be augmented with zero's, so $P_{2a} = P$ and similarly $q_{2a} = 0_{p+q}$.

We ran our `Matlab` implementation of the above version of (SVM$_{s4}$) on the data set of Section 18.12. Since the value of ν is irrelevant, we picked $\nu = 1$. First we ran our program with $K = 190$; see Figure 18.22. We have $p_m = 23$ and $q_m = 18$. The program does not converge for $K \geq 200$.

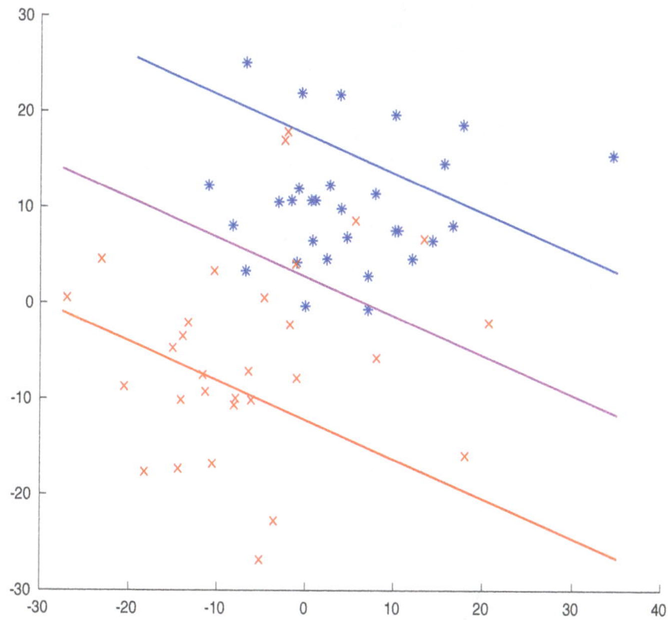

Fig. 18.22 Running (SVM$_{s4}$) on two sets of 30 points; $K = 190$.

Our second run was made with $K = 1/12000$; see Figure 18.23. We have $p_m = 30$ and $q_m = 30$ and we see that the width of the slab is a bit excessive. This example demonstrates that the margin lines need not contain data points.

18.15 Soft Margin SVM; (SVM$_{s5}$)

In this section we consider the version of Problem (SVM$_{s4}$) in which we add the term $(1/2)b^2$ to the objective function. We also drop the constraint $\eta \geq 0$ which is redundant.

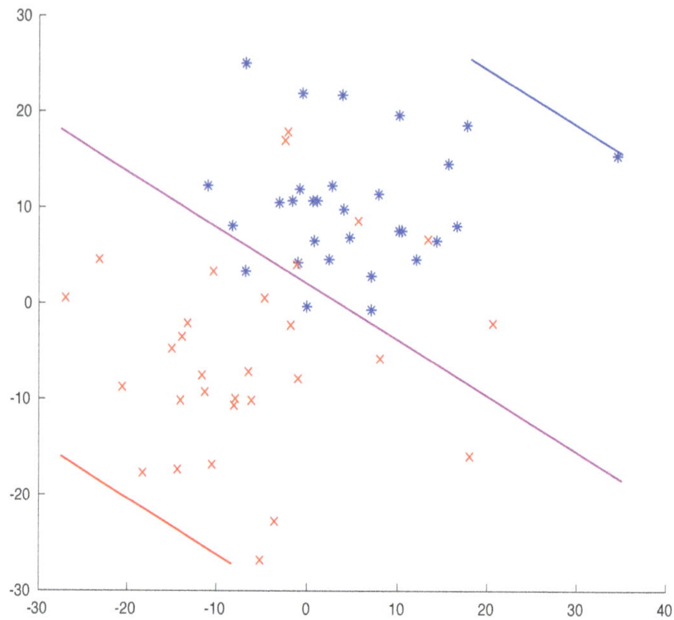

Fig. 18.23 Running (SVM_{s4}) on two sets of 30 points; $K = 1/12000$.

Soft margin SVM (SVM_{s5}):

$$\text{minimize} \quad \frac{1}{2}w^\top w + \frac{1}{2}b^2 + (p+q)K_s\left(-\nu\eta + \frac{1}{p+q}(\epsilon^\top\epsilon + \xi^\top\xi)\right)$$

subject to

$$\begin{aligned} w^\top u_i - b &\geq \eta - \epsilon_i, & i &= 1, \ldots, p \\ -w^\top v_j + b &\geq \eta - \xi_j, & j &= 1, \ldots, q, \end{aligned}$$

where ν and K_s are two given positive constants. As we saw earlier, it is convenient to pick $K_s = 1/(p+q)$. When writing a computer program, it is preferable to assume that K_s is arbitrary. In this case ν must be replaced by $(p+q)K_s\nu$ in all the formulae.

One of the advantages of this methods is that ϵ is determined by λ, ξ is determined by μ (as in (SVM_{s4})), and both η and b determined by λ and μ. As the previous method, this method *does not* require support vectors to compute b. We can omit the constraint $\eta \geq 0$, because for an optimal solution it can be shown using duality that $\eta \geq 0$.

A drawback of Program (SVM$_{s5}$) is that for fixed K_s, the quantity $\delta = \eta / \|w\|$ and the hyperplanes $H_{w,b}, H_{w,b+\eta}$ and $H_{w,b-\eta}$ are independent of ν. This will be shown in Theorem 18.4. Thus this method is less flexible than (SVM$_{s2'}$) and (SVM$_{s3}$).

The Lagrangian is given by

$$\begin{aligned} L(w, \epsilon, \xi, b, \eta, \lambda, \mu) &= \frac{1}{2} w^\top w + \frac{1}{2} b^2 - \nu\eta + K_s(\epsilon^\top \epsilon + \xi^\top \xi) + w^\top X \begin{pmatrix} \lambda \\ \mu \end{pmatrix} \\ &\quad - \epsilon^\top \lambda - \xi^\top \mu + b(\mathbf{1}_p^\top \lambda - \mathbf{1}_q^\top \mu) + \eta(\mathbf{1}_p^\top \lambda + \mathbf{1}_q^\top \mu) \\ &= \frac{1}{2} w^\top w + w^\top X \begin{pmatrix} \lambda \\ \mu \end{pmatrix} + \eta(\mathbf{1}_p^\top \lambda + \mathbf{1}_q^\top \mu - \nu) + \frac{1}{2} b^2 \\ &\quad + K_s(\epsilon^\top \epsilon + \xi^\top \xi) - \epsilon^\top \lambda - \xi^\top \mu + b(\mathbf{1}_p^\top \lambda - \mathbf{1}_q^\top \mu). \end{aligned}$$

To find the dual function $G(\lambda, \mu)$ we minimize $L(w, \epsilon, \xi, b, \eta, \lambda, \mu)$ with respect to w, ϵ, ξ, b, and η. Since the Lagrangian is convex and $(w, \epsilon, \xi, b, \eta) \in \mathbb{R}^n \times \mathbb{R}^p \times \mathbb{R}^q \times \mathbb{R} \times \mathbb{R}$, a convex open set, by Theorem 4.5, the Lagrangian has a minimum in $(w, \epsilon, \xi, b, \eta)$ iff $\nabla L_{w,\epsilon,\xi,b,\eta} = 0$, so we compute $\nabla L_{w,\epsilon,\xi,b,\eta}$. The gradient $\nabla L_{w,\epsilon,\xi,b,\eta}$ is given by

$$\nabla L_{w,\epsilon,\xi,b,\eta} = \begin{pmatrix} w + X \begin{pmatrix} \lambda \\ \mu \end{pmatrix} \\ 2K_s \epsilon - \lambda \\ 2K_s \xi - \mu \\ b + \mathbf{1}_p^\top \lambda - \mathbf{1}_q^\top \mu \\ \mathbf{1}_p^\top \lambda + \mathbf{1}_q^\top \mu - \nu \end{pmatrix}.$$

By setting $\nabla L_{w,\epsilon,\xi,b,\eta} = 0$ we get the equations

$$w = -X \begin{pmatrix} \lambda \\ \mu \end{pmatrix}, \qquad (*_w)$$

$$\begin{aligned} 2K_s \epsilon &= \lambda \\ 2K_s \xi &= \mu \\ b &= -(\mathbf{1}_p^\top \lambda - \mathbf{1}_q^\top \mu) \\ \mathbf{1}_p^\top \lambda + \mathbf{1}_q^\top \mu &= \nu. \end{aligned}$$

As we said earlier, both w an b are determined by λ and μ. We can use the equations to obtain the following expression for the dual function $G(\lambda, \mu, \gamma)$,

$$\begin{aligned} G(\lambda, \mu, \gamma) &= -\frac{1}{4K_s}(\lambda^\top \lambda + \mu^\top \mu) - \frac{1}{2} \begin{pmatrix} \lambda^\top & \mu^\top \end{pmatrix} X^\top X \begin{pmatrix} \lambda \\ \mu \end{pmatrix} - \frac{b^2}{2} \\ &= -\frac{1}{2} \begin{pmatrix} \lambda^\top & \mu^\top \end{pmatrix} \left(X^\top X + \begin{pmatrix} \mathbf{1}_p \mathbf{1}_p^\top & -\mathbf{1}_p \mathbf{1}_q^\top \\ -\mathbf{1}_q \mathbf{1}_p^\top & \mathbf{1}_q \mathbf{1}_q^\top \end{pmatrix} + \frac{1}{2K_s} I_{p+q} \right) \begin{pmatrix} \lambda \\ \mu \end{pmatrix}. \end{aligned}$$

Consequently the dual program is equivalent to the minimization program

Dual of the Soft margin SVM (SVM_{s5}):

$$\text{minimize} \quad \frac{1}{2}\begin{pmatrix}\lambda^\top & \mu^\top\end{pmatrix}\left(X^\top X + \begin{pmatrix}\mathbf{1}_p\mathbf{1}_p^\top & -\mathbf{1}_p\mathbf{1}_q^\top \\ -\mathbf{1}_q\mathbf{1}_p^\top & \mathbf{1}_q\mathbf{1}_q^\top\end{pmatrix} + \frac{1}{2K_s}I_{p+q}\right)\begin{pmatrix}\lambda \\ \mu\end{pmatrix}$$

subject to

$$\sum_{i=1}^p \lambda_i + \sum_{j=1}^q \mu_j = \nu$$
$$\lambda_i \geq 0, \quad i = 1, \ldots, p$$
$$\mu_j \geq 0, \quad j = 1, \ldots, q.$$

It is shown in Section 18.16 how the dual program is solved using ADMM from Section 16.6. If the primal problem is solvable, this yields solutions for λ and μ.

The constraint

$$\sum_{i=1}^p \lambda_i + \sum_{j=1}^q \mu_j = \nu$$

implies that either there is some i_0 such that $\lambda_{i_0} > 0$ or there is some j_0 such that $\mu_{j_0} > 0$, so we have $\epsilon_{i_0} > 0$ or $\xi_{j_0} > 0$, which means that at least one point is misclassified. Thus Problem (SVM_{s5}) should only be used when the sets $\{u_i\}$ and $\{v_j\}$ are *not* linearly separable.

We can use the fact that the duality gap is 0 to find η. We have

$$\begin{aligned}&\frac{1}{2}w^\top w + \frac{b^2}{2} - \nu\eta + K_s(\epsilon^\top\epsilon + \xi^\top\xi) \\ &\quad = -\frac{1}{2}\begin{pmatrix}\lambda^\top & \mu^\top\end{pmatrix}\left(X^\top X + \begin{pmatrix}\mathbf{1}_p\mathbf{1}_p^\top & -\mathbf{1}_p\mathbf{1}_q^\top \\ -\mathbf{1}_q\mathbf{1}_p^\top & \mathbf{1}_q\mathbf{1}_q^\top\end{pmatrix} + \frac{1}{2K_s}I_{p+q}\right)\begin{pmatrix}\lambda \\ \mu\end{pmatrix},\end{aligned}$$

so we get

$$\begin{aligned}\nu\eta &= K_s(\epsilon^\top\epsilon + \xi^\top\xi) \\ &\quad + \begin{pmatrix}\lambda^\top & \mu^\top\end{pmatrix}\left(X^\top X + \begin{pmatrix}\mathbf{1}_p\mathbf{1}_p^\top & -\mathbf{1}_p\mathbf{1}_q^\top \\ -\mathbf{1}_q\mathbf{1}_p^\top & \mathbf{1}_q\mathbf{1}_q^\top\end{pmatrix} + \frac{1}{4K_s}I_{p+q}\right)\begin{pmatrix}\lambda \\ \mu\end{pmatrix} \\ &= \begin{pmatrix}\lambda^\top & \mu^\top\end{pmatrix}\left(X^\top X + \begin{pmatrix}\mathbf{1}_p\mathbf{1}_p^\top & -\mathbf{1}_p\mathbf{1}_q^\top \\ -\mathbf{1}_q\mathbf{1}_p^\top & \mathbf{1}_q\mathbf{1}_q^\top\end{pmatrix} + \frac{1}{2K_s}I_{p+q}\right)\begin{pmatrix}\lambda \\ \mu\end{pmatrix}.\end{aligned}$$

The above confirms that at optimality we have $\eta \geq 0$.

Remark: If we do not assume that $K_s = 1/(p+q)$, then the above formula must be replaced by

$$(p+q)K_s\nu\eta = (\lambda^\top\ \mu^\top)\left(X^\top X + \begin{pmatrix} \mathbf{1}_p\mathbf{1}_p^\top & -\mathbf{1}_p\mathbf{1}_q^\top \\ -\mathbf{1}_q\mathbf{1}_p^\top & \mathbf{1}_q\mathbf{1}_q^\top \end{pmatrix} + \frac{1}{2K_s} I_{p+q}\right)\begin{pmatrix}\lambda \\ \mu\end{pmatrix}.$$

There is a version of Theorem 18.3 stating that for a fixed K_s, the solution to Problem (SVM$_{s5}$) is unique and independent of the value of ν.

Theorem 18.4. *For K_s and ν fixed, if Problem* (SVM$_{s5}$) *succeeds then it has a unique solution. If Problem* (SVM$_{s5}$) *succeeds and returns $(\lambda, \mu, \eta, w, b)$ for the value ν and $(\lambda^\kappa, \mu^\kappa, \eta^\kappa, w^\kappa, b^\kappa)$ for the value $\kappa\nu$ with $\kappa > 0$, then*

$$\lambda^\kappa = \kappa\lambda,\ \mu^\kappa = \kappa\mu,\ \eta^\kappa = \kappa\eta,\ w^\kappa = \kappa w,\ b^\kappa = \kappa b.$$

As a consequence, $\delta = \eta/\left\|w\right\| = \eta^\kappa/\left\|w^\kappa\right\| = \delta^\kappa$, and the hyperplanes $H_{w,b}, H_{w,b+\eta}$ and $H_{w,b-\eta}$ are independent of ν.

Proof. The proof is an easy adaptation of the proof of Theorem 18.3 so we only give a sketch. The two crucial points are that the matrix

$$P = X^\top X + \begin{pmatrix} \mathbf{1}_p\mathbf{1}_p^\top & -\mathbf{1}_p\mathbf{1}_q^\top \\ -\mathbf{1}_q\mathbf{1}_p^\top & \mathbf{1}_q\mathbf{1}_q^\top \end{pmatrix} + \frac{1}{2K_s} I_{p+q}$$

is symmetric positive definite and that we have the single equational constraint

$$\mathbf{1}_p^\top\lambda + \mathbf{1}_q^\top\mu = (p+q)K_s\nu$$

defining the convex set

$$U = \left\{\begin{pmatrix}\lambda \\ \mu\end{pmatrix} \in \mathbb{R}_+^{p+q} \mid \mathbf{1}_p^\top\lambda + \mathbf{1}_q^\top\mu = (p+q)K_s\nu\right\}.$$

The proof is essentially the proof of 18.3 using the above SPD matrix and convex set. ☐

The "kernelized" version of Problem (SVM$_{s5}$) is the following:

Soft margin kernel SVM (SVM$_{s5}$):

$$\begin{aligned} &\text{minimize} \quad \frac{1}{2}\langle w, w\rangle + \frac{1}{2}b^2 - \nu\eta + K_s(\epsilon^\top\epsilon + \xi^\top\xi) \\ &\text{subject to} \\ &\qquad \langle w, \varphi(u_i)\rangle - b \geq \eta - \epsilon_i, \qquad i = 1, \ldots, p \\ &\qquad -\langle w, \varphi(v_j)\rangle + b \geq \eta - \xi_j, \qquad j = 1, \ldots, q, \end{aligned}$$

with $K_s = 1/(p+q)$.

Tracing through the derivation of the dual program, we obtain

Dual of the Soft margin kernel SVM (SVM_{s5}):

$$\begin{aligned}&\text{minimize} \quad \frac{1}{2}\begin{pmatrix}\lambda^\top & \mu^\top\end{pmatrix}\left(\mathbf{K} + \begin{pmatrix}\mathbf{1}_p\mathbf{1}_p^\top & -\mathbf{1}_p\mathbf{1}_q^\top \\ -\mathbf{1}_q\mathbf{1}_p^\top & \mathbf{1}_q\mathbf{1}_q^\top\end{pmatrix} + \frac{1}{2K_s}I_{p+q}\right)\begin{pmatrix}\lambda \\ \mu\end{pmatrix}\\ &\text{subject to}\\ &\qquad\sum_{i=1}^p \lambda_i + \sum_{j=1}^q \mu_j = \nu\\ &\qquad\lambda_i \geq 0, \quad i = 1, \ldots, p\\ &\qquad\mu_j \geq 0, \quad j = 1, \ldots, q,\end{aligned}$$

where $\mathbf{K}$ is the kernel matrix of Section 18.1. Then w, b, and $f(x)$ are obtained exactly as in Section 18.13.

18.16 Solving SVM (SVM_{s5}) Using ADMM

In order to solve (SVM_5) using ADMM we need to write the matrix corresponding to the constraints in equational form,

$$\sum_{i=1}^p \lambda_i + \sum_{j=1}^q \mu_j = K_m,$$

with $K_m = (p+q)K_s\nu$. This is the $1 \times (p+q)$ matrix A given by

$$A = \begin{pmatrix}\mathbf{1}_p^\top & \mathbf{1}_q^\top\end{pmatrix}.$$

Obviously, A has rank 1. The right-hand side is

$$c = K_m.$$

The symmetric positive definite $(p+q) \times (p+q)$ matrix P defining the quadratic functional is

$$P = X^\top X + \begin{pmatrix}\mathbf{1}_p\mathbf{1}_p^\top & -\mathbf{1}_p\mathbf{1}_q^\top \\ -\mathbf{1}_q\mathbf{1}_p^\top & \mathbf{1}_q\mathbf{1}_q^\top\end{pmatrix} + \frac{1}{2K_s}I_{p+q},$$

with

$$X = \begin{pmatrix}-u_1 & \cdots & -u_p & v_1 & \cdots & v_q\end{pmatrix} \quad \text{and} \quad q = 0_{p+q}.$$

Since there are $p+q$ Lagrange multipliers (λ, μ), the $(p+q) \times (p+q)$ matrix P does not have to be augmented with zero's.

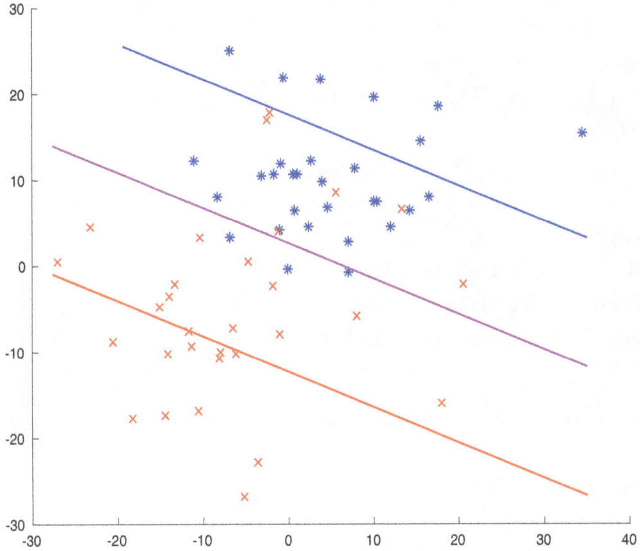

Fig. 18.24 Running (SVM_{s5}) on two sets of 30 points; $K = 190$.

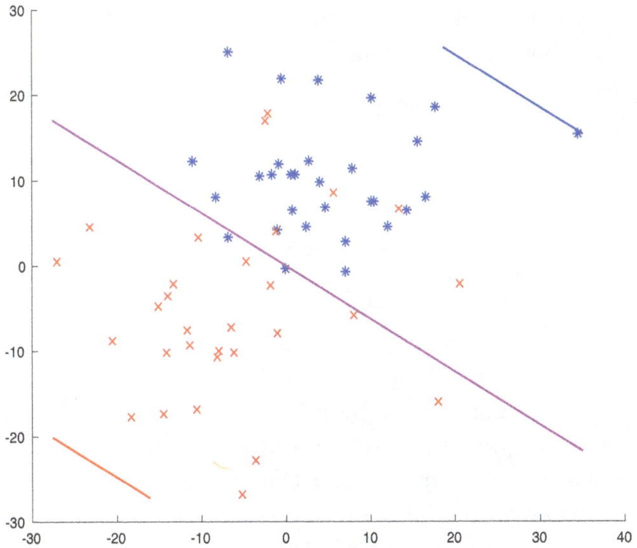

Fig. 18.25 Running (SVM_{s5}) on two sets of 30 points; $K = 1/13000$.

We ran our `Matlab` implementation of the above version of (SVM_{s5}) on the data set of Section 18.14. Since the value of ν is irrelevant, we picked $\nu = 1$. First we ran our program with $K = 190$; see Figure 18.24. We have $p_m = 23$ and $q_m = 18$. The program does not converge for $K \geq 200$.

Our second run was made with $K = 1/13000$; see Figure 18.25. We have $p_m = 30$ and $q_m = 30$ and we see that the width of the slab is a bit excessive. This example demonstrates that the margin lines need not contain data points.

Method (SVM_{s5}) always returns a value for b and η smaller than the value returned by (SVM_{s4}) (because of the term $(1/2)b^2$ added to the objective function) but in this example the difference is too small to be noticed.

18.17 Summary and Comparison of the SVM Methods

In this chapter we considered six variants for solving the soft margin binary classification problem for two sets of points $\{u_i\}_{i=1}^p$ and $\{v_j\}_{j=1}^q$ using support vector classification methods. The objective is to find a separating hyperplane $H_{w,b}$ of equation $w^\top x - b = 0$. We also try to find two "margin hyperplanes" $H_{w,b+\delta}$ of equation $w^\top x - b - \delta = 0$ (the blue margin hyperplane) and $H_{w,b-\delta}$ of equation $w^\top x - b + \delta = 0$ (the red margin hyperplane) such that δ is as big as possible and yet the number of misclassified points is minimized, which is achieved by allowing an error $\epsilon_i \geq 0$ for every point u_i, in the sense that the constraint

$$w^\top u_i - b \geq \delta - \epsilon_i$$

should hold, and an error $\xi_j \geq 0$ for every point v_j, in the sense that the constraint

$$-w^\top v_j + b \geq \delta - \xi_j$$

should hold.

The goal is to design an objective function that minimizes ϵ and ξ and maximizes δ. The optimization problem should also solve for w and b, and for this some constraint has to be placed on w. Another goal is to try to use the dual program to solve the optimization problem, because the solutions involve inner products, and thus the problem is amenable to a generalization using kernel functions.

The first attempt, which is to use the objective function

$$J(w, \epsilon, \xi, b, \delta) = -\delta + K \left(\epsilon^\top \; \xi^\top\right) \mathbf{1}_{p+q}$$

and the constraint $w^\top w \leq 1$, does not work very well because this constraint needs to be guarded by a Lagrange multiplier $\gamma \geq 0$, and as a result, minimizing the Lagrangian L to find the dual function G gives an equation for solving w of the form

$$2\gamma w = -X^\top \begin{pmatrix} \lambda \\ \mu \end{pmatrix},$$

but if the sets $\{u_i\}_{i=1}^p$ and $\{v_j\}_{j=1}^q$ are not linearly separable, then an optimal solution may occurs for $\gamma = 0$, in which case it is impossible to determine w. This is Problem (SVM_{s1}) considered in Section 18.1.

Soft margin SVM (SVM_{s1}):

$$\begin{aligned} &\text{minimize} \quad -\delta + K\Big(\sum_{i=1}^p \epsilon_i + \sum_{j=1}^q \xi_j\Big) \\ &\text{subject to} \\ &\qquad w^\top u_i - b \geq \delta - \epsilon_i, \quad \epsilon_i \geq 0 \qquad i = 1, \ldots, p \\ &\qquad - w^\top v_j + b \geq \delta - \xi_j, \quad \xi_j \geq 0 \qquad j = 1, \ldots, q \\ &\qquad w^\top w \leq 1. \end{aligned}$$

It is customary to write $\ell = p + q$.

It is shown in Section 18.1 that the dual program is equivalent to the following minimization program:

Dual of the Soft margin SVM (SVM_{s1}):

$$\begin{aligned} &\text{minimize} \quad \begin{pmatrix} \lambda^\top & \mu^\top \end{pmatrix} X^\top X \begin{pmatrix} \lambda \\ \mu \end{pmatrix} \\ &\text{subject to} \\ &\qquad \sum_{i=1}^p \lambda_i - \sum_{j=1}^q \mu_j = 0 \\ &\qquad \sum_{i=1}^p \lambda_i + \sum_{j=1}^q \mu_j = 1 \\ &\qquad 0 \leq \lambda_i \leq K, \quad i = 1, \ldots, p \\ &\qquad 0 \leq \mu_j \leq K, \quad j = 1, \ldots, q. \end{aligned}$$

The points u_i and v_j are naturally classified in terms of the values of λ_i and μ_j. The numbers of points in each category have a direct influence on the choice of the parameter K. Let us summarize some of the keys items from Definition 18.1.

The vectors u_i on the blue margin $H_{w,b+\delta}$ and the vectors v_j on the red margin $H_{w,b-\delta}$ are called *support vectors*. Support vectors correspond to vectors u_i for which $w^\top u_i - b - \delta = 0$ (which implies $\epsilon_i = 0$), and vectors v_j for which $w^\top v_j - b + \delta = 0$ (which implies $\xi_j = 0$). Support vectors u_i such that $0 < \lambda_i < K$ and support vectors v_j such that $0 < \mu_j < K$ are *support vectors of type 1*. Support vectors of type 1 play a special role so we denote the sets of indices associated with them by

$$\begin{aligned} I_\lambda &= \{i \in \{1, \ldots, p\} \mid 0 < \lambda_i < K\} \\ I_\mu &= \{j \in \{1, \ldots, q\} \mid 0 < \mu_j < K\}. \end{aligned}$$

We denote their cardinalities by $numsvl_1 = |I_\lambda|$ and $numsvm_1 = |I_\mu|$.

The vectors u_i for which $\lambda_i = K$ and the vectors v_j for which $\mu_j = K$ are said to *fail the margin*. The sets of indices associated with the vectors failing the margin are denoted by

$$\begin{aligned} K_\lambda &= \{i \in \{1, \ldots, p\} \mid \lambda_i = K\} \\ K_\mu &= \{j \in \{1, \ldots, q\} \mid \mu_j = K\}. \end{aligned}$$

We denote their cardinalities by $p_f = |K_\lambda|$ and $q_f = |K_\mu|$.

Vectors u_i such that $\lambda_i > 0$ and vectors v_j such that $\mu_j > 0$ are said to *have margin at most* δ. The sets of indices associated with these vectors are denoted by

$$\begin{aligned} I_{\lambda>0} &= \{i \in \{1, \ldots, p\} \mid \lambda_i > 0\} \\ I_{\mu>0} &= \{j \in \{1, \ldots, q\} \mid \mu_j > 0\}. \end{aligned}$$

We denote their cardinalities by $p_m = |I_{\lambda>0}|$ and $q_m = |I_{\mu>0}|$.

Obviously, $p_f \leq p_m$ and $q_f \leq q_m$. There are $p - p_m$ points u_i classified correctly on the blue side and outside the δ-slab and there are $q - q_m$ points v_j classified correctly on the red side and outside the δ-slab. Intuitively a blue point that fails the margin is on the wrong side of the blue margin and a red point that fails the margin is on the wrong side of the red margin.

It can be shown that that K must be chosen so that

$$\max\left\{\frac{1}{2p_m}, \frac{1}{2q_m}\right\} \leq K \leq \min\left\{\frac{1}{2p_f}, \frac{1}{2q_f}\right\}.$$

If the optimal value is 0, then $\gamma = 0$ and $X\begin{pmatrix}\lambda \\ \mu\end{pmatrix} = 0$, so in this case it is not possible to determine w. However, if the optimal value is > 0, then once a solution for λ and μ is obtained, we have

$$\gamma = \frac{1}{2}\left(\begin{pmatrix}\lambda^\top & \mu^\top\end{pmatrix} X^\top X \begin{pmatrix}\lambda \\ \mu\end{pmatrix}\right)^{1/2}$$

$$w = \frac{1}{2\gamma}\left(\sum_{i=1}^{p} \lambda_i u_i - \sum_{j=1}^{q} \mu_j v_j\right),$$

so we get

$$w = \frac{\displaystyle\sum_{i=1}^{p} \lambda_i u_i - \sum_{j=1}^{q} \mu_j v_j}{\left(\begin{pmatrix}\lambda^\top & \mu^\top\end{pmatrix} X^\top X \begin{pmatrix}\lambda \\ \mu\end{pmatrix}\right)^{1/2}},$$

If the following mild hypothesis holds, then b and δ can be found.

Standard Margin Hypothesis for (SVM_{s1}). There is some index i_0 such that $0 < \lambda_{i_0} < K$ and there is some index j_0 such that $0 < \mu_{j_0} < K$. This means that some u_{i_0} is a support vector of type 1 on the blue margin, and some v_{j_0} is a support vector of type 1 on the red margin.

If the **Standard Margin Hypothesis** for (SVM_{s1}) holds, then $\epsilon_{i_0} = 0$ and $\mu_{j_0} = 0$, and we have the active equations

$$w^\top u_{i_0} - b = \delta \quad \text{and} \quad -w^\top v_{j_0} + b = \delta,$$

and we obtain the value of b and δ as

$$b = \frac{1}{2} w^\top (u_{i_0} + v_{j_0})$$

$$\delta = \frac{1}{2} w^\top (u_{i_0} - v_{j_0}).$$

The second more successful approach is to add the term $(1/2)w^\top w$ to the objective function and to drop the constraint $w^\top w \leq 1$. There are several variants of this method, depending on the choice of the regularizing term involving ϵ and ξ (linear or quadratic), how the margin is dealt with (implicitly with the term 1 or explicitly with a term η), and whether the term $(1/2)b^2$ is added to the objective function or not.

These methods all share the property that if the primal problem has an optimal solution with $w \neq 0$, then the dual problem always determines w, and then under mild conditions which we call standard margin hypotheses, b and η can be determined. Then ϵ and ξ can be determined using the constraints that are active. When $(1/2)b^2$ is added to the objective function, b is determined by the equation

$$b = -(\mathbf{1}_p^\top \lambda - \mathbf{1}_q^\top \mu).$$

All these problems are convex and the constraints are qualified, so the duality gap is zero, and if the primal has an optimal solution with $w \neq 0$, then it follows that $\eta \geq 0$.

We now consider five variants in more details.

(1) **Basic Soft margin SVM**: (SVM_{s2}).
This is the optimization problem in which the regularization term $K\left(\epsilon^\top \; \xi^\top\right) \mathbf{1}_{p+q}$ is linear and the margin δ is given by $\delta = 1/\left\|w\right\|$:

$$\begin{aligned}
&\text{minimize} \quad \frac{1}{2} w^\top w + K\left(\epsilon^\top \; \xi^\top\right) \mathbf{1}_{p+q} \\
&\text{subject to} \\
&\qquad w^\top u_i - b \geq 1 - \epsilon_i, \quad \epsilon_i \geq 0 \qquad i = 1, \ldots, p \\
&\qquad - w^\top v_j + b \geq 1 - \xi_j, \quad \xi_j \geq 0 \qquad j = 1, \ldots, q.
\end{aligned}$$

This problem is the classical one discussed in all books on machine learning or pattern analysis, for instance Vapnik [Vapnik (1998)], Bishop [Bishop (2006)], and Shawe-Taylor and Christianini [Shawe-Taylor and Cristianini (2004)]. It is shown in Section 18.3 that the dual program is

Dual of the Basic Soft margin SVM: (SVM_{s2}):

$$\begin{aligned}
&\text{minimize} \quad \frac{1}{2}\left(\lambda^\top \; \mu^\top\right) X^\top X \begin{pmatrix} \lambda \\ \mu \end{pmatrix} - \left(\lambda^\top \; \mu^\top\right) \mathbf{1}_{p+q} \\
&\text{subject to} \\
&\qquad \sum_{i=1}^p \lambda_i - \sum_{j=1}^q \mu_j = 0 \\
&\qquad 0 \leq \lambda_i \leq K, \quad i = 1, \ldots, p \\
&\qquad 0 \leq \mu_j \leq K, \quad j = 1, \ldots, q.
\end{aligned}$$

We can use the dual program to solve the primal. Once $\lambda \geq 0, \mu \geq 0$ have been found, w is given by

$$w = -X \begin{pmatrix} \lambda \\ \mu \end{pmatrix} = \sum_{i=1}^p \lambda_i u_i - \sum_{j=1}^q \mu_j v_j,$$

but b is not determined by the dual.

The complementary slackness conditions imply that if $\epsilon_i > 0$, then $\lambda_i = K$, and if $\xi_j > 0$, then $\mu_j = K$. Consequently, if $\lambda_i < K$, then $\epsilon_i = 0$ and u_i is correctly classified, and similarly if $\mu_j < K$, then $\xi_j = 0$ and v_j is correctly classified.

A priori nothing prevents the situation where $\lambda_i = K$ for all nonzero λ_i or $\mu_j = K$ for all nonzero μ_j. If this happens, we can rerun the optimization method with a larger value of K. If the following mild hypothesis holds then b can be found.

Standard Margin Hypothesis for (SVM_{s2}). There is some support vector u_{i_0} of type 1 on the blue margin, and some support vector v_{j_0} of type 1 on the red margin.

If the **Standard Margin Hypothesis** for (SVM_{s2}) holds then $\epsilon_{i_0} = 0$ and $\mu_{j_0} = 0$, and then we have the active equations

$$w^\top u_{i_0} - b = 1 \quad \text{and} \quad -w^\top v_{j_0} + b = 1,$$

and we obtain

$$b = \frac{1}{2} w^\top (u_{i_0} + v_{j_0}).$$

(2) **Basic Soft margin ν-SVM Problem** ($\text{SVM}_{s2'}$).

This a generalization of Problem (SVM_{s2}) for a version of the soft margin SVM coming from Problem (SVM_{h2}), obtained by adding an extra degree of freedom, namely instead of the margin $\delta = 1/\|w\|$, we use the margin $\delta = \eta/\|w\|$ where η is some positive constant that we wish to maximize. To do so, we add a term $-K_m\eta$ to the objective function. We have the following optimization problem:

$$\begin{aligned}
&\text{minimize} \quad \frac{1}{2} w^\top w - K_m \eta + K_s \left(\epsilon^\top \ \xi^\top\right) \mathbf{1}_{p+q} \\
&\text{subject to} \\
&\qquad w^\top u_i - b \geq \eta - \epsilon_i, \quad \epsilon_i \geq 0 \qquad i = 1, \ldots, p \\
&\qquad -w^\top v_j + b \geq \eta - \xi_j, \quad \xi_j \geq 0 \qquad j = 1, \ldots, q \\
&\qquad \eta \geq 0,
\end{aligned}$$

where $K_m > 0$ and $K_s > 0$ are fixed constants that can be adjusted to determine the influence of η and the regularizing term.

This version of the SVM problem was first discussed in Schölkopf, Smola, Williamson, and Bartlett [Schölkopf *et al.* (2000)] under the name of *ν-SVC*, and also used in Schölkopf, Platt, Shawe-Taylor, and Smola [Schölkopf *et al.* (2001)].

In order for the problem to have a solution we must pick K_m and K_s so that

$$K_m \leq \min\{2pK_s, 2qK_s\}.$$

It is shown in Section 18.5 that the dual program is

Dual of the Basic Soft margin ν-SVM Problem ($\mathrm{SVM}_{s2'}$):

$$\begin{aligned}
&\text{minimize} \quad \frac{1}{2}\begin{pmatrix}\lambda^\top & \mu^\top\end{pmatrix} X^\top X \begin{pmatrix}\lambda \\ \mu\end{pmatrix} \\
&\text{subject to} \\
&\qquad \sum_{i=1}^p \lambda_i - \sum_{j=1}^q \mu_j = 0 \\
&\qquad \sum_{i=1}^p \lambda_i + \sum_{j=1}^q \mu_j \geq K_m \\
&\qquad 0 \leq \lambda_i \leq K_s, \quad i = 1, \ldots, p \\
&\qquad 0 \leq \mu_j \leq K_s, \quad j = 1, \ldots, q.
\end{aligned}$$

If the primal problem has an optimal solution with $w \neq 0$, then using the fact that the duality gap is zero we can show that $\eta \geq 0$. Thus constraint $\eta \geq 0$ could be omitted. As in the previous case w is given by

$$w = -X\begin{pmatrix}\lambda \\ \mu\end{pmatrix} = \sum_{i=1}^p \lambda_i u_i - \sum_{j=1}^q \mu_j v_j,$$

but b and η are not determined by the dual.
If we drop the constraint $\eta \geq 0$, then the inequality

$$\sum_{i=1}^p \lambda_i + \sum_{j=1}^q \mu_j \geq K_m$$

is replaced by the equation

$$\sum_{i=1}^p \lambda_i + \sum_{j=1}^q \mu_j = K_m.$$

It convenient to define $\nu > 0$ such that

$$\nu = \frac{K_m}{(p+q)K_s},$$

so that the objective function $J(w, \epsilon, \xi, b, \eta)$ is given by

$$J(w, \epsilon, \xi, b, \eta) = \frac{1}{2} w^\top w + (p+q) K_s \left(-\nu\eta + \frac{1}{p+q} \left(\epsilon^\top \; \xi^\top\right) \mathbf{1}_{p+q} \right).$$

Since we obtain an equivalent problem by rescaling by a common positive factor, theoretically it is convenient to normalize K_s as

$$K_s = \frac{1}{p+q},$$

in which case $K_m = \nu$. This method is called the *ν-support vector machine.*

Under the **Standard Margin Hypothesis** for $(\mathrm{SVM}_{s2'})$, there is some support vector u_{i_0} of type 1 and some support vector v_{j_0} of type 1, and by the complementary slackness conditions $\epsilon_{i_0} = 0$ and $\xi_{j_0} = 0$, so we have the two active constraints

$$w^\top u_{i_0} - b = \eta, \quad -w^\top v_{j_0} + b = \eta,$$

and we can solve for b and η and we get

$$\begin{aligned} b &= \frac{w^\top (u_{i_0} + v_{j_0})}{2} \\ \eta &= \frac{w^\top (u_{i_0} - v_{j_0})}{2}. \end{aligned}$$

Due to numerical instability, when writing a computer program it is preferable to compute the lists of indices I_λ and I_μ given by

$$I_\lambda = \{i \in \{1, \ldots, p\} \mid 0 < \lambda_i < K_s\}, I_\mu = \{j \in \{1, \ldots, q\} \mid 0 < \mu_j < K_s\}.$$

Then b and η are given by the following averaging formulae:

$$\begin{aligned} b &= w^\top \left(\left(\sum_{i \in I_\lambda} u_i \right) / |I_\lambda| + \left(\sum_{j \in I_\mu} v_j \right) / |I_\mu| \right) / 2 \\ \eta &= w^\top \left(\left(\sum_{i \in I_\lambda} u_i \right) / |I_\lambda| - \left(\sum_{j \in I_\mu} v_j \right) / |I_\mu| \right) / 2. \end{aligned}$$

Proposition 18.1 yields bounds on ν for the method to converge, namely

$$\max\left\{ \frac{2p_f}{p+q}, \frac{2q_f}{p+q} \right\} \le \nu \le \min\left\{ \frac{2p_m}{p+q}, \frac{2q_m}{p+q} \right\}.$$

In Section 18.7 we investigate conditions on ν that ensure that some point u_{i_0} and some point v_{j_0} is a support vector. Theorem 18.1 shows that for every optimal solution $(w, b, \eta, \epsilon, \xi)$ of Problem $(\mathrm{SVM}_{s2'})$ with $w \neq 0$ and $\eta > 0$, if

$$\max\{2p_f/(p+q), 2q_f/(p+q)\} < \nu < \min\{2p/(p+q), 2q/(p+q)\},$$

then some u_{i_0} and some v_{j_0} is a support vector. Under the same conditions on ν Proposition 18.3 shows that η and b can always be determined in terms of (λ, μ) using a single support vector.

(3) **Soft margin ν-SVM Problem** (SVM_{s3}). This is the variation of Problem $(\mathrm{SVM}_{s2'})$ obtained by adding the term $(1/2)b^2$ to the objective function. The result is that in minimizing the Lagrangian to find the dual function G, not just w but also b is determined. We also suppress the constraint $\eta \geq 0$ which turns out to be redundant. If $\nu > (p_f + q_f)/(p+q)$, then η is also determined. The fact that b and η are determined by the dual seems to be an advantage of Problem (SVM_{s3}). The optimization problem is

$$\begin{aligned}
&\text{minimize} \quad \frac{1}{2}w^\top w + \frac{1}{2}b^2 + (p+q)K_s\left(-\nu\eta + \frac{1}{p+q}\begin{pmatrix}\epsilon^\top & \xi^\top\end{pmatrix}\mathbf{1}_{p+q}\right)\\
&\text{subject to}\\
&\qquad w^\top u_i - b \geq \eta - \epsilon_i, \quad \epsilon_i \geq 0 \qquad i = 1, \ldots, p\\
&\qquad -w^\top v_j + b \geq \eta - \xi_j, \quad \xi_j \geq 0 \qquad j = 1, \ldots, q.
\end{aligned}$$

Theoretically it is convenient to assume that $K_s = 1/(p+q)$. Otherwise, ν needs to be replaced by $(p+q)K_s\nu$ in all the formulae below.
It is shown in Section 18.13 that the dual is given by

Dual of the Soft margin ν-SVM Problem (SVM_{s3}):

$$\begin{aligned}
&\text{minimize} \quad \frac{1}{2}\begin{pmatrix}\lambda^\top & \mu^\top\end{pmatrix}\left(X^\top X + \begin{pmatrix}\mathbf{1}_p\mathbf{1}_p^\top & -\mathbf{1}_p\mathbf{1}_q^\top\\ -\mathbf{1}_q\mathbf{1}_p^\top & \mathbf{1}_q\mathbf{1}_q^\top\end{pmatrix}\right)\begin{pmatrix}\lambda\\ \mu\end{pmatrix}\\
&\text{subject to}\\
&\qquad \sum_{i=1}^{p}\lambda_i + \sum_{j=1}^{q}\mu_j = \nu\\
&\qquad 0 \leq \lambda_i \leq K_s, \quad i = 1, \ldots, p\\
&\qquad 0 \leq \mu_j \leq K_s, \quad j = 1, \ldots, q.
\end{aligned}$$

Once a solution for λ and μ is obtained, we have

$$w = -X\begin{pmatrix}\lambda \\ \mu\end{pmatrix} = \sum_{i=1}^{p} \lambda_i u_i - \sum_{j=1}^{q} \mu_j v_j$$

$$b = -\sum_{i=1}^{p} \lambda_i + \sum_{j=1}^{q} \mu_j.$$

Note that the constraint

$$\sum_{i=1}^{p} \lambda_i - \sum_{j=1}^{q} \mu_j = 0$$

occurring in the dual of Program ($\text{SVM}_{s2'}$) has been traded for the equation

$$b = -\sum_{i=1}^{p} \lambda_i + \sum_{j=1}^{q} \mu_j$$

determining b.

If $\nu > (p_f + q_f)/(p+q)$, then η is determined by expressing that the duality gap is zero. We obtain

$$\begin{aligned}((p+q)\nu - p_f - q_f)\eta &= (p_f - q_f)b + w^\top\Bigg(\sum_{j\in K_\mu} v_j - \sum_{i\in K_\lambda} u_i\Bigg) \\ &\quad + \frac{1}{K_s}\begin{pmatrix}\lambda^\top & \mu^\top\end{pmatrix}\left(X^\top X + \begin{pmatrix}\mathbf{1}_p\mathbf{1}_p^\top & -\mathbf{1}_p\mathbf{1}_q^\top \\ -\mathbf{1}_q\mathbf{1}_p^\top & \mathbf{1}_q\mathbf{1}_q^\top\end{pmatrix}\right)\begin{pmatrix}\lambda \\ \mu\end{pmatrix}.\end{aligned}$$

In practice another way to compute η is to assume the Standard Margin Hypothesis for (SVM_{s3}). Under the **Standard Margin Hypothesis** for (SVM_{s3}), either some u_{i_0} is a support vector of type 1 *or* some v_{j_0} is a support vector of type 1. By the complementary slackness conditions $\epsilon_{i_0} = 0$ or $\xi_{j_0} = 0$, so we have

$$w^\top u_{i_0} - b = \eta, \quad \text{or} \quad -w^\top v_{j_0} + b = \eta,$$

and we can solve for η. As in ($\text{SVM}_{s2'}$) we get more numerically stable formulae by averaging over the sets I_λ and I_μ.

Proposition 18.4 gives bounds ν, namely

$$\frac{p_f + q_f}{p+q} \le \nu \le \frac{p_m + q_m}{p+q}.$$

In Section 18.11 we investigate conditions on ν that ensure that either there is some blue support vector u_{i_0} *or* there is some red support vector v_{j_0}. Theorem 18.2 shows that for every optimal solution $(w, b, \eta, \epsilon, \xi)$ of Problem (SVM_{s3}) with $w \neq 0$ and $\eta > 0$, if

$$(p_{sf} + q_{sf})/(p+q) < \nu < 1,$$

then some u_{i_0} or some v_{j_0} is a support vector.

(4) **Basic Quadratic Soft margin ν-SVM Problem** (SVM_{s4}). This is the version of Problem ($\mathrm{SVM}_{s2'}$) in which instead of using the linear function $K_s \left(\epsilon^\top \ \xi^\top\right) \mathbf{1}_{p+q}$ as a regularizing function we use the quadratic function $K(\|\epsilon\|_2^2 + \|\xi\|_2^2)$. The optimization problem is

$$\begin{aligned} &\text{minimize} \quad \frac{1}{2} w^\top w + (p+q)K_s \left(-\nu\eta + \frac{1}{p+q}(\epsilon^\top \epsilon + \xi^\top \xi)\right) \\ &\text{subject to} \\ &\qquad w^\top u_i - b \geq \eta - \epsilon_i, \qquad i = 1, \ldots, p \\ &\qquad - w^\top v_j + b \geq \eta - \xi_j, \qquad j = 1, \ldots, q \\ &\qquad \eta \geq 0, \end{aligned}$$

where ν and K_s are two given positive constants. As we saw earlier, theoretically, it is convenient to pick $K_s = 1/(p+q)$. When writing a computer program, it is preferable to assume that K_s is arbitrary. In this case ν needs to be replaced by $(p+q)K_s\nu$ in all the formulae obtained with $K_s = 1/(p+q)$.

In this method, it is no longer necessary to require $\epsilon \geq 0$ and $\xi \geq 0$, because an optimal solution satisfies these conditions.

One of the advantages of this methods is that ϵ is determined by λ, ξ is determined by μ, and η and b are determined by λ and μ. We can omit the constraint $\eta \geq 0$, because for an optimal solution it can be shown using duality that $\eta \geq 0$; see Section 18.14. For K_s and ν fixed, if Program (SVM_{s4}) has an optimal solution, then it is unique; see Theorem 18.3.

A drawback of Program (SVM_{s4}) is that for fixed K_s, the quantity $\delta = \eta / \|w\|$ and the hyperplanes $H_{w,b}, H_{w,b+\eta}$ and $H_{w,b-\eta}$ are *independent* of ν. This is shown in Theorem 18.3. Thus this method is less flexible than ($\mathrm{SVM}_{s2'}$) and (SVM_{s3}).

It is shown in Section 18.9 that the dual is given by

Dual of the Basic Quadratic Soft margin ν-SVM Problem (SVM_{s4}):

$$\begin{aligned} &\text{minimize} \quad \frac{1}{2} \left(\lambda^\top \ \mu^\top\right) \left(X^\top X + \frac{1}{2K} I_{p+q}\right) \begin{pmatrix} \lambda \\ \mu \end{pmatrix} \\ &\text{subject to} \end{aligned}$$

$$\begin{aligned}
&\sum_{i=1}^{p} \lambda_i - \sum_{j=1}^{q} \mu_j = 0 \\
&\sum_{i=1}^{p} \lambda_i + \sum_{j=1}^{q} \mu_j \geq \nu \\
&\lambda_i \geq 0, \quad i = 1, \ldots, p \\
&\mu_j \geq 0, \quad j = 1, \ldots, q.
\end{aligned}$$

The above program is similar to the program that was obtained for Problem ($\mathrm{SVM}_{s2'}$) but the matrix $X^\top X$ is replaced by the matrix $X^\top X + (1/2K)I_{p+q}$, which is positive definite since $K > 0$, and also the inequalities $\lambda_i \leq K$ and $\mu_j \leq K$ no longer hold. If the constraint $\eta \geq 0$ is dropped, then the inequality

$$\sum_{i=1}^{p} \lambda_i + \sum_{j=1}^{q} \mu_j \geq \nu$$

is replaced by the equation

$$\sum_{i=1}^{p} \lambda_i + \sum_{j=1}^{q} \mu_j = \nu.$$

We obtain w from λ and μ, and γ, as in Problem ($\mathrm{SVM}_{s2'}$); namely,

$$w = -X \begin{pmatrix} \lambda \\ \mu \end{pmatrix} = \sum_{i=1}^{p} \lambda_i u_i - \sum_{j=1}^{q} \mu_j v_j$$

and η is given by

$$(p+q)K_s \nu \eta = \begin{pmatrix} \lambda^\top & \mu^\top \end{pmatrix} \left(X^\top X + \frac{1}{2K_s} I_{p+q} \right) \begin{pmatrix} \lambda \\ \mu \end{pmatrix}.$$

The constraints imply that there is some i_o such that $\lambda_{i_0} > 0$ *and* some j_0 such that $\mu_{j_0} > 0$, which means that at least two points are misclassified, so Problem (SVM_{s4}) should only be used when the sets $\{u_i\}$ and $\{v_j\}$ are *not* linearly separable. We can solve for b using the active constraints corresponding to any i_0 such that $\lambda_{i_0} > 0$ and any j_0 such that $\mu_{j_0} > 0$. To improve numerical stability we average over the sets of indices I_λ and I_μ.

(5) **Quadratic Soft margin ν-SVM Problem** (SVM_{s5}). This is the variant of Problem (SVM_{s4}) in which we add the term $(1/2)b^2$ to the objective function. We also drop the constraint $\eta \geq 0$ which is redundant. We have the following optimization problem:

$$\text{minimize} \quad \frac{1}{2}w^\top w + \frac{1}{2}b^2 + (p+q)K_s\left(-\nu\eta + \frac{1}{p+q}(\epsilon^\top\epsilon + \xi^\top\xi)\right)$$

subject to

$$\begin{aligned} w^\top u_i - b &\geq \eta - \epsilon_i, & i &= 1, \ldots, p \\ -w^\top v_j + b &\geq \eta - \xi_j, & j &= 1, \ldots, q, \end{aligned}$$

where ν and K_s are two given positive constants. As we saw earlier, it is convenient to pick $K_s = 1/(p+q)$. When writing a computer program, it is preferable to assume that K_s is arbitrary. In this case ν must be replaced by $(p+q)K_s\nu$ in all the formulae.

One of the advantages of this methods is that ϵ is determined by λ, ξ is determined by μ (as in (SVM_{s4})), and both η and b determined by λ and μ. We can omit the constraint $\eta \geq 0$, because for an optimal solution it can be shown using duality that $\eta \geq 0$. For K_s and ν fixed, if Program (SVM_{s5}) has an optimal solution, then it is unique; see Theorem 18.4.

A drawback of Program (SVM_{s5}) is that for fixed K_s, the quantity $\delta = \eta/\|w\|$ and the hyperplanes $H_{w,b}, H_{w,b+\eta}$ and $H_{w,b-\eta}$ are *independent* of ν. This is shown in Theorem 18.4. Thus this method is less flexible than ($\mathrm{SVM}_{s2'}$) and (SVM_{s3}).

It is shown in Section 18.15 that the dual of Program (SVM_{s5}) is given by

Dual of the Quadratic Soft margin ν-SVM Problem (SVM_{s5}):

$$\text{minimize} \quad \frac{1}{2}\begin{pmatrix}\lambda^\top & \mu^\top\end{pmatrix}\left(X^\top X + \begin{pmatrix}\mathbf{1}_p\mathbf{1}_p^\top & -\mathbf{1}_p\mathbf{1}_q^\top \\ -\mathbf{1}_q\mathbf{1}_p^\top & \mathbf{1}_q\mathbf{1}_q^\top\end{pmatrix} + \frac{1}{2K}I_{p+q}\right)\begin{pmatrix}\lambda \\ \mu\end{pmatrix}$$

subject to

$$\begin{aligned} &\sum_{i=1}^p \lambda_i + \sum_{j=1}^q \mu_j = \nu \\ &\lambda_i \geq 0, \quad i = 1, \ldots, p \\ &\mu_j \geq 0, \quad j = 1, \ldots, q. \end{aligned}$$

This time we obtain w, b, η, ϵ and ξ from λ and μ:

$$\begin{aligned} w &= \sum_{i=1}^{p} \lambda_i u_i - \sum_{j=1}^{q} \mu_j v_j \\ b &= -\sum_{i=1}^{p} \lambda_i + \sum_{j=1}^{q} \mu_j \\ \epsilon &= \frac{\lambda}{2K} \\ \xi &= \frac{\mu}{2K}, \end{aligned}$$

and

$$(p+q)K_s\nu\eta = \begin{pmatrix} \lambda^\top & \mu^\top \end{pmatrix} \left(X^\top X + \begin{pmatrix} \mathbf{1}_p\mathbf{1}_p^\top & -\mathbf{1}_p\mathbf{1}_q^\top \\ -\mathbf{1}_q\mathbf{1}_p^\top & \mathbf{1}_q\mathbf{1}_q^\top \end{pmatrix} + \frac{1}{2K_s} I_{p+q} \right) \begin{pmatrix} \lambda \\ \mu \end{pmatrix}.$$

The constraint

$$\sum_{i=1}^{p} \lambda_i + \sum_{j=1}^{q} \mu_j = \nu$$

implies that either there is some i_0 such that $\lambda_{i_0} > 0$ or there is some j_0 such that $\mu_{j_0} > 0$, we have $\epsilon_{i_0} > 0$ or $\xi_{j_0} > 0$, which means that at least one point is misclassified, so Problem (SVM$_{s5}$) should only be used when the sets $\{u_i\}$ and $\{v_j\}$ are *not* linearly separable.

These methods all have a kernelized version.

We implemented all these methods in `Matlab`, solving the dual using ADMM.

From a theoretical point of view, Problems (SVM$_{s4}$) and (SVM$_{s5}$) seem to have more advantages than the others since they determine w, b, η and b without requiring any condition about support vectors of type 1. However, from a practical point of view, Problems (SVM$_{s4}$) and (SVM$_{s5}$) are less flexible that (SVM$_{s2'}$) and (SVM$_{s3}$), and we have observed that (SVM$_{s4}$) and (SVM$_{s5}$) are unable to produce as small a margin δ as (SVM$_{s2'}$) and (SVM$_{s3}$).

18.18 Problems

Problem 18.1. Prove the following inequality

$$\max\left\{\frac{1}{2p_m}, \frac{1}{2q_m}\right\} \leq K \leq \min\left\{\frac{1}{2p_f}, \frac{1}{2q_f}\right\}$$

stated just after Definition 18.1.

Problem 18.2. Prove the averaging formulae

$$b = w^\top \left(\Big(\sum_{i \in I_\lambda} u_i \Big) / |I_\lambda| + \Big(\sum_{j \in I_\mu} v_j \Big) / |I_\mu| \right) /2$$

$$\delta = w^\top \left(\Big(\sum_{i \in I_\lambda} u_i \Big) / |I_\lambda| - \Big(\sum_{j \in I_\mu} v_j \Big) / |I_\mu| \right) /2$$

stated at the end of Section 18.1.

Problem 18.3. Prove that the matrix

$$A = \begin{pmatrix} \mathbf{1}_p^\top & -\mathbf{1}_q^\top & 0_p^\top & 0_q^\top \\ \mathbf{1}_p^\top & \mathbf{1}_q^\top & 0_p^\top & 0_q^\top \\ I_p & 0_{p,q} & I_q & 0_{p,q} \\ 0_{q,p} & I_q & 0_{q,p} & I_q \end{pmatrix}$$

has rank $p + q + 2$.

Problem 18.4. Prove that the dual program of the kernel version of (SVM_{s1}) is given by:

Dual of Soft margin kernel SVM (SVM_{s1}):

$$\begin{aligned} &\text{minimize} \quad \begin{pmatrix} \lambda^\top & \mu^\top \end{pmatrix} \mathbf{K} \begin{pmatrix} \lambda \\ \mu \end{pmatrix} \\ &\text{subject to} \\ &\qquad \sum_{i=1}^p \lambda_i = \sum_{j=1}^q \mu_j = \frac{1}{2} \\ &\qquad 0 \leq \lambda_i \leq K, \quad i = 1, \ldots, p \\ &\qquad 0 \leq \mu_j \leq K, \quad j = 1, \ldots, q, \end{aligned}$$

where $\mathbf{K}$ is the $\ell \times \ell$ kernel symmetric matrix (with $\ell = p + q$) given by

$$\mathbf{K}_{ij} = \begin{cases} \kappa(u_i, u_j) & 1 \leq i \leq p,\ 1 \leq j \leq q \\ -\kappa(u_i, v_{j-p}) & 1 \leq i \leq p,\ p+1 \leq j \leq p+q \\ -\kappa(v_{i-p}, u_j) & p+1 \leq i \leq p+q,\ 1 \leq j \leq p \\ \kappa(v_{i-p}, v_{j-q}) & p+1 \leq i \leq p+q,\ p+1 \leq j \leq p+q. \end{cases}$$

Problem 18.5. Prove the averaging formula

$$b = w^\top \left(\left(\sum_{i \in I_\lambda} u_i \right) / |I_\lambda| + \left(\sum_{j \in I_\mu} v_j \right) / |I_\mu| \right) / 2$$

stated in Section 18.3.

Problem 18.6. Prove that the kernel version of Program (SVM_{s2}) is given by:

Dual of Soft margin kernel SVM (SVM_{s2}):

$$\begin{aligned} &\text{minimize} \quad \frac{1}{2} \begin{pmatrix} \lambda^\top & \mu^\top \end{pmatrix} \mathbf{K} \begin{pmatrix} \lambda \\ \mu \end{pmatrix} - \begin{pmatrix} \lambda^\top & \mu^\top \end{pmatrix} \mathbf{1}_{p+q} \\ &\text{subject to} \\ &\qquad \sum_{i=1}^p \lambda_i - \sum_{j=1}^q \mu_j = 0 \\ &\qquad 0 \leq \lambda_i \leq K, \quad i = 1, \ldots, p \\ &\qquad 0 \leq \mu_j \leq K, \quad j = 1, \ldots, q, \end{aligned}$$

where $\mathbf{K}$ is the $\ell \times \ell$ kernel symmetric matrix (with $\ell = p + q$) given at the end of Section 18.1.

Problem 18.7. Prove that the matrix

$$A = \begin{pmatrix} \mathbf{1}_p^\top & -\mathbf{1}_q^\top & 0_p^\top & 0_q^\top \\ I_p & 0_{p,q} & I_q & 0_{p,q} \\ 0_{q,p} & I_q & 0_{q,p} & I_q \end{pmatrix}$$

has rank $p + q + 1$.

Problem 18.8. Prove that the matrices

$$A = \begin{pmatrix} \mathbf{1}_p^\top & -\mathbf{1}_q^\top & 0_p^\top & 0_q^\top & 0 \\ \mathbf{1}_p^\top & \mathbf{1}_q^\top & 0_p^\top & 0_q^\top & -1 \\ I_p & 0_{p,q} & I_q & 0_{p,q} & 0_p \\ 0_{q,p} & I_q & 0_{q,p} & I_q & 0_q \end{pmatrix} \quad \text{and} \quad A_2 = \begin{pmatrix} \mathbf{1}_p^\top & -\mathbf{1}_q^\top & 0_p^\top & 0_q^\top \\ \mathbf{1}_p^\top & \mathbf{1}_q^\top & 0_p^\top & 0_q^\top \\ I_p & 0_{p,q} & I_q & 0_{p,q} \\ 0_{q,p} & I_q & 0_{q,p} & I_q \end{pmatrix}$$

have rank $p + q + 2$.

Problem 18.9. Prove that the kernel version of Program ($\text{SVM}_{s2'}$) is given by:

Dual of the Soft margin kernel SVM ($\text{SVM}_{s2'}$):

$$\begin{aligned}
&\text{minimize} \quad \frac{1}{2}\begin{pmatrix}\lambda^\top & \mu^\top\end{pmatrix}\mathbf{K}\begin{pmatrix}\lambda \\ \mu\end{pmatrix} \\
&\text{subject to} \\
&\qquad \sum_{i=1}^p \lambda_i - \sum_{j=1}^q \mu_j = 0 \\
&\qquad \sum_{i=1}^p \lambda_i + \sum_{j=1}^q \mu_j \geq K_m \\
&\qquad 0 \leq \lambda_i \leq K_s, \quad i = 1, \ldots, p \\
&\qquad 0 \leq \mu_j \leq K_s, \quad j = 1, \ldots, q,
\end{aligned}$$

where $\mathbf{K}$ is the kernel matrix of Section 18.1.

Problem 18.10. Prove the formulae determining b in terms of η stated just before Theorem 18.3.

Problem 18.11. Prove that the matrix

$$A = \begin{pmatrix} \mathbf{1}_p^\top & \mathbf{1}_q^\top & 0_p^\top & 0_q^\top \\ I_p & 0_{p,q} & I_q & 0_{p,q} \\ 0_{q,p} & I_q & 0_{q,p} & I_q \end{pmatrix}$$

has rank $p + q + 1$.

Problem 18.12. Prove that the kernel version of Program (SVM_{s3}) is given by:

Dual of the Soft margin kernel SVM (SVM_{s3}):

$$\begin{aligned}
&\text{minimize} \quad \frac{1}{2}\begin{pmatrix}\lambda^\top & \mu^\top\end{pmatrix}\left(\mathbf{K} + \begin{pmatrix}\mathbf{1}_p\mathbf{1}_p^\top & -\mathbf{1}_p\mathbf{1}_q^\top \\ -\mathbf{1}_q\mathbf{1}_p^\top & \mathbf{1}_q\mathbf{1}_q^\top\end{pmatrix}\right)\begin{pmatrix}\lambda \\ \mu\end{pmatrix} \\
&\text{subject to} \\
&\qquad \sum_{i=1}^p \lambda_i + \sum_{j=1}^q \mu_j = \nu \\
&\qquad 0 \leq \lambda_i \leq K_s, \quad i = 1, \ldots, p \\
&\qquad 0 \leq \mu_j \leq K_s, \quad j = 1, \ldots, q,
\end{aligned}$$

where $\mathbf{K}$ is the kernel matrix of Section 18.1.

Problem 18.13. Prove that the matrices

$$A = \begin{pmatrix} \mathbf{1}_p^\top & -\mathbf{1}_q^\top & 0 \\ \mathbf{1}_p^\top & \mathbf{1}_q^\top & -1 \end{pmatrix} \quad \text{and} \quad A_2 = \begin{pmatrix} \mathbf{1}_p^\top & -\mathbf{1}_q^\top \\ \mathbf{1}_p^\top & \mathbf{1}_q^\top \end{pmatrix}$$

have rank 2.

Problem 18.14. Implement Program (SVM_{s4}) in `Matlab`. You may adapt the programs given in Section B.2 and Section B.3.

Problem 18.15. Prove that the kernel version of Program (SVM_{s4}) is given by:

Dual of the Soft margin kernel SVM (SVM_{s4}):

$$\begin{aligned} &\text{minimize} \quad \frac{1}{2}\begin{pmatrix}\lambda^\top & \mu^\top\end{pmatrix}\left(\mathbf{K} + \frac{p+q}{2}I_{p+q}\right)\begin{pmatrix}\lambda \\ \mu\end{pmatrix} \\ &\text{subject to} \\ &\qquad \sum_{i=1}^p \lambda_i - \sum_{j=1}^q \mu_j = 0 \\ &\qquad \sum_{i=1}^p \lambda_i + \sum_{j=1}^q \mu_j \geq \nu \\ &\qquad \lambda_i \geq 0, \quad i = 1, \ldots, p \\ &\qquad \mu_j \geq 0, \quad j = 1, \ldots, q, \end{aligned}$$

where $\mathbf{K}$ is the kernel matrix of Section 18.1.

Problem 18.16. Implement Program (SVM_{s5}) in `Matlab`. You may adapt the programs given in Section B.2 and Section B.3.

Problem 18.17. Prove that the kernel version of Program (SVM_{s5}) is given by:

Dual of the Soft margin kernel SVM (SVM_{s5}):

$$\begin{aligned} &\text{minimize} \quad \frac{1}{2}\begin{pmatrix}\lambda^\top & \mu^\top\end{pmatrix}\left(\mathbf{K} + \begin{pmatrix} \mathbf{1}_p\mathbf{1}_p^\top & -\mathbf{1}_p\mathbf{1}_q^\top \\ -\mathbf{1}_q\mathbf{1}_p^\top & \mathbf{1}_q\mathbf{1}_q^\top \end{pmatrix} + \frac{p+q}{2}I_{p+q}\right)\begin{pmatrix}\lambda \\ \mu\end{pmatrix} \\ &\text{subject to} \\ &\qquad \sum_{i=1}^p \lambda_i + \sum_{j=1}^q \mu_j = \nu \\ &\qquad \lambda_i \geq 0, \quad i = 1, \ldots, p \\ &\qquad \mu_j \geq 0, \quad j = 1, \ldots, q, \end{aligned}$$

where $\mathbf{K}$ is the kernel matrix of Section 18.1.

Chapter 19

Ridge Regression, Lasso, Elastic Net

In this chapter we discuss linear regression. This problem can be cast as a learning problem. We observe a sequence of (distinct) pairs $((x_1, y_1), \ldots, (x_m, y_m))$ called a *set of training data* (or *predictors*), where $x_i \in \mathbb{R}^n$ and $y_i \in \mathbb{R}$, viewed as input-output pairs of some unknown function f that we are trying to infer. The simplest kind of function is a linear function $f(x) = x^\top w$, where $w \in \mathbb{R}^n$ is a vector of coefficients usually called a *weight vector*. Since the problem is overdetermined and since our observations may be subject to errors, we can't solve for w exactly as the solution of the system $Xw = y$, where X is the $m \times n$ matrix

$$X = \begin{pmatrix} x_1^\top \\ \vdots \\ x_m^\top \end{pmatrix},$$

where the row vectors $x_i^\top$ are the rows of X, and thus the $x_i \in \mathbb{R}^n$ are column vectors. So instead we solve the least-squares problem of minimizing $\|Xw - y\|_2^2$. In general there are still infinitely many solutions so we add a regularizing term. If we add the term $K \|w\|_2^2$ to the objective function $J(w) = \|Xw - y\|_2^2$, then we have *ridge regression*. This problem is discussed in Section 19.1 where we derive the dual program. The dual has a unique solution which yields a solution of the primal. However, the solution of the dual is given in terms of the matrix $XX^\top$ (whereas the solution of the primal is given in terms of $X^\top X$), and since our data points x_i are represented by the rows of the matrix X, we see that this solution only involves inner products of the x_i. This observation is the core of the idea of kernel functions, which were discussed in Chapter 17. We also explain how to solve the problem of learning an affine function $f(x) = x^\top w + b$.

In general the vectors w produced by ridge regression have few zero entries. In practice it is highly desirable to obtain sparse solutions, that is vectors w with many components equal to zero. This can be achieved by replacing the regularizing term $K \|w\|_2^2$ by the regularizing term $K \|w\|_1$; that is, to use the ℓ^1-norm instead of the ℓ^2-norm; see Section 19.4. This method has the exotic name of *lasso regression*. This time there is no closed-form solution, but this is a convex optimization problem and there are efficient iterative methods to solve it. One of the best methods relies on ADMM (see Section 16.8) and is discussed in Section 19.4. The lasso method has some limitations, in particular when the number m of data is smaller than the dimension n of the data. This happens in some applications in genetics and medicine. Fortunately there is a way to combine the best features of ridge regression and lasso, which is to use *two* regularizing terms:

(1) An ℓ^2-term $(1/2)K \|w\|_2^2$ as in ridge regression (with $K > 0$).
(2) An ℓ^1-term $\tau \|w\|_1$ as in lasso.

This method is known as *elastic net regression* and is discussed in Section 19.6. It retains most of the desirable features of ridge regression and lasso, and eliminates some of their weaknesses. Furthermore, it is effectively solved by ADMM.

19.1 Ridge Regression

The problem of solving an overdetermined or underdetermined linear system $Aw = y$, where A is an $m \times n$ matrix, arises as a "learning problem" in which we observe a sequence of data $((a_1, y_1), \ldots, (a_m, y_m))$, viewed as input-output pairs of some unknown function f that we are trying to infer, where the a_i are the *rows* of the matrix A and $y_i \in \mathbb{R}$. The values y_i are sometimes called *labels* or *responses*. The simplest kind of function is a linear function $f(x) = x^\top w$, where $w \in \mathbb{R}^n$ is a vector of coefficients usually called a *weight vector*, or sometimes an *estimator*. In the statistical literature w is often denoted by β. Since the problem is overdetermined and since our observations may be subject to errors, we can't solve for w exactly as the solution of the system $Aw = y$, so instead we solve the least-square problem of minimizing $\|Aw - y\|_2^2$.

In Section 21.1 (Vol. I) we showed that this problem can be solved using the pseudo-inverse. We know that the minimizers w are solutions of the normal equations $A^\top A w = A^\top y$, but when $A^\top A$ is not invertible, such a solution is not unique so some criterion has to be used to choose among these solutions.

One solution is to pick the unique vector w^+ of smallest Euclidean norm $\|w^+\|_2$ that minimizes $\|Aw - y\|_2^2$. The solution w^+ is given by $w^+ = A^+b$, where A^+ is the pseudo-inverse of A. The matrix A^+ is obtained from an SVD of A, say $A = V\Sigma U^\top$. Namely, $A^+ = U\Sigma^+V^\top$, where Σ^+ is the matrix obtained from Σ by replacing every nonzero singular value σ_i in Σ by σ_i^{-1}, leaving all zeros in place, and then transposing. The difficulty with this approach is that it requires knowing whether a singular value is zero or very small but nonzero. A very small nonzero singular value σ in Σ yields a very large value σ^{-1} in Σ^+, but $\sigma = 0$ remains 0 in Σ^+.

This discontinuity phenomenon is not desirable and another way is to control the size of w by adding a regularization term to $\|Aw - y\|^2$, and a natural candidate is $\|w\|^2$.

It is customary to rename each column vector $a_i^\top$ as x_i (where $x_i \in \mathbb{R}^n$) and to rename the input data matrix A as X, so that the row vector $x_i^\top$ are the *rows* of the $m \times n$ matrix X

$$X = \begin{pmatrix} x_1^\top \\ \vdots \\ x_m^\top \end{pmatrix}.$$

Our optimization problem, called *ridge regression*, is

Program (RR1):

$$\text{minimize} \quad \|y - Xw\|^2 + K\|w\|^2,$$

which by introducing the new variable $\xi = y - Xw$ can be rewritten as

Program (RR2):

$$\begin{aligned} &\text{minimize} \quad \xi^\top\xi + Kw^\top w \\ &\text{subject to} \\ &\qquad\qquad y - Xw = \xi, \end{aligned}$$

where $K > 0$ is some constant determining the influence of the regularizing term $w^\top w$, and we minimize over ξ and w.

The objective function of the first version of our minimization problem can be expressed as

$$\begin{aligned} J(w) &= \|y - Xw\|^2 + K\|w\|^2 \\ &= (y - Xw)^\top(y - Xw) + Kw^\top w \\ &= y^\top y - 2w^\top X^\top y + w^\top X^\top Xw + Kw^\top w \\ &= w^\top(X^\top X + KI_n)w - 2w^\top X^\top y + y^\top y. \end{aligned}$$

The matrix $X^\top X$ is symmetric positive semidefinite and $K > 0$, so the matrix $X^\top X + KI_n$ is positive definite. It follows that

$$J(w) = w^\top (X^\top X + KI_n)w - 2w^\top X^\top y + y^\top y$$

is strictly convex, so by Theorem 4.5(2)–(4), it has a unique minimum iff $\nabla J_w = 0$. Since

$$\nabla J_w = 2(X^\top X + KI_n)w - 2X^\top y,$$

we deduce that

$$w = (X^\top X + KI_n)^{-1} X^\top y. \qquad (*_{wp})$$

There is an interesting connection between the matrix $(X^\top X + KI_n)^{-1}X^\top$ and the pseudo-inverse X^+ of X.

Proposition 19.1. *The limit of the matrix $(X^\top X+KI_n)^{-1}X^\top$ when $K > 0$ goes to zero is the pseudo-inverse X^+ of X.*

Proof. To show this let $X = V\Sigma U^\top$ be a SVD of X. Then

$$(X^\top X + KI_n) = U\Sigma^\top V^\top V\Sigma U^\top + KI_n = U(\Sigma^\top\Sigma + KI_n)U^\top,$$

so

$$\begin{aligned}(X^\top X + KI_n)^{-1}X^\top &= U(\Sigma^\top\Sigma + KI_n)^{-1}U^\top U\Sigma^\top V^\top \\ &= U(\Sigma^\top\Sigma + KI_n)^{-1}\Sigma^\top V^\top.\end{aligned}$$

The diagonal entries in the matrix $(\Sigma^\top\Sigma + KI_n)^{-1}\Sigma^\top$ are

$$\frac{\sigma_i}{\sigma_i^2 + K}, \quad \text{if } \sigma_i > 0,$$

and zero if $\sigma_i = 0$. All nondiagonal entries are zero. When $\sigma_i > 0$ and $K > 0$ goes to 0,

$$\lim_{K \mapsto 0} \frac{\sigma_i}{\sigma_i^2 + K} = \sigma_i^{-1},$$

so

$$\lim_{K \mapsto 0} (\Sigma^\top\Sigma + KI_n)^{-1}\Sigma^\top = \Sigma^+,$$

which implies that

$$\lim_{K \mapsto 0} (X^\top X + KI_n)^{-1}X^\top = X^+. \qquad \square$$

The dual function of the first formulation of our problem is a constant function (with value the minimum of J) so it is not useful, but the second formulation of our problem yields an interesting dual problem. The Lagrangian is

$$\begin{aligned} L(\xi, w, \lambda) &= \xi^\top \xi + K w^\top w + (y - Xw - \xi)^\top \lambda \\ &= \xi^\top \xi + K w^\top w - w^\top X^\top \lambda - \xi^\top \lambda + \lambda^\top y, \end{aligned}$$

with $\lambda, \xi, y \in \mathbb{R}^m$. The Lagrangian $L(\xi, w, \lambda)$, as a function of ξ and w with λ held fixed, is obviously convex, in fact strictly convex.

To derive the dual function $G(\lambda)$ we minimize $L(\xi, w, \lambda)$ with respect to ξ and w. Since $L(\xi, w, \lambda)$ is (strictly) convex as a function of ξ and w, by Theorem 4.5(4), it has a minimum iff its gradient $\nabla L_{\xi,w}$ is zero (in fact, by Theorem 4.5(2), a unique minimum since the function is strictly convex). Since

$$\nabla L_{\xi,w} = \begin{pmatrix} 2\xi - \lambda \\ 2Kw - X^\top \lambda \end{pmatrix},$$

we get

$$\begin{aligned} \lambda &= 2\xi \\ w &= \frac{1}{2K} X^\top \lambda = X^\top \frac{\xi}{K}. \end{aligned}$$

The above suggests defining the variable α so that $\xi = K\alpha$, so we have $\lambda = 2K\alpha$ and $w = X^\top \alpha$. Then we obtain the dual function as a function of α by substituting the above values of ξ, λ and w back in the Lagrangian and we get

$$\begin{aligned} G(\alpha) &= K^2 \alpha^\top \alpha + K\alpha^\top X X^\top \alpha - 2K\alpha^\top X X^\top \alpha - 2K^2 \alpha^\top \alpha + 2K\alpha^\top y \\ &= -K\alpha^\top (XX^\top + KI_m)\alpha + 2K\alpha^\top y. \end{aligned}$$

This is a strictly concave function so by Theorem 4.5(4), its maximum is achieved iff $\nabla G_\alpha = 0$, that is,

$$2K(XX^\top + KI_m)\alpha = 2Ky,$$

which yields

$$\alpha = (XX^\top + KI_m)^{-1} y.$$

Putting everything together we obtain

$$\begin{aligned} \alpha &= (XX^\top + KI_m)^{-1} y \\ w &= X^\top \alpha \\ \xi &= K\alpha, \end{aligned}$$

which yields

$$w = X^\top (XX^\top + KI_m)^{-1} y. \qquad (*_{wd})$$

Earlier in $(*_{wp})$ we found that

$$w = (X^\top X + KI_n)^{-1} X^\top y,$$

and it is easy to check that

$$(X^\top X + KI_n)^{-1} X^\top = X^\top (XX^\top + KI_m)^{-1}.$$

If $n < m$ it is cheaper to use the formula on the left-hand side, but if $m < n$ it is cheaper to use the formula on the right-hand side.

19.2 Ridge Regression; Learning an Affine Function

It is easy to adapt the above method to learn an affine function $f(x) = x^\top w + b$ instead of a linear function $f(x) = x^\top w$, where $b \in \mathbb{R}$. We have the following optimization program

Program (RR3):

$$\begin{aligned} &\text{minimize} \quad \xi^\top \xi + K w^\top w \\ &\text{subject to} \\ &\qquad\qquad y - Xw - b\mathbf{1} = \xi, \end{aligned}$$

with $y, \xi, \mathbf{1} \in \mathbb{R}^m$ and $w \in \mathbb{R}^n$. Note that in Program (**RR3**) minimization is performed over ξ, w and b, but b is not penalized in the objective function. As in Section 19.1, the objective function is strictly convex.

The Lagrangian associated with this program is

$$L(\xi, w, b, \lambda) = \xi^\top \xi + K w^\top w - w^\top X^\top \lambda - \xi^\top \lambda - b\mathbf{1}^\top \lambda + \lambda^\top y.$$

Since L is (strictly) convex as a function of ξ, b, w, by Theorem 4.5(4), it has a minimum iff $\nabla L_{\xi,b,w} = 0$. We get

$$\begin{aligned} \lambda &= 2\xi \\ \mathbf{1}^\top \lambda &= 0 \\ w &= \frac{1}{2K} X^\top \lambda = X^\top \frac{\xi}{K}. \end{aligned}$$

As before, if we set $\xi = K\alpha$ we obtain $\lambda = 2K\alpha$, $w = X^\top \alpha$, and

$$G(\alpha) = -K\alpha^\top (XX^\top + KI_m)\alpha + 2K\alpha^\top y.$$

Since $K > 0$ and $\lambda = 2K\alpha$, the dual to ridge regression is the following program

Program (DRR3):

$$\begin{aligned} &\text{minimize} \quad \alpha^\top (XX^\top + KI_m)\alpha - 2\alpha^\top y \\ &\text{subject to} \\ &\qquad\qquad \mathbf{1}^\top \alpha = 0, \end{aligned}$$

where the minimization is over α.

Observe that up to the factor $1/2$, this problem satisfies the conditions of Proposition 6.3 with

$$\begin{aligned} A &= (XX^\top + KI_m)^{-1} \\ b &= y \\ B &= \mathbf{1}_m \\ f &= 0, \end{aligned}$$

and x renamed as α. Therefore, it has a unique solution (α, μ) (beware that $\lambda = 2K\alpha$ is **not** the λ used in Proposition 6.3, which we rename as μ), which is the unique solution of the KKT-equations

$$\begin{pmatrix} XX^\top + KI_m & \mathbf{1}_m \\ \mathbf{1}_m^\top & 0 \end{pmatrix} \begin{pmatrix} \alpha \\ \mu \end{pmatrix} = \begin{pmatrix} y \\ 0 \end{pmatrix}.$$

Since the solution given by Proposition 6.3 is

$$\mu = (B^\top AB)^{-1}(B^\top Ab - f), \quad \alpha = A(b - B\mu),$$

we get

$$\begin{aligned} \mu &= (\mathbf{1}^\top (XX^\top + KI_m)^{-1}\mathbf{1})^{-1}\mathbf{1}^\top (XX^\top + KI_m)^{-1} y \\ \alpha &= (XX^\top + KI_m)^{-1}(y - \mu\mathbf{1}). \end{aligned}$$

Note that the matrix $B^\top AB$ is the scalar $\mathbf{1}^\top (XX^\top + KI_m)^{-1}\mathbf{1}$, which is the negative of the Schur complement of $XX^\top + KI_m$.

Interestingly $b = \mu$, which is not obvious a priori.

Proposition 19.2. *We have $b = \mu$.*

Proof. To prove this result we need to express α differently. Since μ is a scalar, $\mu\mathbf{1} = \mathbf{1}\mu$, so

$$\mu\mathbf{1} = \mathbf{1}\mu = (\mathbf{1}^\top (XX^\top + KI_m)^{-1}\mathbf{1})^{-1}\mathbf{1}\mathbf{1}^\top (XX^\top + KI_m)^{-1} y,$$

and we obtain

$$\alpha = (XX^\top + KI_m)^{-1}(I_m - (\mathbf{1}^\top (XX^\top + KI_m)^{-1}\mathbf{1})^{-1}\mathbf{1}\mathbf{1}^\top \\ (XX^\top + KI_m)^{-1})y. \quad (*_{\alpha_3})$$

Since $w = X^\top \alpha$, we have

$$w = X^\top (XX^\top + KI_m)^{-1}(I_m - (\mathbf{1}^\top (XX^\top + KI_m)^{-1}\mathbf{1})^{-1}\mathbf{1}\mathbf{1}^\top (XX^\top + KI_m)^{-1})y. \quad (*_{w_3})$$

From $\xi = K\alpha$, we deduce that b is given by the equation

$$b\mathbf{1} = y - Xw - K\alpha.$$

Since $w = X^\top \alpha$, using $(*_{\alpha_3})$ we obtain

$$\begin{aligned}
b\mathbf{1} &= y - Xw - K\alpha \\
&= y - (XX^\top + KI_m)\alpha \\
&= y - (I_m - (\mathbf{1}^\top (XX^\top + KI_m)^{-1}\mathbf{1})^{-1}\mathbf{1}\mathbf{1}^\top (XX^\top + KI_m)^{-1})y \\
&= (\mathbf{1}^\top (XX^\top + KI_m)^{-1}\mathbf{1})^{-1}\mathbf{1}\mathbf{1}^\top (XX^\top + KI_m)^{-1})y \\
&= \mu\mathbf{1},
\end{aligned}$$

and thus

$$b = \mu = (\mathbf{1}^\top (XX^\top + KI_m)^{-1}\mathbf{1})^{-1}\mathbf{1}^\top (XX^\top + KI_m)^{-1}y, \quad (*_{b_3})$$

as claimed. □

In summary the KKT-equations determine both α and μ, and so $w = X^\top \alpha$ and b as well.

There is also a useful expression of b as an average.

Since $\mathbf{1}^\top \mathbf{1} = m$ and $\mathbf{1}^\top \alpha = 0$, we get

$$b = \frac{1}{m}\mathbf{1}^\top y - \frac{1}{m}\mathbf{1}^\top Xw - \frac{1}{m}K\mathbf{1}^\top \alpha = \overline{y} - \sum_{j=1}^{n} \overline{X^j} w_j,$$

where $\overline{y}$ is the mean of y and $\overline{X^j}$ is the mean of the jth column of X. Therefore,

$$b = \overline{y} - \sum_{j=1}^{n} \overline{X^j} w_j = \overline{y} - (\overline{X^1} \cdots \overline{X^n})w,$$

where $(\overline{X^1} \cdots \overline{X^n})$ is the $1 \times n$ row vector whose jth entry is $\overline{X^j}$.

We will now show that solving the dual (**DRR3**) for α and obtaining $w = X^\top \alpha$ is equivalent to solving our previous ridge regression Problem (**RR2**) applied to the centered data $\widehat{y} = y - \overline{y}\mathbf{1}_m$ and $\widehat{X} = X - \overline{X}$, where $\overline{X}$ is the $m \times n$ matrix whose jth column is $\overline{X^j}\mathbf{1}_m$, the vector whose coordinates are all equal to the mean $\overline{X^j}$ of the jth column X^j of X.

The expression

$$b = \overline{y} - (\overline{X^1} \cdots \overline{X^n})w$$

suggests looking for an intercept term b (also called bias) of the above form, namely

Program (RR4):

$$\begin{aligned} &\text{minimize} \quad \xi^\top \xi + K w^\top w \\ &\text{subject to} \\ &\qquad y - Xw - b\mathbf{1} = \xi \\ &\qquad b = \widehat{b} + \overline{y} - (\overline{X^1} \cdots \overline{X^n})w, \end{aligned}$$

with $\widehat{b} \in \mathbb{R}$. Again, in Program (**RR4**), minimization is performed over ξ, w, b and $\widehat{b}$, but b and $\widehat{b}$ are not penalized.

Since

$$b\mathbf{1} = \widehat{b}\mathbf{1} + \overline{y}\mathbf{1} - (\overline{X^1}\mathbf{1} \cdots \overline{X^n}\mathbf{1})w,$$

if $\overline{X} = (\overline{X^1}\mathbf{1} \cdots \overline{X^n}\mathbf{1})$ is the $m \times n$ matrix whose jth column is the vector $\overline{X^j}\mathbf{1}$, then the above program is equivalent to the program

Program (RR5):

$$\begin{aligned} &\text{minimize} \quad \xi^\top \xi + K w^\top w \\ &\text{subject to} \\ &\qquad y - Xw - \overline{y}\mathbf{1} + \overline{X}w - \widehat{b}\mathbf{1} = \xi, \end{aligned}$$

where minimization is performed over ξ, w and $\widehat{b}$. If we write $\widehat{y} = y - \overline{y}\mathbf{1}$ and $\widehat{X} = X - \overline{X}$, then the above program becomes

Program (RR6):

$$\begin{aligned} &\text{minimize} \quad \xi^\top \xi + K w^\top w \\ &\text{subject to} \\ &\qquad \widehat{y} - \widehat{X}w - \widehat{b}\mathbf{1} = \xi, \end{aligned}$$

minimizing over ξ, w and $\widehat{b}$. If the solution to this program is $\widehat{w}$, then $\widehat{b}$ is given by

$$\widehat{b} = \overline{\widehat{y}} - (\overline{\widehat{X}^1} \cdots \overline{\widehat{X}^n})\widehat{w} = 0,$$

since the data $\widehat{y}$ and $\widehat{X}$ are centered. Therefore (**RR6**) *is equivalent to ridge regression without an intercept term applied to the centered data* $\widehat{y} = y - \overline{y}\mathbf{1}$ *and* $\widehat{X} = X - \overline{X}$,

Program (RR6′):

$$\begin{aligned}
&\text{minimize} \quad \xi^\top \xi + K w^\top w \\
&\text{subject to} \\
&\qquad\qquad \widehat{y} - \widehat{X} w = \xi,
\end{aligned}$$

minimizing over ξ and w.

If $\widehat{w}$ is the optimal solution of this program given by

$$\widehat{w} = \widehat{X}^\top (\widehat{X}\widehat{X}^\top + K I_m)^{-1} \widehat{y}, \qquad (*_{w_6})$$

then b is given by

$$b = \overline{y} - (\overline{X^1} \ \cdots \ \overline{X^n})\widehat{w}.$$

Remark: Although this is not obvious a priori, the optimal solution w^* of the Program (**RR3**) given by $(*_{w_3})$ is equal to the optimal solution $\widehat{w}$ of Program (**RR6′**) given by $(*_{w_6})$. We believe that it should be possible to prove the equivalence of these formulae but a proof eludes us at this time. We leave this as an open problem. In practice the Program (**RR6′**) involving the centered data appears to be the preferred one.

Example 19.1. Consider the data set (X, y_1) with

$$X = \begin{pmatrix} -10 & 11 \\ -6 & 5 \\ -2 & 4 \\ 0 & 0 \\ 1 & 2 \\ 2 & -5 \\ 6 & -4 \\ 10 & -6 \end{pmatrix}, \quad y_1 = \begin{pmatrix} 0 \\ -2.5 \\ 0.5 \\ -2 \\ 2.5 \\ -4.2 \\ 1 \\ 4 \end{pmatrix}$$

as illustrated in Figure 19.1. We find that $\overline{y} = -0.0875$ and $(\overline{X^1}, \overline{X^2}) = (0.125, 0.875)$. For the value $K = 5$, we obtain

$$w = \begin{pmatrix} 0.9207 \\ 0.8677 \end{pmatrix}, \quad b = -0.9618,$$

for $K = 0.1$, we obtain

$$w = \begin{pmatrix} 1.1651 \\ 1.1341 \end{pmatrix}, \quad b = -1.2255,$$

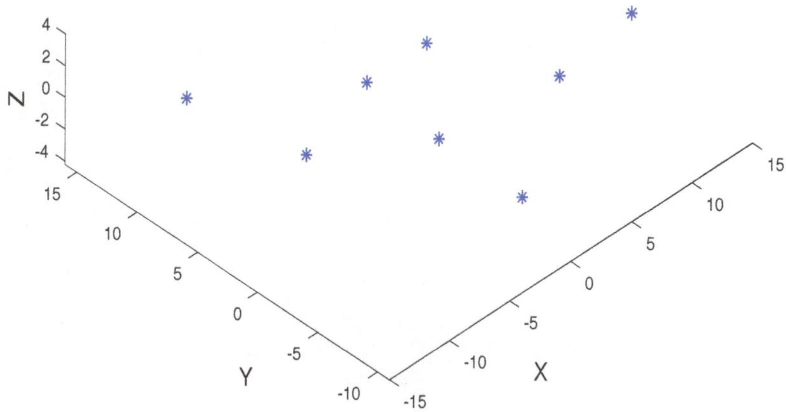

Fig. 19.1 The data set (X, y_1) of Example 19.1.

and for $K = 0.01$,

$$w = \begin{pmatrix} 1.1709 \\ 1.1405 \end{pmatrix}, \quad b = -1.2318.$$

See Figure 19.2.

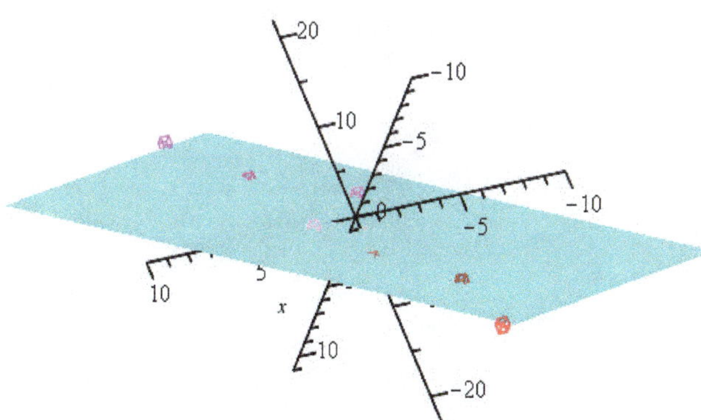

Fig. 19.2 The graph of the plane $f(x, y) = 1.1709x + 1.1405y - 1.2318$ as an approximate fit to the data (X, y_1) of Example 19.1.

We conclude that the points (X_i, y_i) (where X_i is the ith row of X) almost lie on the plane of equation

$$x + y - z - 1 = 0,$$

and that f is almost the function given by $f(x, y) = 1.1x + 1.1y - 1.2$. See Figures 19.3 and 19.4.

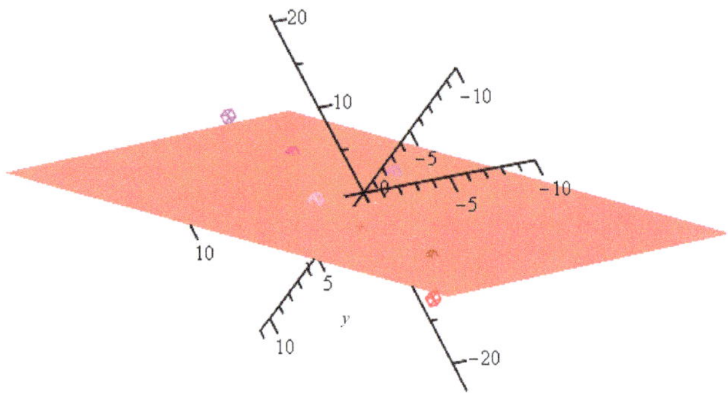

Fig. 19.3 The graph of the plane $f(x, y) = 1.1x + 1.1y - 1.2$ as an approximate fit to the data (X, y_1) of Example 19.1.

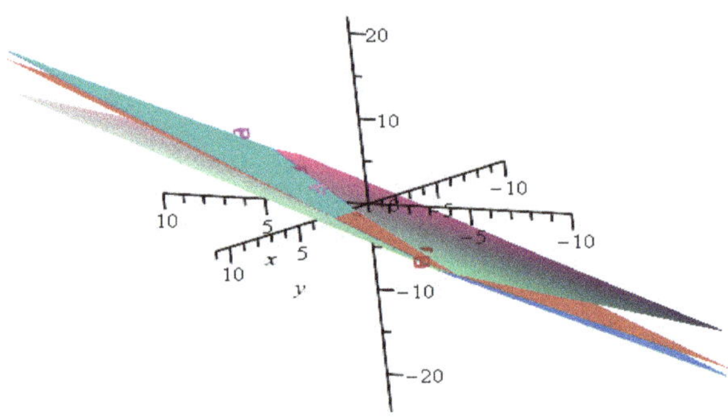

Fig. 19.4 A comparison of how the graphs of the planes corresponding to $K = 1, 0.1, 0.01$ and the salmon plane of equation $f(x, y) = 1.1x + 1.1y - 1.2$ approximate the data (X, y_1) of Example 19.1.

If we change y_1 to

$$y_2 = \begin{pmatrix} 0 & -2 & 1 & -1 & 2 & -4 & 1 & 3 \end{pmatrix}^\top,$$

as evidenced by Figure 19.5, the exact solution is

$$w = \begin{pmatrix} 1 \\ 1 \end{pmatrix}, \quad b = -1,$$

and for $K = 0.01$, we find that

$$w = \begin{pmatrix} 0.9999 \\ 0.9999 \end{pmatrix}, \quad b = -0.9999.$$

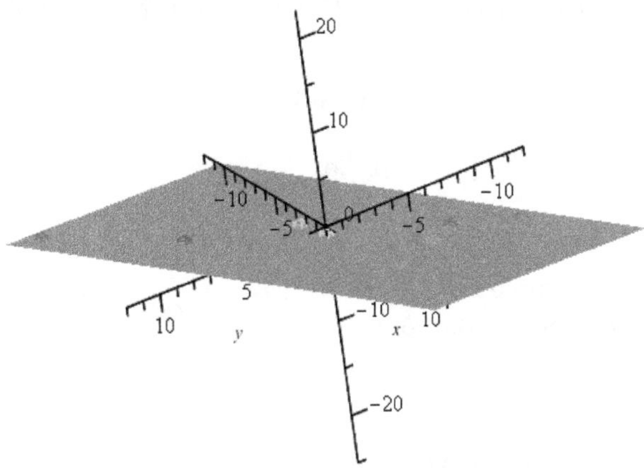

Fig. 19.5 The data (X, y_2) of Example 19.1 is contained within the graph of the plane $f(x, y) = x + y - 1$.

We can see how the choice of K affects the quality of the solution (w, b) by computing the norm $\|\xi\|_2$ of the error vector $\xi = \widehat{y} - \widehat{X}w$. We notice that the smaller K is, the smaller is this norm.

It is natural to wonder what happens *if we also penalize* b in program (**RR3**). Let us add the term Kb^2 to the objective function. Then we obtain the program

$$\begin{aligned} &\text{minimize} \quad \xi^\top \xi + K w^\top w + K b^2 \\ &\text{subject to} \\ &\qquad\qquad y - Xw - b\mathbf{1} = \xi, \end{aligned}$$

minimizing over ξ, w and b.

This suggests treating b an an extra component of the weight vector w and by forming the $m \times (n+1)$ matrix $[X\ \mathbf{1}]$ obtained by adding a column of 1's (of dimension m) to the matrix X, we obtain

Program (RR3b):

$$\begin{aligned}&\text{minimize} \quad \xi^\top \xi + K w^\top w + K b^2\\&\text{subject to}\\&\qquad y - [X\ \mathbf{1}]\begin{pmatrix} w \\ b \end{pmatrix} = \xi,\end{aligned}$$

minimizing over ξ, w and b.

This program is solved just as Program (**RR2**). In terms of the dual variable α, we get

$$\begin{aligned}\alpha &= ([X\ \mathbf{1}][X\ \mathbf{1}]^\top + K I_m)^{-1} y\\ \begin{pmatrix} w \\ b \end{pmatrix} &= [X\ \mathbf{1}]^\top \alpha\\ \xi &= K\alpha.\end{aligned}$$

Thus $b = \mathbf{1}^\top \alpha$. Observe that $[X\ \mathbf{1}][X\ \mathbf{1}]^\top = XX^\top + \mathbf{1}\mathbf{1}^\top$.

If $n < m$, it is preferable to use the formula

$$\begin{pmatrix} w \\ b \end{pmatrix} = ([X\ \mathbf{1}]^\top [X\ \mathbf{1}] + K I_{n+1})^{-1} [X\ \mathbf{1}]^\top y.$$

Since we also have the equation

$$y - Xw - b\mathbf{1} = \xi,$$

we obtain

$$\frac{1}{m}\mathbf{1}^\top y - \frac{1}{m}\mathbf{1}^\top X w - \frac{1}{m} b \mathbf{1}^\top \mathbf{1} = \frac{1}{m}\mathbf{1}^\top K\alpha,$$

so

$$\overline{y} - (\overline{X^1} \ \cdots \ \overline{X^n}) w - b = \frac{1}{m} K b,$$

which yields

$$b = \frac{m}{m+K}(\overline{y} - (\overline{X^1} \ \cdots \ \overline{X^n}) w).$$

Remark: As a least squares problem, the solution is given in terms of the pseudo-inverse $[X\ \mathbf{1}]^+$ of $[X\ \mathbf{1}]$ by

$$\begin{pmatrix} w \\ b \end{pmatrix} = [X\ \mathbf{1}]^+ y.$$

Example 19.2. Applying Program (**RR3**b) to the data set of Example 19.1 with $K = 0.01$ yields

$$w = \begin{pmatrix} 1.1706 \\ 1.1401 \end{pmatrix}, \quad b = -1.2298.$$

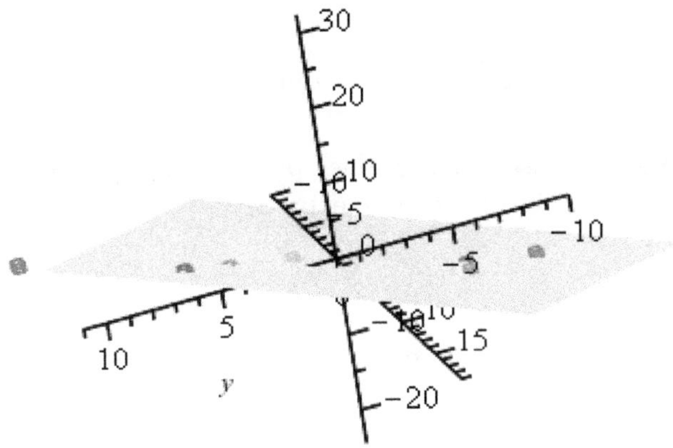

Fig. 19.6 The graph of the plane $f(x, y) = 1.1706x + 1.1401y - 1.2298$ as an approximate fit to the data (X, y_1) of Example 19.1.

See Figure 19.6. We can see how the choice of K affects the quality of the solution (w, b) by computing the norm $\|\xi\|_2$ of the error vector $\xi = y - Xw - b\mathbf{1}_m$. As in Example 19.1 we notice that the smaller K is, the smaller is this norm. We also observe that for a given value of K, Program (**RR6**$'$) gives a slightly smaller value of $\|\xi\|_2$ than (**RR3**b) does.

As pointed out by Hastie, Tibshirani, and Friedman [Hastie *et al.* (2009)] (Section 3.4), a defect of the approach where b is also penalized is that the solution for b is not invariant under adding a constant c to each value y_i. This is not the case for the approach using Program (**RR6**$'$).

19.3 Kernel Ridge Regression

One interesting aspect of the dual (of either (**RR2**) or (**RR3**)) is that it shows that the solution w being of the form $X^\top \alpha$, is a linear combination

$$w = \sum_{i=1}^{m} \alpha_i x_i$$

of the data points x_i, with the coefficients α_i corresponding to the dual variable $\lambda = 2K\alpha$ of the dual function, and with

$$\alpha = (XX^\top + KI_m)^{-1}y.$$

If m is smaller than n, then it is more advantageous to solve for α. But what really makes the dual interesting is that with our definition of X as

$$X = \begin{pmatrix} x_1^\top \\ \vdots \\ x_m^\top \end{pmatrix},$$

the matrix $XX^\top$ consists of the inner products $x_i^\top x_j$, and similarly the function learned $f(x) = x^\top w$ can be expressed as

$$f(x) = \sum_{i=1}^m \alpha_i x_i^\top x,$$

namely that both w and $f(x)$ are given *in terms of the inner products* $x_i^\top x_j$ and $x_i^\top x$.

This fact is the key to a generalization to ridge regression in which the input space $\mathbb{R}^n$ is embedded in a larger (possibly infinite dimensional) Euclidean space F (with an inner product $\langle -, - \rangle$) usually called a *feature space*, using a function

$$\varphi\colon \mathbb{R}^n \to F.$$

The problem becomes (*kernel ridge regression*)

Program (KRR2):

$$\begin{aligned} &\text{minimize} \quad \xi^\top \xi + K\langle w, w\rangle \\ &\text{subject to} \\ &\qquad\qquad y_i - \langle w, \varphi(x_i)\rangle = \xi_i, \quad i = 1, \ldots, m, \end{aligned}$$

minimizing over ξ and w. Note that $w \in F$. This problem is discussed in Shawe-Taylor and Christianini [Shawe-Taylor and Cristianini (2004)] (Section 7.3).

We will show below that the solution is exactly the same:

$$\begin{aligned} \alpha &= (\mathbf{G} + KI_m)^{-1}y \\ w &= \sum_{i=1}^m \alpha_i \varphi(x_i) \\ \xi &= K\alpha, \end{aligned}$$

where $\mathbf{G}$ is the Gram matrix given by $\mathbf{G}_{ij} = \langle \varphi(x_i), \varphi(x_j) \rangle$. This matrix is also called the *kernel matrix* and is often denoted by $\mathbf{K}$ instead of $\mathbf{G}$.

In this framework we have to be a little careful in using gradients since the inner product $\langle -, - \rangle$ on F is involved and F could be infinite dimensional, but this causes no problem because we can use derivatives, and by Proposition 3.5 we have

$$d\langle -, - \rangle_{(u,v)}(x, y) = \langle x, v \rangle + \langle u, y \rangle.$$

This implies that the derivative of the map $u \mapsto \langle u, u \rangle$ is

$$d\langle -, - \rangle_u(x) = 2\langle x, u \rangle. \tag{d_1}$$

Since the map $u \mapsto \langle u, v \rangle$ (with v fixed) is linear, its derivative is

$$d\langle -, v \rangle_u(x) = \langle x, v \rangle. \tag{d_2}$$

The derivative of the Lagrangian

$$L(\xi, w, \lambda) = \xi^\top \xi + K\langle w, w \rangle - \sum_{i=1}^m \lambda_i \langle \varphi(x_i), w \rangle - \xi^\top \lambda + \lambda^\top y$$

with respect to ξ and w is

$$dL_{\xi,w}(\widetilde{\xi}, \widetilde{w}) = 2(\widetilde{\xi})^\top \xi - (\widetilde{\xi})^\top \lambda + \left\langle 2Kw - \sum_{i=1}^m \lambda_i \varphi(x_i), \widetilde{w} \right\rangle,$$

where we used (d_1) to calculate the derivative of $\xi^\top \xi + K\langle w, w \rangle$ and (d_2) to calculate the derivative of $-\sum_{i=1}^m \lambda_i \langle \varphi(x_i), w \rangle - \xi^\top \lambda$. We have $dL_{\xi,w}(\widetilde{\xi}, \widetilde{w}) = 0$ for all $\widetilde{\xi}$ and $\widetilde{w}$ iff

$$\begin{aligned} 2Kw &= \sum_{i=1}^m \lambda_i \varphi(x_i) \\ \lambda &= 2\xi. \end{aligned}$$

Again we define $\xi = K\alpha$, so we have $\lambda = 2K\alpha$, and

$$w = \sum_{i=1}^m \alpha_i \varphi(x_i).$$

Plugging back into the Lagrangian we get

$$\begin{aligned} G(\alpha) &= K^2 \alpha^\top \alpha + K \sum_{i,j=1}^m \alpha_i \alpha_j \langle \varphi(x_i), \varphi(x_j) \rangle - 2K \sum_{i,j=1}^m \alpha_i \alpha_j \langle \varphi(x_i), \varphi(x_j) \rangle \\ &\quad - 2K^2 \alpha^\top \alpha + 2K\alpha^\top y \\ &= -K^2 \alpha^\top \alpha - K \sum_{i,j=1}^m \alpha_i \alpha_j \langle \varphi(x_i), \varphi(x_j) \rangle + 2K\alpha^\top y. \end{aligned}$$

If $\mathbf{G}$ is the matrix given by $\mathbf{G}_{ij} = \langle \varphi(x_i), \varphi(x_j) \rangle$, then we have

$$G(\alpha) = -K\alpha^\top (\mathbf{G} + KI_m)\alpha + 2K\alpha^\top y.$$

The function G is strictly concave, so by Theorem 4.5(4) it has a maximum for

$$\alpha = (\mathbf{G} + KI_m)^{-1} y,$$

as claimed earlier.

As in the standard case of ridge regression, if $F = \mathbb{R}^n$ (but the inner product $\langle -, - \rangle$ is arbitrary), we can adapt the above method to learn an affine function $f(w) = x^\top w + b$ instead of a linear function $f(w) = x^\top w$, where $b \in \mathbb{R}$. This time we assume that b is of the form

$$b = \overline{y} - \langle w, (\overline{X^1} \ \cdots \ \overline{X^n}) \rangle,$$

where X^j is the j column of the $m \times n$ matrix X whose ith row is the transpose of the column vector $\varphi(x_i)$, and where $(\overline{X^1} \ \cdots \ \overline{X^n})$ is viewed as a column vector. We have the minimization problem

Program (KRR6′):

$$\begin{aligned} &\text{minimize} \quad \xi^\top \xi + K\langle w, w \rangle \\ &\text{subject to} \\ &\qquad \widehat{y_i} - \langle w, \widehat{\varphi(x_i)} \rangle = \xi_i, \quad i = 1, \dots, m, \end{aligned}$$

minimizing over ξ and w, where $\widehat{\varphi(x_i)}$ is the n-dimensional vector $\varphi(x_i) - (\overline{X^1} \ \cdots \ \overline{X^n})$.

The solution is given in terms of the matrix $\widehat{\mathbf{G}}$ defined by

$$\widehat{\mathbf{G}}_{ij} = \langle \widehat{\varphi(x_i)}, \widehat{\varphi(x_j)} \rangle,$$

as before. We get

$$\alpha = (\widehat{\mathbf{G}} + KI_m)^{-1} \widehat{y},$$

and according to a previous computation, b is given by

$$b = \overline{y} - \frac{1}{m} \mathbf{1} \widehat{\mathbf{G}} \alpha.$$

We explained in Section 17.4 how to compute the matrix $\widehat{\mathbf{G}}$ from the matrix $\mathbf{G}$.

Since the dimension of the feature space F may be very large, one might worry that computing the inner products $\langle \varphi(x_i), \varphi(x_j) \rangle$ might be very expensive. This is where kernel functions come to the rescue. A *kernel function* κ for an embedding $\varphi\colon \mathbb{R}^n \to F$ is a map $\kappa\colon \mathbb{R}^n \times \mathbb{R}^n \to \mathbb{R}$ with the property that

$$\kappa(u, v) = \langle \varphi(u), \varphi(v) \rangle \quad \text{for all } u, v \in \mathbb{R}^n.$$

If $\kappa(u, v)$ can be computed in a reasonably cheap way, and if $\varphi(u)$ can be computed cheaply, then the inner products $\langle \varphi(x_i), \varphi(x_j) \rangle$ (and $\langle \varphi(x_i), \varphi(x) \rangle$) can be computed cheaply; see Chapter 17. Fortunately there are good kernel functions. Two very good sources on kernel methods are Schölkopf and Smola [Schölkopf and Smola (2002)] and Shawe-Taylor and Christianini [Shawe-Taylor and Cristianini (2004)].

19.4 Lasso Regression (ℓ^1-Regularized Regression)

The main weakness of ridge regression is that the estimated weight vector w usually has many nonzero coefficients. As a consequence, ridge regression does not scale up well. In practice we need methods capable of handling millions of parameters, or more. A way to encourage sparsity of the vector w, which means that many coordinates of w are zero, is to replace the quadratic penalty function $\tau w^\top w = \tau \left\| w \right\|_2^2$ by the penalty function $\tau \left\| w \right\|_1$, with the ℓ^2-norm replaced by the ℓ^1-norm.

This method was first proposed by Tibshirani arround 1996, under the name *lasso*, which stands for "least absolute selection and shrinkage operator." This method is also known as *ℓ^1-regularized regression*, but this is not as cute as "lasso," which is used predominantly.

Given a set of training data $\{(x_1, y_1), \ldots, (x_m, y_m)\}$, with $x_i \in \mathbb{R}^n$ and $y_i \in \mathbb{R}$, if X is the $m \times n$ matrix

$$X = \begin{pmatrix} x_1^\top \\ \vdots \\ x_m^\top \end{pmatrix},$$

in which the row vectors $x_i^\top$ are the rows of X, then *lasso regression* is the following optimization problem

Program (lasso1):

$$\begin{aligned} &\text{minimize} \quad \frac{1}{2}\xi^\top \xi + \tau \left\| w \right\|_1 \\ &\text{subject to} \\ &\qquad\qquad y - Xw = \xi, \end{aligned}$$

minimizing over ξ and w, where $\tau > 0$ is some constant determining the influence of the regularizing term $\left\| w \right\|_1$.

The difficulty with the regularizing term $\|w\|_1 = |w_1| + \cdots + |w_n|$ is that the map $w \mapsto \|w\|_1$ is not differentiable for all w. This difficulty can be overcome by using subgradients, but the dual of the above program can also be obtained in an elementary fashion by using a trick that we already used, which is that if $x \in \mathbb{R}$, then

$$|x| = \max\{x, -x\}.$$

Using this trick, by introducing a vector $\epsilon \in \mathbb{R}^n$ of nonnegative variables, we can rewrite lasso minimization as follows:

Program lasso regularization (**lasso2**):

$$\begin{aligned} &\text{minimize} \quad \frac{1}{2}\xi^\top \xi + \tau \mathbf{1}_n^\top \epsilon \\ &\text{subject to} \\ &\qquad\qquad y - Xw = \xi \\ &\qquad\qquad w \leq \epsilon \\ &\qquad\qquad -w \leq \epsilon. \end{aligned}$$

minimizing over ξ, w and ϵ, with $y, \xi \in \mathbb{R}^m$, and $w, \epsilon, \mathbf{1}_n \in \mathbb{R}^n$.

The constraints $w \leq \epsilon$ and $-w \leq \epsilon$ are equivalent to $|w_i| \leq \epsilon_i$ for $i = 1, \ldots, n$, so for an optimal solution we must have $\epsilon \geq 0$ and $|w_i| = \epsilon_i$, that is, $\|w\|_1 = \epsilon_1 + \cdots + \epsilon_n$.

The Lagrangian $L(\xi, w, \epsilon, \lambda, \alpha_+, \alpha_-)$ is given by

$$\begin{aligned} L(\xi, w, \epsilon, \lambda, \alpha_+, \alpha_-) &= \frac{1}{2}\xi^\top \xi + \tau \mathbf{1}_n^\top \epsilon + \lambda^\top (y - Xw - \xi) \\ &\quad + \alpha_+^\top (w - \epsilon) + \alpha_-^\top (-w - \epsilon) \\ &= \frac{1}{2}\xi^\top \xi - \xi^\top \lambda + \lambda^\top y \\ &\quad + \epsilon^\top (\tau \mathbf{1}_n - \alpha_+ - \alpha_-) + w^\top (\alpha_+ - \alpha_- - X^\top \lambda), \end{aligned}$$

with $\lambda \in \mathbb{R}^m$ and $\alpha_+, \alpha_- \in \mathbb{R}^n_+$. Since the objective function is convex and the constraints are affine (and thus qualified), the Lagrangian L has a minimum with respect to the primal variables, ξ, w, ϵ iff $\nabla L_{\xi,w,\epsilon} = 0$. Since the gradient $\nabla L_{\xi,w,\epsilon}$ is given by

$$\nabla L_{\xi,w,\epsilon} = \begin{pmatrix} \xi - \lambda \\ \alpha_+ - \alpha_- - X^\top \lambda \\ \tau \mathbf{1}_n - \alpha_+ - \alpha_- \end{pmatrix},$$

we obtain the equations

$$\begin{aligned} \xi &= \lambda \\ \alpha_+ - \alpha_- &= X^\top \lambda \\ \alpha_+ + \alpha_- &= \tau \mathbf{1}_n. \end{aligned}$$

Using these equations, the dual function $G(\lambda, \alpha_+, \alpha_-) = \min_{\xi,w,\epsilon} L$ is given by

$$\begin{aligned} G(\lambda, \alpha_+, \alpha_-) &= \frac{1}{2}\xi^\top\xi - \xi^\top\lambda + \lambda^\top y = \frac{1}{2}\lambda^\top\lambda - \lambda^\top\lambda + \lambda^\top y \\ &= -\frac{1}{2}\lambda^\top\lambda + \lambda^\top y = -\frac{1}{2}\left(\|y-\lambda\|_2^2 - \|y\|_2^2\right), \end{aligned}$$

so

$$G(\lambda, \alpha_+, \alpha_-) = -\frac{1}{2}\left(\|y-\lambda\|_2^2 - \|y\|_2^2\right).$$

Since $\alpha_+, \alpha_- \geq 0$, for any $i \in \{1, \ldots, n\}$ the minimum of $(\alpha_+)_i - (\alpha_-)_i$ is $-\tau$, and the maximum is τ. If we recall that for any $z \in \mathbb{R}^n$,

$$\|z\|_\infty = \max_{1 \leq i \leq n} |z_i|,$$

it follows that the constraints

$$\begin{aligned} \alpha_+ + \alpha_- &= \tau\mathbf{1}_n \\ X^\top\lambda &= \alpha_+ - \alpha_- \end{aligned}$$

are equivalent to

$$\left\|X^\top\lambda\right\|_\infty \leq \tau.$$

The above is equivalent to the $2n$ constraints

$$-\tau \leq (X^\top\lambda)_i \leq \tau, \quad 1 \leq i \leq n.$$

Therefore, the dual lasso program is given by

$$\begin{aligned} &\text{maximize} \quad -\frac{1}{2}\left(\|y-\lambda\|_2^2 - \|y\|_2^2\right) \\ &\text{subject to} \\ &\qquad\qquad \left\|X^\top\lambda\right\|_\infty \leq \tau, \end{aligned}$$

which (since $\|y\|_2^2$ is a constant term) is equivalent to

Program (Dlasso2):

$$\begin{aligned} &\text{minimize} \quad \frac{1}{2}\|y-\lambda\|_2^2 \\ &\text{subject to} \\ &\qquad\qquad \left\|X^\top\lambda\right\|_\infty \leq \tau, \end{aligned}$$

minimizing over $\lambda \in \mathbb{R}^m$.

One way to solve lasso regression is to use the dual program to find $\lambda = \xi$, and then to use linear programming to find w by solving the linear program arising from the lasso primal by holding ξ constant. The best way is to use ADMM as explained in Section 16.8(4). There are also a number of variations of gradient descent; see Hastie, Tibshirani, and Wainwright [Hastie *et al.* (2015)].

In theory, if we know the support of w and the signs of its components, then w is determined as we now explain.

In view of the constraint $y - Xw = \xi$ and the fact that for an optimal solution we must have $\xi = \lambda$, the following condition must hold:

$$\left\|X^\top (Xw - y)\right\|_\infty \leq \tau. \tag{$*$}$$

Also observe that for an optimal solution, we have

$$\begin{aligned}\frac{1}{2}\left\|y - Xw\right\|_2^2 + w^\top X^\top (y - Xw) &= \frac{1}{2}\left\|y\right\|^2 - w^\top X^\top y + \frac{1}{2} w^\top X^\top X w \\ &\quad + w^\top X^\top y - w^\top X^\top X w \\ &= \frac{1}{2}\left(\left\|y\right\|_2^2 - \left\|Xw\right\|_2^2\right) \\ &= \frac{1}{2}\left(\left\|y\right\|_2^2 - \left\|y - \lambda\right\|_2^2\right) = G(\lambda).\end{aligned}$$

Since the objective function is convex and the constraints are qualified, by Theorem 14.7(2) the duality gap is zero, so for optimal solutions of the primal and the dual, $G(\lambda) = L(\xi, w, \epsilon)$, that is

$$\frac{1}{2}\left\|y - Xw\right\|_2^2 + w^\top X^\top (y - Xw) = \frac{1}{2}\left\|\xi\right\|_2^2 + \tau \left\|w\right\|_1 = \frac{1}{2}\left\|y - Xw\right\|_2^2 + \tau \left\|w\right\|_1,$$

which yields the equation

$$w^\top X^\top (y - Xw) = \tau \left\|w\right\|_1. \tag{$**_1$}$$

The above is the inner product of w and $X^\top (y - Xw)$, so whenever $w_i \neq 0$, since $\left\|w\right\|_1 = |w_1| + \cdots + |w_n|$, in view of ($*$), we must have $(X^\top (y - Xw))_i = \tau \mathrm{sgn}(w_i)$. If

$$S = \{i \in \{1, \ldots, n\} \mid w_i \neq 0\}, \tag{$\dagger$}$$

if X_S denotes the matrix consisting of the columns of X indexed by S, and if w_S denotes the vector consisting of the nonzero components of w, then we have

$$X_S^\top (y - X_S w_S) = \tau \mathrm{sgn}(w_S). \tag{$**_2$}$$

We also have

$$\left\|X_{\overline{S}}^\top (y - X_S w_S)\right\|_\infty \leq \tau, \tag{$**_3$}$$

where $\overline{S}$ is the complement of S.

Equation $(**_2)$ yields

$$X_S^\top X_S w_S = X_S^\top y - \tau \mathrm{sgn}(w_S),$$

so if $X_S^\top X_S$ is invertible (which will be the case if the columns of X are linearly independent), we get

$$w_S = (X_S^\top X_S)^{-1}(X_S^\top y - \tau \mathrm{sgn}(w_S)). \qquad (**_4)$$

In theory, if we know the support of w and the signs of its components, then w_S is determined, but in practice this is useless since the problem is to find the support and the sign of the solution.

19.5 Lasso Regression; Learning an Affine Function

In the preceding section we made the simplifying assumption that we were trying to learn a linear function $f(x) = x^\top w$. To learn an affine function $f(x) = x^\top w + b$, we solve the following optimization problem

Program (lasso3):

$$\begin{aligned} \text{minimize} \quad & \frac{1}{2}\xi^\top \xi + \tau \mathbf{1}_n^\top \epsilon \\ \text{subject to} \quad & \\ & y - Xw - b\mathbf{1}_m = \xi \\ & w \leq \epsilon \\ & -w \leq \epsilon. \end{aligned}$$

Observe that as in the case of ridge regression, minimization is performed over ξ, w, ϵ and b, but b is not penalized in the objective function.

The Lagrangian associated with this optimization problem is

$$\begin{aligned} L(\xi, w, \epsilon, b, \lambda, \alpha_+, \alpha_-) &= \frac{1}{2}\xi^\top \xi - \xi^\top \lambda + \lambda^\top y - b\mathbf{1}_m^\top \lambda \\ &\quad + \epsilon^\top(\tau \mathbf{1}_n - \alpha_+ - \alpha_-) + w^\top(\alpha_+ - \alpha_- - X^\top \lambda), \end{aligned}$$

so by setting the gradient $\nabla L_{\xi, w, \epsilon, b}$ to zero we obtain the equations

$$\begin{aligned} \xi &= \lambda \\ \alpha_+ - \alpha_- &= X^\top \lambda \\ \alpha_+ + \alpha_- &= \tau \mathbf{1}_n \\ \mathbf{1}_m^\top \lambda &= 0. \end{aligned}$$

Using these equations, we find that the dual function is also given by

$$G(\lambda, \alpha_+, \alpha_-) = -\frac{1}{2}\left(\|y - \lambda\|_2^2 - \|y\|_2^2\right),$$

and the dual lasso program is given by

$$\begin{aligned}\text{maximize} \quad & -\frac{1}{2}\left(\|y - \lambda\|_2^2 - \|y\|_2^2\right) \\ \text{subject to} \quad & \\ & \left\|X^\top \lambda\right\|_\infty \leq \tau \\ & \mathbf{1}_m^\top \lambda = 0,\end{aligned}$$

which is equivalent to

Program (Dlasso3):

$$\begin{aligned}\text{minimize} \quad & \frac{1}{2}\|y - \lambda\|_2^2 \\ \text{subject to} \quad & \\ & \left\|X^\top \lambda\right\|_\infty \leq \tau \\ & \mathbf{1}_m^\top \lambda = 0,\end{aligned}$$

minimizing over $\lambda \in \mathbb{R}^m$.

Once $\lambda = \xi$ and w are determined, we obtain b using the equation

$$b\mathbf{1}_m = y - Xw - \xi,$$

and since $\mathbf{1}_m^\top \mathbf{1}_m = m$ and $\mathbf{1}_m^\top \xi = \mathbf{1}_m^\top \lambda = 0$, the above yields

$$b = \frac{1}{m}\mathbf{1}_m^\top y - \frac{1}{m}\mathbf{1}_m^\top Xw - \frac{1}{m}\mathbf{1}_m^\top \xi = \overline{y} - \sum_{j=1}^n \overline{X^j} w_j,$$

where $\overline{y}$ is the mean of y and $\overline{X^j}$ is the mean of the jth column of X.

The equation

$$b = \widehat{b} + \overline{y} - \sum_{j=1}^n \overline{X^j} w_j = \widehat{b} + \overline{y} - (\overline{X^1} \cdots \overline{X^n})w,$$

can be used as in ridge regression, (see Section 19.2), to show that the Program (**lasso3**) is equivalent to applying lasso regression (**lasso2**) without an intercept term to the centered data, by replacing y by $\widehat{y} = y - \overline{y}\mathbf{1}$ and X by $\widehat{X} = X - \overline{X}$. Then b is given by

$$b = \overline{y} - (\overline{X^1} \cdots \overline{X^n})\widehat{w},$$

where $\widehat{w}$ is the solution given by (**lasso2**). This is the method described by Hastie, Tibshirani, and Wainwright [Hastie *et al.* (2015)] (Section 2.2).

Example 19.3. We can create a data set (X, y) where X a 100×5 matrix and y is a 100×1 vector using the following `Matlab` program in which the command `randn` creates an array of normally distributed numbers.

```
X = randn(100,5);
ww = [0; 2; 0; -3; 0];
y = X*ww + randn(100,1)*0.1;
```

The purpose of the third line is to add some small noise to the "output" $X * ww$. The first five rows of X are

$$\begin{pmatrix} -1.1658 & -0.0679 & -1.6118 & 0.3199 & 0.4400 \\ -1.1480 & -0.1952 & -0.0245 & -0.5583 & -0.6169 \\ 0.1049 & -0.2176 & -1.9488 & -0.3114 & 0.2748 \\ 0.7223 & -0.3031 & 1.0205 & -0.5700 & 0.6011 \\ 2.5855 & 0.0230 & 0.8617 & -1.0257 & 0.0923 \end{pmatrix},$$

and the first five rows of y are

$$y = \begin{pmatrix} -1.0965 \\ 1.2155 \\ 0.4324 \\ 1.1902 \\ 3.1346 \end{pmatrix}.$$

We ran the program for lasso using ADMM (see Problem 16.7) with various values of ρ and τ, including $\rho = 1$ and $\rho = 10$. We observed that the program converges a lot faster for $\rho = 10$ than for $\rho = 1$. We plotted the values of the five components of $w(\tau)$ for values of τ from $\tau = 0$ to $\tau = 0.5$ by increment of 0.02, and observed that the first, third, and fifth coordinate drop basically linearly to zero (a value less that 10^{-4}) around $\tau = 0.2$. See Figures 19.7, 19.8, and 19.9. This behavior is also observed in Hastie, Tibshirani, and Wainwright [Hastie *et al.* (2015)] (Section 2.2).

For $\tau = 0.02$, we have

$$w = \begin{pmatrix} 0.00003 \\ 2.01056 \\ -0.00004 \\ -2.99821 \\ 0.00000 \end{pmatrix}, \quad b = 0.00135.$$

This weight vector w is very close to the original vector $ww = [0; 2; 0; -3; 0]$ that we used to create y. For large values of τ, the weight vector is essentially the zero vector. This happens for $\tau = 235$, where every component of w is less than 10^{-5}.

Another way to find b is to add the term $(C/2)b^2$ to the objective function, for some positive constant C, obtaining the program

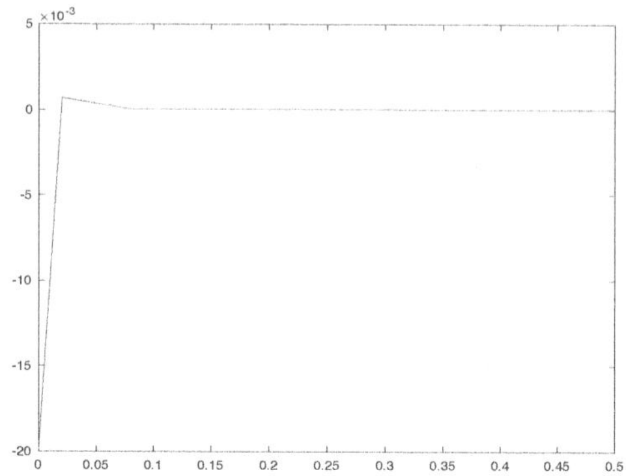

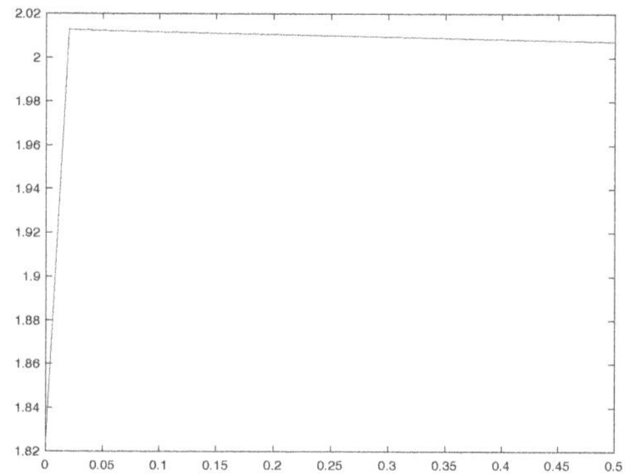

Fig. 19.7 First and second component of w.

Program (lasso4):

$$\begin{aligned} \text{minimize} \quad & \frac{1}{2}\xi^\top \xi + \tau \mathbf{1}_n^\top \epsilon + \frac{1}{2} C b^2 \\ \text{subject to} \quad & \\ & y - Xw - b\mathbf{1}_m = \xi \\ & w \leq \epsilon \\ & -w \leq \epsilon, \end{aligned}$$

minimizing over ξ, w, ϵ and b.

This time the Lagrangian is

$$\begin{aligned} L(\xi, w, \epsilon, b, \lambda, \alpha_+, \alpha_-) &= \frac{1}{2}\xi^\top \xi - \xi^\top \lambda + \lambda^\top y + \frac{C}{2} b^2 - b\mathbf{1}_m^\top \lambda \\ &\quad + \epsilon^\top (\tau \mathbf{1}_n - \alpha_+ - \alpha_-) + w^\top (\alpha_+ - \alpha_- - X^\top \lambda), \end{aligned}$$

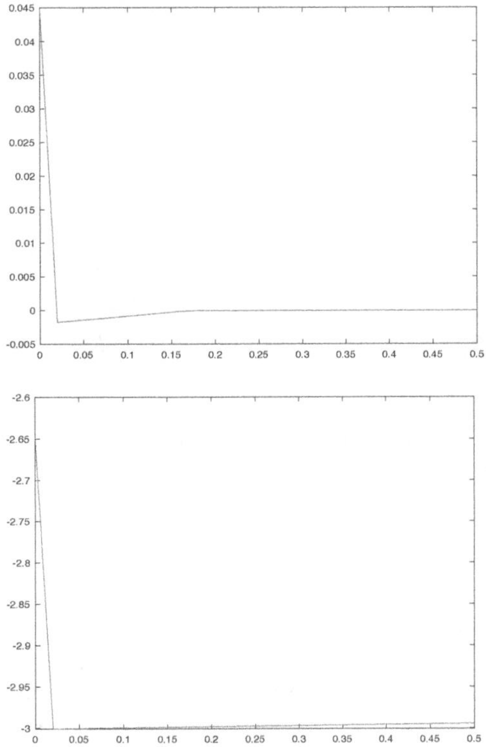

Fig. 19.8 Third and fourth component of w.

so by setting the gradient $\nabla L_{\xi,w,\epsilon,b}$ to zero we obtain the equations

$$\begin{aligned} \xi &= \lambda \\ \alpha_+ - \alpha_- &= X^\top \lambda \\ \alpha_+ + \alpha_- &= \tau \mathbf{1}_n \\ Cb &= \mathbf{1}_m^\top \lambda. \end{aligned}$$

Thus b is also determined, and the dual lasso program is identical to the first lasso dual (**Dlasso2**), namely

$$\text{minimize} \quad \frac{1}{2} \|y - \lambda\|_2^2$$

subject to

$$\left\|X^\top \lambda\right\|_\infty \leq \tau,$$

minimizing over λ.

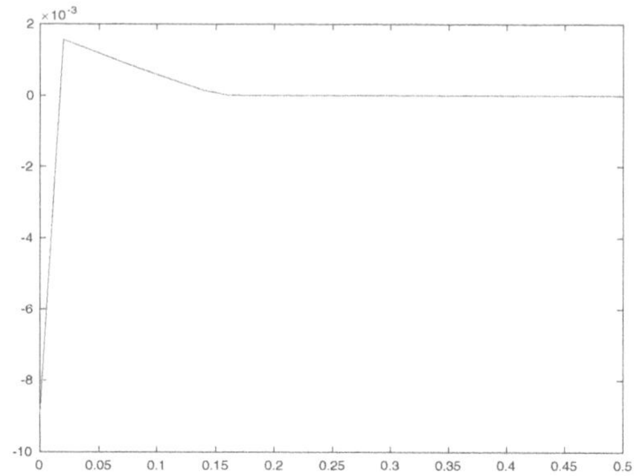

Fig. 19.9 Fifth component of w.

Since the equations $\xi = \lambda$ and

$$y - Xw - b\mathbf{1}_m = \xi$$

hold, from $Cb = \mathbf{1}_m^\top \lambda$ we get

$$\frac{1}{m}\mathbf{1}_m^\top y - \frac{1}{m}\mathbf{1}_m^\top Xw - b\frac{1}{m}\mathbf{1}_m^\top \mathbf{1} = \frac{1}{m}\mathbf{1}_m^\top \lambda,$$

that is

$$\overline{y} - (\overline{X^1} \cdots \overline{X^n})w - b = \frac{C}{m}b,$$

which yields

$$b = \frac{m}{m+C}(\overline{y} - (\overline{X^1} \cdots \overline{X^n})w).$$

As in the case of ridge regression, a defect of the approach where b is also penalized is that the solution for b is not invariant under adding a constant c to each value y_i

It is interesting to compare the behavior of the methods:

(1) Ridge regression (**RR6′**) (which is equivalent to (**RR3**)).
(2) Ridge regression (**RR3**b), with b penalized (by adding the term Kb^2 to the objective function).
(3) Least squares applied to $[X \;\; \mathbf{1}]$.
(4) (**lasso3**).

When $n \leq 2$ and K and τ are small and of the same order of magnitude, say 0.1 or 0.01, there is no noticeable difference. We ran out programs on the data set of 200 points generated by the following `Matlab` program:

```
X14 = 15*randn(200,1);
ww14 = 1;
y14 = X14*ww14 + 10*randn(200,1) + 20;
```

The result is shown in Figure 19.10, with the following colors: Method (1) in magenta, Method (2) in red, Method (3) in blue, and Method (4) in cyan. All four lines are identical.

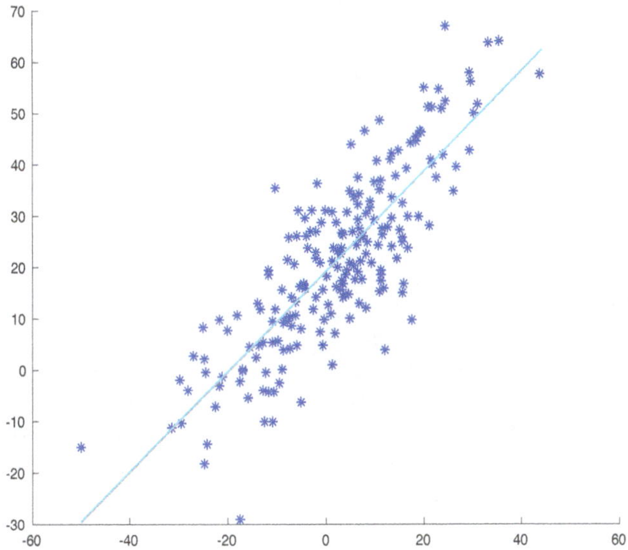

Fig. 19.10 Comparison of the four methods with $K = \tau = 0.1$.

In order to see a difference we also ran our programs with $K = 1000$ and $\tau = 10000$; see Figure 19.11.

As expected, due to the penalization of b, Method (3) yields a significantly lower line (in red), and due to the large value of τ, the line corresponding to lasso (in cyan) has a smaller slope. Method (1) (in magenta) also has a smaller slope but still does not deviate that much from least squares (in blue). It is also interesting to experiment on data sets where n is larger (say $20, 50$).

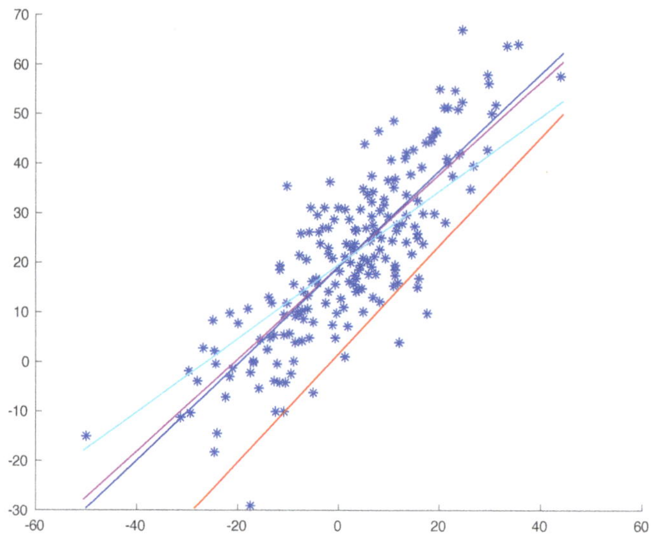

Fig. 19.11 Comparison of the four methods with $K = 1000, \tau = 10000$.

19.6 Elastic Net Regression

The lasso method is unsatisfactory when n (the dimension of the data) is much larger than the number m of data, because it only selects m coordinates and sets the others to values close to zero. It also has problems with groups of highly correlated variables. A way to overcome this problem is to add a "ridge-like" term $(1/2)Kw^\top w$ to the objective function. This way we obtain a hybrid of lasso and ridge regression called the *elastic net method* and defined as follows:

Program (elastic net):

$$\begin{aligned} &\text{minimize} \quad \frac{1}{2}\xi^\top\xi + \frac{1}{2}Kw^\top w + \tau\mathbf{1}_n^\top\epsilon \\ &\text{subject to} \\ &\qquad\qquad y - Xw - b\mathbf{1}_m = \xi \\ &\qquad\qquad w \leq \epsilon \\ &\qquad\qquad -w \leq \epsilon, \end{aligned}$$

where $K > 0$ and $\tau \geq 0$ are two constants controlling the influence of the ℓ^2-regularization and the ℓ^1-regularization.[1] Observe that as in the case of ridge regression, minimization is performed over ξ, w, ϵ and b, but b is not penalized in the objective function. The objective function is strictly convex so if an optimal solution exists, then it is unique; the proof is left as an exercise.

The Lagrangian associated with this optimization problem is

$$\begin{aligned} L(\xi, w, \epsilon, b, \lambda, \alpha_+, \alpha_-) &= \frac{1}{2}\xi^\top \xi - \xi^\top \lambda + \lambda^\top y - b\mathbf{1}_m^\top \lambda + \frac{1}{2}K w^\top w \\ &\quad + \epsilon^\top(\tau \mathbf{1}_n - \alpha_+ - \alpha_-) + w^\top(\alpha_+ - \alpha_- - X^\top \lambda), \end{aligned}$$

so by setting the gradient $\nabla L_{\xi,w,\epsilon,b}$ to zero we obtain the equations

$$\begin{aligned} \xi &= \lambda \\ Kw &= -(\alpha_+ - \alpha_- - X^\top \lambda) \qquad (*_w) \\ \alpha_+ + \alpha_- - \tau\mathbf{1}_n &= 0 \\ \mathbf{1}_m^\top \lambda &= 0. \end{aligned}$$

We find that $(*_w)$ determines w. Using these equations, we can find the dual function but in order to solve the dual using ADMM, since $\lambda \in \mathbb{R}^m$, it is more convenient to write $\lambda = \lambda_+ - \lambda_-$, with $\lambda_+, \lambda_- \in \mathbb{R}^m_+$ (recall that $\alpha_+, \alpha_- \in \mathbb{R}^n_+$). As in the derivation of the dual of ridge regression, we rescale our variables by defining $\beta_+, \beta_-, \mu_+, \mu_-$ such that

$$\alpha_+ = K\beta_+, \ \ \alpha_- = K\beta_-, \ \ \mu_+ = K\lambda_+, \ \ \mu_- = K\lambda_-.$$

We also let $\mu = \mu_+ - \mu_-$ so that $\lambda = K\mu$. Then $\mathbf{1}_m^\top \lambda = 0$ is equivalent to

$$\mathbf{1}_m^\top \mu_+ - \mathbf{1}_m^\top \mu_- = 0,$$

and since $\xi = \lambda = K\mu$, we have

$$\begin{aligned} \xi &= K(\mu_+ - \mu_-) \\ \beta_+ + \beta_- &= \frac{\tau}{K}\mathbf{1}_n. \end{aligned}$$

Using $(*_w)$ we can write

$$\begin{aligned} w &= -(\beta_+ - \beta_- - X^\top \mu) = -\beta_+ + \beta_- + X^\top \mu_+ - X^\top \mu_- \\ &= \begin{pmatrix} -I_n & I_n & X^\top & -X^\top \end{pmatrix} \begin{pmatrix} \beta_+ \\ \beta_- \\ \mu_+ \\ \mu_- \end{pmatrix}. \end{aligned}$$

[1] Some of the literature denotes K by λ_2 and τ by λ_1, but we prefer not to adopt this notation since we use λ, μ *etc.* to denote Lagrange multipliers.

Then we have

$$(-I_n\ I_n\ X^\top\ -X^\top)^\top(-I_n\ I_n\ X^\top\ -X^\top) = \begin{pmatrix} -I_n \\ I_n \\ X \\ -X \end{pmatrix}(-I_n\ I_n\ X^\top\ -X^\top)$$

$$= \begin{pmatrix} I_n & -I_n & -X^\top & X^\top \\ -I_n & I_n & X^\top & -X^\top \\ -X & X & XX^\top & -XX^\top \\ X & -X & -XX^\top & XX^\top \end{pmatrix}.$$

If we define the symmetric positive semidefinite $2(n+m)\times 2(n+m)$ matrix Q as

$$Q = \begin{pmatrix} I_n & -I_n & -X^\top & X^\top \\ -I_n & I_n & X^\top & -X^\top \\ -X & X & XX^\top & -XX^\top \\ X & -X & -XX^\top & XX^\top \end{pmatrix},$$

then

$$\frac{1}{2}w^\top w = \frac{1}{2}\left(\beta_+^\top\ \beta_-^\top\ \mu_+^\top\ \mu_-^\top\right) Q \begin{pmatrix} \beta_+ \\ \beta_- \\ \mu_+ \\ \mu_- \end{pmatrix}.$$

As a consequence, using $(*_w)$ and the fact that $\xi = K\mu$, we find that the dual function is given by

$$\begin{aligned} G(\mu,\beta_+,\beta_-) &= \frac{1}{2}\xi^\top\xi - \xi^\top\lambda + \lambda^\top y + w^\top(\alpha_+ - \alpha_- - X^\top\lambda) + \frac{1}{2}Kw^\top w \\ &= \frac{1}{2}\xi^\top\xi - K\xi^\top\mu + K\mu^\top y + Kw^\top(\beta_+ - \beta_- - X^\top\mu) + \frac{1}{2}Kw^\top w \\ &= \frac{1}{2}K^2\mu^\top\mu - K^2\mu^\top\mu + Ky^\top\mu - Kw^\top w + \frac{1}{2}Kw^\top w \\ &= -\frac{1}{2}K^2\mu^\top\mu - \frac{1}{2}Kw^\top w + Ky^\top\mu. \end{aligned}$$

But

$$\mu = (I_m\ -I_m)\begin{pmatrix} \mu_+ \\ \mu_- \end{pmatrix},$$

so

$$\frac{1}{2}\mu^\top\mu = \frac{1}{2}\left(\mu_+^\top\ \mu_-^\top\right)\begin{pmatrix} I_m & -I_m \\ -I_m & I_m \end{pmatrix}\begin{pmatrix} \mu_+ \\ \mu_- \end{pmatrix},$$

so we get

$$G(\beta_+, \beta_-, \mu_+, \mu_-) = -\frac{1}{2}K\left(\beta_+^\top\ \beta_-^\top\ \mu_+^\top\ \mu_-^\top\right) P \begin{pmatrix} \beta_+ \\ \beta_- \\ \mu_+ \\ \mu_- \end{pmatrix} - Kq^\top \begin{pmatrix} \beta_+ \\ \beta_- \\ \mu_+ \\ \mu_- \end{pmatrix}$$

with

$$\begin{aligned} P &= Q + K \begin{pmatrix} 0_{n,n} & 0_{n,n} & 0_{n,m} & 0_{n,m} \\ 0_{n,n} & 0_{n,n} & 0_{n,m} & 0_{n,m} \\ 0_{m,n} & 0_{m,n} & I_m & -I_m \\ 0_{m,n} & 0_{m,n} & -I_m & I_m \end{pmatrix} \\ &= \begin{pmatrix} I_n & -I_n & -X^\top & X^\top \\ -I_n & I_n & X^\top & -X^\top \\ -X & X & XX^\top + KI_m & -XX^\top - KI_m \\ X & -X & -XX^\top - KI_m & XX^\top + KI_m \end{pmatrix}, \end{aligned}$$

and

$$q = \begin{pmatrix} 0_n \\ 0_n \\ -y \\ y \end{pmatrix}.$$

The constraints are the equations

$$\begin{aligned} \beta_+ + \beta_- &= \frac{\tau}{K}\mathbf{1}_n \\ \mathbf{1}_m^\top \mu_+ - \mathbf{1}_m^\top \mu_- &= 0, \end{aligned}$$

which correspond to the $(n+1) \times 2(n+m)$ matrix

$$A = \begin{pmatrix} I_n & I_n & 0_{n,m} & 0_{n,m} \\ 0_n^\top & 0_n^\top & \mathbf{1}_m^\top & -\mathbf{1}_m^\top \end{pmatrix}$$

and the right-hand side

$$c = \begin{pmatrix} \frac{\tau}{K}\mathbf{1}_n \\ 0 \end{pmatrix}.$$

Since $K > 0$, the dual of elastic net is equivalent to

Program (Dual Elastic Net):

$$\text{minimize} \quad \frac{1}{2}\begin{pmatrix}\beta_+^\top & \beta_-^\top & \mu_+^\top & \mu_-^\top\end{pmatrix} P \begin{pmatrix}\beta_+ \\ \beta_- \\ \mu_+ \\ \mu_-\end{pmatrix} + q^\top \begin{pmatrix}\beta_+ \\ \beta_- \\ \mu_+ \\ \mu_-\end{pmatrix}$$

subject to

$$A \begin{pmatrix}\beta_+ \\ \beta_- \\ \mu_+ \\ \mu_-\end{pmatrix} = c,$$

$\beta_+, \beta_- \in \mathbb{R}^n_+, \mu_+, \mu_- \in \mathbb{R}^m_+$.

Once $\xi = K\mu = K(\mu_+ - \mu_-)$ and w are determined by $(*_w)$, we obtain b using the equation

$$b\mathbf{1}_m = y - Xw - \xi,$$

which as in Section 19.5 yields

$$b = \overline{y} - \sum_{j=1}^{n} \overline{X^j} w_j,$$

where $\overline{y}$ is the mean of y and $\overline{X^j}$ is the mean of the jth column of X.

We leave it as an easy exercise to show that A has rank $n+1$. Then we can solve the above program using ADMM, and we have done so. This very similar to what is done in Section 20.3, and hence the details are left as an exercise.

Observe that when $\tau = 0$, the elastic net method reduces to ridge regression. As K tends to 0 the elastic net method tends to lasso, but $K = 0$ is not an allowable value since $\tau/0$ is undefined. Anyway, if we get rid of the constraint

$$\beta_+ + \beta_- = \frac{\tau}{K}\mathbf{1}_n$$

the corresponding optimization program not does determine w. Experimenting with our program we found that convergence becomes very slow for $K < 10^{-3}$. What we can can do if K is small, say $K < 10^{-3}$, is to run lasso. A nice way to "blend" ridge regression and lasso is to call the elastic net method with $K = C(1-\theta)$ and $\tau = C\theta$, where $0 \leq \theta < 1$ and $C > 0$.

Running the elastic net method on the data set $(X14, y14)$ of Section 19.5 with $K = \tau = 0.5$ shows absolutely no difference, but the reader should conduct more experiments to see how elastic net behaves as K and τ are varied (the best way to do this is to use θ as explained above). Here is an example involving a data set $(X20, y20)$ where $X20$ is a 200×30 matrix generated as follows:

```
X20 = randn(50,30);
ww20 = [0; 2; 0; -3; 0; -4; 1; 0; 2; 0; 2; 3; 0; -5; 6; 0; 1;
2; 0; 10; 0; 0; 3; 4; 5; 0; 0; -6; -8; 0];
y20 = X20*ww20 + randn(50,1)*0.1 + 5;
```

Running our program with $K = 0.01$ and $K = 0.99$, and then with $K = 0.99$ and $K = 0.01$, we get the following weight vectors (in the left column is the weight vector corresponding to $K = 0.01$ and $K = 0.99$):

```
0.0254        0.2007
1.9193        2.0055
0.0766        0.0262
-3.0014       -2.8008
0.0512        0.0089
-3.8815       -3.7670
0.9591        0.8552
-0.0086       -0.3243
1.9576        1.9080
-0.0077       -0.1041
1.9881        2.0566
2.9223        2.8346
-0.0046       -0.0832
-4.9989       -4.8332
5.8640        5.4598
-0.0207       -0.2141
0.8285        0.8585
1.9310        1.8559
0.0046        0.0413
9.9232        9.4836
-0.0216       0.0303
0.0453        -0.0193
2.9384        3.0004
4.0525        3.9753
4.8723        4.6530
0.0767        0.1192
```

```
0.0132          -0.0203
-5.9750         -5.7537
-7.9764         -7.7594
-0.0054         0.0528
```

Generally, the numbers in the left column, which are more "lasso-like," have clearer zeros and nonzero values closer to those of the weight vector $ww20$ that was used to create the data set. The value of b corresponding to the first call is $b = 5.1372$, and the value of b corresponding to the second call is $b = 5.208$.

We have observed that lasso seems to converge much faster than elastic net when $K < 10^{-3}$. For example, running the above data set with $K = 10^{-3}$ and $\tau = 0.999$ requires 140520 steps to achieve primal and dual residuals less than 10^{-7}, but lasso only takes 86 steps to achieve the same degree of convergence. We observed that the larger K is the faster is the convergence. This could be attributed to the fact that the matrix P becomes more "positive definite." Another factor is that ADMM for lasso solves an $n \times n$ linear system, but ADMM for elastic net solves a $2(n+m) \times 2(n+m)$ linear system. So even though elastic net does not suffer from some of the undesirable properties of ridge regression and lasso, it appears to have a slower convergence rate, in fact much slower when K is small (say $K < 10^{-3}$). It also appears that elastic net may be too expensive a choice if m is much larger than n. Further investigations are required to gain a better understanding of the convergence issue.

19.7 Summary

The main concepts and results of this chapter are listed below:

- Ridge regression.
- Kernel ridge regression.
- Kernel functions.
- Lasso regression.
- Elastic net regression.

19.8 Problems

Problem 19.1. Check the formula

$$(X^\top X + KI_n)^{-1}X^\top = X^\top(XX^\top + KI_m)^{-1},$$

stated in Section 19.1.

Problem 19.2. Implement the ridge regression method described in Section 19.1 in `Matlab`. Also implement ridge regression with intercept and compare solving Program (**DRR3**) and Program (**RR6′**) using centered data.

Problem 19.3. Implement the ridge regression with intercept method (**RR3b**) in `Matlab` and compare it with solving (**RR6′**) using centered data.

Problem 19.4. Verify that (**lasso3**) is equivalent to (**lasso2**) applied to the centered data $\widehat{y} = y - \overline{y}\mathbf{1}$ and $\widehat{X} = X - \overline{X}$.

Problem 19.5. Verify the formulae obtained for the kernel ridge regression program (**KRR6′**).

Problem 19.6. Implement in `Matlab` and test (**lasso3**) for various values of ρ and τ. Write a program to plot the coordinates of w as a function of τ. Compare the behavior of lasso with ridge regression (**RR6′**), (**RR3**b) (b penalized), and with least squares.

Problem 19.7. Check the details of the derivation of the dual of elastic net.

Problem 19.8. Write a `Matlab` program, solving the dual of elastic net; use inspiration from Section 20.3. Run tests to compare the behavior of ridge regression, lasso, and elastic net.

Problem 19.9. Prove that if an optimal solution exists for the elastic net method, then it is unique.

Problem 19.10. Prove that the matrix

$$P = \begin{pmatrix} I_n & -I_n & -X^\top & X^\top \\ -I_n & I_n & X^\top & -X^\top \\ -X & X & XX^\top + KI_m & -XX^\top - KI_m \\ X & -X & -XX^\top - KI_m & XX^\top + KI_m \end{pmatrix}$$

is almost positive definite, in the sense that

$$\begin{pmatrix} \beta_+^\top & \beta_-^\top & \mu_+^\top & \mu_-^\top \end{pmatrix} P \begin{pmatrix} \beta_+ \\ \beta_- \\ \mu_+ \\ \mu_- \end{pmatrix} = 0$$

if and only if $\beta_+ = \beta_-$ and $\mu_+ = \mu_-$, that is, $\beta = 0$ and $\mu = 0$.

Chapter 20

ν-SV Regression

20.1 ν-SV Regression; Derivation of the Dual

Let $\{(x_1, y_1), \ldots, (x_m, y_m)\}$ be a set of observed data usually called a set of *training data*, with $x_i \in \mathbb{R}^n$ and $y_i \in \mathbb{R}$. As in Chapter 19, we form the $m \times n$ matrix X where the row vectors $x_i^\top$ are the *rows* of X. Our goal is to learn an affine function f of the form $f(x) = x^\top w + b$ that fits the set of training data, but does not penalize errors below some given $\epsilon \geq 0$. Geometrically, we view the pairs (x_i, y_i) are points in $\mathbb{R}^{n+1}$, and we try to fit a hyperplane $H_{w,b}$ of equation

$$(w^\top \;\; -1)\begin{pmatrix} x \\ z \end{pmatrix} + b = w^\top x - z + b = 0$$

that best fits the set of points (x_i, y_i) (where $(x, z) \in \mathbb{R}^{n+1}$). We seek an $\epsilon > 0$ such that most points (x_i, y_i) are inside the slab (or tube) of width 2ϵ bounded by the hyperplane $H_{w,b-\epsilon}$ of equation

$$(w^\top \;\; -1)\begin{pmatrix} x \\ z \end{pmatrix} + b - \epsilon = w^\top x - z + b - \epsilon = 0$$

and the hyperplane $H_{w,b+\epsilon}$ of equation

$$(w^\top \;\; -1)\begin{pmatrix} x \\ z \end{pmatrix} + b + \epsilon = w^\top x - z + b + \epsilon = 0.$$

Observe that the hyperplanes $H_{w,b-\epsilon}, H_{w,b}$ and $H_{w,b+\epsilon}$ intersect the z-axis when $x = 0$ for the values $(b - \epsilon, b, b + \epsilon)$. Since $\epsilon \geq 0$, the hyperplane $H_{w,b-\epsilon}$ is below the hyperplane $H_{w,b}$ which is below the hyperplane $H_{w,b+\epsilon}$. We refer to the lower hyperplane $H_{w,b-\epsilon}$ as the *blue margin*, to the upper hyperplane $H_{w,b+\epsilon}$ as the *red margin*, and to the hyperplane $H_{w,b}$ as the *best fit hyperplane*. Also note that since the term $-z$ appears in the equations of these hyperplanes, points for which $w^\top x - z + b \leq 0$ are *above* the hyperplane $H_{w,b}$, and points for which $w^\top x - z + b \geq 0$ are *below* the hyperplane $H_{w,b}$ (and similarly for $H_{w,b-\epsilon}$ and $H_{b+\epsilon}$). The region bounded by the hyperplanes $H_{w,b-\epsilon}$ and $H_{b+\epsilon}$ (which contains the best fit hyperplane $H_{w,b}$) is called the *ϵ-slab*.

We also allow *errors* by allowing the point (x_i, y_i) to be outside of the ϵ-slab but in the slab between the hyperplane $H_{w,b-\epsilon-\xi_i}$ of equation

$$(w^\top \ -1)\begin{pmatrix} x \\ z \end{pmatrix} + b - \epsilon - \xi_i = w^\top x - z + b - \epsilon - \xi_i = 0$$

for some $\xi_i > 0$ (which is below the blue margin hyperplane $H_{w,b-\epsilon}$) and the hyperplane $H_{w,b+\epsilon+\xi'_i}$ of equation

$$(w^\top \ -1)\begin{pmatrix} x \\ z \end{pmatrix} + b + \epsilon + \xi'_i = w^\top x_i - z + b + \epsilon + \xi'_i = 0$$

for some $\xi'_i > 0$ (which is above the red margin hyperplane $H_{w,b+\epsilon}$), so that $w^\top x_i - y_i + b - \epsilon - \xi_i \leq 0$ and $w^\top x_i - y_i + b + \epsilon + \xi'_i \geq 0$, that is,

$$\begin{aligned} f(x) - y_i &= w^\top x_i + b - y_i \leq \epsilon + \xi_i, \\ -(f(x) - y_i) &= -w^\top x_i - b + y_i \leq \epsilon + \xi'_i. \end{aligned}$$

Our goal is to minimize ϵ and the errors ξ_i and ξ'_i. See Figure 20.1. The trade off between the size of ϵ and the size of the slack variables ξ_i and ξ'_i is achieved by using two constants $\nu \geq 0$ and $C > 0$. The method of *ν-support vector regression*, for short *ν-SV regression*, is specified by the following minimization problem:

Program ν-SV Regression:

$$\text{minimize} \quad \frac{1}{2} w^\top w + C\left(\nu\epsilon + \frac{1}{m}\sum_{i=1}^{m}(\xi_i + \xi'_i)\right)$$

subject to

$$\begin{aligned} & w^\top x_i + b - y_i \leq \epsilon + \xi_i, \quad \xi_i \geq 0 \qquad i = 1, \ldots, m \\ & - w^\top x_i - b + y_i \leq \epsilon + \xi'_i, \quad \xi'_i \geq 0 \qquad i = 1, \ldots, m \\ & \epsilon \geq 0, \end{aligned}$$

minimizing over the variables w, b, ϵ, ξ, and ξ'. The constraints are affine. The problem is to minimize ϵ and the errors ξ_i, ξ'_i so that the ℓ^1-error is "squeezed down" to zero as much as possible, in the sense that the right-hand side of the inequality

$$\sum_{i=1}^{m} |y_i - x_i^\top w - b| \leq m\epsilon + \sum_{i=1}^{m} \xi_i + \sum_{i=1}^{m} \xi'_i$$

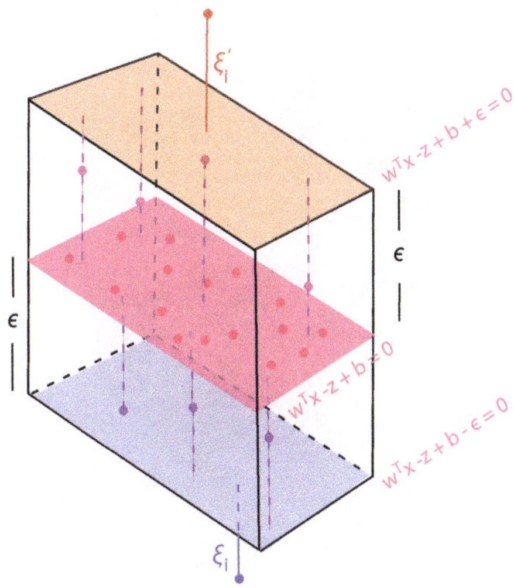

Fig. 20.1 The ϵ-slab around the graph of the best fit affine function $f(x) = x^\top w + b$.

is as small as possible. As shown by Figure 20.2, the region associated with the constraint $w^\top x_i - z + b \leq \epsilon$ contains the ϵ-slab.

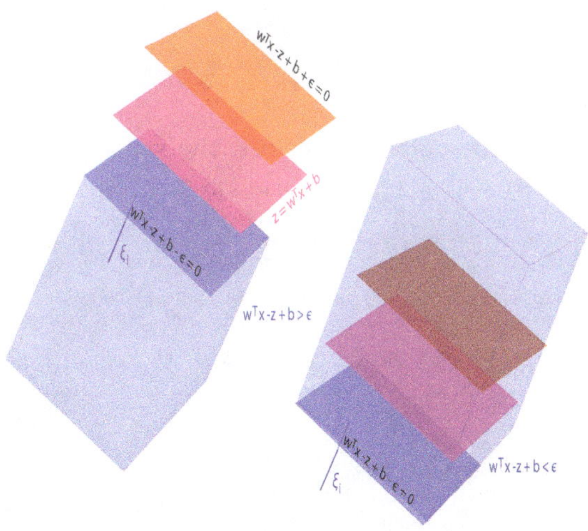

Fig. 20.2 The two blue half spaces associated with the hyperplane $w^\top x_i - z + b = \epsilon$.

Fig. 20.3 The two red half spaces associated with the hyperplane $w^\top x_i - z + b = -\epsilon$.

Similarly, as illustrated by Figure 20.3, the region associated with the constraint $w^\top x_i - z + b \geq -\epsilon$, equivalently $-w^\top x_i + z - b \leq \epsilon$, also contains the ϵ-slab.

Observe that if we require $\epsilon = 0$, then the problem is equivalent to minimizing

$$\|y - Xw - b\mathbf{1}\|_1 + \frac{1}{2} w^\top w.$$

Thus it appears that the above problem is the version of Program (**RR3**) (see Section 19.2) in which the ℓ^2-norm of $y - Xw - b\mathbf{1}$ is replaced by its ℓ^1-norm. This a sort of "dual" of lasso (see Section 19.5) where $(1/2)w^\top w = (1/2)\|w\|_2^2$ is replaced by $\tau \|w\|_1$, and $\|y - Xw - b\mathbf{1}\|_1$ is replaced by $\|y - Xw - b\mathbf{1}\|_2^2$.

Proposition 20.1. *For any optimal solution, the equations*

$$\xi_i \xi_i' = 0, \quad i = 1, \ldots, m \tag{$\xi\xi'$}$$

hold. If $\epsilon > 0$, then the equations

$$\begin{aligned} w^\top x_i + b - y_i &= \epsilon + \xi_i \\ -w^\top x_i - b + y_i &= \epsilon + \xi_i' \end{aligned}$$

cannot hold simultaneously.

Proof. For an optimal solution we have

$$-\epsilon - \xi_i' \leq w^\top x_i + b - y_i \leq \epsilon + \xi_i.$$

If $w^\top x_i + b - y_i \geq 0$, then $\xi_i' = 0$ since the inequality

$$-\epsilon - \xi_i' \leq w^\top x_i + b - y_i$$

is trivially satisfied (because $\epsilon, \xi_i' \geq 0$), and if $w^\top x_i + b - y_i \leq 0$, then similarly $\xi_i = 0$. See Figure 20.4.

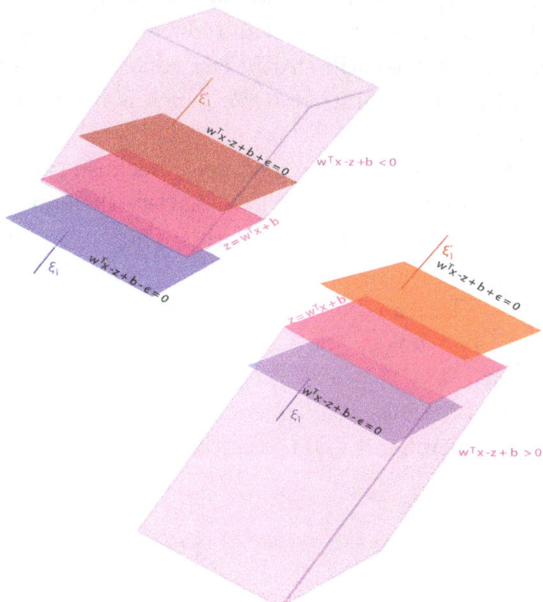

Fig. 20.4 The two pink open half spaces associated with the hyperplane $w^\top x_i - z + b = 0$. Note, $\xi_i > 0$ is outside of the half space $w^\top x_i - z + b - \epsilon < 0$, and $\xi_i' > 0$ is outside of the half space $w^\top x_i - z + b + \epsilon > 0$.

Observe that the equations

$$\begin{aligned} w^\top x_i + b - y_i &= \epsilon + \xi_i \\ -w^\top x_i - b + y_i &= \epsilon + \xi_i' \end{aligned}$$

can only hold simultaneously if

$$\epsilon + \xi_i = -\epsilon - \xi',$$

that is,

$$2\epsilon + \xi_i + \xi_i' = 0,$$

and since $\epsilon, \xi_i, \xi_i' \geq 0$, this can happen only if $\epsilon = \xi_i = \xi_i' = 0$, and then

$$w^\top x_i + b = y_i.$$

In particular, if $\epsilon > 0$, then the equations

$$\begin{aligned} w^\top x_i + b - y_i &= \epsilon + \xi_i \\ -w^\top x_i - b + y_i &= \epsilon + \xi_i' \end{aligned}$$

cannot hold simultaneously. □

Observe that if $\nu > 1$, then an optimal solution of the above program must yield $\epsilon = 0$. Indeed, if $\epsilon > 0$, we can reduce it by a small amount $\delta > 0$ and increase $\xi_i + \xi_i'$ by δ to still satisfy the constraints, but the objective function changes by the amount $-\nu\delta + \delta$, which is negative since $\nu > 1$, so $\epsilon > 0$ is not optimal.

Driving ϵ to zero is not the intended goal, because typically the data is not noise free so very few pairs (x_i, y_i) will satisfy the equation $w^\top x_i + b = y_i$, and then many pairs (x_i, y_i) will correspond to an error ($\xi_i > 0$ or $\xi_i' > 0$). Thus, *typically we assume that* $0 < \nu \leq 1$.

To construct the Lagrangian, we assign Lagrange multipliers $\lambda_i \geq 0$ to the constraints $w^\top x_i + b - y_i \leq \epsilon + \xi_i$, Lagrange multipliers $\mu_i \geq 0$ to the constraints $-w^\top x_i - b + y_i \leq \epsilon + \xi_i'$, Lagrange multipliers $\alpha_i \geq 0$ to the constraints $\xi_i \geq 0$, Lagrange multipliers $\beta_i \geq 0$ to the constraints $\xi_i' \geq 0$, and the Lagrange multiplier $\gamma \geq 0$ to the constraint $\epsilon \geq 0$. The Lagrangian is

$$\begin{aligned} L(w, b, \lambda, \mu, \gamma, \xi, \xi', \epsilon, \alpha, \beta) &= \frac{1}{2} w^\top w + C\bigg(\nu\epsilon + \frac{1}{m}\sum_{i=1}^m (\xi_i + \xi_i')\bigg) - \gamma\epsilon \\ &\quad - \sum_{i=1}^m (\alpha_i \xi_i + \beta_i \xi_i') + \sum_{i=1}^m \lambda_i (w^\top x_i + b - y_i - \epsilon - \xi_i) \\ &\quad + \sum_{i=1}^m \mu_i (-w^\top x_i - b + y_i - \epsilon - \xi_i'). \end{aligned}$$

The Lagrangian can also be written as

$$\begin{aligned} L(w, b, \lambda, \mu, \gamma, \xi, \xi', \epsilon, \alpha, \beta) &= \frac{1}{2} w^\top w + w^\top \bigg(\sum_{i=1}^m (\lambda_i - \mu_i) x_i\bigg) \\ &\quad + \epsilon\bigg(C\nu - \gamma - \sum_{i=1}^m (\lambda_i + \mu_i)\bigg) + \sum_{i=1}^m \xi_i \bigg(\frac{C}{m} - \lambda_i - \alpha_i\bigg) \\ &\quad + \sum_{i=1}^m \xi_i' \bigg(\frac{C}{m} - \mu_i - \beta_i\bigg) + b\bigg(\sum_{i=1}^m (\lambda_i - \mu_i)\bigg) - \sum_{i=1}^m (\lambda_i - \mu_i) y_i. \end{aligned}$$

To find the dual function $G(\lambda, \mu, \gamma, \alpha, \beta)$, we minimize $L(w, b, \lambda, \mu, \gamma, \xi, \xi', \epsilon, \alpha, \beta)$ with respect to the primal variables w, ϵ, b, ξ and ξ'. Observe that the Lagrangian is convex, and since $(w, \epsilon, \xi, \xi', b) \in \mathbb{R}^n \times \mathbb{R} \times \mathbb{R}^m \times \mathbb{R}^m \times \mathbb{R}$, a convex open set, by Theorem 4.5, the Lagrangian has a minimum iff $\nabla L_{w,\epsilon,b,\xi,\xi'} = 0$, so we compute the gradient $\nabla L_{w,\epsilon,b,\xi,\xi'}$. We obtain

$$\nabla L_{w,\epsilon,b,\xi,\xi'} = \begin{pmatrix} w + \sum_{i=1}^m (\lambda_i - \mu_i) x_i \\ C\nu - \gamma - \sum_{i=1}^m (\lambda_i + \mu_i) \\ \sum_{i=1}^m (\lambda_i - \mu_i) \\ \frac{C}{m} - \lambda - \alpha \\ \frac{C}{m} - \mu - \beta \end{pmatrix},$$

where

$$\left(\frac{C}{m} - \lambda - \alpha\right)_i = \frac{C}{m} - \lambda_i - \alpha_i, \quad \text{and} \quad \left(\frac{C}{m} - \mu - \beta\right)_i = \frac{C}{m} - \mu_i - \beta_i.$$

Consequently, if we set $\nabla L_{w,\epsilon,b,\xi,\xi'} = 0$, we obtain the equations

$$w = \sum_{i=1}^m (\mu_i - \lambda_i) x_i = X^\top (\mu - \lambda), \tag{$*_w$}$$

$$C\nu - \gamma - \sum_{i=1}^m (\lambda_i + \mu_i) = 0$$

$$\sum_{i=1}^m (\lambda_i - \mu_i) = 0$$

$$\frac{C}{m} - \lambda - \alpha = 0, \quad \frac{C}{m} - \mu - \beta = 0.$$

Substituting the above equations in the second expression for the Lagrangian, we find that the dual function G is independent of the variables γ, α, β and is given by

$$G(\lambda, \mu) = -\frac{1}{2} \sum_{i,j=1}^m (\lambda_i - \mu_i)(\lambda_j - \mu_j) x_i^\top x_j - \sum_{i=1}^m (\lambda_i - \mu_i) y_i$$

if

$$\sum_{i=1}^m \lambda_i - \sum_{i=1}^m \mu_i = 0$$

$$\sum_{i=1}^m \lambda_i + \sum_{i=1}^m \mu_i + \gamma = C\nu$$

$$\lambda + \alpha = \frac{C}{m}, \quad \mu + \beta = \frac{C}{m},$$

and $-\infty$ otherwise.

The dual program is obtained by maximizing $G(\alpha, \mu)$ or equivalently by minimizing $-G(\alpha, \mu)$, over $\alpha, \mu \in \mathbb{R}^m_+$. Taking into account the fact that $\alpha, \beta \geq 0$ and $\gamma \geq 0$, we obtain the following dual program:

Dual Program for ν-SV Regression:

$$\text{minimize} \quad \frac{1}{2} \sum_{i,j=1}^{m} (\lambda_i - \mu_i)(\lambda_j - \mu_j) x_i^\top x_j + \sum_{i=1}^{m} (\lambda_i - \mu_i) y_i$$

subject to

$$\sum_{i=1}^{m} \lambda_i - \sum_{i=1}^{m} \mu_i = 0$$
$$\sum_{i=1}^{m} \lambda_i + \sum_{i=1}^{m} \mu_i \leq C\nu$$
$$0 \leq \lambda_i \leq \frac{C}{m}, \quad 0 \leq \mu_i \leq \frac{C}{m}, \quad i = 1, \ldots, m,$$

minimizing over α and μ.

Solving the dual program (for example, using ADMM, see Section 20.3) does not determine b, and for this we use the KKT conditions. The KKT conditions (for the primal program) are

$$\begin{aligned}
\lambda_i(w^\top x_i + b - y_i - \epsilon - \xi_i) &= 0, \quad i = 1, \ldots, m \\
\mu_i(-w^\top x_i - b + y_i - \epsilon - \xi'_i) &= 0, \quad i = 1, \ldots, m \\
\gamma\epsilon &= 0 \\
\alpha_i \xi_i &= 0, \quad i = 1, \ldots, m \\
\beta_i \xi'_i &= 0, \quad i = 1, \ldots, m.
\end{aligned}$$

If $\epsilon > 0$, since the equations

$$\begin{aligned}
w^\top x_i + b - y_i &= \epsilon + \xi_i \\
-w^\top x_i - b + y_i &= \epsilon + \xi'_i
\end{aligned}$$

cannot hold simultaneously, we must have

$$\lambda_i \mu_i = 0, \quad i = 1, \ldots, m. \tag{$\lambda\mu$}$$

From the equations

$$\lambda_i + \alpha_i = \frac{C}{m}, \quad \mu_i + \beta_i = \frac{C}{m}, \quad \alpha_i \xi_i = 0, \quad \beta_i \xi'_i = 0,$$

we get the equations

$$\left(\frac{C}{m} - \lambda_i\right) \xi_i = 0, \quad \left(\frac{C}{m} - \mu_i\right) \xi'_i = 0, \quad i = 1, \ldots, m. \tag{$*$}$$

Suppose we have optimal solution with $\epsilon > 0$. Using the above equations and the fact that $\lambda_i \mu_i = 0$ we obtain the following classification of the points x_i in terms of λ and μ.

(1) $0 < \lambda_i < C/m$. By $(*)$, $\xi_i = 0$, so the equation $w^\top x_i + b - y_i = \epsilon$ holds and x_i is on the blue margin hyperplane $H_{w,b-\epsilon}$. See Figure 20.5.
(2) $0 < \mu_i < C/m$. By $(*)$, $\xi'_i = 0$, so the equation $-w^\top x_i - b + y_i = \epsilon$ holds and x_i is on the red margin hyperplane $H_{w,b+\epsilon}$. See Figure 20.5.
(3) $\lambda_i = C/m$. By $(\lambda\mu)$, $\mu_i = 0$, and by $(*)$, $\xi'_i = 0$. Thus we have

$$\begin{aligned} w^\top x_i + b - y_i &= \epsilon + \xi_i \\ -w^\top x_i - b + y_i &\leq \epsilon. \end{aligned}$$

The second inequality is equivalent to $-\epsilon \leq w^\top x_i + b - y_i$, and since $\epsilon > 0$ and $\xi_i \geq 0$ it is trivially satisfied. If $\xi_i = 0$, then x_i is on the blue margin $H_{w,b-\epsilon}$, else x_i is an error and it lies in the open half-space bounded by the blue margin $H_{w,b-\epsilon}$ and not containing the best fit hyperplane $H_{w,b}$ (it is outside of the ϵ-slab). See Figure 20.5.
(4) $\mu_i = C/m$. By $(\lambda\mu)$, $\lambda_i = 0$, and by $(*)$, $\xi_i = 0$. Thus we have

$$\begin{aligned} w^\top x_i + b - y_i &\leq \epsilon \\ -w^\top x_i - b + y_i &= \epsilon + \xi'_i. \end{aligned}$$

The second equation is equivalent to $w^\top x_i + b - y_i = -\epsilon - \xi'_i$, and since $\epsilon > 0$ and $\xi'_i \geq 0$, the first inequality it is trivially satisfied. If $\xi'_i = 0$, then x_i is on the red margin $H_{w,b+\epsilon}$, else x_i is an error and it lies in the open half-space bounded by the red margin $H_{w,b-\epsilon}$ and not containing the best fit hyperplane $H_{w,b}$ (it is outside of the ϵ-slab). See Figure 20.5.
(5) $\lambda_i = 0$ and $\mu_i = 0$. By $(*)$, $\xi_i = 0$ and $\xi'_i = 0$, so we have

$$\begin{aligned} w^\top x_i + b - y_i &\leq \epsilon \\ -w^\top x_i - b + y_i &\leq \epsilon, \end{aligned}$$

that is

$$-\epsilon \leq w^\top x_i + b - y_i \leq \epsilon.$$

If $w^\top x_i + b - y_i = \epsilon$, then x_i is on the blue margin, and if $w^\top x_i + b - y_i = -\epsilon$, then x_i is on the red margin. If $-\epsilon < w^\top x_i + b - y_i < \epsilon$, then x_i is strictly inside of the ϵ-slab (bounded by the blue margin and the red margin). See Figure 20.6.

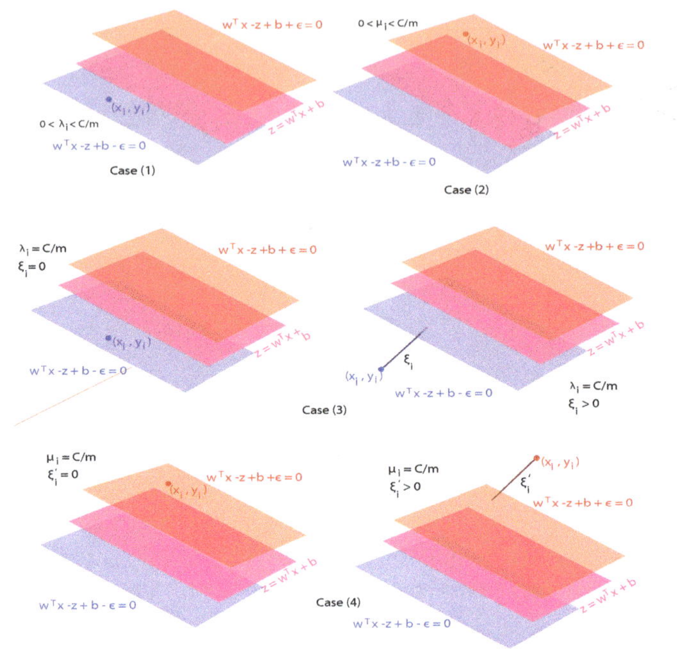

Fig. 20.5 Classifying x_i in terms of nonzero λ and μ.

The above classification shows that the point x_i is an error iff $\lambda_i = C/m$ and $\xi_i > 0$ or or $\mu_i = C/m$ and $\xi_i' > 0$.

As in the case of SVM (see Section 14.6) we define support vectors as follows.

Definition 20.1. A vector x_i such that either $w^\top x_i + b - y_i = \epsilon$ (which implies $\xi_i = 0$) or $-w^\top x_i - b + y_i = \epsilon$ (which implies $\xi_i' = 0$) is called a *support vector*. Support vectors x_i such that $0 < \lambda_i < C/m$ and support vectors x_j such that $0 < \mu_j < C/m$ are *support vectors of type 1*. Support vectors of type 1 play a special role so we denote the sets of indices associated with them by

$$
\begin{aligned}
I_\lambda &= \{i \in \{1, \ldots, m\} \mid 0 < \lambda_i < C/m\} \\
I_\mu &= \{j \in \{1, \ldots, m\} \mid 0 < \mu_j < C/m\}.
\end{aligned}
$$

We denote their cardinalities by $numsvl_1 = |I_\lambda|$ and $numsvm_1 = |I_\mu|$. Support vectors x_i such that $\lambda_i = C/m$ and support vectors x_j such that

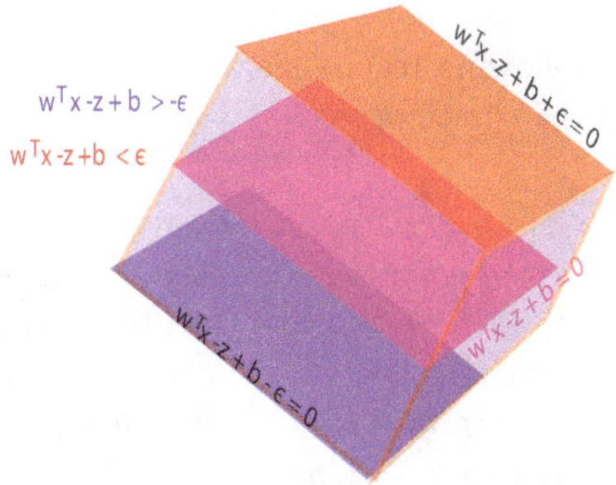

Fig. 20.6 The closed ϵ- tube associated with zero multiplier classification, namely $\lambda_i = 0$ and $\mu_i = 0$.

$\mu_j = C/m$ are *support vectors of type 2*. Support vectors for which $\lambda_i = \mu_i = 0$ are called *exceptional support vectors*.

The following definition also gives a useful classification criterion.

Definition 20.2. A point x_i such that either $\lambda_i = C/m$ or $\mu_i = C/m$ is said to *fail the margin*. The sets of indices associated with the vectors failing the margin are denoted by

$$K_\lambda = \{i \in \{1, \ldots, m\} \mid \lambda_i = C/m\}$$
$$K_\mu = \{j \in \{1, \ldots, m\} \mid \mu_j = C/m\}.$$

We denote their cardinalities by $p_f = |K_\lambda|$ and $q_f = |K_\mu|$.

Vectors u_i such that $\lambda_i > 0$ and vectors v_j such that $\mu_j > 0$ are said to *have margin at most* ϵ. A point x_i such that either $\lambda_i > 0$ or $\mu_i > 0$ is said to have *margin at most* ϵ. The sets of indices associated with these vectors are denoted by

$$I_{\lambda>0} = \{i \in \{1, \ldots, m\} \mid \lambda_i > 0\}$$
$$I_{\mu>0} = \{j \in \{1, \ldots, m\} \mid \mu_j > 0\}.$$

We denote their cardinalities by $p_m = |I_{\lambda>0}|$ and $q_m = |I_{\mu>0}|$.

Points that fail the margin and are not on the boundary of the ϵ-slab lie outside the closed ϵ-slab, so they are errors, also called *outliers*; they correspond to $\xi_i > 0$ or $\xi'_i > 0$.

Observe that we have the equations $I_\lambda \cup K_\lambda = I_{\lambda>0}$ and $I_\mu \cup K_\mu = I_{\mu>0}$, and the inequalities $p_f \leq p_m$ and $q_f \leq q_m$.

We also have the following results showing that p_f, q_f, p_m and q_m have a direct influence on the choice of ν.

Proposition 20.2.

(1) Let p_f be the number of points x_i such that $\lambda_i = C/m$, and let q_f be the number of points x_i such that $\mu_i = C/m$. We have $p_f, q_f \leq (m\nu)/2$.
(2) Let p_m be the number of points x_i such that $\lambda_i > 0$, and let q_m be the number of points x_i such that $\mu_i > 0$. We have $p_m, q_m \geq (m\nu)/2$.
(3) If $p_f \geq 1$ or $q_f \geq 1$, then $\nu \geq 2/m$.

Proof. (1) Recall that for an optimal solution with $w \neq 0$ and $\epsilon > 0$ we have $\gamma = 0$, so we have the equations

$$\sum_{i=1}^{m} \lambda_i = \frac{C\nu}{2} \quad \text{and} \quad \sum_{j=1}^{m} \mu_j = \frac{C\nu}{2}.$$

If there are p_f points such that $\lambda_i = C/m$, then

$$\frac{C\nu}{2} = \sum_{i=1}^{m} \lambda_i \geq p_f \frac{C}{m},$$

so

$$p_f \leq \frac{m\nu}{2}.$$

A similar reasoning applies if there are q_f points such that $\mu_i = C/m$, and we get

$$q_f \leq \frac{m\nu}{2}.$$

(2) If $I_{\lambda>0} = \{i \in \{1, \ldots, m\} \mid \lambda_i > 0\}$ and $p_m = |I_{\lambda>0}|$, then

$$\frac{C\nu}{2} = \sum_{i=1}^{m} \lambda_i = \sum_{i \in I_{\lambda>0}} \lambda_i,$$

and since $\lambda_i \leq C/m$, we have

$$\frac{C\nu}{2} = \sum_{i \in I_{\lambda>0}} \lambda_i \leq p_m \frac{C}{m},$$

which yields

$$p_m \geq \frac{\nu m}{2}.$$

A similar reasoning applies if $\mu_i > 0$.

(3) This follows immediately from (1). ◻

Proposition 20.2 yields bounds on ν, namely

$$\max\left\{\frac{2p_f}{m}, \frac{2q_f}{m}\right\} \leq \nu \leq \min\left\{\frac{2p_m}{m}, \frac{2q_m}{m}\right\},$$

with $p_f \leq p_m$, $q_f \leq q_m$, $p_f + q_f \leq m$ and $p_m + q_m \leq m$. Also, $p_f = q_f = 0$ means that the ϵ-slab is wide enough so that there are no errors (no points strictly outside the slab).

Observe that a small value of ν keeps p_f and q_f small, which is achieved if the ϵ-slab is wide. A large value of ν allows p_m and q_m to be fairly large, which is achieved if the ϵ-slab is narrow. Thus the smaller ν is, the wider the ϵ-slab is, and the larger ν is, the narrower the ϵ-slab is.

20.2 Existence of Support Vectors

We now consider the issue of the existence of support vectors. We will show that in the generic case, for any optimal solution for which $\epsilon > 0$, there is some support vector on the blue margin *and* some support vector on the red margin. Here generic means that there is an optimal solution for some $\nu < (m-1)/m$.

If the data set (X, y) is well fit by some affine function $f(x) = w^\top x + b$, in the sense that for many pairs (x_i, y_i) we have $y_i = w^\top x_i + b$ and the ℓ^1-error

$$\sum_{i=1}^{m} |w^\top x_i + b - y_i|$$

is small, then an optimal solution may have $\epsilon = 0$. Geometrically, many points (x_i, y_i) belong to the hyperplane $H_{w,b}$. The situation in which $\epsilon = 0$ corresponds to minimizing the ℓ^1-error with a quadratic penalization of b. This is a sort of dual of lasso. The fact that the affine function $f(x) = w^\top x + b$ fits perfectly many points corresponds to the fact that an ℓ^1-minimization tends to encourage sparsity. In this case, if C is chosen too small, it is possible that all points are errors (although "small") and there are no support vectors. But if C is large enough, the solution will be sparse and there will be *many* support vectors on the hyperplane $H_{w,b}$.

Let $E_\lambda = \{i \in \{1, \ldots, m\} \mid \xi_i > 0\}$, $E_\mu = \{j \in \{1, \ldots, m\} \mid \xi'_j > 0\}$, $p_{sf} = |E_\lambda|$ and $q_{sf} = |E_\mu|$. Obviously, E_λ and E_μ are disjoint.

Given any real numbers u, v, x, y, if $\max\{u, v\} < \min\{x, y\}$, then $u < x$ and $v < y$. This is because $u, v \leq \max\{u, v\} < \min\{x, y\} \leq x, y$.

Proposition 20.3. *If $\nu < (m-1)/m$, then $p_f < \lfloor m/2 \rfloor$ and $q_f < \lfloor m/2 \rfloor$.*

Proof. By Proposition 20.2, $\max\{2p_f/m, 2q_f/m\} \leq \nu$. If m is even, say $m = 2k$, then

$$2p_f/m = 2p_f/(2k) \leq \nu < (m-1)/m = (2k-1)/2k,$$

so $2p_f < 2k - 1$, which implies $p_f < k = \lfloor m/2 \rfloor$. A similar argument shows that $q_f < k = \lfloor m/2 \rfloor$.

If m is odd, say $m = 2k + 1$, then

$$2p_f/m = 2p_f/(2k+1) \leq \nu < (m-1)/m = 2k/(2k+1),$$

so $2p_f < 2k$, which implies $p_f < k = \lfloor m/2 \rfloor$. A similar argument shows that $q_f < k = \lfloor m/2 \rfloor$. □

Since $p_{sf} \leq p_f$ and $q_{sf} \leq q_f$, we also have $p_{sf} < \lfloor m/2 \rfloor$ and $q_{sf} < \lfloor m/2 \rfloor$. This implies that $\{1, \ldots, m\} - (E_\lambda \cup E_\mu)$ contains at least two elements and there are constraints corresponding to at least two $i \notin (E_\lambda \cup E_\mu)$ (in which case $\xi_i = \xi'_i = 0$), of the form

$$\begin{aligned} w^\top x_i + b - y_i &\leq \epsilon && i \notin (E_\lambda \cup E_\mu) \\ -w^\top x_i - b + y_i &\leq \epsilon && i \notin (E_\lambda \cup E_\mu). \end{aligned}$$

If $w^\top x_i + b - y_i = \epsilon$ for some $i \notin (E_\lambda \cup E_\mu)$ and $-w^\top x_j - b + y_j = \epsilon$ for some $j \notin (E_\lambda \cup E_\mu)$ with $i \neq j$, then we have a blue support vector and a red support vector. Otherwise, we show how to modify b and ϵ to obtain an optimal solution with a blue support vector and a red support vector.

Proposition 20.4. *For every optimal solution $(w, b, \epsilon, \xi, \xi')$ with $w \neq 0$ and $\epsilon > 0$, if*

$$\nu < (m-1)/m$$

and if either no x_i is a blue support vector or no x_i is a red support vector, then there is another optimal solution (for the same w) with some i_0 such that $\xi_{i_0} = 0$ and $w^\top x_{i_0} + b - y_{i_0} = \epsilon$, and there is some j_0 such that $\xi'_{j_0} = 0$ and $-w^\top x_{j_0} - b + y_{j_0} = \epsilon$; in other words, some x_{i_0} is a blue support vector and some x_{j_0} is a red support vector (with $i_0 \neq j_0$). If all points (x_i, y_i) that are not errors lie on one of the margin hyperplanes, then there is an optimal solution for which $\epsilon = 0$.

Proof. By Proposition 20.3 if $\nu < (m-1)/m$, then $p_f < \lfloor m/2 \rfloor$ and $q_f < \lfloor m/2 \rfloor$, so the following constraints hold:

$$\begin{aligned} w^\top x_i + b - y_i &= \epsilon + \xi_i && \xi_i > 0 && i \in E_\lambda \\ -w^\top x_j - b + y_j &= \epsilon + \xi'_j && \xi'_j > 0 && j \in E_\mu \\ w^\top x_i + b - y_i &\leq \epsilon && && i \notin (E_\lambda \cup E_\mu) \\ -w^\top x_i - b + y_i &\leq \epsilon && && i \notin (E_\lambda \cup E_\mu), \end{aligned}$$

where $|\{1, \ldots, m\} - (E_\lambda \cup E_\mu)| \geq 2$.

If our optimal solution does not have a blue support vector and a red support vector, then either $w^\top x_i + b - y_i < \epsilon$ for all $i \notin (E_\lambda \cup E_\mu)$ or $-w^\top x_i - b + y_i < \epsilon$ for all $i \notin (E_\lambda \cup E_\mu)$.

Case 1. We have

$$\begin{aligned} w^\top x_i + b - y_i &< \epsilon & i &\notin (E_\lambda \cup E_\mu) \\ -w^\top x_i - b + y_i &\leq \epsilon & i &\notin (E_\lambda \cup E_\mu). \end{aligned}$$

There are two subcases.

Case 1a. Assume that there is some $j \notin (E_\lambda \cup E_\mu)$ such that $-w^\top x_j - b + y_j = \epsilon$. Our strategy is to decrease ϵ and increase b by a small amount θ in such a way that some inequality $w^\top x_i + b - y_i < \epsilon$ becomes an equation for some $i \notin (E_\lambda \cup E_\mu)$. Geometrically, this amounts to raising the separating hyperplane $H_{w,b}$ and decreasing the width of the slab, keeping the red margin hyperplane unchanged. See Figure 20.7.

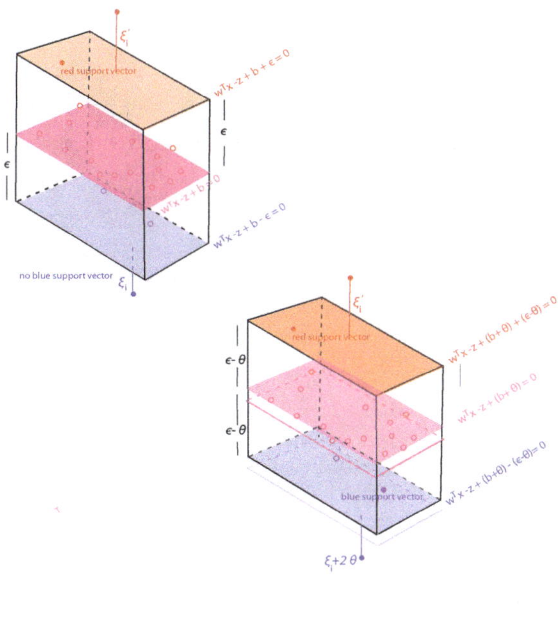

Fig. 20.7 In this illustration points within the ϵ-tube are denoted by open circles. In the original, upper left configuration, there is no blue support vector. By raising the pink separating hyperplane and decreasing the width of the slab, we end up with a blue support vector.

The inequalities imply that

$$-\epsilon \leq w^\top x_i + b - y_i < \epsilon.$$

Let us pick θ such that

$$\theta = (1/2)\min\{\epsilon - w^\top x_i - b + y_i \mid i \notin (E_\lambda \cup E_\mu)\}.$$

Our hypotheses imply that $\theta > 0$, and we have $\theta \leq \epsilon$, because $(1/2)(\epsilon - w^\top x_i - b + y_i) \leq \epsilon$ is equivalent to $\epsilon - w^\top x_i - b + y_i \leq 2\epsilon$ which is equivalent to $-w^\top x_i - b + y_i \leq \epsilon$, which holds for all $i \notin (E_\lambda \cup E_\mu)$ by hypothesis.

We can write

$$\begin{aligned}
w^\top x_i + b + \theta - y_i &= \epsilon - \theta + \xi_i + 2\theta & \xi_i > 0 \quad & i \in E_\lambda \\
-w^\top x_j - (b + \theta) + y_j &= \epsilon - \theta + \xi'_j & \xi'_j > 0 \quad & j \in E_\mu \\
w^\top x_i + b + \theta - y_i &\leq \epsilon - \theta & & i \notin (E_\lambda \cup E_\mu) \\
-w^\top x_i - (b + \theta) + y_i &\leq \epsilon - \theta & & i \notin (E_\lambda \cup E_\mu).
\end{aligned}$$

By hypothesis

$$-w^\top x_j - (b + \theta) + y_j = \epsilon - \theta \quad \text{for some } j \notin (E_\lambda \cup E_\mu)$$

and by the choice of θ,

$$w^\top x_i + b + \theta - y_i = \epsilon - \theta \quad \text{for some } i \notin (E_\lambda \cup E_\mu).$$

The value of $C > 0$ is irrelevant in the following argument so we may assume that $C = 1$. The new value of the objective function is

$$\begin{aligned}
\omega(\theta) &= \frac{1}{2} w^\top w + \nu(\epsilon - \theta) + \frac{1}{m}\left(\sum_{i \in E_\lambda} (\xi_i + 2\theta) + \sum_{j \in E_\mu} \xi'_j\right) \\
&= \frac{1}{2} w^\top w + \nu\epsilon + \frac{1}{m}\left(\sum_{i \in E_\lambda} \xi_i + \sum_{j \in E_\mu} \xi'_j\right) - \left(\nu - \frac{2p_{sf}}{m}\right)\theta.
\end{aligned}$$

By Proposition 20.2 we have

$$\max\left\{\frac{2p_f}{m}, \frac{2q_f}{m}\right\} \leq \nu$$

and $p_{sf} \leq p_f$ and $q_{sf} \leq q_f$, which implies that

$$\nu - \frac{2p_{sf}}{m} \geq 0, \tag{$*_1$}$$

and so $\omega(\theta) \leq \omega(0)$. If inequality $(*_1)$ is strict, then this contradicts the optimality of the original solution. Therefore, $\nu = 2p_{sf}/m$, $\omega(\theta) = \omega(0)$ and $(w, b + \theta, \epsilon - \theta, \xi + 2\theta, \xi')$ is an optimal solution such that

$$\begin{aligned}
w^\top x_i + b + \theta - y_i &= \epsilon - \theta \\
-w^\top x_j - (b + \theta) + y_j &= \epsilon - \theta
\end{aligned}$$

for some $i, j \notin (E_\lambda \cup E_\mu)$ with $i \neq j$.

Observe that the exceptional case in which $\theta = \epsilon$ may arise. In this case all points (x_i, y_i) that are not errors (strictly outside the ϵ-slab) are on the red margin hyperplane. This case can only arise if $\nu = 2p_{sf}/m$.

Case 1b. We have $-w^\top x_i - b + y_i < \epsilon$ for all $i \notin (E_\lambda \cup E_\mu)$. Our strategy is to decrease ϵ and increase the errors by a small θ in such a way that some inequality becomes an equation for some $i \notin (E_\lambda \cup E_\mu)$. Geometrically, this corresponds to decreasing the width of the slab, keeping the separating hyperplane unchanged. See Figures 20.8 and 20.9. Then we are reduced to *Case 1a* or *Case 2a*.

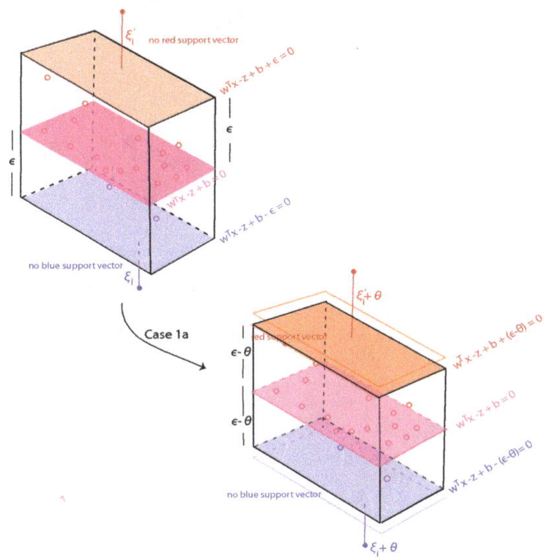

Fig. 20.8 In this illustration points within the ϵ-tube are denoted by open circles. In the original, upper left configuration, there is no blue support vector and no red support vector. By decreasing the width of the slab, we end up with a red support vector and reduce to Case 1a.

We have

$$
\begin{aligned}
w^\top x_i + b - y_i &= \epsilon + \xi_i & \xi_i &> 0 & i &\in E_\lambda \\
-w^\top x_j - b + y_j &= \epsilon + \xi_j' & \xi_j' &> 0 & j &\in E_\mu \\
w^\top x_i + b - y_i &< \epsilon & & & i &\notin (E_\lambda \cup E_\mu) \\
-w^\top x_i - b + y_i &< \epsilon & & & i &\notin (E_\lambda \cup E_\mu).
\end{aligned}
$$

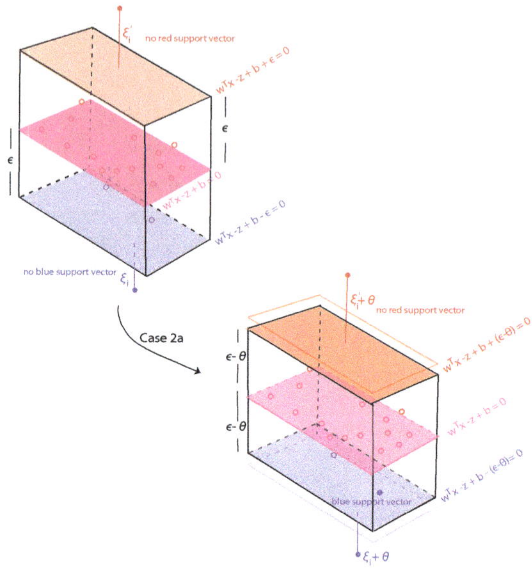

Fig. 20.9 In this illustration points within ϵ-tube are denoted by open circles. In the original, upper left configuration, there is no blue support vector and no red support vector. By decreasing the width of the slab, we end up with a blue support vector and reduce to Case 2a.

Let us pick θ such that

$$\theta = \min\{\epsilon - (w^\top x_i + b - y_i),\ \epsilon + w^\top x_i + b - y_i \mid i \notin (E_\lambda \cup E_\mu)\},$$

Our hypotheses imply that $0 < \theta < \epsilon$. We can write

$$\begin{aligned} w^\top x_i + b - y_i &= \epsilon - \theta + \xi_i + \theta & \xi_i &> 0 & i &\in E_\lambda \\ -w^\top x_j - b + y_j &= \epsilon - \theta + \xi'_j + \theta & \xi'_j &> 0 & j &\in E_\mu \\ w^\top x_i + b - y_i &\leq \epsilon - \theta & & & i &\notin (E_\lambda \cup E_\mu) \\ -w^\top x_i - b + y_i &\leq \epsilon - \theta & & & i &\notin (E_\lambda \cup E_\mu), \end{aligned}$$

and by the choice of θ, either

$$w^\top x_i + b - y_i = \epsilon - \theta \quad \text{for some } i \notin (E_\lambda \cup E_\mu)$$

or

$$-w^\top x_i - b + y_i = \epsilon - \theta \quad \text{for some } i \notin (E_\lambda \cup E_\mu).$$

The new value of the objective function is

$$\begin{aligned}\omega(\theta) &= \frac{1}{2}w^\top w + \nu(\epsilon - \theta) + \frac{1}{m}\bigg(\sum_{i\in E_\lambda}(\xi_i + \theta) + \sum_{j\in E_\mu}(\xi'_j + \theta)\bigg)\\ &= \frac{1}{2}w^\top w + \nu\epsilon + \frac{1}{m}\bigg(\sum_{i\in E_\lambda}\xi_i + \sum_{j\in E_\mu}\xi'_j\bigg) - \bigg(\nu - \frac{p_{sf}+q_{sf}}{m}\bigg)\theta.\end{aligned}$$

Since $\max\{2p_f/m, 2q_f/m\} \leq \nu$ implies that $(p_f+q_f)/m \leq \nu$ and $p_{sf} \leq p_f$, $q_{sf} \leq q_f$, we have

$$\nu - \frac{p_{sf}+q_{sf}}{m} \geq 0, \qquad (*_2)$$

and so $\omega(\theta) \leq \omega(0)$. If inequality $(*_2)$ is strict, then this contradicts the optimality of the original solution. Therefore, $\nu = (p_{sf}+q_{sf})/m$, $\omega(\theta) = \omega(0)$ and $(w, b, \epsilon - \theta, \xi + \theta, \xi' + \theta)$ is an optimal solution such that either

$$w^\top x_i + b - y_i = \epsilon - \theta \quad \text{for some } i \notin (E_\lambda \cup E_\mu)$$

or

$$-w^\top x_i - b + y_i = \epsilon - \theta \quad \text{for some } i \notin (E_\lambda \cup E_\mu).$$

We are now reduced to *Case 1a* or or *Case 2a*.

Case 2. We have

$$\begin{aligned} w^\top x_i + b - y_i &\leq \epsilon & i &\notin (E_\lambda \cup E_\mu)\\ -w^\top x_i - b + y_i &< \epsilon & i &\notin (E_\lambda \cup E_\mu).\end{aligned}$$

Again there are two subcases.

Case 2a. Assume that there is some $i \notin (E_\lambda \cup E_\mu)$ such that $w^\top x_i + b - y_i = \epsilon$. Our strategy is to decrease ϵ and decrease b by a small amount θ in such a way that some inequality $-w^\top x_j - b + y_j < \epsilon$ becomes an equation for some $j \notin (E_\lambda \cup E_\mu)$. Geometrically, this amounts to lowering the separating hyperplane $H_{w,b}$ and decreasing the width of the slab, keeping the blue margin hyperplane unchanged. See Figure 20.10.

The inequalities imply that

$$-\epsilon < w^\top x_i + b - y_i \leq \epsilon.$$

Let us pick θ such that

$$\theta = (1/2)\min\{\epsilon - (-w^\top x_i - b + y_i) \mid i \notin (E_\lambda \cup E_\mu)\}.$$

Our hypotheses imply that $\theta > 0$, and we have $\theta \leq \epsilon$, because $(1/2)(\epsilon - (-w^\top x_i - b + y_i)) \leq \epsilon$ is equivalent to $\epsilon - (-w^\top x_i - b + y_i) \leq 2\epsilon$ which

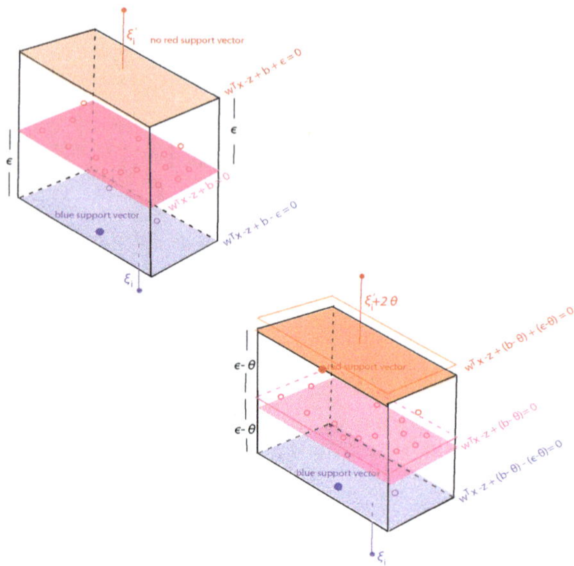

Fig. 20.10 In this illustration points within the ϵ-tube are denoted by open circles. In the original, upper left configuration, there is no red support vector. By lowering the pink separating hyperplane and decreasing the width of the slab, we end up with a red support vector.

is equivalent to $w^\top x_i + b - y_i \leq \epsilon$ which holds for all $i \notin (E_\lambda \cup E_\mu)$ by hypothesis.

We can write

$$\begin{aligned} w^\top x_i + b - \theta - y_i &= \epsilon - \theta + \xi_i & \xi_i > 0 \quad & i \in E_\lambda \\ -w^\top x_j - (b - \theta) + y_j &= \epsilon - \theta + \xi'_j + 2\theta & \xi'_j > 0 \quad & j \in E_\mu \\ w^\top x_i + b - \theta - y_i &\leq \epsilon - \theta & & i \notin (E_\lambda \cup E_\mu) \\ -w^\top x_i - (b - \theta) + y_i &\leq \epsilon - \theta & & i \notin (E_\lambda \cup E_\mu). \end{aligned}$$

By hypothesis

$$w^\top x_i + (b - \theta) - y_i = \epsilon - \theta \quad \text{for some } i \notin (E_\lambda \cup E_\mu),$$

and by the choice of θ,

$$-w^\top x_j - (b - \theta) + y_j = \epsilon - \theta \quad \text{for some } j \notin (E_\lambda \cup E_\mu).$$

The new value of the objective function is

$$\begin{aligned}\omega(\theta) &= \frac{1}{2}w^\top w + \nu(\epsilon - \theta) + \frac{1}{m}\Bigg(\sum_{i\in E_\lambda} \xi_i + \sum_{j \in E_\mu} (\xi'_j + 2\theta)\Bigg)\\ &= \frac{1}{2}w^\top w + \nu\epsilon + \frac{1}{m}\Bigg(\sum_{i\in E_\lambda} \xi_i + \sum_{j \in E_\mu} \xi'_j\Bigg) - \left(\nu - \frac{2q_{sf}}{m}\right)\theta.\end{aligned}$$

The rest of the proof is similar except that $2p_{sf}/m$ is replaced by $2q_{sf}/m$. Observe that the exceptional case in which $\theta = \epsilon$ may arise. In this case all points (x_i, y_i) that are not errors (strictly outside the ϵ-slab) are on the blue margin hyperplane. This case can only arise if $\nu = 2q_{sf}/m$.

Case 2b. We have $w^\top x_i + b - y_i < \epsilon$ for all $i \notin (E_\lambda \cup E_\mu)$. Since we also assumed that $-w^\top x_i - b + y_i < \epsilon$ for all $i \notin (E_\lambda \cup E_\mu)$, Case 2b is identical to Case 1b and we are done. □

The proof of Proposition 20.4 reveals that there are three critical values for ν:

$$\frac{2p_{sf}}{m}, \ \frac{2q_{sf}}{m}, \ \frac{p_{sf} + q_{sf}}{m}.$$

These values can be avoided by requiring the strict inequality

$$\max\left\{\frac{2p_{sf}}{m}, \frac{2q_{sf}}{m}\right\} < \nu.$$

Then the following corollary holds.

Theorem 20.1. *For every optimal solution $(w, b, \epsilon, \xi, \xi')$ with $w \neq 0$ and $\epsilon > 0$, if*

$$\max\left\{\frac{2p_{sf}}{m}, \frac{2q_{sf}}{m}\right\} < \nu < (m-1)/m,$$

then some x_{i_0} is a blue support vector and some x_{j_0} is a red support vector (with $i_0 \neq j_0$).

Proof. We proceed by contradiction. Suppose that for every optimal solution with $w \neq 0$ and $\epsilon > 0$ no x_i is a blue support vector or no x_i is a red support vector. Since $\nu < (m-1)/m$, Proposition 20.4 holds, so there is another optimal solution. But since the critical values of ν are avoided, the proof of Proposition 20.4 shows that the value of the objective function for this new optimal solution is strictly smaller than the original optimal value, a contradiction. □

Remark: If an optimal solution has $\epsilon = 0$, then depending on the value of C there may not be any support vectors, or many.

If the primal has an optimal solution with $w \neq 0$ and $\epsilon > 0$, then by $(*_w)$ and since

$$\sum_{i=1}^{m} \lambda_i - \sum_{i=1}^{m} \mu_i = 0 \quad \text{and} \quad \lambda_i \mu_i = 0,$$

there is i_0 such that $\lambda_{i_0} > 0$ and some $j_0 \neq i_0$ such that $\mu_{j_0} > 0$.

Under the mild hypothesis called the **Standard Margin Hypothesis** that there is some i_0 such that $0 < \alpha_{i_0} < \frac{C}{m}$ *and* there is some $j_0 \neq i_0$ such that $0 < \mu_{j_0} < \frac{C}{m}$, in other words there is a blue support vector of type 1 and there is a red support vector of type 1, then by $(*)$ we have $\xi_{i_0} = 0, \xi'_{j_0} = 0$, and we have the two equations

$$\begin{aligned} w^\top x_{i_0} + b - y_{i_0} &= \epsilon \\ -w^\top x_{j_0} - b + y_{j_0} &= \epsilon, \end{aligned}$$

so b and ϵ can be computed. In particular,

$$\begin{aligned} b &= \frac{1}{2}\left(y_{i_0} + y_{j_0} - w^\top (x_{i_0} + x_{j_0})\right) \\ \epsilon &= \frac{1}{2}\left(y_{j_0} - y_{i_0} + w^\top (x_{i_0} - x_{j_0})\right). \end{aligned}$$

The function $f(x) = w^\top x + b$ (often called *regression estimate*) is given by

$$f(x) = \sum_{i=1}^{m} (\mu_i - \lambda_i) x_i^\top x + b.$$

In practice, due to numerical inaccuracy, it is complicated to write a computer program that will select two *distinct indices* as above. It is preferable to compute the list I_λ of indices i such that $0 < \lambda_i < C/m$ and the list I_μ of indices j such that $0 < \mu_j < C/m$. Then it is easy to see that

$$\begin{aligned} b = {} & \left(\left(\sum_{i_0 \in I_\lambda} y_{i_0}\right)/|I_\lambda| + \left(\sum_{j_0 \in I_\mu} y_{j_0}\right)/|I_\mu|\right)/2 \\ & - w^\top \left(\left(\sum_{i_0 \in I_\lambda} x_{i_0}\right)/|I_\lambda| + \left(\sum_{j_0 \in I_\mu} x_{j_0}\right)/|I_\mu|\right)/2 \\ \epsilon = {} & \left(\left(\sum_{j_0 \in I_\mu} y_{j_0}\right)/|I_\mu|\right) - \left(\sum_{i_0 \in I_\lambda} y_{i_0}\right)/|I_\lambda|\right)/2 \\ & + w^\top \left(\left(\sum_{i_0 \in I_\lambda} x_{i_0}\right)/|I_\lambda| - \left(\sum_{j_0 \in I_\mu} x_{j_0}\right)/|I_\mu|\right)/2. \end{aligned}$$

These formulae are numerically a lot more stable, but we still have to be cautious to set suitable tolerance factors to decide whether $\lambda_i > 0$ and $\lambda_i < C/m$ (and similarly for μ_i).

The following result gives sufficient conditions for expressing ϵ in terms of a single support vector.

Proposition 20.5. *For every optimal solution $(w, b, \epsilon, \xi, \xi')$ with $w \neq 0$ and $\epsilon > 0$, if*

$$\max\left\{\frac{2p_{sf}}{m}, \frac{2q_{sf}}{m}\right\} < \nu < (m-1)/m,$$

then ϵ and b are determined from a solution (λ, μ) of the dual in terms of a single support vector.

Proof sketch. If we express that the duality gap is zero we obtain the following equation expressing ϵ in terms of b:

$$\begin{aligned} C\left(\nu - \frac{p_f + q_f}{m}\right)\epsilon &= -\begin{pmatrix}\lambda^\top & \mu^\top\end{pmatrix} P \begin{pmatrix}\lambda \\ \mu\end{pmatrix} - \begin{pmatrix}y^\top & -y^\top\end{pmatrix}\begin{pmatrix}\lambda \\ \mu\end{pmatrix} \\ &\quad - \frac{C}{m}\left(w^\top\left(\sum_{i\in K_\lambda} x_i - \sum_{j\in K_\mu} x_j\right) - \sum_{i\in K_\lambda} y_i + \sum_{j\in K_\mu} y_j + (p_f - q_f)b\right). \end{aligned}$$

The proof is very similar to the proof of the corresponding formula in Section 20.5. By Theorem 20.1, there is some support vector x_i, say

$$w^\top x_{i_0} + b - y_{i_0} = \epsilon \quad \text{or} \quad -w^\top x_{j_0} - b + y_{j_0} = \epsilon.$$

Then we find an equation expressing ϵ in terms of λ, μ and w, provided that $\nu \neq 2p_f/m$ and $\nu \neq 2q_f/m$. The proof is analogous to the proof of Proposition 18.3 and is left as an exercise. □

20.3 Solving ν-Regression Using ADMM

The quadratic functional $F(\lambda, \mu)$ occurring in the dual program given by

$$F(\lambda, \mu) = \frac{1}{2}\sum_{i,j=1}^m (\lambda_i - \mu_i)(\lambda_j - \mu_j)x_i^\top x_j + \sum_{i=1}^m (\lambda_i - \mu_i)y_i$$

is not of the form $\frac{1}{2}\begin{pmatrix}\lambda^\top & \mu^\top\end{pmatrix} P \begin{pmatrix}\lambda \\ \mu\end{pmatrix} + q^\top \begin{pmatrix}\lambda \\ \mu\end{pmatrix}$, but it can be converted in such a form using a trick. First, if we let $\mathbf{K}$ be the $m \times m$ symmetric matrix $\mathbf{K} = XX^\top = (x_i^\top x_j)$, then we have

$$F(\lambda, \mu) = \frac{1}{2}(\lambda^\top - \mu^\top)\mathbf{K}(\lambda - \mu) + y^\top\lambda - y^\top\mu.$$

Consequently, if we define the $2m \times 2m$ symmetric matrix P by

$$P = \begin{pmatrix} XX^\top & -XX^\top \\ -XX^\top & XX^\top \end{pmatrix} = \begin{pmatrix} \mathbf{K} & -\mathbf{K} \\ -\mathbf{K} & \mathbf{K} \end{pmatrix}$$

and the $2m \times 1$ matrix q by

$$q = \begin{pmatrix} y \\ -y \end{pmatrix},$$

it is easy to check that

$$\begin{aligned} F(\lambda, \mu) &= \frac{1}{2} \begin{pmatrix} \lambda^\top & \mu^\top \end{pmatrix} P \begin{pmatrix} \lambda \\ \mu \end{pmatrix} + q^\top \begin{pmatrix} \lambda \\ \mu \end{pmatrix} \\ &= \frac{1}{2}\lambda^\top \mathbf{K}\lambda + \frac{1}{2}\mu^\top \mathbf{K}\mu - \lambda^\top \mathbf{K}\mu + y^\top \lambda - y^\top \mu. \qquad (*_q) \end{aligned}$$

Since

$$\frac{1}{2} \begin{pmatrix} \lambda^\top & \mu^\top \end{pmatrix} P \begin{pmatrix} \lambda \\ \mu \end{pmatrix} = \frac{1}{2}(\lambda^\top - \mu^\top)\mathbf{K}(\lambda - \mu)$$

and the matrix $\mathbf{K} = XX^\top$ is symmetric positive semidefinite, the matrix P is also symmetric positive semidefinite. Thus we are in a position to apply ADMM since the constraints are

$$\begin{aligned} \sum_{i=1}^m \lambda_i - \sum_{i=1}^m \mu_i &= 0 \\ \sum_{i=1}^m \lambda_i + \sum_{i=1}^m \mu_i + \gamma &= C\nu \\ \lambda + \alpha = \frac{C}{m}, \quad \mu + \beta &= \frac{C}{m}, \end{aligned}$$

namely affine. We need to check that the $(2m+2) \times (4m+1)$ matrix A corresponding to this system has rank $2m+2$. Let us clarify this point. The matrix A corresponding to the above equations is

$$A = \begin{pmatrix} \mathbf{1}_m^\top & -\mathbf{1}_m^\top & 0_m^\top & 0_m^\top & 0 \\ \mathbf{1}_m^\top & \mathbf{1}_m^\top & 0_m^\top & 0_m^\top & 1 \\ I_m & 0_{m,m} & I_m & 0_{m,m} & 0_m \\ 0_{m,m} & I_m & 0_{m,m} & I_m & 0_m \end{pmatrix}.$$

For example, for $m = 3$ we have the 8×13 matrix

$$\begin{pmatrix} 1 & 1 & 1 & -1 & -1 & -1 & 0 & 0 & 0 & 0 & 0 & 0 & 0 \\ 1 & 1 & 1 & 1 & 1 & 1 & 0 & 0 & 0 & 0 & 0 & 0 & 1 \\ 1 & 0 & 0 & 0 & 0 & 0 & 1 & 0 & 0 & 0 & 0 & 0 & 0 \\ 0 & 1 & 0 & 0 & 0 & 0 & 0 & 1 & 0 & 0 & 0 & 0 & 0 \\ 0 & 0 & 1 & 0 & 0 & 0 & 0 & 0 & 1 & 0 & 0 & 0 & 0 \\ 0 & 0 & 0 & 1 & 0 & 0 & 0 & 0 & 0 & 1 & 0 & 0 & 0 \\ 0 & 0 & 0 & 0 & 1 & 0 & 0 & 0 & 0 & 0 & 1 & 0 & 0 \\ 0 & 0 & 0 & 0 & 0 & 1 & 0 & 0 & 0 & 0 & 0 & 1 & 0 \end{pmatrix}.$$

We leave it as an exercise to show that A has rank $2m + 2$. Recall that

$$q = \begin{pmatrix} y \\ -y \end{pmatrix}$$

and we also define the vector c (of dimension $2m + 2$) as

$$c = \begin{pmatrix} 0 \\ C\nu \\ \frac{C}{m}\mathbf{1}_{2m} \end{pmatrix}.$$

The constraints are given by the system of affine equations $Ax = c$, where

$$x = \begin{pmatrix} \lambda^\top & \mu^\top & \alpha^\top & \beta^\top & \gamma \end{pmatrix}^\top.$$

Since there are $4m + 1$ Lagrange multipliers $(\lambda, \mu, \alpha, \beta, \gamma)$, we need to pad the $2m \times 2m$ matrix P with zeros to make it into a $(4m+1) \times (4m+1)$ matrix

$$P_a = \begin{pmatrix} P & 0_{2m,2m+1} \\ 0_{2m+1,2m} & 0_{2m+1,2m+1} \end{pmatrix}.$$

Similarly, we pad q with zeros to make it a vector q_a of dimension $4m + 1$,

$$q_a = \begin{pmatrix} q \\ 0_{2m+1} \end{pmatrix}.$$

In order to solve our dual program, we apply ADMM to the quadractic functional

$$\frac{1}{2} x^\top P_a x + q_a^\top x,$$

subject to the constraints

$$Ax = c, \quad x \geq 0,$$

with P_a, q_a, A, b and x, as above.

Since for an optimal solution with $\epsilon > 0$ we must have $\gamma = 0$ (from the KKT conditions), we can solve the dual problem with the following set of constraints only involving the Lagrange multipliers $(\lambda, \mu, \alpha, \beta)$,

$$\sum_{i=1}^{m} \lambda_i - \sum_{i=1}^{m} \mu_i = 0$$

$$\sum_{i=1}^{m} \lambda_i + \sum_{i=1}^{m} \mu_i = C\nu$$

$$\lambda + \alpha = \frac{C}{m}, \quad \mu + \beta = \frac{C}{m},$$

which corresponds to the $(2m+2) \times 4m$ A_2 given by

$$A_2 = \begin{pmatrix} \mathbf{1}_m^\top & -\mathbf{1}_m^\top & 0_m^\top & 0_m^\top \\ \mathbf{1}_m^\top & \mathbf{1}_m^\top & 0_m^\top & 0_m^\top \\ I_m & 0_{m,m} & I_m & 0_{m,m} \\ 0_{m,m} & I_m & 0_{m,m} & I_m \end{pmatrix}.$$

We leave it as an exercise to show that A_2 has rank $2m+2$. We define the vector c_2 (of dimension $2m+2$) as

$$c_2 = c = \begin{pmatrix} 0 \\ C\nu \\ \frac{C}{m}\mathbf{1}_{2m} \end{pmatrix}.$$

Since there are $4m$ Lagrange multipliers $(\lambda, \mu, \alpha, \beta)$, we need to pad the $2m \times 2m$ matrix P with zeros to make it into a $4m \times 4m$ matrix

$$P_{2a} = \begin{pmatrix} P & 0_{2m,2m} \\ 0_{2m,2m} & 0_{2m,2m} \end{pmatrix}.$$

Similarly, we pad q with zeros to make it a vector q_{2a} of dimension $4m$,

$$q_{2a} = \begin{pmatrix} q \\ 0_{2m} \end{pmatrix}.$$

We implemented the above methods in `Matlab`; see Appendix B, Section B.4. Choosing $C = m$ is typically a good choice because then the values of λ_i and μ_j are not too small $(C/m = 1)$. If C is chosen too small, we found that numerical instability increases drastically and very poor results are obtained. Increasing C tends to encourage sparsity.

We ran our `Matlab` implementation of the above method on the set of 50 points generated at random by the program shown below with $C = 50$ and various values of ν starting with $\nu = 0.03$:

```
X13 = 15*randn(50,1);
ww13 = 1;
y13 = X13*ww13 + 10*randn(50,1) + 20;
[~,~,~,~,~,~,~,~,w1] = runuregb(rho,0.03,X13,y13,50)
```

Figure 20.11 shows the result of running the program with $\nu = 0.03$. We have $p_f = 0, q_f = 0, p_m = 2$ and $q_m = 1$. There are 47 points strictly inside the slab. The slab is large enough to contain all the data points, so none of them is considered an error.

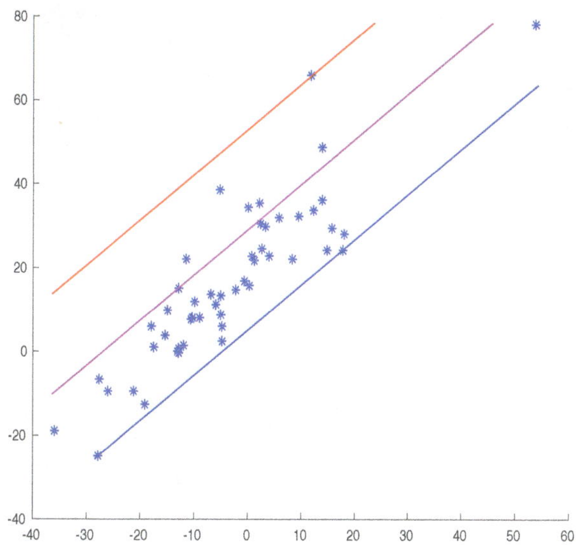

Fig. 20.11 Running ν-SV regression on a set of 50 points; $\nu = 0.03$.

The next value of ν is $\nu = 0.21$, see Figure 20.12. We have $p_f = 4, q_f = 5, p_m = 6$ and $q_m = 6$. There are 38 points strictly inside the slab.

The next value of ν is $\nu = 0.5$, see Figure 20.13. We have $p_f = 12, q_f = 12, p_m = 13$ and $q_m = 14$. There are 23 points strictly inside the slab.

The next value of ν is $\nu = 0.7$, see Figure 20.14. We have $p_f = 17, q_f = 17, p_m = 18$ and $q_m = 19$. There are 13 points strictly inside the slab.

The last value of ν is $\nu = 0.97$, see Figure 20.15. We have $p_f = 23, q_f = 24, p_m = 25$ and $q_m = 25$. There are 0 points strictly inside the slab. The slab is so narrow that it does not contain any of the points x_i in it.

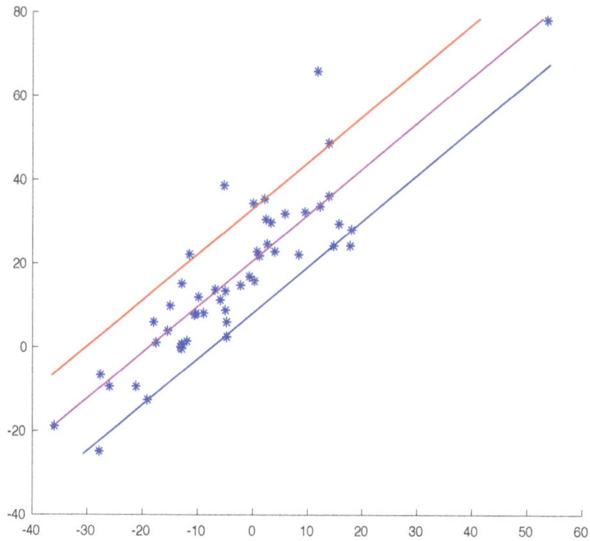

Fig. 20.12 Running ν-SV regression on a set of 50 points; $\nu = 0.21$.

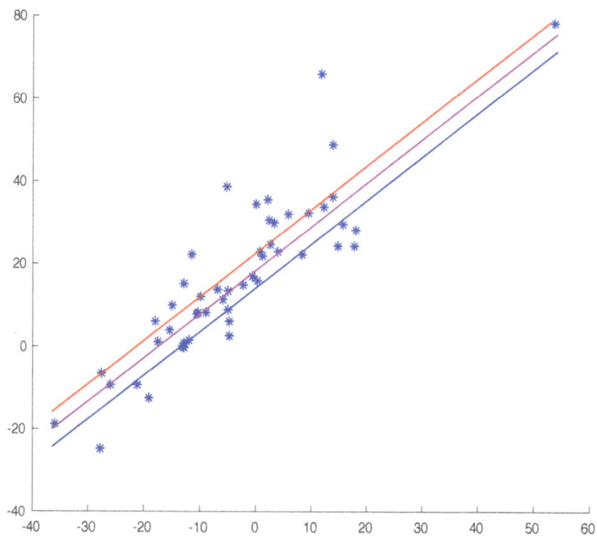

Fig. 20.13 Running ν-SV regression on a set of 50 points; $\nu = 0.5$.

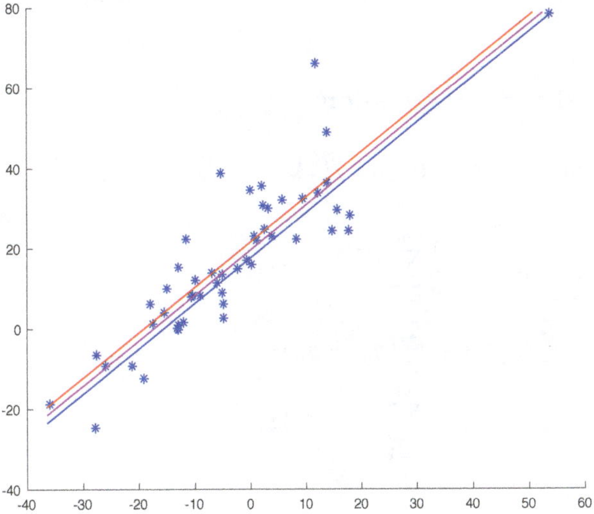

Fig. 20.14 Running ν-SV regression on a set of 50 points; $\nu = 0.7$.

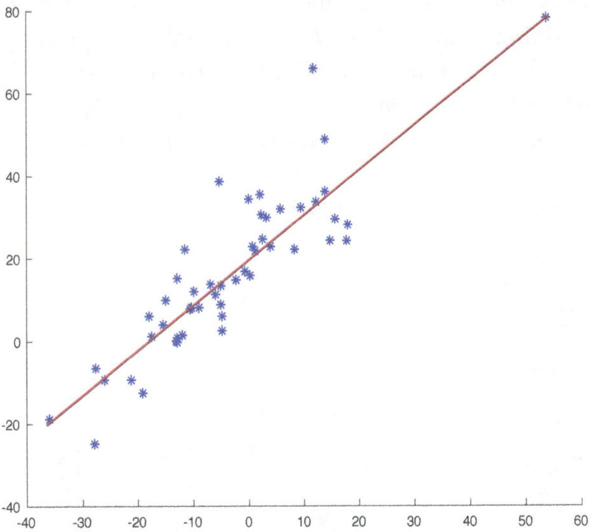

Fig. 20.15 Running ν-SV regression on a set of 50 points; $\nu = 0.97$.

Running the program with any value $\nu > 0.97$ yields $\epsilon = 0$.

20.4 Kernel ν-SV Regression

Since the formulae for w, b, and $f(x)$,

$$\begin{aligned} w &= \sum_{i=1}^{m} (\mu_i - \lambda_i) x_i \\ b &= \frac{1}{2}\left(y_{i_0} + y_{j_0} - w^\top (x_{i_0} + x_{j_0})\right) \\ f(x) &= \sum_{i=1}^{m} (\mu_i - \lambda_i) x_i^\top x + b, \end{aligned}$$

only involve inner products among the data points x_i and x, and since the objective function $-G(\alpha, \mu)$ of the dual program also only involves inner products among the data points x_i, we can kernelize the ν-SV regression method.

As in the previous section, we assume that our data points $\{x_1, \ldots, x_m\}$ belong to a set $\mathcal{X}$ and we pretend that we have feature space $(F, \langle -, - \rangle)$ and a feature embedding map $\varphi \colon \mathcal{X} \to F$, but we only have access to the kernel function $\kappa(x_i, x_j) = \langle \varphi(x_i), \varphi(x_j) \rangle$. We wish to perform ν-SV regression in the feature space F on the data set $\{(\varphi(x_1), y_1), \ldots, (\varphi(x_m), y_m)\}$. Going over the previous computation, we see that the primal program is given by

Program kernel ν-SV Regression:

$$\text{minimize} \quad \frac{1}{2}\langle w, w \rangle + C\left(\nu\epsilon + \frac{1}{m}\sum_{i=1}^{m} (\xi_i + \xi_i')\right)$$

subject to

$$\begin{aligned} \langle w, \varphi(x_i) \rangle + b - y_i &\le \epsilon + \xi_i, \quad \xi_i \ge 0 \qquad i = 1, \ldots, m \\ -\langle w, \varphi(x_i) \rangle - b + y_i &\le \epsilon + \xi_i', \quad \xi_i' \ge 0 \qquad i = 1, \ldots, m \\ \epsilon &\ge 0, \end{aligned}$$

minimizing over the variables w, ϵ, b, ξ, and ξ'.

The Lagrangian is given by

$$L(w,b,\lambda,\mu,\gamma,\xi,\xi',\epsilon,\alpha,\beta) = \frac{1}{2}\langle w,w\rangle + \left\langle w, \sum_{i=1}^m (\lambda_i - \mu_i)\varphi(x_i)\right\rangle$$
$$+ \epsilon\left(C\nu - \gamma - \sum_{i=1}^m (\lambda_i + \mu_i)\right) + \sum_{i=1}^m \xi_i\left(\frac{C}{m} - \lambda_i - \alpha_i\right)$$
$$+ \sum_{i=1}^m \xi_i'\left(\frac{C}{m} - \mu_i - \beta_i\right) + b\left(\sum_{i=1}^m (\lambda_i - \mu_i)\right) - \sum_{i=1}^m (\lambda_i - \mu_i)y_i.$$

Setting the gradient $\nabla L_{w,\epsilon,b,\xi,\xi'}$ of the Lagrangian to zero, we also obtain the equations

$$w = \sum_{i=1}^m (\mu_i - \lambda_i)\varphi(x_i), \tag{$*_w$}$$
$$\sum_{i=1}^m \lambda_i - \sum_{i=1}^m \mu_i = 0$$
$$\sum_{i=1}^m \lambda_i + \sum_{i=1}^m \mu_i + \gamma = C\nu$$
$$\lambda + \alpha = \frac{C}{m}, \quad \mu + \beta = \frac{C}{m}.$$

Using the above equations, we find that the dual function G is independent of the variables β, α, β, and we obtain the following dual program:

Dual Program kernel ν-SV Regression:

$$\text{minimize} \quad \frac{1}{2}\sum_{i,j=1}^m (\lambda_i - \mu_i)(\lambda_j - \mu_j)\kappa(x_i, x_j) + \sum_{i=1}^m (\lambda_i - \mu_i)y_i$$

subject to

$$\sum_{i=1}^m \lambda_i - \sum_{i=1}^m \mu_i = 0$$
$$\sum_{i=1}^m \lambda_i + \sum_{i=1}^m \mu_i \leq C\nu$$
$$0 \leq \lambda_i \leq \frac{C}{m}, \quad 0 \leq \mu_i \leq \frac{C}{m}, \quad i = 1, \ldots, m,$$

minimizing over α and μ.

Everything we said before also applies to the kernel ν-SV regression method, except that x_i is replaced by $\varphi(x_i)$ and that the inner product $\langle -, - \rangle$ must be used, and we have the formulae

$$\begin{aligned} w &= \sum_{i=1}^{m} (\mu_i - \lambda_i)\varphi(x_i) \\ b &= \frac{1}{2}\left(y_{i_0} + y_{j_0} - \sum_{i=1}^{m} (\mu_i - \lambda_i)(\kappa(x_i, x_{i_0}) + \kappa(x_i, x_{j_0})) \right) \\ f(x) &= \sum_{i=1}^{m} (\mu_i - \lambda_i)\kappa(x_i, x) + b, \end{aligned}$$

expressions that only involve κ.

Remark: There is a variant of ν-SV regression obtained by setting $\nu = 0$ and holding $\epsilon > 0$ fixed. This method is called *ϵ-SV regression* or *(linear) ϵ-insensitive SV regression*. The corresponding optimization program is

Program ϵ-SV Regression:

$$\text{minimize} \quad \frac{1}{2} w^\top w + \frac{C}{m} \sum_{i=1}^{m} (\xi_i + \xi'_i)$$

subject to

$$\begin{aligned} w^\top x_i + b - y_i &\leq \epsilon + \xi_i, \quad \xi_i \geq 0 \qquad i = 1, \ldots, m \\ -\, w^\top x_i - b + y_i &\leq \epsilon + \xi'_i, \quad \xi'_i \geq 0 \qquad i = 1, \ldots, m, \end{aligned}$$

minimizing over the variables w, b, ξ, and ξ', holding ϵ fixed.

It is easy to see that the dual program is

Dual Program ϵ-SV Regression:

$$\text{minimize} \quad \frac{1}{2} \sum_{i,j=1}^{m} (\lambda_i - \mu_i)(\lambda_j - \mu_j) x_i^\top x_j + \sum_{i=1}^{m} (\lambda_i - \mu_i) y_i + \epsilon \sum_{i=1}^{m} (\lambda_i + \mu_i)$$

subject to

$$\begin{aligned} &\sum_{i=1}^{m} \lambda_i - \sum_{i=1}^{m} \mu_i = 0 \\ &0 \leq \lambda_i \leq \frac{C}{m}, \quad 0 \leq \mu_i \leq \frac{C}{m}, \quad i = 1, \ldots, m, \end{aligned}$$

minimizing over α and μ.

The constraint

$$\sum_{i=1}^{m} \lambda_i + \sum_{i=1}^{m} \mu_i \leq C\nu$$

is gone but the extra term $\epsilon \sum_{i=1}^{m} (\lambda_i + \mu_i)$ has been added to the dual function, to prevent λ_i and μ_i from blowing up.

There is an obvious kernelized version of ϵ-SV regression. It is easy to show that ν-SV regression subsumes ϵ-SV regression, in the sense that if ν-SV regression succeeds and yields $w, b, \epsilon > 0$, then ϵ-SV regression with the same C and the same value of ϵ also succeeds and returns the same pair (w, b). For more details on these methods, see Schölkopf, Smola, Williamson, and Bartlett [Schölkopf *et al.* (2000)].

Remark: The linear penalty function $\sum_{i=1}^{m} (\xi_i + \xi'_i)$ can be replaced by the quadratic penalty function $\sum_{i=1}^{m} (\xi_i^2 + \xi_i'^2)$; see Shawe-Taylor and Christianini [Shawe-Taylor and Cristianini (2004)] (Chapter 7). In this case, it is easy to see that for an optimal solution we must have $\xi_i \geq 0$ and $\xi'_i \geq 0$, so we may omit the constraints $\xi_i \geq 0$ and $\xi'_i \geq 0$. We must also have $\gamma = 0$ so we omit the variable γ as well. It can be shown that $\xi = (m/2C)\lambda$ and $\xi' = (m/2C)\mu$. This problem is very similar to the Soft Margin SVM (SVM_{s4}) discussed in Section 18.13.

20.5 ν-Regression Version 2; Penalizing b

Yet another variant of ν-SV regression is to add the term $\frac{1}{2}b^2$ to the objective function. We will see that solving the dual not only determines w but also b and ϵ (provided a mild condition on ν). We wish to solve the following program:

Program ν-SV Regression Version 2

$$\text{minimize} \quad \frac{1}{2} w^\top w + \frac{1}{2} b^2 + C\left(\nu\epsilon + \frac{1}{m} \sum_{i=1}^{m} (\xi_i + \xi'_i)\right)$$

subject to

$$\begin{aligned} w^\top x_i + b - y_i &\leq \epsilon + \xi_i, \quad \xi_i \geq 0 \qquad i = 1, \ldots, m \\ - w^\top x_i - b + y_i &\leq \epsilon + \xi'_i, \quad \xi'_i \geq 0 \qquad i = 1, \ldots, m, \end{aligned}$$

minimizing over the variables w, b, ϵ, ξ, and ξ'. The constraint $\epsilon \geq 0$ is omitted since the problem has no solution if $\epsilon < 0$.

We leave it as an exercise to show that the new Lagrangian is

$$\begin{aligned}
L(w, b, \lambda, \mu, \xi, \xi', \epsilon, \alpha, \beta) &= \frac{1}{2} w^\top w + w^\top \left(\sum_{i=1}^m (\lambda_i - \mu_i) x_i \right) \\
&+ \epsilon \left(C\nu - \sum_{i=1}^m (\lambda_i + \mu_i) \right) + \sum_{i=1}^m \xi_i \left(\frac{C}{m} - \lambda_i - \alpha_i \right) \\
+ \sum_{i=1}^m \xi'_i \left(\frac{C}{m} - \mu_i - \beta_i \right) &+ \frac{1}{2} b^2 + b \left(\sum_{i=1}^m (\lambda_i - \mu_i) \right) - \sum_{i=1}^m (\lambda_i - \mu_i) y_i.
\end{aligned}$$

If we set the Laplacian $\nabla L_{w,\epsilon,b,\xi,\xi'}$ to zero we obtain the equations

$$\begin{aligned}
w &= \sum_{i=1}^m (\mu_i - \lambda_i) x_i = X^\top (\mu - \lambda) && (*_w) \\
C\nu - \sum_{i=1}^m (\lambda_i + \mu_i) &= 0 \\
b + \sum_{i=1}^m (\lambda_i - \mu_i) &= 0 \\
\frac{C}{m} - \lambda - \alpha = 0, &\quad \frac{C}{m} - \mu - \beta = 0.
\end{aligned}$$

We obtain the new equation

$$b = -\sum_{i=1}^m (\lambda_i - \mu_i) = -(\mathbf{1}_m^\top \lambda - \mathbf{1}_m^\top \mu) \qquad (*_b)$$

determining b, which replaces the equation

$$\sum_{i=1}^m \lambda_i - \sum_{i=1}^m \mu_i = 0.$$

Plugging back w from $(*_w)$ and b from $(*_b)$ into the Lagrangian we get

$$\begin{aligned}
G(\lambda, \mu, \alpha, \beta) &= -\frac{1}{2} \begin{pmatrix} \lambda^\top & \mu^\top \end{pmatrix} P \begin{pmatrix} \lambda \\ \mu \end{pmatrix} - q^\top \begin{pmatrix} \lambda \\ \mu \end{pmatrix} + \frac{1}{2} b^2 - b^2 \\
&= -\frac{1}{2} \begin{pmatrix} \lambda^\top & \mu^\top \end{pmatrix} P \begin{pmatrix} \lambda \\ \mu \end{pmatrix} - q^\top \begin{pmatrix} \lambda \\ \mu \end{pmatrix} - \frac{1}{2} b^2 \\
&= -\frac{1}{2} \begin{pmatrix} \lambda^\top & \mu^\top \end{pmatrix} \left(P + \begin{pmatrix} \mathbf{1}_m \mathbf{1}_m^\top & -\mathbf{1}_m \mathbf{1}_m^\top \\ -\mathbf{1}_m \mathbf{1}_m^\top & \mathbf{1}_m \mathbf{1}_m^\top \end{pmatrix} \right) \begin{pmatrix} \lambda \\ \mu \end{pmatrix} - q^\top \begin{pmatrix} \lambda \\ \mu \end{pmatrix},
\end{aligned}$$

with

$$P = \begin{pmatrix} XX^\top & -XX^\top \\ -XX^\top & XX^\top \end{pmatrix} = \begin{pmatrix} \mathbf{K} & -\mathbf{K} \\ -\mathbf{K} & \mathbf{K} \end{pmatrix}$$

and

$$q = \begin{pmatrix} y \\ -y \end{pmatrix}.$$

The new dual program is

Dual Program ν-SV Regression Version 2

$$\text{minimize} \quad \frac{1}{2}\begin{pmatrix}\lambda^\top & \mu^\top\end{pmatrix}\left(P + \begin{pmatrix} \mathbf{1}_m\mathbf{1}_m^\top & -\mathbf{1}_m\mathbf{1}_m^\top \\ -\mathbf{1}_m\mathbf{1}_m^\top & \mathbf{1}_m\mathbf{1}_m^\top \end{pmatrix}\right)\begin{pmatrix}\lambda \\ \mu\end{pmatrix} + q^\top \begin{pmatrix}\lambda \\ \mu\end{pmatrix}$$

subject to

$$\sum_{i=1}^m \lambda_i + \sum_{i=1}^m \mu_i = C\nu$$

$$0 \leq \lambda_i \leq \frac{C}{m}, \quad 0 \leq \mu_i \leq \frac{C}{m}, \quad i = 1, \ldots, m.$$

Definition 20.1 and Definition 20.2 are unchanged. We have the following version of Proposition 20.2 showing that p_f, q_f, p_m an q_m have direct influence on the choice of ν.

Proposition 20.6.

(1) Let p_f be the number of points x_i such that $\lambda_i = C/m$, and let q_f be the number of points x_i such that $\mu_i = C/m$. We have $p_f + q_f \leq m\nu$.
(2) Let p_m be the number of points x_i such that $\lambda_i > 0$, and let q_m be the number of points x_i such that $\mu_i > 0$. We have $p_m + q_m \geq m\nu$.
(3) If $p_f \geq 1$ or $q_f \geq 1$, then $\nu \geq 1/m$.

Proof. (1) Let K_λ and K_μ be the sets of indices corresponding to points failing the margin,

$$K_\lambda = \{i \in \{1, \ldots, m\} \mid \lambda_i = C/m\}$$
$$K_\mu = \{i \in \{1, \ldots, m\} \mid \mu_i = C/m\}.$$

By definition $p_f = |K_\lambda|, q_f = |K_\mu|$. Since the equation

$$\sum_{i=1}^m \lambda_i + \sum_{j=1}^m \mu_j = C\nu$$

holds, by definition of K_λ and K_μ we have

$$(p_f + q_f)\frac{C}{m} = \sum_{i \in K_\lambda} \lambda_i + \sum_{j \in K_\mu} \mu_j \leq \sum_{i=1}^m \lambda_i + \sum_{j=1}^m \mu_j = C\nu,$$

which implies that

$$p_f + q_f \leq m\nu.$$

(2) Let $I_{\lambda>0}$ and $I_{\mu>0}$ be the sets of indices

$$\begin{aligned} I_{\lambda>0} &= \{i \in \{1, \ldots, m\} \mid \lambda_i > 0\} \\ I_{\mu>0} &= \{i \in \{1, \ldots, m\} \mid \mu_i > 0\}. \end{aligned}$$

By definition $p_m = |I_{\lambda>0}|, q_m = |I_{\mu>0}|$. We have

$$\sum_{i=1}^{m} \lambda_i + \sum_{j=1}^{m} \mu_j = \sum_{i \in I_{\lambda>0}} \lambda_i + \sum_{j \in I_{\mu>0}} \mu_j = C\nu.$$

Since $\lambda_i \leq C/m$ and $\mu_j \leq C/m$, we obtain

$$C\nu \leq (p_m + q_m)\frac{C}{m},$$

that is, $p_m + q_m \geq m\nu$.

(3) follows immediately from (1). □

Proposition 20.6 yields the following bounds on ν:

$$\frac{p_f + q_f}{m} \leq \nu \leq \frac{p_m + q_m}{m}.$$

Again, the smaller ν is, the wider the ϵ-slab is, and the larger ν is, the narrower the ϵ-slab is.

Remark: It can be shown that for any optimal solution with $w \neq 0$ and $\epsilon > 0$, if the inequalities $(p_f + q_f)/m < \nu < 1$ hold, then some point x_i is a support vector. The proof is essentially Case 1b in the proof of Proposition 20.4. We leave the details as an exercise.

The new dual program is solved using ADMM. The $(2m + 1) \times 4m$ matrix A_3 corresponding to the equational constraints

$$\sum_{i=1}^{m} \lambda_i + \sum_{i=1}^{m} \mu_i = C\nu$$

$$\lambda + \alpha = \frac{C}{m}, \quad \mu + \beta = \frac{C}{m},$$

is given by

$$A_3 = \begin{pmatrix} \mathbf{1}_m^\top & \mathbf{1}_m^\top & 0_m^\top & 0_m^\top \\ I_m & 0_{m,m} & I_m & 0_{m,m} \\ 0_{m,m} & I_m & 0_{m,m} & I_m \end{pmatrix}.$$

We leave it as an exercise to show that A_3 has rank $2m+1$. We define the vector c_3 (of dimension $2m+1$) as

$$c_3 = \begin{pmatrix} C\nu \\ \frac{C}{m}\mathbf{1}_{2m} \end{pmatrix}.$$

Since there are $4m$ Lagrange multipliers $(\lambda, \mu, \alpha, \beta)$, we need to pad the $2m \times 2m$ matrix $P_3 = P + \begin{pmatrix} \mathbf{1}_m\mathbf{1}_m^\top & -\mathbf{1}_m\mathbf{1}_m^\top \\ -\mathbf{1}_m\mathbf{1}_m^\top & \mathbf{1}_m\mathbf{1}_m^\top \end{pmatrix}$ with zeros to make it into a $4m \times 4m$ matrix

$$P_{3a} = \begin{pmatrix} P_3 & 0_{2m,2m} \\ 0_{2m,2m} & 0_{2m,2m} \end{pmatrix}.$$

Similarly, we pad q with zeros to make it a vector q_{3a} of dimension $4m$,

$$q_{3a} = \begin{pmatrix} q \\ 0_{2m} \end{pmatrix}.$$

It remains to compute ϵ. There are two methods to do this.

The first method assumes the **Standard Margin Hypothesis**, which is that there is some i_0 such that $0 < \lambda_{i_0} < C/m$ *or* there is some j_0 such that $0 < \mu_{j_0} < C/m$; in other words, there is some support vector of type 1. By the complementary slackness conditions, $\xi_{i_0} = 0$ or $\xi'_{j_0} = 0$, so we have either $w^\top x_{i_0} + b - y_{i_0} = \epsilon$ or $-w^\top x_{j_0} - b + y_{j_0} = \epsilon$, which determines ϵ.

Due to numerical instability, when writing a computer program it is preferable to compute the lists of indices I_λ and I_μ given by

$$\begin{aligned} I_\lambda &= \{i \in \{1, \dots, m\} \mid 0 < \lambda_i < C/m\} \\ I_\mu &= \{j \in \{1, \dots, m\} \mid 0 < \mu_j < C/m\}. \end{aligned}$$

Then it is easy to see that we can compute ϵ using the following averaging formulae: if $I_\lambda \neq \emptyset$, then

$$\epsilon = w^\top \Big(\sum_{i \in I_\lambda} x_i\Big)/|I_\lambda| + b - \Big(\sum_{i \in I_\lambda} y_i\Big)/|I_\lambda|,$$

and if $I_\mu \neq \emptyset$, then

$$\epsilon = -w^\top \Big(\sum_{j \in I_\mu} x_j\Big)/|I_\mu| - b + \Big(\sum_{i \in I_\mu} y_i\Big)/|I_\mu|.$$

The second method uses duality. Under a mild condition, expressing that the duality gap is zero, we can determine ϵ in terms of λ, μ and b. This is because points x_i that fail the margin, which means that $\lambda_i = C/m$

or $\mu_i = C/m$, are the only points for which $\xi_i > 0$ or $\xi_i' > 0$. But in this case we have an active constraint

$$w^\top x_i + b - y_i = \epsilon + \xi_i \qquad (*_\xi)$$

or

$$-w^\top x_i - b + y_i = \epsilon + \xi_i', \qquad (*_{\xi'})$$

so ξ_i and ξ_i' can be expressed in terms of w and b. Since the duality gap is zero for an optimal solution, the optimal value of the primal is equal to the optimal value of the dual. Using the fact that

$$\begin{aligned} w &= X^\top(\mu - \lambda) \\ b &= -(\mathbf{1}_m^\top \lambda - \mathbf{1}_m^\top \mu) = \begin{pmatrix} \lambda^\top & \mu^\top \end{pmatrix} \begin{pmatrix} -\mathbf{1}_m \\ \mathbf{1}_m \end{pmatrix} \end{aligned}$$

we obtain an expression for the optimal value of the primal. First we have

$$\begin{aligned} \frac{1}{2}w^\top w + \frac{1}{2}b^2 &= \frac{1}{2}(\lambda^\top - \mu^\top)XX^\top(\lambda - \mu) \\ &\quad + \frac{1}{2}\begin{pmatrix} \lambda^\top & \mu^\top \end{pmatrix} \begin{pmatrix} \mathbf{1}_m\mathbf{1}_m^\top & -\mathbf{1}_m\mathbf{1}_m^\top \\ -\mathbf{1}_m\mathbf{1}_m^\top & \mathbf{1}_m\mathbf{1}_m^\top \end{pmatrix} \begin{pmatrix} \lambda \\ \mu \end{pmatrix} \\ &= \frac{1}{2}\begin{pmatrix} \lambda^\top & \mu^\top \end{pmatrix} \left(P + \begin{pmatrix} \mathbf{1}_m\mathbf{1}_m^\top & -\mathbf{1}_m\mathbf{1}_m^\top \\ -\mathbf{1}_m\mathbf{1}_m^\top & \mathbf{1}_m\mathbf{1}_m^\top \end{pmatrix} \right) \begin{pmatrix} \lambda \\ \mu \end{pmatrix}, \end{aligned}$$

with

$$P = \begin{pmatrix} XX^\top & -XX^\top \\ -XX^\top & XX^\top \end{pmatrix}.$$

Let K_λ and K_μ be the sets of indices corresponding to points failing the margin,

$$\begin{aligned} K_\lambda &= \{i \in \{1, \ldots, m\} \mid \lambda_i = C/m\} \\ K_\mu &= \{i \in \{1, \ldots, m\} \mid \mu_i = C/m\}. \end{aligned}$$

Because $\lambda_i \mu_i = 0$, the sets K_λ and K_μ are disjoint. Observe that from Definition 20.2 we have $p_f = |K_\lambda|$ and $q_f = |K_\mu|$. Then by $(*_\xi)$ and $(*_{\xi'})$, we have

$$\begin{aligned} \sum_{i=1}^m (\xi_i + \xi_i') &= \sum_{i \in K_\lambda} \xi_i + \sum_{j \in K_\mu} \xi_j' \\ &= \sum_{i \in K_\lambda} (w^\top x_i + b - y_i - \epsilon) + \sum_{j \in K_\mu} (-w^\top x_j - b + y_j - \epsilon) \\ &= w^\top \left(\sum_{i \in K_\lambda} x_i - \sum_{j \in K_\mu} x_j \right) - \sum_{i \in K_\lambda} y_i + \sum_{j \in K_\mu} y_j \\ &\quad + (p_f - q_f)b - (p_f + q_f)\epsilon. \end{aligned}$$

The optimal value of the dual is given by

$$-\frac{1}{2}\begin{pmatrix}\lambda^\top & \mu^\top\end{pmatrix}\left(P + \begin{pmatrix}\mathbf{1}_m\mathbf{1}_m^\top & -\mathbf{1}_m\mathbf{1}_m^\top \\ -\mathbf{1}_m\mathbf{1}_m^\top & \mathbf{1}_m\mathbf{1}_m^\top\end{pmatrix}\right)\begin{pmatrix}\lambda \\ \mu\end{pmatrix} - q^\top\begin{pmatrix}\lambda \\ \mu\end{pmatrix},$$

with

$$q = \begin{pmatrix}y \\ -y\end{pmatrix}.$$

Expressing that the duality gap is zero we have

$$\begin{aligned}&\frac{1}{2}w^\top w + \frac{1}{2}b^2 + C\nu\epsilon + \frac{C}{m}\sum_{i=1}^m(\xi_i + \xi_i') \\ &\qquad = -\frac{1}{2}\begin{pmatrix}\lambda^\top & \mu^\top\end{pmatrix}\left(P + \begin{pmatrix}\mathbf{1}_m\mathbf{1}_m^\top & -\mathbf{1}_m\mathbf{1}_m^\top \\ -\mathbf{1}_m\mathbf{1}_m^\top & \mathbf{1}_m\mathbf{1}_m^\top\end{pmatrix}\right)\begin{pmatrix}\lambda \\ \mu\end{pmatrix} - q^\top\begin{pmatrix}\lambda \\ \mu\end{pmatrix},\end{aligned}$$

that is,

$$\begin{aligned}&\frac{1}{2}\begin{pmatrix}\lambda^\top & \mu^\top\end{pmatrix}\left(P + \begin{pmatrix}\mathbf{1}_m\mathbf{1}_m^\top & -\mathbf{1}_m\mathbf{1}_m^\top \\ -\mathbf{1}_m\mathbf{1}_m^\top & \mathbf{1}_m\mathbf{1}_m^\top\end{pmatrix}\right)\begin{pmatrix}\lambda \\ \mu\end{pmatrix} + C\nu\epsilon \\ &\qquad + \frac{C}{m}\bigg(w^\top\Big(\sum_{i\in K_\lambda} x_i - \sum_{j\in K_\mu} x_j\Big) - \sum_{i\in K_\lambda} y_i + \sum_{j\in K_\mu} y_j \\ &\qquad\qquad + (p_f - q_f)b - (p_f + q_f)\epsilon\bigg) \\ &\qquad = -\frac{1}{2}\begin{pmatrix}\lambda^\top & \mu^\top\end{pmatrix}\left(P + \begin{pmatrix}\mathbf{1}_m\mathbf{1}_m^\top & -\mathbf{1}_m\mathbf{1}_m^\top \\ -\mathbf{1}_m\mathbf{1}_m^\top & \mathbf{1}_m\mathbf{1}_m^\top\end{pmatrix}\right)\begin{pmatrix}\lambda \\ \mu\end{pmatrix} - q^\top\begin{pmatrix}\lambda \\ \mu\end{pmatrix}.\end{aligned}$$

Solving for ϵ we get

$$\begin{aligned}C\left(\nu - \frac{p_f + q_f}{m}\right)\epsilon &= -\begin{pmatrix}\lambda^\top & \mu^\top\end{pmatrix}\left(P + \begin{pmatrix}\mathbf{1}_m\mathbf{1}_m^\top & -\mathbf{1}_m\mathbf{1}_m^\top \\ -\mathbf{1}_m\mathbf{1}_m^\top & \mathbf{1}_m\mathbf{1}_m^\top\end{pmatrix}\right)\begin{pmatrix}\lambda \\ \mu\end{pmatrix} \\ &\quad - \begin{pmatrix}y^\top & -y^\top\end{pmatrix}\begin{pmatrix}\lambda \\ \mu\end{pmatrix} - \frac{C}{m}\bigg(w^\top\Big(\sum_{i\in K_\lambda} x_i - \sum_{j\in K_\mu} x_j\Big) \\ &\quad - \sum_{i\in K_\lambda} y_i + \sum_{j\in K_\mu} y_j + (p_f - q_f)b\bigg),\end{aligned}$$

so we get

$$\begin{aligned}&(m\nu - p_f - q_f)\epsilon = \\ &-\Bigg(\frac{m}{C}\bigg(\begin{pmatrix}\lambda^\top & \mu^\top\end{pmatrix}\left(P + \begin{pmatrix}\mathbf{1}_m\mathbf{1}_m^\top & -\mathbf{1}_m\mathbf{1}_m^\top \\ -\mathbf{1}_m\mathbf{1}_m^\top & \mathbf{1}_m\mathbf{1}_m^\top\end{pmatrix}\right)\begin{pmatrix}\lambda \\ \mu\end{pmatrix} + \begin{pmatrix}y^\top & -y^\top\end{pmatrix}\begin{pmatrix}\lambda \\ \mu\end{pmatrix}\bigg) \\ &\qquad + w^\top\Big(\sum_{i\in K_\lambda} x_i - \sum_{j\in K_\mu} x_j\Big) - \sum_{i\in K_\lambda} y_i + \sum_{j\in K_\mu} y_j + (p_f - q_f)b\Bigg).\end{aligned}$$

Using the equations

$$w = X^\top(\mu - \lambda)$$

$$b = -(\mathbf{1}_m^\top \lambda - \mathbf{1}_m^\top \mu) = \begin{pmatrix}\lambda^\top & \mu^\top\end{pmatrix} \begin{pmatrix} -\mathbf{1}_m \\ \mathbf{1}_m \end{pmatrix},$$

we see that ϵ is determined by λ and μ provided that $(p_f + q_f)/m < \nu$.

By Proposition 20.6(1),

$$\frac{p_f + q_f}{m} \leq \nu,$$

therefore the condition $(p_f + q_f)/m < \nu$ is very natural.

We have implemented this method in `Matlab`, and we have observed that for some examples the choice of ν caused the equation $\nu(p_f + q_f) = m$ to hold. In such cases, running the program again with a slightly perturbed value of ν always succeeded.

The other observation we made is that b tends to be smaller and ϵ tends to be bigger in ν-SV Regression Version 2, so the fit is actually not as good as in ν-SV Regression without penalizing b. Figure 20.16 shows the result of running our program on the data set of Section 20.3. Compare with Figure 20.13.

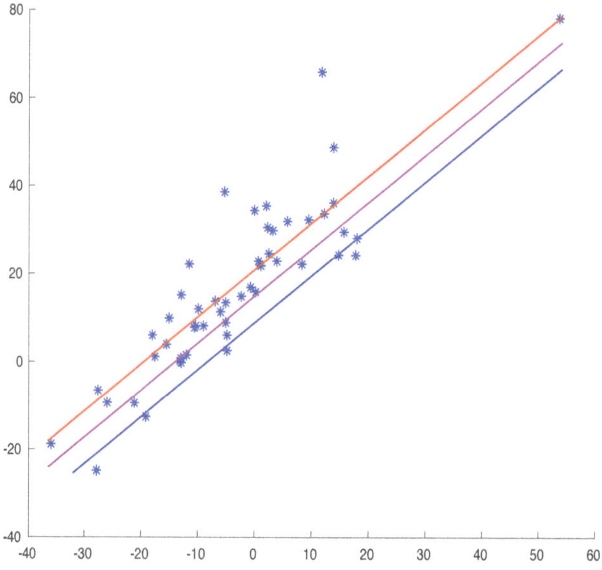

Fig. 20.16 Running ν-SV regression version 2 on a set of 50 points; $\nu = 0.5$.

20.6 Summary

The main concepts and results of this chapter are listed below:

- ν-support vector regression (ν-SV regression).
- Regression estimate.
- Kernel ν-SV regression.
- ϵ-SV regression, ϵ-insensitive SV regression.
- ν-SV regression Version 2; penalizing b.

20.7 Problems

Problem 20.1. Prove that if ν-SV regression succeeds and yields $w, b, \epsilon > 0$, then ϵ-SV regression with the same C and the same value of ϵ also succeeds and returns the same pair (w, b).

Problem 20.2. Prove the formulae

$$\begin{aligned} b &= \Bigg(\Big(\sum_{i_0 \in I_\lambda} y_{i_0}\Big)/|I_\lambda| + \Big(\sum_{j_0 \in I_\mu} y_{j_0}\Big)/|I_\mu|\Bigg)/2 \\ &\quad - w^\top \Bigg(\Big(\sum_{i_0 \in I_\lambda} x_{i_0}\Big)/|I_\lambda| + \Big(\sum_{j_0 \in I_\mu} x_{j_0}\Big)/|I_\mu|\Bigg)/2 \\ \epsilon &= \Bigg(\Big(\sum_{j_0 \in I_\mu} y_{j_0}\Big)/|I_\mu|\Bigg) - \Big(\sum_{i_0 \in I_\lambda} y_{i_0}\Big)/|I_\lambda|\Bigg)/2 \\ &\quad + w^\top \Bigg(\Big(\sum_{i_0 \in I_\lambda} x_{i_0}\Big)/|I_\lambda| - \Big(\sum_{j_0 \in I_\mu} x_{j_0}\Big)/|I_\mu|\Bigg)/2 \end{aligned}$$

stated just before Proposition 20.5.

Problem 20.3. Give the details of the proof of Proposition 20.5. In particular, prove that

$$\begin{aligned} C\left(\nu - \frac{p_f + q_f}{m}\right)\epsilon &= -\begin{pmatrix}\lambda^\top & \mu^\top\end{pmatrix} P \begin{pmatrix}\lambda \\ \mu\end{pmatrix} - \begin{pmatrix}y^\top & -y^\top\end{pmatrix}\begin{pmatrix}\lambda \\ \mu\end{pmatrix} \\ &\quad - \frac{C}{m}\left(w^\top\Big(\sum_{i \in K_\lambda} x_i - \sum_{j \in K_\mu} x_j\Big) - \sum_{i \in K_\lambda} y_i + \sum_{j \in K_\mu} y_j + (p_f - q_f)b\right). \end{aligned}$$

Problem 20.4. Prove that the matrices

$$A = \begin{pmatrix} \mathbf{1}_m^\top & -\mathbf{1}_m^\top & 0_m^\top & 0_m^\top & 0 \\ \mathbf{1}_m^\top & \mathbf{1}_m^\top & 0_m^\top & 0_m^\top & 1 \\ I_m & 0_{m,m} & I_m & 0_{m,m} & 0_m \\ 0_{m,m} & I_m & 0_{m,m} & I_m & 0_m \end{pmatrix}, \quad A_2 = \begin{pmatrix} \mathbf{1}_m^\top & -\mathbf{1}_m^\top & 0_m^\top & 0_m^\top \\ \mathbf{1}_m^\top & \mathbf{1}_m^\top & 0_m^\top & 0_m^\top \\ I_m & 0_{m,m} & I_m & 0_{m,m} \\ 0_{m,m} & I_m & 0_{m,m} & I_m \end{pmatrix}$$

have rank $2m + 2$.

Problem 20.5. Derive the version of ν-SV regression in which the linear penalty function $\sum_{i=1}^m (\xi_i + \xi_i')$ is replaced by the quadratic penalty function $\sum_{i=1}^m (\xi_i^2 + \xi_i'^2)$. Derive the dual program.

Problem 20.6. The linear penalty function $\sum_{i=1}^m (\xi_i + \xi_i')$ can be replaced by the quadratic penalty function $\sum_{i=1}^m (\xi_i^2 + \xi_i'^2)$. Prove that for an optimal solution we must have $\xi_i \geq 0$ and $\xi_i' \geq 0$, so we may omit the constraints $\xi_i \geq 0$ and $\xi_i' \geq 0$. We must also have $\gamma = 0$ so we may omit the variable γ as well. Prove that $\xi = (m/2C)\lambda$ and $\xi' = (m/2C)\mu$. This problem is very similar to the Soft Margin SVM (SVM_{s4}) discussed in Section 18.13.

Problem 20.7. Consider the version of ν-SV regression in Section 20.5. Prove that for any optimal solution with $w \neq 0$ and $\epsilon > 0$, if the inequalities $(p_f + q_f)/m < \nu < 1$ hold, then some point x_i is a support vector.

Problem 20.8. Prove that the matrix

$$A_3 = \begin{pmatrix} \mathbf{1}_m^\top & \mathbf{1}_m^\top & 0_m^\top & 0_m^\top \\ I_m & 0_{m,m} & I_m & 0_{m,m} \\ 0_{m,m} & I_m & 0_{m,m} & I_m \end{pmatrix}$$

has rank $2m + 1$.

Problem 20.9. Consider the version of ν-SV regression in Section 20.5. Prove the following formulae: If $I_\lambda \neq \emptyset$, then

$$\epsilon = w^\top \Big(\sum_{i \in I_\lambda} x_i \Big) / |I_\lambda| + b - \Big(\sum_{i \in I_\lambda} y_i \Big) / |I_\lambda|,$$

and if $I_\mu \neq \emptyset$, then

$$\epsilon = -w^\top \Big(\sum_{j \in I_\mu} x_j \Big) / |I_\mu| - b + \Big(\sum_{i \in I_\mu} y_i \Big) / |I_\mu|.$$

Problem 20.10. Implement ν-Regression Version 2 described in Section 20.5. Run examples using both the original version and version 2 and compare the results.

Appendix A

Total Orthogonal Families in Hilbert Spaces

A.1 Total Orthogonal Families (Hilbert Bases), Fourier Coefficients

We conclude our quick tour of Hilbert spaces by showing that the notion of orthogonal basis can be generalized to Hilbert spaces. However, the useful notion is not the usual notion of a basis, but a notion which is an abstraction of the concept of Fourier series. Every element of a Hilbert space is the "sum" of its Fourier series.

Definition A.1. Given a Hilbert space E, a family $(u_k)_{k\in K}$ of nonnull vectors is an *orthogonal family* iff the u_k are pairwise orthogonal, i.e., $\langle u_i, u_j\rangle = 0$ for all $i \neq j$ $(i, j \in K)$, and an *orthonormal family* iff $\langle u_i, u_j\rangle = \delta_{i,j}$, for all $i, j \in K$. A *total orthogonal family (or system)* or *Hilbert basis* is an orthogonal family that is dense in E. This means that for every $v \in E$, for every $\epsilon > 0$, there is some finite subset $I \subseteq K$ and some family $(\lambda_i)_{i\in I}$ of complex numbers, such that

$$\left\| v - \sum_{i\in I} \lambda_i u_i \right\| < \epsilon.$$

Given an orthogonal family $(u_k)_{k\in K}$, for every $v \in E$, for every $k \in K$, the scalar $c_k = \langle v, u_k\rangle / \|u_k\|^2$ is called the *k-th Fourier coefficient of v over* $(u_k)_{k\in K}$.

Remark. The terminology Hilbert basis is misleading because a Hilbert basis $(u_k)_{k\in K}$ is not necessarily a basis in the algebraic sense. Indeed, in general, $(u_k)_{k\in K}$ does not span E. Intuitively, it takes linear combinations of the u_k's with infinitely many nonnull coefficients to span E. Technically, this is achieved in terms of limits. In order to avoid the confusion between bases in the algebraic sense and Hilbert bases, some authors refer to algebraic bases as *Hamel bases* and to total orthogonal families (or Hilbert bases) as *Schauder bases*.

Given an orthogonal family $(u_k)_{k\in K}$, for any finite subset I of K, we often call sums of the form $\sum_{i\in I}\lambda_i u_i$ *partial sums of Fourier series*, and if these partial sums converge to a limit denoted as $\sum_{k\in K} c_k u_k$, we call $\sum_{k\in K} c_k u_k$ a *Fourier series*.

However, we have to make sense of such sums! Indeed, when K is unordered or uncountable, the notion of limit or sum has not been defined. This can be done as follows (for more details, see Dixmier [Dixmier (1984)]):

Definition A.2. Given a normed vector space E (say, a Hilbert space), for any nonempty index set K, we say that a family $(u_k)_{k\in K}$ of vectors in E is *summable with sum* $v \in E$ iff for every $\epsilon > 0$, there is some finite subset I of K, such that,

$$\left\| v - \sum_{j\in J} u_j \right\| < \epsilon$$

for every finite subset J with $I \subseteq J \subseteq K$. We say that the family $(u_k)_{k\in K}$ is *summable* iff there is some $v \in E$ such that $(u_k)_{k\in K}$ is summable with sum v. A family $(u_k)_{k\in K}$ is a *Cauchy family* iff for every $\epsilon > 0$, there is a finite subset I of K, such that,

$$\left\| \sum_{j\in J} u_j \right\| < \epsilon$$

for every finite subset J of K with $I \cap J = \emptyset$,

If $(u_k)_{k\in K}$ is summable with sum v, we usually denote v as $\sum_{k\in K} u_k$. The following technical proposition will be needed:

Proposition A.1. *Let E be a complete normed vector space (say, a Hilbert space).*

(1) For any nonempty index set K, a family $(u_k)_{k\in K}$ is summable iff it is a Cauchy family.
(2) Given a family $(r_k)_{k\in K}$ of nonnegative reals $r_k \geq 0$, if there is some real number $B > 0$ such that $\sum_{i\in I} r_i < B$ for every finite subset I of K, then $(r_k)_{k\in K}$ is summable and $\sum_{k\in K} r_k = r$, where r is least upper bound of the set of finite sums $\sum_{i\in I} r_i$ $(I \subseteq K)$.

Proof. (1) If $(u_k)_{k\in K}$ is summable, for every finite subset I of K, let

$$u_I = \sum_{i\in I} u_i \quad \text{and} \quad u = \sum_{k\in K} u_k$$

For every $\epsilon > 0$, there is some finite subset I of K such that

$$\|u - u_L\| < \epsilon/2$$

for all finite subsets L such that $I \subseteq L \subseteq K$. For every finite subset J of K such that $I \cap J = \emptyset$, since $I \subseteq I \cup J \subseteq K$ and $I \cup J$ is finite, we have

$$\|u - u_{I \cup J}\| < \epsilon/2 \quad \text{and} \quad \|u - u_I\| < \epsilon/2,$$

and since

$$\|u_{I \cup J} - u_I\| \leq \|u_{I \cup J} - u\| + \|u - u_I\|$$

and $u_{I \cup J} - u_I = u_J$ since $I \cap J = \emptyset$, we get

$$\|u_J\| = \|u_{I \cup J} - u_I\| < \epsilon,$$

which is the condition for $(u_k)_{k \in K}$ to be a Cauchy family.

Conversely, assume that $(u_k)_{k \in K}$ is a Cauchy family. We define inductively a decreasing sequence (X_n) of subsets of E, each of diameter at most $1/n$, as follows: For $n = 1$, since $(u_k)_{k \in K}$ is a Cauchy family, there is some finite subset J_1 of K such that

$$\|u_J\| < 1/2$$

for every finite subset J of K with $J_1 \cap J = \emptyset$. We pick some finite subset J_1 with the above property, and we let $I_1 = J_1$ and

$$X_1 = \{u_I \mid I_1 \subseteq I \subseteq K,\ I \text{ finite}\}.$$

For $n \geq 1$, there is some finite subset J_{n+1} of K such that

$$\|u_J\| < 1/(2n + 2)$$

for every finite subset J of K with $J_{n+1} \cap J = \emptyset$. We pick some finite subset J_{n+1} with the above property, and we let $I_{n+1} = I_n \cup J_{n+1}$ and

$$X_{n+1} = \{u_I \mid I_{n+1} \subseteq I \subseteq K,\ I \text{ finite}\}.$$

Since $I_n \subseteq I_{n+1}$, it is obvious that $X_{n+1} \subseteq X_n$ for all $n \geq 1$. We need to prove that each X_n has diameter at most $1/n$. Since J_n was chosen such that

$$\|u_J\| < 1/(2n)$$

for every finite subset J of K with $J_n \cap J = \emptyset$, and since $J_n \subseteq I_n$, it is also true that

$$\|u_J\| < 1/(2n)$$

for every finite subset J of K with $I_n \cap J = \emptyset$ (since $I_n \cap J = \emptyset$ and $J_n \subseteq I_n$ implies that $J_n \cap J = \emptyset$). Then for every two finite subsets J, L such that $I_n \subseteq J, L \subseteq K$, we have

$$\|u_{J-I_n}\| < 1/(2n) \quad \text{and} \quad \|u_{L-I_n}\| < 1/(2n),$$

and since

$$\|u_J - u_L\| \leq \|u_J - u_{I_n}\| + \|u_{I_n} - u_L\| = \|u_{J-I_n}\| + \|u_{L-I_n}\|,$$

we get

$$\|u_J - u_L\| < 1/n,$$

which proves that $\delta(X_n) \leq 1/n$. Now if we consider the sequence of closed sets $(\overline{X_n})$, we still have $\overline{X_{n+1}} \subseteq \overline{X_n}$, and by Proposition 12.3, $\delta(\overline{X_n}) = \delta(X_n) \leq 1/n$, which means that $\lim_{n\to\infty} \delta(\overline{X_n}) = 0$, and by Proposition 12.3, $\bigcap_{n=1}^{\infty} \overline{X_n}$ consists of a single element u. We claim that u is the sum of the family $(u_k)_{k\in K}$.

For every $\epsilon > 0$, there is some $n \geq 1$ such that $n > 2/\epsilon$, and since $u \in \overline{X_m}$ for all $m \geq 1$, there is some finite subset J_0 of K such that $I_n \subseteq J_0$ and

$$\|u - u_{J_0}\| < \epsilon/2,$$

where I_n is the finite subset of K involved in the definition of X_n. However, since $\delta(X_n) \leq 1/n$, for every finite subset J of K such that $I_n \subseteq J$, we have

$$\|u_J - u_{J_0}\| \leq 1/n < \epsilon/2,$$

and since

$$\|u - u_J\| \leq \|u - u_{J_0}\| + \|u_{J_0} - u_J\|,$$

we get

$$\|u - u_J\| < \epsilon$$

for every finite subset J of K with $I_n \subseteq J$, which proves that u is the sum of the family $(u_k)_{k\in K}$.

(2) Since every finite sum $\sum_{i\in I} r_i$ is bounded by the uniform bound B, the set of these finite sums has a least upper bound $r \leq B$. For every $\epsilon > 0$, since r is the least upper bound of the finite sums $\sum_{i\in I} r_i$ (where I finite, $I \subseteq K$), there is some finite $I \subseteq K$ such that

$$\left| r - \sum_{i\in I} r_i \right| < \epsilon,$$

and since $r_k \geq 0$ for all $k \in K$, we have

$$\sum_{i\in I} r_i \leq \sum_{j\in J} r_j$$

whenever $I \subseteq J$, which shows that

$$\left| r - \sum_{j\in J} r_j \right| \leq \left| r - \sum_{i\in I} r_i \right| < \epsilon$$

for every finite subset J such that $I \subseteq J \subseteq K$, proving that $(r_k)_{k\in K}$ is summable with sum $\sum_{k\in K} r_k = r$. □

Remark. The notion of summability implies that the sum of a family $(u_k)_{k\in K}$ is independent of any order on K. In this sense it is a kind of "commutative summability." More precisely, it is easy to show that for every bijection $\varphi\colon K \to K$ (intuitively, a reordering of K), the family $(u_k)_{k\in K}$ is summable iff the family $(u_l)_{l\in\varphi(K)}$ is summable, and if so, they have the same sum.

The following proposition gives some of the main properties of Fourier coefficients. Among other things, at most countably many of the Fourier coefficient may be nonnull, and the partial sums of a Fourier series converge. Given an orthogonal family $(u_k)_{k\in K}$, we let $U_k = \mathbb{C}u_k$, and $p_{U_k}\colon E \to U_k$ is the projection of E onto U_k.

Proposition A.2. *Let E be a Hilbert space, $(u_k)_{k\in K}$ an orthogonal family in E, and V the closure of the subspace generated by $(u_k)_{k\in K}$. The following properties hold:*

(1) For every $v \in E$, for every finite subset $I \subseteq K$, we have

$$\sum_{i\in I} |c_i|^2 \leq \|v\|^2,$$

where the c_k are the Fourier coefficients of v.

(2) For every vector $v \in E$, if $(c_k)_{k\in K}$ are the Fourier coefficients of v, the following conditions are equivalent:

(2a) $v \in V$

(2b) The family $(c_k u_k)_{k\in K}$ is summable and $v = \sum_{k\in K} c_k u_k$.

(2c) The family $(|c_k|^2)_{k\in K}$ is summable and $\|v\|^2 = \sum_{k\in K} |c_k|^2$;

(3) The family $(|c_k|^2)_{k\in K}$ is summable, and we have the Bessel inequality:

$$\sum_{k\in K} |c_k|^2 \leq \|v\|^2.$$

As a consequence, at most countably many of the c_k *may be nonzero. The family* $(c_k u_k)_{k \in K}$ *forms a Cauchy family, and thus, the Fourier series* $\sum_{k \in K} c_k u_k$ *converges in* E *to some vector* $u = \sum_{k \in K} c_k u_k$. *Furthermore,* $u = p_V(v)$.

See Figure A.1.

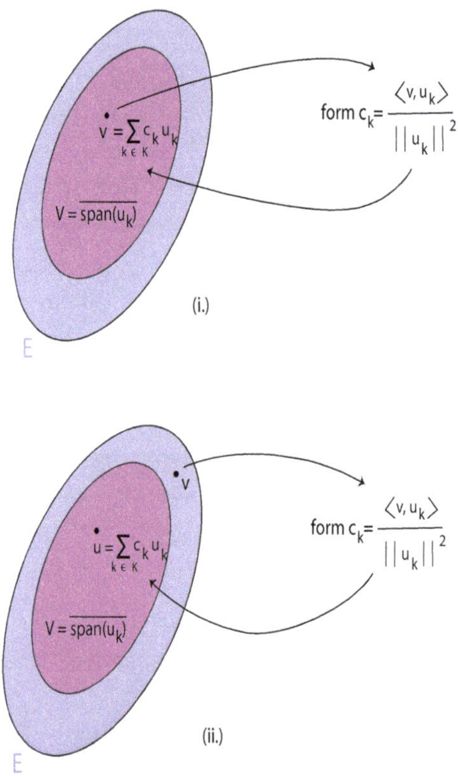

Fig. A.1 A schematic illustration of Proposition A.2. Figure (i.) illustrates Condition (2b), while Figure (ii.) illustrates Condition (3). Note E is the purple oval and V is the magenta oval. In both cases, take a vector of E, form the Fourier coefficients c_k, then form the Fourier series $\sum_{k \in K} c_k u_k$. Condition (2b) ensures v equals its Fourier series since $v \in V$. However, if $v \notin V$, the Fourier series does not equal v. Eventually, we will discover that $V = E$, which implies that that Fourier series converges to its vector v.

Proof. (1) Let

$$u_I = \sum_{i \in I} c_i u_i$$

for any finite subset I of K. We claim that $v - u_I$ is orthogonal to u_i for every $i \in I$. Indeed,

$$\begin{aligned}
\langle v - u_I, u_i \rangle &= \left\langle v - \sum_{j \in I} c_j u_j, u_i \right\rangle \\
&= \langle v, u_i \rangle - \sum_{j \in I} c_j \langle u_j, u_i \rangle \\
&= \langle v, u_i \rangle - c_i \|u_i\|^2 \\
&= \langle v, u_i \rangle - \langle v, u_i \rangle = 0,
\end{aligned}$$

since $\langle u_j, u_i \rangle = 0$ for all $i \neq j$ and $c_i = \langle v, u_i \rangle / \|u_i\|^2$. As a consequence, we have

$$\begin{aligned}
\|v\|^2 &= \left\| v - \sum_{i \in I} c_i u_i + \sum_{i \in I} c_i u_i \right\|^2 \\
&= \left\| v - \sum_{i \in I} c_i u_i \right\|^2 + \left\| \sum_{i \in I} c_i u_i \right\|^2 \\
&= \left\| v - \sum_{i \in I} c_i u_i \right\|^2 + \sum_{i \in I} |c_i|^2,
\end{aligned}$$

since the u_i are pairwise orthogonal, that is,

$$\|v\|^2 = \left\| v - \sum_{i \in I} c_i u_i \right\|^2 + \sum_{i \in I} |c_i|^2.$$

Thus,

$$\sum_{i \in I} |c_i|^2 \leq \|v\|^2,$$

as claimed.

(2) We prove the chain of implications $(a) \Rightarrow (b) \Rightarrow (c) \Rightarrow (a)$.

$(a) \Rightarrow (b)$: If $v \in V$, since V is the closure of the subspace spanned by $(u_k)_{k \in K}$, for every $\epsilon > 0$, there is some finite subset I of K and some family $(\lambda_i)_{i \in I}$ of complex numbers, such that

$$\left\| v - \sum_{i \in I} \lambda_i u_i \right\| < \epsilon.$$

Now for every finite subset J of K such that $I \subseteq J$, we have

$$\begin{aligned}\Big\|v - \sum_{i\in I}\lambda_i u_i\Big\|^2 &= \Big\|v - \sum_{j\in J} c_j u_j + \sum_{j\in J} c_j u_j - \sum_{i\in I}\lambda_i u_i\Big\|^2 \\ &= \Big\|v - \sum_{j\in J} c_j u_j\Big\|^2 + \Big\|\sum_{j\in J} c_j u_j - \sum_{i\in I}\lambda_i u_i\Big\|^2,\end{aligned}$$

since $I \subseteq J$ and the u_j (with $j \in J$) are orthogonal to $v - \sum_{j\in J} c_j u_j$ by the argument in (1), which shows that

$$\Big\|v - \sum_{j\in J} c_j u_j\Big\| \leq \Big\|v - \sum_{i\in I}\lambda_i u_i\Big\| < \epsilon,$$

and thus, that the family $(c_k u_k)_{k\in K}$ is summable with sum v, so that

$$v = \sum_{k\in K} c_k u_k.$$

(b) $\Rightarrow$ *(c)*: If $v = \sum_{k\in K} c_k u_k$, then for every $\epsilon > 0$, there some finite subset I of K, such that

$$\Big\|v - \sum_{j\in J} c_j u_j\Big\| < \sqrt{\epsilon},$$

for every finite subset J of K such that $I \subseteq J$, and since we proved in (1) that

$$\|v\|^2 = \Big\|v - \sum_{j\in J} c_j u_j\Big\|^2 + \sum_{j\in J} |c_j|^2,$$

we get

$$\|v\|^2 - \sum_{j\in J} |c_j|^2 < \epsilon,$$

which proves that $(|c_k|^2)_{k\in K}$ is summable with sum $\|v\|^2$.

(c) $\Rightarrow$ *(a)*: Finally, if $(|c_k|^2)_{k\in K}$ is summable with sum $\|v\|^2$, for every $\epsilon > 0$, there is some finite subset I of K such that

$$\|v\|^2 - \sum_{j\in J} |c_j|^2 < \epsilon^2$$

for every finite subset J of K such that $I \subseteq J$, and again, using the fact that

$$\|v\|^2 = \Big\|v - \sum_{j\in J} c_j u_j\Big\|^2 + \sum_{j\in J} |c_j|^2,$$

we get

$$\left\| v - \sum_{j \in J} c_j u_j \right\| < \epsilon,$$

which proves that $(c_k u_k)_{k \in K}$ is summable with sum $\sum_{k \in K} c_k u_k = v$, and $v \in V$.

(3) Since $\sum_{i \in I} |c_i|^2 \leq \|v\|^2$ for every finite subset I of K, by Proposition A.1(2), the family $(|c_k|^2)_{k \in K}$ is summable. The Bessel inequality

$$\sum_{k \in K} |c_k|^2 \leq \|v\|^2$$

is an obvious consequence of the inequality $\sum_{i \in I} |c_i|^2 \leq \|v\|^2$ (for every finite $I \subseteq K$). Now for every natural number $n \geq 1$, if K_n is the subset of K consisting of all c_k such that $|c_k| \geq 1/n$, the number of elements in K_n is at most

$$\sum_{k \in K_n} |nc_k|^2 \leq n^2 \sum_{k \in K} |c_k|^2 \leq n^2 \|v\|^2,$$

which is finite, and thus, at most a countable number of the c_k may be nonzero.

Since $(|c_k|^2)_{k \in K}$ is summable with sum c, by Proposition A.1(1) we know that for every $\epsilon > 0$, there is some finite subset I of K such that

$$\sum_{j \in J} |c_j|^2 < \epsilon^2$$

for every finite subset J of K such that $I \cap J = \emptyset$. Since

$$\left\| \sum_{j \in J} c_j u_j \right\|^2 = \sum_{j \in J} |c_j|^2,$$

we get

$$\left\| \sum_{j \in J} c_j u_j \right\| < \epsilon.$$

This proves that $(c_k u_k)_{k \in K}$ is a Cauchy family, which, by Proposition A.1(1), implies that $(c_k u_k)_{k \in K}$ is summable since E is complete. Thus, the Fourier series $\sum_{k \in K} c_k u_k$ is summable, with its sum denoted $u \in V$.

Since $\sum_{k \in K} c_k u_k$ is summable with sum u, for every $\epsilon > 0$, there is some finite subset I_1 of K such that

$$\left\| u - \sum_{j \in J} c_j u_j \right\| < \epsilon$$

for every finite subset J of K such that $I_1 \subseteq J$. By the triangle inequality, for every finite subset I of K,

$$\left\|u - v\right\| \leq \left\|u - \sum_{i \in I} c_i u_i\right\| + \left\|\sum_{i \in I} c_i u_i - v\right\|.$$

By (2), every $w \in V$ is the sum of its Fourier series $\sum_{k \in K} \lambda_k u_k$, and for every $\epsilon > 0$, there is some finite subset I_2 of K such that

$$\left\|w - \sum_{j \in J} \lambda_j u_j\right\| < \epsilon$$

for every finite subset J of K such that $I_2 \subseteq J$. By the triangle inequality, for every finite subset I of K,

$$\left\|v - \sum_{i \in I} \lambda_i u_i\right\| \leq \|v - w\| + \left\|w - \sum_{i \in I} \lambda_i u_i\right\|.$$

Letting $I = I_1 \cup I_2$, since we showed in (2) that

$$\left\|v - \sum_{i \in I} c_i u_i\right\| \leq \left\|v - \sum_{i \in I} \lambda_i u_i\right\|$$

for every finite subset I of K, we get

$$\begin{aligned}
\|u - v\| &\leq \left\|u - \sum_{i \in I} c_i u_i\right\| + \left\|\sum_{i \in I} c_i u_i - v\right\| \\
&\leq \left\|u - \sum_{i \in I} c_i u_i\right\| + \left\|\sum_{i \in I} \lambda_i u_i - v\right\| \\
&\leq \left\|u - \sum_{i \in I} c_i u_i\right\| + \|v - w\| + \left\|w - \sum_{i \in I} \lambda_i u_i\right\|,
\end{aligned}$$

and thus

$$\|u - v\| \leq \|v - w\| + 2\epsilon.$$

Since this holds for every $\epsilon > 0$, we have

$$\|u - v\| \leq \|v - w\|$$

for all $w \in V$, i.e., $\|v - u\| = d(v, V)$, with $u \in V$, which proves that $u = p_V(v)$. ☐

A.2 The Hilbert Space $\ell^2(K)$ and the Riesz–Fischer Theorem

Proposition A.2 suggests looking at the space of sequences $(z_k)_{k\in K}$ (where $z_k \in \mathbb{C}$) such that $(|z_k|^2)_{k\in K}$ is summable. Indeed, such spaces are Hilbert spaces, and it turns out that every Hilbert space is isomorphic to one of those. Such spaces are the infinite-dimensional version of the spaces $\mathbb{C}^n$ under the usual Euclidean norm.

Definition A.3. Given any nonempty index set K, the space $\ell^2(K)$ is the set of all sequences $(z_k)_{k\in K}$, where $z_k \in \mathbb{C}$, such that $(|z_k|^2)_{k\in K}$ is summable, i.e., $\sum_{k\in K} |z_k|^2 < \infty$.

Remarks.

(1) When K is a finite set of cardinality n, $\ell^2(K)$ is isomorphic to $\mathbb{C}^n$.
(2) When $K = \mathbb{N}$, the space $\ell^2(\mathbb{N})$ corresponds to the space ℓ^2 of Example 2 in Section 13.1 (Vol. I). In that example, we claimed that ℓ^2 was a Hermitian space, and in fact, a Hilbert space. We now prove this fact for any index set K.

Proposition A.3. *Given any nonempty index set K, the space $\ell^2(K)$ is a Hilbert space under the Hermitian product*

$$\langle (x_k)_{k\in K}, (y_k)_{k\in K}\rangle = \sum_{k\in K} x_k\overline{y_k}.$$

The subspace consisting of sequences $(z_k)_{k\in K}$ such that $z_k = 0$, except perhaps for finitely many k, is a dense subspace of $\ell^2(K)$.

Proof. First we need to prove that $\ell^2(K)$ is a vector space. Assume that $(x_k)_{k\in K}$ and $(y_k)_{k\in K}$ are in $\ell^2(K)$. This means that $(|x_k|^2)_{k\in K}$ and $(|y_k|^2)_{k\in K}$ are summable, which, in view of Proposition A.1(2), is equivalent to the existence of some positive bounds A and B such that $\sum_{i\in I} |x_i|^2 < A$ and $\sum_{i\in I} |y_i|^2 < B$, for every finite subset I of K. To prove that $(|x_k + y_k|^2)_{k\in K}$ is summable, it is sufficient to prove that there is some $C > 0$ such that $\sum_{i\in I} |x_i + y_i|^2 < C$ for every finite subset I of K. However, the parallelogram inequality implies that

$$\sum_{i\in I} |x_i + y_i|^2 \leq \sum_{i\in I} 2(|x_i|^2 + |y_i|^2) \leq 2(A + B),$$

for every finite subset I of K, and we conclude by Proposition A.1(2). Similarly, for every $\lambda \in \mathbb{C}$,

$$\sum_{i\in I} |\lambda x_i|^2 \leq \sum_{i\in I} |\lambda|^2|x_i|^2 \leq |\lambda|^2 A,$$

and $(\lambda_k x_k)_{k\in K}$ is summable. Therefore, $\ell^2(K)$ is a vector space.

By the Cauchy-Schwarz inequality,

$$\sum_{i\in I} |x_i\overline{y_i}| \leq \sum_{i\in I} |x_i||y_i| \leq \Big(\sum_{i\in I} |x_i|^2\Big)^{1/2}\Big(\sum_{i\in I} |xy_i|^2\Big)^{1/2}$$
$$\leq \sum_{i\in I}(|x_i|^2 + |y_i|^2)/2 \leq (A+B)/2,$$

for every finite subset I of K. For the third inequality we used the fact that

$$4CD \leq (C+D)^2,$$

(with $C = \sum_{i\in I} |x_i|^2$ and $D = \sum_{i\in I} |y_i|^2$) which is equivalent to

$$(C-D)^2 \geq 0.$$

By Proposition A.1(2), $(|x_k\overline{y_k}|)_{k\in K}$ is summable. The customary language is that $(x_k\overline{y_k})_{k\in K}$ is absolutely summable. However, it is a standard fact that this implies that $(x_k\overline{y_k})_{k\in K}$ is summable (For every $\epsilon > 0$, there is some finite subset I of K such that

$$\sum_{j\in J} |x_j\overline{y_j}| < \epsilon$$

for every finite subset J of K such that $I \cap J = \emptyset$, and thus

$$|\sum_{j\in J} x_j\overline{y_j}| \leq \sum_{i\in J} |x_j\overline{y_j}| < \epsilon,$$

proving that $(x_k\overline{y_k})_{k\in K}$ is a Cauchy family, and thus summable.) We still have to prove that $\ell^2(K)$ is complete.

Consider a sequence $((\lambda_k^n)_{k\in K})_{n\geq 1}$ of sequences $(\lambda_k^n)_{k\in K} \in \ell^2(K)$, and assume that it is a Cauchy sequence. This means that for every $\epsilon > 0$, there is some $N \geq 1$ such that

$$\sum_{k\in K} |\lambda_k^m - \lambda_k^n|^2 < \epsilon^2$$

for all $m, n \geq N$. For every fixed $k \in K$, this implies that

$$|\lambda_k^m - \lambda_k^n| < \epsilon$$

for all $m, n \geq N$, which shows that $(\lambda_k^n)_{n\geq 1}$ is a Cauchy sequence in $\mathbb{C}$. Since $\mathbb{C}$ is complete, the sequence $(\lambda_k^n)_{n\geq 1}$ has a limit $\lambda_k \in \mathbb{C}$. We claim that $(\lambda_k)_{k\in K} \in \ell^2(K)$ and that this is the limit of $((\lambda_k^n)_{k\in K})_{n\geq 1}$.

Given any $\epsilon > 0$, the fact that $((\lambda_k^n)_{k\in K})_{n\geq 1}$ is a Cauchy sequence implies that there is some $N \geq 1$ such that for every finite subset I of K, we have

$$\sum_{i\in I} |\lambda_i^m - \lambda_i^n|^2 < \epsilon/4$$

for all $m, n \geq N$. Let $p = |I|$. Then

$$|\lambda_i^m - \lambda_i^n| < \frac{\sqrt{\epsilon}}{2\sqrt{p}}$$

for every $i \in I$. Since λ_i is the limit of $(\lambda_i^n)_{n\geq 1}$, we can find some n large enough so that

$$|\lambda_i^n - \lambda_i| < \frac{\sqrt{\epsilon}}{2\sqrt{p}}$$

for every $i \in I$. Since

$$|\lambda_i^m - \lambda_i| \leq |\lambda_i^m - \lambda_i^n| + |\lambda_i^n - \lambda_i|,$$

we get

$$|\lambda_i^m - \lambda_i| < \frac{\sqrt{\epsilon}}{\sqrt{p}},$$

and thus,

$$\sum_{i\in I} |\lambda_i^m - \lambda_i|^2 < \epsilon,$$

for all $m \geq N$. Since the above holds for every finite subset I of K, by Proposition A.1(2), we get

$$\sum_{k\in K} |\lambda_k^m - \lambda_k|^2 < \epsilon,$$

for all $m \geq N$. This proves that $(\lambda_k^m - \lambda_k)_{k\in K} \in \ell^2(K)$ for all $m \geq N$, and since $\ell^2(K)$ is a vector space and $(\lambda_k^m)_{k\in K} \in \ell^2(K)$ for all $m \geq 1$, we get $(\lambda_k)_{k\in K} \in \ell^2(K)$. However,

$$\sum_{k\in K} |\lambda_k^m - \lambda_k|^2 < \epsilon$$

for all $m \geq N$, means that the sequence $(\lambda_k^m)_{k\in K}$ converges to $(\lambda_k)_{k\in K} \in \ell^2(K)$. The fact that the subspace consisting of sequences $(z_k)_{k\in K}$ such that $z_k = 0$ except perhaps for finitely many k is a dense subspace of $\ell^2(K)$ is left as an easy exercise. ☐

Remark: The subspace consisting of all sequences $(z_k)_{k \in K}$ such that $z_k = 0$, except perhaps for finitely many k, provides an example of a subspace which is not closed in $\ell^2(K)$. Indeed, this space is strictly contained in $\ell^2(K)$, since there are countable sequences of nonnull elements in $\ell^2(K)$ (why?).

We just need two more propositions before being able to prove that every Hilbert space is isomorphic to some $\ell^2(K)$.

Proposition A.4. *Let E be a Hilbert space, and $(u_k)_{k \in K}$ an orthogonal family in E. The following properties hold:*

(1) For every family $(\lambda_k)_{k \in K} \in \ell^2(K)$, the family $(\lambda_k u_k)_{k \in K}$ is summable. Furthermore, $v = \sum_{k \in K} \lambda_k u_k$ is the only vector such that $c_k = \lambda_k$ for all $k \in K$, where the c_k are the Fourier coefficients of v.
(2) For any two families $(\lambda_k)_{k \in K} \in \ell^2(K)$ and $(\mu_k)_{k \in K} \in \ell^2(K)$, if $v = \sum_{k \in K} \lambda_k u_k$ and $w = \sum_{k \in K} \mu_k u_k$, we have the following equation, also called Parseval identity:

$$\langle v, w \rangle = \sum_{k \in K} \lambda_k \overline{\mu_k}.$$

Proof. (1) The fact that $(\lambda_k)_{k \in K} \in \ell^2(K)$ means that $(|\lambda_k|^2)_{k \in K}$ is summable. The proof given in Proposition A.2 (3) applies to the family $(|\lambda_k|^2)_{k \in K}$ (instead of $(|c_k|^2)_{k \in K}$), and yields the fact that $(\lambda_k u_k)_{k \in K}$ is summable. Letting $v = \sum_{k \in K} \lambda_k u_k$, recall that $c_k = \langle v, u_k \rangle / \|u_k\|^2$. Pick some $k \in K$. Since $\langle -, - \rangle$ is continuous, for every $\epsilon > 0$, there is some $\eta > 0$ such that

$$| \langle v, u_k \rangle - \langle w, u_k \rangle | < \epsilon \|u_k\|^2$$

whenever

$$\|v - w\| < \eta.$$

However, since for every $\eta > 0$, there is some finite subset I of K such that

$$\left\| v - \sum_{j \in J} \lambda_j u_j \right\| < \eta$$

for every finite subset J of K such that $I \subseteq J$, we can pick $J = I \cup \{k\}$ and letting $w = \sum_{j \in J} \lambda_j u_j$ we get

$$\left| \langle v, u_k \rangle - \left\langle \sum_{j \in J} \lambda_j u_j, u_k \right\rangle \right| < \epsilon \|u_k\|^2.$$

However,

$$\langle v, u_k \rangle = c_k \|u_k\|^2 \quad \text{and} \quad \left\langle \sum_{j \in J} \lambda_j u_j, u_k \right\rangle = \lambda_k \|u_k\|^2,$$

and thus, the above proves that $|c_k - \lambda_k| < \epsilon$ for every $\epsilon > 0$, and thus, that $c_k = \lambda_k$.

(2) Since $\langle -, - \rangle$ is continuous, for every $\epsilon > 0$, there are some $\eta_1 > 0$ and $\eta_2 > 0$, such that

$$|\langle x, y \rangle| < \epsilon$$

whenever $\|x\| < \eta_1$ and $\|y\| < \eta_2$. Since $v = \sum_{k \in K} \lambda_k u_k$ and $w = \sum_{k \in K} \mu_k u_k$, there is some finite subset I_1 of K such that

$$\left\| v - \sum_{j \in J} \lambda_j u_j \right\| < \eta_1$$

for every finite subset J of K such that $I_1 \subseteq J$, and there is some finite subset I_2 of K such that

$$\left\| w - \sum_{j \in J} \mu_j u_j \right\| < \eta_2$$

for every finite subset J of K such that $I_2 \subseteq J$. Letting $I = I_1 \cup I_2$, we get

$$\left| \left\langle v - \sum_{i \in I} \lambda_i u_i, w - \sum_{i \in I} \mu_i u_i \right\rangle \right| < \epsilon.$$

Furthermore,

$$\begin{aligned}
\langle v, w \rangle &= \left\langle v - \sum_{i \in I} \lambda_i u_i + \sum_{i \in I} \lambda_i u_i, w - \sum_{i \in I} \mu_i u_i + \sum_{i \in I} \mu_i u_i \right\rangle \\
&= \left\langle v - \sum_{i \in I} \lambda_i u_i, w - \sum_{i \in I} \mu_i u_i \right\rangle + \sum_{i \in I} \lambda_i \overline{\mu_i},
\end{aligned}$$

since the u_i are orthogonal to $v - \sum_{i \in I} \lambda_i u_i$ and $w - \sum_{i \in I} \mu_i u_i$ for all $i \in I$. This proves that for every $\epsilon > 0$, there is some finite subset I of K such that

$$\left| \langle v, w \rangle - \sum_{i \in I} \lambda_i \overline{\mu_i} \right| < \epsilon.$$

We already know from Proposition A.3 that $(\lambda_k \overline{\mu_k})_{k \in K}$ is summable, and since $\epsilon > 0$ is arbitrary we get

$$\langle v, w \rangle = \sum_{k \in K} \lambda_k \overline{\mu_k}.$$

$\square$

The next proposition states properties characterizing Hilbert bases (total orthogonal families).

Proposition A.5. *Let E be a Hilbert space, and let $(u_k)_{k\in K}$ be an orthogonal family in E. The following properties are equivalent:*

(1) The family $(u_k)_{k\in K}$ is a total orthogonal family.
(2) For every vector $v \in E$, if $(c_k)_{k\in K}$ are the Fourier coefficients of v, then the family $(c_k u_k)_{k\in K}$ is summable and $v = \sum_{k\in K} c_k u_k$.
(3) For every vector $v \in E$, we have the Parseval identity:

$$\|v\|^2 = \sum_{k\in K} |c_k|^2.$$

(4) For every vector $u \in E$, if $\langle u, u_k\rangle = 0$ for all $k \in K$, then $u = 0$.

See Figure A.2.

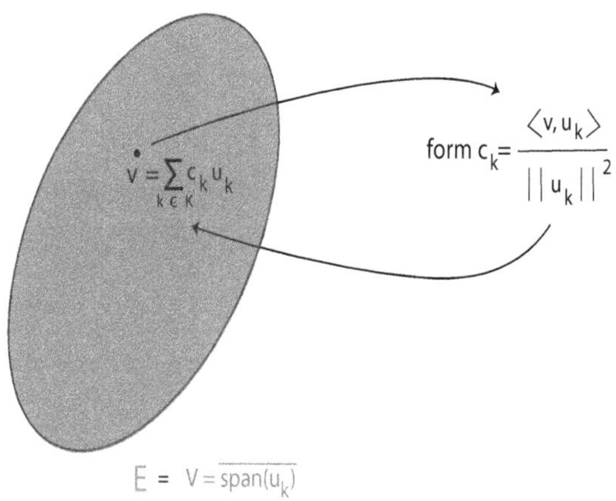

Fig. A.2 A schematic illustration of Proposition A.5. Since $(u_k)_{k\in K}$ is a Hilbert basis, $V = E$. Then given a vector of E, if we form the Fourier coefficients c_k, then form the Fourier series $\sum_{k\in K} c_k u_k$, we are ensured that v is equal to its Fourier series.

Proof. The equivalence of (1), (2), and (3) is an immediate consequence of Proposition A.2 and Proposition A.4.

(4) If $(u_k)_{k\in K}$ is a total orthogonal family and $\langle u, u_k\rangle = 0$ for all $k \in K$, since $u = \sum_{k\in K} c_k u_k$ where $c_k = \langle u, u_k\rangle/\|u_k\|^2$, we have $c_k = 0$ for all $k \in K$, and $u = 0$.

Conversely, assume that the closure V of $(u_k)_{k\in K}$ is different from E. Then by Proposition 12.6, we have $E = V \oplus V^\perp$, where $V^\perp$ is the orthogonal complement of V, and $V^\perp$ is nontrivial since $V \neq E$. As a consequence, there is some nonnull vector $u \in V^\perp$. But then u is orthogonal to every vector in V, and in particular,

$$\langle u, u_k \rangle = 0$$

for all $k \in K$, which, by assumption, implies that $u = 0$, contradicting the fact that $u \neq 0$. □

Remarks:

(1) If E is a Hilbert space and $(u_k)_{k\in K}$ is a total orthogonal family in E, there is a simpler argument to prove that $u = 0$ if $\langle u, u_k \rangle = 0$ for all $k \in K$ based on the continuity of $\langle -, - \rangle$. The argument is to prove that the assumption implies that $\langle v, u \rangle = 0$ for all $v \in E$. Since $\langle -, - \rangle$ is positive definite, this implies that $u = 0$. By continuity of $\langle -, - \rangle$, for every $\epsilon > 0$, there is some $\eta > 0$ such that for every finite subset I of K, for every family $(\lambda_i)_{i\in I}$, for every $v \in E$,

$$\left| \langle v, u \rangle - \left\langle \sum_{i\in I} \lambda_i u_i, u \right\rangle \right| < \epsilon$$

whenever

$$\left\| v - \sum_{i\in I} \lambda_i u_i \right\| < \eta.$$

Since $(u_k)_{k\in K}$ is dense in E, for every $v \in E$, there is some finite subset I of K and some family $(\lambda_i)_{i\in I}$ such that

$$\left\| v - \sum_{i\in I} \lambda_i u_i \right\| < \eta,$$

and since by assumption, $\left\langle \sum_{i\in I} \lambda_i u_i, u \right\rangle = 0$, we get

$$|\langle v, u \rangle| < \epsilon.$$

Since this holds for every $\epsilon > 0$, we must have $\langle v, u \rangle = 0$

(2) If V is any nonempty subset of E, the kind of argument used in the previous remark can be used to prove that $V^\perp$ is closed (even if V is not), and that $V^{\perp\perp}$ is the closure of V.

We will now prove that every Hilbert space has some Hilbert basis. This requires using a fundamental theorem from set theory known as *Zorn's lemma*, which we quickly review.

Given any set X with a partial ordering $\leq$, recall that a nonempty subset C of X is a *chain* if it is totally ordered (i.e., for all $x, y \in C$, either $x \leq y$ or $y \leq x$). A nonempty subset Y of X is *bounded* iff there is some $b \in X$ such that $y \leq b$ for all $y \in Y$. Some $m \in X$ is *maximal* iff for every $x \in X$, $m \leq x$ implies that $x = m$. We can now state Zorn's lemma. For more details, see Rudin [Rudin (1987)], Lang [Lang (1993)], or Artin [Artin (1991)].

Proposition A.6. *(Zorn's lemma) Given any nonempty partially ordered set X, if every (nonempty) chain in X is bounded, then X has some maximal element.*

We can now prove the existence of Hilbert bases. We define a partial order on families $(u_k)_{k \in K}$ as follows: for any two families $(u_k)_{k \in K_1}$ and $(v_k)_{k \in K_2}$, we say that

$$(u_k)_{k \in K_1} \leq (v_k)_{k \in K_2}$$

iff $K_1 \subseteq K_2$ and $u_k = v_k$ for all $k \in K_1$. This is clearly a partial order.

Proposition A.7. *Let E be a Hilbert space. Given any orthogonal family $(u_k)_{k \in K}$ in E, there is a total orthogonal family $(u_l)_{l \in L}$ containing $(u_k)_{k \in K}$.*

Proof. Consider the set $\mathcal{S}$ of all orthogonal families greater than or equal to the family $B = (u_k)_{k \in K}$. We claim that every chain in $\mathcal{S}$ is bounded. Indeed, if $C = (C_l)_{l \in L}$ is a chain in $\mathcal{S}$, where $C_l = (u_{k,l})_{k \in K_l}$, the union family

$$(u_k)_{k \in \bigcup_{l \in L} K_l}, \text{ where } u_k = u_{k,l} \text{ whenever } k \in K_l,$$

is clearly an upper bound for C, and it is immediately verified that it is an orthogonal family. By Zorn's Lemma A.6, there is a maximal family $(u_l)_{l \in L}$ containing $(u_k)_{k \in K}$. If $(u_l)_{l \in L}$ is not dense in E, then its closure V is strictly contained in E, and by Proposition 12.6, the orthogonal complement $V^{\perp}$ of V is nontrivial since $V \neq E$. As a consequence, there is some nonnull vector $u \in V^{\perp}$. But then u is orthogonal to every vector in $(u_l)_{l \in L}$, and we can form an orthogonal family strictly greater than $(u_l)_{l \in L}$ by adding u to this family, contradicting the maximality of $(u_l)_{l \in L}$. Therefore, $(u_l)_{l \in L}$ is dense in E, and thus it is a Hilbert basis. ☐

Remark: It is possible to prove that all Hilbert bases for a Hilbert space E have index sets K of the same cardinality. For a proof, see Bourbaki [Bourbaki (1981)].

Definition A.4. A Hilbert space E is *separable* if its Hilbert bases are countable.

At last, we can prove that every Hilbert space is isomorphic to some Hilbert space $\ell^2(K)$ for some suitable K.

Theorem A.1. *(Riesz–Fischer) For every Hilbert space E, there is some nonempty set K such that E is isomorphic to the Hilbert space $\ell^2(K)$. More specifically, for any Hilbert basis $(u_k)_{k\in K}$ of E, the maps $f\colon \ell^2(K) \to E$ and $g\colon E \to \ell^2(K)$ defined such that*

$$f\left((\lambda_k)_{k\in K}\right) = \sum_{k\in K} \lambda_k u_k \quad \text{and} \quad g(u) = \left(\langle u, u_k\rangle / \|u_k\|^2\right)_{k\in K} = (c_k)_{k\in K},$$

are bijective linear isometries such that $g \circ f = \mathrm{id}$ and $f \circ g = \mathrm{id}$.

Proof. By Proposition A.4 (1), the map f is well defined, and it is clearly linear. By Proposition A.2 (3), the map g is well defined, and it is also clearly linear. By Proposition A.2 (2b), we have

$$f(g(u)) = u = \sum_{k\in K} c_k u_k,$$

and by Proposition A.4 (1), we have

$$g(f\left((\lambda_k)_{k\in K}\right)) = (\lambda_k)_{k\in K},$$

and thus $g \circ f = \mathrm{id}$ and $f \circ g = \mathrm{id}$. By Proposition A.4 (2), the linear map g is an isometry. Therefore, f is a linear bijection and an isometry between $\ell^2(K)$ and E, with inverse g. ☐

Remark: The surjectivity of the map $g\colon E \to \ell^2(K)$ is known as the *Riesz–Fischer* theorem.

Having done all this hard work, we sketch how these results apply to Fourier series. Again we refer the readers to Rudin [Rudin (1987)] or Lang [Lang (1996, 1997)] for a comprehensive exposition.

Let $\mathcal{C}(T)$ denote the set of all periodic continuous functions $f\colon [-\pi, \pi] \to \mathbb{C}$ with period 2π. There is a Hilbert space $\mathrm{L}^2(T)$ containing $\mathcal{C}(T)$ and such that $\mathcal{C}(T)$ is dense in $\mathrm{L}^2(T)$, whose inner product is given by

$$\langle f, g\rangle = \int_{-\pi}^{\pi} f(x)\overline{g(x)}dx.$$

The Hilbert space $\mathrm{L}^2(T)$ is the space of *Lebesgue square-integrable periodic functions* (of period 2π).

It turns out that the family $(e^{ikx})_{k\in\mathbb{Z}}$ is a total orthogonal family in $\mathrm{L}^2(T)$, because it is already dense in $\mathcal{C}(T)$ (for instance, see Rudin [Rudin (1987)]). Then the Riesz–Fischer theorem says that for every family $(c_k)_{k\in\mathbb{Z}}$ of complex numbers such that

$$\sum_{k\in\mathbb{Z}} |c_k|^2 < \infty,$$

there is a unique function $f \in \mathrm{L}^2(T)$ such that f is equal to its Fourier series

$$f(x) = \sum_{k\in\mathbb{Z}} c_k e^{ikx},$$

where the Fourier coefficients c_k of f are given by the formula

$$c_k = \frac{1}{2\pi}\int_{-\pi}^{\pi} f(t)e^{-ikt}dt.$$

The Parseval theorem says that

$$\sum_{k=-\infty}^{+\infty} c_k\overline{d_k} = \frac{1}{2\pi}\int_{-\pi}^{\pi} f(t)\overline{g(t)}dt$$

for all $f, g \in \mathrm{L}^2(T)$, where c_k and d_k are the Fourier coefficients of f and g.

Thus, there is an isomorphism between the two Hilbert spaces $\mathrm{L}^2(T)$ and $\ell^2(\mathbb{Z})$, which is the deep reason why the Fourier coefficients "work." Theorem A.1 implies that the Fourier series $\sum_{k\in\mathbb{Z}} c_k e^{ikx}$ of a function $f \in \mathrm{L}^2(T)$ converges to f in the L^2-sense, i.e., in the mean-square sense. This does not necessarily imply that the Fourier series converges to f pointwise! This is a subtle issue, and for more on this subject, the reader is referred to Lang [Lang (1996, 1997)] or Schwartz [Schwartz (1993a,b)].

We can also consider the set $\mathcal{C}([-1,1])$ of continuous functions $f\colon [-1,1] \to \mathbb{C}$. There is a Hilbert space $\mathrm{L}^2([-1,1])$ containing $\mathcal{C}([-1,1])$ and such that $\mathcal{C}([-1,1])$ is dense in $\mathrm{L}^2([-1,1])$, whose inner product is given by

$$\langle f, g\rangle = \int_{-1}^{1} f(x)\overline{g(x)}dx.$$

The Hilbert space $\mathrm{L}^2([-1,1])$ is the space of *Lebesgue square-integrable functions* over $[-1,1]$. The Legendre polynomials $P_n(x)$ defined in Example 5 of Section 11.2 (Chapter 11, Vol. I) form a Hilbert basis of $\mathrm{L}^2([-1,1])$. Recall that if we let f_n be the function

$$f_n(x) = (x^2 - 1)^n,$$

$P_n(x)$ is defined as follows:

$$P_0(x) = 1, \quad \text{and} \quad P_n(x) = \frac{1}{2^n n!} f_n^{(n)}(x),$$

where $f_n^{(n)}$ is the nth derivative of f_n. The reason for the leading coefficient is to get $P_n(1) = 1$. It can be shown with much efforts that

$$P_n(x) = \sum_{0 \le k \le n/2} (-1)^k \frac{(2(n-k))!}{2^n (n-k)! k! (n-2k)!} x^{n-2k}.$$

A.3 Summary

The main concepts and results of this chapter are listed below:

- Hilbert space
- Orthogonal family, total orthogonal family.
- Hilbert basis.
- Fourier coefficients.
- Hamel bases, Schauder bases.
- Fourier series.
- Cauchy family, summable family.
- Bessel inequality.
- The Hilbert space $\ell^2(K)$.
- Parseval identity.
- Zorn's lemma.
- Riesz–Fischer theorem.
- Legendre polynomials.

A.4 Problems

Problem A.1. Prove that the subspace consisting of sequences $(z_k)_{k \in K}$ such that $z_k = 0$ except perhaps for finitely many k is a dense suspace of $\ell^2(K)$.

Problem A.2. If V is any nonempty subset of E, prove that $V^{\perp}$ is closed (even if V is not) and that $V^{\perp\perp}$ is the closure of V (see the remarks following Proposition A.5).

Appendix B

Matlab Programs

B.1 Hard Margin (SVM_{h2})

The following Matlab programs implement the method described in Section 16.7.

The first program is the heart of the method; it implements ADMM for quadratic programming.

```
function [x,u,nr,ns,k] = qsolve1(P, q, A, b, rho, tolr, tols,
                                          iternum)
%  Solve a quadratic programming problem
%  min (1/2) x^T P x + x^T q + r
%  subject to Ax = b, x >= 0  using ADMM
%  P n x n,   q, r, in R^n, A m x n,  b in R^m
%  A of rank m

m = size(A,1); fprintf('m =  %d ',m)
n = size(P,1); fprintf('    n =  %d \n',n)
u = ones(n,1); u(1,1) = 0; % to initialize u
z = ones(n,1);             % to initialize z
% iternum = maximum number of iterations;
% iternum = 80000 works well
k = 0; nr= 1; ns = 1;
%  typically tolr = 10^(-10); tols = 10^(-10);
%  Convergence is controlled by the norm nr of the primal
%  residual r
%  and the norm ns of the dual residual s

while (k <= iternum) && (ns > tols || nr > tolr)
```

```
    z0 = z;
    k = k+1;
    %  Makes KKT matrix
    KK = [P + rho* eye(n) A'; A zeros(m,m)];
    % Makes right hand side of KKT equation
    bb = [-q + rho*(z - u); b];
    % Solves KKT equation
    xx = KK\bb;
    %  update x, z, u (ADMM update steps)
    x = xx(1:n);
    z = poslin(x + u);
    u = u + x - z;
    %  to test stopping criterion
    r = x - z;                      % primal residual
    nr = sqrt(r'*r);                % norm of primal residual
    s = rho*(z - z0);               % dual residual
    ns = sqrt(s'*s);                % norm of dual residual
end
end
```

The second program SBVMhard2 implements hard margin SVM (v2).

```
function [lamb,mu,w] = SVMhard2(rho,u,v)
%
%   Runs hard margin SVM version 2
%
%   p green vectors u_1, ..., u_p in n x p array u
%   q red   vectors v_1, ..., v_q in n x q array v
%
%   First builds the matrices for the dual program
%
p = size(u,2); q = size(v,2); n = size(u,1);
[A,c,X,Pa,qa] = buildhardSVM2(u,v);
%
%  Runs quadratic solver
%
tolr = 10^(-10); tols = 10^(-10); iternum = 80000;
[lam,U,nr,ns,kk] = qsolve1(Pa, qa, A, c, rho, tolr, tols,
                           iternum);
```

```
fprintf('nr =  %d ',nr)
fprintf('   ns =  %d \n',ns)
fprintf('kk =  %d \n',kk)
if kk > iternum
   fprintf('** qsolve did not converge. Problem
           not solvable ** \n')
end
w = -X*lam;
nw = sqrt(w'*w);   % norm of w
fprintf('nw =  %.15f \n',nw)
delta = 1/nw;
fprintf('delta =  %.15f \n',delta)
if   delta < 10^(-9)
     fprintf('** Warning, delta too small, program
              does not converge ** \n')
end
%
lamb = lam(1:p,1);
mu = lam(p+1:p+q,1);
b = 0;
tols = 10^(-10);
% tols < lambda_i; finds the nonzero lambda_i
[lambnz,numsvl1] = countmlu2(lamb,tols);
% tols < mu_i; finds the nonzero mu_j
[munz,numsvm1] = countmlv2(mu,tols);
fprintf('numsvl1 = %d ',numsvl1)
fprintf('  numsvm1 = %d \n',numsvm1)

if numsvl1 > 0 && numsvm1 > 0
   sx1 = zeros(n,1);  num1 = 0;
   sx2 = zeros(n,1);  num2 = 0;
   for i = 1:p
       if lambnz(i) > 0
          sx1 = sx1 + u(:,i);
          num1 = num1 + 1;
       end
   end
   for j = 1:q
       if munz(j) > 0
```

```
            sx2 = sx2 + v(:,j);
            num2 = num2 + 1;
         end
      end
      b =  (w'*(sx1/num1 + sx2/num2))/2;
      fprintf('b =  %.15f \n',b)
else
   fprintf('** Not enough support vectors ** \n')
end

if n == 2
   [ll,mm] = showdata(u,v);
   if numsvl1 > 0 && numsvm1 > 0
      showSVMs2(w,b,1,ll,mm,nw)
   end
end
end
```

The function `buildhardSVM2` builds the constraint matrix and the matrices defining the quadratic functional.

```
function [A,c,X,Xa,q] = buildhardSVM2(u,v)
%   builds the matrix of constraints A for
%   hard SVM h2, and the right hand side c
%   Aso builds X and Xa = X'*X, and the vector q = -1_{p+q}
%   for the linear part of the quadratic function
%   The right-hand side is c = 0 (Ax = 0).
p = size(u,2); q = size(v,2);
A = [ones(1,p) -ones(1,q)];
c = 0;
X = [-u v];
Xa = X'*X;
q = -ones(p+q,1);
end
```

The function `countmlu2` returns a vector consisting of those λ_i such that $\lambda_i > 0$, and the number of such λ_i.

```
function [lambnz, mlu] = countmlu2(lambda,tols)
%  Counts the number of points u_i (in u)
```

```
%  such that lambda_i > 0 and returns a vector
%  of these lambda_i

% tols = 10^(-11);
p = size(lambda,1); lambnz = zeros(p,1);
mlu = 0;
for i = 1:p
    if lambda(i) > tols
        mlu = mlu + 1;
        lambnz(i) = lambda(i);
    end
end
end
```

The function countmlv2 returns a vector consisting of those μ_j such that $\mu_j > 0$, and the number of such μ_j. It is similar to countmlu2. Here a judicious choice of tols is crucial and one has to experiment with various values.

The function showdata displays the data points (the u_i and the v_j) and the function showSVMs2 displays the separating line and the two margin lines.

```
function  showSVMs2(w,b,eta,ll,mm,nw)
%
%  Function to display the result of running SVM
%  on p blue points u_1, ..., u_p in u
%  and q red points v_1, ..., v_q in v

l = makeline(w,b,ll,mm,nw);          %  makes separating line
lm1 = makeline(w,b+eta,ll,mm,nw);  %  makes blue margin line
lm2 = makeline(w,b-eta,ll,mm,nw);  %  makes red margin line

%  plots separating line
plot(l(1,:),l(2,:),'-m','LineWidth',1.2)
%  plots blue margin line
plot(lm1(1,:),lm1(2,:),'-b','LineWidth',1.2)
%  plots red margin line
plot(lm2(1,:),lm2(2,:),'-r','LineWidth',1.2)
hold off
end
```

Actually, implementing the above function is not entirely trivial. It is necessary to write a function **makeline** to plot the line segment which is part of the line of equation $w_1x + w_2y = b$ inside a box containing the data points. We leave the details an exercises.

B.2 Soft Margin SVM ($\mathrm{SVM}_{s2'}$)

The following Matlab programs implement the method described in Section 18.8.

The function doSVMs2pbv3 calls the function solve1 given in Section 16.7.

```
function [lamb,mu,alpha,beta,lambnz,munz,numsvl1,numsvm1,badnu,
w,nw,b,eta] = doSVMs2pbv3(nu,rho,u,v,K)
%
%   Best version
%   Uses the duality gap to compute eta
%   In principle, needs a single support vector of type 1
%
%   Soft margin nu-SVM version s2'
%   with the constraint
%   \sum_{i = 1}^p + \sum_{j = 1}^q mu_j  = K_m
%   (without the variable gamma)
%
%   p green vectors u_1, ..., u_p in n x p array u
%   q red   vectors v_1, ..., v_q in n x q array v
%
%   First builds the matrices for the dual program
%    K is a scale factor
%
p = size(u,2); q = size(v,2); n = size(u,1);
[A,c,X,Pa,qa] = buildSVMs2pb(nu,u,v,K);
%
% Runs quadratic solver
%
tolr = 10^(-10); tols = 10^(-10); iternum = 80000;
[x,U,nr,ns,kk] = qsolve1(Pa, qa, A, c, rho, tolr, tols, iternum);
fprintf('nr =  %d ',nr)
fprintf('   ns =  %d \n',ns)
```

```
fprintf('kk =  %d \n',kk)
noconv = 0;
if kk > iternum
   noconv = 1;
   fprintf('** qsolve did not converge. Problem
           not solvable ** \n')
end
lam = x(1:(p+q),1);
alpha = x((p+q+1):2*p+q,1);
beta = x(2*p+q+1:2*(p+q),1);
w = -X*lam;
nw = sqrt(w'*w);   % norm of w
fprintf('nw =  %d \n',nw)
%
lamb = x(1:p,1);
mu = x(p+1:p+q,1);
tols = 10^(-10); tolh = 10^(-9);
% tols < lambda_i < K - tolh
[lambnz,numsvl1] = findpsv2(lamb,K,tols,tolh);
% tols < mu_i < K - tolh
[munz,numsvm1] = findpsv2(mu,K,tols,tolh);
fprintf('numsvl1 = %d ',numsvl1)
fprintf('   numsvm1 = %d \n',numsvm1)
%  lambda_i >= K - tolh
% number of blue margin failures
[lamK,pf] = countumf2(lamb,K,tolh);
%  mu_j >= K - tolh
% number of red margin failures
[muK,qf] = countvmf2(mu,K,tolh);
fprintf('pf =  %d ',pf)
fprintf('   qf =  %d \n',qf)
% number of  points such that lambda_i > tols
[~,pm] = countmlu2(lamb,tols);
% number of  points such that mu_i > tols
[~,qm] = countmlv2(mu,tols);
fprintf('pm =  %d ',pm)
fprintf('   qm =  %d \n',qm)
fprintf('p - pm =  %d ',p - pm)
fprintf('   q - qm =  %d \n',q - qm)
```

```
lnu = max(2*pf/(p+q),2*qf/(p+q));
unu = min(2*pm/(p+q),2*qm/(p+q));
fprintf('lnu =  %d ',lnu)
fprintf('    unu =  %d \n',unu)
if nu < lnu
      fprintf('** Warning; nu is too small ** \n')
else
      if nu > unu
        fprintf('** Warning; nu is too big ** \n')
      end
end

sx1 = zeros(n,1);  num1 = 0;
sKu = zeros(n,1);  Knum1 = 0;
for i = 1:p
       if lambnz(i) > 0
          sx1 = sx1 + u(:,i);
          num1 = num1 + 1;
       end
       if lamK(i) > 0
          sKu = sKu + u(:,i);
          Knum1 = Knum1 + 1;
       end
end
% Knum1
sx2 = zeros(n,1);  num2 = 0;
sKv = zeros(n,1);  Knum2 = 0;
for j = 1:q
       if munz(j) > 0
          sx2 = sx2 + v(:,j);
          num2 = num2 + 1;
       end
       if muK(j) > 0
          sKv = sKv + v(:,j);
          Knum2 = Knum2 + 1;
       end
end
% Knum2
```

```
b = 0; eta = 0;
epsilon = 0; xi = 0;
P2 = X'*X;
badnu = 0;
if numsvl1 > 0
   if numsvm1 > 0
      b =  (w'*(sx1/num1 + sx2/num2))/2;
      fprintf('b =  %.15f \n',b)
      eta =  (w'*(sx1/num1 - sx2/num2))/2;
      fprintf('eta =  %.15f \n',eta)
   else
       errterm = w'*(sKv - sKu) + (pf - qf)*w'*(sx1/num1);
       Pterm = (1/K)*(lam'*P2*lam);
       denomqf = (p+q)*nu  -2*qf;
       fprintf('denomqf =  %.15f \n',denomqf)
       if denomqf > 0
          eta = (errterm + Pterm)/denomqf;
          fprintf('eta =  %.15f \n',eta)
          b = -eta + w'*sx1/num1;
       else
           badnu = 1;
           fprintf('** Warning: numsvl1 > 0,
                    numsvm1 = 0 and nu = 2*qf/(p+q) **  \n')
       end
   end
else
    if numsvm1 > 0
       errterm = w'*(sKv - sKu) + (pf - qf)*w'*(sx2/num2);
       Pterm = (1/K)*(lam'*P2*lam);
       denompf = (p+q)*nu  -2*pf;
       fprintf('denompf =  %.15f \n',denompf)
       if denompf > 0
          eta = (errterm + Pterm)/denompf;
          fprintf('eta =  %.15f \n',eta)
          b = eta + w'*sx2/num2;
       else
           badnu = 1;
           fprintf('** Warning: numsvm1 > 0,
                    numsvl1 = 0 and nu = 2*pf/(p+q) **  \n')
```

```
      end
   else
      fprintf('** Not enough support vectors ** \n')
   end
end

Km = (p+q)*nu*K;
fprintf('K  =  %.15f ',K)
fprintf('    (p+q)*nu*Ks/2 =  %.15f \n',Km/2)
fprintf('sum(lambda) =  %.15f ',sum(lamb))
fprintf('    sum(mu) =  %.15f \n',sum(mu))

if (numsvl1 > 0 || numsvm1 > 0) && badnu == 0
   if eta < 10^(-9)
      fprintf('** Warning, eta too small or negative ** \n')
      eta = 0;
   end
   delta = eta/nw;
   fprintf('delta =  %.15f \n',delta)
   tolxi = 10^(-10);
   % tols < lambda_i < K - tolh or K - tolh <= lambda_i
   %  and epsilon_i < tolxi
   [lamsv,psf,epsilon] = findsvl2(lamb,w,b,u,eta,K,tols,
    tolh,tolxi);
   % tols < mu_i < K - tolh or K - tolh <= mu_i
   % and xi_i < tolxi
   [musv,qsf,xi] = findsvm2(mu,w,b,v,eta,K,tols,tolh,tolxi);
   fprintf('psf =  %d ',psf)
   fprintf('   qsf =  %d \n',qsf)
   fprintf('pf - psf =  %d ',pf - psf)
   fprintf('   qf - qsf =  %d \n',qf - qsf)

   % computes eta from the duality gap
   errterm = w'*(sKv - sKu) + (pf - qf)*b;
   Pterm = (1/K)*(lam'*P2*lam);
   denom = (p+q)*nu - pf -qf;
   fprintf('denom =  %.15f \n',denom)
   if denom > 0
      eta1 = (errterm + Pterm)/denom;
```

```
        fprintf('eta1 =  %.15f \n',eta1)
    end
end
end
```

The constraint matrix and the matrices defining the quadratic program are constructed by the function `buildSVMs2pb`.

```
function [A,c,X,Pa,q] = buildSVMs2pb(nu,u,v,K)
%  builds the matrix of constraints A for
%  soft margin nu-SVM s2'
%  with the constraint
%  \sum_{i = 1}^p + \sum_{j = 1}^q mu_j  = K_m
%  (without the  variable gamma) and the right-hand side c
%  u: vector of p blue points (each an n-dim vector)
%  v: vector of q red points (each an n-dim vector)
%  builds the matrix X = [-u_1 ... -u_p v1 .... v_q]
%  and the matrix Pa as 2(p+q) matrix obtained
%  by augmenting X'*X with zeros
%  K is a scale factor (K = Ks)

p = size(u,2); q = size(v,2);
% Ks = 1/(p+q);
Ks = K; Km = (p+q)*K*nu;
A = [ones(1,p) -ones(1,q) zeros(1,p+q);
     ones(1,p) ones(1,q)  zeros(1,p+q) ;
     eye(p) zeros(p,q) eye(p) zeros(p,q);
     zeros(q,p) eye(q)  zeros(q,p) eye(q) ];
c = [0; Km; Ks*ones(p+q,1)];
X = [-u v];
XX = X'*X;
Pa = [XX zeros(p+q,p+q); zeros(p+q, 2*(p+q))];
q = zeros(2*(p+q),1);
end
```

The function `findpsv2` makes a vector of λ_i (and μ_j) corresponding to support vectors of type 1.

```
function [lampsv,num] = findpsv2(lambda,K,tols,tolh)
%
%   This function find the vector of
%   lambda_i's such that 0 < lambda_i < K
%   and the number of such  lambda_i.
%
% tols = 10^(-11);  %  the smaller this is, the larger
% the number of points on the margin
% tolh = 10^(-9);  %
m = size(lambda,1); lampsv = zeros(m,1);
num = 0;
for i = 1:m
    if lambda(i) > tols && lambda(i) < K - tolh
       lampsv(i) = lambda(i);
       num = num + 1;
    end
end
end
```

The function countumf2 finds those λ_i such that $\lambda_i = K$.

```
function [lamK,mf] = countumf2(lambda,K,tolh)
%  Counts the number of margin failures, that is,
%  points u_i (in u) such that lambda_i = K

p = size(lambda,1);
mf = 0; lamK = zeros(p,1);
for i = 1:p
    if lambda(i) >= K - tolh
       mf = mf + 1;
       lamK(i) = lambda(i);
    end
end
end
```

Similarly, the function countvmf2 finds those μ_j such that $\mu_j = K$.

The function countmlu2 finds those λ_i such that $\lambda_i > 0$.

```
function [lambnz, mlu] = countmlu2(lambda,tols)
%  Counts the number of points u_i (in u)
```

```
%  such that lambda_i > 0 and returns a vector
%  of these lambda_i

% tols = 10^(-11);
p = size(lambda,1); lambnz = zeros(p,1);
mlu = 0;
for i = 1:p
    if lambda(i) > tols
        mlu = mlu + 1;
        lambnz(i) = lambda(i);
    end
end
end
```

Similarly, the function **countmlv2** finds those μ_j such that $\mu_j > 0$. The function **findsvl2** finds the λ_i corresponding to blue support vectors of type 1 and 2 and the error vector ϵ. The number of blue errors is *psf* (the u_i for which $\epsilon_i > 0$). Similarly the function **findsvm2** finds the μ_j corresponding to red support vectors of type 1 and 2 and the error vector ξ. The number of red errors is *qsf* (the v_j for which $\xi_j > 0$).

The main function **runSVMs2pbv3** calls **doSVMs2pbv3** and displays the separating line (or plane) and the two margin lines (or planes).

```
function [lamb,mu,alpha,beta,lambnz,munz,w]
              = runSVMs2pbv3(nu,rho,u,v,K)
%
%   Best version
%   Uses the duality gap to compute eta
%   In principle, needs a single support vector of type 1
%
%   Runs soft margin nu-SVM version s2'
%   with the constraint
%   \sum_{i = 1}^p + \sum_{j = 1}^q mu_j  = K_m
%   (without the variable gamma)
%
%   p green vectors u_1, ..., u_p in n x p array u
%   q red   vectors v_1, ..., v_q in n x q array v
%
%   First builds the matrices for the dual program
```

```
%     K is a scale factor
%
p = size(u,2); q = size(v,2); n = size(u,1);

[lamb,mu,alpha,beta,lambnz,munz,numsvl1,numsvm1,badnu,w,
          nw,b,eta] = doSVMs2pbv3(nu,rho,u,v,K);

   if n == 2
       [ll,mm] = showdata(u,v);
       if (numsvl1 > 0 || numsvm1 > 0) && badnu == 0
          showSVMs2(w,b,eta,ll,mm,nw)
       end
   else
      if n == 3
         showpointsSVM(u,v)
         if (numsvl1 > 0 || numsvm1 > 0) && badnu == 0
            offset = 10;
            C1 = [1 0 1];  % magenta
            plotplaneSVM(u,v,w,b,offset,C1)
            C2 = [0 0 1];  % blue
            plotplaneSVM(u,v,w,b+eta,offset,C2)
            C3 = [1,0,0];  % red
            plotplaneSVM(u,v,w,b-eta,offset,C3)
         end
         axis equal
         view([-1 -1 1]);
         xlabel('X','fontsize',14);ylabel('Y','fontsize',14);
               zlabel('Z','fontsize',14);
         hold off
      end
   end
end
```

B.3 Soft Margin SVM (SVM_{s3})

The following Matlab programs implement the method described in Section 18.12. The main function doSVMs3b is given below.

```
function [lamb,mu,alpha,beta,lambnz,munz,lamK,muK,w,b,eta,
```

```
          nw,fail] = doSVMs3b (nu,rho,u,v,K)
%
% Soft margin nu-SVM version s3
%
% Computes eta using the duality gap
% Needs a single support vector of type 1
%
%  p green vectors u_1, ..., u_p in n x p array u
%  q red   vectors v_1, ..., v_q in n x q array v
%
%  First builds the matrices for the dual program
%   K is a scale factor
%
p = size(u,2); q = size(v,2); n = size(u,1);
[A,c,X,P2,Pa,qa] = buildSVMs3b (nu,u,v,K);
%
% Runs quadratic solver
%
tolr = 10^(-10); tols = 10^(-10); iternum = 80000;
[x,U,nr,ns,kk] = qsolve1(Pa, qa, A, c, rho, tolr, tols,
                         iternum);
fprintf('nr =  %d ',nr)
fprintf('    ns =  %d ',ns)
fprintf('    kk =  %d \n',kk)
noconv = 0;
if kk > iternum
   noconv = 1;
   fprintf('** qsolve did not converge. Problem
            not solvable ** \n')
end
lam = x(1:(p+q),1);
alpha = x((p+q+1):2*p+q,1);
beta = x(2*p+q+1:2*(p+q),1);
w = -X*lam;
nw = sqrt(w'*w);   % norm of w
fprintf('nw =  %d \n',nw)
lamb = x(1:p,1);
mu = x(p+1:p+q,1);
b = -(sum(lamb) - sum(mu));
```

```
fprintf('b =  %.15f \n',b)
%

tols = 10^(-10); tolh = 10^(-9);
% tols < lambda_i < K - tolh
[lambnz,numsvl1] = findpsv2(lamb,K,tols,tolh);
% tols < mu_i < K - tolh
[munz,numsvm1] = findpsv2(mu,K,tols,tolh);
fprintf('numsvl1 = %d ',numsvl1)
fprintf('    numsvm1 = %d \n',numsvm1)
%  lambda_i >= K - tolh
% number of blue margin failures
[lamK,pf] = countumf2(lamb,K,tolh);
%  mu_j >= K - tolh
% number of red margin failures
[muK,qf] = countvmf2(mu,K,tolh);
fprintf('pf =  %d ',pf)
fprintf('   qf =  %d \n',qf)
% number of  points such that lambda_i > tols
[~,pm] = countmlu2(lamb,tols);
% number of  points such that mu_i > tols
[~,qm] = countmlv2(mu,tols);
fprintf('pm =  %d ',pm)
fprintf('   qm =  %d \n',qm)
fprintf('p - pm =  %d ',p - pm)
fprintf('   q - qm =  %d \n',q - qm)
lnu =  (pf + qf)/(p+q); unu = (pm + qm)/(p+q);
fprintf('lnu =  %d ',lnu)
   fprintf('    unu =  %d \n',unu)
if nu < lnu
      fprintf('** Warning; nu is too small ** \n')
else
      if nu > unu
         fprintf('** Warning; nu is too big ** \n')
      end
end

sx1 = zeros(n,1);  num1 = 0;
sKu = zeros(n,1);  Knum1 = 0;
```

```
for i = 1:p
       if lambnz(i) > 0
          sx1 = sx1 + u(:,i);
          num1 = num1 + 1;
       end
       if lamK(i) > 0
          sKu = sKu + u(:,i);
          Knum1 = Knum1 + 1;
       end
end
% Knum1
sx2 = zeros(n,1);  num2 = 0;
sKv = zeros(n,1);  Knum2 = 0;
for j = 1:q
       if munz(j) > 0
          sx2 = sx2 + v(:,j);
          num2 = num2 + 1;
       end
       if muK(j) > 0
          sKv = sKv + v(:,j);
          Knum2 = Knum2 + 1;
       end
end
% Knum2
% computes eta from the duality gap
errterm = w'*(sKv - sKu) + (pf - qf)*b;
Pterm = (1/K)*(lam'*P2*lam);
denom = (p+q)*nu - pf -qf;
fprintf('denom =  %.15f \n',denom)
epsilon = 0; xi = 0;
if denom > 0
   eta = (errterm + Pterm)/denom;
   fprintf('eta =  %.15f \n',eta)
   if eta < 10^(-10)
      fprintf('** Warning; eta is too small
               or negative ** \n')
   end

   tolxi = 10^(-10);
```

```
   % tols < lambda_i < K - tolh or K - tolh <= lambda_i
   % and epsilon_i < tolxi
   [lamsv,psf,epsilon] = findsvl2(lamb,w,b,u,eta,K,tols,
                                  tolh,tolxi);
    % tols < mu_i < K - tolh or K - tolh <= mu_i
    % and xi_i < tolxi
   [musv,qsf,xi] = findsvm2(mu,w,b,v,eta,K,tols,tolh,tolxi);
   fprintf('psf =  %d ',psf)
   fprintf('   qsf =  %d \n',qsf)
   fprintf('pf - psf =  %d ',pf - psf)
   fprintf('   qf - qsf =  %d \n',qf - qsf)
else
    eta = 0;
    denom = 0;
    fprintf('** Warning, nu = (pf + qf)/(p+q) **  \n')
end

Km = (p+q)*nu*K;
fprintf('K  =  %.15f ',K)
fprintf('    (p+q)*nu*Ks =  %.15f \n',Km)
fprintf('sum(lambda) + sum(mu)=  %.15f \n',sum(lamb) + sum(mu))

eta1 = 0;
if numsvl1 > 0 || numsvm1 > 0
   if numsvl1 > numsvm1
      eta1 =  w'*sx1/num1 - b;
   else
      eta1 =  b - w'*sx2/num2;
   end
   fprintf('eta1 =  %.15f \n',eta1)
else
      fprintf('** Warning: not enough support vectors ** \n')
end
if denom == 0
   if numsvl1 > 0 || numsvm1 > 0
      eta = eta1;
      fail = 0;
   else
      fail = 1;
```

```
      fprintf('** Warning, denom = 0 and
              not enough support vectors **  \n')
   end
else
   fail = 0;
end
end
```

The main function doSVMs3b is executed by the following function:

```
function [lamb,mu,alpha,beta,lambnz,munz,w]
          = runSVMs3b(nu,rho,u,v,K)
%
%   Runs soft margin nu-SVM version s3
%
%  Computes eta using the duality gap
%  Needs a single support vector of type 1
%
%   p green vectors u_1, ..., u_p in n x p array u
%   q red   vectors v_1, ..., v_q in n x q array v
%
%   First builds the matrices for the dual program
%    K is a scale factor
%
p = size(u,2); q = size(v,2); n = size(u,1);

[lamb,mu,alpha,beta,lambnz,munz,lamK,muK,w,b,eta,nw,fail]
   = doSVMs3b(nu,rho,u,v,K);

   if n == 2
       [ll,mm] = showdata(u,v);
       if fail == 0
          showSVMs2(w,b,eta,ll,mm,nw)
       end
   else
      if n == 3
         showpointsSVM(u,v)
         if fail == 0
            offset = 10;
            C1 = [1 0 1];   % magenta
```

```
            plotplaneSVM(u,v,w,b,offset,C1)
            C2 = [0 0 1];  % blue
            plotplaneSVM(u,v,w,b+eta,offset,C2)
            C3 = [1,0,0];  % red
            plotplaneSVM(u,v,w,b-eta,offset,C3)
         end
         axis equal
  %      axis([ll(1) mm(1) ll(2) mm(2)]);
         view([-1 -1 1]);
         xlabel('X','fontsize',14);ylabel('Y','fontsize',14);
          zlabel('Z','fontsize',14);
         hold off
      end
   end
end
```

The function buildSVMs3b builds the constraint matrix and the matrices defining the quadratic program.

```
function [A,c,X,P2,Pa,q] = buildSVMs3b(nu,u,v,K)
%  builds the matrix of constraints A for
%  soft margin nu-SVM s3 and the right-hand side c
%  u: vector of p blue points (each an n-dim vector)
%  v: vector of q red points (each an n-dim vector)
%  builds the matrix X = [-u_1 ... -u_p v1 .... v_q]
%  and the matrix Xa as 2(p+q)  matrix obtained
%  by augmenting X'*X with zeros
%  K is a scale factor (K = Ks)

p = size(u,2); q = size(v,2);
% Ks = 1/(p+q);
Ks = K; Km = (p+q)*K*nu;
A = [ones(1,p) ones(1,q)  zeros(1,p+q) ;
     eye(p) zeros(p,q) eye(p) zeros(p,q);
     zeros(q,p) eye(q)  zeros(q,p) eye(q) ];
c = [Km; Ks*ones(p+q,1)];
X = [-u v];
XX1 = X'*X;
XX2 = [ones(p,1)*ones(p,1)' -ones(p,1)*ones(q,1)';
      -ones(q,1)*ones(p,1)' ones(q,1)*ones(q,1)'];
```

```
P2 = XX1 + XX2;
Pa = [P2 zeros(p+q,p+q); zeros(p+q, 2*(p+q))];
q = zeros(2*(p+q),1);
end
```

B.4 ν-SV Regression

The main function donuregb is given below.

```
function
[lamb,mu,alpha,beta,lambnz,munz,lamK,muK,numsvl1,numsvm1,w,
        epsilon,b] = donuregb (rho,nu,X,y,C)
%
%   Soft margin nu-regression
%   with the constraint
%   \sum_{i = 1}^m + \sum_{j = 1}^m mu_j = C nu
%   (Without the variable gamma)
%
%  Input: an m x n matrix of data points represented as
%  as the rows of X, and y a vector in R^n
%
%   First builds the matrices for the dual program
%    C is a scale factor
%
m = size(X,1); n = size(X,2);
[A,c,P,Pa,qa] = buildnuregb(nu,X,y,C);
%
%  Runs quadratic solver
%
tolr = 10^(-10); tols = 10^(-10); iternum = 80000;
[x,U,nr,ns,kk] = qsolve1(Pa, qa, A, c, rho, tolr, tols,
                           iternum);
% fprintf('nr =  %d ',nr)
% fprintf('   ns =  %d \n',ns)
fprintf('nr =  %d',nr)
fprintf('  ns =  %d',ns)
fprintf('  kk =  %d \n',kk)
noconv = 0;
if kk > iternum
```

```
   noconv = 1;
   fprintf('** qsolve did not converge. Problem
           not solvable ** \n')
end
lamb = x(1:m,1);
mu = x(m+1:2*m,1);
alpha = x((2*m+1):3*m,1);
beta = x(3*m+1:4*m,1);
w = X'*(mu - lamb);
%
b = 0; epsilon = 0;
tols = 10^(-10); tolh = 10^(-9);
% tols < lambda_i < C/m - tolh
[lambnz,numsvl1] = findpsv2(lamb,C/m,tols,tolh);
% tols < mu_i < C/m - tolh
[munz,numsvm1] = findpsv2(mu,C/m,tols,tolh);
fprintf('numsvl1 = %d',numsvl1)
fprintf('   numsvm1 = %d \n',numsvm1)
%  lambda_i >= C/m - tolh
% number of blue margin failures
[lamK,pf] = countumf2(lamb,C/m,tolh);
%  mu_j >= C/m - tolh
% number of red margin failures
[muK,qf] = countvmf2(mu,C/m,tolh);
fprintf('pf =  %d',pf)
fprintf('    qf =  %d \n',qf)
% number of  points such that lambda_i > tols
[~,pm] = countmlu2(lamb,tols);
% number of  points such that mu_i > tols
[~,qm] = countmlv2(mu,tols);
fprintf('pm =  %d',pm)
fprintf('    qm =  %d \n',qm)
%  lambda_i <= tols
[lmz,nz] = countLzero(lamb,mu,tols);
pm2 = numsvl1 + pf; qm2 = numsvm1 + qf;
fprintf('pm2 =  %d',pm2)
fprintf('   qm2 =  %d \n',qm2)
lnu = max(2*pf/m,2*qf/m); unu = min(2*pm/m,2*qm/m);
fprintf('lnu =  %d',lnu)
```

```
fprintf('  unu =  %d \n',unu)
fprintf('nz =  %d \n',nz)
if nu < lnu
      fprintf('** Warning; nu is too small ** \n')
else
      if nu > unu
           fprintf('** Warning; nu is too big ** \n')
      end
end

fprintf('C/m  =  %.15f ',C/m)
fprintf('    (C nu)/2 =  %.15f \n',(C*nu)/2)
fprintf('sum(lambda) =  %.15f ',sum(lamb))
fprintf('    sum(mu) =  %.15f \n',sum(mu))
lamsv = 0; musv = 0; xi = 0; xip = 0;
if numsvl1 > 0 && numsvm1 > 0
   sx1 = zeros(n,1); sy1 = 0; num1 = 0;
   sx2 = zeros(n,1); sy2 = 0; num2 = 0;
   for i = 1:m
       if lambnz(i) > 0
          sx1 = sx1 + X(i,:)'; sy1 = sy1 + y(i);
          num1 = num1 + 1;
       end
       if munz(i) > 0
          sx2 = sx2 + X(i,:)'; sy2 = sy2 + y(i);
          num2 = num2 + 1;
       end
   end
   % num1
   % num2
   b =  (sy1/num1 + sy2/num2 - w'*(sx1/num1 + sx2/num2))/2;
   fprintf('b =  %.15f \n',b)
   epsilon =  (w'*(sx1/num1 - sx2/num2)
             + sy2/num2 - sy1/num1)/2;
   fprintf('epsilon =  %.15f \n',epsilon)
   if epsilon < 10^(-10)
      fprintf('** Warning; epsilon is too small
              or negative ** \n')
   end
```

```
   nw = sqrt(w'*w);   % norm of w
   fprintf('nw =  %.15f \n',nw)
   %
   tolxi = 10^(-10);
   % tols < lambda_i < C/m - tolh or C/m - tolh <= lambda_i
   % and xi_i < tolxi
   [lamsv,psf,xi] = findnuregsvl2(lamb,w,b,X,y,epsilon,C/m,
                    tols,tolh,tolxi);
    % tols < mu_i < C/m - tolh or C/m - tolh <= mu_i
   % and xi_i' < tolxi
   [musv,qsf,xip] = findnuregsvm2(mu,w,b,X,y,epsilon,C/m,tols,
                    tolh,tolxi);
   fprintf('psf =  %d ',psf)
   fprintf('   qsf =  %d \n',qsf)
else
   fprintf('** Not enough support vectors ** \n')
end
end
```

To run donuregb use the function runuregb listed below.

```
function [lamb,mu,alpha,beta,lambnz,munz,lamK,muK,w]
          = runuregb (rho,nu,X,y,C)
%
%   Runs soft margin nu-regression
%   with the constraint
%   \sum_{i = 1}^m + \sum_{j = 1}^m mu_j = C nu
%  (Without the variable gamma)
%
%  Input: an m x n matrix of data points represented as
%  as the rows of X, and y a vector in R^n
%
%   First builds the matrices for the dual program
%    C is a scale factor
%

m = size(X,1); n = size(X,2);
[lamb,mu,alpha,beta,lambnz,munz,lamK,muK,numsvl1,numsvm1,w,
           epsilon,b] = donuregb(rho,nu,X,y,C);
```

```
    if n == 1
       [ll,mm] = showgraph(X,y);
       ww = [w;-1]; n1 = sqrt(ww'*ww);
       if numsvl1 > 0 && numsvm1 > 0
          showSVMs2(ww,-b,epsilon,ll,mm,n1)
       end
    else
        if n == 2
           offset = 10;
           [ll,mm] = showpoints(X,y,offset);
           if numsvl1 > 0 && numsvm1 > 0
              showplanes(w,b,ll,mm,epsilon)
           end
           axis equal
           axis([ll(1) mm(1) ll(2) mm(2)]);
           view([-1 -1 1]);
           xlabel('X','fontsize',14);ylabel('Y','fontsize',14);
           zlabel('Z','fontsize',14);
        end
    end
end
```

The function buildnuregb creates the constraint matrix and the matrices defining the quadratic functional.

```
function [A,c,P,Pa,qa] = buildnuregb (nu,X,y,C)
%  builds the matrix of constraints A for
%  soft margin nu-regression
%   with the constraint
%   \sum_{i = 1}^m + \sum_{j = 1}^m mu_j = C nu
%  (without the  variable gamma)
%  and the right-hand side c.
%  Input: an m x n matrix X of data points represented as
%  as the rows of X, and y a vector in R^n.
%  builds the m x m matrix X*X^T, the 2m x 2m matrix
%  P = [X*X^T -X*X^T;  -X*X^T X*X^T],
%  and the matrix Pa as the 4m  x 4m   matrix obtained
%  by augmenting  with zeros.
```

```
% Also builds the vector q_a (q augmented with zeros).
% C is a scale factor.

m = size(X,1); n = size(X,2);
% Ks = 1/(p+q);
Ks = C; Km = C*nu;
A = [ones(1,m) -ones(1,m) zeros(1,2*m);
     ones(1,m) ones(1,m)  zeros(1,2*m) ;
     eye(m) zeros(m,m) eye(m) zeros(m,m);
     zeros(m,m) eye(m)  zeros(m,m) eye(m)];
c = [0; Km; (Ks/m)*ones(2*m,1)];
XX1 = X*X';
P = [XX1 -XX1; -XX1 XX1];
Pa = [P zeros(2*m,2*m); zeros(2*m, 4*m )];
qa = [y; -y; zeros(2*m,1)];
end
```

Bibliography

Abraham, R. and Marsden, J. E. (1978). *Foundations of Mechanics*, 2nd edn. (Addison Wesley).

Apostol, T. (1974). *Analysis*, 2nd edn. (Addison Wesley).

Arnold, V. (1989). *Mathematical Methods of Classical Mechanics*, 2nd edn., GTM No. 102 (Springer Verlag).

Artin, M. (1991). *Algebra*, 1st edn. (Prentice Hall).

Avez, A. (1991). *Calcul Différentiel*, 1st edn. (Masson).

Berger, M. (1990a). *Géométrie 1* (Nathan), english edition: Geometry 1, Universitext, Springer Verlag.

Berger, M. (1990b). *Géométrie 2* (Nathan), english edition: Geometry 2, Universitext, Springer Verlag.

Berger, M. and Gostiaux, B. (1992). *Géométrie différentielle: variétés, courbes et surfaces*, 2nd edn., Collection Mathématiques (Puf), english edition: Differential geometry, manifolds, curves, and surfaces, GTM No. 115, Springer Verlag.

Bertsekas, D. P. (2009). *Convex Optimization Theory*, 1st edn. (Athena Scientific).

Bertsekas, D. P. (2015). *Convex Optimization Algorithms*, 1st edn. (Athena Scientific).

Bertsekas, D. P. (2016). *Nonlinear Programming*, 3rd edn. (Athena Scientific).

Bertsekas, D. P., Nedić, A., and Ozdaglar, A. E. (2003). *Convex Analysis and Optimization*, 1st edn. (Athena Scientific).

Bertsekas, D. P. and Tsitsiklis, J. N. (1997). *Parallel and Distributed Computation: Numerical Methods*, 1st edn. (Athena Scientific).

Bertsimas, D. and Tsitsiklis, J. N. (1997). *Introduction to Linear Optimization*, 3rd edn. (Athena Scientific).

Bishop, C. M. (2006). *Pattern Recognition and Machine Learning*, 1st edn., Information Science and Statistics (Springer).

Bourbaki, N. (1981). *Espaces Vectoriels Topologiques*, Eléments de Mathématiques (Masson).

Boyd, S., Parikh, N., Chu, E., Peleato, B., and Eckstein, J. (2010). Distributed optimization and statistical learning via the alternating direction method of multiplier, *Foundations and Trends in Machine Learning* **3(1)**, pp. 1–122.

Boyd, S. and Vandenberghe, L. (2004). *Convex Optimization*, 1st edn. (Cambridge University Press).

Bredon, G. E. (1993). *Topology and Geometry*, 1st edn., GTM No. 139 (Springer Verlag).

Brezis, H. (2011). *Functional Analysis, Sobolev Spaces and Partial Differential Equations*, 1st edn., Universitext (Springer-Verlag).

Cartan, H. (1990). *Cours de Calcul Différentiel*, Collection Méthodes (Hermann).

Chang, C.-C. and Chih-Jen, L. (2001). Training ν-support vector classifiers: Theory and algorithms, *Neural Computation* **13**, pp. 2119–2147.

Choquet-Bruhat, Y., DeWitt-Morette, C., and Dillard-Bleick, M. (1982). *Analysis, Manifolds, and Physics, Part I: Basics*, 1st edn. (North-Holland).

Chvatal, V. (1983). *Linear Programming*, 1st edn. (W.H. Freeman).

Ciarlet, P. (1989). *Introduction to Numerical Matrix Analysis and Optimization*, 1st edn. (Cambridge University Press), french edition: Masson, 1994.

Cour, T. and Shi, J. (2007). Solving markov random fields with spectral relaxation, in M. Meila and X. Shen (eds.), *Artifical Intelligence and Statistics* (Society for Artificial Intelligence and Statistics).

Demmel, J. W. (1997). *Applied Numerical Linear Algebra*, 1st edn. (SIAM Publications).

Dixmier, J. (1984). *General Topology*, 1st edn., UTM (Springer Verlag).

do Carmo, M. P. (1976). *Differential Geometry of Curves and Surfaces* (Prentice Hall).

do Carmo, M. P. (1992). *Riemannian Geometry*, 2nd edn. (Birkhäuser).

Faugeras, O. (1996). *Three-Dimensional Computer Vision, A geometric Viewpoint*, 1st edn. (the MIT Press).

Foley, J., van Dam, A., Feiner, S., and Hughes, J. (1993). *Computer Graphics. Principles and Practice*, 2nd edn. (Addison-Wesley).

Gabay, D. (1983). Applications of the method of multipliers to variational inequalities, *Studies in Mathematics and Applications* **15(C)**, pp. 299–331.

Gallier, J. H. (2011). *Geometric Methods and Applications, For Computer Science and Engineering*, 2nd edn., TAM, Vol. 38 (Springer).

Gallier, J. H. (2016). Notes on Convex Sets, Polytopes, Polyhedra, Combinatorial Topology, Voronoi Diagrams, and Delaunay Triangulations, Tech. rep., University of Pennsylvania, CIS Department, Philadelphia, PA 19104, book in Preparation.

Gander, W., Golub, G. H., and von Matt, U. (1989). A constrained eigenvalue problem, *Linear Algebra and its Applications* **114/115**, pp. 815–839.

Golub, G. H. (1973). Some modified eigenvalue problems, *SIAM Review* **15(2)**, pp. 318–334.

Gray, A. (1997). *Modern Differential Geometry of Curves and Surfaces*, 2nd edn. (CRC Press).

Hastie, T., Tibshirani, R., and Friedman, J. (2009). *The Elements of Statistical Learning: Data Mining, Inference, and Prediction*, 2nd edn. (Springer).

Hastie, T., Tibshirani, R., and Wainwright, M. (2015). *Statistical Learning with Sparsity. The Lasso and Generalizations*, 1st edn. (CRC Press).

Helgason, S. (2000). *Groups and Geometric Analysis. Integral Geometry, Invariant Differential Operators and Spherical Functions*, 1st edn., MSM, Vol. 83 (AMS).

Higham, N. J. (2008). *Functions of Matrices. Theory and Computation*, 1st edn. (SIAM).

Horn, R. A. and Johnson, C. R. (1990). *Matrix Analysis*, 1st edn. (Cambridge University Press).

Jain, R., Katsuri, R., and Schunck, B. G. (1995). *Machine Vision*, 1st edn. (McGraw-Hill).

Kolmogorov, A. and Fomin, S. (1975). *Introductory Real Analysis*, 1st edn. (Dover).

Kreyszig, E. (1991). *Differential Geometry*, 1st edn. (Dover).

Lang, S. (1993). *Algebra*, 3rd edn. (Addison Wesley).

Lang, S. (1995). *Differential and Riemannian Manifolds*, 3rd edn., GTM No. 160 (Springer Verlag).

Lang, S. (1996). *Real and Functional Analysis*, 3rd edn., GTM 142 (Springer Verlag).

Lang, S. (1997). *Undergraduate Analysis*, 2nd edn., UTM (Springer Verlag).

Lax, P. (2007). *Linear Algebra and Its Applications*, 2nd edn. (Wiley).

Luenberger, D. G. (1997). *Optimization by Vector Space Methods*, 1st edn. (Wiley).

Luenberger, D. G. and Ye, Y. (2016). *Linear and Nonlinear Programming*, 4th edn. (Verlag).

Matousek, J. and Gartner, B. (2007). *Understanding and Using Linear Programming*, 1st edn., Universitext (Springer Verlag).

Metaxas, D. N. (1997). *Physics-Based Deformable Models*, 1st edn. (Kluwer Academic Publishers).

Milnor, J. W. (1969). *Topology from the Differentiable Viewpoint*, 2nd edn. (The University Press of Virginia).

Molla, T. (2015). Class notes, math 588 example 5, Tech. rep., http://myweb.usf.edu/molla/2015_spring_math588/example5.pdf.

Munkres, J. R. (1991). *Analysis on Manifolds* (Addison Wesley).

Munkres, J. R. (2000). *Topology*, 2nd edn. (Prentice Hall).

Papadimitriou, C. H. and Steiglitz, K. (1998). *Combinatorial Optimization. Algorithms and Complexity*, 1st edn. (Dover).

Rockafellar, R. T. (1970). *Convex Analysis*, Princeton Landmarks in Mathematics (Princeton University Press).

Rudin, W. (1987). *Real and Complex Analysis*, 3rd edn. (McGraw Hill).

Rudin, W. (1991). *Functional Analysis*, 2nd edn. (McGraw Hill).

Schölkopf, B., Platt, J. C., Shawe-Taylor, J., and Smola, A. J. (2001). Estimating the support of a high-dimensional distribution, *Neural Computation* **13**, pp. 1443–1471.

Schölkopf, B. and Smola, A. J. (2002). *Learning with Kernels*, 1st edn. (MIT Press).

Schölkopf, B., Smola, A. J., Williamson, R. C., and Bartlett, P. L. (2000). New support vector algorithms, *Neural Computation* **12**, pp. 1207–1245.

Schrijver, A. (1999). *Theory of Linear and Integer Programming*, 1st edn. (Wiley).

Schwartz, L. (1980). *Topologie Générale et Analyse Fonctionnelle*, Collection Enseignement des Sciences (Hermann).

Schwartz, L. (1991). *Analyse I. Théorie des Ensembles et Topologie*, Collection Enseignement des Sciences (Hermann).

Schwartz, L. (1992). *Analyse II. Calcul Différentiel et Equations Différentielles*, Collection Enseignement des Sciences (Hermann).

Schwartz, L. (1993a). *Analyse III. Calcul Intégral*, Collection Enseignement des Sciences (Hermann).

Schwartz, L. (1993b). *Analyse IV. Applications à la Théorie de la Mesure*, Collection Enseignement des Sciences (Hermann).

Seifert, H. and Threlfall, W. (1980). *A Textbook of Topology*, 1st edn. (Academic Press).

Shawe-Taylor, J. and Cristianini, N. (2004). *Kernel methods for Pattern Analysis*, 1st edn. (Cambridge University Press).

Stoker, J. (1989). *Differential Geometry*, 1st edn., Wiley Classics (Wiley-Interscience).

Strang, G. (1986). *Introduction to Applied Mathematics*, 1st edn. (Wellesley-Cambridge Press).

Strang, G. (2019). *Linear Algebra and Learning from Data*, 1st edn. (Wellesley-Cambridge Press).

Trefethen, L. and Bau III, D. (1997). *Numerical Linear Algebra*, 1st edn. (SIAM Publications).

Trucco, E. and Verri, A. (1998). *Introductory Techniques for 3D Computer Vision*, 1st edn. (Prentice-Hall).

Vanderbei, R. J. (2014). *Linear Programming: Foundations and Extensions*, 4th edn. (Springer).

Vapnik, V. N. (1998). *Statistical Learning Theory*, 1st edn. (Wiley).

Warner, F. (1983). *Foundations of Differentiable Manifolds and Lie Groups*, 1st edn., GTM No. 94 (Springer Verlag).

Yu, S. X. and Shi, J. (2001). Grouping with bias, in T. G. Dietterich, S. Becker, and Z. Ghahramani (eds.), *Neural Information Processing Systems, Vancouver, Canada, 3–8 Dec. 2001* (MIT Press).

Ziegler, G. (1997). *Lectures on Polytopes*, 1st edn., GTM No. 152 (Springer Verlag).

Index

www.ingramcontent.com/pod-product-compliance
Lightning Source LLC
Chambersburg PA
CBHW071248180926
51613CB00002B/160
9789811216565